Teichmüller theory and moduli problems

Ramanujan Mathematical Society Lectures Notes Series

Vol. 1: Number theory

Vol. 2: The Riemann zeta function and related themes

Vol. 3: Formal language aspects of natural computing

Vol. 4: Commutative algebra and combinatorics

Vol. 5: Convexity in discrete structures

Vol. 6: Number theory and discrete geometry

Vol. 7: Discrete mathematics

Vol. 8: Lectures on operator theory

Vol. 9: Essays on geometric group theory

Vol. 10: Teichmüller theory and moduli problems

Ramanujan Mathematical Society

Lecture Notes Series

Volume 10

Teichmüller theory and moduli problems

Proceedings of the U.S.-India workshop in
Teichmüller theory and moduli problems,
Harish-Chandra Research Institute,
Allahabad, January 2006

Volume editors

Indranil Biswas
Ravi S. Kulkarni
Sudeb Mitra

International Press
www.intlpress.com

Ramanujan Mathematical Society
Lecture Notes Series, Volume 10
Teichmüller theory and moduli problems

Volume Editors:
Indranil Biswas
Ravi S. Kulkarni
Sudeb Mitra

Published in 2010 by International Press, Somerville, Massachusetts, U.S.A. under license from the Ramanujan Mathematical Society.

This work is also published in the Republic of India, exclusively by the Ramanujan Mathematical Society.

ISBN 978-1-57146-195-7

Printed in the United States of America.

15 14 13 12 11 2 3 4 5 6 7 8 9

Members of the Advisory Board

Contents

Teichmüller Theory and Moduli Problem No. 10, 2010, pp. 1–3.

Introduction

The Harish-Chandra Research Institute (HRI), Allahabad, India, hosted the "US-India Workshop: Teichmüller Theory and Moduli Problems" from January 5, 2006 to January 15, 2006. This was the culmination of the "Year in Teichmüller Theory and Moduli Problems." This year-long program was inaugurated in January 2005 by Clifford J. Earle (Cornell University, USA), and Frederick P. Gardiner (City University of New York, USA). They visited HRI during January – February, 2005, and gave a series of introductory lectures on Teichmüller theory. Clifford Earle gave a series of lectures on "Teichmüller theory, past, present, and future." Frederick Gardiner gave some lectures on "A short course on Teichmüller's theorem." These lectures were followed by some survey lectures on "Moduli spaces of vector bundles" by M. S. Narasimhan in March 2005.. William J. Harvey of King's College, London, visited HRI in March 2005 and lectured on "Mapping Class Groups." He surveyed some recent work about these groups which play a pivotal role in low-dimensional topology and geometry of surfaces. John H. Hubbard (Cornell University, USA) visited HRI in August 2005 and gave a course of lectures on "Holomorphic Dynamics."

The principal aim of this Workshop was to bring closer two mathematical traditions, namely the Ahlfors-Bers School of the complex analytic approach to Teichmüller theory that has a strong presence in USA, and the algebro-geometric tradition of the moduli of vector bundles that has developed in India. The idea was to connect scholars from different countries with similar research interests, as an important step towards furthering joint research. The larger objective was to foster increased cooperation between the Indian, American and European scientific communities, working in various areas of moduli problems.

Teichmüller theory originated as a specialized branch of complex analysis. Over the last 30 years, it has blossomed into a big and booming field of research, interacting with many other areas of mathematics, like topology, geometry, and dynamics. A natural question in the study of Riemann surfaces is to parametrize the space of conformal structures on a given topological surface. Riemann mapping theorem already says that there is a unique conformal structure on the Riemann sphere, complex plane and the unit disk, i.e any Riemann surface structure on a simply connected domain is equivalent to one of these three cases. There is a well-studied 1-complex parameter family of Riemann surface structure on a compact surface of genus 1 (elliptic curves).

Riemann asserted that, for a compact Riemann surface of genus $g \geq 2$, the space M_g of distinct conformal structures has complex dimension $3g - 3$. The space M_g is called the Riemann's moduli space. The algebraic geometers have studied this space extensively. During the late 1930s, Teichmüller followed an analytic approach, using quasiconformal mappings, and constructed a new space T_g, now called the Teichmüller space. The Teichmüller space has a canonical complex structure. The automorphism group of this structure is Γ_g, the Teichmüller Modular Group, which is isomorphic the outer automorphism group of the fundamental group of a compact surface of genus g. The quotient space T_g / Γ_g can be canonically identified with M_g.

Broadly speaking, there are three main approaches to the study of moduli of Riemann surfaces and their higher-dimensional analogues:

(i) The analytic approach was initiated by Teichmüller and was further developed by Ahlfors, Bers, and their students and followers. An important aspect of this approach is that it can also be extended to noncompact surfaces including those of infinite type. The methods of quasiconformal mappings and Teichmüller theory also found interesting applications in the study of the dynamics of rational maps. Fundamental work in this area was done by Douady, Hubbard, McMullen, Sullivan, Thurston, and others during the 1980s.

(ii) The algebro-geometric approach was developed by Grothendieck, Deligne, Mumford, and many others.

(iii) The topological and hyperbolic/conformal geometric approach was initiated by Nielsen and Fenchel in the 1930s. Its importance for the study of 3-manifolds and Kleinian groups was recognized by Kulkarni, Marden, Maskit, and was revolutionized by Thurston, and later developed in fundamental ways by McMullen, Penner, Sullivan, and others in the 1980s.

As is well-known, Teichmüller theory has found interesting applications in modern particle physics. In the branch of physics known as String theory, elementary particles are modelled by loops of strings that generate Riemann surfaces as they move through time. These Riemann surfaces have physical interpretations, and Teichmüller theory can be used to study how they vary.

The mapping class group of a genus g surface acts on T_g, and the quotient M_g, as noted before, parametrizes conformal structures on that surface. We can have interesting parameter spaces for many other structures and objects other than the conformal structure. For example, we can have moduli spaces of holomorphic vector bundles on a given base, moduli space of holomorphic maps from surfaces of fixed genus to a given projective variety etc. We refer to the article of Seshadri (in this volume) for a more technical introduction for the moduli spaces.

This workshop was funded by the National Science Foundation, USA, Department of Science and Technology, India, and the Infosys Foundations. There were speakers from USA, UK, Spain, Japan, China, France, Germany, Switzerland, Mexico, Sultanate of Oman, and India. The lectures covered a remarkably broad range, helping the younger participants to get an exposure to current research areas. There was also a special event – John Hubbard (USA), Richard Canary (USA), and

S. Ramanan (India) gave general survey lectures on the historical developments and various ramifications of Teichmüller theory, hyperbolic geometry, 3-manifolds, and moduli of vector bundles.

The editors wish to thank the graduate students and the staff at HRI who worked very hard to make this workshop a success. Many thanks to Professor Amitabha Raychowdhury (Director, HRI), and the graduate students Vikram Aithal, Kuntal Chatterjee, Krishnendu Gangopadhyay, and Siddhartha Sarkar. Special thanks to Prof. Madhav Modak, who took pains to go through every paper, and corrected many typos and ambiguities.

Indranil Biswas, Ravi S. Kulkarni, Sudeb Mitra

Teichmüller Theory and Moduli Problem No. 10, 2010, pp. 5–33.

Teichmüller Spaces as Complex Manifolds: Warwick 1992

Clifford J. Earle

Department of Mathematics, Cornell University, Ithaca, New York 14853
e-mail: cliff@math.cornell.edu

Abstract. We present an introduction to the complex analytic theory of Teichmüller spaces.

Keywords. Quasiconformal maps, Riemann surfaces, Teichmüller spaces.

Prologue

In January 2005, Fred Gardiner and I had the happy experience of offering parallel introductory courses about Teichmüller spaces at the Harish Chandra Research Institute in Allahabad, India. Fred's lecture notes evolved into the Gardiner-Hu paper [7] in this volume. My Allahabad notes are unfit for publication, but the editors suggested that I substitute the notes from a much shorter course of lectures at Warwick in 1992. I am very grateful for the opportunity to publish them here. Although they are rather informal, I have made no alterations except for updating some references and correcting some errors pointed out by the editors.

As the other papers in this lively volume show, Teichmüller spaces are still widely studied and used in a variety of contexts. The supplementary chapters of the second edition of Ahlfors's book [2] contain further information about Teichmüller spaces and related areas of study. All three chapters have extensive bibliographies.

1. Introduction

These notes are based on a series of four lectures given during the Instructional Conference on hyperbolic geometry and Teichmüller spaces at the University of Warwick in September 1992. The conference program consisted of 13 coordinated lectures and a number of problem sessions. A few problems from those sessions are included here as examples or as proofs left to the reader. In the interest of greater clarity the later sections of the notes include some examples and interpretive comments that were not presented at the conference.

The material presented here is well known, and most of it can be found in the books [6], [8], [9], [12], and [15]. A number of important subjects went unmentioned, notably Teichmüller's theorem about the extremal quasiconformal maps between compact Riemann surfaces (see the Gardiner-Hu paper [7]) and the universal property of the Teichmüller curve (see [8] or [15]).

If there is any novelty in the presentation, it is in some parts of Section 5, where the one-dimensional Teichmüller spaces are discussed. This discussion presents very concretely many important conceptual features of the general case.

2. Riemann surfaces

2.1 *Definitions and examples*

By definition, a *surface* is a (nonempty) connected Hausdorff space each point of which has an open neighborhood that is homeomorphic to the complex plane $\mathbb{C}$. An *atlas* for the surface X is a family $\mathcal{A}$ of functions h satisfying the two conditions

(a) each h in $\mathcal{A}$ is a homeomorphism whose domain is an open subset of X and whose image is an open subset of $\mathbb{C}$, and
(b) each point of X belongs to the domain of some h in $\mathcal{A}$.

The functions h in the atlas $\mathcal{A}$ are called *coordinate charts*, or simply *charts*.

Obviously, every surface X has an atlas, and the union of two atlases is an atlas. The maximal atlas for X consists of all the homeomorphisms whose domain and range are open subsets of X and $\mathbb{C}$ respectively.

Definitions. *The atlas $\mathcal{A}$ is holomorphic (smooth) iff the transition mapping $g \circ h^{-1}$ is holomorphic (of class C^∞) whenever $g \in \mathcal{A}$, $h \in \mathcal{A}$, and the domain of $g \circ h^{-1}$ is not empty. Two holomorphic (smooth) atlases are equivalent iff their union is a holomorphic (smooth) atlas. A Riemann surface (smooth surface) is a surface together with an equivalence class of holomorphic (smooth) atlases.*

Observe that every holomorphic atlas is smooth, and every equivalence class of holomorphic atlases is contained in an equivalence class of smooth atlases. Thus, every Riemann surface is also smooth.

To determine a Riemann surface structure (or smooth structure) on X, it is necessary only to specify a holomorphic (or smooth) atlas, and it is often efficient to specify a small atlas. (See the examples below.) However, it is also useful to have charts with small domains. This requires large atlases, so it is customary to equip a Riemann surface with its maximal holomorphic atlas, which consists of all the charts in all the holomorphic atlases in the given equivalence class. We shall follow that custom.

Examples.

1) Any region (connected open set) V in $\mathbb{C}$ has a standard Riemann surface structure determined by the atlas consisting of the identity map of V onto itself.
2) The Riemann sphere $\hat{\mathbb{C}}$ has a standard Riemann surface structure. It is determined by the holomorphic atlas that consists of the identity map of $\mathbb{C}$ onto itself and the homeomorphism $z \mapsto 1/z$ of $\hat{\mathbb{C}} - \{0\}$ onto $\mathbb{C}$.
3) Let h be any homeomorphism of $\mathbb{C}$ onto the open unit disk $\Delta = \{z \in \mathbb{C}; |z| < 1\}$. Then $\mathcal{A} = \{h\}$ is a holomorphic atlas for $\mathbb{C}$. This atlas is not equivalent to the standard holomorphic atlas defined in Example 1, for Liouville's theorem implies that h is not a holomorphic function.

4) Any subregion V of a Riemann surface X inherits a Riemann surface structure. It is determined by the holomorphic atlas consisting of all the charts for X whose domains are contained in V. (Here it is helpful to provide X with its maximal holomorphic atlas.)
5) Let $Z[i] = \{m + ni;\, m \text{ and } n \in Z\}$ be the additive group of Gaussian integers, let X be the quotient group $\mathbb{C}/Z[i]$, and let $\pi : \mathbb{C} \to X$ be the quotient map. The quotient topology makes X a compact surface (homeomorphic to the torus $S^1 \times S^1$) and π a universal covering map. Let $\mathcal{S}$ be the set of all regions V in $\mathbb{C}$ such that $\pi|V$ is injective. The set

$$\mathcal{A} = \{(\pi|V)^{-1} : \pi(V) \to V;\, V \in \mathcal{S}\}$$

is a holomorphic atlas for X. More generally, for each t in Δ, let $L_t : \mathbb{C} \to \mathbb{C}$ be the real-linear map

$$L_t(z) = (z + t\bar{z})/(1 + t).$$

The family of homeomorphisms $\mathcal{A}(t) = \{L_t \circ h;\, h \in \mathcal{A}\}$ is a holomorphic atlas for X and is not equivalent to $\mathcal{A}$ unless $t = 0$.

2.2 *Holomorphic maps and isomorphisms*

Let X and Y be Riemann surfaces. By definition, a *holomorphic map* (*smooth map*) $f : X \to Y$ is a continuous map such that the composite function $h \circ f \circ g^{-1}$ is holomorphic (of class C^∞) whenever $g \in \mathcal{A}(X)$, $h \in \mathcal{A}(Y)$, and the domain of $h \circ f \circ g^{-1}$ is not empty. Here (and always) $\mathcal{A}(X)$ and $\mathcal{A}(Y)$ are holomorphic atlases on X and Y determining their respective Riemann surface structures.

A holomorphic map of X into $\mathbb{C}$ is called a *holomorphic function*.

Examples. If V is a subregion of $\mathbb{C}$ (with its standard Riemann surface structure), the definition of holomorphic function given above agrees with the usual definition. In addition, the holomorphic maps of V into the Riemann sphere $\hat{\mathbb{C}}$ are the meromorphic functions on V (together with the constant map to the point at infinity). The holomorphic maps of the Riemann sphere into itself are the rational functions and the constant maps.

Holomorphic maps between Riemann surfaces enjoy many of the basic properties of holomorphic functions in $\mathbb{C}$. Here are some examples.

Proposition 1 (Open Mapping Theorem). *A holomorphic map $f : X \to Y$ is either constant or an open map.*

Proof. Given any x in X, choose a chart h in $\mathcal{A}(Y)$ whose domain V contains $f(x)$ and a chart g in $\mathcal{A}(X)$ whose domain U contains x and is contained in $f^{-1}(V)$. The holomorphic function $h \circ f \circ g^{-1}$ is either constant or an open map, so $f|U$ is either constant or open. Thus X is the union of the disjoint open sets

$$A = \{x \in X;\, f|U \text{ is open for some neighborhood } U \text{ of } x\},$$

$$B = \{x \in X;\, f|U \text{ is constant for some neighborhood } U \text{ of } x\}.$$

If $X = A$, f is an open map. If $X = B$, f is locally constant, hence constant. □

Corollary 1. *Suppose X is compact and $f : X \to Y$ is holomorphic. Then either f is constant or f is surjective. In particular, if Y is not compact then f is constant.*

Proof. If f is not constant, then $f(X)$ is a nonempty compact open subset of Y. □

Corollary 2 (Maximum Principle). *Let f be a holomorphic function on the Riemann surface X. If $|f|$ has a local maximum, then f is constant.*

Here are a few additional elementary facts, the proofs of which are left as exercises for the reader.

Proposition 2 (Identity Theorem). *Suppose f and g are holomorphic maps of a Riemann surface X into a Riemann surface Y. If $f = g$ on a subset of X that has a limit point in X, then $f = g$.*

Proposition 3. *Let $f : X \to Y$ be continuous. If $x_0 \in X$ and $f|(X - \{x_0\})$ is holomorphic, then f is holomorphic.*

Proposition 4. *Let $f : X \to Y$ be a bijective holomorphic map. Then f^{-1} is holomorphic.*

Definitions. *A bijective holomorphic map $f : X \to Y$ is called an isomorphism or a conformal equivalence. The Riemann surfaces X and Y are isomorphic iff there is an isomorphism $f : X \to Y$. An isomorphism of X onto itself is an automorphism.*

Example. Return to Example 5 in Section 2.1, and for each t in Δ let X_t be the torus X with the Riemann surface structure determined by the holomorphic atlas $\mathcal{A}(t)$ defined in that example. For any complex number c, the translation $z \mapsto z+c$ covers a homeomorphism $f_c : X_t \to X_t$. Each of these homeomorphisms is an automorphism.

Now suppose that $f : \mathbb{C} \to \mathbb{C}$ is a homeomorphism that fixes the origin and covers an isomorphism of X_t onto X_s. Since the induced map from X_t to X_s is holomorphic, the homeomorphism $L_s \circ f \circ L_t^{-1}$ of $\mathbb{C}$ onto itself is holomorphic, hence complex linear (since it fixes the origin). Thus f is real linear. Since $L_s \circ f \circ L_t^{-1}$ is holomorphic, we can compute s in terms of t and f. Since f maps $Z[i]$ onto itself and preserves orientation, its matrix (with respect to the standard basis $\{1, i\}$ for the real vector space $\mathbb{C}$) belongs to the group $SL(2, Z)$. We conclude that $SL(2, Z)$ acts on Δ in such a way that the Riemann surfaces X_t and X_s are isomorphic iff t and s are in the same $SL(2, Z)$ orbit. We shall see later that every Riemann surface homeomorphic to X is isomorphic to one of the Riemann surfaces X_t, so the quotient space $\Delta/SL(2, Z)$ represents the space of isomorphism classes of Riemann surfaces homeomorphic to X.

2.3 *Uniformization*

Appropriate maps between surfaces can be used to transport Riemann surface structures from one surface to another. The following results are sufficient for our purposes.

Definition. *A local homeomorphism is an open, continuous, and locally injective map.*

Proposition 5. *Let X and Y be surfaces and $f : X \to Y$ a local homeomorphism. If Y is a Riemann surface, there is exactly one Riemann surface structure on X such that the map f is holomorphic.*

Proof. Suppose $\mathcal{A}$ is a maximal holomorphic atlas on X such that f is holomorphic. Let V be any region in X such that $f|V$ is injective and there is an injective holomorphic function $h : f(V) \to \mathbb{C}$. Then the map $h \circ (f|V)$ belongs to $\mathcal{A}$. But the collection of all such maps is a holomorphic atlas for X that makes f holomorphic. □

Definitions. *If x is a point in the Hausdorff space X and G is a group of homeomorphisms of X, the orbit (or G-orbit) of x is the set $G(x) = \{g(x); g \in G\}$. The quotient space (or orbit space) X/G is the set of all G-orbits, and is given the quotient topology induced by the quotient map $x \mapsto G(x)$. We say G acts freely if for every x in X the map $g \mapsto g(x)$ is injective. We say G acts freely and properly discontinuously iff G acts freely, X/G is Hausdorff, and the quotient map from X to X/G is locally injective (and hence a local homeomorphism).*

Proposition 6. *Let G be a group of homeomorphisms of a surface X, let $Y = X/G$, and let $f : X \to Y$ be the quotient map. Suppose G acts freely and properly discontinuously.*

(a) *If Y is a Riemann surface and X is given the unique Riemann surface structure that makes f holomorphic, then G is a group of automorphisms of X.*
(b) *Conversely, if X is a Riemann surface and G is a group of automorphisms (still acting freely and properly discontinuously), there is a unique Riemann surface structure on Y that makes f holomorphic.*

Proof.

(a) Let $\mathcal{A}$ be the maximal holomorphic atlas on X that makes f holomorphic and choose g in G. The atlas $\mathcal{B} = \{x \mapsto h(g(x)); h \in \mathcal{A}\}$ is holomorphic and makes f holomorphic (since $f \circ g = f$), so $\mathcal{B} \subset \mathcal{A}$ and g is holomorphic.
(b) The required Riemann surface structure on Y is determined by the holomorphic atlas consisting of all the maps $h \circ (f|V)^{-1}$, where h is in the maximal holomorphic atlas of X, V is a subregion of the domain of h, and $f|V$ is injective. □

Corollary 3. *Let Y be a Riemann surface, X its universal covering surface, and $f : X \to Y$ the universal covering map. If X is given the unique Riemann surface structure such that f is holomorphic, then every cover transformation is an automorphism of X and Y is isomorphic to X/G (with the Riemann surface structure that makes the quotient map holomorphic).*

This corollary is very useful because of

Theorem 1 (Uniformization Theorem). *Every simply connected Riemann surface is isomorphic to $\hat{\mathbb{C}}$, $\mathbb{C}$, or Δ.*

Proofs of this deep and important result can be found in [3], [4], [5], [8], and many other places.

Corollary 4. *Every Riemann surface is isomorphic to $\hat{\mathbb{C}}$, $\mathbb{C}$ modulo a discrete group of translations, or Δ / Γ, where Γ is a group of automorphisms of Δ, acting freely and properly discontinuously.*

Proof. The universal covering surface X of the Riemann surface Y is (isomorphic to) one of the three simply connected Riemann surfaces listed above. If X is the Riemann sphere, then Y is isomorphic to $\hat{\mathbb{C}}$ modulo a group G of Möbius transformations. Since the group of cover transformations acts freely, G is trivial and Y is isomorphic to the Riemann sphere. The remainder of the proof is left to the reader. □

2.4 *The ideal boundary and the hyperbolic metric*

There are many ways to put a boundary on a Riemann surface. The ideal boundary for our purposes is defined as follows. If the universal covering surface of X (with its induced Riemann surface structure) is not (isomorphic to) Δ, the *ideal boundary* δX is empty. If the universal cover is Δ, the group Γ of cover transformations consists of Möbius transformations and therefore acts on all of $\hat{\mathbb{C}}$. The limit points of the orbit $\Gamma(0)$ are a closed Γ-invariant subset $L(\Gamma)$ of the unit circle S^1. If $L(\Gamma) = S^1$, the ideal boundary δX is again empty. If $L(\Gamma)$ is not all of S^1, then Γ acts freely and properly discontinuously on the complementary region $R(\Gamma) = \hat{\mathbb{C}} - L(\Gamma)$. The quotient Riemann surface $\hat{X} = R(\Gamma)/\Gamma$ is called the *double* of X, and the ideal boundary δX is the boundary of X in $\hat{X}$. Each connected component of δX is a Jordan curve or arc.

If the universal covering surface of X is Δ, then every cover transformation is an isometry with respect to the hyperbolic (Poincaré) metric of Δ. Therefore X inherits a Riemannian metric of constant negative curvature so that the universal covering map from Δ to X is a local isometry. This is called the hyperbolic or Poincaré metric of X.

3. Quasiconformal maps, Teichmüller spaces, and the modular group

3.1 *Quasiconformal maps in the plane*

Let V be a plane region. By definition, a *quasiconformal map* (a *qc map* for short) of V into the plane is an orientation-preserving homeomorphism f such that the function $H_f : V \to [1, \infty]$ defined by

$$H_f(z) = \limsup_{r \to 0+} \frac{\max\{|f(w) - f(z)|;\ |w - z| = r\}}{\min\{|f(w) - f(z)|;\ |w - z| = r\}}$$

has a finite upper bound in V. The qc map f is *K-quasiconformal* (*K-qc*) iff $H_f \leq K$ almost everywhere in V. Thus f is a qc map iff f is K-qc for some K in the interval $[1, \infty)$. We define $K(f)$ as the smallest number K such that f is K-qc.

Example. The homeomorphisms $L_t : \mathbb{C} \to \mathbb{C}$ in Example 1 of Section 2.1 are qc maps with $K(L_t) = (1 + |t|)/(1 - |t|)$.

The definition above is picturesque but is not the most useful characterization of qc-maps. There are at least two other important characterizations. The first uses the distributional derivatives of f. These are defined as follows. Recall first that if f is a C^1 map of V into $\mathbb{C}$ (so that $f'(z) : \mathbb{C} \to \mathbb{C}$ is a real-linear map for every z in V), the complex derivatives $f_z : V \to \mathbb{C}$ and $f_{\bar{z}} : V \to \mathbb{C}$ are defined by the requirement that

$$f'(z)h = f_z(z)h + f_{\bar{z}}(z)\bar{h} \quad \text{whenever } z \in V \text{ and } h \in \mathbb{C}.$$

The locally integrable (and measurable) complex-valued functions g and h in V are called the *distributional derivatives* of the measurable function $f : V \to \mathbb{C}$ with respect to z and $\bar{z}$ respectively iff

$$\iint (f\varphi_z + g\varphi) = \iint (f\varphi_{\bar{z}} + h\varphi) = 0$$

for all smooth functions $\varphi : V \to \mathbb{C}$ whose support is a compact subset of V. When this condition is satisfied we of course denote g and h by f_z and $f_{\bar{z}}$ respectively.

Theorem 2. *Suppose f is a homeomorphism of a plane region V into $\mathbb{C}$ and $1 \leq K < \infty$. Then f is a K-qc map iff f has locally integrable distributional derivatives that satisfy the inequality*

$$|f_{\bar{z}}| \leq \frac{K-1}{K+1}|f_z| \quad \textit{almost everywhere in } V. \tag{1}$$

Moreover, if f is a K-qc map then, at almost all points of V, f is differentiable, its Jacobian is positive, and its ordinary complex derivatives equal its distributional derivatives. In addition, there is a number $p > 2$, depending only on K, such that f_z is locally an L^p function.

Corollary 5. *The 1-qc maps are precisely the injective holomorphic functions.*

Proof. If f is holomorphic and injective, then f is a homeomorphism onto an open subset of $\mathbb{C}$. In addition, the Cauchy-Riemann equations say that $f_{\bar{z}} = 0$ everywhere in V, so (1) holds with $K = 1$, and f is 1-qc. Conversely, if f is 1-qc, then f is certainly injective, and $f_{\bar{z}} = 0$ almost everywhere in V. Hence, Weyl's Lemma (see for instance [4]) implies that f is holomorphic. □

For a proof of the theorem above (and for a proof of the corollary that does not require Weyl's Lemma) see the book [13].

The remaining important characterization of qc-maps uses the notion of extremal length, for which we refer to [6], [16], or the Gardiner-Hu article [7] in this volume. If Γ is a path family (set of paths) in the plane, we denote its extremal length by $\lambda(\Gamma)$. The following theorem is proved in [16].

Theorem 3. *Let f be an orientation-preserving homeomorphism of the plane region V into $\mathbb{C}$. Then f is K-qc iff for every path family Γ in V we have*

$$\frac{1}{K}\lambda(\Gamma) \leq \lambda(f(\Gamma)) \leq K\lambda(\Gamma).$$

Corollary 6. *The inverse of a K-qc map is K-qc. If f_1 and f_2 are K_1-qc and K_2-qc maps, respectively, then $f_1 \circ f_2$ is $K_1 K_2$-qc.*

The following facts will be important to us. For proofs we again refer to [2] or [13].

Theorem 4. *Every qc map of Δ onto itself extends to a homeomorphism of the closed unit disk onto itself.*

Theorem 5. *The set of all K-qc maps of the plane onto itself that fix the points 0 and 1 is compact with respect to the compact-open topology.*

Remarks.

1) Theorem 2 guarantees that for many purposes the distributional derivatives of a qc map can be treated like ordinary complex derivatives. For instance, the derivatives of inverse or composite maps obey the chain rule.
2) If V is an open set and $f : V \to \mathbb{C}$ is a qc map, the area of the image $f(V)$ is equal to the integral over V of the Jacobian of f. In particular, the image of a set of zero area under a qc map has zero area. (See [2] or [13].)
3) At a point where f is differentiable its Jacobian equals $|f_z|^2 - |f_{\bar{z}}|^2$. If the Jacobian is positive at the point z it is easy to verify that
$$H_f(z) = \frac{|f_z(z)| + |f_{\bar{z}}(z)|}{|f_z(z)| - |f_{\bar{z}}(z)|}.$$
This observation proves that a diffeomorphism f of V into $\mathbb{C}$ is K-qc iff the inequality (1) holds.
4) A *quadrilateral* in the region V consists of a closed Jordan domain Q in V and an ordered quadruple of distinct boundary points of Q, arranged in cyclic (anticlockwise) order. The path family $\Gamma(Q(z_1, z_2, z_3, z_4))$ consists of all the arcs in Q that join a point of the boundary arc $[z_1, z_2]$ to a point of the (disjoint) boundary arc $[z_3, z_4]$. Theorem 3 remains true if the set of all path families in V is replaced by the set of path families of this special form. (See [2] and [13].)

3.2 *Quasiconformal maps of Riemann surfaces*

Let X and Y be Riemann surfaces and let K be a real number ≥ 1. By definition, the homeomorphism $f : X \to Y$ is *K-quasiconformal* (*K-qc*) iff $h \circ f \circ g^{-1}$ is K-qc whenever g and h are charts on X and Y such that the composed map makes sense. The homeomorphism f is a *qc map* iff f is K-qc for some K. Again we denote by $K(f)$ the least K such that f is K-qc.

It is easy to verify that Corollaries 5 and 6 remain valid for qc maps between Riemann surfaces. (Use the fact that the transition map between two charts on a Riemann surface is 1-qc.) The counterpart of Theorem 4 is

Theorem 6. *Every qc map of X onto Y extends to a homeomorphism of $X \cup \delta X$ onto $Y \cup \delta Y$.*

We leave the proof to the reader. At one point in the argument it is necessary to show that there is no qc map of $\mathbb{C}$ onto Δ. That fact can be deduced from Theorem 3 or from Theorem 8 and Proposition 7 below.

3.3 *The Teichmüller space of a Riemann surface*

Let X be a Riemann surface whose universal covering surface is Δ. The Teichmüller space $Teich(X)$ of X is the set of equivalence classes of qc maps with domain X. By definition, two such maps $f_1 : X \to Y_1$ and $f_2 : X \to Y_2$ are equivalent iff there is an isomorphism $g : Y_1 \to Y_2$ such that the qc map $f_2^{-1} \circ g \circ f_1$ of X onto itself is homotopic to the identity rel the ideal boundary. (That condition means that if δX is not empty, the extended homeomorphism of $X \cup \delta X$ onto itself is homotopic to the identity through a homotopy that maps X into X and holds δX pointwise fixed at each stage.) The equivalence class of f will be denoted by $[f]$.

Definition. *Let $f_1 : X \to Y_1$ and $f_2 : X \to Y_2$ be qc maps. Let $\mathcal{F}(f_1, f_2)$ be the set of all qc maps $g : Y_1 \to Y_2$ such that $[g \circ f_1] = [f_2]$. The Teichmüller distance*

$$\tau([f_1], [f_2]) = \inf\left\{\frac{1}{2}\log K(g);\ g \in \mathcal{F}(f_1, f_2)\right\}. \tag{2}$$

The compactness properties of qc mappings lead to the following theorem, whose proof is in Chapter 5 of [6].

Theorem 7. *The Teichmüller metric on $Teich(X)$ is complete. Moreover, for all $[f_1]$ and $[f_2]$ in $Teich(X)$ the infimum in (2) is attained by at least one g in $\mathcal{F}(f_1, f_2)$.*

Remark. The maps g that achieve the minimum in (2) are called *extremal* qc maps and play a central role in certain aspects of the theory.

3.4 *The Teichmüller modular group*

A qc map g of Y onto X induces a map $g^* : Teich(X) \to Teich(Y)$ by the obvious rule

$$g^*([f]) = [f \circ g]$$

for all $[f]$ in $Teich(X)$. This rule produces a (right) action on $Teich(X)$ by the group $QC(X)$ of all qc maps of X onto itself:

$$[f] \cdot g = g^*([f]) = [f \circ g] \tag{3}$$

if $[f] \in Teich(X)$ and $g \in QC(X)$. These maps g^* are isometries in the Teichmüller metric.

Let $QC_0(X)$ be the normal subgroup of $QC(X)$ consisting of the qc maps that are homotopic to the identity rel δX. It is clear that the induced map g^* on $Teich(X)$ is the identity whenever g belongs to $QC_0(X)$. Therefore the right action (3) of $QC(X)$ induces a right action of the quotient group $Mod(X) = QC(X)/QC_0(X)$ on $Teich(X)$. The group $Mod(X)$ is called the *modular group* (or *Teichmüller modular group*) of X.

4. Beltrami forms

4.1 *The Beltrami equation*

Let f be a qc map whose domain is a subregion V of the plane. By definition, the *complex dilatation* of f is the function

$$\mu_f = f_{\bar{z}}/f_z.$$

This function is measurable and defined almost everywhere in V. It therefore extends to a measurable function on V, and (1) says that

$$\|\mu_f\|_\infty = (K(f) - 1)/(K(f) + 1) \; (< 1). \tag{4}$$

It is a remarkable fact that every measurable complex valued function μ on V with $\|\mu\|_\infty < 1$ has the form μ_f for some qc-map f defined in V. The following theorem was proved by C.B. Morrey in 1938, although it was not until 1957 that Lipman Bers noticed that Morrey's solutions of (5) are qc maps. For modern proofs of this theorem see [2] or [13].

Theorem 8. *Given any function μ in $L^\infty(\mathbb{C}, \mathbb{C})$ with $\|\mu\|_\infty < 1$ there is a qc map f of $\mathbb{C}$ onto itself such that the distributional derivatives of f are locally L^2 functions satisfying Beltrami's equation*

$$f_{\bar{z}} = \mu f_z. \tag{5}$$

Corollary 7. *If V is a plane region, every function in the open unit ball of $L^\infty(V, \mathbb{C})$ is the complex dilatation of some qc map defined in V.*

Proof. Given ν in $L^\infty(V, \mathbb{C})$, define μ in $L^\infty(\mathbb{C}, \mathbb{C})$ by setting $\mu = \nu$ in V and $\mu = 0$ in the complement of V. Let f be a qc map of $\mathbb{C}$ onto itself satisfying the Beltrami equation (5). Then ν is the complex dilatation of the restricted qc map $f|V$. □

If f and g are qc maps, the chain rule gives the useful formula

$$\mu_{f \circ g} = \frac{\mu_g + (\mu_f \circ g)\alpha_g}{1 + \bar{\mu}_g(\mu_f \circ g)\alpha_g} \tag{6}$$

with $\alpha_g = (|g_z|/g_z)^2$. Appropriate use of this formula shows that if $\mu_f = \mu_g$ in a region V, then $\mu_{f \circ g^{-1}} = 0$ in $g(V)$. This proves

Proposition 7. *If f and g are qc maps in V with $\mu_f = \mu_g$, then $f \circ g^{-1}$ is holomorphic in $g(V)$.*

Remark. There are two classical ways to interpret the complex dilatation and the Beltrami equation when μ is a smooth function. The first is from Riemannian geometry. Any Riemannian metric on the plane region V can be written uniquely in the complex form

$$ds = \lambda(z)\,|dz + \mu(z)\,d\bar{z}| \tag{7}$$

with $\lambda(z) > 0$ and $|\mu(z)| < 1$. Two Riemannian metrics on V are said to be *conformally equivalent* iff they are positive multiples of each other, and this is true iff their complex forms have the same function μ. The smooth maps μ from V into Δ thus parametrize the conformal equivalence classes of smooth Riemannian metrics on V.

An orientation-preserving diffeomorphism $w = f(z)$ of V into the plane is said to define *isothermal coordinates* for the metric (7) on V iff the metrics ds and $f^*(|dw|)$ are conformally equivalent. This will happen iff f satisfies the Beltrami equation (5). Thus Theorem 8 provides us with isothermal coordinates for very general metrics.

A second interpretation is complex analytic. A smooth map μ of V into Δ defines at each point z of V a real-linear map $L_{\mu(z)}(h) = h + \mu(z)\bar{h}$ from the tangent space $V_z(= \mathbb{C})$ onto $\mathbb{C}$. There is a unique real-linear automorphism $J_{\mu(z)}$ of V_z such that

$$L_{\mu(z)} \circ J_{\mu(z)} = iL_{\mu(z)}.$$

The automorphisms $J_{\mu(z)}$ determine an *almost complex structure* J_μ on V, and a function f from V into $\mathbb{C}$ is J_μ-holomorphic iff it satisfies equation (5). From this point of view, Theorem 8 provides injective J_μ-holomorphic functions.

These two interpretations of (5) are essentially equivalent since the orientation-preserving conformal maps in the plane are precisely the injective holomorphic functions.

4.2 *Beltrami forms on a Riemann surface*

The complex dilatation of a qc map $f : X \to Y$ between Riemann surfaces is defined as follows. Given a chart g defined in a neighborhood of x in X, choose a chart h defined in a neighborhood of $f(x)$ on Y. Since $h \circ f \circ g^{-1}$ is qc in a neighborhood of $g(x)$, it has a well defined complex dilatation, *which does not depend on the choice of* h, because of equation (6). To study the dependence on g we replace g by $\varphi \circ g$ for some injective holomorphic function $w = \varphi(z)$ defined in a neighborhood of $g(x)$, and we compute using (6). We find that $\mu = \mu_{h \circ f \circ g^{-1}}$ is transformed to $\tilde{\mu}$, where

$$\tilde{\mu}(\varphi(z))\overline{\varphi'(z)}/\varphi'(z) = \mu(z). \tag{8}$$

In other words μ is well defined (almost everywhere on X) as a tensor such that the expression

$$\mu(z)\, d\bar{z}/dz$$

remains invariant under changes of coordinates on X. Such a tensor is called a *Beltrami form* on X. The Beltrami form μ determined as above by the qc map f is called the complex dilatation of f and is denoted by μ_f.

If μ is a Beltrami form on X, equation (8) shows that $|\mu|$ is a well defined function on X. We denote by *Belt*(X) the complex Banach space consisting of the Beltrami forms μ that are measurable, defined almost everywhere on X, and satisfy $|\mu| \leq k$ almost everywhere for some finite number k. By definition, the least such k is the *norm* $\|\mu\|_\infty$ of μ. The open unit ball in *Belt*(X) is denoted by $M(X)$.

Theorem 9. *The complex dilatation of any qc map $f : X \to Y$ belongs to $M(X)$. In fact $\|\mu_f\|_\infty$ satisfies equation (4). Conversely, given any μ in $M(X)$ there is a unique Riemann surface X_μ such that its underlying topological space is X and the identity map from X to X_μ has complex dilatation μ. Moreover, if μ is the complex dilatation of $f : X \to Y$, then f is an isomorphism of X_μ onto Y.*

Here is an outline of the proof. The first two statements are obvious. For the third, a Beltrami form μ and a chart g determine a function $\mu(z)$ on the image V of g. By Corollary 7, there is a qc map $f(z)$ on V with complex dilatation $\mu(z)$. The map $f \circ g$ is a homeomorphism into $\mathbb{C}$, and the set of all such homeomorphisms is a defining atlas for X_μ. By construction, the complex dilatation of the identity map from X to X_μ is μ. To prove the final statement we observe more generally that if $f_1 : X \to Y_1$ and $f_2 : X \to Y_2$ have the same complex dilatation, then $f_2 \circ f_1^{-1}$ is an isomorphism of Y_1 onto Y_2.

4.3 *Teich*(X) *as a quotient of* $M(X)$

The theorem above shows that there is a well defined surjective map Φ from $M(X)$ to *Teich*(X) that sends μ to the equivalence class of the identity map from X to X_μ. Equivalently we can write

$$\Phi(\mu_f) = [f] \tag{9}$$

for all qc maps f with domain X.

Proposition 8. *The group $QC(X)$ acts from the right on $M(X)$ by the rule*

$$\mu_f \cdot g = \mu_{f \circ g} \quad \text{if } \mu_f \in M(X) \text{ and } g \in QC(X). \tag{10}$$

Moreover, $\Phi(\mu) = \Phi(\nu)$ iff $\nu = \mu \cdot g$ for some g in $QC_0(X)$. Thus Φ induces a bijection between the quotient space $M(X)/QC_0(X)$ and Teich(X).

We leave the proof as an exercise. The importance of this proposition lies in the fact that we can use the map Φ to induce quotient structures on *Teich*(X) from natural structures on $M(X)$. For instance $M(X)$ is the open ball in the complex Banach space *Belt*(X), so it has a natural complex manifold structure. One of the fundamental theorems about Teichmüller spaces is

Theorem 10. *The space Teich(X) has a unique complex manifold structure such that the map Φ from $M(X)$ to Teich(X) is holomorphic and every point of $M(X)$ is in the image of some locally defined holomorphic cross section of Φ.*

Proofs are given in [8], [12], and [15]. See also [6] and [9] for the finite dimensional case.

Remark. The group *Diff*$_0(X)$ of diffeomorphisms of X homotopic to the identity is a subgroup of $QC_0(X)$ when X is compact. In addition, two smooth Beltrami forms in $M(X)$ are in the same $QC_0(X)$-orbit iff they are in the same *Diff*$_0(X)$-orbit, so we can identify *Teich*(X) with the quotient space $\mathcal{M}(X)/\mathit{Diff}_0(X)$, where $\mathcal{M}(X)$ is the set of smooth forms in $M(X)$. That set can be interpreted as the set of smooth conformal structures on X or as the set of smooth almost complex structures on X that are compatible with its orientation. Since each conformal equivalence class of smooth Riemannian metrics on X contains exactly one metric of constant curvature -1, we can also interpret $\mathcal{M}(X)$ as the set of such metrics. In any case, $Teich(X)$ is the base space of a very natural principal fibre bundle with structure group $Diff_0(X)$.

4.4 *Dependence on parameters*

The Beltrami equation is an important tool in the study of $Teich(X)$, and it is especially important to understand how its solutions depend on the parameter μ. The major results were proved by Ahlfors and Bers. (See [15] or [12] for references and some details.) The following result is sufficient for most purposes. We denote the open unit ball of the complex Banach space $L^\infty(\mathbb{C}, \mathbb{C})$ by $M(\mathbb{C})$. For any positive number r we denote the complex Banach space of continuous functions on the closed ball $\{z \in \mathbb{C}; |z| \leq r\}$ by $C(B_r)$ and we denote by $L^2(B_r)$ the complex Hilbert space of square integrable functions on B_r.

Theorem 11. *For each μ in $M(\mathbb{C})$ let w^μ be the unique qc map of $\mathbb{C}$ onto itself that fixes the points* 0 *and* 1 *and has complex dilatation μ. For every positive number r the map $\mu \mapsto w^\mu$ is holomorphic, viewed as a map from $M(\mathbb{C})$ to $C(B_r)$. In addition, $\mu \mapsto w^\mu_z$ is a holomorphic map from $M(\mathbb{C})$ to $L^2(B_r)$.*

5. The one-dimensional Teichmüller spaces

5.1 *The punctured torus*

We consider $Teich(X')$ when X' is the torus $X = \mathbb{C}/Z[i]$ with the coset $[0] = Z[i]$ deleted. The quotient map from $\mathbb{C}$ to X restricts to a holomorphic covering $\pi : \mathbb{C}' \to X'$, where $\mathbb{C}' = \mathbb{C} - Z[i]$. The group L of cover transformations consists of the translations $z \mapsto z + \omega$, $\omega \in Z[i]$, and is transitive on the fibres of π. As usual, $Aut(L)$ is the group of automorphisms of the group L.

We shall use the map π to lift the space $M(X')$ and the groups $QC(X')$ and $QC_0(X')$ to $\mathbb{C}'$, where we shall obtain a quite explicit model for $Teich(X')$. Put

$$M(L, \mathbb{C}) = \{\mu \in M(\mathbb{C}); \mu(z + \omega) = \mu(z) \text{ if } \omega \in Z[i] \text{ and } z \in \mathbb{C}\},$$

and let $QC(L)$ be the group of g in $QC(\mathbb{C})$ such that $g(0) = 0$ and there is an automorphism α_g in $Aut(L)$ such that

$$g \circ \lambda = \alpha_g(\lambda) \circ g \quad \text{whenever } \lambda \in L.$$

The map $g \mapsto \alpha_g$ is a homomorphism of $QC(L)$ into $Aut(L)$. We denote its kernel by $QC_0(L)$. Thus

$$QC_0(L) = \{g \in QC(L); g(\omega) = \omega \text{ for all } \omega \in Z[i]\}.$$

Observe that $SL(2, Z)$, acting on the real vector space $\mathbb{C}$ as a group of real-linear transformations (not as Möbius transformations), is a subgroup of $QC(L)$. It is easy to see that each g in $QC(L)$ can be written uniquely in the form

$$g = A_g \circ g_0$$

with A_g in $SL(2, Z)$ and g_0 in $QC_0(L)$. (A_g is determined by the property that $A_g(\omega) = g(\omega)$ for all ω in $Z[i]$.) The map $g \mapsto A_g$ induces an isomorphism of $QC(L)/QC_0(L)$ onto $SL(2, Z)$.

Lemma 1. *If $\mu \in M(L, \mathbb{C})$, the Beltrami form $\mu(z)\, d\bar{z}/dz$ on $\mathbb{C}$ is L-invariant and induces a Beltrami form $\pi_*(\mu)$ on X'. The map π_* from $M(L, \mathbb{C})$ to $M(X')$ is a holomorphic isomorphism.*

Lemma 2. *Suppose $\mu \in M(\mathbb{C})$. Then $\mu \in M(L, \mathbb{C})$ iff there exist a discrete group L^μ of translations and an isomorphism θ^μ of L onto L^μ such that*

$$w^\mu \circ \lambda = \theta^\mu(\lambda) \circ w^\mu \qquad \text{for all } \lambda \in L. \tag{11}$$

Proof. If (11) holds, application of the chain rule formula (6) to both sides of (11) shows that $\mu \in M(L, \mathbb{C})$. Conversely, if $\mu \in M(L, \mathbb{C})$, the chain rule formula shows that the qc map $\theta^\mu(\lambda)$ defined by equation (11) is holomorphic. Since it is conjugate to λ (in the group of homeomorphisms of $\mathbb{C}$) it has no fixed points and is a translation. The group L^μ is discrete because it is conjugate to L. □

Lemma 3. *For any qc map $f : X' \to Y$ there is a torus $\overline{Y}$ containing Y so that f extends to a qc map of X onto $\overline{Y}$. In particular the restriction map is an isomorphism from the group $QC(X, [0])$ of f in $QC(X)$ that fix the point $[0]$ to the group $QC(X')$.*

Proof. The complex dilatation of f equals $\pi_*(\mu)$ for some $\mu \in M(L, \mathbb{C})$. By Lemma 2, the qc map w^μ covers a qc map f^μ from X onto the torus $X^\mu = \mathbb{C}/L^\mu$. The map $f^\mu \circ f^{-1}$ is an isomorphism of Y onto $X^\mu - \{f^\mu([0])\}$. □

Lemma 4. *Every g in $QC(L)$ covers a qc map $\pi_*(g)$ in $QC(X')$. The map*

$$\pi_* : QC(L) \to QC(X')$$

is a group isomorphism, and the image of $QC_0(L)$ is $QC_0(X')$.

Proof. Every map in $QC(X')$ extends uniquely to a map in $QC(X, [0])$. The extended map has a unique lift that fixes zero and therefore belongs to $QC(L)$. So π_* is an isomorphism of $QC(L)$ onto $QC(X')$. To see that the image of $QC_0(L)$ is contained in $QC_0(X')$, choose g in $QC_0(L)$ and put $\mu = \mu_g$. Note that $\mu \in M(L, \mathbb{C})$ and that $g = w^\mu$ since every point of $Z[i]$ is fixed by g. For t in the unit interval $[0, 1]$ put

$$g_t(z) = \frac{\alpha(t) w^{t\mu}(z) - \beta(t)\overline{w^{t\mu}(z)}}{\alpha(t) - \beta(t)}, \tag{12}$$

with $\alpha(t) = \overline{w^{t\mu}(i)} - i$ and $\beta(t) = w^{t\mu}(i) - i$. We leave it to the reader to verify that each g_t is a qc map of $\mathbb{C}$ onto itself and belongs to $QC_0(L)$. In addition, g_0 is the identity map and $g_1 = g$. It follows that $\pi_*(g) \in QC_0(X')$.

Finally, we must verify that every map in $QC_0(X')$ is in the image of $QC_0(L)$. Suppose $g \in QC(L)$ and $\pi_*(g) \in QC_0(X')$. To see that $g \in QC_0(L)$, choose a homotopy H of $\pi_*(g)$ to the identity on X'. The covering homotopy theorem provides a continuous map $\hat{H} : [0, 1] \times \mathbb{C}' \to \mathbb{C}'$ such that $\hat{H}(0, z) = g(z)$ and $\pi(\hat{H}(t, z)) = H(t, \pi(z))$ for all z in $\mathbb{C}'$ and t in $[0, 1]$. It follows that for all ω in $Z[i]$, z in $\mathbb{C}'$, and t in $[0, 1]$ the complex number $\hat{H}(t, z + \omega) - \hat{H}(t, z)$ belongs to $Z[i]$ and therefore depends only on ω. Since the map $\hat{H}(1, \cdot)$ of $\mathbb{C}'$ into itself covers the identity map on X', we conclude that

$$\hat{H}(t, z + \omega) - \hat{H}(t, z) = \omega$$

when $t = 1$, hence also when $t = 0$. □

Remarks.

1) Lemma 4 and the observations that precede Lemma 1 provide us with an explicit isomorphism of the Teichmüller modular group $Mod(X')$ onto $SL(2, Z)$.
2) Let $\mathit{Diff}_0(L)$ be the group of C^∞ diffeomorphisms in $QC_0(L)$, endowed with the C^∞ topology. If g_t is defined by equation (12), the map

$$(1 - t, w^\mu) \mapsto g_t$$

from $[0, 1] \times \mathit{Diff}_0(L)$ to $\mathit{Diff}_0(L)$ is continuous and provides an explicit contraction of $\mathit{Diff}_0(L)$ to a point. This provides a very explicit proof that the group $\mathit{Diff}_0(X, [0])$ of diffeomorphisms of X that fix the point $[0]$ and are homotopic to the identity is contractible.

Lemma 5. *If $\mu \in M(L, \mathbb{C})$ and $g \in QC(L)$, then*

(a) $\mu \cdot g \in M(L, \mathbb{C})$
(b) $\pi_*(\mu \cdot g) = \pi_*(\mu) \cdot \pi_*(g)$.

Proof. Fix g in $QC(L)$ and μ in $M(L, \mathbb{C})$. Put $\nu = \mu \cdot g$. This means there is a bijective holomorphic map $h : \mathbb{C} \to \mathbb{C}$ such that $w^\nu = h \circ w^\mu \circ g$.

To prove (a), fix λ in L and note that $w^\nu \circ \lambda = h \circ w^\mu \circ g \circ \lambda = h \circ w^\mu \circ \alpha_g(\lambda) \circ g$. Using Lemma 2 we can write this as $h \circ \theta^\mu(\alpha_g(\lambda)) \circ w^\mu \circ g$. Therefore

$$w^\nu \circ \lambda = (h \circ \theta^\mu(\alpha_g(\lambda)) \circ h^{-1}) \circ w^\nu.$$

It follows from Lemma 2 that $\nu \in M(L, \mathbb{C})$, as required.

To prove (b) we observe that

$$\pi_*(\mu) \cdot \pi_*(g) = \mu_{\pi_*(w^\mu)} \cdot \pi_*(g) = \mu_{\pi_*(w^\mu \circ g)} = \mu_{\pi_*(w^\nu)} = \pi_*(\nu).$$ □

Corollary 8. *The holomorphic isomorphism $\pi_* : M(L, \mathbb{C}) \to M(X')$ induces a bijective map from $M(L, \mathbb{C})/QC_0(L)$ to $Teich(X')$.*

By combining Corollary 8 with the following lemma we shall identify $Teich(X')$ with the upper half plane $\mathbb{H}^+ = \{\tau \in \mathbb{C};\ \mathrm{Im}(\tau) > 0\}$.

Lemma 6. *The map $P : M(L, \mathbb{C}) \to \mathbb{H}^+$ defined by $P(\mu) = w^\mu(i)$ has these properties*

(a) *$P(\mu) = P(\nu)$ iff $\nu = \mu \cdot g$ for some g in $QC_0(L)$.*
(b) *P is surjective. In fact $P(\sigma(\tau)) = \tau$ for all τ in $\mathbb{H}^+$, where $\sigma : \mathbb{H}^+ \to M(L, \mathbb{C})$ is the holomorphic map such that $\sigma(\tau)$ is the constant function $-(\tau - i)/(\tau + i)$.*
(c) *For any μ_0 in $M(L, \mathbb{C})$ there is a holomorphic map $s : \mathbb{H}^+ \to M(L, \mathbb{C})$ such that $P \circ s$ is the identity map and $s(P(\mu_0)) = \mu_0$.*

Proof. Let $\mu \in M(L, \mathbb{C})$. Since the translations $z \mapsto z + 1$ and $z \mapsto z + P(\mu)$ generate the group L^μ, $P(\mu)$ cannot be a real number. Since $P(0) = i$ and P is a continuous map, $P(\mu) \in \mathbb{H}^+$.

To prove (a), assume first that $\nu = \mu \cdot g$ for some g in $QC_0(L)$. That means $w^\nu = w^\mu \circ g$, so $P(\nu) = w^\mu(g(i)) = w^\mu(i)$ as required. Conversely, if $P(\nu) = P(\mu)$, put $g = (w^\mu)^{-1} \circ w^\nu$. We must show that $g \in QC_0(L)$. Lemma 2 implies that for any integers m and n

$$w^\mu(m + ni) = w^\mu(m) + w^\mu(ni) = m + nP(\mu) = m + nP(\nu) = w^\nu(m + ni).$$

Therefore $w^\mu(\omega) = w^\nu(\omega)$ for all ω in $Z[i]$, so $\theta^\mu = \theta^\nu$. Now for any λ in L,

$$g \circ \lambda = (w^\mu)^{-1} \circ w^\nu \circ \lambda = (w^\mu)^{-1} \circ \theta^\nu(\lambda) \circ w^\nu = \lambda \circ g,$$

so $g \in QC_0(L)$, as required.

To prove (b) consider any τ in $\mathbb{H}^+$. We look for a constant function t in $M(L, \mathbb{C})$, in other words for a number t in Δ, such that $\tau = P(t) = w^t(i)$. Since

$$w^t(z) = (z + t\bar{z})/(1 + t),$$

we need to have $\tau = (i - ti)/(1 + t)$, or $t = -(\tau - i)/(\tau + i) = \sigma(\tau)$.

To prove (c) choose μ_0 in $M(L, \mathbb{C})$, and let $t_0 = \sigma(P(\mu_0))$. Then $P(t_0) = P(\mu_0)$, so $\mu_0 = t_0 \cdot g$ for some g in $QC_0(L)$. The map s defined by $s(\tau) = \sigma(\tau) \cdot g$ has the required properties. (The fact that s is holomorphic is a consequence of equation (6).) □

Corollary 9. *The map P induces a bijective map P^* from $M(L, \mathbb{C})/QC_0(L)$ to $\mathbb{H}^+$. The restriction of P to the set $\sigma(\mathbb{H}^+)$ of constant functions in $M(L, \mathbb{C})$ is a bijective map to $\mathbb{H}^+$.*

Lemma 6, Corollary 8, and Theorem 10 imply

Theorem 12. *The holomorphic map $P \circ \pi_*^{-1} : M(X') \to \mathbb{H}^+$ induces a holomorphic isomorphism of Teich(X') onto $\mathbb{H}^+$.*

Remarks.

1) Theorem 10 asserts that every Teichmüller space has a unique complex structure with certain properties. The uniqueness of such a structure is obvious but its existence requires proof. Corollary 8 and Lemma 6 above provide that existence proof for the Teichmüller space of a punctured torus.

2) A right action of the group $Mod(X')$ on $Teich(X')$ was defined in Section 3.4. Our identifications of $Mod(X')$ and $Teich(X')$ with $SL(2, Z)$ and $\mathbb{H}^+$ allow us to interpret that action as a right action of $SL(2, Z)$ on $\mathbb{H}^+$. It is determined by the formula

$$P(\mu) \cdot A = P(\mu \cdot A)$$

for all μ in $M(L, \mathbb{C})$ and A in $SL(2, Z)$. Here we are again thinking of $SL(2, Z)$ as a group of real-linear transformations of $\mathbb{C}$. Thus, if

$$A = \begin{pmatrix} a & b \\ c & d \end{pmatrix}$$

then $A(z) = A(x + yi) = (ax + by) + (cx + dy)i$. If $\nu = \mu \cdot A$, then

$$w_\mu(A(z)) = w_\mu(A(1))w_\nu(z) = w_\mu(a + ci)w_\nu(z) = (a + cP(\mu))w_\nu(z),$$

so $(a + cP(\mu))P(\nu) = (a + cP(\mu))w_\nu(i) = w_\mu(A(i)) = w_\mu(b + di) = b + dP(\mu)$, and

$$P(\mu) \cdot A = P(\nu) = \frac{dP(\mu) + b}{cP(\mu) + a}. \tag{13}$$

The corresponding left action of $SL(2, Z)$ on $\mathbb{H}^+$ is

$$A(P(\mu)) = P(\mu) \cdot A^{-1} = \frac{aP(\mu) - b}{-cP(\mu) + d},$$

which is not quite what we might have expected.

3) The action (13) of $SL(2, Z)$ on $\mathbb{H}^+$ lifts to an action on the *Teichmüller curve* $V(X')$, which we shall now define. (We were unable to include this material in the lectures but decided to include it here, in part to show how natural the action (13) is.) The first goal is to construct a complex manifold $V(X')$ and a holomorphic projection map $\varpi : V(X') \to \mathbb{H}^+$ so that for each τ in $\mathbb{H}^+$, $\varpi^{-1}(\tau)$ is the punctured torus that corresponds to τ. That is easy. We define an action of $Z \oplus Z$ on $\mathbb{H}^+ \times \mathbb{C}$ by

$$(m, n) \cdot (\tau, z) = (\tau, z + m + n\tau).$$

By definition, $V(X)$ is the quotient complex manifold $(\mathbb{H}^+ \times \mathbb{C})/(Z \oplus Z)$, and $V(X')$ is the open complex submanifold obtained by deleting from $V(X)$ the projection of the set $\{0\} \times \mathbb{C}$. The projection $\varpi : V(X') \to \mathbb{H}^+$ is induced by the map $(\tau, z) \mapsto \tau$.

Now the punctured tori $\varpi^{-1}(\tau)$ and $\varpi^{-1}(\tau \cdot A)$ are isomorphic for any τ in $\mathbb{H}^+$ and A in $SL(2, Z)$. Moreover, the homotopy class of an isomorphism between them is determined by A, interpreted as an element of $Mod(X')$. It is therefore reasonable to hope that the action (13) can be lifted to a fiber-preserving action on $V(X')$ whose restrictions to the fibers are the appropriate isomorphisms of punctured tori. The desired action is induced by the right action of $SL(2, Z)$ on $\mathbb{H}^+ \times \mathbb{C}$ defined by

$$(\tau, z) \cdot A = ((d\tau + b)/(c\tau + a), z/(c\tau + a))$$

if $\tau \in \mathbb{H}^+$, $z \in \mathbb{C}$, and

$$A = \begin{pmatrix} a & b \\ c & d \end{pmatrix}$$

belongs to $SL(2, Z)$.

5.2 *The sphere with four punctures*

Let Γ be the group of Möbius transformations generated by $z \mapsto z + 2$, $z \mapsto z + 2i$, and $z \mapsto -z$. The set of fixed points of elliptic transformations in Γ is precisely $Z[i]$. The group Γ acts properly discontinuously on $\mathbb{C}' = \mathbb{C} - Z[i]$, and the quotient surface $X' = \mathbb{C}'/\Gamma$ is (isomorphic to) the Riemann sphere with the set $\{-1, 0, 1, \infty\}$ deleted. (See Remark 1 below.) The quotient surface $X = \mathbb{C}/\Gamma$ is of course isomorphic to $\hat{\mathbb{C}}$. Let $\pi : \mathbb{C}' \to X'$ be the quotient map, and let $Aut(\Gamma)$ be the automorphism group of Γ.

By analogy with Section 5.1, put

$$M(\Gamma, \mathbb{C}) = \{\mu \in M(\mathbb{C});\ \mu \circ \gamma = \mu \text{ if } \gamma \in \Gamma\},$$

and let $QC(\Gamma)$ be the group of g in $QC(\mathbb{C})$ such that there is an automorphism α_g in $Aut(\Gamma)$ satisfying

$$g \circ \gamma = \alpha_g(\gamma) \circ g \quad \text{whenever } \gamma \in \Gamma.$$

Let $QC_0(\Gamma)$ be the kernel of the homomorphism $g \mapsto \alpha_g$ from $QC(\Gamma)$ into $Aut(\Gamma)$. As before,

$$QC_0(\Gamma) = \{g \in QC(\Gamma);\ g(\omega) = \omega \text{ for all } \omega \in Z[i]\}.$$

We leave the proofs of the following results to the reader. They use the same ideas as the proofs of the corresponding statements in Section 5.1.

Lemma 1′. *If $\mu \in M(\Gamma, \mathbb{C})$, the Beltrami form $\mu(z)\, d\bar{z}/dz$ on $\mathbb{C}$ is L-invariant and induces a Beltrami form $\pi_*(\mu)$ on X'. The map π_* from $M(\Gamma, \mathbb{C})$ to $M(X')$ is a holomorphic isomorphism.*

Lemma 2′. *Suppose $\mu \in M(\mathbb{C})$. Then $\mu \in M(\Gamma, \mathbb{C})$ iff there exist a group Γ^μ of Möbius transformations and an isomorphism θ^μ of Γ onto Γ^μ such that*

$$w^\mu \circ \gamma = \theta^\mu(\gamma) \circ w^\mu \qquad \text{for all } \gamma \in \Gamma.$$

Lemma 3′. *For any qc map $f : X' \to Y$ there is a compact Riemann surface $\overline{Y}$ containing Y so that f extends to a qc map of X onto $\overline{Y}$. (Thus $\overline{Y}$ is in fact isomorphic to the Riemann sphere.)*

Lemma 4′. *Every g in $QC(\Gamma)$ covers a qc map $\pi_*(g)$ in $QC(X')$. The map*

$$\pi_* : QC(\Gamma) \to QC(X')$$

is a surjective group homomorphism. Its kernel is Γ, and the restriction of π_* to $QC_0(\Gamma)$ is an isomorphism onto $QC_0(X')$.

Lemma 5′. *If* $\mu \in M(\Gamma, \mathbb{C})$ *and* $g \in QC(\Gamma)$, *then*

(a) $\mu \cdot g \in M(\Gamma, \mathbb{C})$
(b) $\pi_*(\mu \cdot g) = \pi_*(\mu) \cdot \pi_*(g)$.

The proof of the next result will require a bit of explanation.

Lemma 6′. *The map* $P : M(\Gamma, \mathbb{C}) \to \mathbb{H}^+$ *defined by* $P(\mu) = w^\mu(i)$ *has these properties.*

(a) $P(\mu) = P(\nu)$ *iff* $\nu = \mu \cdot g$ *for some* g *in* $QC_0(\Gamma)$.
(b) P *is surjective. In fact* $P(\sigma(\tau)) = \tau$ *for all* τ *in* $\mathbb{H}^+$, *where* $\sigma : \mathbb{H}^+ \to M(\Gamma, \mathbb{C})$ *is the holomorphic map such that* $\sigma(\tau)$ *is the constant function* $-(\tau - i)/(\tau + i)$.
(c) *For any* μ_0 *in* $M(\Gamma, \mathbb{C})$ *there is a holomorphic map* $s : \mathbb{H}^+ \to M(\Gamma, \mathbb{C})$ *such that* $P \circ s$ *is the identity map and* $s(P(\mu_0)) = \mu_0$.

Sketch of Proof. Let $\mu \in M(\Gamma, \mathbb{C})$. For any ω in $Z[i]$, let γ_ω be the elliptic transformation $z \mapsto -z + 2\omega$ in Γ that fixes ω. Then $\theta^\mu(\gamma_\omega)$ is the elliptic transformation of order two that fixes the points $w^\mu(\omega)$ and ∞, so

$$w^\mu(-z + 2\omega) = \theta^\mu(\gamma_\omega)(w^\mu(z)) = -w^\mu(z) + 2w^\mu(\omega)$$

for all z in $\mathbb{C}$. Therefore $w^\mu(2\omega) = 2w^\mu(\omega)$ for all ω in $Z[i]$. In particular $w^\mu(2) = 2$ and $w^\mu(2i) = 2P(\mu)$.

Now let Γ_0 be the subgroup of Γ consisting of the translations $z \mapsto z + 2\omega$, $\omega \in Z[i]$. Since $\theta^\mu(\Gamma_0)$ is a discrete group generated by $z \mapsto z + 2$ and $z \mapsto z + 2P(\mu)$, $P(\mu)$ cannot be real. We conclude as before that $P(\mu) \in \mathbb{H}^+$.

To prove (a), assume first that $\nu = \mu \cdot g$ for some g in $QC_0(\Gamma)$. Then $w^\nu = w^\mu \circ g$, so $P(\nu) = w^\mu(g(i)) = w^\mu(i)$, as before. Conversely, if $P(\nu) = P(\mu)$, then $\theta^\nu(\gamma_\omega) = \theta^\mu(\gamma_\omega)$ when $\omega = 0,\ 1,$ or i. These three transformations generate Γ, so $\theta^\nu = \theta^\mu$. Put $g = (w^\mu)^{-1} \circ w^\nu$. Then g commutes with Γ, so $g \in QC_0(\Gamma)$ and $\nu = \mu \cdot g$ as required.

The proofs of (b) and (c) are left to the reader. □

Once again these lemmas lead to an explicit representation of *Teich*(X'), given by

Theorem 13. *The holomorphic map* $P \circ \pi_*^{-1} : M(X') \to \mathbb{H}^+$ *induces a holomorphic isomorphism of Teich*(X') *onto* $\mathbb{H}^+$.

Remarks.

1) Let $\wp(z)$ be the Weierstrass $\wp$-function with fundamental periods 2 and $2i$. Since $\wp(z) = z^{-2} + O(z^2)$ near $z = 0$, the elliptic function $\wp(z) + \wp(iz)$ is identically zero. In particular,

$$\wp(1 + i) = -\wp(i - 1) = -\wp(i + 1) = 0 \quad \text{and} \quad \wp(i) = -\wp(1) \neq 0,$$

It follows readily that the elliptic function

$$f(z) = \frac{\wp(z)}{\wp(1)}$$

induces an isomorphism of X' onto $\hat{\mathbb{C}} - \{-1, 0, 1, \infty\}$

2) We can again represent $Mod(X')$ explicitly. The homomorphism π_* in Lemma 4′ induces an isomorphism of the quotient group $QC(\Gamma)/\Gamma \cdot QC_0(\Gamma)$ onto $Mod(X')$. Every element of $QC(\Gamma)$ can be written uniquely in the form

$$\lambda \circ A \circ g,$$

where $g \in QC_0(\Gamma)$, $A \in SL(2, Z)$, and λ belongs to the group L of translations $z \mapsto z + \omega$, $\omega \in Z[i]$. The subgroup $Aff(\Gamma)$ of $QC(\Gamma)$ generated by L and $SL(2, Z)$ is the normalizer of Γ in the group of orientation-preserving real-affine transformations of $\mathbb{C}$, and π_* induces an isomorphism of $Aff(\Gamma)/\Gamma$ onto $Mod(X')$. The reader can verify that $Mod(X')$ is a semidirect product of $PSL(2, Z)$ $(= SL(2, Z)/\{\pm I\})$ and Klein's group of order four $(= Z[i]/2Z[i])$, which is a normal subgroup of $Mod(X')$.
The discussion above shows that $X' = \hat{\mathbb{C}} - \{-1, 0, 1, \infty\}$ has an affine structure such that every qc homeomorphism of X' onto itself can be written uniquely as the composition of an affine homeomorphism and a homeomorphism that is homotopic to the identity. The modular group $Mod(X')$ is isomorphic to the group of orientation-preserving affine homeomorphisms of X'.

6. The infinitesimal theory

6.1 *The variational formula*

Let μ in $L^\infty(\mathbb{C}, \mathbb{C})$ and z in $\mathbb{C}$ be given. It is sometimes useful to have an explicit formula for the derivative at $t = 0$ of the map $t \mapsto w^{t\mu}(z)$. A heuristic derivation of the formula (which can be made rigorous) goes as follows. (We denote by $\dot{w}[\mu]$ the derivative of $t \mapsto w^{t\mu}$ at $t = 0$.) According to Theorem 11,

$$w^{t\mu}(z) = z + t\dot{w}[\mu](z) + o(t),$$

$$(w^{t\mu})_z = 1 + t\dot{w}[\mu]_z + o(t), \qquad \text{and}$$

$$(w^{t\mu})_{\bar{z}} = t\dot{w}[\mu]_{\bar{z}} + o(t).$$

Since $(w^{t\mu})_{\bar{z}} = t\mu(w^{t\mu})_z$, we must have $\dot{w}[\mu]_{\bar{z}} = \mu$. Since $w^{t\mu}$ fixes 0, 1, and ∞, $\dot{w}[\mu]$ must vanish at 0, 1, and ∞. (The last condition means that $\dot{w}[\mu](z)/z^2$ approaches zero as z approaches ∞.) These conditions uniquely determine the function $\dot{w}[\mu]$. Explicitly we have

$$\dot{w}[\mu](z) = \frac{1}{2\pi i} \iint_{\mathbb{C}} \mu(\zeta) \frac{z(z-1)}{\zeta(\zeta-1)(\zeta-z)} d\zeta \wedge d\bar{\zeta}. \tag{14}$$

Example. Let $\mu_n(z) = (\bar{z}/|z|)^n$ for $n \geq 0$. If $|t| < 1$, then direct computation shows that

$$w^{t\mu_n}(z) = [z^{\frac{n+2}{2}} + t\bar{z}^{\frac{n+2}{2}}]^{\frac{2}{n+2}}[1+t]^{-\frac{2}{n+2}} = z[1 + t(\bar{z}/|z|)^{n+2}]^{\frac{2}{n+2}}[1+t]^{-\frac{2}{n+2}}$$

for all z in $\mathbb{C} - \{0\}$. It follows that

$$\dot{w}[\mu_n](z) = \frac{2z[(\bar{z}/|z|)^{n+2} - 1]}{n+2}. \tag{15}$$

The fact that $\dot{w}[\mu_n]_{\bar{z}} = \mu_n$ is easily verified.

Remark. The equation $\dot{w}[\mu]_{\bar{z}} = \mu$ shows that in a sense the equation $f_{\bar{z}} = \mu$ is an infinitesimal version of the Beltrami equation.

6.2 *Infinitesimal variation of complex structure*

Any Beltrami form μ in $Belt(X)$ defines a tangent vector to $Teich(X)$. Indeed $t \mapsto t\mu$, for t real, is a path in $Belt(X)$ mapping a neighborhood $J = (-\epsilon, \epsilon)$ of $t = 0$ into $M(X)$. Therefore $t \mapsto \Phi(t\mu)$, t in J, is a path through the *origin* $\Phi(0)$ in $Teich(X)$. Its tangent vector at the origin is $\Phi'(0)\mu$, the derivative of Φ at 0 in the direction μ. This vector is zero iff $\mu \in ker(\Phi'(0))$, and Theorem 10 guarantees that the quotient Banach space $Belt(X)/ker(\Phi'(0))$ is the tangent space to $Teich(X)$ at the origin.

Example. Let $X' = \mathbb{C}'/L$ be the punctured torus in Section 5.1. Lifting to $\mathbb{C}$ as we did in Section 5.1, we find that the roles of $\Phi'(0)$ and $Belt(X')$ are played by $P'(0)$ and $Belt(L, \mathbb{C})$ respectively, where

$$Belt(L, \mathbb{C}) = \{\mu \in L^\infty(\mathbb{C}, \mathbb{C});\ \mu(z + \omega) = \mu(z) \text{ if } \omega \in Z[i] \text{ and } z \in \mathbb{C}\}.$$

The tangent space to $Teich(X')$ at the origin is $Belt(L, \mathbb{C})/ker(P'(0))$, and we can compute $P'(0)$ and its kernel explicitly by using formula (14).

Let S be the interior of the square in $\mathbb{C}$ with corners 0, 1, $1 + i$, and i. Since $P(\mu) = w^\mu(i)$, (14) gives

$$\begin{aligned} P'(0)\mu = \dot{w}[\mu](i) &= c \iint_{\mathbb{C}} \frac{\mu(\zeta)}{\zeta(\zeta - 1)(\zeta - i)} d\zeta \wedge d\bar{\zeta} \\ &= c \sum_{\omega \in Z[i]} \iint_{S+\omega} \frac{\mu(\zeta)}{\zeta(\zeta - 1)(\zeta - i)} d\zeta \wedge d\bar{\zeta} \\ &= c \sum_{\omega \in Z[i]} \iint_{S} \frac{\mu(z + \omega)}{(z + \omega)(z + \omega - 1)(z + \omega - i)} dz \wedge d\bar{z}, \end{aligned}$$

with $c = (i - 1)/2\pi$. Since $\mu \in Belt(L, \mathbb{C})$, we find that

$$P'(0)\mu = \iint_S \mu(z)\varphi(z) dz \wedge d\bar{z},$$

where $\varphi(z)$ is the elliptic function

$$\varphi(z) = \sum_{\omega \in Z[i]} \frac{c}{(z + \omega)(z + \omega - 1)(z + \omega - i)}.$$

The only possible poles of φ are at the points of $Z[i]$. They are all simple, so if φ is not constant the holomorphic branched covering map from $X = \mathbb{C}/Z[i]$ to $\hat{\mathbb{C}}$ that φ induces is a homeomorphism. That is impossible, so φ is a constant function, and

$$P'(0)\mu = \hat{c} \iint_S \mu(z) dz \wedge d\bar{z},$$

for some constant $\hat{c}$. Equation (15) with $n = 0$ gives $P'(0)\mu = \dot{w}[\mu](i) = -2i$ if $\mu(z)$ is the constant function $\mu(z) = 1$. Therefore $\hat{c} = 1$ and

$$P'(0)\mu = \iint_S \mu(z) dz \wedge d\bar{z}. \tag{16}$$

In particular the kernel of $P'(0)$ consists of the μ in $Belt(L, \mathbb{C})$ whose integral over S is zero.

Remark. We cannot resist giving an application of equation (16) that was omitted from the lectures for lack of time. We wish to determine which Beltrami forms in $M(L, \mathbb{C})$ are the complex dilatations of extremal qc maps from $X' = \mathbb{C}'/L$ to other punctured tori. By definition, these *extremal* Beltrami forms ν are characterized by the property that $\|\nu\|_\infty \leq \|\mu\|_\infty$ if $\Phi(\nu) = \Phi(\mu)$.

Consider any μ in $M(L, \mathbb{C})$. By Lemma 6, there is a unique constant function t in $M(L, \mathbb{C})$ with $\Phi(t) = \Phi(\mu)$, and $t = \sigma(P(\mu))$. (In particular, the restriction of Φ to the set of constant functions maps bijectively to $Teich(X')$.) By the compactness properties of qc maps, there is at least one extremal Beltrami form ν with $\Phi(\nu) = \Phi(\mu)$. We shall prove that every extremal form is constant; it will follow that the extremal forms are precisely the constant functions. This is the result predicted by Teichmüller's general theorem about extremal qc maps (see [6], [7], [8], [9], or [12]).

For the proof we fix any μ in $M(L, \mathbb{C}) - \{0\}$ and consider the holomorphic function

$$f(\zeta) = \iint_S \sigma(P(\zeta\mu/\|\mu\|_\infty))(z)\, dx\, dy$$

on Δ. Using the chain rule, the definition of σ in Lemma 6, and (16) we find that

$$f'(0) = \iint_S (\mu(z)/\|\mu\|_\infty)\, dx\, dy. \tag{17}$$

Since the function $\sigma(P(\zeta\mu/\|\mu\|_\infty))$ is constant and S has area one,

$$|f(\zeta)| = \|\sigma(P(\zeta\mu/\|\mu\|_\infty))\|_\infty < 1$$

for all ζ in Δ. Since $f(0) = 0$, Schwarz's lemma implies that

$$\|\mu\|_\infty \geq |f(\|\mu\|_\infty)| = \|\sigma(P(\mu))\|_\infty. \tag{18}$$

If μ is extremal, then equality must occur in (18), so Schwarz's lemma implies that $|f'(0)| = 1$. By (17), this can happen only if μ is constant almost everywhere (in S, hence also in $\mathbb{C}$), which is what we were required to prove.

6.3 *Infinitesimally trivial μ and the dimension of Teich(X)*

A Beltrami form on X is said to be infinitesimally trivial if it projects to the zero tangent vector at the origin in *Teich*(X); thus μ is infinitesimally trivial iff it belongs to the kernel of $\Phi'(0)$. There is an important characterization of that kernel, which we shall now describe. By definition, a *holomorphic quadratic differential* φ is a tensor on X defined by assigning to every chart h a function $\varphi(z)$, holomorphic in the image of h, in such a way that the expression

$$\varphi(z)dz^2 \tag{19}$$

is invariant under changes of coordinates.

It is clear that $|\varphi|$ is a differential two-form on X. Let $Q(X)$ be the Banach space of holomorphic quadratic differentials whose norm

$$\|\varphi\| = \iint_X |\varphi| \tag{20}$$

is finite. If $\mu \in$ *Belt*(X) and $\varphi \in Q(X)$, then $\mu\varphi$ is an integrable differential two-form on X. We obtain a linear map L from *Belt*(X) onto the dual space $Q(X)^*$ of $Q(X)$ by setting

$$L(\mu)(\varphi) = \iint_X \mu\varphi \tag{21}$$

for all $\mu \in$ *Belt*(X) and $\varphi \in Q(X)$.

The following theorem owes its name to the fact that it was proved by Teichmüller in the case that *Teich*(X) is finite dimensional. A proof in the general case can be found in [9] or [15].

Theorem 14 (Teichmüller's Lemma). *If $\Phi : M(X) \to Teich(X)$ is the quotient map, then the linear maps $\Phi'(0)$ and L have the same kernel. Therefore L induces an isomorphism from the tangent space at the origin of Teich(X) onto $Q(X)^*$.*

Definition. *A Riemann surface X is said to be analytically finite iff there is a compact Riemann surface $\hat{X}$ such that X is isomorphic to the complement of a finite subset of $\hat{X}$.*

Corollary 10. *The spaces Teich(X) and $Q(X)^*$ have equal dimension. In particular, Teich(X) is finite dimensional iff X is analytically finite.*

Proof. The second statement of the corollary follows from the first since $Q(X)$ is finite dimensional iff X is analytically finite. (See pp. 324–335 of [10] for a detailed proof of this classical fact.) □

Remarks.

1) Theorem 14 says that a Beltrami form is infinitesimally trivial iff it is orthogonal to the integrable holomorphic quadratic differentials. Equation (16) in Section 6.2 illustrates this statement for Beltrami forms on the punctured torus $X' = \mathbb{C}'/Z[i]$. In fact, the integral in (16) can be written as

$$\iint_{X'} \mu\varphi,$$

where φ is the holomorphic quadratic differential on X' that lifts to the quadratic differential dz^2 on $\mathbb{C}'$. Observe that φ spans the one-dimensional space $Q(X')$.

2) Let V be any closed subspace of *Belt*(X) that is complementary to the kernel of L. According to the inverse function theorem some neighborhood of zero in V is biholomorphically equivalent to a neighborhood of the origin in *Teich*(X). When X is analytically finite we can choose V so that its members are all supported in any given nonempty open subset S of X. To this end let ρ be the hyperbolic metric $\rho(z)|dz|$ on X (induced by the Poincaré metric on its universal covering surface Δ), let f be a smooth function that maps X onto the closed interval $[0, 1]$ and has compact support contained in S, and observe that $f\bar{\varphi}\rho^{-2} \in$ *Belt*(X) whenever $\varphi \in Q(X)$. The space V of all such Beltrami forms is complementary to the kernel of L, for it has the same finite dimension as $Q(X)^*$ and its intersection with the kernel of L is $\{0\}$.

3) We remarked in Section 6.1 that in the plane we can think of the equation $f_{\bar{z}} = \mu$ as the infinitesimal version of Beltrami's equation. Extending that idea to a Riemann surface X, we may expect a Beltrami form on X to be infinitesimally trivial iff it is in some sense the $\bar{z}$-derivative of an infinitesimal qc map in $QC_0(X)$. To make a precise statement we must identify the appropriate class of "infinitesimal maps". Once again let ρ be the hyperbolic metric $\rho(z)|dz|$ on X. Let f be a tensor on X such that the expression $f(z)/dz$ is invariant under change of coordinates on X. Observe that $\rho|f|$ is a well defined function on X. We call f a *qc vectorfield that vanishes on* δX provided that the functions $f(z)$ are continuous and satisfy the following two regularity and boundedness conditions. (The first guarantees that f vanishes on δX.)

The function $\rho|f|$ is bounded on X.

The tensor $f_{\bar{z}}d\bar{z}/dz$ belongs to *Belt*(X).

It is a rather easy consequence of formula (14) that μ in *Belt*(X) is infinitesimally trivial iff it has the form $f_{\bar{z}}d\bar{z}/dz$ for some qc vectorfield f that vanishes on δX.

7. The Teichmüller space of a Fuchsian group

In Section 5 we saw that explicit representations of the one-dimensional Teichmüller spaces can be obtained from appropriate covering spaces of the Riemann surface. A somewhat less explicit model of an arbitrary space *Teich*(X) was given by Lipman Bers, using the universal covering of X by Δ. We shall give a brief description of this Bers embedding. As usual, more details can be found in the books [6], [8], [9], [12], and [15].

Let $\pi : \Delta \to X$ be a holomorphic covering map, and let Γ be the group of cover transformations. As in Section 5, π induces identifications of $M(X)$ and $QC_0(X)$ with the space $M(\Gamma, \Delta)$ and group $QC_0(\Gamma)$ defined as follows:

$$M(\Gamma, \Delta) = \{\mu \in M(\Delta); (\mu \circ \gamma)\overline{\gamma'}/\gamma' = \mu \text{ if } \gamma \in \Gamma\},$$

$$QC_0(\Gamma) = \{g \in QC(\Delta); g \circ \gamma = \gamma \circ g \text{ if } \gamma \in \Gamma, \text{ and } g|S^1 \text{ is the identity}\}.$$

The right action of $QC_0(\Gamma)$ on $M(\Gamma, \Delta)$ is defined so that $\nu = \mu \cdot g$ iff $g : \Delta_\nu \to \Delta_\mu$ is an isomorphism of Riemann surfaces, and the *Teichmüller space* $T(\Gamma)$ is, by definition, the quotient space $M(\Gamma, \Delta)/QC_0(\Gamma)$. The quotient map from $M(\Gamma, \Delta)$ to $T(\Gamma)$ is again denoted by Φ. As in Section 5, the covering map π induces an isomorphism between *Teich*(X) and $T(\Gamma)$.

For each μ in $M(\Gamma, \Delta)$ let f^μ be the unique qc homeomorphism f of $\mathbb{C}$ onto itself such that $f|\Delta$ has complex dilatation μ and the restriction of f to the exterior Δ^* of Δ is holomorphic, with the behavior

$$f(z) = z + O(1/z)$$

at infinity. It is rather easy to verify that $\Phi(\mu) = \Phi(\nu)$ iff $f^\mu - f^\nu$ is identically zero on S^1 and hence (by the maximum principle) in Δ^*.

Recall that the Schwarzian derivative of the locally injective meromorphic function f in a plane region D is the holomorphic function

$$(f''/f')' - \frac{1}{2}(f''/f')^2.$$

With Bers, let φ^μ be the Schwarzian derivative of $f^\mu|\Delta^*$. Classical properties of Schwarzian derivatives and conformal maps imply that φ^μ belongs to the complex Banach space $B(\Gamma, \Delta^*)$ of holomorphic functions $\varphi : \Delta^* \to \mathbb{C}$ with the following two properties:

$$\|\varphi\|_* = \sup\{|\varphi(z)|(|z|^2 - 1)^2; z \in \Delta^*\} < \infty, \tag{22}$$

$$(\varphi \circ \gamma)(\gamma')^2 = \varphi \quad \text{for all } \gamma \in \Gamma. \tag{23}$$

(To make (23) true when ∞ is not a fixed point of the group Γ we should add the point ∞ to Δ^* and put $\varphi(\infty) = 0$. This is consistent with (22), which implies that $|\varphi(z)| < 2\|\varphi\|_*/|z|^4$ if $|z| > 2$.)

A remarkable and fundamental theorem of Bers states that the map $\mu \mapsto \varphi^\mu$ from $M(\Gamma)$ to $B(\Gamma)$ is holomorphic and can be factored through the quotient map $\Phi : M(\Gamma) \to T(\Gamma)$, yielding a holomorphic isomorphism $\Phi(\mu) \mapsto \varphi^\mu$ between $T(\Gamma)$ and a bounded open subset of $B(\Gamma)$. (That isomorphism is known as the "Bers embedding" of $T(\Gamma)$ in $B(\Gamma)$.) The most difficult part of this deep theorem is the proof that the inverse map $\varphi^\mu \mapsto \Phi(\mu)$ is holomorphic.

Remarks.

1) For μ in $M(\Gamma)$ the qc map f^μ induces an isomorphism

$$\theta^\mu(\gamma) = f^\mu \circ \gamma \circ (f^\mu)^{-1}$$

of Γ onto a group Γ^μ of Möbius transformations. Since it is quasiconformally conjugate (as a transformation group on $\mathbb{C}$) to a Fuchsian group, Γ^μ is called a *quasifuchsian* group. The Möbius transformation $\theta^\mu(\gamma)$ is completely determined by its values at three points, so the isomorphism θ^μ depends only on the equivalence class $\Phi(\mu)$. Conversely, if Γ is a Fuchsian group "*of the first kind*" (which means that the limit set $L(\Gamma)$ equals S^1), θ^μ determines the restriction of f^μ to S^1, which in turn determines $\Phi(\mu)$. Thus, for groups of the first kind the map $\Phi(\mu) \mapsto \theta^\mu$ identifies $T(\Gamma)$ with the set of isomorphisms θ^μ.

If X is analytically finite and $\pi : \Delta \to X$ is a holomorphic covering map, the group Γ of cover transformations is finitely generated and of the first kind. If the transformations $\gamma_1, \gamma_2, \dots, \gamma_n$ generate Γ and if we identify the group of all Möbius transformations with $PSL(2, \mathbb{C})$ in the usual way, then the map

$$\Phi(\mu) \mapsto (\theta^\mu(\gamma_1), \dots, \theta^\mu(\gamma_n))$$

is a holomorphic embedding of $T(\Gamma)$ in the complex manifold $PSL(2, \mathbb{C})^n$.

2) The definitions of $M(\Gamma, \Delta)$, $QC_0(\Gamma)$, $T(\Gamma)$, and the Bers embedding $\Phi(\mu) \mapsto \varphi^\mu$ make sense even if the Fuchsian group Γ contains elliptic transformations. In this case the quotient space Δ/Γ is an "orbifold" and $T(\Gamma)$ is its Teichmüller space. A celebrated theorem of Bers and Greenberg says that the Teichmüller space of the orbifold Δ/Γ is isomorphic to the Teichmüller space of the Riemann surface obtained from Δ/Γ by deleting the images of the elliptic fixed points.

8. The Teichmüller space of a Kleinian group

For our purposes a Kleinian group is a finitely generated torsion-free group Γ of Möbius transformations that acts properly discontinuously on a dense open subset $\Omega(\Gamma)$ of $\hat{\mathbb{C}}$ whose complement $L(\Gamma)$ is perfect, nowhere dense, and contained in every closed Γ-invariant subset of $\hat{\mathbb{C}}$. According to the Ahlfors finiteness theorem, the quotient space $\Omega(\Gamma)/\Gamma$ has finitely many connected components, each of which is an analytically finite Riemann surface. (See [10] for a proof of this theorem.) Since conjugation by a Möbius transformation does not alter the geometry of Γ we shall assume that the points 0, 1, and ∞ belong to $L(\Gamma)$.

The Teichmüller space $T(\Gamma)$ of a Kleinian group Γ is defined once again as the quotient space of an open ball $M(\Gamma)$ by a group $QC_0(\Gamma)$. This time

$$M(\Gamma) = \{\mu \in M(\hat{\mathbb{C}});\ (\mu \circ \gamma)\overline{\gamma'}/\gamma' = \mu \text{ if } \gamma \in \Gamma\},$$

$$QC_0(\Gamma) = \{g \in QC(\hat{\mathbb{C}});\ g \circ \gamma = \gamma \circ g \text{ if } \gamma \in \Gamma\},$$

and the right action of $QC_0(\Gamma)$ on $M(\Gamma)$ is defined so that $\nu = \mu \cdot g$ iff $w^\nu = w^\mu \circ g$. Thus μ and ν are in the same $QC_0(\Gamma)$-orbit iff the restrictions of w^μ and w^ν to $L(\Gamma)$ coincide. This happens iff

$$\theta^\mu(\gamma) = w^\mu \circ \gamma \circ (w^\mu)^{-1} = w^\nu \circ \gamma \circ (w^\nu)^{-1} = \theta^\nu(\gamma)$$

for all γ in Γ. (See [11]. In that paper the reader should assume that the set $L(G)$ contains the points 0, 1, and ∞ unless the contrary is explicitly stated.)

Once again the quotient space $T(\Gamma)$ is a complex manifold and (as for a finitely generated Fuchsian group of the first kind) evaluation of the isomorphisms θ^μ on a set of generators produces a holomorphic embedding of $T(\Gamma)$ in a product of copies of $PSL(2, \mathbb{C})$. The complex manifold $T(\Gamma)$ depends only on the quotient space $\Omega(\Gamma)/\Gamma$ and the toplogy of the components of $\Omega(\Gamma)$. We summarize the results as follows.

Theorem 15. *Let Γ be a Kleinian group satisfying the conditions stated at the beginning of this section, and let $X_1, X_2, \ldots, X_n$ be analytically finite Riemann surfaces whose disjoint union is isomorphic to $\Omega(\Gamma)/\Gamma$. If every connected component of $\Omega(\Gamma)$ is simply connected, then $T(\Gamma)$ is isomorphic to the product of the spaces Teich(X_j). If not, $T(\Gamma)$ is isomorphic to the product of quotient spaces of the Teich(X_j) and the quotient map from the product of the Teich(X_j) to $T(\Gamma)$ is a universal covering.*

A more precise statement can be found in [11].

Examples.

1) If $L(\Gamma)$ is contained in the extended real axis $\hat{\mathbb{R}}$ and the upper half plane is Γ-invariant, then Γ is Fuchsian, and $T(\Gamma)$ has been defined in two different ways, according as Γ is viewed as a Fuchsian or a Kleinian group. The resulting spaces are quite different, especially when $L(\Gamma)$ is a proper subset of $\hat{\mathbb{R}}$. When Γ is of the first kind (i.e. $L(\Gamma) = \hat{\mathbb{R}}$), its Fuchsian Teichmüller space is isomorphic to *Teich*(X), where X is the analytically finite Riemann surface $\mathbb{H}^+/\Gamma$. (We are still requiring Γ to be finitely generated and to contain no elliptic transformations.) According to Theorem 15 the Kleinian Teichmüller space of Γ is isomorphic to *Teich*$(X) \times$ *Teich*$(\overline{X})$, where $\overline{X}$, the "conjugate surface" of X, is isomorphic to the quotient of the *lower* half plane by Γ. The Kleinian groups $\theta^\mu(\Gamma)$ in this Kleinian Teichmüller space are quasifuchsian, as were the groups that arose in the construction of the Bers embedding in Section 7. A quasifuchsian group (of the first kind) represents two Riemann surfaces. In the Fuchsian construction of Section 7 we varied the complex structure of one of them, leaving the other surface unchanged. In the Kleinian Teichmüller space, the two surfaces are varied independently.
2) A *Schottky group* is a Kleinian group that is free on $p \geq 2$ generators and contains no transformations with only one fixed point. If Γ is a Schottky group, then $\Omega(\Gamma)$ is connected, $L(\Gamma)$ is a Cantor set, and $X = \Omega(\Gamma)/\Gamma$ is a compact Riemann surface of genus p. (For these and many other interesting facts see Maskit's book [14].) By Theorem 15, *Teich*(X) is the universal covering space of the "Schottky space" $T(\Gamma)$, so the (complex) dimension of Schottky space is $3p - 3$. If we choose Γ to be freely generated by transformations γ_j, $1 \leq j \leq p$, so that γ_1 fixes 0 and ∞ and γ_2 fixes 1, the transformations $\theta^\mu(\gamma_j)$ will depend on $3p - 3$ complex parameters that provide a rather natural global coordinate system for $T(\Gamma)$.

9. Opening a node

The "moduli space" of isomorphism classes of compact Riemann surfaces of genus $p \geq 2$ is the quotient space *Teich*$(X)/$*Mod*(X), where X is any compact Riemann surface of genus p. Understanding the structure of this moduli space is one of the major goals of the theory. Deligne and Mumford obtained a natural compactification of the moduli space by adjoining certain "noded Riemann surfaces of genus p" to the space. The "$zw = t$" construction for opening up a node plays a fundamental role in this compactification process. (See [1] and [9].) We shall conclude with an example to show how this construction is represented by an appropriate Kleinian group. We have chosen our example so that the group is particularly easy to construct. Similar constructions in the general case use slightly more elaborate groups. Groups of the sort constructed here play an important role in our unpublished work with Al Marden on the compactification of the moduli space.

Let X and Y be compact Riemann surfaces of genus one, and choose points P in X and Q in Y. The space S_0 formed from the disjoint union of X and Y by identifying the points P and Q is a noded Riemann surface of genus two; the point P $(= Q)$ is the node. Choose charts z on X and w on Y so that z and w both have range Δ and $z(P) = w(Q) = 0$. For each nonzero t in Δ set

$$X_t = X - z^{-1}(\Delta_t) \text{ and } Y_t = Y - w^{-1}(\Delta_t),$$

where $\Delta_t = \{\zeta; |\zeta| \leq |t|\}$. The Riemann surface S_t is formed from the disjoint union of X_t and Y_t by identifying A in X with B in Y iff $z(A)$ and $w(B)$ are defined and their product $z(A)w(B) = t$. This is the $zw = t$ construction for "opening the node" P of S_0.

To represent this construction by a Kleinian group we choose a universal covering of $X' = X - P$ by $\mathbb{H}^+$ so that P corresponds to the point at ∞. This means the group G of cover transformations contains a cyclic group of translations whose generator $\zeta \mapsto \zeta + c$, $c > 0$, corresponds to a simple loop around P. We may assume that $c = 1$. Under these circumstances we can define a chart z on X by putting $z(P) = 0$ and

$$z(\pi^+(\zeta)) = e^{2\pi i(\zeta - i)} \text{ if } \operatorname{Im}(\zeta) > 1,$$

where $\pi^+ : \mathbb{H}^+ \to X'$ is our universal covering map. Now G acts also on the lower half plane $\mathbb{H}^-$. We obtain a compact Riemann surface Y from the quotient Riemann surface $Y' = \mathbb{H}^-/G$ by adding to Y' a point Q that corresponds to the point at ∞. (By construction Y is isomorphic to $\overline{X}$.) A chart w on Y is defined by putting $w(Q) = 0$ and

$$w(\pi^-(\zeta)) = e^{-2\pi i(\zeta + i)} \text{ if } \operatorname{Im}(\zeta) < -1,$$

where π^- has the obvious meaning.

Choose any τ in $\mathbb{H}^+$ and put $e^{2\pi i\tau} = t$. The $zw = t$ construction requires us to discard from X and Y the respective sets

$$\pi^+(\{\zeta; \operatorname{Im}(\zeta) \geq \operatorname{Im}(\tau) + 1\}) \text{ and } \pi^-(\{\zeta; \operatorname{Im}(\zeta) \leq -(\operatorname{Im}(\tau) + 1)\})$$

and to identify $\pi^-(\zeta_1)$ with $\pi^+(\zeta_2)$ when

$$w(\pi^-(\zeta_1)) \times z(\pi^+(\zeta_2)) = e^{2\pi i\tau},$$

which happens when $\zeta_2 - \zeta_1 - \tau - 2i$ is an integer. Combination theorems for Kleinian groups allow us to make precisely these identifications. The theorems in Chapter 7 of [14] guarantee that if $\tau \in \mathbb{H}^+$ the group Γ_τ generated by G and the additional translation $A_\tau(\zeta) = \zeta + \tau + 2i$ is Kleinian and that the quotient $\Omega(\Gamma)/\Gamma$ is the

Riemann surface S_t (with $t = e^{2\pi i \tau}$). Each component of $\Omega(\Gamma_\tau)$ is simply connected; in fact the subgroup of Γ_τ that stabilizes any given component is a Maskit B-group (see [14]). Therefore, by Theorem 15, $T(\Gamma_\tau)$ is isomorphic to $Teich(S_t)$. This model of $Teich(S_t)$ is useful for studying the properties of Teichmüller space "near" the noded surface S_0.

References

[1] W. Abikoff, *The Real Analytic Theory of Teichmüller Space*, Lecture Notes in Mathematics, vol. 820, Springer-Verlag (1980).

[2] L. V. Ahlfors, *Lectures on Quasiconformal Mappings (second edition)*, University Lecture Series, vol. 38, American Mathematical Society (2006).

[3] A. F. Beardon, *A Primer on Riemann Surfaces*, London Mathematical Society Lecture Notes, vol. 78, Cambridge University Press (1984).

[4] H. M. Farkas and I. Kra, *Riemann Surfaces (second edition)*, Graduate Texts in Mathematics, vol. 71, Springer-Verlag (1992).

[5] O. Forster, *Lectures on Riemann Surfaces*, Graduate Texts in Mathematics, vol. 81, Springer-Verlag (1981).

[6] F. P. Gardiner, *Teichmüller Theory and Quadratic Differentials*, Wiley-Interscience (1987).

[7] F. P. Gardiner and Jun Hu, *A short course on Teichmüller's theorem*, in this volume.

[8] J. H. Hubbard, *Teichmüller Theory and Applications to Geometry, Topology, and Dynamics, vol. 1: Teichmüller Theory*, Matrix Editions, Ithaca, NY (2006).

[9] Y. Imayoshi and M. Taniguchi, *An Introduction to Teichmüller Spaces*, Springer-Verlag (1992).

[10] I. Kra, *Automorphic Forms and Kleinian Groups*, Benjamin (1972).

[11] I. Kra, *Deformation spaces*, A Crash Course on Kleinian Groups, Lecture Notes in Mathematics, vol. 400, Springer-Verlag, (1974) pp. 48–70.

[12] O. Lehto, *Univalent Functions and Teichmüller Spaces*, Graduate Texts in Mathematics, vol. 109, Springer-Verlag (1987).

[13] O. Lehto and K. I. Virtanen, *Quasiconformal Mappings in the Plane*, Springer-Verlag (1973).

[14] B. Maskit, *Kleinian Groups*, Springer-Verlag (1988).

[15] S. Nag, *The Complex Analytic Theory of Teichmüller Space*, Wiley-Interscience (1988).

[16] J. Väisälä, *Lectures on n-Dimensional Quasiconformal Mappings*, Lecture Notes in Mathematics, vol. 229, Springer-Verlag (1971).

Teichmüller Theory and Moduli Problem No. 10, 2010, pp. 35–39.

Moduli

C. S. Seshadri

Chennai Mathematical Institute, H1, SIPCOT IT Park, Siruseri 603 103, India
e-mail: css@cmi.ac.in

The basic notion for moduli is that of deformation say of schemes, vector bundles, coherent sheaves etc. Let $f : X \longrightarrow T$ be a flat morphism and say T is connected. Then if $t_1, t_2 \in T$, we say that X_{t_1} is a *deformation* of X_{t_2}. If X, T are smooth, f is proper and we are working over the field $\mathbb{C}$ of complex numbers, then X_{t_1} is topologically isomorphic to X_{t_2} i.e. the underlying topology of X_t is a deformation invariant. If F is a coherent sheaf on X which is flat over T, we get a deformation of coherent sheaves. In particular if $f : X \times T \longrightarrow T$ is the canonical projection and V is a vector bundle on $X \times T$, V is flat over T and V_t is the restriction of V to $X \times t \approx X$, we get deformations of vector bundles on X and one simply calls $\{V_t\}$ a *family* of vector bundles on X parametrized by T. If X is a smooth projective curve, deg V_t is a deformation invariant. Similarly the Hilbert polynomial $\chi(V_t(m))$ (when X is projective with ample $\mathcal{O}_X(1)$) is a deformation invariant. One can say that the moduli problem is the study of objects like varieties, vector bundles etc. with fixed deformation invariants (see [21] for an introduction to the subject).

There has been a great progress in the subject of moduli during the last fifty years. A basic factor which has contributed to this is Grothendieck's theory of schemes, providing a powerful language for dealing with moduli problems (flat morphisms, infinitesimal deformations, formal schemes, representable functors etc.). Another one is the work of Mumford on Geometric Invariant Theory (GIT) (see [19]). Moduli problems reduce to the study of orbit spaces under the actions of algebraic groups. This is a subtle affair in algebraic geometry and GIT has provided a systematic approach to the study of orbit spaces which arise in moduli problems. Using GIT Mumford gave the first construction of the moduli spaces of smooth projective curves of genus g and that of principally polarized abelian varieties of dimension g, as quasi-projective algebraic schemes in arbitrary characteristic. Around the same time, the moduli space $U(n, d)$ of semi-stable vector bundles of rank n and degree d on a smooth projective X of genus g ($g \geq 2$) under a certain equivalence (S-equivalence) was constructed as a normal projective variety in arbitrary characteristic; further, if the ground field is $\mathbb{C}$ it was shown that $U(n, 0)$ gets identified with the set of isomorphism classes of unitary representations of rank n of the fundamental group of X with a suitable generalization for $d \neq 0$ (see [20] and [27]). A lot of work has been done in generalizing these results and studying the moduli spaces. I shall not attempt here to give a survey of these developments but limit myself to making a few remarks related to moduli of vector bundles.

Let X be a smooth projective curve. Let $\mathcal{O}_X(1)$ be an ample line bundle on X. Then if V is a vector bundle on X, we see that V is stable (resp. semistable) if and only if for every proper sub-bundle W of V, we have

$$\frac{h^0(W(m))}{\operatorname{rk} W} < \frac{h^0(V(m))}{\operatorname{rk} V} \quad \text{(resp. } \leq \text{) for } m \gg 0.$$

If X is now a projective variety with an ample $\mathcal{O}_X(1)$, we can define stable and semi-stable torsion free coherent sheaves in a similar manner. We call this Gieseker stability or χ-stability (resp. semi-stability) ([11]). If we now define the *degree* of a torsion free sheaf V as the highest coefficient of the Hilbert polynomial $h^0(V(m)), m \gg 0$, and imitate the definition of stability as for curves, we get the notion of μ-stability (and semi-stability).

In the early eighties the paper of Atiyah and Bott ([1]) gave a new impetus to the study of moduli spaces of vector bundles. This paper gave a differential geometric formulation of the notion of μ-stability of vector bundles in terms of Yang-Mills functionals. The approach of Atiyah-Bott and the subsequent spectacular work of Donaldson ([8]) revealed deep interconnections between the moduli spaces of vector bundles, the topology of 4-manifolds and gauge theories in mathematical physics (see [9] for a comprehensive treatment of the subject).

Recall that the moduli spaces of χ-semi-stable torsion free sheaves on X have been constructed when the characteristic of the ground field is zero, in fact, the best existence theorem is due to Simpson (see [28]) who constructs these moduli schemes on arbitrary projective schemes with a suitable notion of a "pure" coherent sheaf (again the ground field is of characteristic zero). These schemes are projective when we fix the Hilbert polynomial of the semistable torsion free sheaves. One of the long standing problems in this area was to prove this result in arbitrary characteristic. The difficulty lies in proving the *boundedness* of semi-stable torsion free sheaves with a fixed Hilbert polynomial. Recently Adrian Langer has settled this question in the affirmative (see [16]) so that all these moduli spaces exist as projective schemes in arbitrary characteristics (presumably in the same generality as Simpson's). See ([15]) for a comprehensive account of these questions and the recent developments.

There has been a lot of work on the moduli spaces of principal bundles i.e. bundles with reductive groups as structure groups. Recall that these moduli spaces have been constructed over smooth projective curves in characteristic zero by Ramanathan with a suitable definition of stability ([23]). However, generalizing his results in arbitrary characteristic has presented non-trivial difficulties. A principal reason for this is the failure of the tensor product theorem for semistable bundles over fields of positive characteristic. More precisely, let P be a principal G-bundle on X (in characteristic zero). Then, as an easy consequence of the unitary characterization of stable vector bundles, it follows that P is polystable (resp. semi-stable) in Ramanathan's sense if and only if the vector bundle V associated to a faithful representation of G is polystable (resp. semi-stable) so that the moduli spaces of principal bundles get related to that of vector bundles (this can be formulated in a more precise manner using the

Tannakian language [4]). However, this fact is not true in char p, $p > 0$. In this context the first construction of the moduli space of principal bundles as a projective variety has been given in [5], under the hypothesis that p is large with an explicit bound (called the separability index). However, the boundedness of these moduli spaces has been proved with no assumption on the characteristic [14], as well as the valuative criterion for properness ([13]). But it is not clear whether the moduli spaces can be constructed as projective varieties without some hypothesis on p, the problem being the non-emptiness of these spaces.

The construction of moduli spaces of principal G-bundles on higher dimensional projective varieties has been taken up in recent years. To have projective moduli spaces, one cannot restrict only to principal bundles (e.g. we have to take torsion free sheaves to compactify moduli spaces of vector bundles). For this purpose with the aid of a fixed representation ρ of G, concepts such as *quasi-G bundles* and their semi-stability have been given. Moduli spaces of such bundles have been constructed as projective schemes, which however depend on the choice of the representation ρ of G. A substantial survey of this development appears in this volume (Moduli Spaces for Principal Bundles in Large Characteristic by Tomás L. Gómez, Adrian Langer, Alexander H. W. Schmitt, Ignacio Sols), see also [12], [24] and [25]. In the case of a smooth projective surface an intrinsic moduli space of principal G-bundles (rather quasi G-bundles) has been constructed using a generalization of μ-stability ([2] and [3]). This gave a generalization of the Donaldson-Uhlenbeck compactification for arbitrary reductive G. This paper also proves the semistable reduction theorem for μ-semistable quasi bundles over arbitrary smooth projective varieties, generalizing the paper of Langton ([18]) to arbitrary structure groups.

An interesting problem is the construction of moduli spaces of principal G-bundles and their compactifications on a projective curve with only nodes as singularities (even in characteristic zero). This moduli space should have good specialization properties i.e. if a smooth projective curve Y specializes to X, then the moduli space on Y should specialize to the one on X. Further, the description of this moduli space should be sufficiently explicit so that, for example, one can deduce the Verlinde formula for the moduli space on Y. There has been some work and good candidates for moduli spaces (see for example [7], [26] and [29]). Equivariant compactifications of G seem to play an important role in this problem. The paper (Stable Maps into the Classifying Space of the General Linear Group by Ivan Kausz) in this volume is related to this problem.

Madhav Nori in ([22]) constructed what is called a "fundamental group scheme" of a smooth projective curve in characteristic p, $p > 0$. The motivation for this goes back to an earlier work of A. Weil ([30]). Suppose first that we work in characteristic zero. Let us call a vector bundle V to be *finite* if the element defined by V in the Grothendieck ring formed by direct sums and tensor products of vector bundles, is integral over $\mathbb{Z}$. Then $V \approx V_p$ where $\rho : \pi \longrightarrow GL(n, \mathbb{C})$ is a representation of the fundamental group such that $\operatorname{Ker} \rho$ is of finite index in π (A. Weil proved such a result purely in the group theoretic context). Now finite bundles form a Tannaka category (i.e. preserved under direct sums and tensor products) and the Tannaka group turns out to be the algebraic fundamental group of X. Now in characteristic p, $p > 0$, Nori defines the notion of *essentially finite* bundles which again form a Tannaka Category. He calls the Tannaka group of this category the *fundamental group scheme*. Now direct sums of stable bundles of degree zero (in characteristic zero) form a Tannaka category. Recently there has been interesting developments ([6]) in the study of the Tannaka group of this category and its generalisations.

The papers in this volume concerning moduli are mostly devoted to vector bundles. Besides the papers in this volume which have been mentioned above, there is a fairly complete account of the moduli spaces of vector bundles on curves in (Notes on Vector Bundles on Curves by Usha N. Bhosle and Indranil Biswas), an expository talk on stacks whose language has become indispensable in moduli problems (The moduli stack of vector bundles on a curve by Norbert Hoffmann). There is also an expository account on "variations of Hodge structures" which is a very important aspect of the subject of moduli (Rigidity, Past and Present by Chris Peters).

References

[1] M. F. Atiyah and R. Bott, The Yang-Mills equations over Riemann Surfaces, *Phil. Tr. R. Soc. Lond. A.*, **308** (1982) 523–615.

[2] V. Balaji, Principal bundles on projective varieties and the Donaldson-Uhlenbeck compactification, *J. Differential Geometry*, **76** (2007) pp. 351–398. (arXiv:math.AG/0505106).

[3] V. Balaji, *Addendum to* Principal bundles on projective varieties and the Donaldson-Uhlenbeck compactification, to appear in *J. Differential Geometry*.

[4] V. Balaji and C. S. Seshadri, Semistable Principal Bundles-I (in characteristic zero), *Journal of Algebra*, Special issue in celebration of Claudio Procesi's 60th birthday, **258** (2002) 321–347.

[5] V. Balaji and A. J. Parameswaran, Semistable principal bundles-II (in positive characteristics), *Transformation Groups*, **l8** (2003) no. 1, pp. 3–36.

[6] V. Balaji and A. J. Parameswaran, An analogue of the Narasimhan-Seshadri theorem and some applications (preprint) (2009).

[7] Usha Bhosle, Tensor fields and singular principal bundles Bhosle, *Int Math Res Notices*, (2004) 3057–3077.

[8] S. K. Donaldson, Polynomial invariants for smooth four manifolds, *Topology*, **29** (1990) 257–315.

[9] S. K. Donaldson and P. B. Kronheimer, The Geometry of Four-Manifolds, Oxford (1990).

[10] G. Faltings, Stable G-bundles and projective connections, *J. Alg. Geom.*, **2** (1993) 507–568.

[11] D. Gieseker, A construction of stable bundles on an algebraic surface, *J. Differential. Geom.*, **27** (1988) 137–154.

[12] T. Gómez and I. Sols, Moduli spaces of principal sheaves over projective varieties, *Annals of Math.*, **161** (2005) no. 2, 1037–1092.

[13] J. Heinloth, Semistable reduction for G-bundles on curves, *J. Algebraic Geom.*, **17** (2008) 167–183.

[14] Y. Holla and M. S. Narasimhan, A generalisation of Nagata's theorem on ruled surfaces, *Compositio Mathematica*, **127** (2001) 32132.

[15] D. Huybrechts and M. Lehn, The Geometry of moduli spaces of sheaves, Aspects of Mathematics E31, Vieweg, Braunschweig/Wiesbaden, (1997).

[16] A. Langer, Semistable sheaves in positive characteristic, *Annals of Math.*, **159** (2004) no. 1, 251–276.

[17] S. Ilangovan, V. B. Mehta and A. J. Parameswaran, Semistability and semisimplicity in representations of low height in positive characteristic, A Tribute to C. S. Seshadri, Birkhauser (2004).

[18] S. Langton, Valuative criterion for families of vector bundles on algebraic varieties, *Annals of Mathematics (2)* **101** (1975) pp. 88–110.

[19] D. Mumford, J. Fogarty and F. Kirwan, Geometric Invariant theory, Ergebnisse 34, Springer 3rd Edition.

[20] M. S. Narasimhan and C. S. Seshadri, Stable and unitary vector bundles on a compact Riemann surface, *Annals of Mathematics (2)*, **82** (1965) pp. 540–567.

[21] P. Newstead, Introduction to moduli problems, TIFR Lecture Notes in Mathematics, (1978).

[22] M. V. Nori, The fundamental group scheme, *Proc. Ind. Acad. Sci (Math. Sci.)*, **91** (1982) 73–122.

[23] A. Ramanathan, Stable principal bundles on a compact Riemann surface – Construction of moduli space (Thesis, Bombay University 1976), *Proc. Ind. Acad. Sci.*, **106** (1996) 301–328 and 421–449.

[24] A. Schmitt, Singular Principal Bundles over higher dimensional manifolds and their moduli spaces, *Int Math Res Notices*, **23** (2002) 1183–1209.

[25] A. Schmitt, A closer look at semistability for singular principal bundles, *Int. Math. Res. Notices*, **62** (2004) 3327–3366.

[26] A. Schmitt, Moduli spaces for semistable honest singular principal bundles on a nodal curve which are compatible with degenerationa remark on Bhosle's paper tensor fields and singular principal bundles, *Int Math Res Notices*, (2005) 1427–1437.

[27] C. S. Seshadri, *Fibrés vectoriels sur un courbes algébriques*, Asterisque., **96** (1982).

[28] C. Simpson, Moduli of representations of the fundamental group of a smooth projective variety, I, *Inst. Hautes Etudes Sci. Publ. Math.*, **79** (1994) 47–129.

[29] C. S. Seshadri, Moduli spaces of torsion free sheaves on nodal curves and generalisations – I, (to appear in the volume dedicated to P. E. Newstead).

[30] A. Weil, Généralisation des fonctiones abeliennes, *J. Math. pures. appl.*, **17** (1938) 47–87.

Teichmüller Theory and Moduli Problem No. 10, 2010, pp. 41–51.

On the Divergence of Certain Geodesic Rays in Teichmüller Space

Abdelhadi Belkhirat

The American University of the Middle East, P. O. Box. 220 Dasman, 15453 Kuwait
e-mail: hbelkhirat@yahoo.fr; Abdelhadi.Belkhirat@aum.edu.kw

Abstract. We prove the divergence of geodesic rays with respect to the Teichmüller metric in certain direction compared with stretch lines which are geodesics with respect to Thurston's weak metric in the same direction.

1. Introduction

Let (X, d) be a metric space. A geodesic segment joining the points x and y (resp. geodesic ray emanating from x) is an isometry $r : [a, b] \subset \mathbb{R} \longrightarrow X$ (resp. $r : [a, +\infty[\longrightarrow X$) such that $r(0) = x, r(1) = y$ (resp. $r(0) = x$.) The metric space X is called a geodesic metric space if any two distinct points can be joined by a geodesic segment. We know that the Teichmüller space of closed surface or more generally of surface of finite anayltical type, is a geodesic metric space. Two geodesic rays are called divergent if the distance between them does not remain bounded above by any positive constant. More precisely, let $r, r' : [0, +\infty[\longrightarrow X$ be two geodesic rays in X. We say that r and r' are divergent geodesic rays iff

$$d(r(K), r'(K)) \longrightarrow +\infty, \quad \text{whenever} \quad K \longrightarrow +\infty.$$

The main result of this paper is to prove a result about the divergence of geodesic rays with respect to two different distances (Teichmüller metric and Thurston's weak metric) which are defined in the same direction.

1.1 *Teichmüller space, metric, maps*

Let S be closed C^∞ surface of genus $\xi \geq 2$; everthing in this note extends to punctured surfaces with only additionnal notation, so we concentrate on the closed surface case. We recall that points in the Teichmüller space $\mathbb{T}(S)$ are equivalence classes of Riemann surface structures M on S, the structure M_1 is equivalent to the structure M_2 if there is a homeomorphism $f : S \longrightarrow S$, homotopic to the identity, which is conformal map of the structures M_1 and M_2. We define the Teichmüller distance $d(\{M_1\}, \{M_2\})$ by

$$d(\{M_1\}, \{M_2\}) = \frac{1}{2} \log \inf_f K[f],$$

where $f : M_1 \longrightarrow M_2$ is a quasiconformal homeomorphism homotopic to the identity on S and $K[f]$ is the maximal dilatation of f. This metric is well defined, so we may unambiguously write M_1 for $\{M_1\}$. For a good general introduction to the Teichmüller theory, the reader should consult the excellent books by Gardiner [7], Gardiner and Lakic [8], Lehto [10], Abikoff [1] and the beautiful paper by Lipman Bers [4].

1.2 *Modulus, extremal length*

The *modulus* of a flat cylinder C of circumference l and height h is

$$\mathrm{Mod}(C) = h/l.$$

For a simple closed curve $\gamma \subset S$, we define the modulus $\mathrm{Mod}_S(\gamma)$ of γ to be the supremum of the moduli of all cylinders embedded in S with core curve isotopic to γ. This definition was called "geometric definition" by Kerckhoff [9].

The *extremal length* ext_{M_0} of a curve γ on a surface M_0 is defined (see Ahlfors [2], for more details) to be

$$\sup(l_\rho([\gamma]))^2/\mathrm{A}_\rho,$$

where ρ ranges over all conformal metrics on M_0 with area A_ρ satisfying $0 < \mathrm{A}_\rho < \infty$, and where $l_\rho([\gamma])$ denotes the infimum of lengths of simple closed curves homotopic to γ. One can show that

$$\mathrm{ext}_{M_0}(\gamma) = 1/\,\mathrm{Mod}_{M_0}(\gamma).$$

Kerckhoff [9] gave a characterization of the Teichmüller metric d in terms of the extremal length of corresponding curves on the surfaces M_1 and M_2. He proved

$$d(M_1, M_2) = \frac{1}{2}\log\left(\sup_{\gamma\in\mathcal{S}} \frac{\mathrm{ext}_{M_2}(\gamma)}{\mathrm{ext}_{M_1}(\gamma)}\right), \tag{1}$$

where $\mathcal{S}$ is the set of isotopic classes of non-trivial simple closed curves non isotopic to a puncture in the surface S.

Any two distinct points M_1 and M_2 in $\mathbb{T}(S)$ are joined by a unique geodesic segment (i.e. an isometric image of a Euclidean interval), $[M_1, M_2]$, and lie in a unique geodesic line (i.e. an isometric image of $\mathbb{R}$), $r(M_1, M_2)$. Now, fix a conformal structure M on S and let $QD(M)$ be the space of holomorphic quadratic differentials on M. The geodesic rays (i.e. isometric images of $[0, \infty)$) which emanate from the point M in $\mathbb{T}(S)$ are described in terms of $QD(M)$. If q is a holmorphic quadratic differential on M, p is point on M and z is local parameter on M defined in a neighborhood U of p, then q may be written in the form $\varphi(z)dz^2$ for some holomorphic function φ in U. If $\varphi(p) \neq 0$ and $z_0 = z(p)$, then on a sufficiently small neighborhood V of p contained in U, we may define a branch $\sqrt{\varphi(z)}$ of the square root of φ. The integral $w = \int_{z_0}^{z} \sqrt{\varphi(z)}dz$ is conformal function of z and determines a local parameter for M on a sufficently small neighborhood W of p in V. This parameter is called *natural parameter* for q at the regular point p. In terms of this parameter w, q may be written in the form dw^2. For each nonzero quadratic differential q on M, there is one-parameter family M_K of conformal structures on S and quadratic differentials $\{q_K\}$ obtained by replacing the natural parameters w for q on M by natural parameters w_K for q_K on M_K. The relationship between w_K and w is given by the rule:

$$\text{Re } w_K = K^{1/2}\text{Re } w \qquad \text{and} \qquad \text{Im } w_K = K^{-1/2}\text{Im } w,$$

The Teichmüller distance from M_K to M is equal to $\frac{1}{2}\log K$. The map $t \mapsto M_{e^{2t}}$ is the Teichmüller geodesic ray emanating from M in the direction of the quadratic differential q. Two quadratic differentials determine the same direction of a geodesic ray emanating from M if and only if they are positive multiples of one another.

We shall be particularly interested in the Jenkins-Strebel differentials. These are the quadratic differentials all of whose noncritical horizontal trajectories are closed.

Jenkins [6] and Strebel [15] proved the existence of quadratic differentials $q \in QD(M)$ with some prescribed trajectory topology. Specifically, they showed that one could specify m disjoint simple closed loops $\gamma_1, \ldots, \gamma_m$, with $1 \leq m \leq 3\xi - 3$, on M representing distinct non-trivial free homotopy classes, with m positive numbers $N_1, \ldots, N_m$, and then one could find a unique (up to scalar multiple) quadratic differential $q(\gamma_1, \ldots, \gamma_m)dz^2 \in QD(M)$ with the following property: if M' is the result of removing the critical trajectory of $q(\gamma_1, \ldots, \gamma_m)dz^2$ from M, then M' is the union of annuli $A_1, \ldots, A_m$ with A_j homotopically equivalent to γ_j and the moduli of A_j was N_j, up to some fixed (independent of j) scalar multiple. Further $M - M'$ is the union of a finite number of analytical arcs, the smooth pieces of the critical trajectories. Note that two Jenkins-Strebel differentials on M determine the same Teichmüller geodesic in $\mathbb{T}(S)$ emanating from M if and only if the horizontal foliations of these Jenkins-Strebel differentials are projectively equivalent.

Following Masur [12], we define a *Strebel ray* in $\mathbb{T}(S)$ emanating from M to be a Teichmüller geodesic ray in the direction of a Jenkins-Strebel differential on M.

By assuming that $1 \leq p < 3\xi - 3$, we may choose a simple and nontrivial closed curve in S disjoint from the closed curves $\alpha_1, \ldots, \alpha_p$.

Proposition 1.1. *The modulus (resp. extremal length) of the closed simple curve β along the Teichmüller geodesic ray $(h_K)_{K\geq 1}$ is bounded below (resp. above) by a constant independent of the parameter K.*

Proof. We know from [12], that we can represent the Teichmüller map between h_0 and h_K as a radial expansion of the annuli defined by the horizontal foliation of the Jenkins-Strebel $F(\alpha_1, \ldots, \alpha_p)$. Thus we can build a model which is conformal to h_K by an operation of conformal plumbing. We cut the conformal cylinders along the core curves $\alpha_1, \ldots, \alpha_p$. We then glue in one cylinder each edge of the cut, again leaving a pair of boundary components. Finally, we glue these free edges together without twisting.

As the closed curve β is disjoint from the curves $\alpha_1, \ldots, \alpha_p$, then its homotopy class is represented by a union of geodesic segments on the boundary of the annuli. Therefore, we can find an annulus embedded around β, and also disjoint from the core curves along which our initial cut is made. That annulus will be unaffected by the plumbing, and we can conclude that for all h_K we can find an embedded annulus about β of modulus bounded uniformly away from zero, independently of K.

Thus the extremal length of β is then uniformly bounded above, independently of K, concluding the proof of the proposition. □

Lemma 1.1. *The hyperbolic length* $l_{h_K}(\beta)$ *of* β *along the geodesic ray* $(h_K)_{K\geq 1}$ *is bounded above by a constant indepedent of the parameter* K.

Proof. The proof is a direct consequence of the proposition and Maskit's comparison between hyperbolic and extremal lengths see ([11]). □

1.3 *Thurston's nonsymmetric metric*

We shall adopt here this definition of *weak metric*, in the sense of Ribeiro [14].

Definition 1.1. *A weak metric* $\mathcal{K}$ *on a set* X *is a map* $\mathcal{K} : X \times X \longrightarrow \mathbb{R}$ *such that for all* x, y *and* $z \in X$, *we have*

1. $\mathcal{K}(x, x) = 0$;
2. $\mathcal{K}(x, y) \geq 0$;
3. $\mathcal{K}(x, z) \leq \mathcal{K}(x, y) + \mathcal{K}(y, z)$.

According to the uniformization theorem, we can also see the Teichmüller space as the set $\mathcal{H}(S)$ of hyperbolic metrics on S modulo the action of the group $\mathrm{Diff}_0(S)$ of diffeomorphisms of S isotopic to the identity by taking the pull back of the metric. Thus, we can write

$$\mathbb{T}(S) := \mathrm{Diff}_0(S)/\mathcal{H}(S)$$

In [16], Thurston introduced and studied an alternative distance between two hyperbolic metrics g and h, which is based on the comparison of the length of closed curves with respect to the hyperbolic metrics g and h. More precisely, for every isotopy class γ of closed curves, we denote by $l_g(\gamma)$ the infimum of the length of closed curves $a : [0, 1] \rightarrow S$ in the isotopy class γ computed in the metric g:

$$l_g(\gamma) = \inf_{\gamma} \left\{ \int_0^1 \sqrt{g(\dot{a}, \dot{a})} | a \in \gamma \right\}$$

Thurston defined the distance $\mathcal{K}$ by setting

$$\mathcal{K}(g, h) = \log\left(\sup_{\gamma \in \mathcal{S}} \frac{l_h(\gamma)}{l_g(\gamma)} \right), \tag{2}$$

This distance is a weak metric. Furthermore, this weak metric separates points in $\mathbb{T}(S)$, that is, it satisfies

$$\mathcal{K}(g, h) = 0 \Longleftrightarrow g = h.$$

However, it is not a metric in the usual sense since it is not symmetric.

Remark. In the case of the flat torus, we can define, after a suitable normalization, a similar weak metric, but we can prove (see [3]) that this weak metric does not separate points.

Figure 1.

1.4 *A parametrization of Teichmüller space (following Thurston), Stretch lines*

This section contains some terminology and a construction made by Thurston which provides the global parameters of Teichmüller space.

For every maximal geodesic lamination μ on the hyperbolic surface S (see [17] for definitions and further developments,) Thurston has constructed in [16] a measured foliation, denoted by $F_\mu(h)$, and which we shall call the horocyclic foliation (associated to μ and to the hyperbolic structure h on S). The construction is as follows:

By assumption, the complementary components of μ are all isometric to ideal triangles. In each of these components, we define a partial foliation (i.e, a foliation whose support is a subsurface), whose leaves are obtained by intersecting the ideal triangle with horocycles centered at the ideal vertices. These horocycles meet the edges of the triangle at right angles. The partial foliation is chosen in such a way that the non-foliated region is equal to a little triangle whose edges are subarcs of horocycles which meet tangentially at their endpoints. That this foliation exists and is uniquely defined can be seen by making the construction in a particular triangle in the upper half plane (or the unit disk) like the one represented in Fig. 1, and using the fact that any two ideal triangles are isometric.

These partial foliations fit together on the surface and define a partial foliation on this surface, which has a well-defined invariant transverse measure, which is uniquely specified by the fact that on the leaves of μ, this transverse measure coincides with the hyperbolic distance. We obtain in this way a measured foliation, $F_\mu(h)$, which has a well-defined class in the space of measured foliations on S up to isotopy and Whitehead moves $\mathcal{MF}(S)$ (see [5] for properties of this space). Note that by construction, this class has the property that it can be represented by a measured foliation that is transverse to μ. Conversely, Thurston proves in [16] that the elements of $\mathcal{MF}(S)$ which possess this last property are classes of horocyclic foliations which arise as above, for the same μ and for some hyperbolic metric. More precisely, let $\mathcal{MF}(\mu)$ denote the subset of $\mathcal{MF}(S)$ which consists of equivalence classes which admit representatives transverse to μ. We have the following theorem.

Theorem 1.1 (Thurston [16]). *For any maximal geodesic lamination μ, the map φ_μ which associates to each hyperbolic metric the equivalence class of its horocyclic foliation is a homeomorphism from the Teichmüller space to the subset $\mathcal{MF}(\mu)$ of $\mathcal{MF}(S)$.*

1.5 *Geometric intersection functions*

We shall also make use of the following few facts about the geometric intersection function, which are easy to prove. First, we make a definition.

Let F_i, for $i = 1$ and 2 be either a measured foliation, a partial measured foliation (i.e. a measured foliation supported on a subsurface of S, like for example the horocyclic foliation associated to a maximal geodesic lamination) or a measured lamination on the surface S, and suppose that F_1 is transverse to F_2 (transverse at each point where they intersect; we do not suppose that their supports are the same).

Definition 1.2. *We define the quantity $I(F_1, F_2)$ as the total mass on the surface of the product measure $dx_1 \times dx_2$, where for $i = 1$ and 2, dx_i denotes the transverse measure of F_i, i.e.*

$$I(F_1, F_2) = \int_S dx_1 \times dx_2$$

We can define also the geometric intersection function which is the extension of the geometric intersection function defined on the simple closed curves as follows:

Definition 1.3. *Let $[F_1]$, $[F_2]$ be two equivalence classes of measured foliation with full support, we define the geometric intersection number by the formula:*

$$i([F_1], [F_2]) = \inf\{I(F_1', F_2') | F_1' \in [F_1], F_2' \in [F_2]\}$$

In [13] Papadopoulos proved the following theorem, relating these two functions

Theorem 1.2. *Suppose that there is no Whitney disk for the couple (F_1, F_2), that is, a disk on the surface whose boundary is the union of a segement in F_1 and a segment in F_2. Then, we have:*

$$I(F_1, F_2) = i([F_1], [F_2]).$$

Remark. The function i defined above is the extension of the classical geometric intersection function defined on the set $\mathcal{S}$ of isotopic classes of non-trivial simple closed curves non isotopic to a puncture in the surface S.

1.6 *Stretch lines*

To define these lines, recall that the set $\mathcal{MF}(\mu)$ of equivalence classes of measured foliation transverse to μ has a natural cone structure, and define a positive ray in that space to be a one-parameter family of measured foliation classes of the form $(k \cdot F)_{k \geq 0}$, where F is an element of $\mathcal{MF}(\mu)$, and where this ray is equipped with the positive orientation indexed by the real numbers. The image of such ray by the map φ_μ, is called *stretch line* in the direction of the measured foliation F in Teichmüller space, and is a geodesic for Thurston's weak metric. Note that it is important to specify the orientation in the definition of stretch line, because this line equipped with the opposite orientation is not a stretch line, that it is not a geodesic.

1.6.1 *Jenkins-Strebel stretch lines*

We define by analogy, according to Masur [12], the Jenkins-Strebel Stretch lines.

Definition 1.4. *Let $(h_K)_{K \geq 1} = \varphi^{-1}(K \cdot F(\alpha_1, \dots, \alpha_p))$ be a Stretch line in the direction of the Jenkins-Strebel measured foliation $F(\alpha_1, \dots .\alpha_p)$. Then the stretch line $(h_K)_{K \geq 1}$ is called Jenkins-Strebel stretch line.*

Remark. As the geodesic lamination μ is complete, then it is transverse to any Jenkins-Strebel measured foliation $F(\alpha_1, \dots, \alpha_p)$, for $1 \leq p < 3\xi - 3$ where ξ is the genus of the surface S.

Proposition 1.2. *The hyperbolic length $l_{h'_K}(\beta)$ of the closed simple curve β, along the Jenkins-Strebel stretch line $(h'_K)_{K \geq 1}$ goes to infinity when the parameter K goes to infinity.*

Proof. The proof follows closely the proof of the fundamental lemma done by Papadopoulos [13]. Given the hyperbolic structure (h'_K) on the surface S, let β^* denote the closed geodesic reprensentig the class β, (β^* depends on the hyperbolic metric h'_K, but to save notations, we do not put an index to β). The closed geodesic curve β^* is transverse to μ since μ has no closed leaves.

Define N to be the subset of the surface S equal to the complement of the lamination μ in the support of $(K \cdot F(\alpha_1, \dots, \alpha_p))$. This subset is equal to the union of foliated parts of ideal triangles (where the boundary is not included).

Consider a connected component of the intersection of β^* with N and let k denote its closure, k is a segment which can be one of the following 3 types

- **Type 1:** the two endpoints of k are on μ;
- **Type 2:** there is exactly one endpoint of k that is on μ;
- **Type 3:** no endpoint of k is on μ.

By elementary hyperbolic geometry, the segment k can have at most one point of tangency with the foliation $F_{h'_K}(\mu)$. We can therefore modify k in the following manner:

We replace k by a segment k' having the same endpoints as k, and obtained by pushing k in a direction transverse to the leaves of the foliation, so that the resulting segment k' is transverse to the leaves.

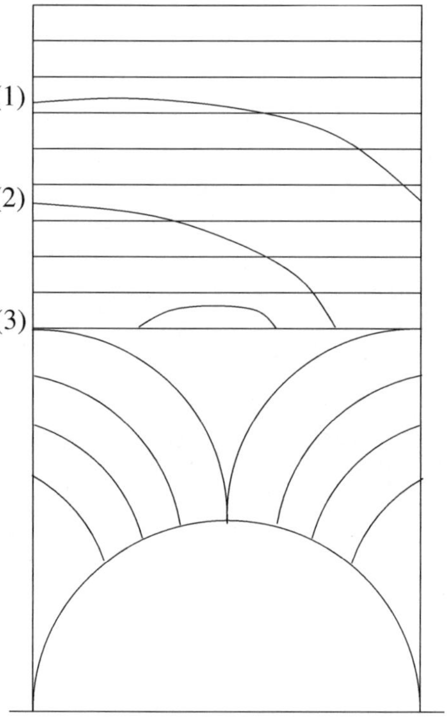

Figure 2.

The transverse measure $I(F_{h'_K}(\mu), k')$ of the segment k' is equal to the hyperbolic length of the projection (which is contained on one side of the ideal triangle) along the leaves of horocyclic foliation. Again, by hyperbolic geometry, the length of the projection is not greater than of k. Therefore, we have:

$$I(F_{h'_K}(\mu), k') \leq l_{h'_K}(k) \tag{3}$$

Let β^{**} denote the closed curve obtained out of β^* by applying the above modification to each connected component of the intersection of β^* with N_0. By (3), we have $I(F_{h'_K}(\mu), \beta^{**}) \leq l_{h'_K}(\beta)$. Therefore, $I(F_{h'_K}(\mu), \beta) \leq l_{h'_K}(\beta)$, but along a stretch line the transverse measure of $F_{h_K}(\mu)$ is multiplied by K times the transverse measure of $F_h(\mu)$, which means that:

$$I(F_{h'_K}(\mu), \beta) = K \cdot I(F_{h'}(\mu), \beta)$$

When the parameter K goes to infinity $I(F_{h'_K}(\mu), \beta)$ goes to infinity, because $I(F_{h'}(\mu), \beta) \neq 0$ and according to the inequality (3) we may conclude the proposition. □

Remark. Such curves β exist because we have assumed that $1 \leq p < 3\xi - 3$.

Lemma 1.2. *The extremal length of β with resepect to the conformal structure h'_K goes to infinity when the parameter K goes also to infinity.*

Proof. According to Maskit's inequality (see [11]), we have

$$l_{h'_K}(\beta) \leq \pi \operatorname{ext}_{h'_K}(\beta)$$

Thus, the lemma follows directly from this inequality and the result of the last proposition. □

2. Divergence of geodesic rays

In this last section we prove the main theorem of this paper. We fix a maximal measured geodesic lamination μ, and let $F = F(\alpha_1, \ldots, \alpha_p)$, where $1 \leq p < 3\xi - 3$ be a the horizontal foliation associated with the Jenkins-Strebel quadratic differential $q(\alpha_1, \ldots, \alpha_p)$. As F is always transverse to μ (F has closed trajectories and μ is complete geodesic lamination), we denote by $F_\mu(h') = \varphi_\mu(h')$, the horocyclic foliation in the same class of F, then we can take the inverse image of the ray $(K \cdot F_\mu(h'))_{K\geq 1}$ in $\mathcal{MF}(\mu)$ by Thurston's homeomorphism φ_μ. We obtain the stretch line $(h'_K)_{K\geq 1}$ emanating from the point h'.

Let $(h_K)_{K\geq 1}$ be the geodesic ray with respect to the Teichmüller metric d emanating from a point h, in the direction of the Jenkins-Strebel differential $q(\alpha_1, \ldots, \alpha_p)$, where $1 \leq p < 3\xi - 3$.

Theorem 2.1. *We consider the two geodesic rays $(h_K)_{K\geq 1}$, $(h'_K)_{K\geq 1}$ for the Teichmüller metric and for Thurston weak metric respectively. Then, these two geodesic rays are divergent either with respect to the Teichmüller metric d or with respect to the Thurston's weak metric $\mathcal{K}$.*

Proof. We distinguish two cases, up to the metric used to compare between the geodesic rays.

With respect to Teichmüller's metric

According to the proposition (1.1), the extremal length of β is bounded above by a constant c_1 independent of the parameter K along the ray $(h_K)_{K\geq 1}$, but the extremal length β goes to infinity along the stretch line $(h'_K)_{K\geq 1}$ when K tends to infinity. By Kerckhoff's characterization of the Teichmüller metric, we have:

$$d(h_K, h'_K) = \frac{1}{2}\log\left(\sup_{\gamma\in\mathcal{S}} \frac{\operatorname{ext}_{h'_K}(\gamma)}{\operatorname{ext}_{h_K}(\gamma)}\right),$$

thus

$$d(h_K, h'_K) \geq \frac{1}{2}\log\left(\frac{\operatorname{ext}_{h'_K}(\beta)}{\operatorname{ext}_{h_K}(\beta)}\right).$$

$$d(h_K, h'_K) \geq \frac{1}{2}\log\left(\frac{\operatorname{ext}_{h'_K}(\beta)}{\operatorname{ext}_{h_K}(\beta)}\right) \geq \frac{1}{2}\log\left(\frac{\operatorname{ext}_{h'_K}(\beta)}{c_1}\right).$$

Therefore the distance $d(h_K, h'_K)$ goes to infinity. □

With respect to Thurston's weak metric

According to the proposition (1.2), the hyperbolic length of β tends to infinity when K goes to infinity along the stretch line $(h'_K)_{K\geq 1}$ and by the lemma (1.1) the hyperbolic length β is bounded above by a constant c_2 independent of K along $(h_K)_{K\geq 1}$, thus

$$\mathcal{K}(h_K, h'_K) = \log\left(\sup_{\gamma\in\mathcal{S}} \frac{l_{h'_K}(\gamma)}{l_{h_K}(\gamma)}\right),$$

$$\mathcal{K}(h_K, h'_K) = \log\left(\sup_{\gamma\in\mathcal{S}} \frac{l_{h'_K}(\gamma)}{l_{h_K}(\gamma)}\right) \geq \log\left(\frac{l_{h'_K}(\beta)}{l_{h_K}(\beta)}\right),$$

therefore

$$\mathcal{K}(h_K, h'_K) \geq \log\left(\frac{l_{h'_K}(\beta)}{l_{h_K}(\beta)}\right) \geq \log\left(\frac{l_{h'_K}(\beta)}{c_2}\right).$$

The distance $\mathcal{K}(h_K, h'_K)$ goes to infinity when K goes to infinity.

The author would like to thank the organizers of the workshop for their kind invitation and their financial support.

References

[1] W. Abikoff, The Real-Analytic Theory of Teichmüller space, *Lecture notes in mathematics*, **820** (1980) Springer-Verlag.

[2] L. V. Ahlfors, Conformal invariants, Topics in Geometric Theory, MacGraw-Hill Book company (1973).

[3] A. Belkhirat, A. Papadopoulos and M. Troyanov, Thurston weak metric on the Teichmüller space of the torus, *Trans. Amer. Math. Soc.*, **357** (2005) 3311–3324.

[4] L. Bers, Quasiconformal mappings and Teichmüller's theorem, Analytic functions, Princeton University Press (1960).

[5] A. Fathi, F. Laudenbach and V. Poénaru V, Travaux de Thurston sur les surfaces, Astérique, 66–67 (1979).

[6] J. A. Jenkins, On the existence of certain extremal metrics, *Ann. of. Math.*, **66** (1957) no. 3, 440–453.

[7] F. Gardiner, Teichmüller Theory and Quadratic Differentials, A Wiley-Interscience Publication, (1987).

[8] F. Gardiner and N. Lakic, Quasiconformal Teichmüller Theory, American Mathematical Society, (2000).

[9] S. Kerckhoff, The asymptotic geometry of Teichmüller space, *Topology*, **19** (1980) 23–41.

[10] O. Lehto, Univalent Functions and Teichmüller space, Springer-Verlag, **109** (1987).

[11] A. Maskit, Comparison of hyperbolic and extremal lengths, *Ann. Acad. Sci. Fenn. Ser. A I Math.*, **10** (1985) 381–386.

[12] H. Masur, On a class of geodesics in Teichmüller space, *Annals of Mathematics*, **102** (1975) 205–221.

[13] A. Papadopoulos, On Thurston's boundary of the Teichmüller space and the extention of earthquakes, *Topology Appl.*, **41** (1991) 147–177.

[14] H. Ribeiro, Sur les espaces à métrique faible, *Portugaliae Math.*, **4** (1943) 21–40.

[15] K. Strebel, Quadratic Differentials, Springer-Verlag, Berlin (1984).

[16] W. P. Thurston, Minimal Stretch maps between hyperbolic surfaces, Preprint (1985) available at http://arxiv.org/abs/math.GT/9801039.
[17] W. P. Thurston, The geometry and topology of 3-manifolds, Princeton Academic Press (1979), available at: http://www.msri.org/publications/books/gt3m/.
[18] S. Wolpert, The length spectra as moduli for compact hyperbolic surfaces, *Ann. of. Math.*, **109** (1979) 323–351.

Teichmüller Theory and Moduli Problem No. 10, 2010, pp. 53–60.

On the Non-Quasi-Isometry of the Teichmüller Metric and Thurston's Weak Metric

Abdelhadi Belkhirat

The American University of the Middle East, P. O. Box. 220, Dasman 15453, Kuwait
e-mail: hbelkhirat@yahoo.fr; Abdelhadi.Belkhirat@aum.edu.kw

Abstract. We prove that the identity map of the Teichmüller space of a punctured torus, where the domain space is equipped with Teichmüller's metric and the range space is equipped with Thurston's weak metric, is not a quasi-isometry.

1. Introduction

1.1 *Teichmüller space, metric*

Let S be a closed orientable surface of genus g with b distinguished points, regarded as punctures. We are interested in conformal structures on S minus punctures that can be extended across the punctures, i.e., in the conformal structures of finite type on S minus punctures. Let $\mathbb{T}(S)$ be the Teichmüller space of such conformal structures. Two conformal structures on S define the same point in $\mathbb{T}(S)$ if one can be transformed to the other by a conformal homeomorphism $S \longrightarrow S$ fixing the distinguished points and isotopic to the identity.
The Teichmüller metric d between the points S_1 and S_2 is defined by

$$d(S_1, S_2) = \frac{1}{2} \log \inf K(f)$$

where $f : S_1 \longrightarrow S_2$ is quasiconformal homeomorphism isotopic to the identity on S, and $K(f)$ is the maximal dilatation of f.

1.2 *Modulus, extremal length*

The *modulus* of a flat cylinder C of circumference l and height h is

$$\mathrm{Mod}(C) = h/l.$$

For a simple closed curve $\gamma \subset S$, we define the modulus $\mathrm{Mod}_S(\gamma)$ of γ to be the supremum of the moduli of all cylinders embedded in S with core curve isotopic to γ. This definition was called "geometric definition" by Kerckhoff [3].

The *extremal length* ext_{S_0} of a curve γ on a surface S_0 is defined to be

$$\sup(l_\rho([\gamma]))^2/\mathrm{A}_\rho,$$

where ρ ranges over all conformal metrics on S_0 with area A_ρ satisfying $0 < A_\rho < \infty$, and where $l_\rho([\gamma])$ denotes the infimum of lengths of simple closed curves homotopic to γ. One can show that

$$\mathrm{ext}_{S_0}(\gamma) = 1/\mathrm{Mod}_{S_0}(\gamma).$$

Kerckhoff [3] gave a characterization of the Teichmüller metric d in terms of the extremal length of corresponding curves on the surfaces S_1 and S_2. He proved

$$d(S_1, S_2) = \frac{1}{2}\log\left(\sup_{\gamma \in \mathcal{S}} \frac{\mathrm{ext}_{S_2}(\gamma)}{\mathrm{ext}_{S_1}(\gamma)}\right), \tag{1}$$

where $\mathcal{S}$ is the set of isotopic classes of non-trivial simple closed curves non isotopic to a puncture in the surface S.

1.3 *Thurston's weak metric*

We shall adopt here this definition of *weak metric*, in the sense of Ribeiro [5].

Definition 1.1. *A weak metric $\mathcal{K}$ on a set X is a map $\mathcal{K} : X \times X \longrightarrow \mathbb{R}$ such that for all x, y and $z \in X$, we have*

1. $\mathcal{K}(x, x) = 0$;
2. $\mathcal{K}(x, y) \geq 0$;
3. $\mathcal{K}(x, z) \leq \mathcal{K}(x, y) + \mathcal{K}(y, z)$.

According to the uniformization theorem, we can also see the Teichmüller space as the set $\mathcal{H}(S)$ of hyperbolic metrics on S modulo the action of the group $\mathrm{Diff}_0(S)$ of diffeomorphisms of S isotopic to the identity by taking the pull back of the metric. Thus, we can write

$$\mathbb{T}(S) := \mathrm{Diff}_0(S)/\mathcal{H}(S)$$

In [6], Thurston introduced and studied an alternative distance between two hyperbolic metrics g and h, which is based on the comparison of the length of closed curves with respect to the hyperbolic metrics g and h. More precisely, for every isotopy class γ of closed curves, we denote by $l_g(\gamma)$ the infimum of the length of closed curves $a : [0, 1] \to S$ in the isotopy class γ computed in the metric g:

$$l_g(\gamma) = \inf_{\gamma}\left\{\int_0^1 \sqrt{g(\dot{a}, \dot{a})} \mid a \in \gamma\right\}$$

Thurston defined the distance $\mathcal{K}$ by setting

$$\mathcal{K}(g, h) = \log\left(\sup_{\gamma \in \mathcal{S}} \frac{l_h(\gamma)}{l_g(\gamma)}\right), \tag{2}$$

This distance is a weak metric. Furthermore, this weak metric separates points in $\mathbb{T}(S)$, that is, it satisfies

$$\mathcal{K}(g, h) = 0 \Longleftrightarrow g = h.$$

However, it is not a metric in the usual sense since it is not symmetric.

Remark. In the case of the flat torus, we can define, after a suitable normalization, a similar weak metric, but we can prove (see [1]) that this weak metric does not separate points.

Definition 1.2. *Two weak metric spaces* (X, d) *and* $(X', \mathcal{K})$ *are* quasi-isometric *if there exists a map* $f : X \longrightarrow X'$ *and two constants* $c \geq 0$ *and* $k \geq 1$ *such that for all* x, y *in* X, *we have*

$$\frac{1}{k}\mathcal{K}(f(x), f(y)) - c \leq d(x, y) \leq k\mathcal{K}(f(x), f(y)) + c \tag{3}$$

and for all x' *in* X', *there exists* x *in* X *such that*

$$\mathcal{K}(x', f(x)) \leq c.$$

Such a map is then called a quasi-isometry.

2. Maskit's estimates

Maskit [4] has compared the extremal and hyperbolic lengths of curves on any surface S, whose Euler-Poincaré characteristic $\chi(S)$ is negative

Theorem 2.3 (Maskit). *Let* M *be a conformal structure defined on* S, *and* h *be the unique hyperbolic structure on* M. *Then*

$$l_h(\gamma) \leq \pi \operatorname{ext}_M(\gamma) \tag{4}$$

$$\operatorname{ext}_M(\gamma) \leq \frac{1}{2} l_h(\gamma) \exp\left(\frac{l_h(\gamma)}{2}\right) \tag{5}$$

2.1 *Maskit's example*

In [4], Maskit has given an example on a punctured torus $S_{1,1}$ showing that the estimate (5) is not very far from being sharp.

For each α, $0 < \alpha < \frac{\pi}{2}$, he considers a Fuchsian group G_α generated by

$$A_\alpha = \begin{pmatrix} \csc\alpha & \cot\alpha \\ \cot\alpha & \csc\alpha \end{pmatrix} \quad \text{and} \quad B_\alpha = \begin{pmatrix} \sec\alpha & i\tan\alpha \\ -i\tan\alpha & \sec\alpha \end{pmatrix}$$

and he has concluded that for any $\varepsilon > 0$, we can find an $\alpha \in]0, \frac{\pi}{2}[$ so that

$$\operatorname{ext}_{h_\alpha}(\pi(A_\alpha)) \geq \left(\frac{1}{2} - \varepsilon\right) \exp\left(\frac{l_{h_\alpha}(\pi(A_\alpha))}{2}\right) \tag{6}$$

where $\pi : \mathbb{D}^2 \longrightarrow \mathbb{D}^2/G_\alpha$ is the canonical projection.

Remark. To simplify the notations, $l_{h_\alpha}(A_\alpha)$ (resp. $\operatorname{ext}_{h_\alpha}(A_\alpha)$) denotes the length measured with the metric h_α (resp. the extremal length) of the projection of the axis of A_α.

We shall prove the following theorem

Theorem 2.2. *The application*

$$\mathrm{id} : (\mathbb{T}(S_{1,1}), d) \longrightarrow (\mathbb{T}(S_{1,1}), \mathcal{K}),$$

is not a quasi-isometry.

Remark. In [1], we prove that the Teichmüller space of the flat torus with the Teichmüller metric $(\mathbb{T}(T^2), d)$ is not quasi-isometric to the Teichmüller space with Thurston's weak metric $(\mathbb{T}(T^2), \mathcal{K})$, a result which is stronger than the result of Theorem 2.2, where we consider only the identity map.

3. Proof of theorem 2.2

3.1 *A parametrization of Teichmüller space (following Thurston)*

This section contains some terminology and a construction made by Thurston which provides the global parameters of Teichmüller space.

For every maximal geodesic lamination μ on the hyperbolic surface S (see [7] for definitions and further developments,) Thurston has constructed in [6] a measured foliation, denoted by $F_\mu(h)$, and which we shall call the horocyclic foliation (associated to μ and to the hyperbolic structure h on S). The construction is as follows:

By assumption, the complementary components of μ are all isometric to ideal triangles. In each of these components, we define a partial foliation (i.e, a foliation whose support is a subsurface), whose leaves are obtained by intersecting the ideal triangle with horocycles centered at the ideal vertices. These horocycles meet the edges of the triangle at right angles. The partial foliation is chosen in such a way that the nonfoliated region is equal to a little triangle whose edges are subarcs of horocycles which meet tangentially at their endpoints. That this foliation exists and is uniquely defined can be seen by making the construction in a particular triangle in the upper half plane (or the unit disk) like the one represented in Fig. 1, and using the fact that any two ideal triangles are isometric.

These partial foliations fit together on the surface and define a partial foliation on this surface, which has a well-defined invariant transverse measure, which is uniquely specified by the fact that on the leaves of μ, this transverse measure coincides with the hyperbolic distance. We obtain in this way a measured foliation, $F_\mu(h)$, which has a well-defined class in the space of measured foliations on S up to isotopy and Whitehead moves $\mathbf{MF}(S)$ (see [2] for proprieties of this space). Note that by construction, this class has the property that it can be represented by a measured foliation that is transverse to μ. Conversely, Thurston proves in [6] that the elements of $\mathbf{MF}(S)$ which possess this last property are classes of horocyclic foliations which arise as above, for the same μ and for some hyperbolic metric. More precisely, let $\mathbf{MF}(\mu)$ denote the subset of $\mathbf{MF}(S)$ which consists of equivalence classes which admit representatives transverse to μ. We have the following

Figure 1.

Theorem 3.1 (Thurston [6]). *For any maximal geodesic lamination μ, the map φ_μ which associates to each hyperbolic metric the equivalence class of its horocyclic foliation is a homeomorphism from the Teichmüller space to the subset* $\mathbf{MF}(\mu)$ *of* $\mathbf{MF}(S)$.

3.2 *Construction of hyperbolic structures*

We will construct a hyperbolic structure g on $S_{1,1}$ that verifies certain proprieties. But before that, we recall another construction by Thurston [6]. Let μ be a maximal lamination, and h a hyperbolic structure. By the homeomorphism φ_μ, we can associate a horocyclic foliation $F_\mu(h)$. On each neighbourhood V of an ideal vertex υ, the foliation is composed of horocycles. We can define on each V a sharpness function $f_{h,\upsilon}(.)$ by

$$f_{h,\upsilon}(x) = \log(l_h(C_x)),$$

where x is point in V and $l_h(C_x))$ denotes the length measured with the hyperbolic metric h, of the horocycle C_x passing through x.

Proposition 3.1. *For every $K > 1$, there exists a hyperbolic structure g_α on the punctured torus $S_{1,1}$, and a K-Lipschitz mapping*

$$f : (S_{1,1}, g_\alpha) \longrightarrow (S_{1,1}, h_\alpha) = \mathbb{D}^2/G_\alpha,$$

such that:

$$K = K(f) = \frac{l_h(A_\alpha)}{l_g(A_\alpha)}. \tag{7}$$

Proof. We recall that the projection of the axis of the hyperbolic isometry A_α is simple closed geodesic in $\mathbb{D}^2/G_\alpha$. We extend this geodesic to a maximal lamination μ. According to the Thurston's homeomorphism φ_μ, there is a unique class of horocyclic foliation $F_{h_\alpha}(\mu)$ which is transverse to μ and associated to the hyperbolic structure h_α. On each neighbourhood V of an ideal vertex v of μ, we multiply the transverse measure of $F_{h_\alpha}(\mu)$ by the factor $1/K$, thus we multiply the arclength of each horocycle (which are the leaves of the foliation $F_{h_\alpha}(\mu)$) by $\exp(1/K)$. We define on each neighborhood of an ideal vertex, a new sharpness function $\exp(1/K) \cdot f_{h_\alpha, v}$. These functions parametrize the hyperbolic structures on each V, and as the lamination μ is maximal, then the sharpness functions define a new hyperbolic structure on the whole surface which we denote by g_α (see [6] corollary 4.2, p.17). The lamination μ and the hyperbolic structure g_α being fixed, we construct the extremal Lipschitz homeomorphism

$$f : (S_{1,1}, g_\alpha) \longrightarrow (S_{1,1}, h_\alpha) = S_\alpha,$$

as explained in [6] corollary 4.2, p17. □

We have a relation between the trace of the matrix A_α and the hyperbolic length of the geodesic $l_{h_\alpha}(A_\alpha)$ which is the projection of the A_α-axis.

$$|\mathrm{tr}(A_\alpha)| = 2\cosh(l_{h_\alpha}(A_\alpha)/2) \tag{8}$$

For a proof of this fact see Douady [2] (exposé 7, lemma 1, p 135)

Lemma 3.1. *We have*

$$\lim_{\alpha\to 0} l_{h_\alpha}(A_\alpha) = \lim_{\alpha\to 0} l_{g_\alpha}(A_\alpha) = +\infty$$

Proof. A simple computation from formula (8). □

Lemma 3.2. *The function defined from* $\left]0, \frac{\pi}{2}\right[$ *to* $\mathbb{R}_+^*$ *by*

$$\alpha \mapsto \frac{\mathrm{ext}_{h_\alpha}(A_\alpha)}{\mathrm{ext}_{g_\alpha}(A_\alpha)},$$

in not bounded above.

Proof. We proceed by contradiction. Suppose that our function is bounded above. Using Maskit's estimates (5) and (6), then there exists a constant $M_0 \geq 0$, such that for all $\alpha \in \left]0, \frac{\pi}{2}\right[$

$$\frac{\left(\frac{1}{2} - \varepsilon\right)\exp(l_{h_\alpha}(A_\alpha)/2)}{\frac{1}{2}\cdot l_{g_\alpha}(A_\alpha)\exp(l_{g_\alpha}(A_\alpha)/2)} \leq \frac{\mathrm{ext}_{h_\alpha}(A_\alpha)}{\mathrm{ext}_{g_\alpha}(A_\alpha)} \leq M_0,$$

but g_α and h_α are related by the following relation

$$l_{h_\alpha}(A_\alpha) = K l_{g_\alpha}(A_\alpha).$$

We obtain

$$\frac{(1-2\varepsilon)\exp((K-1)\cdot l_{g_\alpha}(A_\alpha)/2)}{l_{g_\alpha}(A_\alpha)} \leq M_0.$$

According to lemma 3.1, for α small, $l_g(A_\alpha)$ is a huge quantity, thus there exists $\xi > 1$, such that

$$((K-1)\cdot l_{g_\alpha}(A_\alpha)/2)^{\xi} \leq \exp((K-1)\cdot l_{g_\alpha}(A_\alpha)/2)$$

hence

$$(1-2\epsilon)((K-1)/2)^{\xi}\cdot(l_{g_\alpha}(A_\alpha))^{\xi-1} \leq M_0.$$

This proves that

$$\lim_{\alpha\to 0} l_{g_\alpha}(A_\alpha) < \infty$$

which contradicts lemma 3.1. □

Proof of Theorem 2.2. We know that $\mathcal{K}(g,h) \leq d(g,h)$ (see Wolpert [8]). If id : $(\mathbb{T}(S_{1,1}), d) \longrightarrow (\mathbb{T}(S_{1,1}), \mathcal{K})$, is a quasi-isometry, then there exist two constants $k \geq 1$ and $c \geq 0$, such that

$$d(g,h) \leq k\cdot\mathcal{K}(g,h) + c \tag{9}$$

for all $g, h \in \mathbb{T}(S_{1,1})$.

We proceed by contradiction. Suppose that inequality (9) is true. Using the formulas (1), (2) and the fact that $p(A_\alpha) \in \mathcal{S}$, we obtain

$$\frac{\mathrm{ext}_{h_\alpha}(A_\alpha)}{\mathrm{ext}_{g_\alpha}(A_\alpha)} \leq \sup_{\gamma\in\mathcal{S}} \frac{\mathrm{ext}_{h_\alpha}(\gamma)}{\mathrm{ext}_{g_\alpha}(\gamma)} \leq \sup_{\gamma\in\mathcal{S}} \left(\frac{l_{h_\alpha}(\gamma)}{l_{g_\alpha}(\gamma)}\right)^{2k} e^c = K^{2k}e^c. \tag{10}$$

Since the metrics g and h are constructed such that

$$\sup_{\gamma\in\mathcal{S}} \frac{l_{h_\alpha}(\gamma)}{l_{g_\alpha}(\gamma)} = \frac{l_{h_\alpha}(A_\alpha)}{l_{g_\alpha}(A_\alpha)} = K,$$

for all $0 < \alpha < \frac{\pi}{2}$, inequality (10) implies that the function

$$\alpha \longmapsto \frac{\mathrm{ext}_{h_\alpha}(A_\alpha)}{\mathrm{ext}_{g_\alpha}(A_\alpha)},$$

is bounded above which contradicts the result of lemma 3.2. □

Aknowledgement

The author would like to thank Professor Athanase Papodopoulos for his constant encouragement.

References

[1] A. Belkhirat, A. Papadopoulos and M. Troyanov, Thurston weak metric on the Teichmüller space of the torus, *Trans. Amer. Math. Soc.*, **357** (2005) 3311–3324.

[2] A. Fathi, F. Laudenbach and V. Poénaru, Travaux de Thurston sur les surfaces, *Astérique* (1979) 66–67.

[3] S. Kerckhoff, The asymptotic geometry of Teichmüller space, *Topology*, **19** (1980) 23–41.

[4] A. Maskit, Comparison of hyperbolic and extremal lengths, *Ann. Acad. Sci. Fenn. Ser. A I Math.*, **10** (1985) 381–386.

[5] H. Ribeiro, Sur les espaces à métrique faible, *Portugaliae Math.*, **4** (1943) 21–40.

[6] W. P. Thurston, Minimal Stretch maps between hyperbolic surfaces, Preprint (1985), available at http://arxiv.org/abs/math.GT/9801039.

[7] W. P. Thurston, The geometry and topology of 3-manifolds, Princeton Academic Press (1979), available at: http://www.msri.org/publications/books/gt3m/.

[8] S. Wolpert, The length spectra as moduli for compact hyperbolic surfaces, *Ann. of. Math.*, **109** (1979) 323–351.

Teichmüller Theory and Moduli Problem No. 10, 2010, pp. 61–93.

Notes on Vector Bundles on Curves

Usha N. Bhosle and Indranil Biswas

School of Mathematics, Tata Institute of Fundamental Research, Homi Bhabha Road, Mumbai 400 005, India
e-mail: usha@math.tifr.res.in; indranil@math.tifr.res.in

Introduction

Vector bundles over algebraic curves is an active area of research in algebraic geometry for more than four decades. It has connections to many subjects — algebraic geometry, differential geometry, number theory, topology, representation theory and quantum field theory. Last two decades have witnessed a remarkable renaissance in the interaction between geometry and physics. Moduli spaces of vector bundles have played an important role in it. In this expository article we give a few glimpses in the vast development in the theory of vector bundles, mainly from the algebraic geometric point of view.

We start with Grothendieck's classification of vector bundles on the projective line. We then describe in detail Atiyah's classification of vector bundles on elliptic curves. The story is much more complicated in higher genera, it unfolds in the rest of the sections. We explain the concept of moduli spaces, the rigors of the construction of moduli spaces are omitted. However, we give explicit descriptions of moduli spaces in lower ranks and genera. They give valuable insight in these moduli spaces and are very useful in studying their geometry.

Some important properties of moduli spaces are studied in detail. Universal moduli spaces over moduli spaces of curves are described. The Narasimhan-Seshadri theorem, which is a bridge between the algebraic and topological points of view, is described in some detail. Vector bundles with additional structures on them are also of great interest. We study the most interesting among them, viz. the Higgs bundles. In some places we have given complete arguments and some places we haven't. The details can be filled in from the references given. The proofs here are mostly algebraic geometric. For a more differential geometric approach, see [Ga].

1. Preliminaries

Let X denote an irreducible smooth projective curve defined over a field k. Thus, for $k = \mathbb{C}$, the curve X is a compact connected Riemann surface. To each smooth projective curve there is an associated integral invariant known as *genus*. We recall that the genus of X is the number of independent differentials on X. The projective line $\mathbb{P}^1_k$ is the unique, up to isomorphism, smooth curve of genus zero.

If the genus of X is one, then X is an elliptic curve. An elliptic curve has very rich structures including that of an abelian variety of dimension one.

The space of isomorphism classes of smooth curves of genus g defined over k, where $g \geq 2$, is parametrized by a moduli space of dimension $3g - 3$. This moduli space is an irreducible normal quasiprojective variety defined over k.

A particularly interesting class of curves is the hyperelliptic curves. Under the assumption that the characteristic of k is different from 2, they are defined by

$$y^2 = (x - \omega_1)(x - \omega_2) \cdots (x - \omega_{2g+2}), \omega_i \in k.$$

The map $(x, y) \longmapsto x$ makes X a two sheeted cover over $\mathbb{P}^1_k$. All curves of genus 1 and 2 are hyperelliptic curves, but if $g \geq 3$, then the general curve of genus g is not hyperelliptic.

Divisors on X are elements of the free group generated by the closed points of X. Thus a divisor D on X is given by a finite sum of the form

$$D = \sum_i n_i x_i,$$

where n_i are integers and x_i are points of X. The degree of the divisor D is defined to be the integer $\sum_i n_i$. Any line bundle L over X corresponds to some divisor D on X. Isomorphism classes of line bundles correspond to linearly equivalence classes of divisors. We recall that two divisors are called linearly equivalent if their difference is the divisor of a rational function on X.

The *degree* of a line bundle L is the degree of any divisor defining L. The *degree* of a vector bundle E over X of rank r is defined to be the degree of the line bundle $\wedge^r E$. For any vector bundle E, this line bundle $\wedge^{\mathrm{rk}(E)} E$ is called the *determinant line bundle* of E, and it is denoted by $\det E$.

We shall denote the rank and degree of E by $r(E)$ and $d(E)$ respectively. For a non-negative integer i, we will denote by $H^i(X, E)$, or simply $H^i(E)$, the i-th cohomology of E and $h^i(E) = h^i(X, E) := \dim H^i(E)$. The dual vector bundle of E will be denoted by E^*. Let K denote the canonical line bundle of X, which is also the cotangent bundle of X.

For the convenience of readers, we list here a few results frequently used in the theory of vector bundles. For an introduction to vector bundles, the reader may refer to [At3], Chapter I; [Gu].

Theorem 1.1 (Riemann-Roch Theorem). *Let E be a vector bundle on X. Then*

$$h^0(X, E) - h^1(X, E) = d(E) + r(E)(1 - g).$$

Theorem 1.2 (Serre duality).

$$H^1(X, E) \cong H^0(X, E^* \otimes K)^*,$$

and hence

$$H^0(X, E) \cong H^1(X, E^* \otimes K)^*.$$

In particular, $h^1(E) = h^0(E^* \otimes K)$, and $h^0(E) = h^1(E^* \otimes K)$.

2. Atiyah-Krull-Schmidt theorem

Let k be any field. Let M be a projective variety defined over k. A vector bundle E over M is called *decomposable* if there are vector bundles E_1 and E_2 over M of positive ranks such that

$$E \cong E_1 \oplus E_2.$$

A vector bundle over M is called *indecomposable* if it is not decomposable. Therefore, a vector bundle E is indecomposable if and only if whenever

$$E \cong E_1 \oplus E_2$$

then either $E = E_1$ or $E = E_2$.

From the above definition of indecomposable vector bundles it follows immediately that any vector bundle over M is isomorphic to a direct sum of indecomposable vector bundles.

The following theorem is due to Atiyah (see [At1, page 315, Theorem 3]):

Theorem 2.1. *Let F be a vector bundle over a projective variety M defined over a field k. Let*

$$F = \bigoplus_{i=1}^{m} E_i$$

and

$$F = \bigoplus_{i=1}^{n} E'_i$$

be two decompositions of F into direct sum of indecomposable vector bundles; so all E_i and E'_i are indecomposable vector bundles. Then $m = n$, and furthermore, there is a permutation τ of $\{1, \ldots, m\}$ such that the vector bundle E'_i is isomorphic to $E_{\tau(i)}$.

Proof. Fix an isomorphism

$$\rho : F \longrightarrow \bigoplus_{i=1}^{m} E_i. \tag{2.1}$$

For any $j \in [1, m]$, set

$$f_j := \rho^{-1} \circ \iota_j \circ q_j \circ \rho \in H^0(M, \mathrm{End}(F)), \tag{2.2}$$

where $q_j : \oplus_{i=1}^{m} E_i \longrightarrow E_j$ is the natural projection and

$$\iota_j : E_j \longrightarrow \bigoplus_{i=1}^{m} E_i$$

is the natural inclusion. Similarly, fix an isomorphism

$$\rho' : F \longrightarrow \bigoplus_{i=1}^{n} E'_i$$

and for any $j \in [1, n]$ set

$$f'_j := (\rho')^{-1} \circ \iota'_j \circ q'_j \circ \rho' \in H^0(M, \mathrm{End}(F)), \tag{2.3}$$

where $q'_j : \oplus_{i=1}^n E'_i \longrightarrow E'_j$ is the natural projection and ι'_j the natural inclusion of E'_j in $\oplus_{i=1}^n E'_i$. So we have

$$f_j^2 - f_j = 0 = (f'_j)^2 - f'_j \tag{2.4}$$

for all j.

Consider the k-algebra $R := H^0(M, \mathrm{End}(F))$. For any $j \in [1, m]$ (respectively, $j \in [1, n]$) let I_j (respectively, I'_j) denote the right ideal of R generated by f_j (respectively, f'_j) defined in (2.2) (respectively, (2.3)). Therefore, I_j and I'_j are right modules over R. We will prove that all $\{I_j\}_{j=1}^m$ and $\{I'_j\}_{j=1}^n$ are indecomposable right R-modules.

For that take the right R-module I_j. Assume that I_j is decomposable. Take any decomposition

$$I_j = J_1 \oplus J_2 \tag{2.5}$$

of the right R-module, where $J_1 \neq 0 \neq J_2$. Therefore, J_1 and J_2 are right ideals of the algebra R. Recall that I_j is generated by f_j. Consider

$$f_j = a_1 + a_2 \in J_1 \oplus J_2 \subset R \tag{2.6}$$

in terms of (2.5), where $a_i \in J_i$, $i = 1, 2$.

From (2.6) we have

$$a_1 a_2 = f_j a_2 - a_2^2. \tag{2.7}$$

On the other hand, as $a_2 \in I_j$, we have $a_2 = f_j a$ for some $a \in R$. Hence from (2.7) and (2.4) we have

$$a_1 a_2 = f_j a_2 - a_2^2 = f_j^2 a - a_2^2 = f_j a - a_2^2 = a_2 - a_2^2 \in J_2.$$

But $a_1 a_2 \in J_1$, as a_1 is an element of the right ideal J_1. Therefore, we have

$$a_1 a_2 \in J_1 \cap J_2 = 0. \tag{2.8}$$

Similarly, we have

$$a_2 a_1 \in J_1 \cap J_2 = 0. \tag{2.9}$$

Since $(a_1 + a_2)^2 = a_1 + a_2$ (see (2.4) and (2.6)), using (2.8) and (2.9) we conclude that

$$0 = a_1 a_2 = a_2 a_1 = a_1^2 - a_1 = a_2^2 - a_2 \tag{2.10}$$

(note that $a_1^2 - a_1 = a_2^2 - a_2 \in J_1 \cap J_2 = 0$). The restriction to the subbundle $E_j \subset F$ of the endomorphism f_j of F constructed in (2.2) clearly coincides with the identity map of E_j. Therefore, from (2.10) and (2.6) we have

$$E_j = \mathrm{image}(a_1) \oplus \mathrm{image}(a_2).$$

But this contradicts the given condition that E_j is indecomposable. Indeed, such a decomposition implies that both image(a_1) and image(a_2) are subbundles of E_j, and also $a_1 \neq 0 \neq a_2$. Therefore, we conclude that the right R-module I_j is indecomposable.

Similarly, the right R-module I'_j is indecomposable, where $j \in [1, n]$.

Since ρ in (2.1) is a decomposition of F, from the construction of f_j (done in (2.2)) it follows immediately that

$$\sum_{j=1}^{m} f_j = \mathrm{Id}_F.$$

It further follows that

$$R = \bigoplus_{j=1}^{m} I_j. \tag{2.11}$$

Similarly, we have $\sum_{j=1}^{n} f'_j = \mathrm{Id}_F$ and

$$R = \bigoplus_{j=1}^{n} I'_j. \tag{2.12}$$

Since M is a projective variety, the dimension of the k-vector space R is finite. Therefore, R as a right R-module is both Noetherian and Artinian. As $\{I_j\}_{j=1}^{m}$ and $\{I'_j\}_{j=1}^{n}$ are irreducible right R-modules, using the Krull–Schmidt theorem for decompositions (2.11) and (2.12) we conclude that

(1) $n = m$, and
(2) there is an automorphism G of the right R-module R and a permutation τ of $\{1, \ldots, m\}$ such that for each $j \in [1, m]$ the right R-submodule $G(I_j) \subset R$ coincides with the right R-submodule $I'_{\tau(j)} \subset R$.

See [Ja, page 115] for the Krull–Schmidt theorem.

An automorphism G as above coincides with the left multiplication of R with the element $G(\mathrm{Id}_F) \in R$. Therefore, G is given by the automorphism $G(\mathrm{Id}_F) \in H^0(M, \mathrm{End}(F))$ of the vector bundle F. Now the above condition (2) on the pair (G, τ) says that for each $j \in [1, m]$ the automorphism $G(\mathrm{Id}_F)$ of the vector bundle F takes the subbundle E_j isomorphically to $E'_{\tau(j)}$; here E_j and $E'_{\tau(j)}$ are realized as subbundles of F using ρ and ρ' respectively. This completes the proof of the theorem. □

3. Grothendieck's theorem

As before, k is any field. Let $\mathcal{O}_{\mathbb{P}^1_k}(1)$ be the tautological line bundle over the projective line $\mathbb{P}^1_k$ defined over k. We recall that any line bundle over $\mathbb{P}^1_k$ is isomorphic to a line bundle of the form $\mathcal{O}_{\mathbb{P}^1_k}(n)$, where

$$\mathcal{O}_{\mathbb{P}^1_k}(n) = \mathcal{O}_{\mathbb{P}^1_k}(1)^{\otimes n}$$

if $n \geq 0$ and

$$\mathcal{O}_{\mathbb{P}^1_k}(n) = \mathcal{O}_{\mathbb{P}^1_k}(-n)^*$$

if $n < 0$.

Let E be any vector bundle over the projective line $\mathbb{P}^1_k$. Grothendieck proved that E decomposes into a direct sum of line bundles – we will prove it below. Therefore, vector bundles over $\mathbb{P}^1_k$ have a rather simple description.

Theorem 3.1. *Let E be a vector bundle of rank r over $\mathbb{P}^1_k$. Then E is isomorphic to a direct sum of line bundles, or in other words,*

$$E \cong \bigoplus_{i=1}^{r} \mathcal{O}_{\mathbb{P}^1_k}(n_i),$$

where $n_i \in \mathbb{Z}$. Furthermore, the integers $\{n_i\}_{i=1}^r \in \mathbb{Z}^{\oplus r}$ are determined by E uniquely up to a permutation of the index set $\{1, \ldots, r\}$.

Proof. We will prove the first part using induction on rank. Assume that any vector bundle over $\mathbb{P}^1_k$, whose rank is at most $r - 1$, decomposes as a direct sum of line bundles.

Take a vector bundle E over $\mathbb{P}^1_k$ of rank r. For any integer n, the vector bundle $E \otimes_{\mathcal{O}_{\mathbb{P}^1_k}} \mathcal{O}_{\mathbb{P}^1_k}(n)$ over $\mathbb{P}^1_k$ will be denoted by $E(n)$. Using Riemann–Roch it follows immediately that $H^0(\mathbb{P}^1_k, E(n)) \neq 0$ for all sufficiently large positive integers n.

Since $\dim \mathbb{P}^1_k = 1$, any vector bundle over $\mathbb{P}^1_k$ admits a filtration of subbundles such that each successive quotient is of rank one. Applying this to E we conclude that $H^0(\mathbb{P}^1_k, E(-n)) = 0$ for all sufficiently large positive integers n. As $H^0(\mathbb{P}^1_k, \mathcal{O}_{\mathbb{P}^1_k}(1)) \neq 0$, using these observations we conclude that there is a unique integer ℓ such that the following two conditions hold:

$$H^0(\mathbb{P}^1_k, E(\ell)) \neq 0$$

and

$$H^0(\mathbb{P}^1_k, E(\ell - 1)) = 0. \tag{3.1}$$

Note that as the line bundle $\mathcal{O}_{\mathbb{P}^1_k}(1)$ is ample, for any sufficiently large positive integer m, the vector bundle $E(m)$ is generated by its global sections, and also, the dual vector bundle $E(-m)^*$ is ample. Therefore, $E(-m)$ does not admit any nonzero sections for all m sufficiently large.

Fix any nonzero section $\sigma \in H^0(\mathbb{P}^1_k, E(\ell))\setminus\{0\}$. It gives the following exact sequence of coherent sheaves on $\mathbb{P}^1_k$:

$$0 \longrightarrow \mathcal{O}_{\mathbb{P}^1_k} \longrightarrow E(\ell) \longrightarrow W \longrightarrow 0, \tag{3.2}$$

where the homomorphism $\mathcal{O}_{\mathbb{P}^1_k} \longrightarrow E(\ell)$ is multiplication by the above section σ. Since this homomorphism is induced by a homomorphism $\mathcal{O}_{\mathbb{P}^1_k}(D) \longrightarrow E(\ell)$, where D is the divisor over which σ vanishes, from (3.1) it follows immediately that $D = \emptyset$, or in other words, the image of $\mathcal{O}_{\mathbb{P}^1_k}$ in $E(\ell)$ is a line subbundle of $E(\ell)$. Consequently,

W is a vector bundle of rank $r-1$. Therefore, by the induction hypothesis, the vector bundle W decomposes as a direct sum of line bundles.

Since $H^i(\mathcal{O}_{\mathbb{P}^1_k}(-1)) = 0$ for $i = 0, 1$, we conclude that

$$H^0(W(-1)) = H^0(E(\ell - 1)) = 0$$

(see (3.1)). This and the above observation that W is a direct sum of line bundles together imply that

$$W \cong \bigoplus_{i=1}^{r-1} \mathcal{O}_{\mathbb{P}^1_k}(m_i)$$

with $m_i \leq 0$ for all $i \in [1, r-1]$. Hence we have

$$H^1(\mathbb{P}^1_k, W^*) = 0,$$

where W^* is the dual of W. Consequently, the exact sequence of vector bundles in (3.2) splits. Therefore, E is isomorphic to a direct sum of line bundles.

The uniqueness of $\{n_i\}_{i=1}^r \in \mathbb{Z}^{\oplus r}$ up to a permutation follows from Theorem 2.1. □

4. Vector bundles on elliptic curves

Now-onwards, the field k will be assumed to be algebraically closed.

4.1 *Indecomposable vector bundles*

Since every vector bundle on a curve is a direct sum of indecomposable vector bundles (see Theorem 2.1), the problem is to classify indecomposable vector bundles. The latter was solved by Atiyah for vector bundles on elliptic curves.

Henceforth, in this section, X will denote an elliptic curve defined over the algebraically closed field k; the characteristic of k is arbitrary. Let $\mathcal{E}(r, d)$ denote the set of isomorphism classes of indecomposable vector bundles of rank r and degree d on X. For $r = 1$, the set $\mathcal{E}(1, 0)$ is the set of isomorphism classes of line bundles of degree 0 on X. It has the structure of an abelian variety called the Jacobian of X. Fixing a point $p \in X$ and associating to any point $x \in X$ the line bundle $\mathcal{O}(x - p)$ gives an isomorphism of X with its Jacobian.

Let I_n denote the trivial vector bundle over X of rank n.

Definition 4.1. *A* maximal splitting *of a vector bundle E is an ordered sequence of finitely many line bundles $(L_1, \ldots, L_r)$ such that*

- *L_1 is a line subbundle of maximal degree of E, and*
- *$(L_2, \ldots, L_r)$ is a maximal splitting of E/L_1.*

A line subbundle of E of maximal degree will be called a maximal line subbundle *of E.*

Lemma 4.2. *Let E be an indecomposable vector bundle on the elliptic curve X with $H^0(X, E) \neq 0$. Then E has a maximal splitting $(L_1, \dots, L_r)$ with $L_i \supseteq L_1 \supseteq I, i = 1, \dots, r$ (here we identify a line bundle with the coherent sheaf of rank one associated with it).*

Proof. Since X has genus 1 and $h^0(E) \neq 0$, the vector bundle E has a maximal line subbundle $E_1 \supseteq I$. Suppose that we have constructed a sequence

$$0 = E_0 \subset E_1 \subset \cdots \subset E_i,$$

where $L_j = E_j/E_{j-1}$ is a maximal line subbundle of $E'_j = E/E_{j-1}$ with $L_j \supseteq L_1 \supseteq I$ for all $j = 1, \dots, i$. Since E is an extension of E'_i by E_i, one has $H^1(X, E'^*_i \otimes E_i) \neq 0$. Since X has trivial canonical bundle, by Serre duality, this implies that there is a nonzero $f \in \mathrm{Hom}\,(E_i, E'_i)$. Let $j \in [1, i]$ be the smallest integer such that $f(E_j) \neq 0$. Then f induces a non-zero homomorphism

$$\widehat{f} : L_j \longrightarrow E'_i.$$

If $\widehat{f}(L_j)$ has maximal degree, let $L_{i+1} = \widehat{f}(L_j)$. Otherwise let L_{i+1} be a maximal line subbundle of E'_i, then $\deg(L_{i+1}) \geq \deg(\widehat{f}(L_j)) + 1$. Since X is elliptic, this implies that $L_{i+1} \supset \widehat{f}(L_j)$. Hence $L_{i+1} \supseteq L_j$ and $L_j \supseteq L_1 \supseteq I$ by induction. Thus L_{i+1} is a maximal line subbundle of E/E_i with $L_{i+1} \supseteq L_1 \supseteq I$. Define $E_{i+1} = p^{-1}(L_{i+1})$, $p : E \to E'_i$ being the projection. Thus the maximal splitting with required properties can be constructed inductively. □

Lemma 4.3. *Let $E \in \mathcal{E}(r, r)$ with $r \geq 2$. Then E has a maximal splitting $(L, \dots, L)$ with* $\deg(L) = 1$.

Proof. Let $(L_1, \dots, L_r)$ with $L_i \supseteq L_1 \supseteq I, i = 1, \dots, r$, be a maximal splitting given by Lemma 4.2. If the maximal line subbundle L_1 has degree zero, no section of E can vanish at any point i.e., the evaluation map $H^0(X, E) \longrightarrow E_x$ is injective for all $x \in X$. It follows that $H^0(E)$ generates a trivial subbundle I_s of E, in particular $s \leq r$. On the other hand, by Riemann-Roch theorem, $s \geq d(E) = r$. Hence $s = r$ and $E = I_r$ contradicting the indecomposability of E. Therefore we must have $d(L_1) \geq 1$. Then

$$r = \deg(E) = \Sigma_i \deg(L_i) \geq r\deg(L_1)$$

with equality if and only $L_i \cong L_1$ for all i. Thus $L_i \cong L_1 \cong L, d(L) = 1$ for all i. □

Lemma 4.4.

(1) *Let $E \in \mathcal{E}(r, d)$, $0 \leq d < r$ and $s = h^0(E) > 0$. Then E has a trivial subbundle I_s of rank s.*

(2) *Let $E \in \mathcal{E}(r, d)$, $d \geq 0$ and $s = h^0(E)$. Then one has:*

- *$s = d$ if $d > 0$, and $s \in \{0, 1\}$ if $d = 0$.*
- *If $d < r$ then E has a trivial subbundle I_s and $E' = E/I_s$ is indecomposable with $h^0(E') = s$.*

Proof.

(1) Lemma 4.2 gives a maximal splitting $(L_1, \ldots, L_r)$ with $L_i \supseteq L_1 \supseteq I, i = 1, \ldots, r$. We claim that $d(L_1) = 0$ and hence $L_1 = I$. If $d(L_1) \geq 1$, then

$$d = d(E) = \Sigma_i d(L_i) \geq rd(L_1) \geq r$$

which is a contradiction as $d < r$. Since a maximal line subbundle L_1 has degree zero, it follows that $H^0(E)$ generates a trivial subbundle I_s of E.

(2) If $d \geq r$, then as in the proof of Lemma 4.3 we can show that if L_1 is a maximal line subbundle, then $d(L_1) = 0$ implies that $E = I_r$, so we must have $d(L_1) > 0$. By Lemma 4.2, one then has $d(L_i) > 0$ so that $h^1(E) \leq \Sigma h^1(L_i) = 0$. Hence $h^1(E) = 0$ and by Riemann-Roch theorem $h^0(E) = d$. This proves the result for $d \geq r$.

Suppose that $d < r$. The result clearly holds for $r = 1$. We proceed by induction on r. If $s \neq 0$, by part (1), E has a trivial subbundle I_s. We check that $E' = E/I_s \in \mathcal{E}(r-s, d)$. Let $p : E^* \to I_s^* = I_s$ be the quotient map. If there is a $\sigma \in H^0(E^*)$ such that $p \circ \sigma \neq 0$ then $q_i \circ p \circ \sigma \in H^0(I)$ is nonzero for some direct summand I of $I_s, q_i : I_s \to I$ projection. This implies that I is a direct summand of E^* contradicting the indecomposability of E^*. Thus $H^0(p) : H^0(E^*) \to H^0(I_s)$ is a zero map. From the cohomology exact sequence associated with the exact sequence

$$0 \to E'^* \to E^* \to I_s \to 0$$

it follows that $h^0(E'^*) = h^0(E^*)$. By Riemann-Roch theorem and Serre duality, this implies that $h^0(E') = h^0(E) = s$ and $h^1(E'^*) = s$. The extensions of I_s by E'^* correspond to $H^1(I_s^* \otimes E'^*) = \text{Hom}\,(H^0(I_s), H^1(E'^*))$ and $H^0(p) = 0$ implies that the corresponding element $\delta : H^0(I_s) \to H^1(E'^*)$ is a monomorphism, in fact an isomorphism as $h^1(E'^*) = s$. If E' is decomposable, one has $E' = F_1 \oplus F_2$ with F_1, F_2 nonzero. Then $\delta = \delta_1 \oplus \delta_2, \delta_i : H^0(I_{f_i}) \to H^1(F_i^*), f_i = h^1(F_i), i = 1, 2$. These δ_i give vector bundles E_i^* obtained by extension of I_{f_i} by F_i^* and $E^* = E_1^* \oplus E_2^*$, a contradiction. Thus E' is indecomposable. By induction one has $h^0(E') = 0$ or 1 if $d = 0$ and $h^0(E') = d$ if $d > 0$. We have already seen that $h^0(E') = h^0(E)$, hence the result follows. □

Theorem 4.5 (At2, Theorem 5).

(1) *There exists a vector bundle $F_r \in \mathcal{E}(r, 0)$, unique up to isomorphism, with $h^0(F_r) \neq 0$. Moreover one has an exact sequence*

$$0 \to I \to F_r \to F_{r-1} \to 0.$$

(2) *Let $E \in \mathcal{E}(r, 0)$, then $E \cong L \otimes F_r$ where L is a line bundle of degree zero, unique up to isomorphism, with $L^r \cong \det E$.*

Proof.

(1) Let $\mathcal{E}'(r, 0)$ be the set of isomorphism classes of $E \in \mathcal{E}(r, 0)$ with $h^0(E) \neq 0$. For $r = 1$, this set consists of the unique element I, i.e., $F_1 = I$. Assume that

the result is true for all $r' < r$. By Lemma 4.4, any $E \in \mathcal{E}'(r, 0)$ has a trivial subbundle I with $E' = E/I \in \mathcal{E}'(r-1, 0)$. By induction, $E' \cong F_{r-1}$. Since $h^1(F^*_{r-1}) = h^0(F_{r-1}) = 1$ (Lemma 4.4), the vector bundle E given by the short exact sequence

$$0 \longrightarrow I \longrightarrow E \longrightarrow F_{r-1} \longrightarrow 0$$

is unique up to isomorphism, proving (1).

(2) Fix a line bundle A of degree 1. If $E \in \mathcal{E}(r, 0)$ then $E \otimes A \in \mathcal{E}(r, r)$. By Lemma 4.2, $E \otimes A$ has a maximal splitting $(L_1, \ldots, L_1)$ with $d(L_1) = 1$. It follows that $E \otimes A \otimes L_1^* \in \mathcal{E}'(r, 0)$ and by (1), $E \otimes A \otimes L_1^* \cong F_r$. Setting $L = L_1 \otimes A^*$, one has $E \cong F_r \otimes L$. Taking determinants we have $det E = L^r$.

For uniqueness of L, it suffices to show that $F_r \otimes L \cong F_r$ implies that $L = I$. Note that $F_r \otimes L$ has maximal splitting type $(L, \ldots, L)$. One has $1 = h^0(F_r) = h^0(F_r \otimes L) \leq rh^0(L) = 0$ if $L \neq I$. This implies that $L = I$. □

Theorem 4.6 (At2, Theorem 6). *Let $h = (r, d)$. Fix a line bundle A of degree 1. Then A determines a one to one correspondence*

$$f_{r,d} : \mathcal{E}(h, 0) \longrightarrow \mathcal{E}(r, d).$$

The correspondence $f_{r,d}$ is defined uniquely by the following properties:

(i) *$f_{r,0} = Id$, the identity,*
(ii) *$f_{r,d+r}(E) \cong f_{r,d}(E) \otimes A$ for $E \in \mathcal{E}(h, 0)$, and*
(iii) *if $0 < d < r$, we have a short exact sequence*

$$0 \longrightarrow I_d \longrightarrow f_{r,d}(E) \longrightarrow f_{r-d,d}(E) \longrightarrow 0.$$

Finally, $\det f_{r,d} \cong \det E \otimes A^d$.

Proof. We briefly sketch the proof. Tensoring with A gives a $1-1$ correspondence between $\mathcal{E}(r, d)$ and $\mathcal{E}(r, d+r)$ corresponding to (ii). Hence we may assume that $0 \leq d < r$. If $d = 0$ then $h = r$ and we take $f_{r,0} = Id$. If $d > 0$, one checks that there is a $1-1$ correspondence between $\mathcal{E}(r, d)$ and $\mathcal{E}(r-d, d)$ given by the short exact sequence

$$0 \longrightarrow I_d \longrightarrow E \longrightarrow E' \longrightarrow 0,$$

where $E \in \mathcal{E}(r, d)$, $E' \in \mathcal{E}(r-d, d)$. Using this with (ii) one gets a $1-1$ correspondence $f_{r,d}$. The Euclidean algorithm to determine the greatest common divisor h of r and d determines the number and order of the steps (ii) and (iii) to be performed. □

Theorems 4.5 and 4.6 together give the following.

Theorem 4.7 (At2, Theorem 7). *Let p be a fixed base point of an elliptic curve X (any characteristic), it may be regarded as the zero element of the abelian variety X. Then the set $\mathcal{E}(r, d)$ of isomorphism classes of indecomposable vector bundles over X of rank r, degree d may be identified with X in such a way that the determinant map*

$$\det : \mathcal{E}(r, d) \longrightarrow \mathcal{E}(1, d)$$

corresponds to the map $X \to X$ defined by $x \mapsto hx = x + \cdots + x$ (h times), h being the highest common factor of r and d.

In fact, every $E \in \mathcal{E}(r, d)$ is of the form $E = L \otimes f_{r,d}(F_r)$, and $E \cong f_{r,d}(F_r)$ if and only if $L^{r/h} \cong \mathcal{O}$. In particular there are only r classes of indecomposable projective bundles of rank $r - 1$ on X.

4.2 *The ring $\tilde{\mathcal{E}}(X)$*

Let $\mathcal{E}(X)$ be the set of isomorphism classes of all vector bundles on X. Since Krull-Schmitt theorem holds in $\mathcal{E}(X)$, it is a free abelian semi-group with respect to direct sum. Let $\tilde{\mathcal{E}}(X)$ be the free abelian group generated by $\mathcal{E}(X)$. The elements of $\tilde{\mathcal{E}}(X)$ are finite sums of the form $\Sigma n_i[E_i]$ where $n_i \in \mathbb{Z}$, and $[E_i]$ are isomorphism classes of indecomposable vector bundles on X. The tensor product induces a multiplication on $\tilde{\mathcal{E}}(X)$ making it a commutative ring with unit I (the trivial line bundle). Atiyah determined the structure of this ring completely in characteristic 0.

Assume that the characteristic $p > 0$. The kernel $X[p]$ of the pth power map $x \mapsto x^p, x \in X$ is the group of p-torsion points of X. It is known that $X[p] = 0$ or $X[p] \cong \mathbf{Z}/p\mathbf{Z}$. If $X[p]$ is nonzero, then X is called ordinary and its Hasse invariant is said to be 1. If $X[p] = 0$, then X is called supersingular and its Hasse invariant is said to be zero. The multiplicative structure of the ring $\tilde{\mathcal{E}}(X)$ depends on the characteristic of the base field and Hasse invariant of X.

Indecomposable vector bundles on elliptic curves have also been studied by T. Oda using abelian varieties techniques [O]. In addition to recovering Atiyah's results, Oda also studies the pullback $E^{[p]}$ of an indecomposable vector bundle E under the Frobenius morphism F of X. We recall that $F : X \to X$ is the unique morphism which is identity on the points of X and is the pth power map on functions on X. Oda shows that $E^{[p]}$ is indecomposable if X is ordinary. If X is supersingular, then $E^{[p]}$ is indecomposable if either $(r, d) = 1$ or $(r, d) \neq 1$ with r/h divisible by p. Otherwise $E^{[p]}$ decomposes into m indecomposable components, $m = \min\{h, p\}$. The injectivity of the morphism $H^1(E) \to H^1(E^{[p]})$ is also studied.

5. Moduli spaces of vector bundles

The situation for $g \geq 2$ is much more complicated, while the lower genera $g = 0, 1$ are somewhat special. The classification of vector bundles for $g \geq 2$ was basically done by David Mumford [M1] and later by C. Seshadri [S] in the sense that they constructed the moduli space of vector bundles. For $g \geq 2$, it is impossible to find a geometric object which parametrizes all vector bundles on X, even the indecomposable ones. Hence one has to restrict to vector bundles which have nicer properties. The right category of objects was discovered by Mumford. They are called the stable bundles and semistable bundles.

The following definition is due to Mumford.

Definition 5.1. *A vector bundle E of rank n and degree d is called* stable *if for every nonzero proper subbundle F of E we have*

$$\frac{\deg(F)}{\operatorname{rank}(F)} < \frac{\deg(E)}{\operatorname{rank}(E)}.$$

If the weaker inequality $\frac{\deg(F)}{\operatorname{rank}(F)} \leq \frac{\deg(E)}{\operatorname{rank}(E)}$ holds, then E is called semistable.

Clearly, every stable vector bundle is semistable. Stable vector bundles are indecomposable, the converse is not true. For example, the vector bundles F_r on elliptic curve are indecomposable but not stable for $r \geq 2$ (Theorem 4.5). Stable vector bundles are simple i.e., any automorphism of a stable vector bundle is a scalar multiplication.

For any vector bundle E, the quotient $\frac{\deg(E)}{\operatorname{rank}(E)}$ will be denoted by $\mu(E)$. This rational number $\mu(E)$ is called the *slope* of E.

Definition 5.2. *A semistable vector bundle is called* polystable *if it is a direct sum of stable vector bundles.*

Thus a polystable vector bundle E is of the form

$$E = \bigoplus_{i=1}^{\ell} E_i,$$

where each E_i is semistable with $\mu(E_i) = \mu(E)$.

Although any polystable vector bundle is semistable, not every semistable vector bundle is polystable. The vector bundles F_r with $r \geq 2$ on an elliptic curve are semistable but not polystable.

If a vector bundle E is semistable but not stable, then it has a proper subbundle E' with $\mu(E') = \mu(E)$. Certainly, E' is semistable. Iterating, we get the following.

Definition 5.3. *If a vector bundle E is semistable but not stable, then there is a unique filtration of subbundles*

$$0 = E_0 \subset E_1 \subset \cdots \subset E_{\ell-1} \subset E_\ell = E \tag{5.1}$$

such that E_i/E_{i-1} is the maximal polystable subbundle of E/E_{i-1} with $\mu(E_i) = \mu(E)$ for all $i = 1, \ldots, \ell$. This filtration is called the Jordan–Hölder filtration *of E.*

For such a semistable vector bundle E, define

$$\operatorname{gr}(E) = \bigoplus_{i=1}^{i=\ell} (E_i/E_{i-1}).$$

The vector bundle $\operatorname{gr}(E)$ is polystable. Although the filtration of E as in (5.1) is not unique, the vector bundle $\operatorname{gr}(E)$ is unique up to an isomorphism.

Definition 5.4. *Two semistable vector bundles E and E' over X are called* S-equivalent *(short form of strongly equivalent) if $gr(E) \cong \operatorname{gr}(E')$.*

The following theorem was proved in [HN]:

Theorem 5.5. *Let E be a vector bundle which is not semistable. Then there is a unique filtration of vector bundles*

$$0 = E_0 \subset E_1 \subset \cdots \subset E_{\ell-1} \subset E_\ell = E$$

such that E_i/E_{i-1} is semistable for all $i \in [1, \ell]$, and furthermore,

$$\mu(E_i/E_{i-1}) > \mu(E_{i+1}/E_i)$$

for all $i \in [1, \ell - 1]$.

Proof. Let V_1 and V_2 be two semistable vector bundles over X with

$$\mu(V_1) > \mu(V_2).$$

We will show that

$$H^0(X, \operatorname{Hom}(V_1, V_2)) = 0. \tag{5.2}$$

To prove this, let $\rho : V_1 \longrightarrow V_2$ be a nonzero homomorphism. Let $\ker(\rho) \subset V_1$ be the coherent subsheaf defined by the kernel of ρ, and let $\operatorname{im}(\rho) \subset V_2$ be the coherent subsheaf defined by the image of ρ. Since $\dim X = 1$, we conclude that $\ker(\rho)$ is a subbundle of V_1. Let $\overline{\operatorname{im}(\rho)} \subset V_2$ be the minimal subbundle generated by $\operatorname{im}(\rho)$. Therefore, $\overline{\operatorname{im}(\rho)}$ is the inverse image of the torsion subsheaf of $V_2/\operatorname{im}(\rho)$. Since both V_1 and V_2 are semistable, we have

$$\mu(V_1) \le \mu(V_1/\ker(\rho)) = \mu(\operatorname{im}(\rho)) \le \mu(\overline{\operatorname{im}(\rho)}) \le \mu(V_2).$$

This contradicts the given condition that $\mu(V_1) > \mu(V_2)$. Therefore,

$$H^0(X, \operatorname{Hom}(V_1, V_2)) = 0.$$

Using induction on rank it follows that any vector bundle contains a semistable vector bundle as a subbundle. Given a vector bundle E, let $\mathcal{S}(E) \subset \mathbb{Q}$ be the subset given by the slopes of all semistable subbundles contained in E. Fixing a filtration of subbundles of E such that each successive quotient is semistable, from (5.2) we conclude that $\mathcal{S}(E)$ contains an upper bound. Let $m(E) \in \mathcal{S}(E)$ be such that $m(E) \ge c$ for all $c \in \mathcal{S}(E)$.

If $V \subset E$ is a coherent subsheaf with $\mu(V) = m(E)$, then V must be a semistable subbundle of E. Indeed, this follows from the observation that the inverse image of the torsion subsheaf of E/V is a subbundle of E, and furthermore, if the torsion subsheaf of E/V is nonzero, then the slope of this subbundle of E is strictly greater than that of V.

Let $V \subset E$ be a subbundle with $\mu(V) = m(E)$. Consider the quotient E/V. If $W \subset E$ is another subbundle with $\mu(W) = m(E)$, then let W' be the image of W in this quotient E/V. Since W is semistable, we have $\mu(W') \ge \mu(W)$. On the other hand, clearly $m(E) \ge m(E/V)$. Therefore, W' is a subbundle of E/V, and $\mu(W') = \mu(W)$. From these it follows immediately that $V + W$ is a subbundle of

E with $\mu(V+W)=m(E)$. Consequently, there is a unique maximal subbundle E_1 of E with $\mu(E_1)=m(E)$. This E_1 is the sum of subbundles of E with slope $m(E)$, hence $m(E)>m(E/E_1)$. Now the proof of the theorem is completed by iterating the argument for E/E_1. □

The filtration of subbundles in Theorem 5.5 is called the *Harder-Narasimhan filtration*.

What is a **moduli space** of vector bundles? Roughly speaking, a moduli space is a geometric object (a manifold, algebraic variety etc.) whose points are in bijective correspondence with classes of objects to be classified. We formulate the precise definition in the following.

Definition 5.6. A family of (semi)stable vector bundles on X parametrized by a variety *S is a vector bundle $\mathcal{E}$ on $S\times X$ such that $\mathcal{E}|_{s\times X}$ is (semi)stable for all $s\in S$. Two families $\mathcal{E}_1$ and $\mathcal{E}_2$ are called equivalent if $\mathcal{E}_1\cong\mathcal{E}_2\otimes p_S^*L$ for some line bundle L on S, p_S being the projection from $S\times X$ to S.*

The moduli functor *$\mathcal{F}$ for (semi)stable vector bundles is the contravariant functor from the category of varieties to the category of sets which associates to S the set $\mathcal{F}(S)$ of equivalence classes of families of (semi)stable vector bundles parametrized by S.*

Definition 5.7. *A* fine moduli space *for the functor $\mathcal{F}$ is a variety M together with an isomorphism of functors $\phi:\mathcal{F}\to Hom\,(-,M)$. In other words, (M,ϕ) represents the functor $\mathcal{F}$.*

In simpler terms, a fine moduli space is a variety M together with a family of vector bundles U parametrized by M (called the universal family) such that, for every variety S and a family F parametrized by it, there is a unique morphism

$$\phi:S\longrightarrow M$$

with F equivalent to ϕ^*U.

Definition 5.8. *A* coarse moduli space *for the functor $\mathcal{F}$ is a variety M together with a natural transformation of functors*

$$\phi:\mathcal{F}\to \mathrm{Hom}\,(-,M)$$

such that

(i) *$\phi(point)$ is bijective;*
(ii) *for any variety N and any natural transformation $\psi:\mathcal{F}\to Hom\,(-,M)$, there exists a unique natural transformation $\eta:Hom\,(-,M)\to Hom\,(-,N)$ such that $\psi=\eta\circ\phi$.*

In simpler terms, M has the following property. The points of M are in bijective correspondence with the vector bundles to be classified. Let F be a family parametrized by S and for any $s\in S$, let $F_s=F|_{s\times X}$. Then the map $f_M:S\to M$ defined by $s\mapsto F_s$ is a morphism.

If N is a variety which has this property for M replaced by N, then there is a morphism $\nu:M\to N$ such that $f_N=\nu\circ f_M$.

A coarse moduli space is unique if it exists. A fine moduli space is also a coarse moduli space.

Mumford showed that the set of isomorphism classes of stable vector bundles of rank n, degree d has the structure of a quasi-projective variety $U^s(n,d)$ [M1], Seshadri constructed a projective compactification $U(n,d)$ of it [S].

Theorem 5.9. *There exists a coarse moduli space $U_X(n,d)$ (or $U(n,d)$) for semistable vector bundles of rank n, degree d on a nonsingular projective curve X. The points of $U(n,d)$ correspond to S-equivalence classes of semistable vector bundles. $U(n,d)$ is a normal projective (irreducible) variety.*

The open subvariety $U^s(n,d)$ of $U(n,d)$ consisting of points corresponding to stable vector bundles is the coarse moduli space for stable vector bundles of rank n, degree d. It is a nonsingular variety.

If $(n,d)=1$, then it is easy to see that E is stable if it is semistable.

Theorem 5.10 (N2).

(1) *If $(n,d)=1$, then $U^s(n,d)=U(n,d)$. In particular, $U^s(n,d)$ is a projective variety if $(n,d)=1$.*
(2) *If $(n,d)=1$, then $U(n,d)$ is a fine moduli space.*

Theorem 5.10(2) says that there exists a universal family $\mathcal{P}$ parametrized by $U(n,d)$ if n,d are coprime. The vector bundle $\mathcal{P}$ on $U(n,d)\times X$ is called the Poincaré bundle, it is unique up to tensoring by (pull-back of) a line bundle on $U(n,d)$. A question arises. If n,d are not coprime, is there a Poincaré bundle at least on $S\times X$ for some open set $S\subset U(n,d)$? This was answered in negative by Ramanan.

Theorem 5.11 (R). *If there is a Poincaré bundle on $S\times X$ for some open set $S\subset U(n,d)$, then n, d are coprime.*

5.1 *Moduli spaces of vector bundles with a fixed determinant*

Fix a line bundle L of rank n, degree d. Let $U_L(n,d)$ be the closed subvariety of the moduli space $U(n,d)$ consisting of vector bundles of rank n, degree d, with a fixed determinant L. It is known that $U_L(n,d)$ is simply connected. It is normal and in case $(n,d)=1$, it is nonsingular. Both the varieties $U(n,d)$ and $U_L(n,d)$ are locally factorial and Fano varieties [DN].

Let J_n denote the subgroup of the Jacobian J consisting of line bundles N such that $N^n=\mathcal{O}$. Since $det(E\otimes N)=det E\otimes N^n=E$ for E of rank n, it follows that J_n acts on $U_L(n,d)$.

Proposition 5.12. *The quotient of $U_L(n,d)\times J$ by J_n, under the action $N(E,\eta)=(E\otimes N, N^{-1}\otimes\eta)$ for $N\in J_n$, $E\in U_L(n,d)$, $\eta\in J$, is isomorphic to $U(n,d)$.*

Proof. There is a natural morphism $U_L(n,d)\times J\to U(n,d)$ defined by $(E,\eta)\mapsto E\otimes\eta$. This induces a morphism $(U_L(n,d)\times J)/J_n\to U(n,d)$. The inverse of this morphism is given by $F\mapsto[(F\otimes N, det F)]$ where $N\in J_n$ is such that $N^n=det F^{-1}\otimes L$. □

In view of this proposition, some problems in the study of $U(n,d)$ are reduced to those for $U_L(n,d)$.

5.2 *The moduli spaces for low ranks and genera*

The moduli spaces are constructed as quotients modulo $PGL(N)$-actions by using geometric invariant theory introduced by D. Mumford [M2]. The construction does not show what the moduli spaces look like. It was later discovered that for lower ranks and genera, these moduli spaces have simple and beautiful explicit geometric descriptions.

(1) **Case** $g = 1$
For $g = 1$, all indecomposable vector bundles are semistable and they are stable if and only if rank and degree are coprime [Tu]. Thus $U^s(n, d)$ is nonempty if and only if n and d are coprime. If $h = gcd(n, d)$, then $U(n, d) \cong S^h(X)$, the hth symmetric power of X [Tu].

(2) **Case** $g = 2$
For $g = 2$, Peter Newstead proved that

$$U_L(2, 1) \cong Q,$$

where Q is a nonsingular intersection of two smooth quadrics in general position in $\mathbb{P}^5(\mathbb{C})$ [N1]. Narasimhan and Ramanan [NR1] showed that

$$U_L(2, 0) \cong \mathbb{P}^3(\mathbb{C}).$$

Here $\mathbb{P}^n(\mathbb{C})$ denotes the n-dimensional complex projective space; it is the quotient of $(\mathbb{C})^{n+1} - \{0\}$ by the diagonal action of $\mathbb{C}^*$.

(3) **Case** $g = 3$
For a non-hyperelliptic curve of $g = 3$, Narasimhan and Ramanan showed that the moduli space $U_L(2, 0)$ is isomorphic to the quartic hypersurface in $\mathbb{P}^7(\mathbb{C})$ known as Coble quartic.

The Coble quartic is given by a single quartic equation in 8 variables. The vanishing of all its partial derivatives gives the singular locus of the Coble quartic. The result of Narasimhan and Ramanan shows that the singular locus of the Coble quartic is the Kummer 3-fold. The Kummer 3-fold is isomorphic to the quotient J/ι, of the Jacobian J of X by an involution ι, embedded in $U_L(2, 0)$. It then follows that the Kummer 3-fold is given by cubic equations. This is an interesting application of the vector bundle theory to classical geometry.

(4) **Moduli of vector bundles on hyperelliptic curves**
Note that curves of genus 1 and 2 are hyperelliptic. More generally we have the following results by Desale and Ramanan.

Theorem 5.13 (Desale-Ramanan [DR2]).

(I) *Let X be a smooth hyperelliptic curve. Then $U_L(2, 1)$ is isomorphic to the variety of $(g - 1)$-dimensional subspaces of $\mathbb{C}^{2g+2}$ which are contained in the intersection of two quadrics $q_1 = \sum_j X_j^2, q_2 = \sum_j a_j X_j^2, a_j$ being all distinct scalars.*

(II) *Let R denote the variety of $g+1$ dimensional subspaces V of $\mathbb{C}^{2g+2}$ which satisfy the following.*

(i) *V is contained in q_1 and belongs to a fixed system of maximum isotropic spaces.*
(ii) *$q_2|V$ has rank ≤ 4.*

Then $U_L(2,0)/i \cong R$, i = being the canonical involution on the moduli space induced by the involution on X.

These results are in fact particular cases of more general results on orthogonal bundles and symplectic bundles on X. Orthogonal bundles and symplectic bundles can be regarded as vector bundles with nondegenerate quadratic forms and symplectic forms (alternating forms) respectively. The results were also generalized to base fields of positive characteristics including characteristic 2 [B1,B2].

6. Universal moduli space, Torelli type theorems

So far we had kept X fixed. If we vary the curve X, how does $U_X(n,d)$ change? Does a moduli space of vector bundles on a curve X determine X? There exists a quasi-projective moduli space $\mathcal{M}_g$ of smooth curves of genus $g \geq 2$. It has a compactification viz., the moduli space $\overline{\mathcal{M}_g}$ of Deligne-Mumford stable curves of genus g. Recall that a stable curve of genus g is a connected, reduced curve of arithmetic genus g (i.e. with $h^1(X,\mathcal{O}) = g$) with at most ordinary nodes as singularities and such that every smooth rational component of X meets the remaining components in at least three points. We may modify our question as: How does $U_X(n,d)$ change as X varies over $\overline{\mathcal{M}_g}$?

If X becomes singular, the moduli space of vector bundles on X is a quasi-projective scheme. To get a projective compactification of this scheme, one needs torsionfree sheaves (i.e. sheaves which are torsionfree on each irreducible component). Let X be a stable curve and X_i, $i \in I$, its irreducible components. Let ω be its canonical sheaf, it is a locally free sheaf of rank 1. Let $w_i = d(\omega|_{X_i})$. For a torsionfree sheaf E on X, let $r_i(E)$ be the rank of $E|_{X_i}$ and $\chi(E) := h^0(E) - h^1(E)$ the Euler characteristic. The slope of E is defined as the rational number

$$\mu(E) = \frac{\chi(E)}{\Sigma_i w_i r_i(E)}.$$

Definition 6.1. *A torsionfree sheaf E is called* (slope) stable *if for every nonzero proper subsheaf F of E we have*

$$\mu(F) < \mu(E).$$

If the weaker inequality $\mu(F) \leq \mu(E)$ holds, then E is called (slope) semistable.

We remark that the condition for (semi)stability of a torsionfree sheaf on a curve with many component depends on the choice of polarization on each component. The preceding definition defines (semi)stability for canonical polarization.

A (slope) semistable sheaf has a Jordon-Holder filtration by stable sheaves. Two semistable sheaves are S-equivalent if they have the same associated graded sheaf for Jordon-Holder filtrations.

Definition 6.2. *Two S-equivalence classes of semistable sheaves are called aut-equivalent if they differ by an automorphism of X.*

If E has a uniform rank $r = r_i(E)$ for all i, then define its total degree

$$d(E) := \chi(E) + r(g-1).$$

The moduli spaces of pairs (X, E) where X is a stable curve and E is a torsionfree sheaf on X were constructed by L. Caporaso [C] in case of uniform rank $r_i = 1$ and by R. Pandharipande [P] for higher ranks by different methods. For rank 1, both constructions give isomorphic moduli spaces. Calculations by Abramovich showed that a generalization of methods of [C] for rank 2 gives a different moduli space than that of [P].

Theorem 6.3. *For integers $g \geq 2, n \geq 1, d$ there exists a projective (irreducible) variety $\boldsymbol{U}_g(n,d)$ and a morphism*

$$\eta : \boldsymbol{U}_g(n,d) \to \overline{\mathcal{M}}_g$$

with the following properties.

(1) *There is a functorial bijective correspondence between the points of $\boldsymbol{U}_g(n,d)$ and aut-equivalence classes of semistable torsionfree sheaves of uniform rank n and total degree d on stable curves of genus g. The map η which associates to an aut-equivalence class the underlying curve is a morphism.*
(2) *For $[X] \in \mathcal{M}_g$, there exists a functorial isomorphism*

$$\eta^{-1}([X]) \cong U_X(n,d)/Aut X,$$

where $Aut X$ denotes the group of automorphisms of X.
(3) *For $r = 1$, $\boldsymbol{U}_g(1,d)$ is normal.*

We now come to the second question: Does a moduli space of vector bundles on X determine X?

Theorem 6.4 (MN, NR2, Tn1). *Let L and L′ be line bundles of degree d on two compact Riemann surfaces X and X′ respectively. If $g \geq 3$, $(n,d) = 1$ and*

$$U_{X,L}(n,d) \cong U_{X',L'}(n,d),$$

then

$$X \cong X'.$$

In the non-coprime case, a similar result is proved by D. Arapura and P. Shastry for large g (dependent on n) [AS]. Kouvidakis and Pantev have a version of Torelli theorem for $U_L(n,d)$ for $n \geq 2$, $g \geq 3$ [KP].

6.1 *Automorphisms of moduli spaces*

Tensoring by an element of order n of the Jacobian defines an automorphism of $U_L(n,d)$. Any automorphism of X also induces an automorphism of $U(n,d)$. In fact, if X has nontrivial automorphisms, then any automorphism of $U_L(n,d)$ is of the form

$$E \mapsto \sigma^* E \otimes \xi,$$

where $\sigma \in Aut(X), \xi \in J, \sigma^* L \otimes \xi^n \cong L$. Any automorphism of $U(n,d)$ decomposes into an automorphism of J and an automorphism of $U_L(n,d)$. More precise results are the following [KP].

Theorem 6.5.

(1) *If* $n \nmid 2d$ *and* X *has no automorphisms, then* $\mathrm{Aut}(U_L(n,d)) \cong J_n$.
(2) *If* $n|2d$, $n \neq 2$, *then* $Aut(U_L(n,d))$ *is isomorphic to the semidirect product of* J_n *and* $\mathbb{Z}/2\mathbb{Z}$.

If $n|2d$, $n \neq 2$, then there is an extra automorphism of the form $\delta : E \mapsto E^* \otimes \nu, \nu^n \cong L^2$.

7. Vector bundles and representations

For X we have its (topological) fundamental group $\pi(X)$. There is a close relation between representations of the fundamental group of X and vector bundles on X.

Definition 7.1. *A linear* representation *of a group* G *on an* n*-dimensional complex vector space* V *is a homomorphism* $\rho : G \to GL(V)$, *where* $GL(V)$ *is the group of invertible linear transformations of* V *onto itself.*

Definition 7.2. *A linear representation* ρ *is called* irreducible *if there is no non-trivial proper subspace* W *of* V *kept invariant by it. That is, if* $\rho(h)W \subset W$ *for all* $h \in G$ *then either* $W = 0$ *or* $W = V$.

A linear representation ρ is called unitary if it maps G into the subgroup $U(n)$, the group of unitary matrices in $GL(V)$.

A linear representation ρ is called completely reducible if it is a direct sum of irreducible representations. This means that $V = \oplus_i V_i$ with each V_i kept invariant by ρ. For example, a unitary representation is completely reducible.

Definition 7.3. *Two representations* ρ *and* ρ' *of* G *in a group* H $(H \subset GL(V))$ *are called* equivalent *if there exists an* $h \in H$ *such that for all* $g \in G$, *one has* $\rho'(g) = h\rho(g)h^{-1}$.

Let X' denote an open smooth curve obtained by removing finitely many points of a smooth projective curve X. The spaces of representations of $\pi(X)$ and $\pi(X')$ have been studied extensively in many subjects including the theory of Lie groups and algebraic groups, differential geometry, algebraic geometry and dynamical systems.

Note that the fundamental group $\pi(X)$ is a generated by $A_1, B_1, \ldots, A_g, B_g$ with the only relation $A_1 B_1 {A_1}^{-1} {B_1}^{-1} \ldots A_g B_g {A_g}^{-1} {B_g}^{-1} = 1$. In particular, for $g = 0$, the fundamental group is trivial. For $g = 1$, the group $\pi(X)$ is an abelian group and hence it has no irreducible representations.

The following theorem was proved in [NS]:

Theorem 7.4. *The equivalence classes of* n*-dimensional irreducible unitary representations of the fundamental group* $\pi(X)$ *of a compact Riemann surface* X *of genus* $g \geq 2$ *form a complex manifold of complex dimension* $n^2(g-1)+1$.

Proof (Idea of the Proof). Let $N = U(n) \times \cdots \times U(n)$, ($2g$ times), and define a map $f : N \to SU(n)$ (the special unitary group) by

$$(A_1, B_1, \ldots, A_g, B_g) \mapsto A_1 B_1 {A_1}^{-1} {B_1}^{-1} \ldots A_g B_g {A_g}^{-1} {B_g}^{-1}.$$

Then the set U of unitary representations of $\pi(X)$ can be identified with $f^{-1}(I)$ where I is the identity matrix. It is shown that f has maximal rank at a point $\rho \in f^{-1}(I)$ if and only if ρ is an irreducible representation. Consequently, the set of n-dimensional irreducible unitary representations has a natural structure of a real analytic manifold. It follows that the equivalence classes of n-dimensional irreducible unitary representations form a real analytic manifold M. It is also shown that the tangent space at a point $m \in M$ can be identified with the cohomology space $H^1(\pi(X), \mathrm{ad}\rho)$ where ρ is a unitary representation in the class m and ad ρ is the composite of ρ with the adjoint representation of $U(n)$ in its Lie algebra (i.e. in the skew symmetric matrices). The complex structure on M is obtained by identifying $H^1(\pi(X), \mathrm{ad}\rho)$ naturally with a complex vector space.

We remark that in the above proof, if we replace $U(n)$ by the special linear group $SL(n, \mathbb{C})$ then $R = f^{-1}(I)$ is the complex space of n-dimensional representations and U is a closed subset of R. □

Representations of type τ – More generally, given a compact Riemann surface X of genus ≥ 1, a point $x_0 \in X$ and an integer $N \geq 1$, there is a simply connected Riemann surface Y with a discrete group π acting holomorphically on it such that X is the quotient of Y by π, the covering $p : Y \to X$ is ramified precisely at x_0 and the isotropy subgroup π_y at $y \in p^{-1}x_0$ is a cyclic group of order N independent of $y \in p^{-1}x_0$. We remark that for $N = 1$, Y is the universal covering of X and $\pi = \pi(X)$. Let us fix a $y_0 \in p^{-1}x_0$ and a character $\tau : \pi_{y_0} \to \mathbb{C}^*$. A representation $\rho : \pi \to GL(n, \mathbb{C})$ is called a *representation of type τ* if the restriction of ρ to π_{y_0} satisfies the condition $\rho(h) = \tau(h)I$, I being the identity matrix.

Theorem 7.5 (Theorem A′). *The space U_0 of n-dimensional irreducible unitary representations of type τ has a natural structure of a real analytic manifold of dimension $2(n^2(g-1)+1) + (n^2-1)$. The equivalence classes of n-dimensional irreducible unitary representations of type τ form a real analytical manifold M of dimension $2(n^2(g-1)+1)$.*

The proof of Theorem A′ is similar to that of Theorem A ([NS], Proposition 9.2). One uses the fact that π is generated by $a_1, b_1, \ldots, a_g, b_g, c$ with the only relations

$$\left(\prod_i (a_i b_i {a_i}^{-1} {b_i}^{-1})\right) c = 1, c^N = 1.$$

Vector bundles associated to representations

Let X be a compact Riemann surface. Let $\rho : \pi(X) \to GL(V)$ be a representation of the fundamental group $\pi(X)$ on a vector space V. Using ρ one defines a vector bundle

E_ρ on X as follows. Let Z be the universal covering of X. The fundamental group acts on Z (by a right action by convention) and X is a quotient of Z by the action of the group $\pi(X)$. There is a left action of $\pi(X)$ on V defined by $gv = (\rho(g))^{-1}v$ for $g \in \pi(X), v \in V$. Define E_ρ as the quotient of $Z \times V$ by the action of $\pi(X)$ given by

$$g(z, v) = (zg, gv)$$

where $g \in \pi(X)$, $z \in Z$ and $v \in V$. The first projection $Z \times V \to Z$ induces a surjective morphism of quotients $E_\rho \to X$. Using the second projection one easily sees that for $x \in X$, the fiber $E_x \approx V$. In fact, E_ρ is a vector bundle on X, we call E_ρ the vector bundle associated to ρ. Clearly, E_ρ has rank n equal to the dimension of V. For $n = 1$, ρ is called a character and the associated bundle is a line bundle. Using cohomological arguments, one can check that a line bundle associated to a character has degree 0. For a representation ρ in an n-dimensional vector space V, define a character $\wedge^n\rho$ by $(\wedge^n\rho)(g) = \wedge^n(\rho(g))$ for all $g \in \pi(X)$.It is easy to see that if E_ρ is associated to ρ then the determinant of E_ρ is associated to $\wedge^n\rho$ and hence the determinant of E_ρ has degree 0. By definition, E_ρ has degree zero.

One has the following classical result relating line bundles to characters of $\pi(X)$.

Theorem 7.6. *A line bundle on X is associated to a (unitary) character of the fundamental group of X if and only if it has degree* 0.

It is natural to ask if every vector bundle of degree 0 is associated to a representation of the fundamental group. An answer to this question was provided by the following theorem of Weil ([We,Gr]):

Theorem 7.7. *A vector bundle is associated to a representation of the fundamental group if and only if all its indecomposable components have degree zero.*

One can further ask if the stability or semistability of E_ρ is related to some property of the representation ρ. Deeper results in this direction were proved by M. S. Narasimhan and C. S. Seshadri in [NS] (given below).

Theorem 7.8.

(1) *A vector bundle of degree zero on X is stable if and only if it is associated to an irreducible unitary representation of the fundamental group of X.*
(2) *A vector bundle is associated to a unitary representation of the fundamental group if and only if it is polystable, i.e., it is a direct sum of stable vector bundles of degree zero.*

Since any unitary representation of the fundamental group is a direct sum of irreducible unitary representations of the fundamental group, the second part follows from the first part.

Let E be a flat vector bundle over X associated to a unitary representation of the fundamental group of X. Given any holomorphic subbundle F of E, considering the second fundamental form of F it can seen that degree$(F) \leq 0$, and furthermore, degree$(F) = 0$ if and only if the second fundamental form vanishes. From this it follows that if E is associated to an irreducible unitary representation of the fundamental group of X, then E is stable. The more technical part of the proof of Theorem 7.8

is to show any stable vector bundle of degree zero admits a unitary flat connection; see [NS] for the details.

One then wonders about vector bundles of non-zero degrees. They are not associated to representations of $\pi(X)$, they are associated to representations of certain Fuchsian groups. In the above construction of E_ρ we may replace Z by a simply connected Riemann surface Y with a discrete group π acting holomorphically on it such that the quotient space $X = Y/\pi$. As above, we can get a vector bundle E_ρ on X associated to a representation ρ of π of type τ. The degree of E_ρ is determined as follows. There exists a generator g_0 of π such that $\tau(g_0) = \zeta^{-d}$, where ζ is a primitive N the root of unity and the integer d depends only on τ ([NS], Remark 6.1). Then the degree of E_ρ is given by $d(E_\rho) = dn/N$. If we take $n = N$, we have $d(E) = d$. We assume henceforth that $n = N$. Given an integer $d, 0 \leq -d < n$ there is a unique character $\tau(d)$ such that $\tau(d)(g_0) = \zeta^{-d}$.

Theorem 7.9 (NS). *A holomorphic vector bundle E of rank n and degree d (with $0 \leq -d(E) < n$) on a compact Riemann surface X is stable if and only if it is associated to an irreducible unitary representation $\rho : \pi \to GL(n, \mathbb{C})$ of type $\tau(d)$. Further, two such stable bundles are isomorphic if and only if the two corresponding unitary representations are equivalent.*

Proof. We give here an outline of the original proof by Narasimhan and Seshadri.

The theorem is proved by Narasimhan and Seshadri in several steps using induction on the rank n [NS], we briefly describe below the steps. The easier part is to show that a vector bundle associated to a unitary (resp. irreducible unitary) representation is stable (resp. polystable), the converse is much harder to prove.

The case $n = 1$ is classical.

(1) The vector bundles E_ρ associated to unitary representations are semistable ([NS], Proposition 10.4).

One first proves this for $\pi(X)$ and deduces the result for π. If F is a subbundle of E_ρ of rank r, then $\wedge^r F$ is a line subbundle of $\wedge^r(E_\rho) = E_{\wedge^r \rho}$, hence we are reduced to the case when F is a line subbundle. In this case one checks that $d(F) \leq 0$ by writing ρ as a direct sum of irreducible representations ρ_j and using the fact that $H^0(X, E_{\rho_j})$ is nonzero if and only if ρ_j is the trivial representation.

(2) The vector bundles E_ρ associated to irreducible representations are stable.

By (1), E_ρ is semistable. If a semistable vector bundle E is not stable, then there exists a proper stable subbundle F such that $d(F)/r(F) = d(E)/r(E)$. Using induction, one shows the existence of an irreducible unitary representation χ such that F is associated to χ. One has Hom $(\chi, \rho) \approx$ Hom (F, E_ρ) ([NS], Proposition 10.1). Thus Hom $(\chi, \rho) \neq 0$, but ρ is irreducible giving a contradiction. Hence E_ρ must be stable.

(3) Every indecomposable vector bundle (and in particular a stable vector bundle) is associated to a representation of type τ.

This is proved using Weil's Theorem ([NS], Proposition 6.2).

(4) The set of isomorphism classes of simple vector bundles of rank n, degree d on X has the structure of a complex manifold S of complex dimension $n^2(g-1)+1$.

A family of (simple) vector bundles parametrized by a complex space T is a vector bundle $\mathcal{E}$ on $T \times X$ (such that for all $t \in T$ the restriction $\mathcal{E}_t = \mathcal{E}|_{t \times X}$ is a simple bundle). Given a family of simple vector bundles, the natural map $T \to X$ sending t to the point corresponding to $\mathcal{E}_t$ is a holomorphic map.

There exists a (universal) family of simple bundles parametrized by S ([NS], Theorem 1).

(5) Given a family of vector bundles parametrized by a complex space T, the subset T_s (respectively T_{ss}) of T corresponding to stable (resp. semistable) bundles is an open set in T ([NS, Lemma 12.1] and [NS, Lemma 12.2]).

In particular, S_s and R_s (or R_{ss}) are open in S and R respectively. Here R is the complex space of n-dimensional representations.

(6) S_s is a connected open subset of S ([NS], Lemma 12.3).

To prove connectedness, one constructs a family of indecomposable vector bundles parametrized by an irreducible smooth algebraic variety (using results of Atiyah [At2]). By (5), A_s is an irreducible subset of A. Since the canonical morphism $A_s \to S_s$ (given by (4)) is surjective, S_s is irreducible, hence connected.

(7) The final step of the proof.

Let $R_1 \subset R_{ss} \times R_{ss}$ consist of points (ρ_1, ρ_2) such that there is a nontrivial homomorphism from E_{ρ_1} to E_{ρ_2}. Let $R_2 = (R_{ss} - R_s) \times (R_{ss} - R_s)$. Then $R' = R_1 \cup R_2$ is a closed subset of $R_{ss} \times R_{ss}$ and defines an equivalence relation on R_{ss} with the properties: (a) two points in R_s are equivalent if and only if the corresponding bundles on X are isomorphic (b) R_s is a saturated open subset of R_{ss}. Let $Q = R_{ss}/R'$ be the quotient and $q : R_{ss} \to Q$ the canonical mapping. By (b), $q(R_s)$ is an open subset of Q. One shows that the natural map $R_s \to S_s$ is a homeomorphism. Hence by (6), $q(R_s)$ is a connected open subset of Q. By (1) (resp. (2)), the subset U (resp. U_0) of R corresponding to unitary (resp. irreducible unitary) representations is contained in R_{ss} (resp. R_s). Since unitary bundles in S form an open subset in S [NS, Proposition 10.3], we conclude that $q(U_0)$ is an open subset in R_s. Since U is compact, $q(U)$ is closed in Q and $q(U_0) = q(U) \cap q(R_s)$ is closed in R_s. Thus $q(U_0)$ is an open and closed subset of the connected space $q(R_s)$, hence $q(U_0) = q(R_s)$. Since every stable bundle belongs to $q(R_s)$ by (3), this completes the proof of the theorem. □

Let $f : U(n)^{2g} \to U(n)$ be given by

$$(A_1, B_1, \ldots, A_g, B_g) \mapsto A_1 B_1 {A_1}^{-1} {B_1}^{-1} \ldots A_g B_g {A_g}^{-1} {B_g}^{-1}.$$

Then the Narasimhan-Seshadri theorem induces a bijection

$$U(n, d) \to f^{-1}(e^{2\pi i d/n})/U(n)$$

where the unitary group $U(n)$ acts by conjugation on all factors.

Proposition 7.10.

(1) *The bijection $U(n, d) \to f^{-1}(e^{2\pi i d/n})/U(n)$ is a homeomorphism.*
(2) *Moreover, the homeomorphism restricts to a diffeomorphism on the stable locus.*

We remark that the Narasimhan-Seshadri Theorems do not hold in positive characteristic. There is an analogue of Narasimhan-Seshadri theorem for vector bundles on a Mumford curve (a class of analytic curves) over a complete non-Archimedian valued field (see [F1,VR]).

There is an analogue of Weil's theorem for vector bundles on nodal curves. Narasimhan-Seshadri Theorem is partially valid over curves with singularities. For example, vector bundles associated to unitary representations of the fundamental group of a nodal curve are polystable, but there exit stable vector bundles on nodal curves which are not associated to any representations of the fundamental group.

8. Desingularizations of the moduli space for rank 2

The moduli space $U_{\mathcal{O}}(2,0)$ of semistable vector bundles of rank 2 with trivial determinant on a smooth curve of genus $g \geq 2$ (except for $g = 2 = n$) is a singular projective variety. Its open subvariety $U^s_{\mathcal{O}}(2,0)$ is nonsingular. Narasimhan and Ramanan constructed a desingularization N of $U_{\mathcal{O}}(2,0)$ using Hecke cycles [NR3]. Seshadri constructed a desingularization S of $U_{\mathcal{O}}(2,0)$ as the moduli space of parabolic vector bundles of rank 4, degree 0. Parabolic vector bundles are vector bundles E together with flags of vector subspaces of the fibers of E at finitely many points. Seshadri considered parabolic vector bundles of rank 4, degree 0 such that the endomorphism algebra of the underlying bundle E is isomorphic to a specialization of the matrix algebra $M(2)$ [S2]. Another desingularization of $U_{\mathcal{O}}(2,0)$ was constructed by Kirwan using successive blow ups of the singular strata [K].

The subsets corresponding to stable points (i.e. $U^s_{\mathcal{O}}(2,0)$) in all these desingularizations are isomorphic. The isomorphism of stable points of K and S can be extended to a morphism $K \to S$ which is a composite of two explicitly described blow downs [KL]. It is conjectured that the desingularization N of $U_{\mathcal{O}}(2,0)$ by Narasimhan and Ramanan lies between K and S as the middle blow up.

The explicit blowdowns are used to show that the moduli space $U_{\mathcal{O}}(2,0)$ has terminal singularities and trivial plurigenera. The stringy E-function of $U_{\mathcal{O}}(2,0)$ is determined. The Poincare polynomial (giving Betti numbers) of Seshadri's desingularization was computed in a compact form by Balaji and Seshadri [BaS].

Let S_X denote the Seshadri desingularisation for the moduli space of semistable vector bundles of rank 2 with trivial determinant on a curve X. Balaji showed that

$$S_{X_1} \cong S_{X_2} \text{ if and only if } X_1 \cong X_2.$$

9. Unirationality and rationality of moduli spaces

In the couple of sections, we study some properties of the moduli spaces. We begin with the birational classification of moduli spaces.

Lemma 9.1. *Let E be a vector bundle of rank n on X generated by sections. Then E has a trivial subbundle of rank $n-1$.*

Proof. Since E is generated by sections, the evaluation at $x \in X$ gives an exact sequence

$$0 \to N_x \to H^0(X, E) \to E_x \to 0.$$

It suffices to show that there is an $(n-1)$-dimensional subspace $W \subset H^0(E)$ such that $W \cap N_x = 0$ for all x. Let Z be the subvariety of $P = \mathbb{P}(H^0(E))$ generated by $\mathbb{P}(N_x), x \in X$. Then $\dim Z \leq 1+ \dim \mathbb{P}(N_x) = 1+\dim P - n$. Hence there is an $(n-2)$-dimensional linear subspace of P which does not intersect Z, i.e. there is an $n-1$ dimensional subspace $W \subset H^0(E)$ such that $W \cap N_x = 0$ for all x. □

Lemma 9.2.

(1) *If E is semistable (resp. stable) then $h^1(E) = 0$ if $\mu(E) > 2g-2$ (resp. $\mu(E) \geq 2g-2$).*
(2) *E is generated by sections if E is semistable and $\mu(E) > 2g-1$ (resp. E is stable and $\mu(E) \geq 2g-1$).*

Proof.

(1) Suppose that E is semistable. If $h^1(E)$ is nonzero, in view of Serre duality, it implies that there is a nonzero morphism $f : E \to K_X$. Let F be the sheaf image of f. Since E is semistable, $\mu(E) \leq \mu(F) \leq d(K_X)$. Thus $\mu(E) \leq 2g-2$. This proves the result in the semistable case. In the case E is stable, the result follows similarly.
(2) For $x \in X$, consider the exact sequence

$$0 \to E(-x) \to E \to E_x \to 0$$

From the cohomology sequence, one has

$$H^0(E) \to E_x \to H^1(E(-x)).$$

Hence E is generated by sections at x if $h^1(E(-x)) = 0$. Since $E(-x)$ is semistable with $\mu(E(-x)) = \mu(E) - 1$, the result follows from Part (1). □

Proposition 9.3. *$U_L(n, d)$ is unirational.*

Proof. Since tensoring by a line bundle gives isomorphism of fixed determinant moduli spaces, we may assume that the degree d is at least $2g-1$. By Lemmas 9.1 and 9.2, any $E \in U_L(n, d)$ can be written as an extension of L by the trivial vector bundle I_{n-1}. Such extensions are parametrized by $P = \mathbb{P}(\oplus_{n-1} H^1(L^*))$. There is an open subset S of P which parametrizes semistable bundles. Thus there is a surjective morphism from $S \to U_L(n, d)$, i.e., $U_L(n, d)$ is unirational. □

Since the intersection of two quadrics is a rational variety, from the explicit descriptions of moduli spaces for $g = n = 2$, it follows that the varieties $U_L(2, d)$ are rational for $g = 2$. Is this true for any n, g? This turned out to be a much more difficult question. Newstead claimed that for $(n, d) = 1$, the variety $U_L(n, d)$ is rational [N3]. He [N4] and others proved this in many particular cases, finally King and Schofield proved the claim for any n, d with $(n, d) = 1$ [KS].

Theorem 9.4.

(1) *For* $(n, d) = 1$, *the variety* $U_L(n, d)$ *is rational.*
(2) *The moduli space* $U(n, d)$ *is birational to* $U(h, 0) \times \mathbf{A}^{(n^2-h^2)(g-1)}$, $h = gcd(n, d)$.

In fact the result is proved in a stronger form. King and Schofield show that there is a dominant rational map $h : U(n, d) \to U(h, 0)$ such that the generic fiber of h is rational and h restricts to a dominant rational map of fixed determinant moduli spaces. Since $U_L(1, 0)$ is a point, this gives rationality of $U_L(n, d)$ in the coprime case. The generic fiber of h is rational means that the function field of $U(n, d)$ is a purely transcendental extension of the function field of $U(h, 0)$. This gives the second part of the theorem.

The question of the rationality of $U_L(n, d)$ in the non-coprime case is still open.

10. Picard groups and generalized theta functions

10.1 *Theta functions*

Let J^{g-1} denote the Jacobian of line bundles of degree $g-1$ on X. It has a canonically defined theta divisor

$$\theta = \{L \in J^{g-1} | h^0(X, L) \geq 1\}.$$

When the base field is $\mathbb{C}$, it is the zero locus of the *Riemann theta function* (a quasi-periodic function on the universal covering $\mathbb{C}^g$ of J^{g-1}). The sections of the line bundle $\mathcal{O}(k\theta)$ can be identified to theta functions of order k. These quasi-periodic functions have been studied in detail classically.

10.2 *The determinant line bundle*

For any $n \geq 1$, one can canonically define a theta divisor in $U_L(n, n(g-1))$ by

$$\theta = \{E \in U_L(n, n(g-1)) | h^0(E) \geq 1\}.$$

This is in fact a Cartier divisor on $U_L(n, n(g-1))$. More generally, let F be a vector bundle with $r(F) = n/h$ and $d(F) = (-d + n(g-1))/h, h = gcd(n, d)$. Then $\mu(E \otimes F) = g - 1$ for $E \in U_L(n, d)$. The subset of $U_L(n, d)$ defined by

$$\theta_F = \{E \in U_L(n, d) | h^0(E \otimes F) \geq 1\},$$

is a proper subset and a Cartier divisor of $U_L(n, d)$. The line bundle $\mathcal{O}(\theta_F)$ is independent of the choice of F [DN]. It is denoted by $\mathcal{L}$ and is called the determinant line bundle on $U_L(n, d)$. It is an ample line bundle. The following theorem by Drezet and Narasimhan shows the crucial role of this line bundle.

Theorem 10.1.

(1) $\mathrm{Pic}(U_L(n, d) \cong \mathbb{Z} \cdot \mathcal{L}$.
(2) $\mathrm{Pic}(U(n, d)) \cong \mathrm{Pic}(J^d) \oplus \mathbb{Z} \cdot \mathcal{L}$.
(3) *The dualising sheaf of* $U_L(n, d)$ *is isomorphic to* $\mathcal{L}^{(-2h)}$.

10.3 *Non-abelian theta functions*

By analogy with the rank 1 case, the global sections of the determinant bundle $\mathcal{L}^{||}$ are called the generalized theta functions or non-abelian theta functions of order k. Let $SU(n) := U_{\mathcal{O}}(n, 0)$. Beauville, Narasimhan and Ramanan found the following interesting relationship.

Theorem 10.2. *There is a canonical isomorphism*

$$i_\theta : H^0(SU(n), \mathcal{L}) \cong H^0(J^{g-1}, \mathcal{O}(n\theta))^*.$$

For $E \in SU(n)$ define

$$D(E) = \{L \in J^{g-1} | h^0(E \otimes L) \geq 1\}.$$

For most E, $D(E)$ is a divisor equivalent to $n\theta$, for some E, $D(E)$ equals J^{g-1}. Thus we have a rational morphism $f_\theta : SU(n) \to |n\theta|$ defined by $E \mapsto D(E)$. Here $|n\theta|$ denotes the linear system of divisors linearly equivalent to $n\theta$. For $n = 2$, the study of f_θ led to explicit descriptions of $U_L(2, d)$ [NR1,DR1,NR2]. The most general result [BV] is the following:

Theorem 10.3. *If X is non-hyperelliptic, then f_θ embeds $SU(2)$ in $|2\theta|$. If X is hyperelliptic, then f_θ embeds $SU(2)/\iota$ in $|2\theta|$, where ι is the involution on $SU(2)$ induced by the hyperelliptic involution.*

10.4 *The Verlinde formula*

From the theory of theta functions, one has

$$h^0(J^{g-1}, \mathcal{O}(k\theta)) = k^g.$$

Hence by Theorem 10.2, we have

$$h^0(SU(n), \mathcal{L}) = n^g.$$

What is the dimension of $h^0(SU(n), \mathcal{L}^k)$? Surprisingly, the answer to this question was provided by mathematical physics. For certain Conformal field theories in mathematical physics, the space of conformal blocks can be identified to the space of generalized theta functions. Verlinde conjectured the following remarkable formula [V].

Theorem 10.4.

$$\dim\ H^0(SU(n)), \mathcal{L}^k) = \sum_{S,T} \prod_{s \in S, t \in T} \left| 2 \sin \pi \frac{s-t}{n+k} \right|^{g-1},$$

where the summation is over all S, T with $S \coprod T = [1, \dots, n+k]$, $|S| = n$.

For $k = 1$, the formula gives $h^0(SU(n), \mathcal{L}) = n^g$ and for $n = 2$, it gives

$$\dim\ H^0(SU(2), \mathcal{L}^k) = \left(\frac{k}{2}+1\right)^{g-1} \sum_{i=1}^{i=k+1} \left(\sin \frac{i\pi}{k+2}\right)^{2-2g}.$$

Note that $H^i(SU(n), \mathcal{L}^k) = 0$ for $i > 0$, by the Kodaira vanishing theorem (or its extension by Grauert and Riemenschneider).

It was a challenge for mathematicians to give a rigorous proof of the Verlinde formula. A wealth of different proofs has appeared in last several years (for rank 2, [BS,Th,NRs,Rs], for general case [TSU, F2]).

11. Cohomology of the moduli spaces

The cohomology of $U_L(2, 1)$ was determined by Newstead [N0]. He computed explicitly the Betti numbers, the generators of the cohomology and the relations among them. The general case turned out far more difficult.

11.1 *Determination of Poincaré polynomials*

The proving of Weil conjectures by Deligne paved way to the determination of Poincaré polynomials (i.e., polynomials whose coefficients are the Betti numbers) of $U_L(n, d)$, with $(n, d) = 1$. Let X denote an irreducible non-singular projective curve defined over a finite field with q elements and of characteristic $p > 0$. Let K be the function field of X and let $\mathcal{A}$ denote the adeles over K. Let ζ_X be the zeta function of X. A vector bundle E of rank n defines a maximal compact subgroup $\mathbf{k}$ of $SL(n, \mathcal{A})$. There is a surjective map from the double cosets $\mathbf{k}|\mathrm{SL}(n, \mathcal{A})/\mathrm{SL}(n, K)$ to the set of isomorphism classes of vector bundles of rank n and with a fixed determinant on X. The fact that the Tamagawa number of the special linear group $SL(n, K)$ is 1 leads to the following Siegel formula

$$\sum_x \frac{1}{|x^{-1}\mathbf{k}x \cap SL(n, k)|} = q^{(n^2-1)(g-1)} \zeta_X(2) \cdots \zeta_X(n),$$

where x runs through the double cosets. Since $|x^{-1}\mathbf{k}x \cap SL(n, k)|$ can be interpreted as the number of automorphisms of the vector bundle corresponding to x and automorphisms of the stable bundles are scalars, this formula can be rewritten as

$$|U_L(n, d)| = q^{(n^2-1)(g-1)} \zeta_X(2) \cdots \zeta_X(n) - \sum_E \frac{1}{|Aut\, E|}, \qquad (11.1)$$

where $|U_L(n, d)|$ denotes the number of F_q-rational points of $U_L(n, d)$ for $(n, d) = 1$, $|Aut\, E|$ denotes the number of automorphisms of E and the summation varies over all unstable vector bundles E. Weil's conjectures give a very simple recipe to write down the Poincaré polynomial of $U_L(n, d)$ from the expression for $|U_L(n, d)|$. Thus the computation of Poincaré polynomial reduces to determination of the sum $\sum_E \frac{1}{|Aut\, E|}$, over non-semistable bundles E. This computation can be done using the Harder-Narasimhan filtrations for non-semistable vector bundles by induction on rank [DR1]. The computation is done in [HN] working entirely with adele group directly from Siegel formula. Carrying on the computation with vector bundles gives greater clarity and better results. Thus one has an inductive formula to determine the Poincaré polynomial of $U_L(n, d)$. There is another proof of this by Atiyah and Bott using Gauge theory, it generalizes these results to principal G-bundles.

Zagier solved the recurrence relation (inductive formula) and gave an explicit closed expression for the right hand side of (11.1) by an ingenious elementary method [Z]. This made it easier to compute Betti numbers using a computer.

11.2 *Cohomology ring of the moduli space in the coprime case*

Let X be a compact Riemann surface. Atiyah and Bott showed that the cohomology of the moduli space $U(n, d)$ on X has no torsion [AB]. The pioneering work of Atiyah and Bott brings together Morse theory and the theory of vector bundles and principal G-bundles on curves. The catalyst that produced this interaction was Yang-Mills equations in theoretical physics. Morse theory is concerned with the relation between (co)homology of a manifold and critical points of a functional on it. Atiyah and Bott observed that the space $\mathcal{A}$ of unitary connections on a fixed C^∞ vector bundle E over a compact Riemann surface X can be viewed as the the space $\mathcal{C}$ of all holomorphic structures on E. The Yang-Mills functional f on $\mathcal{A}$ was defined by associating to a connection the square of the norm of its curvature. The main idea was to apply Morse theory methods to this functional. The connections corresponding to bundles arising from representations of $\pi(X)$ give critical points of f and those corresponding to irreducible representations give Yang-Mills minimum. By Narasimhan-Seshadri theorem, stable holomorphic bundles are associated to irreducible unitary representations of the fundamental group. Hence the moduli space M of stable holomorphic vector bundles is the quotient of an open subset of $\mathcal{A}$ by the gauge group, the group of automorphisms of the fixed C^∞ vector bundle E. Using Morse theory, results on the cohomology of M were deduced. Atiyah and Bott did this in a more general set up for principal G-bundles.

Atiyah and Bott also gave an explicit set of generators of the cohomology ring. These are the Künneth components, in the cohomology $H^*(U(n, d))$, of the Chern classes of a universal bundle V over $U(n, d) \times X$. Ideal of relations satisfied by these generators was studied by many authors. For $n = 2$, the ideal is generated by so called Mumford relations. These relations are obtained by computing, in terms of generators, the expressions $c_r(\pi_*(V))$ for $r > n$ (taking $(2g - 2)n < d < (2g - 1)n$) [KN]. For $n > 2$, the Mumford relations don't suffice. A complete set of generating relations is constructed by suitably generalizing the Mumford relations [KE]. These relations are obtained by taking numbers $\hat{d}, \hat{n}, \hat{d}/\hat{n} > d/n, 0 < \hat{n} < n$ and equating to zero slant products of Chern classes $c_r(-\pi_*(Hom(\hat{V}, V)))$ for $r > n$, $\hat{V}$ being the universal bundle over $U(\hat{n}, \hat{d}) \times X$. Special care has to be taken when $(\hat{n}, \hat{d}) \neq 1$, since $\hat{V}$ is not well-defined in that case. The proof uses Atiyah-Bott stratification by Yang-Mills functional and equivariant cohomology.

The algebraic cohomology ring is the subring of the rational cohomology ring spanned by the fundamental classes of algebraic cycles. Balaji, King and Newstead computed the algebraic cohomology ring of $U_L(2,1)$ in terms of algebraic cohomology ring of the Jacobian of the curve [BKN]. Biswas and Narasimhan defined explicitly a subalgebra I of the cohomology algebra generated entirely by algebraic cohomology classes for $U(n,d)$ and $U_L(n,d)$. For a general curve, they showed that any Hodge class (i.e. a rational cohomology class of type (p,p)) is in I. In particular, any Hodge class is generated by algebraic cycles [BN], thus proving the Hodge conjecture for $U(n,d)$ and $U_L(n,d)$ over a general curve.

12. Higgs bundles over a curve

Let X be a compact connected Riemann surface. Let F be a holomorphic vector bundle over X. A *Higgs field* on F is a $\mathcal{O}_M$-linear homomorphism

$$\theta : F \longrightarrow F \otimes \Omega^1_X,$$

where Ω^1_X is the holomorphic cotangent bundle of X. A pair (F,θ), where θ is a Higgs field on F, is called a *Higgs bundle*.

A Higgs bundle (F,θ) is called *semistable* if for all nonzero holomorphic proper subbundles $F' \subset F$ with

$$\theta(F') \subset F' \otimes \Omega^1_X,$$

the inequality

$$\frac{\text{degree}(F')}{\text{rank}(F')} \leq \frac{\text{degree}(F)}{\text{rank}(F)}$$

holds. If the stronger inequality

$$\frac{\text{degree}(F')}{\text{rank}(F')} < \frac{\text{degree}(F)}{\text{rank}(F)}$$

holds, then (F,θ) is called *stable*. A semistable Higgs bundle (F,θ) is called *polystable* if it is a direct sum of stable Higgs bundles; see [Hi1,Hi2,BNR].

Let $\mathcal{M}_H$ denote the moduli space of semistable Higgs bundles over X of rank r and degree d; see [Hi1,Ni] for the construction of this moduli space. Note that the total space of the cotangent bundle of the moduli space of stable vector bundles of rank r and degree d is contained in $\mathcal{M}_H$ as a Zariski open dense set. This moduli space $\mathcal{M}_H$ has a natural symplectic structure which extends the natural symplectic structure on the cotangent bundle of the moduli space of stable vector bundles.

Let

$$\mathcal{H} : \mathcal{M}_H \longrightarrow \bigoplus_{i=1}^{r} H^0(X, K^{\otimes r})$$

be the morphism defined by

$$(E, \theta) \longmapsto (\mathrm{trace}(\theta), \ldots, \mathrm{trace}(\theta^r)).$$

This morphism $\mathcal{H}$ is called the *Hitchin map*. This Hitchin map is proper [Hi1,Ni]. Furthermore, this map defines an algebraically completely integrable system. In other words, the function $\mathcal{H}$ is Poisson commuting of maximal rank, and the general fiber is a Lagrangian abelian subvariety of $\mathcal{M}_H$ [Hi1,Hi2,BNR].

Let E be a holomorphic vector bundle over X. Let $\overline{\partial}_E$ be the Dolbeault operator on E defining the holomorphic structure of the vector bundle E. For any Hermitian metric h on E, there is a unique connection ∇^h on E such that

- the connection ∇^h preserves the metric h, i.e., the connection ∇^h is unitary, and
- the (01)-part of ∇^h coincides with $\overline{\partial}_E$.

This connection ∇^h is also called the *Chern connection* for the Hermitian metric h on E.

Let (E, θ) be a stable Higgs bundle over X of rank r and degree zero. There is a Hermitian metric h_0 on E, unique up to a constant scalar multiplication, such that $\nabla^{h_0} + \theta + \theta^*$ is a flat connection on E. Here ∇^{h_0} is the Chern connection for the Hermitian metric h_0 on E, and θ^* is the adjoint of θ with respect to h_0 [Hi1,Do]. Furthermore, all irreducible flat connections of rank r on X arise from stable Higgs bundles of rank r and degree zero over X [Do,Hi1].

References

[At1] M. F. Atiyah, On the Krull–Schmidt theorem with application to sheaves, *Bull. Soc. Math. Fr.*, **84** (1956) 307–317.

[At2] M. F. Atiyah, Vector bundles over an elliptic curve, *Proc. London Math. Soc.*, **7** (1957) 412–452.

[At3] M. F. Atiyah, K theory, W. A. Benjamin Inc., Amsterdam (1997).

[AB] M. F. Atiyah and R. Bott, The Yang-Mills equations over Riemann surfaces, *Phil. Trans. Royal Soc. Lond. Ser. A*, **308** (1982) 523–615.

[AS] D. Arapura and P. Sastry, Intermediate Jacobians and Hodge structures of moduli spaces, *Proc. Indian Acad. Sci. (Math. Sci.)*, **110** (2000) no. 1, 1–26.

[Ba] V. Balaji, Intermediate Jacobians of some moduli spaces of vector bundles on curves, *Amer. J. Math.*, **112** (1990) no. 4, 611–629.

[BaS] V. Balaji and C. S. Seshadri, Poincaré Polynomials of some moduli varieties, *Algebraic geometry and analytic geometry, Kyoto (1990), ICM-90 Satell. Conf. Proc., Springer, Tokyo*, (1991) 1–25.

[B1] U. N. Bhosle (Desale), Moduli of orthogonal and spin bundles over hyperelliptic curves, *Compositio Math.*, **51** (1984) 15–40.

[B2] U. N. Bhosle, Pencils of quadrics and hyperelliptic curves in characteristic two, *Jour. Reine Angew. Math.*, **407** (1990) 75–98.

[BKN] V. Balaji, A. D. King and P. E. Newstead, Algebraic cohomology of the moduli space of rank 2 vector bundles on a curve, *Topology*, **36** (1997) no. 2, 567–577.

[BN] I. Biswas and M. S. Narasimhan, Hodge classes of moduli spaces of parabolic bundles over the general curve, *J. algebraic Geom.*, **6** (1997) no. 4, 697–715.

[BNR] A. Beauville, M. S. Narasimhan and S. Ramanan, Spectral curves and the generalised theta divisor, *J. reine angew. Math.*, **398** (1989) 169–179.

[BS] A. Bertram and A. Szenes, Hilbert polynomials of moduli spaces of rank 2 vector bundles II, *Topology*, **32** (1993) 599–609.

[BV] S. Brivio and A. Verra, The Theta divisor of $SU_C(2)$ is very ample if C is not hyperelliptic and Noether-Lefschetz general, *Duke Math. Jour.*, **82** (1996) 503–552.

[C] L. Caporaso, A compactification of the universal Picard variety over the moduli space of stable curves, *J. Amer. Math. Soc.*, **7** (1994) no. 3, 589–660.

[DN] J.-M. Drezet and M. S. Narasimhan, Groupe de Picard des varieties de modules de fibres semi-stables sur les courbes algebraiques, *Invent math.*, **97** (1989) 53–94.

[DR1] U. V. Desale and S. Ramanan, Poincaré Polynomials of the variety of stable bundles, *Math. Ann.*, **216** (1975) 233–244.

[DR2] U. V. Desale and S. Ramanan, Classification of vector bundles of rank 2 on hyperelliptic curves, *Invent. Math.*, **38** (1976) 161–185.

[Do] S. K. Donaldson, Twisted harmonic maps and self–duality equations, *Proc. London Math. Soc.*, **55** (1987) 127–131.

[F1] G. Faltings, Semistable vector bundles on Mumford curves, *Invent. Math.*, **74** (1983) no. 2, 199–212.

[F2] G. Faltings, A proof for Verlinde formula, *Jour. Alg. Geom.*, **3** (1994) 347–374.

[Ga] O. Garcia-Prada, Moduli spaces and geometric structures. To appear as an appendix in the book: Differential analysis on complex manifolds by R. O. Wells, 3rd edition, GTM 65, Springer.

[Gr] A. Grothendieck, Sur le mémoire de Weil-Generalisation des fonctions abéliennes, Séminaire Bourbaki, 9, expose 141 (1956–57).

[Gu] R. C. Gunning, Lectures on vector bundles over Riemann surfaces. Princeton Mathematical notes, 6, Princeton university press, New Jersey; University of Tokyo press, Tokyo, (1967).

[Ha] R. Hartshorne, Algebraic Geometry, Graduate Texts in Mathematics, vol. 52, Springer–Verlag, New York (1977).

[HN] G. Harder and M. S. Narasimhan, On the cohomology groups of moduli spaces of vector bundles on curves, *Math. Ann.*, **212** (1975) 215–248.

[Hi1] N. J. Hitchin, The self-duality equations on a Riemann surface, *Proc. Lond. Math. Soc.*, **55** (1987) 59–126.

[Hi2] N. J. Hitchin, Stable bundles and integrable systems, *Duke Math. Jour.*, **54** (1987) 91–114.

[Ja] N. Jacobson, Basic Algebra, II., W. H. Freeman and Company, San Francisco (1980).

[K] F. Kirwan, Partial desingularisations of quotients of nonsingular varieties and their Betti numbers, *Ann. Math.*, **122** (1985) 41–85.

[Ko] S. Kobayashi, Differential Geometry of Complex Vector Bundles, Publications of the Mathematical Society of Japan 15, Iwanami Shoten Publishers and Princeton University Press (1987).

[KE] F. Kirwan and R. Earl, Complete set of relations in cohomology ring of moduli of holomorphic bundles, *Proc. London Math. Soc.*, **89** (2004) 570–622.

[KL] Y-H. Kiem and J. Li, Desingularisations of the moduli space of rank 2 bundles over a curve, *Math. Ann.*, **330** (2004) 491–518.

[KN] A. D. King and P. E. Newstead, On the cohomology ring of the moduli space of rank 2 vector bundles on a curve, *Topology*, **37** (1998) no. 2, 407–418.

[KP] A. Kouvidakis and T. Pantev, The automorphism group of the moduli space of semistable vector bundles, *Math. Ann.*, **302** (1995) 225–268.

[KS] A. King and A. Schofield, Ratinality of moduli spaces on curves, *Indag. Math. (N.S.)*, **10** (1999) no. 4, 519–535.

[M1] D. Mumford, Projective invariants of projective structures and applications, Proc. Intern. Cong. Math., Stockholm, (1962) 526–530.

[M2] D. Mumford, Geometric invariant theory, Springer-Verlag, (1965).

[MN] D. Mumford and P. E. Newstead, Periods of a moduli space of bundles on curves, *Amer. J. Math.*, **90** (1968) 1200–1208.

[N0] P. E. Newstead, Topological properties of some spaces of stable bundles, *Topology*, **6** (1967) 241–262.

[N1] P. E. Newstead, Stable bundles of rank 2 and odd degree over a curve of genus 2, *Topology*, **7** (1968) 205–215.

[N2] P. E. Newstead, Introduction to moduli problems and orbit spaces, TIFR lecture notes, Springer (1978).

[N3] P. E. Newstead, Rationality of moduli spaces of stable bundles, *Math. Ann.*, **215** (1975) 251–268.

[N4] P. E. Newstead, Correction to 'Rationality of moduli spaces of stable bundles', *Math. Ann.*, **249** (1980) 281–282.

[Ni] N. Nitsure, Moduli space of semistable pairs on a curve, *Proc. Lond. Math. Soc.*, **62** (1991) 275–300.

[NR1] M. S. Narasimhan and S. Ramanan, Moduli of vector bundles on a compact Riemann surface, *Ann. of Math.*, **89** (1969) 19–51.

[NR2] M. S. Narasimhan and S. Ramanan, Deformations of a moduli space of vector bundles, *Ann. of Math.*, **101** (1975) 391–417.

[NR3] M. S. Narasimhan and S. Ramanan, Geometry of Hecke cycles, *C. P. Ramanujam – a tribute, Tata Inst. Fund. Res. Studies in Math.*, **8** Springer (1978) 291–345.

[NRs] M. S. Narasimhan and T. R. Ramadas, Factorization of generalised Theta functions I, *Invent. Math.*, **114** (1993) 565–623.

[NS] M. S. Narasimhan and C. S. Seshadri, Stable and unitary vector bundles on a compact Riemann surface, *Annals of Math.*, **82** (1965) 540–567.

[O] T. Oda, Vector bundles on an elliptic curve, *Nagoya Math. J.*, **43** (1971) 41–72.

[P] R. Pandharipande, A compactification over $\overline{\mathcal{M}}_g$ of the universal moduli space of slope semistable vector bundles, *J. Amer. Math. Soc.*, **9** (1996) no. 2, 425–471.

[R] S. Ramanan, The moduli spaces of vector bundles over an algebraic curve, *Ann. of Math.*, **200** (1973) 69–84.

[Rs] T. R. Ramadas, Factorization of generalised Theta functions II, *Topology*, **35** (1996) no. 3, 641–654.

[S] C. S. Seshadri, Space of unitary vector bundles on a compact Riemann surface, *Ann. of Math.*, **85** (1967) 303–336.

[S2] C. S. Seshadri, Desingularisation of the moduli variety of vector bundles on curves, Proc. Internat. Symposium on Alge. Geom., Kyoto, (1977) 155–184.

[Th] M. Thaddeus, Stable pairs, linear systems and the Verlinde formula, *Invent. Math.*, **117** (1994) no. 2, 317–353.

[Tn] A. N. Tjurin, Classification of vector bundles over an algebraic curve of arbitrary genus, *Amer. Math. Soc. Trans.*, **63** (1967) 245–279.

[Tn1] A. N. Tjurin, The geometry of moduli of vector bundles, *Russ. Math Surveys*, **29** (1974) 57–88.

[Tu] L. Tu, Semistable vector bundles over elliptic curves, *Adv. Math.*, **98** (1993) 1–26.

[TUY] A. Tsuchiya, K. Ueno and Y. Yamada, Conformal field theory on universal family of stable curves with gauge symmetries, *Adv. studies in Pure Math.*, **19** (1989) 459–566.

[V] E. Verlinde, Fusion rules and modular transformations in 2d conformal field theory, *Nuclear Physics*, **B300** (1988) 360–376.

[VR] M. van der Put and M. Reversat, Fibr'es vectoriels semi-stables sur une courbe de Mumford, *Math. Ann.*, **273** (1986) 573–600.

[We] A. Weil, Generalisation des fonctions abeliennes, *Jour. Math. Pure Appl.*, **17** (1938), 47–87.

[Z] D. Zagier, Elementary aspects of the Verlinde formula and of Harder-Narasimhan-Atiyah-Bott formula, *Proc. Hirzebruch 65 Conference on Algebraic Geometry, Ramat Gan*, (1993), 445–462; *Israel Math. Conf. Proc.*, **9** Bar-Elan Univ., Ramat Gan (1996).

Teichmüller Theory and Moduli Problem No. 10, 2010, pp. 95–111.

Moduli Spaces of Connections on a Riemann Surface

Indranil Biswas[1] and Vicente Muñoz[2]

[1]School of Mathematics, Tata Institute of Fundamental Research, Homi Bhabha Road, Bombay 400 005, India
[2]Instituto de Ciencias Matemáticas CSIC-UAM-UC3M-UCM, Consejo Superior de Investigaciones Científicas, Serrano 113 bis, 28006 Madrid, Spain
e-mail: indranil@math.tifr.res.in; vicente.munoz@mat.ucm.es

Abstract. Let E be a holomorphic vector bundle over a compact connected Riemann surface X. The vector bundle E admits a holomorphic projective connection if and only if for every holomorphic direct summand F of E of positive rank, the equality $\text{degree}(E)/\text{rank}(E) = \text{degree}(F)/\text{rank}(F)$ holds. Fix a point x_0 in X. There is a logarithmic connection on E, singular over x_0 with residue $-\frac{d}{n}\,\text{Id}_{E_{x_0}}$ if and only if the equality $\text{degree}(E)/\text{rank}(E) = \text{degree}(F)/\text{rank}(F)$ holds. Fix an integer $n \geq 2$, and also fix an integer d coprime to n. Let $\mathcal{M}(n,d)$ denote the moduli space of logarithmic $\text{SL}(n,\mathbb{C})$-connections on X singular over x_0 with residue $-\frac{d}{n}\,\text{Id}$. The isomorphism class of the variety $\mathcal{M}(n,d)$ determines the isomorphism class of the Riemann surface X.

2000 Subject Classification: 14H60, 14C30

1. Introduction

Let X be a compact connected Riemann surface. Let E be a holomorphic vector bundle over X of rank n. A holomorphic connection on E is given by locally defined holomorphic trivializations of E such that all the transition functions are locally constant maps to $\text{GL}(n,\mathbb{C})$. We recall that if $\{U_i\}_{i\in I}$ is a covering of X by open subsets and

$$\varphi_i : E_{U_i} \longrightarrow U_i \times \mathbb{C}^n$$

are holomorphic isomorphisms, then for an ordered pair $(i,j) \in I \times I$, the transition function

$$g_{i,j} : U_i \cap U_j \longrightarrow \text{GL}(n,\mathbb{C})$$

is the unique function satisfying the identity

$$g_{i,j} \circ \varphi_i = \varphi_j .$$

Therefore, if E is equipped with a holomorphic connection, then it makes sense to talk of locally constant holomorphic sections of E.

A holomorphic projective connection on E is given by locally defined holomorphic trivializations of E such that all the transition functions $g_{i,j}$ project to locally constant maps under the projection $\text{GL}(n,\mathbb{C}) \longrightarrow \text{PGL}(n,\mathbb{C})$. Therefore, if E is equipped with a holomorphic projective connection, then the projective bundle $\mathbb{P}(E)$ has locally defined holomorphic trivializations such that all the transition functions are locally constant maps to $\mathrm{PGL}(n, \mathbb{C})$.

Fix a point $x_0 \in X$. Set $X' := X \setminus \{x_0\}$ to be the complement. A logarithmic connection on E singular at x_0 is a holomorphic connection on $E|_{X'}$ which, locally around x_0, has a holomorphic connection matrix with a single pole at x_0. Given a logarithmic connection on E singular at x_0, the behavior of the flat sections near x_0 are captured by what is called the residue of the connection. The residue is a linear endomorphism of the fiber E_{x_0}. If the residue of a logarithmic connection on E singular at x_0 is a scalar multiple of the identity automorphism of E_{x_0}, then the logarithmic connection gives a holomorphic projective connection on E. Here we shall consider logarithmic connections whose residue is a scalar multiple of the identity automorphism.

We investigate some properties of holomorphic projective connections and logarithmic connections. Especially we address the question of existence of such connections on a given holomorphic vector bundle.

In the last section we explain a recent result of the authors on the moduli spaces of logarithmic connections.

2. Connections and projective connections

Let X be a compact connected Riemann surface; alternatively, X is an irreducible smooth projective curve defined over the field of complex numbers. The complexified tangent bundle $T^{\mathbb{R}}X \otimes_{\mathbb{R}} \mathbb{C}$ decomposes as

$$T^{\mathbb{R}}X \otimes_{\mathbb{R}} \mathbb{C} = T^{1,0}X \oplus T^{0,1}X$$

into $(1, 0)$ and $(0, 1)$ components. The dual line bundle $(T^{1,0}X)^*$ will also be denoted by $\Omega_X^{1,0}$ or K_X, and the complex line bundle $(T^{0,1}X)^*$ will also be denoted by $\Omega_X^{0,1}$.

A holomorphic vector bundle over X is a C^∞ vector bundle of rank n whose transition functions are given by holomorphic maps to $\mathrm{GL}(n, \mathbb{C})$. The usual Dolbeault operator $\bar{\partial}$-operator on locally defined smooth functions over X gives a Dolbeault operator $\bar{\partial}_E$ on a holomorphic vector bundle E. This operator $\bar{\partial}_E$ is a first order differential operator that sends smooth sections of E to those of $E \otimes \Omega_X^{0,1}$.

We note that the line bundles $T^{1,0}X$ and K_X have natural holomorphic structures. The trivial holomorphic line bundle $X \times \mathbb{C}$ will also be denoted by $\mathcal{O}_X$.

Definition 2.1. *Let E be a holomorphic vector bundle over X of rank n. A holomorphic connection on E is a first order holomorphic differential operator*

$$\mathcal{D} : E \longrightarrow E \otimes K_X$$

satisfying the Leibniz identity, which says that $\mathcal{D}(fs) = f\mathcal{D}(s) + \mathrm{d}f \otimes s$, where f (respectively, s) is a locally defined holomorphic function (respectively, holomorphic section of E).

Associated to a holomorphic connection $\mathcal{D}$ there is a C^∞ connection

$$\nabla : C^\infty(E) \longrightarrow C^\infty(E \otimes T^*X),$$

where $C^\infty(E)$ denotes the sheaf of smooth sections of the vector bundle E. Using the isomorphism

$$\Gamma(E) = \mathcal{O}(E) \otimes_{\mathcal{O}} C^\infty,$$

where $\mathcal{O}(E)$ is the sheaf of holomorphic sections of E, we have

$$\nabla(fs) = f\mathcal{D}(s) + \mathrm{d}f \otimes s,$$

for s a holomorphic section of E and $f \in C^\infty(X)$. In other terms, we define $\nabla = \mathcal{D} + \bar{\partial}_E$ on holomorphic sections, and then we extend to C^∞ sections.

Let $\mathcal{D}$ be a holomorphic connection on E. The curvature $\mathcal{D} \circ \mathcal{D}$ of $\mathcal{D}$ is a holomorphic section of the vector bundle $\mathrm{End}(E) \otimes \Omega_X^{2,0}$. Since $\dim_{\mathbb{C}} X = 1$, we have $\Omega_X^{2,0} = 0$. Hence a holomorphic connection on a Riemann surface is automatically flat.

If E has a holomorphic connection, then $\mathrm{degree}(E) = 0$ [At]. Actually, the connection ∇ is also flat, since its curvature is $F_\nabla = \mathcal{D}\bar{\partial}_E + \bar{\partial}_E\mathcal{D}$, which is zero on holomorphic sections, hence $F_\nabla = 0$. Considering a local flat trivialization, the constant sections (in the local trivialization) are automatically holomorphic, hence there are local flat holomorphic trivializations of E.

Note also that any flat connection ∇ on a C^∞ vector bundle E gives rise to a holomorphic structure on E together with a holomorphic connection. For this, we only need to take flat trivializations and declare them to be holomorphic.

Let E be a holomorphic vector bundle over X and F a direct summand of E. This means that F is a holomorphic subbundle of E, and there exists a holomorphic subbundle F' of E such that the natural homomorphism $F \oplus F' \longrightarrow E$ is an isomorphism. Let $\iota : F \hookrightarrow E$ be the inclusion map. Fix a complement F' of F as above. Using the natural isomorphism $F \oplus F' \longrightarrow E$ we get a projection

$$q : E \longrightarrow F.$$

If $\mathcal{D}$ is a holomorphic connection on E, then it is easy to see that the composition

$$F \overset{\iota}{\hookrightarrow} E \overset{\mathcal{D}}{\longrightarrow} E \otimes K_X \overset{q\otimes \mathrm{Id}}{\longrightarrow} F \otimes K_X$$

is a holomorphic connection on F. Hence if E admits a holomorphic connection, then each direct summand of E also admits a holomorphic connection. So, if E admits a holomorphic connection, then the degree of each direct summand of E is zero. A theorem due to Atiyah and Weil says that the converse is also true.

Theorem 2.2 ([At],[We]). *A holomorphic vector bundle E over X admits a holomorphic connection if and only if the degree of each direct summand of E is zero.* □

In particular, any indecomposable bundle E of degree zero admits a holomorphic connection.

A holomorphic vector bundle E of rank n over X admits a holomorphic connection if and only if we can choose (holomorphic) local trivializations of E over a covering of X (by open sets) such that all the transition functions are locally constant functions to $\mathrm{GL}(n, \mathbb{C})$. This follows from the flatness of ∇.

Recall that the Atiyah bundle $\mathrm{At}(E)$ associated to a holomorphic vector bundle E is the holomorphic vector bundle associated to the sheaf of first order holomorphic differential operators on E. The *Atiyah exact sequence* for E is

$$0 \longrightarrow \mathrm{End}(E) \longrightarrow \mathrm{At}(E) \longrightarrow T^{1,0}X \longrightarrow 0 \tag{2.1}$$

where the last arrow is the symbol map. Note that $\mathrm{End}(E)$ is identified with the zeroth order differential operators. Therefore, a holomorphic connection on E corresponds to a holomorphic splitting of the Atiyah exact sequence in (2.1).

Now let P_E be the holomorphic principal $\mathrm{GL}(n,\mathbb{C})$–bundle over X defined by E. So P_E is the space of all bases in the fibers of E. Let $\mathcal{O}_X \subset \mathrm{End}(E)$ be the line subbundle given by the sheaf of endomorphisms of E of the type $s \longmapsto f \cdot s$. Therefore, from (2.1) we have the exact sequence of vector bundles

$$0 \longrightarrow \mathrm{End}(E)/\mathcal{O}_X \longrightarrow \mathrm{At}(E)/\mathcal{O}_X \longrightarrow T^{1,0}X \longrightarrow 0 \tag{2.2}$$

over X. This exact sequence is the Atiyah bundle for the principal $\mathrm{PGL}(n,\mathbb{C})$-bundle over X corresponding to E. We note that this principal $\mathrm{PGL}(n,\mathbb{C})$-bundle is the quotient space $P_E/\mathbb{C}^*$, where $\mathbb{C}^*$ is considered as a subgroup of $\mathrm{GL}(n,\mathbb{C})$ through scalar multiplications of $\mathbb{C}^n$. Also, note that the quotient vector bundle $\mathrm{End}(E)/\mathcal{O}_X$ is identified with the subbundle of $\mathrm{ad}(E) \subset \mathrm{End}(E)$ defined by trace zero endomorphisms. Indeed, the natural projection of the subbundle $\mathrm{ad}(E)$ to the quotient bundle $\mathrm{End}(E)/\mathcal{O}_X$ is an isomorphism.

Definition 2.3. *A holomorphic projective connection on E is a holomorphic splitting of the exact sequence in* (2.2). *By a projective connection we will always mean a holomorphic projective connection.*

It follows that a holomorphic vector bundle E admits a projective connection if we can choose holomorphic local trivializations of E over a covering of X (by open sets) such that images of all the transition functions are locally constant under the projection $\mathrm{GL}(n,\mathbb{C}) \longrightarrow \mathrm{PGL}(n,\mathbb{C})$. This means that they are of the form $f \cdot \mathrm{Id}_{n\times n}$ times a locally constant function, with f being some nonzero holomorphic function.

To see this, consider the splitting of the exact sequence (2.2) given by the projective connection, and take a covering by open sets U_α. Over each U_α, take an arbitrary splitting of $\mathrm{At}(E) \longrightarrow \mathrm{At}(E)/\mathcal{O}_X$. This produces a holomorphic connection $\mathcal{D}_\alpha$ on U_α, and hence a flat trivialization of $E|_{U_\alpha}$. Since $\mathcal{D}_\alpha - \mathcal{D}_\beta = f_{\alpha\beta} \cdot \mathrm{Id}_{n\times n}$ on $U_\alpha \cap U_\beta$, we have that the transition functions of E (with respect to the chosen trivializations) are of the form $f \cdot \mathrm{Id}_{n\times n}$ times a locally constant function.

Assume that the vector bundle E admits a projective connection. Let E_{PGL} be the principal $\mathrm{PGL}(n,\mathbb{C})$-bundle given by E. We noted earlier that $E_{\mathrm{PGL}} = P_E/\mathbb{C}^*$, where P_E is the principal $\mathrm{GL}(n,\mathbb{C})$-bundle given by E. Giving a projective connection on E is equivalent to giving a holomorphic connection on the principal $\mathrm{PGL}(n,\mathbb{C})$-bundle E_{PGL}, that is, a family of flat trivializations of E_{PGL} such that the transition functions are locally constant functions with values in $\mathrm{PGL}(n,\mathbb{C})$. Note that this also can be expressed in terms of giving trivializations of the projective bundle $\mathbb{P}(E)$ whose transition functions are locally constant functions in $\mathrm{PGL}(n,\mathbb{C})$.

Using the commutative diagram of vector bundles

$$\begin{array}{ccccccccc} 0 & \longrightarrow & \mathrm{End}(E) & \longrightarrow & \mathrm{At}(E) & \longrightarrow & T^{1,0}X & \longrightarrow & 0 \\ & & \downarrow & & \downarrow & & \downarrow & & \\ 0 & \longrightarrow & \mathrm{End}(E)/\mathcal{O}_X & \longrightarrow & \mathrm{At}(E)/\mathcal{O}_X & \longrightarrow & T^{1,0}X & \longrightarrow & 0 \end{array}$$

over X, a holomorphic connection on E induces a projective connection on E. However, the converse is not true in the sense that a vector bundle admitting a projective connection need not admit a holomorphic connection.

If E admits a projective connection and $\mathrm{degree}(E) = 0$ then the projective connection is induced by a holomorphic connection. To see this, take a covering by open sets U_α with a holomorphic connection $\mathcal{D}_\alpha$ on each U_α, as before. Since $\mathcal{D}_\alpha - \mathcal{D}_\beta = f_{\alpha\beta} \cdot \mathrm{Id}_{n\times n}$ on $U_\alpha \cap U_\beta$, we have that $f_{\alpha\beta} \in \Omega^1_X(U_\alpha \cap U_\beta)$ form a cocycle, and hence define a class in $H^1(X, K_X) = \mathbb{C}$. Therefore there exists $g_\alpha \in \Omega^1_X(U_\alpha)$ such that if we modify $\mathcal{D}_\alpha$ to $\widetilde{\mathcal{D}}_\alpha = \mathcal{D}_\alpha + g_\alpha \cdot \mathrm{Id}_{n\times n}$, then the corresponding $\widetilde{\mathcal{D}}_\alpha - \widetilde{\mathcal{D}}_\beta$ is a constant multiple of the identity. This gives transition functions for E which are locally constant functions in $\mathrm{GL}(n, \mathbb{C})$.

The following theorem classifies all holomorphic vector bundles over X that admit a projective connection.

Theorem 2.4. *Let E be a holomorphic vector bundle over X of rank n and degree d. Then the following two statements are equivalent:*

(1) *The vector bundle E admits a projective connection.*
(2) *For any direct summand $F \subset E$, the equality*

$$\frac{\mathrm{degree}(F)}{\mathrm{rank}(F)} = \frac{d}{n}$$

holds.

Proof. Assume that the vector bundle E admits a projective connection. Let E_{PGL} be the principal $\mathrm{PGL}(n, \mathbb{C})$-bundle given by E. The projective connection on E gives a holomorphic connection on E_{PGL}. The adjoint vector bundle over X for the principal $\mathrm{PGL}(n, \mathbb{C})$-bundle E_{PGL} is $\mathrm{ad}(E)$. Therefore $\mathrm{ad}(E)$ admits a family of trivializations with transitions functions locally constant functions on $\mathrm{PGL}(n, \mathbb{C})$. This means that $\mathrm{ad}(E)$ admits a projective connection. Since the degree of $\mathrm{ad}(E)$ is zero, the vector bundle $\mathrm{ad}(E)$ admits a holomorphic connection.

Let $\mathcal{D}$ be a holomorphic connection on $\mathrm{ad}(E)$. Assume that

$$E = F_1 \oplus F_2.$$

We will show that $\mathcal{D}$ induces a holomorphic connection on the vector bundle $\mathrm{Hom}(F_1, F_2)$.

For this, first note that $\mathrm{Hom}(F_1, F_2)$ is a direct summand of $\mathrm{ad}(E)$. Indeed, we have a holomorphic decomposition

$$\mathrm{ad}(E) = \mathrm{Hom}(F_1, F_2) \oplus \mathrm{Hom}(F_2, F_1) \oplus (\mathrm{End}(F_1) \oplus \mathrm{End}(F_2))/\mathcal{O}_X \tag{2.3}$$

of $\mathrm{ad}(E)$ into a direct sum of vector bundles; here $\mathcal{O}_X$ is considered as a subbundle of $\mathrm{End}(F_1) \oplus \mathrm{End}(F_2)$ by sending any locally defined holomorphic function f to $f(\mathrm{Id}_{F_1} + \mathrm{Id}_{F_2})$. $\mathrm{Hom}(F_1, F_2)$ is a direct summand of the bundle $\mathrm{ad}(E)$, which admits a holomorphic connection. Hence Theorem 2.2 implies that

$$\mathrm{degree}(\mathrm{Hom}(F_1, F_2)) = n_1 d_2 - d_1 n_2 = 0, \tag{2.4}$$

where n_i (respectively, d_i) is the rank (respectively, degree) of F_i, $i = 1, 2$. Since

$$\mathrm{degree}(E) = d_1 + d_2 = d,$$

from (2.4), it follows immediately that

$$\frac{\mathrm{degree}(F_i)}{\mathrm{rank}(F_i)} = \frac{d}{n}.$$

Therefore, the first statement in the theorem implies the second statement. □

To prove the converse, we begin with a lemma.

Lemma 2.5. *Let E_1 and E_2 are two holomorphic vector bundles over X, both admitting projective connections. If*

$$\frac{\mathrm{degree}(E_1)}{\mathrm{rank}(E_1)} = \frac{\mathrm{degree}(E_2)}{\mathrm{rank}(E_2)},$$

then $E_1 \oplus E_2$ also admits a projective connection.

Proof. Since both E_1 and E_2 admit projective connections, the vector bundle $E_1^* \otimes E_2$ also admits a projective connection. On the other hand, the condition that

$$\frac{\mathrm{degree}(E_1)}{\mathrm{rank}(E_1)} = \frac{\mathrm{degree}(E_2)}{\mathrm{rank}(E_2)}$$

implies that $\mathrm{degree}(E_1^* \otimes E_2) = 0$. This condition together with the condition that $E_1^* \otimes E_2$ admits a projective connection imply that $E_1^* \otimes E_2$ admits a holomorphic connection. Similarly, $E_2^* \otimes E_1$ admits a holomorphic connection. Now using the natural decomposition

$$\mathrm{End}(E_1 \oplus E_2) = \mathrm{End}(E_1) \oplus \mathrm{End}(E_2) \oplus (E_1^* \otimes E_2) \oplus (E_2^* \otimes E_1)$$

it follows that $E_1 \oplus E_2$ also admits a projective connection, and the proof of the lemma is complete. □

Continuing with the proof of the theorem, assume that

$$\frac{\mathrm{degree}(F)}{\mathrm{rank}(F)} = \frac{d}{n}$$

for each direct summand F of E.

We will prove that E admits a projective connection. This will be done by imitating the Atiyah's proof in [At] of the criterion for the existence of a holomorphic connection.

As before, let E_{PGL} denote the holomorphic principal $\mathrm{PGL}(n, \mathbb{C})$-bundle over X defined by E. Let

$$0 \longrightarrow \mathrm{ad}(E) \longrightarrow \mathrm{At}(E_{\mathrm{PGL}}) \longrightarrow T^{1,0}X \longrightarrow 0 \tag{2.5}$$

be the Atiyah exact sequence for the principal $\mathrm{PGL}(n, \mathbb{C})$-bundle E_{PGL} (see [At, page 187, Theorem 1]), where $\mathrm{At}(E_{\mathrm{PGL}})$ is the Atiyah bundle for E_{PGL}. The exact sequence of vector bundles in (2.5) coincides with the exact sequence in (2.2). A projective connection on E is equivalent to a holomorphic connection on the principal $\mathrm{PGL}(n, \mathbb{C})$-bundle E_{PGL}.

Let

$$a(E) \in H^1(X, K_X \otimes \mathrm{ad}(E)) \tag{2.6}$$

be the obstruction class for holomorphic splitting of (2.5). We note that the vector bundle $\mathrm{ad}(E)$ is self-dual, that is, $\mathrm{ad}(E)^* = \mathrm{ad}(E)$. Indeed, the trace map

$$\mathrm{ad}(E) \otimes \mathrm{ad}(E) \longrightarrow \mathcal{O}_X$$

that sends $s \otimes t$ to $\mathrm{trace}(s \circ t)$ identifies $\mathrm{ad}(E)^*$ with $\mathrm{ad}(E)$. Therefore, the Serre duality says that

$$H^1(X, K_X \otimes \mathrm{ad}(E)) = H^0(X, \mathrm{ad}(E))^*.$$

Let

$$b(E) \in H^0(X, \mathrm{ad}(E))^* \tag{2.7}$$

be the linear form corresponding to the cohomology class $a(E)$, constructed in (2.6), by the Serre duality.

In view of Lemma 2.5, it suffices to prove the theorem under the assumption that the vector bundle E is indecomposable.

Assume that E is indecomposable. Then any section

$$s \in H^0(X, \mathrm{ad}(E))$$

is actually nilpotent [At, page 201, Proposition 16]. In other words, the endomorphism s gives a holomorphic filtration of subbundles

$$0 = E_0 \subset E_1 \subset E_2 \cdots \subset E_k = E$$

such that $s(E_i) = E_{i-1}$ for all $1 \leq i \leq k$. More precisely, E_i is the subbundle of E generated by the kernel of the i-fold composition $s^i = s \circ \cdots \circ s$.

Now, for a nilpotent endomorphism s of E we know that

$$b(E)(s) = 0,$$

where $b(E)$ is constructed in (2.7) [At, page 202, Proposition 18(ii)]. Since the functional $b(E)$ vanishes on $H^0(X, \mathrm{ad}(E))$, we have $b(E) = 0$. Therefore, the obstruction class $a(E)$ in (2.6) vanishes, and hence E_{PGL} admits a holomorphic connection. This completes the proof of the theorem. □

3. Logarithmic connections

As before, X will denote a compact connected Riemann surface of genus g, where $g \geq 2$. Fix a point $x_0 \in X$. The holomorphic line bundle over X defined by the divisor x_0 will be denoted by $\mathcal{O}_X(x_0)$.

Definition 3.1. *Let E be a holomorphic vector bundle over X. A logarithmic connection on E singular over x_0 is a holomorphic differential operator*

$$\mathcal{D} : E \longrightarrow E \otimes \mathcal{O}_X(x_0) \otimes K_X$$

satisfying the Leibniz identity

$$\mathcal{D}(fs) = f\mathcal{D}(s) + \mathrm{d}f \otimes s, \tag{3.1}$$

where f (respectively, s) is a locally defined holomorphic function (respectively, holomorphic section of E).

Note that a logarithmic connection on E singular over x_0 produces a holomorphic connection on E over $X \backslash \{x_0\}$. Let U be a chart around x_0 with local holomorphic coordinate z. Then

$$\mathcal{D}(s) = ds + A(z)s\frac{dz}{z}, \tag{3.2}$$

for any local holomorphic section s of $\mathrm{End}(E)$ and a holomorphic connection matrix $A(z)$.

The curvature of a holomorphic connection on E is a holomorphic two–form with values in $\mathrm{End}(E) \otimes \mathcal{O}_X(x_0)$. Since a Riemann surface does not have nonzero holomorphic two forms, any logarithmic connection on a Riemann surface is flat.

A differential operator that satisfies the Leibniz identity (3.1) is clearly of order one. The above condition that a logarithmic connection $\mathcal{D}$ satisfies the Leibniz identity is evidently equivalent to the condition that the symbol of the first order differential operator $\mathcal{D}$ coincides with the section

$$\mathrm{Id}_E \in H^0(X, \mathcal{O}_X(x_0) \otimes \mathrm{End}(E)),$$

where Id_E denotes the identity automorphism of E.

The Poincaré adjunction formula says the following: Let D be a smooth hypersurface on a smooth variety Z, and let L denote the line bundle over Z defined by the divisor D. Then the restriction of the line bundle L to D is canonically identified with the normal bundle of D. See [GH] for a proof.

Therefore, the Poincaré adjunction formula says that the fiber over x_0 of the line bundle $\mathcal{O}_X(x_0)$ is identified with the holomorphic tangent space $T_{x_0}X$. Using this isomorphism between $\mathcal{O}_X(x_0)|_{x_0}$ and $T_{x_0}X$, the fiber $(K_X \otimes \mathcal{O}_X(x_0))_{x_0}$ is identified with $\mathbb{C}$.

We will now recall the definition of the important notion of residue of a logarithmic connection.

Let $\mathcal{D}$ be a logarithmic connection on a vector bundle E over X, which is singular over x_0. Let $v \in E_{x_0}$ be any vector in the fiber of E over the point x_0. Let $\hat{v}$ be a holomorphic section of E defined around x_0 such that $\hat{v}(x_0) = v$. Consider

$$\mathcal{D}(\hat{v})(x_0) \in (K_X \otimes \mathcal{O}_X(x_0))_{x_0} \otimes_{\mathbb{C}} E_{x_0} = \mathbb{C} \otimes_{\mathbb{C}} E_{x_0} = E_{x_0}.$$

Note that if $v = 0$, then $\mathcal{D}(\hat{v})$ is a (locally defined) section of the subsheaf

$$E \otimes K_X \subset E \otimes \mathcal{O}_X(x_0) \otimes K_X.$$

So, in that case the above evaluation $\mathcal{D}(\hat{v})(x_0) \in E_{x_0}$ vanishes. Consequently, we have a well–defined endomorphism

$$\mathrm{Res}(\mathcal{D}, x_0) \in \mathrm{End}(E_{x_0})$$

that sends any $v \in E_{x_0}$ to $\mathcal{D}(\hat{v})(x_0) \in E_{x_0}$. This endomorphism $\mathrm{Res}(\mathcal{D}, x_0)$ is called the *residue* of the logarithmic connection $\mathcal{D}$ at the point x_0. See [De1] for more details.

If $\mathcal{D}$ is a logarithmic connection on E singular over x_0 and $\theta \in H^0(X, \mathrm{End}(E) \otimes K_X)$, then the differential operator $\mathcal{D}+\theta$ is also a logarithmic connection on E singular over x_0. Furthermore, we have

$$\mathrm{Res}(\mathcal{D}, x_0) = \mathrm{Res}(\mathcal{D} + \theta, x_0).$$

Conversely, if $\mathcal{D}$ and $\mathcal{D}'$ are two logarithmic connections on E regular on $X\backslash\{x_0\}$ with

$$\mathrm{Res}(\mathcal{D}, x_0) = \mathrm{Res}(\mathcal{D}', x_0), \tag{3.3}$$

then $\mathcal{D}' = \mathcal{D}+\theta$, where $\theta \in H^0(X, \mathrm{End}(E) \otimes K_X)$. In this way the space of all logarithmic connections $\mathcal{D}'$ on the given vector bundle E, regular on $X\backslash\{x_0\}$ and satisfying (3.3), with $\mathcal{D}$ fixed, is an affine space for the vector space $H^0(X, K_X \otimes \mathrm{End}(E))$.

Let E be a holomorphic vector bundle over X of rank n, and let $\mathcal{D}$ be a logarithmic connection on E, regular on $X\backslash\{x_0\}$, satisfying the residue condition

$$\mathrm{Res}(\mathcal{D}, x_0) = -\frac{d}{n}\,\mathrm{Id}_{E_{x_0}}, \tag{3.4}$$

where $d \in \mathbb{C}$. Consider the nonsingular flat connection on the complement $X\backslash\{x_0\}$ defined by $\mathcal{D}$. The above condition on the residue implies that the monodromy around x_0 of this flat connection is the $n \times n$ diagonal matrix with $\exp(2\pi\sqrt{-1}d/n)$ as the diagonal entries [De1, page 79, Proposition 3.11]. From the expression of the degree of E in terms of the residue of $\mathcal{D}$ it follows immediately that $\mathrm{degree}(E) = d$; the last sentence of Corollary B.3 in [EV, page 186] gives the expression of the first Chern class in terms of the trace of the residue. In particular, d in (3.4) must be an integer.

We will now relate logarithmic connections with projective connections discussed in Section 2.

Let E be a holomorphic vector bundle over X of rank n and degree d. Let $\mathcal{D}$ be a logarithmic connection on E, regular on $X\backslash\{x_0\}$, satisfying the residue condition in (3.4). The logarithmic connection $\mathcal{D}$ gives a holomorphic connection on the vector bundle $E|_{X\setminus\{x_0\}}$ over the open subset $X\setminus\{x_0\} \subset X$. This connection on $E|_{X\setminus\{x_0\}}$ induces a projective connection on the $\mathrm{PGL}(n,\mathbb{C})$-principal bundle $E_{\mathrm{PGL}}|_{X\setminus\{x_0\}}$ over $X\setminus\{x_0\} \subset X$, understood as a family of (holomorphic) trivializations with transition functions which are locally constant in $\mathrm{PGL}(n,\mathbb{C})$. The residue of $\mathcal{D}$ is in the center of $\mathrm{M}(n,\mathbb{C})$ (the Lie algebra of $\mathrm{GL}(n,\mathbb{C})$); its projection to $\mathrm{sl}(n,\mathbb{C})$ vanishes. Therefore, the projective connection of $E_{\mathrm{PGL}}|_{X\setminus\{x_0\}}$ extends across x_0 as a regular projective connection.

Hence we have proved the following lemma:

Lemma 3.2. *Let E be a holomorphic vector bundle over X such that E admits a logarithmic connection $\mathcal{D}$ regular on $X\setminus\{x_0\}$ such that*

$$\mathrm{Res}(\mathcal{D}, x_0) = -\frac{d}{n}\,\mathrm{Id}_{E_{x_0}}.$$

Then the vector bundle E admits a projective connection. □

The following theorem is the analog of Theorem 2.4 for logarithmic connections.

Theorem 3.3. *Let E be a holomorphic vector bundle over X of rank n and degree d. Then the following two statements are equivalent:*

(1) *The vector bundle E admits a logarithmic connection regular on $X\setminus\{x_0\}$ whose residue at x_0 is $-\frac{d}{n}\,\mathrm{Id}_{E_{x_0}}$.*
(2) *For any direct summand $F \subset E$, the equality*

$$\frac{\mathrm{degree}(F)}{\mathrm{rank}(F)} = \frac{d}{n}$$

holds.

Proof. Let $\mathcal{D}$ be a logarithmic connection on the vector bundle E, regular on $X\setminus\{x_0\}$, with residue $\mathrm{Res}(\mathcal{D}, x_0) = -\frac{d}{n}\,\mathrm{Id}_{E_{x_0}}$. From Lemma 3.2 we know that E admits a projective connection. Therefore, from Theorem 2.4 it follows that

$$\frac{\mathrm{degree}(F)}{\mathrm{rank}(F)} = \frac{d}{n}$$

for every direct summand F of E.

To prove the converse, assume that

$$\frac{\mathrm{degree}(F)}{\mathrm{rank}(F)} = \frac{d}{n} \tag{3.5}$$

for every direct summand F of E. Therefore, E admits a projective connection (see Theorem 2.4). Fix a projective connection $\mathcal{D}^P$ on E.

Let

$$\gamma : \widetilde{X} \longrightarrow X \tag{3.6}$$

be a ramified Galois covering of degree n with the property that the map γ is totally ramified over the point $x_0 \in X$. (The map γ is allowed to have ramifications over points in $X\backslash\{x_0\}$.) The condition that γ is totally ramified over x_0 means that there is a unique point $y_0 \in \widetilde{X}$ such that $\gamma(y_0) = x_0$.

Consider the vector bundle

$$V := \mathcal{O}_{\widetilde{X}}(-dy_0) \otimes_{\mathcal{O}_{\widetilde{X}}} \gamma^* E \tag{3.7}$$

over $\widetilde{X}$, where γ is the map in (3.6), and y_0 is the unique point with $\gamma(y_0) = x_0$. The projective bundle $\mathbb{P}(V)$ over $\widetilde{X}$ is canonically identified with the projective bundle $\gamma^*\mathbb{P}(E)$. Therefore, the projective connection $\mathcal{D}^P$ on E gives a projective connection $\gamma^*\mathcal{D}^P$ on V.

Note that degree$(V) = 0$. Therefore, the condition that V admits a projective connection implies that V admits a holomorphic connection. Fix a holomorphic connection $\mathcal{D}'$ on V.

Let Γ denote the Galois group for the Galois covering γ in (3.6). Note that the Galois action of Γ on $\widetilde{X}$ has a canonical lift to the vector bundle V. Consider

$$\mathcal{D}'' = \frac{1}{\#\Gamma} \sum_{h \in \Gamma} h^* \mathcal{D}'$$

which is a holomorphic connection on V (as the space of holomorphic connections on V is an affine space, the average of $\mathcal{D}'$ over the action of the elements of Γ is again a holomorphic connection on V).

It is now straight–forward to check that the Galois invariant connection $\mathcal{D}''$ on V descends to a logarithmic connection on E. This descended logarithmic connection is regular outside x_0, and its residue at x_0 is $-\frac{d}{n}\,\mathrm{Id}_{E_{x_0}}$. This completes the proof of the theorem. □

Theorem 2.4 and Theorem 3.3 together have the following corollary:

Corollary 3.4. *Let E be a holomorphic vector bundle over X of rank n and degree d. Then the following three statements are equivalent:*

(1) *The vector bundle E admits a logarithmic connection regular on $X\backslash\{x_0\}$ whose residue at x_0 is $-\frac{d}{n}\,\mathrm{Id}_{E_{x_0}}$.*
(2) *The vector bundle E admits a projective connection.*
(3) *For any direct summand $F \subset E$, the equality*

$$\frac{\mathrm{degree}(F)}{\mathrm{rank}(F)} = \frac{d}{n}$$

holds. □

4. Moduli space of logarithmic connections

Henceforth, we will assume the rank n and the degree d are mutually coprime.

Let $\mathcal{M}_D(n)$ denote the moduli space of all logarithmic connections $(E, \mathcal{D})$ over X, where E is any holomorphic vector bundle of rank n and degree d and $\mathcal{D}$ is a logarithmic connection on E, regular on $X\backslash\{x_0\}$, satisfying the residue condition

$$\operatorname{Res}(\mathcal{D}, x_0) = -\frac{d}{n} \operatorname{Id}_{E_{x_0}}.$$

(See [Si1], [Si2], [Ni] for the construction of this moduli space.)

So $\mathcal{M}_D(n)$ parametrizes isomorphism classes of pairs of the form $(E, \mathcal{D})$, where E is any rank n holomorphic vector bundle of degree d and $\mathcal{D}$ is a logarithmic connection on E regular on $X\backslash\{x_0\}$ with $\operatorname{Res}(\mathcal{D}, x_0) = -\frac{d}{n} \operatorname{Id}_{E_{x_0}}$. We note that every logarithmic connection over X occurring in the moduli space $\mathcal{M}_D(n)$ is irreducible. Indeed, for $(E, \mathcal{D}) \in \mathcal{M}_D(n)$, if $F \subset E$ is a holomorphic subbundle invariant by the connection $\mathcal{D}$, i.e. $\mathcal{D}(F) \subset F \otimes \mathcal{O}(x_0) \otimes K_X$, then $\mathcal{D}$ induces a logarithmic connection $\mathcal{D}_F$ on F which is regular on $X\backslash\{x_0\}$ and

$$\operatorname{Res}(\mathcal{D}_F, x_0) = -\frac{d}{n} \operatorname{Id}_{F_{x_0}}.$$

Therefore,

$$\frac{\operatorname{degree}(F)}{\operatorname{rank}(F)} = \frac{d}{n}.$$

This contradicts the assumption that d and n are mutually coprime if F is a proper subbundle of E.

Since every logarithmic connection over X occurring in the moduli space $\mathcal{M}_D(n)$ is irreducible, the variety $\mathcal{M}_D(n)$ is smooth; singular points of a moduli space of connections correspond to reducible connections. The variety $\mathcal{M}_D(n)$ is known to be irreducible. The dimension of $\mathcal{M}_D(n)$ is $2(n^2(g-1)+1)$.

Consider the holomorphic line bundle $L := \mathcal{O}_X(d\, x_0)$ over X. The de Rham differential $f \longmapsto df$, defines a logarithmic connection $\mathcal{D}_L$ on L which is regular on $X\backslash\{x_0\}$, and

$$\operatorname{Res}(\mathcal{D}_L, x_0) = -d \operatorname{Id}_{L_{x_0}}. \tag{4.1}$$

Let

$$\mathcal{M}_D(L) \subset \mathcal{M}_D(n) \tag{4.2}$$

be the subvariety parametrizing isomorphism classes of logarithmic connections $(E, \mathcal{D}) \in \mathcal{M}_D(n)$ such that

- $\wedge^n E \cong L$, and
- the logarithmic connection on L induced by $\mathcal{D}$ using an isomorphism $\wedge^n E \longrightarrow L$ coincides with the logarithmic connection $\mathcal{D}_L$ in (4.1).

Note that a logarithmic connection $\mathcal{D}$ on E induces a logarithmic connection on $\wedge^n E$. Since any two holomorphic isomorphisms between $\wedge^n E$ and L differ by a constant scalar, the connection of L given by the induced connection on $\wedge^n E$ using an isomorphism $\wedge^n E \longrightarrow L$ is independent of the choice of the isomorphism.

The subset $\mathcal{M}_D(L)$ in (4.2) is an irreducible smooth closed subvariety of dimension $2(n^2-1)(g-1)$.

We will show that the biholomorphism classes of both $\mathcal{M}_D(L)$ and $\mathcal{M}_D(n)$ are independent of the complex structure of X.

Let $X' := X\backslash\{x_0\}$ be the complement. Fix a point $x' \in X'$. The point x_0 gives a conjugacy class in the fundamental group $\pi_1(X', x')$ as follows. Let

$$f : \mathbb{D} \longrightarrow X$$

be an orientation preserving embedding of the closed unit disk $\mathbb{D}:=\{z\in\mathbb{C} \mid \|z\|^2 \le 1\}$ into the Riemann surface X such that $f(0) = x_0$. The free homotopy class of the map $\partial\mathbb{D} = S^1 \longrightarrow X'$ obtained by restricting f to the boundary of $\mathbb{D}$ is independent of the choice of f. The orientation of $\partial\mathbb{D}$ coincides with the anti–clockwise rotation around x_0. Any free homotopy class of oriented loops in X' gives a conjugacy class in $\pi_1(X', x')$. Let γ denote the orbit in $\pi_1(X', x')$, for the conjugation action of $\pi_1(X', x')$ on itself, defined by the above free homotopy class of oriented loops associated to x_0.

Let

$$\mathrm{Hom}^0(\pi_1(X', x'), \mathrm{GL}(n, \mathbb{C})) \subset \mathrm{Hom}(\pi_1(X', x'), \mathrm{GL}(n, \mathbb{C})) \tag{4.3}$$

be the space of all homomorphisms from the fundamental group $\pi_1(X', x')$ to $\mathrm{GL}(n, \mathbb{C})$ satisfying the condition that the image of γ (the free homotopy class defined above) is $\exp(2\pi\sqrt{-1}d/n) \cdot I_{n\times n}$. It may be noted that since $\exp(2\pi\sqrt{-1}d/n) \cdot I_{n\times n}$ is in the center of $\mathrm{SL}(n, \mathbb{C})$, a homomorphism sends the orbit γ in $\pi_1(X', x')$ (for the adjoint action of $\pi_1(X', x')$ on itself) to $\exp(2\pi\sqrt{-1}d/n) \cdot I_{n\times n}$ if and only if there is an element in the orbit which is mapped to $\exp(2\pi\sqrt{-1}d/n) \cdot I_{n\times n}$.

Take any homomorphism $\rho \in \mathrm{Hom}^0(\pi_1(X', x'), \mathrm{GL}(n, \mathbb{C}))$. Let (V, ∇) be the flat vector bundle or rank n over X' given by ρ. Therefore V is a holomorphic vector bundle on X'. The monodromy of ∇ along the oriented loop γ is $\exp(2\pi\sqrt{-1}d/n) \cdot I_{n\times n}$. Using the logarithm $2\pi\sqrt{-1}d/n \cdot I_{n\times n}$ of the monodromy, the vector bundle V over X' extends to a holomorphic vector bundle $\bar{V}$ over X, and furthermore, the connection ∇ on V extends to a logarithmic connection $\bar{\nabla}$ on the vector bundle $\bar{V}$ over X such that $(\bar{V}, \bar{\nabla}) \in \mathcal{M}_D(n)$, where $\mathcal{M}_D(n)$ is the moduli space of logarithmic connections defined earlier (see [Ma, p. 159, Theorem 4.4]).

Since $\mathrm{GL}(n, \mathbb{C})$ is an algebraic group defined over the field of complex numbers, and $\pi_1(X', x')$ is a finitely presented group, the representation space $\mathrm{Hom}(\pi_1(X', x'), \mathrm{SL}\ (n, \mathbb{C}))$ is a complex algebraic variety in a natural way. The conjugation action of $\mathrm{GL}(n, \mathbb{C})$ on itself induces an action of $\mathrm{GL}(n, \mathbb{C})$ on $\mathrm{Hom}^0(\pi_1(X', x'), \mathrm{GL}(n, \mathbb{C}))$. The action of any $T \in \mathrm{GL}(n, \mathbb{C})$ on $\mathrm{Hom}^0(\pi_1(X', x'), \mathrm{GL}(n, \mathbb{C}))$ sends any homomorphism ρ to the homomorphism $\pi_1(X', x') \longrightarrow \mathrm{GL}(n, \mathbb{C})$ defined by $\beta \longmapsto T^{-1}\rho(\beta)T$. Let

$$\mathcal{R}_g := \mathrm{Hom}^0(\pi_1(X', x'), \mathrm{GL}(n, \mathbb{C}))/\mathrm{GL}(n, \mathbb{C}) \tag{4.4}$$

be the quotient space for this action.

The algebraic structure of $\mathrm{Hom}^0(\pi_1(X', x'), \mathrm{GL}(n, \mathbb{C}))$ induces an algebraic structure on the quotient $\mathcal{R}_g$. The scheme $\mathcal{R}_g$ is an irreducible smooth quasiprojective variety of dimension $2(n^2 - 1)(g - 1) + 2$ defined over $\mathbb{C}$.

The isomorphism class of this variety $\mathcal{R}_g$ is independent of the complex structure of the topological surface X. The isomorphism class depends only on the integers g, n and d. On the other hand, the moduli space $\mathcal{M}_D(n)$ is canonically biholomorphic to $\mathcal{R}_g$. Therefore, the biholomorphism class of the moduli space $\mathcal{M}_D(n)$ is independent of the complex structure of X; the biholomorphism class depends only on the integers g, n and d.

Replacing $\mathrm{GL}(n, \mathbb{C})$ by the algebraic subgroup $\mathrm{SL}(n, \mathbb{C})$ above, we have an algebraic irreducible smooth quasiprojective variety

$$\mathcal{S}_g := \mathrm{Hom}^0(\pi_1(X', x'), \mathrm{SL}(n, \mathbb{C}))/\mathrm{SL}(n, \mathbb{C}),$$

which is biholomorphic to $\mathcal{M}_D(L)$. We conclude that the biholomorphism class of the moduli space $\mathcal{M}_D(L)$ defined in (4.2) is also independent of the complex structure of X.

5. The second intermediate Jacobian of the moduli space

In this section we recall some results of [BM].

A holomorphic vector bundle E over X is called stable if for every nonzero proper subbundle $F \subset E$, the inequality

$$\frac{\mathrm{degree}(F)}{\mathrm{rank}(F)} < \frac{\mathrm{degree}(E)}{\mathrm{rank}(E)}$$

holds.

Let $\mathcal{N}_X$ denote the moduli space parametrizing all stable vector bundles E over X with $\mathrm{rank}(E) = n$ and $\wedge^n E \cong L = \mathcal{O}_X(dx_0)$. The moduli space $\mathcal{N}_X$ is an irreducible smooth projective variety of dimension $(n^2 - 1)(g - 1)$ defined over $\mathbb{C}$.

Let $\mathcal{M}_D(L)$ be the moduli space defined in (4.2). Let

$$\mathcal{U} \subset \mathcal{M}_D(L) \tag{5.1}$$

be the Zariski open subset parametrizing all $(E, \mathcal{D})$ such that the underlying vector bundle E is stable. The openness of this subset follows from [Ma]. Let

$$\Phi : \mathcal{U} \longrightarrow \mathcal{N}_X \tag{5.2}$$

denote the forgetful map that sends any $(E, \mathcal{D})$ to E.

By Theorem 3.3, any $E \in \mathcal{N}_X$ admits a logarithmic connection $\mathcal{D}$ such that $(E, \mathcal{D}) \in \mathcal{M}_D(L)$, since any $E \in \mathcal{N}_X$ is indecomposable. Therefore, the projection Φ in (5.2) is surjective. Furthermore, Φ makes $\mathcal{U}$ an affine bundle over $\mathcal{N}_X$. More precisely, $\mathcal{U}$ is a torsor over $\mathcal{N}_X$ for the holomorphic cotangent bundle $T^*\mathcal{N}_X$. This means that the fibers of the vector bundle $T^*\mathcal{N}_X$ act freely transitively on the fibers of Φ [BR, p. 786]. Since $\mathcal{M}_D(L)$ is irreducible, and $\mathcal{U}$ is nonempty, the subset $\mathcal{U} \subset \mathcal{M}_D(L)$ is Zariski dense.

The following lemma is proved in [BM]:

Lemma 5.1. *Let* $\mathcal{Z} := \mathcal{M}_D(L)\backslash\mathcal{U}$ *be the complement of the Zariski open dense subset. The codimension of the Zariski closed subset* $\mathcal{Z}$ *in* $\mathcal{M}_X$ *is at least* $(n-1)(g-2)+1$. □

For any $i \geq 0$, the i-th cohomology of a complex variety with coefficients in $\mathbb{Z}$ is equipped with a mixed Hodge structure [De2], [De3].

The following proposition is proved in [BM].

Proposition 5.2. *The mixed Hodge structure* $H^3(\mathcal{M}_D(L), \mathbb{Z})$ *is pure of weight three. More precisely, the mixed Hodge structure* $H^3(\mathcal{M}_D(L), \mathbb{Z})$ *is isomorphic to the Hodge structure* $H^3(\mathcal{N}_X, \mathbb{Z})$*, where* $\mathcal{N}_X$ *is the moduli space of stable vector bundles introduced at the beginning of this section.* □

Let

$$J^2(\mathcal{M}_D(L)) := H^3(\mathcal{M}_D(L), \mathbb{C})/(F^2H^3(\mathcal{M}_D(L), \mathbb{C}) + H^3(\mathcal{M}_D(L), \mathbb{Z})) \quad (5.3)$$

be the intermediate Jacobian of the mixed Hodge structure $H^3(\mathcal{M}_D(L))$ (see [Ca, p. 110]). The intermediate Jacobian of any mixed Hodge structure is a generalized torus [Ca, p. 111]. Let

$$J^2(\mathcal{N}_X) := H^3(\mathcal{N}_X, \mathbb{C})/(F^2H^3(\mathcal{N}_X, \mathbb{C}) + H^3(\mathcal{N}_X, \mathbb{Z}))$$

be the intermediate Jacobian for $H^3(\mathcal{N}_X, \mathbb{Z})$, which is a complex torus.

The following proposition is proved in [BM].

Proposition 5.3. *The intermediate Jacobian* $J^2(\mathcal{M}_D(L))$ *is isomorphic to* $J^2(\mathcal{N}_X)$*, which is isomorphic to the Jacobian* $\mathrm{Pic}^0(X)$ *of the Riemann surface* X. □

The homology $H_1(J^2(\mathcal{M}_D(L)), \mathbb{Z})$ has a natural skew–symmetric pairing which we will describe below.

We first note that from Proposition 5.3, and the fact that $H^3(\mathcal{N}_X, \mathbb{Z})$ is torsionfree, it follows that

$$H_1(J^2(\mathcal{M}_D(L)), \mathbb{Z}) = H^3(\mathcal{M}_D(L), \mathbb{Z}) = H^3(\mathcal{N}_X, \mathbb{Z}). \quad (5.4)$$

Also, we have

$$H^2(\mathcal{M}_D(L), \mathbb{Z}) = H^2(\mathcal{N}_X, \mathbb{Z}) = \mathbb{Z}.$$

Fix a generator

$$\alpha \in H^2(\mathcal{M}_D(L), \mathbb{Z}).$$

For any

$$\theta \in H_{2(n^2-1)(g-1)}(\mathcal{M}_D(L), \mathbb{Z}), \quad (5.5)$$

we have a skew–symmetric pairing

$$B_\theta : \wedge^2 H^3(\mathcal{M}_D(L), \mathbb{Z}) \longrightarrow \mathbb{Z} \quad (5.6)$$

defined by

$$B_\theta(\omega_1, \omega_2) = (\omega_1 \cup \omega_2 \cup \alpha^{(n^2-1)(g-1)-3}) \cap \theta \in \mathbb{Z}.$$

The following theorem is proved in [BM].

Theorem 5.4. *There exists a homology class* θ *as in* (5.5) *such that the pairing* B_θ *in* (5.6) *is nonzero.*

Take any θ *such that* B_θ *is nonzero. Using the isomorphism in* (5.4), B_θ *defines a nonzero pairing*

$$\widetilde{B}_\theta : \wedge^2 H_1(J^2(\mathcal{M}_D(L)), \mathbb{Z}) \longrightarrow \mathbb{Z}.$$

The pair $(J^2(\mathcal{M}_D(L)), \widetilde{B}_\theta)$ *is isomorphic to* $\mathrm{Pic}^0(X)$ *equipped with a multiple of the canonical principal polarization on* $\mathrm{Pic}^0(X)$ *given by the class of a theta line bundle.* □

Given a multiple of a principal polarization on an abelian variety, there is a unique way to recover the principal polarization from it. Therefore, from Theorem 5.4 it follows immediately that the isomorphism class of $(\mathrm{Pic}^0(X), \Theta)$, where Θ is the canonical polarization on $\mathrm{Pic}^0(X)$, is determined by the isomorphism class of the variety $\mathcal{M}_D(L)$.

The Torelli theorem says that the isomorphism class of the principally polarized abelian variety $(\mathrm{Pic}^0(X), \Theta)$ determines the curve X up to an isomorphism. Thus Theorem 5.4 gives the following corollary:

Corollary 5.5. *The isomorphism class of the variety* $\mathcal{M}_D(L)$ *uniquely determines the isomorphism class of the curve* X. □

This is in contrast with the fact that the biholomorphism class of the complex manifold $\mathcal{M}_D(L)$ is independent of the complex structure of the Riemann surface X (see the end of Section 4).

References

[At] M. F. Atiyah, Complex analytic connections in fibre bundles, *Trans. Amer. Math. Soc.*, **85** (1957) 181–207.

[BM] I. Biswas and V. Muñoz, Torelli theorem for the moduli spaces of connections on a Riemann surface, *Topology*, **46** (2007) 295–317.

[BR] I. Biswas and N. Raghavendra, Line bundles over a moduli space of logarithmic connections on a Riemann surface, *Geom. Funct. Anal.*, **15** (2005) 780–808.

[Ca] J. Carlson, Extensions of mixed Hodge structures, in: *Journées de géométrie algébrique d'Angers Juillet 1979/Algebraic Geometry*, Angers, 1979, Sijthoff & Noordhoff, Alphen aan den Rijn–Germantown, Md. (1980) pp. 107–127.

[De1] P. Deligne, Equations Différentielles à Points Singuliers Réguliers, *Lecture Notes in Math.*, **163** Springer–Verlag, Berlin (1970).

[De2] P. Deligne, Théorie de Hodge, II, *Inst. Hautes Études Sci. Publ. Math.*, **40** (1971) 5–57.

[De3] P. Deligne, Théorie de Hodge, III, *Inst. Hautes Études Sci. Publ. Math.*, **44** (1974) 5–77.

[EV] H. Esnault and E. Viehweg, Logarithmic de Rham complex and vanishing theorems, *Invent. Math.*, **86** (1986) 161–194.

[GH] P. A. Griffiths and J. Harris, Principles of algebraic geometry, Wiley Classics Library, John Wiley & Sons, Inc., New York (1994).

[Ma] M. Maruyama, Openness of a family of torsion free sheaves, *Jour. Math. Kyoto Univ.*, **16** (1976) 627–637.

[Ni] N. Nitsure, Moduli of semistable logarithmic connections, *Jour. Amer. Math. Soc.*, **6** (1993) 597–609.

[Si1] C. T. Simpson, Moduli of representations of the fundamental group of a smooth projective variety. I, *Inst. Hautes Études Sci. Publ. Math.*, **79** (1994) 47–129.

[Si2] C. T. Simpson, Moduli of representations of the fundamental group of a smooth projective variety. II, *Inst. Hautes Études Sci. Publ. Math.*, **80** (1994) 5–79.

[We] A. Weil, Généralisation des fonctions abéliennes, *Jour. Math. Pures Appl.*, **17** (1938) 47–87.

Teichmüller Theory and Moduli Problem No. 10, 2010, pp. 113–129.

Teichmüller Geodesics, Delaunay Triangulations, and Veech Groups

Joshua P. Bowman

Institute for Mathematical Sciences, Stony Brook University, Stony Brook, NY, USA
e-mail: bowman@math.sunysb.edu; joshua.bowman@gmail.com

Abstract. In this article, we describe a method for computing generators of the Veech group of a flat surface (which we define as a Riemann surface with a non-zero holomorphic quadratic differential). The method employs a cell structure on the (complex) Teichmüller geodesic generated by the surface, using Delaunay triangulations, which are canonically associated to flat surfaces.

Introduction

Since the late 1980s, Veech groups have been an extremely useful tool in the study of both billiard dynamical systems and projections of "Teichmüller discs" to moduli space. It was observed as early as the 1930s [FK] that billiard paths in a rational polygon could be lifted to geodesic paths on a compact surface with a singular locally Euclidean structure. (In the case of simply connected polygons, "rational" is simply the condition that the angles be rational multiples of π.) Following the terminology of Riemann surface theory, R. Fox and R. Kerschner called this surface the "Ueberlagerungsfläche" of the polygon. Indeed, such a locally Euclidean structure is equivalent to a holomorphic quadratic differential on a Riemann surface (this equivalence is sketched in Section 1.1), and so the field of billiard dynamics becomes connected with the study of the moduli space of quadratic differentials.

W. Veech showed that when the group of affine automorphisms of such a *flat surface* (i.e., the *Veech group,* defined in Section 1.3) is "sufficiently large," then the dynamical behavior of the geodesic flow (which projects to the billiard flow in the case that the surface arises from a rational billiard table) satisfies the *Veech dichotomy:*

Theorem 0.1 ([Ve1]). *Suppose the Veech group of a flat surface is a lattice in* $\mathrm{PSL}(2,\mathbb{R})$. *Then for every direction* θ *on the surface, one of the two following (mutually exclusive) possibilities occurs:*

- *The surface decomposes in the direction* θ *into metric cylinders whose moduli are commensurable (in particular, every geodesic either connects two singular points or is closed);*
- *Every trajectory with direction* θ *is dense in* X, *and the geodesic flow in the direction* θ *is uniquely ergodic.*

A flat surface whose Veech group is a lattice in $\mathrm{PSL}(2,\mathbb{R})$ is called a *Veech surface* (also a *lattice surface*). A rational polygon whose Ueberlagerungfläche is a

Veech surface is called a *lattice polygon*. The simplest examples of Veech surfaces are *origami*: surfaces that cover the torus with its flat metric, such that the cover has a single branch point. G. Schmithüsen has developed an algorithm for computing the Veech group of an origami [Sch]. She makes use of the result of E. Gutkin and C. Judge that origamis are precisely the flat surfaces whose Veech groups are finite index subgroups of PSL(2, $\mathbb{Z}$ [GJ]. Veech produced the first examples of flat surfaces to have Veech groups that are lattices but not commensurable to PSL(2, $\mathbb{Z}$); his examples arise from billiards in right triangles that tile regular polygons [Ve2].

There have been concerted efforts both to demonstrate the existence of Veech surfaces with specified properties and to produce examples via explicit constructions. For example, R. Kenyon and J. Smillie classified the acute lattice triangles [KS]. R. Schwartz and P. Hooper have written a Java program called McBilliards that computes periodic paths in triangular billiard tables. Using this program and similar techniques to those of Kenyon and Smillie, they discovered an obtuse triangle that produces a Veech surface of genus 4 [Ho]. C. McMullen [Mc1,Mc2,Mc3,Mc5] and K. Calta [Ca] have independently given criteria for determining when a genus 2 surface, carrying a quadratic differential which is the square of an abelian differential, is Veech.

In this article we develop a general algorithm for computing generating elements of Veech groups. The method depends on keeping track of combinatorial data that is canonically associated to a flat surface—namely, its Delaunay triangulation (Section 2). This produces a natural partition of the orbit of a flat surface under affine equivalence, whose structure is preserved by elements of the Veech group. If the Veech group is in fact a lattice, then this method is guaranteed to determine a full generating set of elements. In many of the cases of the preceding paragraph, the Veech group is known to be finitely generated, so the algorithm we shall describe can be particularly effective in studying these examples.

Figures 5 and 6 show examples of *iso-Delaunay regions,* the type of data we are seeking in the Teichmüller disc of a surface (see Definition 2.11). These images were made with a program written in MATLAB, designed to compute the boundaries of iso-Delaunay regions. A fuller program, which will compute both iso-Delaunay regions and isomorphisms between them, is under construction. More examples of these images, made with the newer program, are available in the author's thesis [Bo].

During preparation of this article, the author became aware of a preprint by Veech [Ve1], which also introduces what we call the *iso-Delaunay complex* (Definition 2.14) under the name of the *tessellation of* $\mathbb{H}$ *subordinate to the quadratic differential.* In future work, we shall call this object the *iso-Delaunay tessellation*.

1. Background

1.1 *Flat surfaces*

Let X be a compact hyperbolic Riemann surface (i.e., of genus $g \geq 2$).

Definition 1.1. *A* holomorphic quadratic differential *on X is a section of the tensor-square of the sheaf of holomorphic one-forms. In symbols, it is an element of* $\Gamma(X, \Omega^{\otimes 2}(X))$.

Let q be a non-zero quadratic differential on X. In a local analytic coordinate z on X, q has the form $q(z) = f(z)\,\mathrm{d}z \otimes \mathrm{d}z$; often one abbreviates $\mathrm{d}z^2 = \mathrm{d}z \otimes \mathrm{d}z$. Let Z denote the set of zeroes of q on X. Given a point $x \in X - Z$ and a simply connected neighborhood $U \subset X - Z$ of x, a chart $U \to \mathbb{C}$ centered at x is

$$y \mapsto \int_\gamma \sqrt{q},$$

where $y \in U$ and $\gamma : [0, 1] \to U$ is a path from x to y. In this chart, q has the form $\mathrm{d}z^2$. The choice of $\sqrt{q}$ is unique up to sign, hence q determines an analytic atlas on $X - Z$ with transition maps all of the form $z \mapsto z + c$ or $z \mapsto -z + c$. Such an atlas is called a *flat structure* on X with singularities in Z.

The Euclidean metric $|\mathrm{d}z| = |\sqrt{q}|$ is invariant under changes of coordinates in the flat structure, and so gives an elementary perspective on the geometry of the quadratic differential. The pair $(X - Z, |\sqrt{q}|)$ is locally isometric to the plane, while each point $z \in Z$ has a *cone angle* of $k\pi$ for some integer $k > 2$; $k - 2$ is called the *order* of the singular point z (this coincides with the analytic definition of the order of a zero).

The reverse construction is also possible: given a flat structure on a compact oriented surface of genus $g \geq 2$, one obtains a unique pair (X', q'), where X' is a Riemann surface and q' is a quadratic differential on X'. The conformal structure on X' is obtained by completing the flat structure to a maximal conformal atlas, and q' is obtained by taking $q'(z) = \mathrm{d}z^2$ in the charts of the flat structure.

Definition 1.2. *Via the above equivalence, a* flat surface *means either a topological surface with a flat structure or a Riemann surface with a chosen non-zero quadratic differential.*

Remark 1.3. One might also consider flat structures with some singularities of cone angle π, i.e., zeroes of order -1; these correspond to *meromorphic* quadratic differentials with at worst simple poles. This class of quadratic differentials is quite important, and much of the material developed in this paper applies also to these flat surfaces. All of our examples, however, will arise as holomorphic quadratic differentials.

For the remainder of the paper, (X, q) will denote a fixed flat surface, and Z the set of zeroes of q.

1.2 *Teichmüller geodesics and the Poincaré half-plane*

Let $\mathscr{T}_X$ denote the *Teichmüller space* of X, i.e., the space of Riemann surfaces marked by X, up to isomorphisms respecting the marking, and $\mathbb{H}$ the open upper half-plane in $\mathbb{C}$.

Definition 1.4. *A (*complex*)* Teichmüller geodesic *(also known as a* Teichmüller disc*) is a complex-analytic embedding $f : \mathbb{H} \to \mathscr{T}_X$ whose image is geodesic for the Teichmüller metric.*

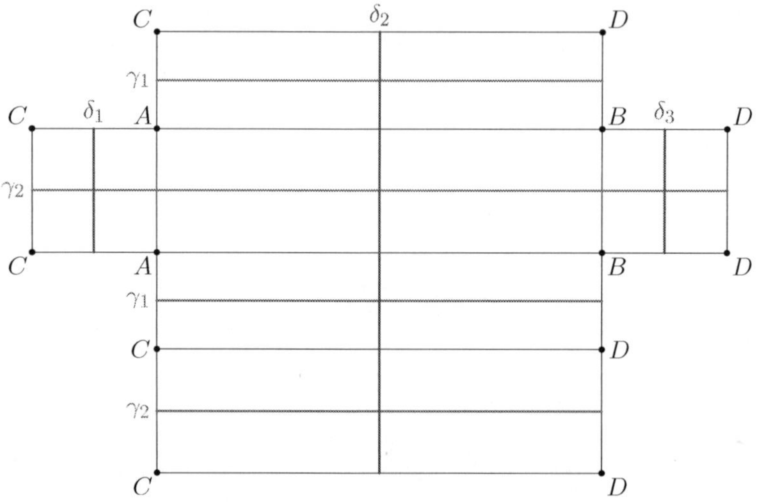

Figure 1. A "bouillabaisse" flat surface, $\lambda = 5 + \sqrt{17}$. (Bouillabaisse surfaces are described in Section 3.) Horizontal exterior edges are identified via vertical translations, and vertical exterior edges are identified via 180° rotations. The singular points are A, B, C, and D; they are *simple* zeroes of the quadratic differential $\mathrm{d}z^2$, each having a cone angle of 3π.

The pull-back of the Teichmüller metric by f coincides with the Poincaré metric on $\mathbb{H}$. It is a fundamental result in Teichmüller theory that the bundle of quadratic differentials $\mathscr{Q}_X \to \mathscr{T}_X$ is the cotangent bundle to $\mathscr{T}_X$, and that there is a canonical pairing between $\mathscr{Q}_X$ and the tangent bundle to $\mathscr{T}_X$. Hence (X, q) generates a complex-analytic embedding of the upper half-plane $\mathbb{H}$ into $\mathscr{Q}_X$, which projects down to a Teichmüller geodesic $f : \mathbb{H} \to \mathscr{T}_X$.

For $t \in \mathbb{H}$, the flat surface $(X_t, q_t) \in \mathscr{Q}_X$ can be described explicitly in terms of the local geometry of q. Take a local coordinate z on X in which $q = \mathrm{d}z^2$, and write $z = x + iy$. Compose with the map $x + iy \mapsto x + ty$ to get a new coordinate z_t. Then X_t is the Riemann surface with local coordinate z_t, and q_t is the quadratic differential with local form $q_t = \mathrm{d}{z_t}^2$. The map $x + iy \mapsto x + ty$ is affine, with derivative given by the matrix $\begin{pmatrix} 1 & \Re t \\ 0 & \Im t \end{pmatrix}$.

Dividing $\begin{pmatrix} 1 & \Re t \\ 0 & \Im t \end{pmatrix}$ by the square root of its determinant $\Im t$, we obtain a representative element in $\mathrm{SL}(2, \mathbb{R})$ for the map $(X, q) \mapsto (X_t, (\Im t)^{-1/2} q_t)$. Normalizing to an element in $\mathrm{SL}(2, \mathbb{R})$ simply yields a quadratic differential having the same area on X_t as the original. As we are interested in the projection to $\mathscr{T}_X$, rather than the complex curve in $\mathscr{Q}_X$, we will in general ignore the difference between $\mathrm{GL}(2, \mathbb{R})$ and $\mathrm{SL}(2, \mathbb{R})$.

Even further, one can consider the action of $\mathrm{SL}(2, \mathbb{R})$ directly on the flat structure of a flat surface, and obtain a picture of the Teichmüller geodesic apart from its embedding into $\mathscr{T}_X$. This description is given in [EG]; we reproduce the salient features for our computational purposes. Let $A \in \mathrm{SL}(2, \mathbb{R})$, and define $A \cdot (X, q)$ by taking each chart in the flat structure of (X, q) and post-composing it with A. Multiplying by an element of $\mathrm{SO}(2,\mathbb{R})$ is equivalent to multiplying q by a unit complex number, which changes neither the conformal nor the metric structure of (X, q). Hence we introduce:

Definition 1.5. *The* $\mathrm{SL}(2,\mathbb{R})$-orbit *of* (X, q) *is the quotient of*

$$\{A \cdot (X, q) \mid A \in \mathrm{SL}(2,\mathbb{R})\}$$

by the left action of $\mathrm{SO}(2,\mathbb{R})$.

We want to systematically identify the $\mathrm{SL}(2,\mathbb{R})$-orbit of (X, q) with the well-understood space $\mathbb{H}$, using Proposition 1.6 below. Define $P : \mathrm{SL}(2,\mathbb{R}) \to \mathbb{C}$ by

$$P : \begin{pmatrix} a & b \\ c & d \end{pmatrix} \mapsto \frac{di + b}{ci + a}$$

Let σ denote the involution on $\mathrm{SL}(2,\mathbb{R})$ given by

$$\sigma : \begin{pmatrix} a & b \\ c & d \end{pmatrix} \mapsto \begin{pmatrix} a & -b \\ -c & d \end{pmatrix}.$$

Recall the standard action of $\mathrm{SL}(2,\mathbb{R})$ on $\mathbb{H}$ by Möbius transformations, which, for $A \in \mathrm{SL}(2,\mathbb{R})$, we will denote by $\zeta \mapsto A\zeta$:

$$A = \begin{pmatrix} a & b \\ c & d \end{pmatrix} : \zeta \mapsto \frac{a\zeta + b}{c\zeta + d};$$

Then $P(A) = (\sigma(A))^{-1} i$. Note that $\mathrm{SO}(2,\mathbb{R})$ is preserved by σ, and that $Ai = i$ for all $A \in \mathrm{SO}(2,\mathbb{R})$. The following result is straightforward computation.

Proposition 1.6. *P descends to a bijection (also denoted P) from the set of right cosets of* $\mathrm{SO}(2,\mathbb{R})$ *to the upper half-plane* $\mathbb{H}$ *in* $\mathbb{C}$. (In particular, $\mathrm{SO}(2,\mathbb{R})$ itself is mapped to $i \in \mathbb{H}$.)

The group $\mathrm{PSL}(2,\mathbb{R}) = \mathrm{SL}(2,\mathbb{R})/\{\pm I\}$ acts faithfully on this coset space via

$$B \cdot A = A(\sigma(B))^{-1},$$

yielding the standard representation of $\mathrm{PSL}(2,\mathbb{R})$ onto the group of conformal automorphisms of $\mathbb{H}$. (That is, $P(B \cdot A) = B(P(A))$.)

1.3 *Veech groups*

Recall that $\mathrm{Mod}(X)$, the Teichmüller modular group of X, (a.k.a. the mapping class group) acts on $\mathcal{T}_X$ by complex-analytic isomorphisms, which are also isometries for the Teichmüller metric, and that the quotient of $\mathcal{T}_X$ by this action is precisely $\mathcal{M}_X$, the moduli space of X.

Definition 1.7. *The* Veech group $\Gamma(X, q)$ *of* (X, q) *is the subgroup of* $\mathrm{PSL}(2,\mathbb{R})$ *that corresponds to mapping classes of X, i.e., elements of* $\mathrm{Mod}(X)$ *that stabilize the image of the geodesic f generated by (X, q). The surface (X, q) is called* Veech *if* $\Gamma(X, q)$ *is a lattice in* $\mathrm{PSL}(2,\mathbb{R})$, *i.e., if* $\mathbb{H}/\Gamma(X, q)$ *has finite hyperbolic area.*

If (X, q) is a Veech surface, the image of the composition $\mathbb{H} \to \mathscr{T}_X \to \mathcal{M}_X$ is an algebraic curve. Veech showed that a Veech surface also has especially "nice" geometric properties [Ve2], as stated in Theorem 0.1.

By Thurston's description [Th] of the action of elements of $\mathrm{Mod}(X)$ on $\mathscr{T}_X$, we can closely relate the topological type of $g \in \Gamma(X, q)$ and the type of isometry of $\mathbb{H}$ that g induces; see also I. Kra's study in [K]. Specifically,

- g is *periodic* if and only if the correponding isometry of $\mathbb{H}$ is *elliptic* (fixes a point in $\mathbb{H}$);
- g is *reducible* if and only if the corresponding isometry of $\mathbb{H}$ is *parabolic* (fixes a point on $\partial\mathbb{H}$ and the class of horocycles through this point);
- g is *pseudo-Anosov* if and only if the corresponding isometry of $\mathbb{H}$ is *hyperbolic* (fixes two points on $\partial\mathbb{H}$ and the Poincaré geodesic connecting them).

2. Delaunay triangulations

Our algorithm for computing generators of the Veech group depends essentially on the existence of a combinatorial structure uniquely determined by the geometry of q. The structure we will use is the Delaunay triangulation of the flat surface. The existence of Delaunay triangulations and a few basic results concerning them are proved by H. Masur and J. Smillie in [MSx]. We will also use several results from I. Rivin's work in [Ri].

2.1 *Triangulations of a flat surface*

Following Vorobets's description of ω-triangles when ω is a holomorphic 1-form on a Riemann surface [Vo], we introduce:

Definition 2.1. *A* q-triangle *on* X *is a* 2*-simplex in* X*, embedded on its interior, whose vertices lie in* Z*, whose sides are geodesic with respect to* $|\sqrt{q}|$*, and whose interior does not contain any points of* Z*.*

On the interior of a q-triangle, the structure is purely Euclidean; for instance, the angles sum to π. Some of the vertices may coincide (for example, if Z contains only a single point). The condition that singular points have a cone angle greater than 2π prevents two edges of a single q-triangle from coinciding, however.

Definition 2.2. *A* q-triangulation *of* X *is a simplicial cell structure on* X*, the closures of whose* 2*-cells are* q*-triangles. A* q*-triangulation on* X *and a* q'*-triangulation on* X' *are* combinatorially equivalent *if there is a bijective map from the* q*-triangles in the triangulation of* X *to the* q'*-triangles in the triangulation of* X' *that preserves edge identifications.*

Remark 2.3. A combinatorial equivalence between (X, q) and (X', q') induces a homeomorphism from X to X', which is piecewise affine for the local coordinates of q and q'. We will use this natural homeomorphism in our algorithm, described in Section 4.

The *Whitehead move* is a natural way to change one q-triangulation into another. Suppose ABC and BCD are two triangles in some q-triangulation of X with the common edge BC. (Some or all of A, B, C, D may actually coincide as elements of Z; as far as the Euclidean structure on $X - Z$ is concerned, we can consider ABC and BCD to simply lie in the plane.) Then no points of Z are contained in the interior of the quadrilateral $ABDC$; provided $ABDC$ is convex, we can obtain a new q-triangulation of X by removing the edge BC and adding the edge AD, forming the triangles ABD and ACD.

Clearly, (X, q) admits many q-triangulations. We wish to describe a particular sort that is unique for an open dense subset of surfaces in the SL(2, $\mathbb{R}$)-orbit of (X, q).

2.2 *Dihedral angles and the Delaunay condition*

Definition 2.4. *Let T_1 and T_2 be a pair of Euclidean triangles with disjoint interiors and a common edge e, and let α_1 and α_2 be the (unsigned) angles opposite e in T_1 and T_2, respectively. The* dihedral angle *of e is $\alpha(e) = \alpha_1 + \alpha_2$.*

Definition 2.5. *A q-triangulation of X is* Delaunay *if $\alpha(e) \leq \pi$ for all edges e of the triangulation. A Delaunay triangulation is* degenerate *if $\alpha(e) = \pi$ for one or more e.*

The following result is a consequence of Theorem 10.2 in [Ri]; the proof is corrected by A. Bobenko and B. Springborn in [BS].

Proposition 2.6. *Given any q-triangulation of X, one obtains a Delaunay triangulation of X via a finite sequence of Whitehead moves, each of which replaces an edge having dihedral angle $> \pi$ with its opposite diagonal.*

Remark 2.7. The Delaunay condition for a triangulation is usually phrased in terms of the circumcircles of the triangles. It is a basic fact of Euclidean geometry that if $\alpha(e) = \pi$, then the vertices of T_1 and T_2 lie on a single circle. If $\alpha(e) > \pi$, then the circumcircle of T_1 contains the remaining vertex of T_2, and vice versa. If $\alpha(e) < \pi$, then the remaining vertex of T_2 lies outside the circumcircle of T_1.

Remark 2.8. The terminology of "dihedral angles" arises from an application to hyperbolic geometry [Ri]. If we use the half-space model of hyperbolic 3-space $\mathbb{H}^3$, then $\partial\mathbb{H}^3$ is the union of the Euclidean plane $\mathbb{E}^2$ and the point ∞. A triangle ABC in $\mathbb{E}^2$ determines an ideal tetrahedron $ABC\infty$ in $\mathbb{H}^3$, with ideal vertices A, B, C, and ∞. A basic fact of hyperbolic geometry is that the opposite edges of an ideal tetrahedron have equal dihedral angles (which measure the "solid" angle between faces of the tetrahedron). The dihedral angle of $\angle ABC$ in $ABC\infty$ equals the angle $\angle ABC$ in the plane, because the faces $AB\infty$ and $BC\infty$ are contained in vertical (Euclidean) planes. Two adjacent triangles ABC and BCD then determine two ideal tetrahedra, joined along the common face $BC\infty$, and the dihedral angle of their common edge BC is the sum of the angles $\angle BAC$ and $\angle BDC$ in $\mathbb{E}^2$.

The Delaunay condition on a triangulation of (X, q) will allow us to introduce additional structure on the SL(2, $\mathbb{R}$)-orbit of (X, q) in the following section.

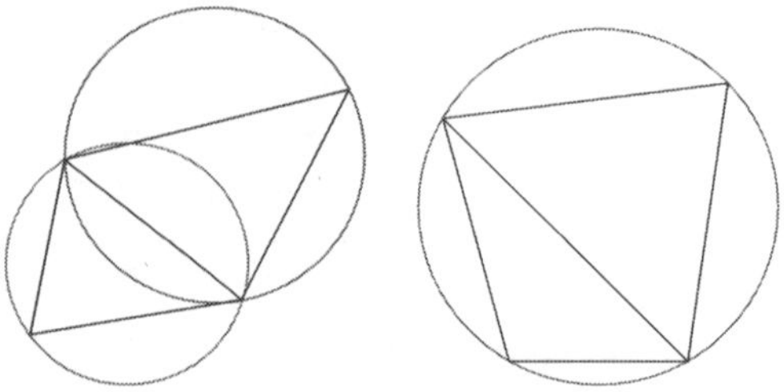

Figure 2. A pair of adjacent triangles in a Delaunay triangulation, and in a degenerate Delaunay triangulation

2.3 *Iso-Delaunay regions in* $\mathbb{H}$

The key observations in our algorithm are the following:

Proposition 2.9. *If a Delaunay triangulation of* $(X', q') \in \mathbb{H}$ *is non-degenerate, then it is unique. If a Delaunay triangulation is degenerate, then it is unique except for exchanges of edges with dihedral angle* π *via Whitehead moves.*

Proposition 2.10. *If* $f \in \Gamma(X, q)$ *and* $\mathcal{T}$ *is a Delaunay triangulation of* (X, q)*, then* $f^*\mathcal{T} = \{f^*(T) \mid T \in \mathcal{T}\}$ *is a Delaunay triangulation of* (f^*X, f^*q)*. I.e., The Delaunay triangulation is independent of the marking of* X*.*

Uniqueness is typically proved via the duality between edges of a Delaunay triangulation and those of the Voronoi cells on the surface (see, e.g., [MSx]; see also [Ri] for a proof using the convexity of a volume function on the triangulation). Identify the $\mathrm{SL}(2, \mathbb{R})$-orbit of (X, q) with $\mathbb{H}$ as in Section 1.2. In this case, the openness of the non-degenerate Delaunay condition implies that (X', q') is contained in a neighborhood of surfaces with combinatorially equivalent Delaunay triangulations.

Definition 2.11. *An* iso-Delaunay region *(IDR) is a maximal connected open subset of* $\mathbb{H}$ *such that all its points have combinatorially equivalent Delaunay triangulations.*

Proposition 2.12. *With respect to the Poincaré metric on* $\mathbb{H}$*, each IDR is convex and has piecewise geodesic boundary.*

To prove this result, we shall use a standard test in computational geometry.

Proposition 2.13 (Incircle test). *Let* $P_1 = (x_1, y_1)$*,* $P_2 = (x_2, y_2)$*, and* $P_3 = (x_3, y_3)$ *be three points in* $\mathbb{R}^2$*, with* O*,* P_1*, and* P_2 *non-colinear. Let* $\mathcal{C}$ *be the circumcircle of* OP_1P_2*. Then*

$$\textit{if} \quad \det\begin{pmatrix} x_1 & y_1 & x_1^2 + y_1^2 \\ x_2 & y_2 & x_2^2 + y_2^2 \\ x_3 & y_3 & x_3^2 + y_3^2 \end{pmatrix} \begin{cases} < 0, & P_3 \text{ is in the interior of } \mathcal{C}. \\ = 0, & P_3 \text{ lies on } \mathcal{C}. \\ > 0, & P_3 \text{ is exterior to } \mathcal{C}. \end{cases} \tag{1}$$

Sketch of proof. Embed $\mathbb{R}^2$ into $\mathbb{R}^3$ as the (x, y)-plane. Project P_1, P_2, and P_3 vertically to Q_1, Q_2, and Q_3, respectively, on the paraboloid $z = x^2 + y^2$. Let $\mathcal{P}$ be the plane determined by O, Q_1, and Q_2. Observe that $\{Q_1, Q_2, Q_3\}$ is a direct basis, an indirect basis, or a dependent set as the above determinant is greater than, less than, or equal to zero; this also corresponds to Q_3 lying above, below, or on $\mathcal{P}$ (where "above" and "below" are determined by the orientation of the basis $\{Q_1, Q_2\}$ for $\mathcal{P}$). Replacing z in the equation $ax + by + cz = 0$ of $\mathcal{P}$ with $x^2 + y^2$, one obtains the equation of a circle in the (x, y)-plane, which must pass through O, P_1, and P_2. One concludes the desired result by comparing the various geometric and algebraic conditions. □

Proof of Proposition 2.12. We will show that each IDR is an intersection of hyperbolic half-planes, which implies the desired results.

Suppose a Delaunay triangulation is given for a certain point in the $SL(2, \mathbb{R})$ orbit of (X, q). For each edge in the Delaunay triangulation, choose a chart as in Section 1.1 such that one endpoint of the edge is at 0. Label the other endpoint P_2, and the remaining vertices P_1 and P_3, so that the ordered bases $\{P_1, P_2\}$ and $\{P_2, P_3\}$ are oriented counterclockwise.

Now let $A \in SL(2, \mathbb{R})$ be a variable matrix, with $u + iv$ the image of A in $\mathbb{H}$. Then we can chose a representative in the same left $SO(2, \mathbb{R})$-coset as A that sends x_i to $x_i + uy_i$ and y_i to vy_i $(i = 1, 2, 3)$. Evaluate the equality in (1) with x_i and y_i replaced by their images under the action of this representative. After some simplification (including dividing by v, which we can do because $v > 0$ in the upper half-plane), this becomes

$$a(u^2 + v^2) + 2bu + c \geq 0 \tag{2}$$

where

$$a = x_1y_2y_3(y_3 - y_2) + x_2y_1y_3(y_1 - y_3) + x_3y_1y_2(y_2 - y_1),$$

$$b = x_1y_1(x_2y_3 - x_3y_2) + x_2y_2(x_3y_1 - x_1y_3) + x_3y_3(x_1y_2 - x_2y_1),$$

$$c = x_1x_2y_3(x_1 - x_2) + x_2x_3y_1(x_2 - x_3) + x_1x_3y_2(x_3 - x_1).$$

If $a \neq 0$, (2) is the equation of a circle whose center lies on the real axis. If $a = 0$ (for example, if $y_1 = 0$ and $y_2 = y_3$), then (2) is the equation of a vertical line. These are precisely the forms of geodesics in $\mathbb{H}$.

Let (X', q') be a surface in the $SL(2, \mathbb{R})$-orbit of (X, q) with nondegenerate Delaunay triangulation. For each edge e, choose a chart in the corresponding flat structure whose domain contains the triangles adjacent to e. We may assume that one of the endpoints of e is at the origin. Then, applying the result of the previous paragraph, we get one inequality for each edge. Because there are only finitely many such inequalities—one for each edge of the Delaunay triangulation—the result follows. □

As noted in Veech [Ve1], the area of each IDR is finite, so it is actually a hyperbolic polygon, possibly with some vertices at infinity.

Observe that, at an edge of an IDR determined by a set of inequalities (1), by performing a Whitehead move on the degenerate edges, we reverse these particular inequalities and obtain a polygonal region on the opposite side of the edge. Hence the Delaunay triangulations of surface in the orbit of (X, q) naturally partition $\mathbb{H}$ into 2-cells (the open IDRs), 1-cells (maximally connected sets of points lying in the closures of exactly two IDRs), and 0-cells (points lying in the closures of at least three IDRs), giving $\mathbb{H}$ the structure of a cell complex.

Definition 2.14. *The above cell complex on* $\mathbb{H}$ *is called the* iso-Delaunay complex *of* (X, q).

Proposition 2.10 implies that the iso-Delaunay complex in the SL(2, $\mathbb{R}$)-orbit of the surface is preserved by the action of $\Gamma(X, q)$. A caveat: the iso-Delaunay complex is not a CW-complex, as some of its cells may not have compact closure in $\mathbb{H}$.

3. Bouillabaisse surfaces

Whenever a flat surface has a direction in which it decomposes into cylinders, and the moduli of these cylinders are commensurable, the Veech group of the surface contains a parabolic element, which consists of appropriate powers of Dehn twists in each of the cylinders. A *bouillabaisse surface* is a flat surface with at least two such directions; in other words, its Veech group contains two transverse parabolic elements. The construction of bouillabaisse surfaces is due to W. Thurston [Th]. In June 2003, J. Hubbard described them at the "bouillabaisse night" lecture during a conference at the Centre International de Rencontres Mathématiques in Luminy, France, whence their name.

3.1 *Basic construction*

The construction begins with an oriented compact surface S of genus $g \geq 2$. A *multi-curve* on S is a collection of pairwise disjoint, non-homotopic simple closed curves. The number of curves is bounded by the genus of the surface. In particular, any maximal multi-curve creates a decomposition of the surface into pairs of pants, and the number of pairs of pants is determined by the genus. Thus the number of elements in a maximal multi-curve is $3g - 3$. A *Dehn twist* around a multi-curve is simply the composition of the Dehn twists around each of the component curves; because the curves are disjoint, all these twists commute, so the order of composition does not matter. We will always assume Dehn twists are "right-handed" with respect to the orientation of S. We will also mildly abuse notation by letting, for example, Γ represent both the set of curves in a multi-curve on S and the set of points on S contained in the images of these curves.

Let Γ and Δ be transverse, not necessarily maximal, multicurves on S such that each component of $S - (\Gamma \cup \Delta)$ is simply connected. Assume further that Γ and Δ are in minimal position, meaning that each pair $(\gamma, \delta) \in \Gamma \times \Delta$ has the minimum number of intersections as γ and δ vary in their respective homotopy classes. Define $M = [m_{\delta\gamma}] : \mathbb{R}^{\Gamma} \to \mathbb{R}^{\Delta}$ by

$$\delta \cdot M(\gamma) = m_{\delta\gamma}, \qquad \gamma \in \Gamma,\ \delta \in \Delta$$

where $m_{\delta\gamma}$ is the geometric intersection number of δ and γ. Let λ be the largest eigenvalue of $M^{\top}M$, and let $v \in \mathbb{R}^{\Gamma}$ be an eigenvector of $M^{\top}M$ corresponding to λ with all entries positive. (Such a v exists because $M^{\top}M$ is a Perron-Frobenius matrix.)

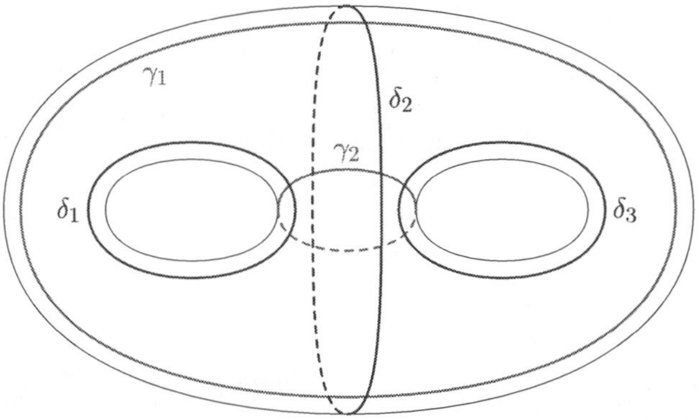

Figure 3. The pair of multi-curves Γ and Δ on a genus 2 surface that produce the flat surface in Figure 1.

Now set $u = Mv$. Then u is an eigenvector of $MM^{\top}$, also corresponding to the eigenvalue λ.

We use the above data to simultaneously construct a Riemann surface $X' \in \mathscr{T}_S$ and a quadratic differential q' on X'. $\Gamma \cup \Delta$ induces a cell structure on S, whose dual structure consists of rectangles $R_{\delta\gamma}$, each containing the intersection of one $\gamma \in \Gamma$ and one $\delta \in \Delta$. For each rectangle $R_{\delta\gamma}$ crossed by γ and δ in the dual cell structure to $\Gamma \cup \Delta$, take a chart on the interior to $\mathbb{C}$ so that γ is horizontal, δ is vertical, the the width of $R_{\delta\gamma}$ is u_δ and the height is v_γ. Assume further that each chart matches the orientation of S with the orientation of the plane. These charts define the complex structure on S. Take $q' = \mathrm{d}z^2$ in each of these charts.

Proposition 3.1. *In these coordinates, the Dehn twists D_Γ and D_Δ, around Γ and Δ respectively, are affine transformations, and their derivatives have the forms*

$$\mathrm{der}(D_\Gamma) = \begin{pmatrix} 1 & \lambda \\ 0 & 1 \end{pmatrix} \quad \textit{and} \quad \mathrm{der}(D_\Delta) = \begin{pmatrix} 1 & 0 \\ -1 & 1 \end{pmatrix}.$$

Proof. Choose a point in the cylinder crossed by $\gamma \in \Gamma$ and a rectangular chart so that one edge of the cylinder lies along the x-axis. In these coordinates, the Dehn twist around Γ maps (x, y) by:

$$\begin{pmatrix} x \\ y \end{pmatrix} \mapsto \begin{pmatrix} x + s_\gamma y \\ y \end{pmatrix}$$

for some s_γ depending a priori *on γ. When $y = tv_\gamma$, x is increased by*

$$\sum_{\delta \in \Delta} m_{\delta\gamma} t u_\delta = (M^{\top} tu)_\gamma = t(M^{\top} Mv)_\gamma = t\lambda v_\gamma .$$

Hence $s_\gamma = \lambda$ independently of γ, and $\mathrm{der}(D_\Gamma)$ *has the form described. The proof for the Dehn twist around Δ is similar.* □

3.2 *Introducing arbitrary rational ratios of moduli*

The above construction admits several generalizations. We remark here on the principal one that applies to our study. As described above, the bouillabaisse construction

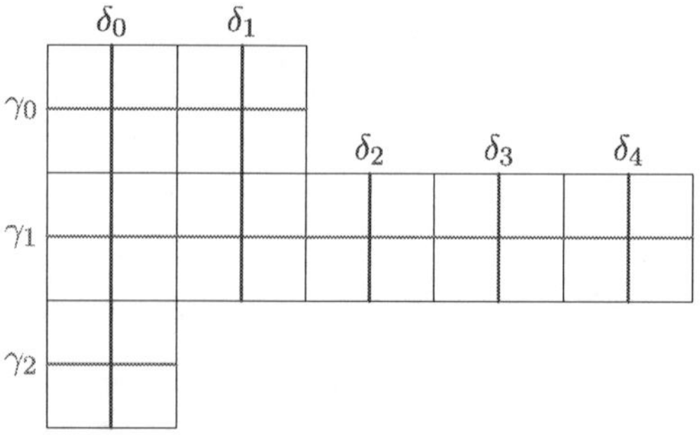

Figure 4. A bouillabaisse surface on which the cylinders have moduli that are rationally related, such as the square-tiled surface in this figure, can be constructed via multi-curves, using the appropriate duplication of each homotopy class of curves.

creates cylinders with *equal* moduli in each of the horizontal and vertical directions. In order to create cylinders which have *commensurable* moduli, one "duplicates" the core curves of cylinders in which higher powers of Dehn twists are necessary. More precisely, fix $\gamma_0 \in \Gamma$, $\delta_0 \in \Delta$, and rational numbers r_γ, r_δ for each $\gamma \in \Gamma$, $\delta \in \Delta$, with $r_{\gamma_0} = r_{\delta_0} = 1$. We will construct a flat structure on S so that r_γ is the ratio of the modulus of the cylinder with core curve γ to that of the cylinder with core curve γ_0 (and likewise for the r_δs).

Let ρ be the least common multiple of the r_γ s (i.e., ρ is the smallest rational number such that ρ/r_γ is an integer for all γ); ρ is an integer because $r_{\gamma_0} = 1$. Set $r'_\gamma = \rho/r_\gamma$, and replace each $\gamma \in \Gamma$ with r'_γ homotopic copies of γ (for example, γ_0 is replaced by ρ copies of itself); denote these copies $\alpha_{\gamma,j}$ for $1 \leq j \leq r'_\gamma$. Then a simultaneous Dehn twist in each $\alpha_{\gamma,j}$ for fixed γ corresponds to the r'_γth power of the Dehn twist in γ. Let $\Gamma' = \{\alpha_{\gamma,j} \mid \gamma \in \Gamma, 1 \leq j \leq r'_\gamma\}$. Note that the r'_γ s depend only on the ratios of the moduli of the cylinders, not on the choice of γ_0.

Now carry out the same process in the vertical direction to get a collection Δ' of $\beta_{\delta,k}$s, and construct $M' : \mathbb{R}^{\Gamma'} \to \mathbb{R}^{\Delta'}$ as before. Because the geometric intersection number is well-defined on homotopy classes, we have $\beta_{\delta,k} \cdot M'(\alpha_{\gamma_j}) = \delta \cdot M(\gamma)$, i.e., M' is obtained from the ordinary $M : \mathbb{R}^\Gamma \to \mathbb{R}^\Delta$ by replacing each column $[m_{\delta\gamma}]_{\delta\in\Delta}$ with r'_γ copies, and each row $[m_{\delta\gamma}]_{\gamma\in\Gamma}$ with r'_δ copies.

By carrying out the same construction with M' as we did with M, we get a flat surface such that the cylinders containing each of $\alpha_{\gamma,j}$ all have the same modulus. The cylinder of which some γ is a core curve is comprised of r'_γ of these cylinders. For $\gamma_1, \gamma_2 \in \Gamma$, the ratio of the moduli of their respective cylinders is $r'_{\gamma_1}/r'_{\gamma_2} = r_{\gamma_2}/r_{\gamma_1}$, as desired.

3.3 *Relevance of bouillabaisse surface*

The following corollary to Theorem 0.1 demonstrates one measure of the importance of bouillabaisse surfaces:

Corollary 3.2. *Every Veech surface is bouillabaisse.*

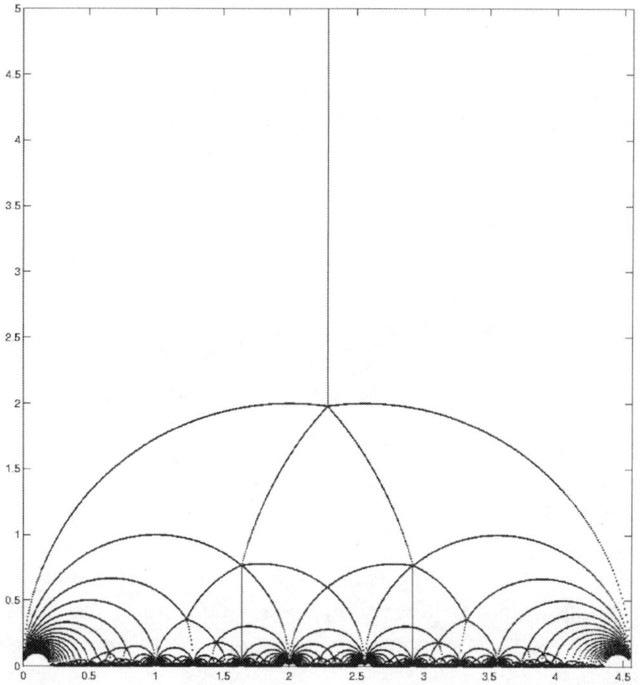

Figure 5. Some IDRs for the flat surface shown in Figures 1 and 3. This is a Veech surface, which explains the extreme regularity of the iso-Delaunay complex.

Proof. Any direction containing a geodesic that connects two singular points decomposes the surface into cylinders whose moduli are commensurable. □

This construction has become an indispensible tool in the study of flat surfaces. For example, McMullen has used it to describe infinite families of Veech surfaces in genus 2, 3, and 4 [Mc4]. Not all flat structures with a non-trivial Veech group arise in this manner, however. The expansion constants for pseudo-Anosov elements of the subgroup $\langle D_\Gamma, D_\Delta\rangle < \Gamma(X, q)$ all lie in the field $\mathbb{Q}(\lambda)$, which is totally real since λ arises as an eigenvalue of a symmetric matrix. P. Arnoux and J.-C. Yoccoz [AY] constructed pseudo-Anosov maps of surfaces in all genera $g \geq 3$ whose expansion constants are *Pisot numbers,* some of whose Galois conjugates lie outside of $\mathbb{R}$. P. Hubert and E. Lanneau [HuLa] showed that this implies the Veech groups of the Arnoux–Yoccoz surfaces contain no parabolic elements; i.e., every stabilizing element of such a surface is either periodic or pseudo-Anosov; hence the Arnoux–Yoccoz examples are not bouillabaisse.

4. The algorithm

The IDRs and iso-Delaunay complex determined by (X, q) partition $\mathbb{H}$ into (not necessarily compact) polygons. A fundamental domain for the action of the Veech group $\Gamma(X, q)$ can be assembled from (possibly infinitely many) IDRs of the surface. In order to turn this observation into an algorithm, we need a method for detecting when two IDRs are equivalent under an element of $\Gamma(X, q)$; we will call such a pair of IDRs *Veech equivalent*.

4.1 *Veech equivalent IDRs*

Combinatorial equivalence is an obvious necessary condition for two IDRs to be Veech equivalent, by the uniqueness of non-degenerate Delaunay triangulations. It is not sufficient, however, so we must also take into account the metric structure of the surfaces. Fortunately, this structure is well-behaved for the action of SL(2, $\mathbb{R}$).

Let R_1 and R_2 be two iso-Delaunay regions with the same combinatorial data, and choose a point (X', q') in one of them, say in R_2. We want to determine if (X', q') is isomorphic to a point in R_1; if it is, then there is an element of $\Gamma(X, q)$ carrying R_1 to R_2.

4.2 *Tree search in* $\mathbb{H}$

To assemble the fundamental domain of $\Gamma(X, q)$, begin with a single point in $\mathbb{H}$ whose Delaunay triangulation is not degenerate. First determine the IDR containing this point; each edge of the IDR corresponds to the degeneration of (a) particular edge(s) in the triangulation.

More systematically, we have the following algorithm:

(1) Begin with a surface in the SL(2, $\mathbb{R}$)-orbit of (X, q) whose Delaunay triangulation is non-degenerate, and compute its iso-Delaunay region $R_i = R_0$ using the formulas in the proof of Proposition 2.12.
(2) Along each (possibly infinite) hyperbolic segment comprising the boundary of R_i, determine which edges have degenerated, and compute the adjacent triangulation that replaces each of these edges by their opposite diagonal. This allows us to "cross" the edge into a new iso-Delaunay region.
(3) For each boundary segment crossed, compute the adjacent IDR from the new triangulation; set R_i equal to this IDR.
(4) Check if the new triangulation in R_i is combinatorially equivalent to that of a previously visited region R_j. If so:
    - check if the new surface is isomorphic to a point in R_j;
    - if such an isomorphism exists, find an isometry that carries R_i to R_j, and eliminate R_i from the tree search; add the isometry to the generating list for $\Gamma(X, q)$.
(5) For each new IDR not found to be Veech equivalent to a previous IDR, return to Step (2).

4.3 *Initial bounds for bouillabaisse surfaces*

From the construction of bouillabaisse surfaces, we automatically get part of the iso-Delaunay complex, and hence initial bounds on the fundamental domain of $\Gamma(X, q)$. The two elements of SL(2, $\mathbb{R}$) described in Proposition 3.1 generate a subgroup

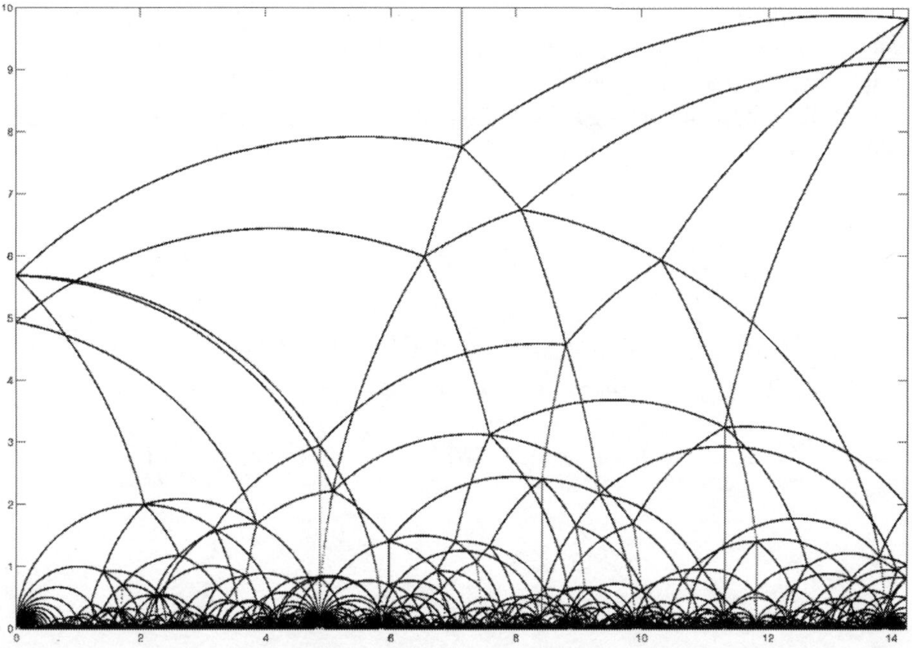

Figure 6. Some IDRs for the flat surface described in Section 4.4. Note the prominent cusp near 4.8595; this is the point ϕ described in the text.

of $\Gamma(X, q)$. This subgroup is in most cases a free group on these two generators; C. Leininger has given the precise conditions for the group to be free [Lei]. In any case, it has a fundamental domain bounded by the geodesics

$$x = 0, \quad x = \lambda, \quad y^2 + \left(x - \frac{1}{2}\right)^2 = \frac{1}{4}, \quad \text{and} \quad y^2 + \left(x - \lambda + \frac{1}{2}\right)^2 = \frac{1}{4}. \quad (3)$$

If $\lambda > 2$, F_2 is not itself a lattice in $\mathrm{SL}(2, \mathbb{R})$.

The trace field of $\Gamma(X, q)$ is exactly $\mathbb{Q}(\lambda)$. This follows from the fact that that the trace field is generated by the trace of a single hyperbolic element ([KS]), and

$$D_\Delta^{-1} \circ D_\Gamma = \begin{pmatrix} 1 & \lambda \\ 1 & 1+\lambda \end{pmatrix}, \quad (4)$$

which has trace $2 + \lambda$, and $\mathbb{Q}(2 + \lambda) = \mathbb{Q}(\lambda)$. Because λ arises from a symmetric matrix, it and all of its Galois conjugates are real, hence the field $\mathbb{Q}(\lambda)$ is *totally real,* i.e., all of its embeddings into $\mathbb{C}$ actually have their image contained in $\mathbb{R}$.

4.4 *Other uses of IDR analysis*

By a slight modification of the surface in Figure 3, we obtain an example where $\lambda \approx 28.475$ is the largest root of the polynomial $x^3 - 30x^2 + 44x - 16$, which is irreducible over $\mathbb{Q}$. A portion of the $\mathrm{SL}(2, \mathbb{R})$-orbit of this surface with its iso-Delaunay complex is shown in Figure 6. One observes immediately that the structure of the iso-Delaunay complex is much more complicated than the one in Figure 5. Indeed, we can demonstrate that this surface is not Veech.

The IDR picture itself suggests a direction on the surface to examine: note the prominent cusp near 4.8595 on the real axis. This is in fact the positive root ϕ of the polynomial $x^2 + x - \lambda$, which is contained in $\mathbb{Q}(\lambda)$:

$$\phi = \frac{-1 + \sqrt{1 + 4\lambda}}{2} = -6 + \frac{15}{2}\lambda - \frac{1}{4}\lambda^2 \approx 4.8595$$

When we apply $\left(\begin{smallmatrix} 1 & \phi \\ 0 & 1 \end{smallmatrix}\right)$ to the surface, we find that the vertical direction decomposes the surface into cylinders whose moduli are:

$$m_1 = \phi + 1, \quad m_2 = -10 + 15\lambda - \frac{1}{2}\lambda^2 = 2(\phi + 1)$$

$$m_3 = -\frac{21}{16} + \frac{73}{32}\lambda - \frac{5}{64}\lambda^2 = \frac{1}{16}(5\phi - \lambda + 9),$$

These moduli are not commensurable:

$$\frac{m_2}{m_1} = 2, \quad \frac{m_3}{m_1} = \frac{1}{16}(5(\phi + 1) - \lambda), \quad \frac{m_3}{m_2} = \frac{1}{32}(5(\phi + 1) - \lambda).$$

Therefore the Veech group of the surface is not a lattice. Further analysis may demonstrate the existence of other cylindrical decompositions, not conjugate to these in $\mathrm{PSL}(2, \mathbb{Q}(\lambda))$.

5. Acknowledgements

I would like to thank my advisor, John Hubbard, for providing me with the problem of computing Veech groups, as well as John Smillie, Bill Thurston, and Cliff Earle for their great encouragement, help, and insights. Thanks to the organizers of the International Workshop on Teichmüller Theory and Moduli Problems for their invitation to speak and to the students and faculty of the Harish-Chandra Research Institute for their great welcome. Thanks also to Melanie Pivarski and my father, Jerry Bowman, who helped with programming through many conversations.

References

[AY] Pierre Arnoux and Jean-Christophe Yoccoz, Construction de difféomorphismes pseudo-Anosov., *C. R. Acad. Sc. Paris Sr. I Math.*, **292** (1981) 75–78.

[BS] I. Alexander Bobenko and A. Boris Springborn, A discrete laplace–beltrami operator for simplicial surfaces, *Discrete Comp. Geom.*, **38** (2007) 740–756.

[Bo] Joshua P. Bowman, Flat structures and complex structures in Teichmüller theory. PhD thesis, Cornell, August (2009).

[Ca] Kariane Calta, Veech surfaces and complete periodicity in genus two, *J. Amer. Math. Soc.*, **17** (2004) 871–908.

[EG] J. Clifford Earle and P. Frederick Gardiner, Teichmüller disks and Veech's $\mathcal{F}$-structures, *Contemporary Mathematics*, **201** (1997) 165–189.

[FK] A. Ralph Fox and B. Richard Kerschner, Concerning the transitive properties of geodesics on a rational polyhedron, *Duke Math. J.*, **2** (1936) 147–150.

[GJ] Eugene Gutkin and Chris Judge, Affine mappings of translation surfaces: Geometry and arithmetic, *Duke Math. J.*, **103** (2000) 191–213.

[Ho] W. Patrick Hooper, Another veech triangle. Preprint, 4/2007. Available online.

[HuLa] Pascal Hubert and Erwan Lanneau, Veech groups without parabolic elements, *Duke Math. J.*, **133** (2006) 335–346.

[KS] Richard Kenyon and John Smillie, Billiards in rational-angled triangles, *Comment. Math. Helv.*, **75** (2000) 65–108.

[K] Irwin Kra, On the nielsen–thurston–bers type of some self-maps of riemann surfaces, *Acta Math.*, **146** (1981) 231–270.

[Lei] J. Christopher Leininger, On groups generated by two positive multi-twists: Teichmüller curves and Lehmer's number, *Geom. and Top.*, **8** (2004) 1301–1359.

[MSx] Howard Masur and John Smillie, Hausdorff dimension of sets of nonergodic measured foliations, *Ann. Math.*, **134** (1991) 455–543.

[Mc1] T. Curtis McMullen, Billiards and Teichmüller curves on Hilbert modular surfaces, *J. Amer. Math. Soc.*, **16** (2003) 857–885.

[Mc2] T. Curtis McMullen, Teichmüller curves in genus two: Discriminant and spin, *Math. Ann.*, **333** (2005) 87–130.

[Mc3] T. Curtis McMullen, Teichmüller curves in genus two: The decagon and beyond, *J. reine angew. Math.*, **582** (2005) 173–200.

[Mc4] T. Curtis McMullen, Prym varieties and Teichmüller curves, *Duke Math. J.*, **133** (2006) 569–590.

[Mc5] T. Curtis McMullen, Teichmüller curves in genus two: Torsion divisors and ratios of sines, *Inv. math.*, **165** (2006) 651–672.

[Ri] Igor Rivin, Euclidean structures on simplicial surfaces and hyperbolic volume, *Ann. Math.*, **139** (1994) 553–580.

[Sch] Gabriela Schmithüsen, An algorithm for finding the veech group of an origami, *Experimental Mathematics*, **13** (2000) 459–472.

[Th] P. William Thurston, On the geometry and dynamics of diffeomorphisms of surfaces, *Bull. A.M.S.*, **19** (1988) 417–431.

[Ve1] A. William Veech, Bicuspid f-structures and Hecke groups. Preprint.

[Ve2] A. William Veech, Teichmüller curves in moduli space, Eisenstein series, and an application to triangular billiards, *Inv. Math.*, **97** (1989) 553–583.

[Vo] B. Yaroslav Vorobets, Planar structures and billiards in rational polygons: the Veech alternative, *Russ. Math. Surv*, **51** (1996) 779–817.

Teichmüller Theory and Moduli Problem No. 10, 2010, pp. 131–150.

Introductory Bumponomics: The Topology of Deformation Spaces of Hyperbolic 3-Manifolds

Richard D. Canary

University of Michigan, Ann Arbor, MI 48109, USA
e-mail: canary@umich.edu

Abstract. We survey work on the topology of the space $AH(M)$ of all (marked) hyperbolic 3-manifolds homotopy equivalent to a fixed compact 3-manifold M with boundary. The interior of $AH(M)$ is quite well-understood, but the topology of the entire space can be quite complicated. However, the topology is well-behaved at many points in the boundary of $AH(M)$.

1. Introduction

In this paper we survey recent work on the topology of the space $AH(M)$ of all hyperbolic 3-manifolds homotopy equivalent to a fixed compact 3-manifold with boundary M. The recent resolution of Thurston's Ending Lamination Conjecture (in Minsky [57] and Brock-Canary-Minsky [14,15]) in combination with the resolution of Marden's Tameness Conjecture (in Agol [1] and Calegari-Gabai [22]), gives a complete classification of the manifolds in $AH(M)$. However, the invariants in this classification vary discontinuously over the space, and we are very far from having a parameterization of $AH(M)$.

The interior $\text{int}(AH(M))$ has been well-understood since the 1970's, due to work of Ahlfors, Bers, Kra, Marden, Maskit, Sullivan, Thurston and others. The components of $\text{int}(AH(M))$ are enumerated by topological data, and each component is a manifold parameterized by natural analytic data. The recent resolution of the Bers-Sullivan-Thurston Density Conjecture assures us that $AH(M)$ is the closure of $\text{int}(AH(M))$.

Since the mid-1990's, there has been a string of results and examples demonstrating that the topology of $AH(M)$ is less well-behaved than originally expected. Anderson and Canary [5] first showed that components of $\text{int}(AH(M))$ can bump, i.e. have intersecting closure, while Anderson, Canary and McCullough [6] characterized exactly which components of $\text{int}(AH(M))$ can bump when M has incompressible boundary. McMullen [55] showed that if $M = S \times I$ (where S is a closed surface), then the only component of $\text{int}(AH(M))$ self-bumps, i.e. there is a point in the boundary such that any sufficiently small neighborhood of the point disconnects the interior. Bromberg and Holt [20] showed that self-bumping occurs whenever M contains a primitive essential annulus. Most recently, Bromberg [19] showed that the space of punctured torus groups is not even locally connected. We will survey these results and

Research supported in part by grants from the National Science Foundation describe the construction which has been responsible for all the pathological behavior discovered so far.

On the other hand, the topology of the deformation space appears to be well-behaved at most points on its boundary and we will describe some recent results establishing this in a variety of settings.

2. Definitions

In this section, we will set up some of the notation and introduce some of the definitions used throughout the remainder of the paper.

Let M be a compact 3-manifold whose interior admits a complete hyperbolic structure. We will assume throughout this paper that all surfaces and 3-manifolds are oriented and have non-abelian fundamental group and that all manifolds are allowed to have boundary. Then $AH(M)$ is the space of (marked) hyperbolic 3-manifolds homotopy equivalent to M. More formally, we define

$$AH(M) = \{\rho : \pi_1(M) \to \mathrm{PSL}_2(\mathbf{C}) \mid \rho \text{ discrete and faithful}\}/\mathrm{PSL}_2(\mathbf{C}).$$

We topologize $AH(M)$ as a subset of the the character variety

$$X(M) = \mathrm{Hom}_T(\pi_1(M), \mathrm{PSL}_2(\mathbf{C}))//\mathrm{PSL}_2(\mathbf{C})$$

where $\mathrm{Hom}_T(\pi_1(M), \mathrm{PSL}_2(\mathbf{C}))$ denotes the space of representations $\rho : \pi_1(M) \to \mathrm{PSL}_2(\mathbf{C})$ with the property that if g is an element of a rank two abelian subgroup of $\pi_1(M)$, then $\rho(g)$ is either parabolic or the identity. (Here, we are taking the Mumford quotient to guarantee that the quotient has the structure of an algebraic variety, see Kapovich [41] for details.)

If $\rho \in AH(M)$, then $N_\rho = \mathbb{H}^3/\rho(\pi_1(M))$ is a hyperbolic 3-manifold homotopy equivalent to M. (The elements of $AH(M)$ are really equivalence classes of representations, but as the manifolds associated to conjugate representations are isometric we will consistently blur this distinction.) There is also a homotopy equivalence $h_\rho : M \to N_\rho$ such that

$$(h_\rho)_* : \pi_1(M) \to \pi_1(N_\rho) = \rho(\pi_1(M))$$

agrees with ρ. The homotopy equivalence h_ρ is the "marking" of N_ρ.

Alternatively, we could have defined $AH(M)$ as the space of pairs (N, h) where N is an oriented hyperbolic 3-manifold and $h : M \to N$ is a homotopy equivalence, where we consider two pairs (N_1, h_1) and (N_2, h_2) to be equivalent if there is an orientation-preserving isometry $j : N_1 \to N_2$ such that $j \circ h_1$ is homotopic to h_2. This alternate definition is reminiscent of the classical definition of Teichmüller space as a space of marked Riemann surfaces of a fixed genus.

If $\rho \in AH(M)$, then the *domain of discontinuity* $\Omega(\rho)$ is the largest open subset of $\hat{\mathbf{C}}$ on which $\rho(\pi_1(M))$ acts properly discontinuously. The *limit set* $\Lambda(\rho) = \hat{\mathbf{C}} - \Omega(\rho)$ is its complement. The *conformal boundary* is defined to be $\partial_c N_\rho = \Omega(\rho)/\rho(\pi_1(M))$ and we let

$$\hat{N}_\rho = N_\rho \cup \partial_c N_\rho = (\mathbb{H}^3 \cup \Omega(\rho))/\rho(\pi_1(M)).$$

We say that N_ρ is *convex cocompact*, if $\hat{N}_\rho$ is compact. We say that N_ρ is *geometrically finite* if $\hat{N}_\rho$ is homeomorphic to $M' - P'$ where M' is a compact 3-manifold and P' is a collection of annuli and tori in $\partial M'$. (These definitions are equivalent to more classical definitions, see Marden [51] and Bowditch [11].)

3. The interior of $AH(M)$

In this section, we survey the classical deformation theory of hyperbolic 3-manifolds which gives a beautiful description of the interior $\text{int}(AH(M))$ of $AH(M)$. If the boundary of M consists entirely of tori, then Mostow-Prasad Rigidity [61,71] implies that any homotopy equivalence between hyperbolic 3-manifolds homotopy equivalent to M is homotopic to an isometry. Therefore, $AH(M)$ has either 0 or 2 points (one gets one point for each orientation of M.) So, we will always assume that the boundary of M has a non-toroidal component.

Marden [51] and Sullivan [74] proved that $\text{int}(AH(M))$ consists exactly of the geometrically finite hyperbolic manifolds $\rho \in AH(M)$ such that every parabolic element of $\rho(\pi_1(M))$ is contained in a free abelian subgroup of rank two. (This is equivalent to requiring that $\hat{N}_\rho$ be homeomorphic to a compact 3-manifold with its toroidal boundary components removed.) The now classical quasiconformal deformation theory of Kleinian groups shows that geometrically finite hyperbolic 3-manifolds are determined by their (marked) homeomorphism type and the conformal structure on the conformal boundary and that every possible conformal structure arises. For an analytically-oriented discussion of quasiconformal deformation theory, see Bers [8]. For a more topological viewpoint on this material, see chapter 7 of Canary-McCullough [24].

In order to formally state the parameterization theorem for $\text{int}(AH(M))$ we need to introduce some more notation.

We define $\mathcal{A}(M)$ to be the set of oriented, compact, irreducible, atoroidal (marked) 3-manifolds homotopy equivalent to M. More formally, $\mathcal{A}(M)$ is the set of pairs (M', h') where M' is an oriented, compact, irreducible, atoroidal 3-manifold and $h' : M \to M'$ is a homotopy equivalence where two pairs (M_1, h_1) and (M_2, h_2) are considered equivalent if there exists an orientation-preserving homeomorphism $j : M_1 \to M_2$ such that $j \circ h_1$ is homotopic to h_2. We recall that M' is said to be *irreducible* if every embedded 2-sphere in M' bounds a ball and is said to be *atoroidal* if every rank two free abelian subgroup of $\pi_1(M')$ is conjugate to a subgroup of $\pi_1(T')$ for some toroidal boundary component T' of M'.

If $[(M', h')] \in \mathcal{A}(M)$, we define $Mod_0(M')$ to be the group of isotopy classes of homeomorphisms of M' which are homotopic to the identity. We define $\partial_{NT} M'$ to be the non-toroidal components of ∂M and we let $\mathcal{T}(\partial_{NT} M')$ denote the Teichmüller space of all (marked) conformal structures on $\partial_{NT} M'$. Recall that the Teichmüller space of a disconnected surface is simply the product of the Teichmüller spaces of its components, so $\mathcal{T}(\partial_{NT} M')$ is always topologically a cell.

Theorem 3.1 (Ahlfors, Bers, Kra, Marden, Maskit, Sullivan,Thurston).

$$\mathrm{int}(AH(M)) \cong \bigcup_{[(M',h')]\in\mathcal{A}(M)} \mathcal{T}(\partial_{NT}M')/Mod_0(M')$$

This identification is quite natural. By the previously mentioned results of Marden and Sullivan, if $\rho \in \mathrm{int}(AH(M))$, there exists a compact, atoroidal, irreducible 3-manifold M_ρ and an orientation-preserving homeomorphism $j_\rho : \hat{N}_\rho \to \mathrm{int}(M_\rho) \cup \partial_{NT}M_\rho$, so we obtain a well-defined marked homeomorphism type $[(M_\rho, j_\rho \circ h_\rho)] \in \mathcal{A}(M)$. If $[(M_\rho, j_\rho \circ h_\rho)] = [(M', h')]$, then we may assume that $M_\rho = M'$ and choose j_ρ so that $j_\rho \circ h_\rho$ is homotopic to h'. With this constraint, j_ρ is well-defined up to post-composition by elements of $Mod_0(M')$ and $\partial_c N_\rho$ is a Riemann surface, so we get a well-defined element of $\mathcal{T}(\partial_{NT}M')/Mod_0(M')$.

Let $\Phi : \mathrm{int}(AH(M)) \to \bigcup_{[(M',h')]\in\mathcal{A}(M)} \mathcal{T}(\partial_{NT}M')/Mod_0(M')$ be the map defined in the previous paragraph. We will discuss the ingredients of the proof that Φ is a homeomorphism. Maskit's Extension Theorem [52] implies that if $\Theta(\rho_1) = \Theta(\rho_2)$, then there exists a quasiconformal map $\varphi : \hat{\mathbb{C}} \to \hat{\mathbb{C}}$ which is conformal on $\Omega(\rho_1)$ and conjugates the action of $\rho_1(\pi_1(M))$ to the action of $\rho_2(\pi_1(M))$. Ahlfors [2] proved that if ρ_1 is geometrically finite, then $\Lambda(\rho_1)$ has measure zero, so we may conclude that φ is conformal. Therefore, Φ is injective. The surjectivity of Φ follows from the Measurable Riemann Mapping Theorem [3] and Thurston's Geometrization Theorem (see Morgan [60]).

If M has incompressible boundary (equivalently if $\pi_1(M)$ is freely indecomposable), then $Mod_0(M')$ is trivial, so each component of $\mathrm{int}(AH(M))$ is topologically a ball. Moreover, Canary and McCullough [24], showed that if M has incompressible boundary, then $\mathcal{A}(M)$ is infinite if and only if M has *double trouble*, i.e. there exist simple closed curves α and β in $\partial_{NT}M$ which are both homotopic to a curve γ in a toroidal boundary component of M but are not homotopic in ∂M. (Alternatively, M has double trouble if and only if there is a thickened torus component of its characteristic submanifold, which intersects the boundary of M in at least two annuli.) We summarize this below.

Corollary 3.2. *If M has incompressible boundary, then* $\mathrm{int}(AH(M))$ *is homeomorphic to a collection of disjoint balls. This collection is infinite if and only if M has double trouble.*

If M has compressible boundary, then typically $\mathcal{A}(M)$ is infinite (see Canary-McCullough [24] for a detailed analysis) and $Mod_0(M)$ is infinitely generated (see McCullough [53].) Maskit [52] showed that $Mod_0(M')$ always acts freely on $\mathcal{T}(\partial_{NT}M')$, so each component of $\mathrm{int}(AH(M))$ is still a manifold.

Examples.

1) If M is homeomorphic to a trivial I-bundle $S \times I$ over a closed orientable surface S of genus $g \geq 2$, then $\mathcal{A}(M)$ has only one element. (Notice that $S \times I$ has the quite rare property that it admits an orientation-reversing self-homeomorphism homotopic to the identity.) The elements of $\mathrm{int}(AH(M))$ are called *quasi-fuchisan* as they are quasiconformally conjugate to Fuchsian groups. In this case,

int$(AH(M))$ is often denoted $QF(S)$ and $QF(S) \cong \mathcal{T}(S) \times \mathcal{T}(S)$. Concretely, if $\rho \in \mathrm{int}(AH(M))$, then N_ρ is determined by the conformal structure on $\partial_c N_\rho$ and given any conformal structure on $S \times \{0, 1\}$ one can construct a complete hyperbolic structure on $S \times (0, 1)$ with appropriate conformal boundary.

2) Let S_i denote a surface of genus i with one (open) disk removed. We consider the I-bundle $J_i = S_i \times [0, 1]$ and let $\partial_r J_i = \partial S_i \times [0, 1]$. Let $\{A_i\}_{i=1}^n$ be a collection of disjoint, parallel, consecutively ordered longitudinal annuli in the boundary of a solid torus V. Then M_n is formed from V and $\{J_1, \ldots, J_n\}$ by attaching $\partial_r J_i$ to A_i. The manifolds M_n are examples of books of I-bundles, see Culler-Shalen [27].

   Any irreducible manifold homotopy equivalent to M_n is formed by attaching the $\{J_i\}$ to the $\{A_i\}$ in a different order. To be more precise, if τ lies in the permutation group Σ_n, then one can form M_n^τ from V and $\{J_1, \ldots, J_n\}$ by attaching $\partial_r J_i$ to $A_{\tau(i)}$. One may extend the identity map on $\{J_1, \ldots, J_n\}$ to a homotopy equivalence $h_\tau : M_n \to M_n^\tau$. It is a consequence, see [5], of Johannson's Deformation Theorem [40], that every $(M', h') \in \mathcal{A}(M_n)$ is equivalent to (M_n^τ, h_τ) for some $\tau \in \Sigma_n$.

   It is easy to check that if $n \geq 3$, (M_n^τ, h_τ) need not be equivalent to (M_n, id). For example, if $n = 4$ and $\tau = (2\ 3)$ one may check that the boundary components of M_4 have genera 3, 5, 5, and 7, while the boundary components of M_4^τ have genera 4, 5, 5, and 6, so M_4 and M_4^τ are clearly not homeomorphic. On the other hand, if τ is any multiple of the rotation $(1\ 2 \cdots n)$, (M_n^τ, h_τ) is clearly equivalent to (M_n, id). Analyzing the situation more carefully, again using Johannson's Deformation Theorem, one shows that the space $\mathcal{A}(M_n)$ of marked homeomorphism types of manifolds homotopy equivalent to M_n may be identified with cosets of the subgroup generated by $(1\ 2 \cdots n)$ in the permutation group Σ_n (see [5]). In particular, $\mathcal{A}(M_n)$ has $(n-1)!$ elements, so int$(AH(M_n))$ is homeomorphic to a disjoint union of $(n-1)!$ balls. Notice that $M_2 = S_3 \times I$ where S_3 is a closed orientable surface of genus 3.

4. The Density Theorem

Much of the recent work on infinite volume hyperbolic 3-manifolds has culminated in the proof of the Bers-Sullivan-Thurston Density Conjecture which asserts that $AH(M)$ is the closure of its interior. More concretely, the Density Conjecture predicts that every hyperbolic 3-manifold with finitely generated fundamental group is an (algebraic) limit of geometrically finite hyperbolic 3-manifolds.

Density Theorem. *If M is a compact hyperbolizable 3-manifold, then $AH(M)$ is the closure of its interior* int$(AH(M))$.

There are two approaches to the proof of the Density Theorem. Both approaches make use of the proof of Marden's Tameness Conjecture by Agol [1] and Calegari-Gabai [22] which asserts that every hyperbolic 3-manifold with finitely generated fundamental group is homeomorphic to the interior of a compact 3-manifold.

In the classical approach one starts with a hyperbolic manifold $N_\rho \in AH(M)$ and constructs a sequence $\{N_{\rho_i}\}$ in int$(AH(M))$ such that the end invariants of $\{N_{\rho_i}\}$ converge to those of N_ρ. One then uses convergence results of Thurston [75], Kleineidam-Souto[45], Lecuire [49] and Kim-Lecuire-Ohshika [44] to show that this sequence converges to a limit $N_{\rho'}$. Arguments of Namazi-Souto [63] or Ohshika [66] can then be used to check that $N_{\rho'}$ has the same end invariants as N_ρ. One then invokes the solution of Thurston's Ending Lamination Conjecture [57,14,15] to show that $N_\rho = N_{\rho'}$ and thus complete the proof that $N_\rho \in \overline{\text{int}(AH(M))}$.

The other approach makes use of the deformation theory of cone-manifolds developed by Hodgson-Kerckhoff [31,32] and Bromberg [16]. These ideas were first used by Bromberg [17] to prove Bers' original Density conjecture for hyperbolic manifolds without cusps. This conjecture asserted that any Kleinian surface group with exactly one invariant component of its domain of discontinuity was in the boundary of the "appropriate" Bers slice (see section 8 for the definition of a Bers slice). Brock and Bromberg [12] showed that if M has incompressible boundary and $N_\rho \in AH(M)$ has no cusps, then $N_\rho \in \overline{\text{int}(AH(M))}$. A complete proof of the Density Theorem using these methods was given by Bromberg and Souto [21].

5. Bumping

It would be reasonable to expect that no two components of int$(AH(M))$ have intersecting closures, since one might expect that any hyperbolic manifold in the closure of a component B would be homeomorphic to a hyperbolic manifold in B. In fact, Jim Anderson and I spent two years attempting to prove this and eventually came up with examples [5] which illustrated that this "bumping" of components can occur. Later, Anderson, Canary and McCullough [6] characterized exactly when two components of int$(AH(M))$ can bump in the case that M has incompressible boundary.

Formally, we say that two components B and C of int$(AH(M))$ *bump at* ρ if $\rho \in \bar{B} \cap \bar{C}$. Notice that whenever two components of int$(AH(M))$ bump, $AH(M)$ is not a manifold.

The first setting in which the phenomenon of bumping was observed, was the books $\{M_n\}$ of I-bundles discussed in section 3.

Theorem 5.1 (Anderson-Canary [5]). *If $n \geq 3$, then any two components of* int$(AH(M_n))$ *bump, where $\{M_n\}$ are the books of I-bundles constructed in example 2 in section 3. In particular, $AH(M_n)$ is connected, while* int$(AH(M_n))$ *has* $(n-1)!$ *components.*

We will outline the construction in the proof of Theorem 5.1 in the section 7. In a finer analysis, Holt [33] showed that there is a single point at which any two components of int$(AH(M_n))$ bump.

Theorem 5.2 (Holt [33]). *If $n \geq 3$, then there exists $\rho \in AH(M_n)$ which lies in the boundary of every component of* int$(AH(M_n))$.

Remark. Holt [33] further observes, that the set of points lying in the closure of every component of int$(AH(M_n))$ contains an open subset of a (complex) codimension 1 subvariety of $X(M_n)$. It is typical that the "bumping locus" is relatively large.

Anderson, Canary and McCullough [6] later gave a complete characterization of which components of $\text{int}(AH(M))$ can bump in the case when M has incompressible boundary. Roughly, the two components can bump if the two (marked) homeomorphism types differ by cutting along a collection of (primitive) solid tori and rearranging the order in which the complementary pieces are attached. One sees that for M_n any two homeomorphism types differ in this specific way. A solid torus V in M is said to be *primitive* if $V \cap \partial M$ is a non-empty collection of annuli, the inclusion of each annulus of $V \cap \partial M$ into V is a homotopy equivalence and the image of $\pi_1(V)$ in $\pi_1(M)$ is a maximal abelian subgroup.

To illustrate the role of primitivity in Anderson, Canary and McCullough's result, we consider a sequence of manifolds $\{M'_n\}_{n=3}^{\infty}$, again obtained from a solid torus V and I-bundles $\{J_1, \ldots, J_n\}$. This time we let $\{A'_1, \ldots, A'_n\}$ be a collection of disjoint, parallel, consecutively ordered annuli in the boundary of a solid torus V such that the inclusion of $\pi_1(A'_i)$ into $\pi_1(V)$ is a subgroup of index 3 (i.e. each annulus wrap 3 times around the longitude of V.) We form M'_n by attaching $\partial_r J_i$ to A'_i. It is again the case that any manifold homotopy equivalent to M'_n is obtained by attaching the J_i in a different order and that $\mathcal{A}(M'_n)$ has $(n-1)!$ elements. However, the results of [6] imply that no two components of $\text{int}(AH(M'_n))$ bump for any n.

In order to give a precise statement of the results of [6] we must introduce the notion of a primitive shuffle equivalence. In what follows, if M is a compact, irreducible 3-manifold with incompressible boundary, then $\Sigma(M)$ will denote its characteristic submanifold. For complete discussions of the theory of characteristic submanifolds see Jaco-Shalen [39] or Johannson [40]. For a discussion in the setting of hyperbolizable 3-manifolds, see Canary-McCullough [24] or Morgan [60].

Given two compact irreducible 3-manifolds M_1 and M_2 with nonempty incompressible boundary, a homotopy equivalence $h\colon M_1 \to M_2$ is a *primitive shuffle equivalence* if there exists a finite collection V_1 of primitive solid torus components of $\Sigma(M_1)$ and a finite collection V_2 of solid torus components of $\Sigma(M_2)$, so that $h^{-1}(V_2) = V_1$ and so that h restricts to an orientation-preserving homeomorphism from the closure of $M_1 - V_1$ to the closure of $M_2 - V_2$.

If M is a compact, hyperbolizable 3-manifold with nonempty incompressible boundary, we say that two elements $[(M_1, h_1)]$ and $[(M_2, h_2)]$ of $\mathcal{A}(M)$ are *primitive shuffle equivalent* if there exists a primitive shuffle equivalence $s\colon M_1 \to M_2$ such that $[(M_2, h_2)] = [(M_2, s \circ h_1)]$.

Theorem 5.3 (Anderson-Canary-McCullough [6]). *Let M be a compact, hyperbolizable 3-manifold with nonempty incompressible boundary, and let $[(M_1, h_1)]$ and $[(M_2, h_2)]$ be two elements of $\mathcal{A}(M)$. The associated components of $\text{int}(AH(M))$ have intersecting closures if and only if $[(M_2, h_2)]$ is primitive shuffle equivalent to $[(M_1, h_1)]$.*

Combining the work of Anderson, Canary and McCullough [6] with the resolution of the Density Conjecture, one obtains a complete enumeration of the components of $AH(M)$ when M has incompressible boundary. In particular, one completely determines exactly when $AH(M)$ has infinitely many components. Primitive shuffle equivalence gives a finite-to-one equivalence relation on $\mathcal{A}(M)$ and we let $\hat{\mathcal{A}}(M)$ be the quotient of $\mathcal{A}(M)$ by this equivalence relation.

Corollary 5.4. *If M has incompressible boundary, then the components of $AH(M)$ are in one-to-one correspondence with $\hat{\mathcal{A}}(M)$. In particular, $AH(M)$ has infinitely many components if and only if M has double trouble.*

Holt [34] refined the analysis of [6] to show that if C_i is a collection of components of $\text{int}(AH(M))$ such that any two components in the collection bump, then then they all bump at a single point.

Theorem 5.5 (Holt [34]). *Let M be a compact, hyperbolizable 3-manifold with non-empty incompressible boundary, and let $\{[(M_i, h_i)]\}_{i=1}^m$ be a collection of elements of $\mathcal{A}(M)$ such that $[(M_i, h_i)]$ is primitive shuffle equivalent to $[(M_1, h_1)]$ for all i. If C_i is the component of $\text{int}(AH(M))$ associated to $[(M_i, h_i)]$, then*

$$\bigcap_{i=1}^{m} \overline{C_i} \neq \emptyset.$$

6. Self bumping

McMullen [55] was the first to observe that individual components of $\text{int}(AH(M))$ can self-bump. A component B of $\text{int}(AH(M))$ *self-bumps* at $\rho \in \bar{B}$ if there is a neighborhood V of ρ such that if $\rho \in W \subset V$ is any sub-neighborhood, then $W \cap B$ is disconnected.

Theorem 6.1 (McMullen [55]). *If S is a closed surface, then $QF(S) = \text{int}(AH(S \times I))$ self-bumps.*

Notice that this implies, in particular, that $AH(S \times I)$ is not a manifold. The self-bumping points in [55] are again obtained using the construction from [5] described in section 7.

McMullen's proof made use, in a crucial manner, of the theory of projective structures on surfaces, so did not generalize immediately to manifolds which are not I-bundles. Bromberg and Holt [20] were able to generalize McMullen's result to all manifolds which contain primitive essential annuli. An embedded annulus A in M is said to be *essential* if it is *incompressible*, i.e. $\pi_1(A)$ injects into $\pi_1(M)$, and is not properly homotopic into the boundary ∂M. It is said to be *primitive* if the image of $\pi_1(A)$ in $\pi_1(M)$ is a maximal abelian subgroup of $\pi_1(M)$.

Theorem 6.2 (Bromberg-Holt [20]). *If M contains a primitive essential annulus, then every component of $\text{int}(AH(M))$ self-bumps.*

Notice that there is no assumption that M has incompressible boundary in this theorem. It implies, in particular, that $AH(M)$ is not a manifold if M contains a primitive essential annulus.

Ito has completed an extensive analysis of related phenomena in the space $P(S)$ of complex projective structures on a surface S. There is a natural map $hol : P(S) \to X(S \times I)$ which takes a complex projective structure to its associated holonomy map. Hejhal [30] showed that the map hol is a local homeomorphism and Goldman [29] showed that the components of $Q(S) = hol^{-1}(QF(S))$ are enumerated by the set of weighted multicurves on S. McMullen [55] showed that $QF(S)$ self-bumps by showing that two components of $Q(S)$ can bump. Ito [36,37] shows that any two components of $Q(S)$ bump, that any component of $Q(S)$ other than the base component self-bumps, and that arbitrarily many components of $Q(S)$ can bump at a single point. (Bromberg and Holt have obtained related results.)

7. The key construction

In this section, we will describe the "wrapping" construction of [5] in the special case of $AH(M_4)$ and show that components of $\mathrm{int}(AH(M_4))$ can bump.

Let $\tau \in \Sigma_4$ be the permutation $(2\ \ 3)$ and let $\hat{M}_4^\tau$ be the manifold obtained from M_4^τ by removing an open neighborhood of the core curve of V. One may construct an infinite cyclic cover $(\hat{M}_4^\tau)'$ of $\hat{M}_4^\tau$ from an infinite thickened annulus $S^1 \times I \times \mathbb{R}$ by attaching infinitely many copies of each J_i to the outer boundary $S^1 \times \{0\} \times \mathbb{R}$ so that they occur repeatedly in the cyclic order $\ldots J_1, J_3, J_2, J_4, J_1, J_3, \ldots\ldots$. (More concretely, for all $n \in \mathbf{Z}$ and $i = 1, 2, 3, 4$ one attaches a copy of J_i to the thickened annulus by identifying $\partial_r J_i$ with $S^1 \times \{0\} \times [12n + 3\tau(i) - 1, 12n + 3\tau(i) + 1]$ by an orientation-reversing homeomorphism. Vertical translation by 12 on the infinite thickened annulus extends to a homeomorphism of $(\hat{M}_4^\tau)'$ which generates the full group of covering transformations of $(\hat{M}_4^\tau)'$ over M_4^τ.) Let $\pi : (\hat{M}_4^\tau)' \to \hat{M}_4^\tau$ be the covering map.

One then constructs an orientation-preserving embedding $\hat{f}_\tau : M_4 \to (\hat{M}_4^\tau)'$ which takes each J_i homeomorphically to a copy of J_i. (More concretely, one may take J_i to the copy of J_i attached to $S^1 \times I \times [12i + 3\tau(i) - 1, 12i + 3\tau(i) + 1]$.) Let $f_\tau = \pi \circ \hat{f}_\tau$.

Let $\tilde{M}_4^\tau$ be the cover of $\hat{M}_4^\tau$ associated to $(f_\tau)_*(\pi_1(M_4))$. One easily checks that f_τ lifts to an embedding $\tilde{f}_\tau : M_4 \to \tilde{M}_4^\tau$ (since it lifts to an embedding in the intermediate cover $(\hat{M}_4^\tau)'$.) Also, notice that if $i_0 : \hat{M}_4^\tau \to M_4^\tau$ denotes the inclusion map, then $i_0 \circ f_\tau$ is a homotopy equivalence and is homotopic to h_τ (where h_τ is the homotopy equivalence defined in example 2 in section 3).

The central tool in the construction is the generalization of Thurston's Hyperbolic Dehn Filling Theorem [75] to the setting of geometrically finite hyperbolic 3-manifolds.

Let T be a toroidal boundary component of compact 3-manifold M and let (m, l) be a choice of meridian and longitude for T. Given a pair (p, q) of relatively prime integers, we may form a new manifold $M(p, q)$ by attaching a solid torus V to M by an orientation-reversing homeomorphism $g\colon \partial V \to T$ so that, if c is the meridian of V, then $g(c)$ is a (p, q) curve on T with respect to the chosen meridian-longitude system. We say that $M(p, q)$ is obtained from M by *(p, q)-Dehn filling along T*.

If $T = \partial M$ and M is hyperbolizable (i.e. its interior admits a complete hyperbolic structure), then Thurston [75] proved that $M(p, q)$ is hyperbolizable for all but finitely many choices of (p, q). Bonahon and Otal [10] were the first to observe that you could generalize this result to the setting of geometrically finite hyperbolic 3-manifolds (see also Comar [26], Hodgson-Kerckhoff [31] and Bromberg [18].) Our statement of the resulting Hyperbolic Dehn Filling Theorem is rather convoluted but it essentially says that given a geometrically finite hyperbolic 3-manifold N homeomorphic to $\text{int}(M)$ and (p, q) near to ∞, then there exists a geometrically finite hyperbolic 3-manifold $N(p, q)$ which is "near" to N. The complications in the statement are largely the result of the need to keep careful track of the marking.

Hyperbolic Dehn Filling Theorem. *Let M be a compact, hyperbolizable 3-manifold and let T be toroidal boundary component of M. Let $N = \mathbb{H}^3/\Gamma$ be a geometrically finite hyperbolic 3-manifold admitting an orientation-preserving homeomorphism $\psi : \text{int}(M) \to N$. Further assume that every parabolic element of Γ lies in a rank-two parabolic subgroup. Let $\{(p_n, q_n)\}$ be an infinite sequence of distinct relatively prime pairs of integers.*

Then, for all sufficiently large n, there exists a (non-faithful) representation $\beta_n : \Gamma \to \text{PSL}_2(\mathbf{C})$ with discrete image such that

(1) *$\beta_n(\Gamma)$ is geometrically finite and every parabolic element of $\beta_n(\Gamma)$ lies in a rank-two parabolic subgroup,*
(2) *$\{\beta_n\}$ converges to the identity representation of Γ, and*
(3) *if $i_n : M \to M(p_n, q_n)$ denotes the inclusion map, then for each n, there exists an orientation-preserving homeomorphism $\psi_n : \text{int}(M(p_n, q_n)) \to \mathbb{H}^3/\beta_n(\Gamma)$ such that $\beta_n \circ \psi_*$ is conjugate to $(\psi_n)_* \circ (i_n)_*$.*

In order to apply the Hyperbolic Dehn Filling Theorem in our setting we need to make a couple more topological observations. We choose a meridian-longitude system for the toroidal boundary component T of $\hat{M}_4^\tau$ so that the meridian bounds a disk in M_4^τ and the longitude bounds an essential annulus A in $\hat{M}_4^\tau$. First, one checks that $\hat{M}_4^\tau(1, n)$ is homeomorphic to M_4^τ for all n. (The easiest way to see this is to note that Dehn twisting n times about A takes the $(1, 0)$ curve on T to the $(1, n)$ curve and observe that $\hat{M}_4^\tau(1, 0) = M_4^\tau$.) Then one similarly notices that $(\hat{M}_4^\tau(1, n), i_n \circ f_\tau)$ is equivalent to (M_4^τ, h_τ) in $\mathcal{A}(M_4)$ for all n, where $i_n : \hat{M}_4^\tau \to M_4^\tau(1, n)$ is the inclusion map and h_τ is the homotopy equivalence defined in example 2 in section 3.

Now let $N = \mathbb{H}^3/\Gamma$ be a geometrically finite hyperbolic 3-manifold admitting an orientation-preserving homeomorphism $\psi : \text{int}(\hat{M}_4^\tau) \to N$. (We further require that all parabolic elements of Γ lie in rank two parabolic subgroups.) In the Hyperbolic Dehn Filling Theorem we choose $(p_n, q_n) = (1, n)$ for all n and let $\beta_n : \Gamma \to \text{PSL}_2(\mathbf{C})$ be the resulting sequence of representations.

For each n, define $\rho_n = \beta_n \circ \psi_* \circ (f_\tau)_*$. The sequence $\{\rho_n\}$ converges to $\rho = \psi \circ (f_\tau)_*$. Since $\beta_n \circ \psi_*$ is conjugate to $(\psi_n)_* \circ (i_n)_*$, we see that ρ_n is conjugate to $(\psi_n)_* \circ (i_n)_* \circ (f_\tau)_*$. Since $i_n \circ f_\tau$ is a homotopy equivalence and ψ_n is a homeomorphism, ρ_n is a discrete faithful representation with image $\pi_1(N_n) = \beta_n(\hat{\Gamma})$. Since $(\hat{M}_4^\tau(1, n), i_n \circ f_\tau)$ is equivalent to (M_4^τ, h_τ), it follows that $(\hat{N}_{\rho_n}, h_{\rho_n})$ is equivalent to (M_4^τ, h_τ) for all n. In particular, ρ_n lies in the component of $\text{int}(AH(M_4))$ associated to (M_4^τ, h_τ) for all n. On the other hand, since f_τ lifts to an embedding, Bonahon'sTameness Theorem [9] and results of McCullough-Miller-Swarup [54] imply that N_ρ is homeomorphic to $\text{int}(M_4)$. So, $\{\rho_n\}$ is an example of a sequence in $AH(M_4)$ where the homeomorphism type changes in the limit.

It remains to show that ρ lies in the closure of the component of $\text{int}(AH(M_4))$ associated to (M_4, id). We may accomplish this by modifying the above construction. We first construct a geometrically finite hyperbolic 3-manifold N' which admits an orientation-preserving homeomorphism $\psi' : \text{int}(\hat{M}_4) \to N'$ (where $\hat{M}_4$ is the manifold obtained from M_4 by removing an open neighborhood of the core curve of V). We further require that there exists an embedding $f : M_4 \to \hat{M}_4$ such that $i'_n \circ f : M_4 \to \text{int}(\hat{M}_4(1,n))$ is a homotopy equivalence which is homotopic to an orientation-preserving homeomorphism for all n (where $i'_n : \hat{M}_4 \to \hat{M}_4(1,n)$ is the inclusion map). Finally, we require that ρ is conjugate to $\psi'_* \circ f_*$. (To construct N' from M_ρ, we may normalize so that there is a maximal abelian subgroup of $\rho(\pi_1(M_4))$ generated by the Möbius transformation $z \to z+1$. If one considers the group Γ' generated by $\rho(\pi_1(M_4))$ and $z \to z + ri$ for a large enough real value of r, then one may take $N' = \mathbb{H}^3/\Gamma'$.) One then applies the Hyperbolic Dehn Filling Theorem just as before to obtain a sequence $\{\rho'_n\}$ of representations lying in the component of $\text{int}(AH(M_4))$ associated to (M_4, id) and converging to ρ. Alternatively, one may simply apply a theorem of Ohshika [64].

It should be fairly clear that the above construction works for all n and all $\tau \in \Sigma_n$. In fact, it works for all primitive shuffle equivalences, see [6].

This construction is also responsible for the phenomenon of self-bumping. Recall that $M_2 = S_3 \times I$ where S_3 is a closed surface of genus 3. We illustrate the modifications of the "wrapping construction" necessary to produce a self-bumping point in the boundary of $QF(S_3)$. Let $\hat{M}_2$ be obtained from M_2 by removing an open neighborhood of the core curve of V and let $(\hat{M}_2)'$ be the infinite cyclic cover obtained by gluing alternating copies of J_1 and J_2 to an infinite thickened annulus. We construct an orientation-preserving embedding $\hat{f} : M_2 \to (\hat{M}_2)'$ which maps J_1 to the "first copy" of J_1 and maps J_2 to the "second" copy of J_2 (so that there are copies of J_2 and J_1 which lie "between" the image of J_1 and the image of J_2). We then let $f = \pi \circ \hat{f}$, where $\pi : (\hat{M}_2)' \to \hat{M}_2$ is the covering map, and apply Thurston's Hyperbolic Dehn Filling Theorem as before, to produce a sequence $\{\rho_n\}$ in $QF(S_3)$ which converges to a self-bumping point $\rho \in \partial AH(M_2)$. (The assertion that ρ is a self-bumping point is not obvious and requires a clever proof, see McMullen [55] or Bromberg-Holt [20]).

8. Relative deformation spaces and the failure of local connectivity

There are various naturally defined subsets of $AH(M)$ which have played a prominent role in the theory of Kleinian groups. Most simply, one may require that certain elements of $\pi_1(M)$ be mapped to parabolic elements. In this setting, it is natural to introduce the language of pared manifolds. We refer the reader to Morgan [60] and Canary-McCullough [24] for a more extensive discussion of pared manifolds.

Let M be a compact, orientable, irreducible 3-manifold with nonempty boundary which is not a 3-ball, and let $P \subseteq \partial M$. We say that (M, P) is a *pared* 3-manifold if

(1) Every component of P is an incompressible torus or annulus,
(2) every noncyclic abelian subgroup of $\pi_1(M)$ is conjugate into the fundamental group of a component of P, and

(3) every map $\varphi\colon (S^1 \times I, S^1 \times \partial I) \to (M, P)$ which induces an injection on fundamental groups is homotopic, as a map of pairs, to a map ψ such that $\psi(S^1 \times I) \subset P$.

We can then define the relative deformation space $AH(M, P)$ to be the set of (conjugacy classes) of representations $\rho \in AH(M)$ such that if Q is any component of P then $\rho(\pi_1(Q))$ consists of parabolic elements. $AH(M, P)$ lies in a relative character variety $X(M, P)$. We may define the set $\mathcal{A}(M, P)$ to be the set of marked, oriented pared manifolds homotopy equivalent to (M, P) up to orientation preserving pared homeomorphisms. Again the interior of $AH(M, P)$ (within $X(M, P)$) can be identified with

$$\bigcup_{(M',P')\in\mathcal{A}(M,P)} \mathcal{T}(\partial M' - P')/Mod_0(M', P')$$

where $Mod_0(M', P')$ is the group of isotopy classes of pared self-homeomorphism of (M', P') which are pared homotopic to the identity (see Canary-McCullough [24] for more details.)

If S is a compact surface with boundary, then the interior $QF(S)$ of $AH(S \times I, \partial S \times I)$ is again the space of quasifuchsian groups which are quasiconformally conjugate to a cofinite area Fuchsian group uniformizing the interior S^0 of S and is naturally identified with $\mathcal{T}(S^0) \times \mathcal{T}(S^0)$. In the case that S^0 is a once-punctured torus, $AH(S \times I, \partial S \times I)$ is known as the space of punctured torus groups. Bromberg [19] proved that the space of punctured torus groups is not locally connected.

Theorem 8.1 (Bromberg [19]). *The space of punctured torus groups is not locally connected.*

Bromberg's proof suggests that many deformation spaces and relative deformation spaces are not locally connected. Magid [50] recently observed that you can use Bromberg's result to prove that an infinite family of relative deformation spaces is not locally connected. The simplest case of his result is the following.

Theorem 8.2 (Magid [50]). *If S is a closed surface, $M = S \times I$, and P is a collection of non-parallel incompressible annuli in $S \times \{1\}$ such that at least one component of $S \times \{1\} - P$ is homeomorphic to a once-punctured torus, then $AH(M, P)$ is not locally connected.*

More generally, Magid's result shows that $AH(M, P)$ fails to be locally connected whenever there is a separating essential annulus A in M with one boundary component lying in a component P_0 of P such that the closure X of one component of $M - A$ is homeomorphic to $S \times I$ where S^0 is a once-punctured torus (by a homeomorphism taking $\partial S \times I$ to A) and that X intersects P only in P_0.

Results of Holt-Souto [35] and Evans-Holt [28] combine to show that the set of self-bumping points, and hence the set of points where the deformation space is not locally connected, is not dense in the boundary of the space of punctured torus groups. In particular, they show that there exists some constant $\epsilon_0 > 0$ such that if ρ is a punctured torus group and every closed geodesic in $\partial_c N_\rho$ which is homotopic into a cusp in $\hat{N}_\rho$ has length at least ϵ_0, then the space of punctured torus groups does not self-bump at ρ.

Ito [38] has quite recently given a complete description of the self-bumping points in the boundary of the space of punctured torus groups. In particular, he shows that all self-bumping points arise from the construction described in section 7. Ohshika [67] has generalized many of Ito's result to the setting of all quasifuchsian spaces. Moreover, he is able to show that there is no self-bumping at many points in the boundary of quasifuchsian space.

Another natural and well-studied class of deformation spaces are the Bers slices which sit inside the quasifuchsian spaces. We recall that if S is a compact surface, then its associated space $QF(S)$ of quasifuchsian groups is naturally identified with $\mathcal{T}(S^0)\times\mathcal{T}(S^0)$. If $\sigma\in\mathcal{T}(S^0)$, then the associated *Bers slice* B_σ is the set $\mathcal{T}(S^0)\times\{\sigma\}$ of quasifuchsian groups, whose bottom conformal boundary component has conformal structure σ. Bers [7] proved that the closure of any Bers slice is compact, so it is natural to study the topology of the closure of such a component. Minsky [56] proved that if S^0 is a once-punctured torus, then the closure of any Bers slice is homeomorphic to a disk.

Theorem 8.3 (Minsky [56]). *If S^0 is a once-punctured torus, then the closure in the space of punctured torus groups of any Bers slice in $QF(S)$ is homeomorphic to a closed disk.*

Remark. Minsky [56] also showed that the closure of the Maskit slice is homeomorphic to a closed disk. Komori [46] has established the same result for the Earle slice and Komori and Parkkonen [47] show that Bers-Maskit slices have the disk as closure.

If the complex dimension of the Bers slice is greater than 1, then very little is known about the topology of the closure of a Bers slice. However, Kerckhoff and Thurston [43] showed that, in this setting, there exist Bers slices B_σ and $B_{\sigma'}$ such that the natural homeomorphism between B_σ and $B_{\sigma'}$ does not extend to a homeomorphism of their closures. They use this to show that there exist Bers slices such that the action of the mapping class group does not extend continuously to the closure. Kerckhoff and Thurston's results were the first indication that the topology of closures of Bers slices must be "complicated."

9. Untouchable points

In this section, we will describe joint work with Jeff Brock, Ken Bromberg and Yair Minsky [13] which shows that the topology of $AH(M)$ is well-behaved at "most" points in the boundary of $AH(M)$ in many cases.

A point $\rho\in\partial AH(M)$ is said to be *untouchable* if there is no bumping or self-bumping at ρ. Notice that $AH(M)$ is locally connected at all untouchable points.

Theorem 9.1 ([13]). *Suppose that M has incompressible boundary. If $\rho\in\partial AH(M)$ and $\rho(\pi_1(M))$ contains no parabolic elements, then ρ is untouchable.*

If ∂M contains no tori, then such points are generic in $\partial AH(M)$.

The proof of the non-bumping portion of this result is a straightforward application of earlier work and we will provide a brief outline of the argument. One first notes that results of Thurston [75], Canary [23] and Anderson-Canary [4] imply that if $\{\rho_i\} \subset \text{int}(AH(M))$ converges to ρ, then $\{N_{\rho_i}\}$ converges geometrically to N_ρ (i.e. larger and larger portions of N_ρ look increasingly like portions of N_{ρ_i}.) Results of Thurston [75], Canary-Minsky [25] and Ohshika [65], then imply that N_ρ is homeomorphic to N_{ρ_i} (by a homeomorphism in the homotopy class determined by $\rho \circ \rho_i^{-1}$) for all large i. Hence the (marked) homeomorphism type is locally constant at ρ, so there is no bumping at ρ.

The proof that there is no self-bumping at ρ is somewhat more involved. One begins by showing that if $\{\rho_i\}$ converges to ρ, then the end invariants of $\{N_{\rho_i}\}$ converge to those of N_ρ. (This is not always the case, so we are also using our restrictions on ρ here.) We then consider any two sequences $\{\rho_i\}$ and $\{\rho_i'\}$ in $\text{int}(AH(M))$ converging to ρ and construct paths $\alpha_i : I \to \text{int}(AH(M))$ joining ρ_i to ρ_i', so that the sequence $\{\alpha_i\}$ converges to the constant path with image ρ. To establish the convergence of these paths, we use the same convergence results and arguments as used in the proof of the Density Theorem as well as the resolution of the Ending Lamination Conjecture.

If we allow N_ρ to have cusps, then $\{N_{\rho_i}\}$ need not converge geometrically to N_ρ in the above argument, so to rule out bumping we must place additional restrictions on ρ. We say that ρ is *quasiconformally rigid*, if every component of $\partial_c N_\rho$ is a thrice-punctured sphere. Notice that this includes the case that $\partial_c N_\rho$ is empty. (We call such representations quasiconformally rigid, since Sullivan's rigidity theorem [73] guarantees that any representation quasiconformally conjugate to ρ is in fact conformally conjugate.)

Theorem 9.2 ([13]). *If $\rho \in \partial AH(M)$ is quasiconformally rigid, then there is no bumping at ρ.*

Remark. There is a related result of Anderson, Canary and McCullough (Corollary 8.2 in [6]) in the setting where M has incompressible boundary which is much stronger.

We provide a brief outline of the argument, which is again largely an exercise in applying previously developed technology. Suppose that $\{\rho_i\} \subset \text{int}(AH(M))$ converges to ρ and $\{N_{\rho_i}\}$ converges geometrically to N_∞. Then there exists a covering map $\pi : N_\rho \to N_\infty$. Results of Anderson-Canary [4] and Canary [23] imply that there exists a compact core C for N_ρ which embeds in N_∞ (via π). For large i, one can pull-back $\pi(C)$ to obtain a compact core C_i for N_{ρ_i} such that C_i is homeomorphic to C (by a homeomorphism in the homotopy class determined by $\rho_i \circ \rho^{-1}$). The main result of McCullough-Miller-Swarup [54] imply that C_i is homeomorphic to N_{ρ_i} (by a homeomorphism homotopic to inclusion). Therefore, N_{ρ_i} is homeomorphic to N_ρ for all large enough i (again by a homeomorphism in the correct homotopy class). So, there is no bumping at ρ.

To rule out self-bumping, we need to further restrict our setting. We recall that a compact, irreducible manifold is said to be *acylindrical* if it does not contain any essential annuli. (We note that there exist manifolds which are not acylindrical, but do not contain any *primitive* essential annuli.)

Theorem 9.3 ([13]). *If M is acylindrical or homeomorphic to $S \times I$ (where S is a closed surface) and $\rho \in \partial AH(M)$ is quasiconformally rigid, then there is no self-bumping at ρ. So, ρ is untouchable.*

The proof of this theorem is much more involved and makes use in a key way of the techniques of proof of the Ending Lamination Conjecture, in particular Minsky's a priori bounds [57], as well as Thurston's bounded image theorem (see Kent [42]).

Remark. In the case that $M \cong S \times I$, our results overlap substantially with results of Ohshika [67]. We refer the reader to Ohshika's paper for the detailed definitions.

Theorem 9.4 (Ohshika [67]). *Suppose that S is a compact hyperbolic surface and $\rho \in \partial AH(S \times I, \partial S \times I)$.*

(1) *If $\partial_c N_\rho$ has one component homeomorphic to S^0 and all other components are thrice-punctured spheres and every rank one cusp of N_ρ (which is not homotopic to a component of ∂S) abuts a geometrically infinite end, then there is no self-bumping at ρ.*
(2) *If ρ is quasiconformally rigid and every rank one cusp of N_ρ (which is not homotopic to a component of ∂S) abuts a geometrically infinite end, then there is no self-bumping at ρ.*

His techniques are quite different and make deep use of Soma's work [72] on geometric limits of quasifuchsian hyperbolic 3-manifolds. Ohshika and Soma [68] have recently extended Soma's work on geometric limits.

10. Questions and conjectures

In this final section, we will collect conjectures and questions about the topology of deformation spaces of hyperbolic 3-manifolds. As we still understand very little, much remains to study and we limit ourselves to a few of the more obvious questions.

10.1 *Local connectivity*

Bromberg has conjectured that the failure of local connectivity is a fairly widespread phenomenon. As a first step, one might hope to show that it holds for all spaces of surface groups.

Conjecture 10.1 (Bromberg [19]). If S is any compact surface, then $AH(S \times I, \partial S \times I)$ is not locally connected.[1]

Bromberg's proof of Theorem 8.1 makes essential use of the wrapping construction described in section 7. However, he expects that local connectivity should fail even in settings where one cannot perform this construction, e.g. for Bers slices.

[1]Conjecture 10.1 has recently been established by Aaron Magid when S is a closed orientable surface of genus at least two.

Conjecture 10.2 (Bromberg [19]). If S is any compact hyperbolic surface whose interior S^0 is not a once-punctured torus, thrice-punctured sphere or 4-times punctured sphere, then the closure of any Bers slice in $QF(S)$ is not locally connected.

As Bromberg points out, these two specific conjectures strongly suggest the following more dramatic conjecture

Conjecture 10.3. If M is any compact hyperbolizable 3-manifold with a non-toroidal boundary component, then $AH(M)$ is not locally connected.

It should be pointed out that if M does not contain a primitive essential annulus, then we don't even know that M is not a manifold, so one might begin with the following easier conjecture.

Conjecture 10.4. If M is any compact hyperbolizable 3-manifold with a non-toroidal boundary component, then $AH(M)$ is not a manifold.

10.2 *The compressible case*

In the case where M has incompressible boundary, Corollary 5.4 gives a complete enumeration of the components of $AH(M)$, but the situation where M is allowed to have compressible boundary is still quite mysterious. It is easy to use the constuction in section 7 to show that various components of int$(AH(M))$ bump. Bromberg and Holt's result, Theorem 6.2, already applies to show that if M has compressible boundary, and is not obtained from one or two thickened tori by adding a single 1-handle, then every component of int$(AH(M))$ self-bumps, since M will contain a primitive essential annulus in these cases.

One would like to give a complete enumeration of the components of $AH(M)$ whenever M has compressible boundary. In the incompressible boundary situation, this is accomplished by showing that the construction in section 7 is responsible for all bumping phenomena.

Problem 10.5. Give a complete enumeration of the components of $AH(M)$ in terms of topological data.

One might hope to begin by establishing the following simpler conjecture, which follows from Theorem 5.3 in the case where M has incompressible boundary.

Conjecture 10.6. $AH(M)$ has finitely many components if and only if its interior has finitely many components.

A first step in the proof of this conjecture might be to show that it is not possible for infinitely many components of int$(AH(M))$ to accumulate at a single point.

10.3 *The fractal nature of deformation spaces*

Closures of Bers slices in punctured torus spaces admit natural embeddings in **C**. The beautiful pictures of these embeddings, see Komori-Sugawa-Wada-Yamashita [48], Yamashita [76], and Mumford-Series-Wright [62], suggest that their boundaries are quite "fractal" in nature. One might guess that the boundaries of these slices have Hausdorff dimension strictly between 1 and 2.

Question 10.7. What can one say about the Hausdorff dimension of the boundary of a Bers slice of a punctured torus?

Miyachi [58,59] has shown that there is a countable dense set of "cusps" in the boundary of a Bers slice of a punctured torus. Parkkonen [69] has further analyzed the shapes of these cusps in the related setting of a Maskit slice. Parkkonen [70] has also studied the shape of Schottky space near certain boundary points. (Schottky space is $\text{int}(AH(H_g))$ where H_g is the handlebody of genus g.)

Of course, one would like to study more general classes of deformation spaces in the future.

10.4 *Components of $AH(M)$ and components of the character variety*

It is known that $AH(M)$ can be disconnected, but it is not known whether or not the different components of $AH(M)$ can lie in different components of the character variety.

Question 10.8. Is it possible that different components of $AH(M)$ lie in different components of the character variety?

Acknowledgements

I would like to thank the organizers of the workshop, Ravi Kulkarni and Sudeb Mitra, for inviting me to visit the Harish-Chandra Research Institute. I enjoyed both the mathematical interaction and the opportunity to visit India for the first time. I would especially like to thank the student organizers, Vikram Aithal, Krishnendu Gongopadhyay, and Siddhartha Sarkar, for their help during my stay.

I would also like to thank the referee for their very careful reading of the original version of this manuscript and their many useful suggestions.

References

[1] I. Agol, Tameness of hyperbolic 3-manifolds, preprint.

[2] L. Ahlfors, Fundamental polyhedrons and limit point sets of Kleinian groups, *Proc. Nat. Acad. Sci. USA*, **55** (1966) 251–254.

[3] L. Ahlfors and L. Bers, Riemann's mapping theorem for variable metrics, *Annals of Math.*, **72** (1960) 385–404.

[4] J. W. Anderson and R. D. Canary, Cores of hyperbolic 3-manifolds and limits of Kleinian groups, *Amer. J. Math.*, **118** (1996) 745–779.

[5] J. W. Anderson and R. D. Canary, Algebraic limits of Kleinian groups which rearrange the pages of a book, *Invent. Math.*, **126** (1996) 205–214.

[6] J. W. Anderson, R. D. Canary and D. McCullough, On the topology of deformation spaces of Kleinian groups, *Annals of Math.*, **152** (2000) 693–741.

[7] L. Bers, On boundaries of Teichmüller spaces and on Kleinian groups, *Annals of Math.*, **91** (1970) 570–600.

[8] L. Bers, On moduli of Kleinian groups, *Russian Math. Surveys*, **29** (1974) 88–102.

[9] F. Bonahon, Bouts des variétés hyperboliques de dimension 3, *Annals of Math.*, **124** (1986) 71–158.

[10] F. Bonahon and J. P. Otal, Variétés hyperboliques à géodésiques arbitrairement courtes, *Bull. L.M.S.*, **20** (1988) 255–261.

[11] B. Bowditch, Geometrical finiteness for hyperbolic groups, *J. Funct. Anal.*, **113** (1993) 245–317.

[12] J. Brock and K. Bromberg, On the density of geometrically finite Kleinian groups, *Acta Math.*, **192** (2004) 33–93

[13] J. Brock, K. Bromberg, R. Canary and Y. Minsky, Local topology in deformation spaces of hyperbolic 3-manifolds, preprint.

[14] J. Brock, R. Canary and Y. Minsky, The Classification of Kleinian Surface Groups II: The Ending Lamination Conjecture, preprint.

[15] J. Brock, R. Canary and Y. Minsky, The Classification of finitely generated Kleinian groups, in preparation.

[16] K. Bromberg, Rigidity of geometrically finite hyperbolic cone-manifolds, *Geom. Ded.*, **105** (2004) 143–170.

[17] K. Bromberg, Projective structures with degenerate holonomy and the Bers density conjecture, *Annals of Math.*, **166** (2007) 77–93.

[18] K. Bromberg, Hyperbolic Dehn surgery on geometrically infinite 3-manifolds, preprint.

[19] K. Bromberg The space of Kleinian punctured torus groups is not locally connected, preprint.

[20] K. Bromberg and J. Holt, Self-bumping of deformation spaces of hyperbolic 3-manifolds, *J. Diff. Geom.*, **57** (2001) 47–65.

[21] K. Bromberg and J. Souto, The density conjecture: A prehistoric approach, in preparation.

[22] D. Calegari and D. Gabai, Shrinkwrapping and the taming of hyperbolic 3-manifolds, *J.A.M.S.*, **19** (2006) 385–446.

[23] R. D. Canary, A covering theorem for hyperbolic 3-manifolds and its applications, *Topology*, **35** (1996) 751–778.

[24] R. D. Canary and D. McCullough, Homotopy equivalences of 3-manifolds and deformation theory of Kleinian groups, *Mem. A.M.S.*, **172** (2004) no. 812.

[25] R. D. Canary and Y. N. Minsky, On limits of tame hyperbolic 3-manifolds, *J. Diff. Geom.*, **43** (1996) 1–41.

[26] T. D. Comar, *Hyperbolic Dehn surgery and convergence of Kleinian groups,* Ph.D. thesis, University of Michigan (1996).

[27] M. Culler and P. B. Shalen, Volumes of hyperbolic Haken manifolds, I *Invent. Math.*, **118** (1994) 285–329.

[28] R. Evans and J. Holt, Non-wrapping of hyperbolic interval bundles, *G.A.F.A.*, **18** (2008) 98–119.

[29] W. M. Goldman, Projective structures with Fuchsian holonomy, *J. Diff. Geom.*, **25** (1987) 297–326.

[30] D. A. Hejhal, Monodromy groups and linearly polymorphic functions, *Acta Math.*, **135** (1975) 1–55.

[31] C. D. Hodgson and S. P. Kerckhoff, Rigidity of hyperbolic cone-manifolds and hyperbolic Dehn surgery, *J. Diff. Geom.*, **48** (1998) 1–59.

[32] C. Hodgson and S. P. Kerckhoff, Universal bounds for hyperbolic Dehn surgery, *Ann. of Math.*, **162** (2005) 367–421.

[33] J. Holt, Some new behaviour in the deformation theory of Kleinian groups, *Comm. Anal. Geom*, **9** (2001) 757–775.

[34] J. Holt, Multiple bumping of components of deformation spaces of hyperbolic 3-manifolds, *Am. J. Math.*, **125** (2003) 691–736.

[35] J. Holt and J. Souto, On the topology of the space of once-punctured torus groups, preprint.

[36] K. Ito, Exotic projective structures and quasifuchsian space, *Duke Math. J.*, **105** (2000) 185–209.
[37] K. Ito, Exotic projective structures and quasifuchsian space, II *Duke Math. J.*, **140** (2007) 85–109.
[38] K. Ito, Convergence and divergence of Kleinian punctured torus groups, preprint.
[39] W. Jaco and P. Shalen, Seifert fibered spaces in 3-manifolds, *Mem. Amer. Math. Soc.*, **220** (1979).
[40] K. Johannson, *Homotopy Equivalences of 3-manifolds with Boundary*, Lecture Notes in Mathematics, vol. 761, Springer-Verlag (1979).
[41] M. Kapovich, *Hyperbolic Manifolds and Discrete groups: Lectures on Thurston's Hyperbolization*, Birkhauser (2000).
[42] R. Kent, Skinning maps, *Duke Math. J.*, preprint.
[43] S. P. Kerckhoff and W. P. Thurston, Noncontinuity of the action of the modular group at Bers' boundary of Teichmüller space, *Invent. Math.*, **100** (1990) 25–47.
[44] I. Kim, C. Lecuire and K. Ohshika, Convergence of freely decomposable Kleinian groups, preprint.
[45] G. Kleineidam and J. Souto, Algebraic convergence of function groups, *Comment. Math. Helv.*, **77** (2002) 244–269.
[46] Y. Komori, On the boundary of the Earle slice for punctured torus groups, in *Kleinian groups and hyperbolic 3-manifolds (Warwick, 2001)*, ed. by Y. Komori, V. Markovic and C. Series, *London Math. Soc. Lecture Notes*, **299** (2003) 293–304.
[47] Y. Komori and J. Parkkonen, On the shape of Bers-Maskit slices, *Ann. Acad. Sci. Fenn. Math.*, **32** (2007) 179–198.
[48] Y. Komori, T. Sugawa, M. Wada and Y. Yamashita, Drawing Bers embeddings of the Teichmüller space of once-punctured tori, *Experiment. Math.*, **15** (2006) 51–60.
[49] C. Lecuire, An extension of Masur domain, in *Spaces of Kleinian groups*, ed. by Y. Minsky, M. Sakuma and C. Series, *London Math. Soc. Lecture Notes*, **329** (2006) 49–73.
[50] A. Magid, Examples of relative deformation spaces that are not locally connected, *Math. Ann.*, **344** (2009) 877–889.
[51] A. Marden, The geometry of finitely generated Kleinian groups, *Annals of Math.*, **99** (1974) 383–462.
[52] B. Maskit, Self-maps of Kleinian groups, *Amer. J. Math.*, **93** (1971) 840–856.
[53] D. McCullough, Twist groups of compact 3-manifolds, *Topology*, **24** (1985) 461–474.
[54] D. McCullough, A. Miller and G. A. Swarup, Uniqueness of cores of noncompact 3-manifolds, *J. London Math. Soc.*, **32** (1985) 548–556.
[55] C. T. McMullen, Complex earthquakes and Teichmüller theory, *J. Amer. Math. Soc.*, **11** (1998) 283–320.
[56] Y. N. Minsky, The classification of punctured torus groups, *Annals of Math.*, **149** (1999) 559–626.
[57] Y. N. Minsky, The classification of Kleinian surface groups I: models and bounds, *Annals of Math.*, to appear.
[58] H. Miyachi, On the horocyclic coordinate for the Teichmüller space of once punctured tori, *Proc. A.M.S.*, **130** (2002) 1019–1029.
[59] H. Miyachi, Cusps in complex boundaries of one-dimensional Teichmüller space, *Conf. Geom. Dyn.*, **7** (2003) 103–151.
[60] J. W. Morgan, On Thurston's uniformization theorem for three-dimensional manifolds, in *The Smith Conjecture*, ed. by J. Morgan and H. Bass, Academic Press (1984) 37–125.
[61] G. D. Mostow, Quasiconformal mappings in n-space and the rigidity of hyperbolic space forms, *Publ. I. H. E. S.*, **34** (1968) 53–104.
[62] D. Mumford, C. Series and D. Wright, *Indra's pearls. The vision of Felix Klein* Cambridge University Press, New York (2002).

[63] H. Namazi and J. Souto, Non-realizability and ending laminations, in preparation.

[64] K. Ohshika, Ending laminations and boundaries for deformation spaces of Kleinian groups, *J. London Math. Soc.*, **42** (1990) 111–121.

[65] K. Ohshika, Kleinian groups which are limits of geometrically finite groups, *Mem. A. M.S.*, **177** (2005) no. 834.

[66] K. Ohshika, Realising end invariants by limits of minimally parabolic, geometrically finite groups, preprint.

[67] K. Ohshika, Divergence, exotic convergence and self-bumping in quasi-Fuchsian spaces, in preparation.

[68] K. Ohshika and T. Soma, Geometry and topology of geometric limits, in preparation.

[69] J. Parkkonen, Conformal Dehn surgery and the shape of Maskit's embedding, *Conform. Geom. Dyn.*, **8** (2004) 143–157.

[70] J. Parkkonen, Geometric limits of cyclic groups and the shape of Schottky space, *Math. Proc. Camb. Phil. Soc.*, **137** (2004) 55–68.

[71] G. Prasad, Strong rigidity of $\mathbb{Q}$-rank 1 lattices, *Invent. Math.*, **21** (1973) 255–286.

[72] T. Soma, Geometric limits of quasi-Fuchsian groups, preprint.

[73] D. Sullivan, On the ergodic theory at infinity of an arbitrary discrete group of hyperbolic motions, in *Riemann Surfaces and Related Topics, Proceedings of the 1978 Stony Brook Conference*, Annals of Math. Studies no. 97 (1980) 465–496.

[74] D. Sullivan, Quasiconformal homeomorphisms and dynamics. II. Structural stability implies hyperbolicity for Kleinian groups, *Acta Mathematica*, **155** (1985) pp. 243–260.

[75] W. P. Thurston, *The Geometry and Topology of Three-Manifolds*, Princeton University course notes, available at http://www.msri.org/publications/books/gt3m/ (1980).

[76] Y. Yamashita, Computer experiments on the discreteness locus in projective structure, in *Spaces of Kleinian groups*, ed. by Y. Minsky, M. Sakuma and C. Series, *London Math. Soc. Lecture Notes*, **329** (2006) 375–390.

Teichmüller Theory and Moduli Problem No. 10, 2010, pp. 151–166.

A Short Note on Nonemptyness of Parabolic Moduli

Arijit Dey

The Institute of Mathematical Sciences, CIT Campus, Taramani, Chennai 600113, India
e-mail: arijit@imsc.res.in

1. Introduction

In this note we continue our study of nonemptiness of moduli of stable parabolic bundles over smooth algebraic surface X defined over $\mathbb{C}$.

In [BDP06b], we proved the following theorem

Theorem I. *The moduli space of rank* 2 *stable parabolic bundles on* X *with* **certain** *parabolic structure is non-empty if* $\operatorname{par} c_2 \gg 0$.

In §4 we use **Theorem I** to prove the following theorem:

Theorem II. *The moduli space of rank* r *stable parabolic bundles on* X *with* **certain** *parabolic structure is nonempty if* $\operatorname{par} c_2 \gg 0$.

The main strategy used in proving **Theorem I** is the *Seshadri–Biswas* correspondence of the category of Γ-bundles of fixed type τ on a certain *Kawamata cover* Y of the surface X with the category of parabolic bundles on X with fixed parabolic datum (see 2 for definitions and terminology). The Kawamata cover Y is non-canonical and is therefore employed only as a stepping stone. Note that by *certain* parabolic structure we mean that parabolic bundle has full quasi–parabolic structure and a condition on highest weight i.e $\alpha_{\max} < \frac{{\Theta_1}^2}{\sum D_i \cdot \Theta_1}$, where Θ_1 is the polarisation on X and D is the parabolic divisor. If for some choices of (X, D) and Θ_1 the right hand side is greater than 1 then we actually get a rank 2 parabolic stable bundle (with respect to Θ_1) of full flag and arbitrary weights.

In §5 we give another approach to tackle this nonemptiness problem. Here we use the concept of a monodromy group for a stable $(\Gamma, SL(n))$-bundle over algebraic surface (see §2 for definitions). This gives rise to another proof of **Theorem II**.

2. Preliminaries

Let X be a smooth projective variety over $\mathbb{C}$ and Θ_1 be a very ample divisor in X. In our proof of nonemptiness (§4) we rely heavily on the correspondence between the category of parabolic bundles on X and the category of Γ-bundles on a suitable Kawamata cover. This strategy has been employed in many papers ([Bis97b],[BDP06b]) but since we need its intricate properties, most of which are scattered in a few papers of Biswas and Seshadri, we recall them briefly. We stress only on those points which are relevant to our purpose.

2.1 *The category of parabolic bundles*

Let D be an effective divisor in X and E be a coherent torsion free sheaf on X. All our degree calculations of a sheaf on X are with respect to a fixed ample divisor Θ_1. The image of $E \otimes_{\mathcal{O}_X} \mathcal{O}_X(-D)$ in E will be denoted by $E(-D)$. The following definition of parabolic sheaf was introduced in [MY92].

Definition 2.1. *Let E be a torsion-free $\mathcal{O}_X$-coherent sheaf on X. A quasi–parabolic structure on E over D is a filtration by $\mathcal{O}_X$-coherent subsheaves*

$$E = F_1(E) \supset F_2(E) \supset \cdots \supset F_l(E) \supset F_{l+1}(E) = E(-D)$$

The integer l is called the length of the filtration. A parabolic structure is a quasi–parabolic structure, as above, together with a system of weights $\{\alpha_1, \ldots, \alpha_l\}$ such that

$$0 \leq \alpha_1 < \alpha_2 < \cdots < \alpha_{l-1} < \alpha_l < 1$$

where the weight α_i corresponds to the subsheaf $F_i(E)$.

In other words a *quasi-parabolic structure* on a sheaf E can also be defined by giving filtration by subsheaves on the restriction $E|_D$ of the sheaf E to each component of the parabolic divisor:

$$E|_D = \mathcal{F}_D^1(E) \supset \mathcal{F}_D^2(E) \supset \cdots \supset \mathcal{F}_D^l(E) \supset \mathcal{F}_D^{l+1}(E) = 0$$

A parabolic sheaf is a sheaf together with a *quasi–parabolic* structure and a system of weights

$$0 \leq \alpha_1 < \alpha_2 < \cdots < \alpha_{l-1} < \alpha_l < 1$$

where each α_i corresponds to $\mathcal{F}^i|_D(E)$.

We shall denote the parabolic sheaf defined above by (E, F_*, α_*). When there is no scope of confusion it will be denoted by E_*.

For a parabolic sheaf (E, F_*, α_*) define the following filtration $\{E_t\}_{t\in\mathbb{R}}$ of coherent sheaves on X parameterized by $\mathbb{R}$:

$$E_t := F_i(E)(-[t]D) \tag{1}$$

where $[t]$ is the integral part of t and $\alpha_{i-1} < t - [t] \leq \alpha_i$, with the convention that $\alpha_0 = \alpha_l - 1$ and $\alpha_{l+1} = 1$.

A *homomorphism* from the parabolic sheaf (E, F_*, α_*) to another parabolic sheaf (E', F'_*, α'_*) is a homomorphism from E to E' which sends any subsheaf E_t into E'_t, where $t \in [0, 1]$ and the filtration are as above. In particular E_* is said to be parabolic subsheaf of E'_* if E_t is a subsheaf of E'_t for every $t \in [0, 1]$.

If the underlying sheaf E is locally free then E_* will be called a parabolic vector bundle.

The notion of *parabolic slope* of a parabolic bundle E_* of rank r is defined as:

$$\mathrm{par}_{\deg}(E_*) := \int_0^1 \deg(E_t)dt + r \cdot \deg(D)$$

Let $\mathrm{par}_\mu(E_*) := \mathrm{par}_{\deg}(E_*)/r$.

Definition 2.2. *A parabolic sheaf* E_* *is called* parabolic semistable (resp parabolic stable) *if for every parabolic subsheaf* V_* *of* E_* *with* $0 < rank(V_*) < rank(E_*)$*, the following holds:*

$$par_\mu(V_*) \leq par_\mu(E_*) \;\; (resp. \; par_\mu(V_*) < par_\mu(E_*)) \tag{2}$$

Remark 2.3. Given any subsheaf F of E with E/F is torsion-free one can give an induced parabolic structure F_* on F defined as $F_t := E_t \cap F$ ([MY92] [Bis97b]). In order to check (semi)stability of a parabolic sheaf E_* it is enough to check the inequality 2 for subsheaves of E with induced parabolic structure.

2.1.1 *Some assumptions* The class of parabolic vector bundles that are dealt with in this note satisfy certain constraints which we will explain now.

(1) The first condition is that all parabolic divisors are assumed to be *divisors with normal crossings*. In other words, any parabolic divisor is assumed to be reduced, its each irreducible component is smooth, and furthermore the irreducible components intersect transversally.
(2) The second condition is that all the parabolic weights are *rational numbers*.
(3) The third and final condition states that on each component of the parabolic divisor the filtration is given by *subbundles*. The precise formulation of the last condition is given in ([Bis97b], Assumptions 3.2 (1)).

Henceforth, all parabolic vector bundles will be assumed to satisfy the above three conditions. Condition 1 and 3 are necessary for the categorical correspondence between parabolic bundles and orbifold bundles. Condition 2 is not a very strong assumption for nonemptiness of stable parabolic moduli (since $\mathbb{Q}$ is dense in $\mathbb{R}$ and a bundle is stable if it satisfies strict inequality of 2.2 [Bis97b]).

Let PVect(X, D) denote the category whose objects are parabolic vector bundles over X with parabolic structure over the divisor D satisfying the above three conditions, and the morphisms of the category are homomorphisms of parabolic vector bundles (which was defined earlier). One can define parabolic direct sum /dual in a natural way. There is also notion of parabolic tensor product which is quite well explained in [Bis97a].

For an integer $N \geq 2$, let PVect$(X, D, N) \subseteq$ PVect(X, D) denote the subcategory consisting of all parabolic vector bundles all of whose parabolic weights are multiples of $1/N$. It is obvious that PVect(X, D) is also closed under direct sum. After little bit of work it can be shown that the subcategory PVect(X, D, N) is also closed under parabolic tensor product, parabolic direct sum and parabolic dual.

2.2 *The Bridge: Kawamata Covering lemma*

The "Covering Lemma" of Y. Kawamata (Theorem 1.1.1 of [KMM87], Theorem 17 of [Kaw81]) says that given a smooth projective variety X over $\mathbb{C}$ and a normal crossing divisor D in X and a natural number $N \geq 1$ there is a connected smooth projective variety Y over $\mathbb{C}$ and a Galois covering morphism

$$p : Y \longrightarrow X \tag{3}$$

such that the reduced divisor $\tilde{D} := (p^*D)_{\text{red}}$ is a normal crossing divisor on Y and furthermore, $p^*D_i = k_i N.(p^*D_i)_{\text{red}}$, where D_i's are irreducible components of D; k_i's, $1 \leq i \leq c$, are positive integers. Let Γ denote the Galois group for the covering map p.

2.3 *The category of* Γ*-bundles*

Let $\Gamma \subseteq \text{Aut}(Y)$ be a finite subgroup of the group of automorphisms of a connected smooth projective variety $Y/\mathbb{C}$ which acts on Y in properly discontinuous way. The natural action of Γ on Y is encoded in a morphism

$$\mu : \Gamma \times Y \longrightarrow Y$$

Denote the projection of $\Gamma \times Y$ to Y by p_2. The projection of $\Gamma \times \Gamma \times Y$ to the i-th factor will be denoted by q_i. A Γ*-linearized vector bundle* on Y is a vector bundle V over Y together with an isomorphism

$$\lambda : p_2^* V \longrightarrow \mu^* V$$

over $\Gamma \times Y$ such that the following diagram of vector bundles over $\Gamma \times \Gamma \times Y$ is commutative:

$$\begin{array}{ccc} q_3^*V & \xrightarrow{(q_2,q_3)^*\lambda} & (\mu \circ (q_2,q_3))^*V \\ & \searrow^{(m \times Id_Y)^*\lambda} & \downarrow{\scriptstyle (Id_\Gamma \times \mu)^*\lambda} \\ & & (\mu \circ (m, Id_Y))^*V \end{array}$$

where m is the multiplication operation on Γ.

The above definition of Γ-linearization is equivalent to giving isomorphisms of vector bundles

$$\bar{g} : V \longrightarrow (g^{-1})^* V$$

for all $g \in \Gamma$, satisfying the condition that $\overline{gh} = \bar{g} \circ \bar{h}$ for any $g, h \in \Gamma$.

A Γ*-homomorphism* between two Γ-linearized vector bundles is a homomorphism between the two underlying vector bundles which commutes with the Γ-linearizations. Clearly direct sum, tensor product of two Γ-linearized vector bundles admits a natural Γ-linearization; so does the dual of a Γ-linearized vector bundle. Let $\text{Vect}_\Gamma(Y)$ denote the additive tensor category of Γ-linearized vector bundles on Y with morphisms being Γ-homomorphisms.

By imitating the above definition of Γ-linearization, a (Γ, G)-bundle is defined to be a principal G-bundle E, together with a lift of the action of Γ on Y to a left action of Γ on E which commutes with the right action of G on P. So a $(\Gamma, Gl(n, \mathbb{C})$-bundle is same as a rank n Γ vector bundle.

2.4 *On local types of* Γ*-bundles*

Recall that since the Γ-action on Y is *properly discontinuous*, for each $y \in Y$, if Γ_y is the isotropy subgroup at y, then there exists an *analytic* neighbourhood $U_y \subset Y$ of y which is Γ_y-invariant and such that for each $g \in G$, $g \cdot U_y \cap U_y \neq \emptyset$.

Definition 2.4. *Let ρ be a representation of Γ in $GL(r, \mathbb{C})$. Then Γ-acts on the trivial bundle $Y \times \mathbb{C}^r$ by $(y, v) \longrightarrow (\gamma y, \rho(\gamma)v)$, $y \in Y, v \in \mathbb{C}^r, \gamma \in \Gamma$. Following [Ses69] we call this Γ-bundle, the Γ-bundle associated to the representation ρ.*

We then have the following equivariant local trivialisation lemma.

Lemma 2.5. *Let E be a Γ-bundle on Y of rank r. Let $y \in Y$ and let Γ_y be the isotropy subgroup of Γ at y. Then there exists a Γ_y-invariant analytic neighbourhood U_y of y such that the Γ_y-bundle $E|_{U_y}$ is associated to a representation $\Gamma_y \to GL(r)$ (in the sense of Def 2.4).*

Remark 2.6. The above Lemma for Γ-bundles with structure group $GL(r)$ can be found in [Ses69, Remark 2, page 162] and [Gro57]. Here the key property that is used is that U_y and U_y/Γ_y are Stein spaces.

2.4.1 Γ*-bundles of fixed local type* We make some general observations on the local structure of Γ-bundles on the Kawamata cover defined in (2.3).

Let $\mathrm{Vect}_\Gamma^D(Y, N)$ denote the subcategory of $\mathrm{Vect}_\Gamma(Y)$ consisting of all Γ-linearized vector bundles W over Y satisfying the following two conditions:

(1) for a general point y of an irreducible component of $(p^* D_i)_{\mathrm{red}}$, the isotropy subgroup Γ_y is *cyclic* of order $|\Gamma_y| = n_y$ which is a divisor of N; the action of the isotropy group Γ_y on the fiber W_y is of order N, which is equivalent to the condition that for any $g \in \Gamma_y$, the action of g^N on W_y is the trivial action;
(2) In fact, the action is given by a representation ρ_y of Γ_y given as follows:

$$\rho_y(\zeta) = \begin{bmatrix} z^{\alpha_1}.I_1 & & 0 \\ & \ddots & \\ 0 & & z^{\alpha_l}.I_l \end{bmatrix} \tag{4}$$

where

- ζ is a generator of the group Γ_y and whose order n_y divides N
- $\alpha_i = \frac{m_j}{N}$ and
- I_j is the identity matrix of order r_j, where r_j is the multiplicity of the weight α_j.
- z is an n_y-th root of unity.
- We have the relation $0 \leq m_1 < m_2 < \cdots < m_l \leq N - 1$.

(3) For a general point y of an irreducible component of a ramification divisor for p not contained in $(p^* D)_{\mathrm{red}}$, the action of Γ_y on W_y is the trivial action.
(4) For a *special point* y contained in $(p^* D)_{\mathrm{red}}$, the isotropy subgroup Γ_y contains the cyclic group Γ_n of order n determined by the irreducible component containing y.

By the rigidity of representations of finite groups, the Γ_y-module structure on W_y (given by Lemma 2.5) when restricted to $\Gamma_n \subset \Gamma_y$ is of type τ.

(5) At special points y of the ramification divisor for p not contained in $(p^* D)_{\text{red}}$, the restriction of the representation to the generic isotropy is trivial.

Definition 2.7. *Following Seshadri [Ses69, page 161] we call the Γ-bundles E in* $\mathrm{Vect}_\Gamma^D(Y, N)$ bundles of fixed local orbifold type τ.

Remark 2.8. The reason for calling it local type τ is that, for a Γ-bundle and a point y the generic point of a divisor as above, the structure of the representation defines the bundle $E|_U$ for a Γ_y-invariant analytic neighbourhood in Y. Seshadri denoted the collection of representations of the cyclic groups which define the local isomorphism type over an analytic neighbourhood by the letter τ; note that the Γ-bundle defines what is known as an orbifold bundle.

Remark 2.9. We remark that this definition of Γ-bundles of fixed local type easily extends to Γ-torsion–free sheaves since the local action is specified only at the generic points of the ramification divisor.

We note that $\mathrm{Vect}_\Gamma^D(Y, N)$ is also an additive tensor category.

2.4.2 *Seshadri-Biswas correspondence* In [Bis97b] an identification between the objects of $\mathrm{PVect}(X, D, N)$ and the objects of $\mathrm{Vect}_\Gamma^D(Y, N)$ has been constructed. Given a Γ-homomorphism between two Γ-linearized vector bundles, there is a naturally associated homomorphisms between the corresponding vector bundles, and this identifies, in a bijective fashion, the space of all Γ-homomorphisms between two objects of $\mathrm{Vect}_\Gamma^D(Y, N)$ and the space of all homomorphisms between the corresponding objects of $\mathrm{PVect}(X, D, N)$. An equivalence between the two additive tensor categories, namely $\mathrm{PVect}(X, D, N)$ and $\mathrm{Vect}_\Gamma^D(Y, N)$, is obtained this way. Since the description of this identification is already given in [Bis97b], and [BBN01], it will not be repeated here.

We observe that an earlier assertion that the parabolic tensor product operation enjoys all the abstract properties of the usual tensor product operation of vector bundles, is a consequence of the fact that the above equivalence of categories indeed preserves the tensor product operation.

The above equivalence of categories has the further property that it takes the parabolic dual of a parabolic vector bundle to the usual dual of the corresponding Γ-linearized vector bundle.

Let $W \in \mathrm{Vect}_\Gamma^D(Y, N)$ be the Γ-linearized vector bundle of rank n on Y that corresponds to the given parabolic vector bundle E_*. The fiber bundle

$$\pi : P \longrightarrow Y$$

whose fiber $\pi^{-1}(y)$ is the space of all $\mathbb{C}$-linear isomorphisms from $\mathbb{C}^n$ to the fiber W_y, has a the structure of a $(\Gamma, GL(n, \mathbb{C}))$-bundle over Y.

Definition 2.10. *A Γ-linearized vector bundle E over Y is called Γ-semistable (resp. Γ-stable) if for any proper nonzero coherent subsheaf $F \subset E$, invariant* *under the action of* Γ *and with* E/F *being torsionfree, the following inequality is valid:*

$$\mu(F) \leq \mu(E) \text{ (resp. } \mu(F) < \mu(E)) \tag{5}$$

where the slope is as usual $\mu(E) = deg(E)/r$ *and* $deg(E)$ *is computed with respect to the* Γ*-linearised very ample divisor* Θ *on* Y.

The Γ*-linearized vector bundle* E *is called* Γ*-polystable if it is a direct sum of* Γ*-stable vector bundles of same slope.*

Remark 2.11. The above correspondence between *parabolic bundles* on X and Γ-bundles on Y preserves the semistable (resp. stable) objects as well, where *parabolic semistability* is as in Definition (2.2). (cf [Bis97b])

Remark 2.12. We remark that it is not hard to check that for Γ-bundles, Γ-semistability (resp. Γ-polystability) is the same as usual semistability (resp. polystability). This can be seen from the fact that the *top term of the Harder-Narasimhan filtration* (resp. *the socle*) is *canonical* and hence invariant under the action of Γ. But we note that a Γ-stable bundle need not be stable, as can be seen by taking a direct sum of Γ-translates of a line bundle.

Remark 2.13. We make some key observations in this remark where we also note the essential nature of assumptions of characteristic zero base fields.

(1) The notion of Γ-cohomology for Γ-sheaves on Y has been constructed and dealt with in great detail in [Gro57]. These can be realised as higher derived functors of the Γ-*fixed points*–sub-functor $(H^0)^\Gamma$ of the section functor H^0. (We use this notation to avoid Γ^Γ, because we have denoted the finite group by the letter Γ!).

   We note immediately that since we work over fields of characteristic zero, the sub-functor $(H^0)^\Gamma \subset H^0$ is in fact a direct summand (by averaging operation). Hence, we see immediately that the higher derived functors of the functor$(H^0)^\Gamma$ are all sub objects of the derived functors of H^0.

(2) When we work with a Kawamata cover as in our case, then we have the following relation between the Γ-cohomology and the usual cohomology on $Y/\Gamma = X$:

$$H^i_\Gamma(Y, \mathcal{F}) = H^i(X, p^\Gamma_*(\mathcal{F}))$$

   $\forall i$.

2.4.3 Γ-*line bundles and parabolic line bundles* A Γ line bundle on Y is a line bundle L on Y together with a lift of action Γ . The Γ line bundle gives a Γ invariant line bundle L^Γ on X. Let D be a divisor of normal crossing on X. Let $D = \sum_{i=1}^d D_i$ be a decomposition into irreducible components. A parabolic line bundle on (X, D) is a pair of the form $(M, \beta_1, \dots, \beta_i, \dots, \beta_d)$ where M is a holomorphic line bundle on X and $0 \leq \beta_i < 1$ is a real number [Bis95]. When we start from a Γ line bundle on Y we get a pair $(L^\Gamma, \beta_1, \dots, \beta_i, \dots, \beta_d)$ where β_i is a rational number and it can be written as $\beta_i = m_i/N$. Let $\tilde{D}_i = (p^* D_i)_{\text{red}}$. Then by following [Bis95] we have $L = p^*(L^\Gamma) \otimes \mathcal{O}_Y(\sum_{i=1}^d k_i m_i \tilde{D}_i)$

Remark 2.14. In our situation, by choice we work with a single weight when we consider Γ-line bundles of fixed local type τ although this may not be absolutely essential.

2.4.4 *Serre duality for Γ-line bundles of fixed local type*

Definition 2.15. *By a line bundle L of fixed local type τ we mean a parabolic line bundle $(L, \alpha_1, \alpha_2, \ldots, \alpha_d)$, where $\alpha_i = \alpha \forall i$. In other words, locally, the generic isotropy on the irreducible components of the inverse image of the parabolic divisor acts by a single character namely α. We will write $L^{(\alpha)}$ to specify the character.*

Let $L = L^{(\alpha)}$ be a Γ line bundle on Y of type τ. Then from above paragraph one knows that $L = p^*(p_*^{\Gamma}(L)) \otimes \mathcal{O}_Y(\sum k_i m_i \tilde{D}_i)$ where all the m_i can be assumed to be equal to m since we have a single weight α. Then if $M = M^{(\alpha)}$ is another Γ-line bundle with the same local character type we have:

$$(p_*^{\Gamma}(L^* \otimes M)) = (p_*^{\Gamma}(L)^* \otimes (p_*^{\Gamma}(M)) \tag{6}$$

Consider the canonical bundles K_X of X and define the Γ-bundle $K_Y^{(\alpha)}$ as follows:

$$K_Y^{(\alpha)} = p^*(K_X) \otimes \mathcal{O}_Y\left(\sum (mk_i \tilde{D}_i)\right) \tag{7}$$

Then, we see as above that $p_*^{\Gamma}(K_Y^{(\alpha)}) = K_X$. We then have the following duality for Γ-line bundles of type τ (for proof see [BDP06b]):

Lemma 2.16. *For Γ-line bundles L of type τ, with local character α, the Γ-line bundle $K_Y^{(\alpha)}$ is the* dualising sheaf. *In other words, we have a canonical isomorphism:*

$$H_{\Gamma}^i(Y, L^* \otimes K_Y^{(\alpha)}) \simeq H_{\Gamma}^{n-i}(Y, L)^*$$

for all i. We have made this statement for Γ-varieties Y of any dimension.

3. Rank 2 case-A short sketch

In [BDP06b] our whole idea was to get a non-split locally free Γ-extension E of the form

$$0 \longrightarrow L^{(\alpha_1)} \longrightarrow E \longrightarrow M^{(\alpha_2)} \otimes \mathcal{I}_Z \longrightarrow 0$$

such that E is Γ-stable with respect to a chosen polarization Θ. We describe the main steps here

- Our first step is to ensure that $\mathrm{Ext}_{\Gamma}^1(M^{(\alpha_2)} \otimes \mathcal{I}_Z, L^{(\alpha_1)})$ is nonzero, for that we need $l(Z) > h_{\Gamma}^0(Y, L^{(\alpha_1)*} \otimes M^{(\alpha_2)} \otimes \mathrm{K}_Y^{(\alpha_1-\alpha_2)})$
- For such OCB tuple with $l(Z) \gg 0$ there exists an extension which is actually Γ-locally free. Note that this is the reason for bundle to have large c_2.
- After that we prove that any such Γ-locally free extension is actually Θ-stable.

4. Rank r case–Using Cayley–Bachrach

Let Y be a smooth projective Γ-surface and $p : Y \longrightarrow X$ be a morphism where $X := Y/\Gamma$ arising from the Kawamata covering lemma (see [KMM87], [Kaw81]).

Let $\mathcal{D}_{Y/X} = \tilde{\mathcal{D}} = \{y \in Y : \Gamma_y \neq e\}$ be the ramification locus in Y, then lemma 2.8 of [Bis97b] tells that $\tilde{\mathcal{D}}$ is a normal crossing divisor. Let R be a codimension two subvariety of X away from $p(\tilde{\mathcal{D}})$ and $Z = p^*(R)$. Then we term the cycle Z in Y a **good** Γ-cycle.

Remark 4.1. Consider Γ-line bundles $L^{(\alpha_1)}$, $M^{(\alpha_2)}$ on Y and let $P = M^{(\alpha_2)} \otimes L^{(\alpha_1)*} \otimes K_Y^{(\alpha_1-\alpha_2)}$. By tensoring the exact sequence

$$0 \longrightarrow \mathcal{I}_Z \longrightarrow \mathcal{O}_Y \longrightarrow \mathcal{O}_Z \longrightarrow 0$$

with P we get the following long exact sequence (see remark 2.13):

$$\begin{aligned} 0 \longrightarrow H^0_\Gamma(P \otimes \mathcal{I}_Z) \longrightarrow H^0_\Gamma(P) \longrightarrow H^0_\Gamma(P \otimes \mathcal{O}_Z) \\ \longrightarrow H^1_\Gamma(P \otimes \mathcal{I}_Z) \longrightarrow H^1_\Gamma(P) \longrightarrow H^1_\Gamma(P \otimes \mathcal{O}_Z) = 0 \end{aligned} \tag{8}$$

Let $\dim H^0_\Gamma(P) = l_1$, then by choosing a reduced 0 dimensional subscheme $Z = p^*(R)$ such that $l(Z) > l_1$ it is easily seen that $H^1_\Gamma(P \otimes \mathcal{I}_Z) \neq 0$. This implies that there is at least one Γ-torsion free sheaf E on Y which is a non-split extension of $M^{(\alpha_2)} \otimes \mathcal{I}_Z$ by $L^{(\alpha_1)}$.

Definition 4.2. *Let $0 \leq \alpha_1 < \alpha_2 < 1$, we say that the Γ-triple $(L^{(\alpha_1)}, M^{(\alpha_2)}, Z)$, satisfies the* Orbifold Cayley Bacharach *(OCB) property if the following holds: for any section $s \in H^0_\Gamma(Y, M^{(\alpha_2)} \otimes L^{(\alpha_1)*} \otimes K_Y^{(\alpha_1-\alpha_2)})$ if the restriction of s to a good Γ-cycle $Z' \subset Z$ is zero implies that $s|_Z = 0$, where $Z' \subset Z$ is a good Γ-cycle such that $l(Z') = l(Z) - d$ and d is the order of the group Γ.*

Lemma 4.3. *Let $(L^{(\alpha_1)}, M^{(\alpha_2)}, Z)$ be a Γ-triple which satisfies OCB. Then for $l(Z) \gg 0$, there exists a Γ-extension*

$$0 \longrightarrow L^{(\alpha_1)} \longrightarrow E \longrightarrow M^{(\alpha_2)} \otimes \mathcal{I}_Z \longrightarrow 0$$

with E locally free.

Proof. Lemma 5.8 of [BDP06b] □

4.0.5 *Choice of $\mathcal{Q}$ and degree bounds*

Remark 4.4. Let $\mathcal{Q} \in \mathrm{Pic}(Y)$ be any Γ-line bundle obtained as follows: Let L be a line bundle on X. Let $Q = L \otimes \mathcal{O}_X(rn\Theta_1)$, we choose n in such a way that

$$c_1(Q) \cdot \Theta_1 < 3\Theta_1^2 \tag{9}$$

This choice of n is made to ensure the resultant rank r bundle (to be constructed) is Θ-stable and has first Chern class $(r-1)c_1(\mathcal{Q})$, where $\mathcal{Q} = p^*(Q) \otimes \mathcal{O}_Y^{(\alpha_2)}$. Note that $\mathcal{Q}$ has same Γ-structure as $\mathcal{O}_Y^{(\alpha_2)}$ (since $p^*(Q)$ has trivial Γ-structure), where by $\mathcal{O}_Y^{(\alpha_2)}$ we mean the trivial line bundle $\mathcal{O}_Y$ with a Γ-structure of type τ given by multiplication by the character corresponding to α_2 (see (2.4.1) for notation).

Let $0 \leq \alpha_1 < \alpha_2 < 1$. Then we claim that for any choice of L on X, there exists a rank 2 Γ-stable bundle whose determinant is $\mathcal{L} = p^*(L) \otimes \mathcal{O}_Y^{(\alpha_2)}$. For that we twist it by a suitable n as in (9) and then produce a triple $(\mathcal{O}_Y^{(\alpha_1)}, \mathcal{Q}, Z)$ which satisfies *OCB* property. Hence there exists a rank 2 bundle which comes as extension of $\mathcal{O}_Y^{\alpha_1}$ by $\mathcal{Q} \otimes \mathcal{I}_Z$. In fact we prove such a bundle is actually Θ-stable. After that we twist it by $\mathcal{O}_Y(-n\Theta)$ to get a Γ-stable bundle whose c_1 is $c_1(\mathcal{L})$. So we need to choose a cycle R such that $(\mathcal{O}_X, Q, R)$ has stronger Cayley-Bacharach property.

Let $\gamma = \alpha_2 \cdot \sum \deg_X(D_i)$, where D_i's are the irreducible components of the *parabolic divisor*. We assume that

$$\gamma < \Theta_1^2 \tag{10}$$

This imposes a condition on the weight α_2 which we therefore have as hypothesis in **Theorem I**.

Let $d = |\Gamma|$. Then we see that by comparing degrees, we have:

$$c_1(\mathcal{Q}) \cdot \Theta = \deg_Y(\mathcal{Q}) = (\text{pardeg}(p_*^\Gamma(\mathcal{Q}))\, d = \{\deg_X(Q) + \gamma\}\, d$$

The non-trivial contribution of γ occurs since $p_*^\Gamma(\mathcal{Q})$ is a parabolic line bundle with underlying line bundle Q but with *non-trivial parabolic structure.*

Again, by (9) and (10) we have;

$$\frac{c_1(\mathcal{Q}) \cdot \Theta}{2} < 2\Theta^2 \tag{11}$$

Let $\mathcal{Q}_i = p^*(Q) \otimes \mathcal{O}_Y^{(\alpha_{i+1})}$, $i = 1 \cdots r-1$; where each α_i's satisfies (10) and are in increasing order. Let $Z_1 \ldots, Z_{r-1}$ are $(r-1)$ Γ-cycles of same length, such that $Z_i \cap Z_j = \emptyset$ for $i \neq j$; where each $Z_i = p^*(R_i)$ and $R_i \cap p(\tilde{\mathcal{D}}) = \emptyset$. Let $W = \oplus_{i=1}^{r-1}(\mathcal{Q}_i \otimes \mathcal{I}_{Z_i})$ and $W_i = \frac{W}{(\mathcal{Q}_i \otimes \mathcal{I}_{Z_i})}$, it is known (from previous section) that there exists an extension $e_i \in \text{Ext}_\Gamma^1(\mathcal{Q}_i, \mathcal{O}_Y^{(\alpha_1)})$ whose corresponding exact sequence

$$0 \longrightarrow \mathcal{O}_Y^{(\alpha_1)} \longrightarrow E_i \longrightarrow \mathcal{Q}_i \longrightarrow 0$$

gives a bundle E_i if and only if $(\mathcal{Q}_i, \mathcal{O}_Y^{(\alpha_1)}, Z_i)$ satisfies orbifold Cayley-Bacharach property. Note that

$$\text{Ext}_\Gamma^1(W, \mathcal{O}_Y^{(\alpha_1)}) \simeq \bigoplus_{i=1}^{r-1} \text{Ext}_\Gamma^1(\mathcal{Q}_i, \mathcal{O}_Y^{(\alpha_1)}) \tag{12}$$

Lemma 4.5. *There exists a reduced Γ-invariant 0-cycle $Z = p^*(R)$ with $l(R) \geq 4(r-1)^2\Theta_1^2$ having the following property: if $\mathcal{Q}$ is a Γ-line bundle on Y such that $h_\Gamma^0(Y, \mathcal{Q} \otimes \mathcal{I}_Z) > 0$, then we have $c_1(\mathcal{Q}) \cdot \Theta \geq 2(r-1)\Theta^2$.*

Proof. Let C_i's are r distinct smooth curves in the linear system Θ_1. For each i choose a set R_i of $2(r-1)\Theta_1^2$ many distinct points away from the ramification divisor $\mathcal{D}$ in the open subset

$$C_i - \left(\bigcup_{j \neq i} C_j)\right)$$

of C_i. Let $R = \cup_i R_i$ and $Z = p^*(R)$. Suppose that $h^0_\Gamma(Y, \mathcal{Q} \otimes I_Z) > 0$, this implies (**??**) $h^0(X, p^\Gamma_*(\mathcal{Q}) \otimes I_R) > 0$. Let $p^\Gamma_*(\mathcal{Q}) = Q$, observe that both $c_1(\mathcal{Q})$ and $c_1(Q)$ are effective. If $c_1(Q)$ contains all the curves C_i's as its irreducible components, then we have

$$c_1(Q) \cdot \Theta_1 \geq 2(r-1)\Theta_1^2$$

If $c_1(\mathcal{Q})$ does not have at least one C_i as its irreducible component for some i, then;

$$c_1(Q) \cdot \Theta_1 \geq 2(r-1)\Theta_1^2$$

Therefore

$$\begin{aligned} c_1(\mathcal{Q}) \cdot \Theta &= \deg_Y(\mathcal{Q}) = (\text{pardeg}(Q))|\Gamma| = (\deg_X(Q) + \gamma)d \\ &\geq 2(r-1)\Theta_1^2 \| \Gamma| = 2(r-1)\Theta^2 \end{aligned}$$

note $\gamma \geq 0$(which we need in the above inequality), since parabolic divisor is effective. □

4.1 *Proof of Theorem II*

Lemma 4.6. *There exists an extension $e \in Ext^1_\Gamma(W, \mathcal{O}_Y^{\alpha_1})$ whose corresponding exact sequence gives a Γ-vector bundle if and only if for each $i = 1, \ldots, (r-1)$ the tuple $(\mathcal{Q}_i, \mathcal{O}_Y^{(\alpha_1)}, Z_i)$, $i = 1 \cdots (r-1)$ satisfies OCB property.*

Proof. Put $e = (e_1, \ldots, e_{r-1})$ where $e_i \in \text{Ext}^1_\Gamma(Y, \mathcal{Q}_i \otimes \mathcal{I}_{Z_i})$. Let E_i be the subsheaf $\phi^{-1}(\mathcal{Q}_i \otimes \mathcal{I}_{Z_i})$ of E. Then E_i is given by the extension e_i

$$0 \longrightarrow \mathcal{O}_Y^{\alpha_1} \longrightarrow E_i \longrightarrow \mathcal{Q}_i \otimes \mathcal{I}_{Z_i} \longrightarrow 0$$

Note that E is locally free outside the zero cycle Z and sits in an exact sequence

$$0 \longrightarrow E_i \longrightarrow E \longrightarrow W_i \longrightarrow 0$$

Since W_i is locally free at the points in Z_i, we see that E is locally free at the points Z_i if and only if E_i is locally free at Z_i. But by choice $(\mathcal{Q}_i, \mathcal{O}_Y^{\alpha_1}, Z_i)$ satisfies OCB, hence our result follows. □

Proposition 4.7. *The rank r Γ-bundle (constructed above) is Θ-stable.*

Proof. Let E be a rank r-vector bundle as before. Let V be a proper sub-bundle of E such that E/V is torsion free. Let U_2 be the image of V in W, and V_1 be the kernel of the surjection $V \longrightarrow U_2 \longrightarrow 0$. Then we have a following commutative diagram of morphisms:

$$\begin{array}{ccccccccc} 0 & \longrightarrow & \mathcal{O}_Y^{(\alpha_1)} & \longrightarrow & E & \longrightarrow & W & \longrightarrow & 0 \\ & & \uparrow & & \uparrow & & \uparrow & & \\ 0 & \longrightarrow & U_1 & \longrightarrow & V & \longrightarrow & U_2 & \longrightarrow & 0 \\ & & \uparrow & & \uparrow & & \uparrow & & \\ & & 0 & & 0 & & 0 & & \end{array}$$

(1) $U_1 \neq 0$: From above diagram we have: $c_1(U_1) = -E_1$, where E_1 is an effective divisor. Let rk(U_2)=$r_2 \leq r-2$, we have $U_2^{**} \hookrightarrow W^{**} = \oplus_{i=1}^{r-1} \mathcal{Q}_i$; $\wedge^{r_2}(U_2^{**}) \hookrightarrow \oplus_{\beta \in S} \mathcal{Q}_\beta$, where $S = \{\beta | \beta = (i_{k_1}, \ldots, i_{k_{r_2}}); 1 \leq k_1 < \cdots < k_{r_2} \leq r-1\}$.

Thus $\wedge^{r_2} U_2 \hookrightarrow \mathcal{Q}_\beta$ for at least one β. Thus $c_1(U_2) = r_2 c_1(\mathcal{Q}) - E_2$.

So $\mu_\Theta(V) = ((c_1(U_1)+c_1(U_2))\cdot\Theta)/(r_2+1)$ $=(r_2 c_1(\mathcal{Q})\cdot\Theta)/(r_2+1) - (E_1 + E_2)\cdot\theta/(r_2+1) \leq r_2(c_1(\mathcal{Q})\cdot\Theta)/r_2+1 < ((r-1)c_1(\mathcal{Q})\cdot\theta)/r = \mu_\Theta(E)$, since $(r-1)/r - r_2/(1+r_2) > 0$.

(2) $U_1 = 0$: We have $V \hookrightarrow W$; let $r_1 = rank(V)$, hence, $\wedge^{r_1}(V) \hookrightarrow \wedge^{r_1}(W) = \oplus_{\beta \in S}(\mathcal{Q}_\beta \otimes \mathcal{I}_{Z_\beta})$ where, $Z_\beta = \cup_{j \in \beta} Z_j$; hence, $H^0_\Gamma(Y, \oplus_\beta(\mathcal{Q}_\beta \otimes (\wedge^{r_1} V)^* \otimes \mathcal{I}_{Z_\beta}) \neq 0$, so $H^0_\Gamma(Y, (\mathcal{Q}^{r_1} \otimes (\wedge^{r_1} V)^* \otimes \mathcal{I}_{Z_\beta}) \neq 0$ for some β.

Observe $H^0_\Gamma(Y, (\mathcal{Q}_\beta \otimes (\wedge^{r_1} V)^* \otimes \mathcal{I}_{Z_\beta}) \hookrightarrow H^0_\Gamma(Y, (\mathcal{Q}_\beta \otimes (\wedge^{r_1} V)^* \otimes \mathcal{I}_{Z_i})$ for all $i \in \beta$. So $H^0_\Gamma(Y, (\mathcal{Q}_\beta \otimes (\wedge^{r_1} V)^* \otimes \mathcal{I}_{Z_i}) \neq 0$ for some i.

We choose Z_i such that $l(Z_i) \geq 4(r-1)\Theta^2$. Now by lemma 4.6 we have $c_1(\mathcal{Q}_\beta \otimes (\wedge^{r_1} V)^*)\cdot\Theta \geq 2(r-1)\Theta^2$. i.e $r_1 \cdot c_1(\mathcal{Q})\cdot\Theta - c_1(V)\cdot\Theta \geq 2(r-1)\cdot\Theta^2$. $\frac{c_1(V)\cdot\Theta)}{r_1} \leq \frac{(r-1)c_1(\mathcal{Q})\cdot\Theta)}{r} + \frac{c_1(\mathcal{Q})\cdot\Theta)}{r} - \frac{2(r-1)\Theta^2}{r_1}$.

Now by choice $c_1(\mathcal{Q})\cdot\Theta < 4\Theta^2$, since $4\Theta^2 < \frac{2r(r-1)\Theta^2}{r_1}$ for any $r \geq 3$, $r_1 < r$; we have $(c_1(\mathcal{Q})\cdot\Theta)/r < (2(r-1)\Theta^2)/r_1$. Hence $\mu(V) < \mu(E)$. □

Remark 4.8. Let W be the Γ-stable bundle as constructed as above, the invariant direct image $p_*^\Gamma(W)$ is parabolic stable and is of following type:

$$\begin{bmatrix} e^{2\pi i \alpha_1} & & 0 \\ & \cdot & \\ & & \cdot \\ 0 & & e^{2\pi i \alpha_r} \end{bmatrix}$$

i.e $p_*^\Gamma W$ has a full flag quasi–parabolic structure with weights $(\alpha_1, \ldots, \alpha_r)$.

5. Nonemptiness through Monodromy

Let $E \in Vect^D_\Gamma(Y, N)$ be a Γ-stable $SL(2)$-bundle of type τ with a Γ-invariant cycle$c_2 \gg 0$ as second Chern class over Y ([BDP06a]). This corresponds to $(p_*^\Gamma(E))_* \in PVect(X, D, N)$, a degree 0 parabolic stable bundle on X. let C be a general high degree Γ-invariant curve such that $E|_C$ is Γ-stable (see §2 for notations). The quotient $C/\Gamma := T$ is also smooth projective curve in $Y/\Gamma := X$. By choosing C sufficiently general, one can make sure that the action of Γ on C is faithful and we can realize Γ as a quotient Γ_1/Γ_2 where $\Gamma_2 = \pi_1(C)$ and Γ_1 acts properly discontinuously on the simply connected cover $\tilde{C}$ of C and $T = \tilde{C}/\Gamma_1$.

Then by proposition 10 of [Ses69] one knows that there exists an irreducible representation $\rho : \Gamma_1 \longrightarrow SU(2)$ such that $E|_C \simeq \tilde{C} \times_\rho \mathbb{C}^2$.

Let $M(E|_C) = \overline{\mathrm{Im}(\rho)}$ be the Zariski closure of $\mathrm{Im}(\rho)$ in $SL(2)$. It is known that $M(E|_C)$ is reductive. In fact it is the smallest reductive subgroup of $SL(2)$ such that E has a reduction to $M(E|_C)$. Since one knows the set of all reductive subgroups of $SL(2)$, the only possibilities for $M(E|_C)$ are the following ([Ser72])

(1) Finite cyclic groups C_n and the dihedral groups D_n.
(2) The alternating groups A_4 and A_5 and the permutation group S_5.
(3) The whole of $SL(2)$ or its maximal torus.

Of these, since $E|_C$ is stable, we can omit the maximal torus and the cyclic groups C_n. So $M(E|_C)$ can either be the alternating groups, S_5, or the dihedral groups apart from the whole of $SL(2)$.

We wish to estimate the set Z_C of representations of Γ_1 in $SL(2)$ which lie entirely in these families of finite groups upto conjugacy by the diagonal action of $SL(2)$.

So we have an exact sequence,

$$0 \longrightarrow \pi_1(C) \longrightarrow \Gamma_1 \longrightarrow \Gamma \longrightarrow 0$$

Since $\pi_1(C)$ has finitely many generators and Γ is a finite subgroup the set Z_C is at most countable. This implies the following lemma.

Lemma 5.1. *The locus of points in the moduli space of Γ-stable vector bundles $M^{\Gamma,s}_{C,SL(2)}$ of rank 2 and trivial determinant on the curve C whose monodromy is among the set of finite subgroups listed above has countable cardinality.*

Now we consider the rational map given by restriction $r_C : M^{\Gamma,s}_{SL(2)}(c) \cdots \longrightarrow M^{\Gamma,s}_{C,SL(2)}$, where $M^{\Gamma,s}_{SL(2)}(c)$ is the moduli space of Γ-stable $SL(2)$ bundles with second chern class c over Y.

Lemma 5.2. *If the Γ-invariant curve C is chosen sufficiently general and high degree then the map r_C is an immersion on a nonempty open set U of $M^{\Gamma,s}_{SL(2)}(c)$.*

Proof. By Γ-Mehta-Ramanathan restriction theorem (Theorem 7.2 of [BDP06b]) there exists a high degree general curve C in X, an open subset $U \subset M^{\Gamma,s}_{SL(2)}(c)$ and a morphism (restriction) $r_C : U \longrightarrow M^{\Gamma,s}_{C,SL(2)}$. To show r_C is an immersion we need to prove r_C is one–one as set map together with it is one–one at tangent space level.

Let E_1 and E_2 are two points in U. Consider the Γ-sheaf $\mathcal{H}om_\Gamma(E_1, E_2)$ and the long exact sequence

$$\mathrm{Hom}_\Gamma(E_1, E_2(-C)) \longrightarrow \mathrm{Hom}_\Gamma(E_1, E_2) \longrightarrow \mathrm{Hom}_\Gamma(E_1|_C, E_2|_C)$$
$$\longrightarrow H^1_\Gamma(\mathcal{H}om(E_1, E_2)(-C))$$

By Γ-Serre duality $\mathrm{Hom}_\Gamma(E_1, E_2(-C)) = 0$ and by Enrique-Severi $H^1_\Gamma(\mathcal{H}om(E_1, E_2)(-C)) = 0$. So $\dim(\mathrm{Hom}_\Gamma(E_1, E_2)) = \dim(\mathrm{Hom}_\Gamma(E_1|_C, E_2|_C)) = 1$, i.e $E_1 \simeq E_2$. This proves r_C is one–one as set map.

Let E be a Γ-stable bundle such that $E|_C$ is Γ-stable. We have the following long exact sequence

$$\longrightarrow H^1_\Gamma(\mathrm{End}(E)(-C)) \longrightarrow H^1_\Gamma(\mathrm{End}(E))$$
$$\longrightarrow H^1_\Gamma(\mathrm{End}(E|_C)) \longrightarrow H^2_\Gamma(\mathrm{End}(E)(-C))$$

By Enrique–Severi we have $H^1_\Gamma(End(E)(-C)) = 0$, Hence r_C is one–one at tangent space level. We choose a common Γ-invariant curve C which satisfies all three conditions and the lemma holds for this common C. □

Proposition 5.3. *There exists a rank 2, Γ-stable bundle E with $c_2(E) \gg 0$ and trivial determinant on the surface Y such that the restriction $E|_C$ to a high degree Γ-invariant general curve $C \subset Y$ has monodromy subgroup to be the whole of $SL(2)$ itself.*

Proof. For $c = c_2(E) \gg 0$, it is known that the moduli space $M_Y^\Gamma(2, \mathcal{O}, c_2)^s = M_{SL(2)}^\Gamma(c)^s$ is non–empty and has non-zero dimension $d(c)$ [BDP06b]. By lemma 5.2 it is clear that $\mathrm{Im}(r_c)$ is not completely contained in the subset Z_c of Γ-stable bundle with finite monodromy groups. Therefore there exists at least one Γ-stable bundle E on Y such that $M(E|_c) = SL(2)$ □

Definition 5.4. *Let K be a maximal compact subgroup of $SL(W)$, where W is an $n+1$ dimensional vector space. A subset $A \subset K$ is said to be* irreducible *if*

$$\{Y \in \mathfrak{sl}(W) | ad\, x(Y) = Y, \ \ \forall\ x \in A\} = \textit{centre of } \mathfrak{sl}(W).$$

A (homomorphism) representation $\delta : G \longrightarrow K$ is said to be irreducible if the image $\delta(G)$ is irreducible, where G is any group and $\mathfrak{sl}(W)$ is the Lie algebra of $SL(W)$.

Consider the natural action of $Sl(2, \mathbb{C})$ on the complex vector space $\mathbb{C}^2$. It induces a representation on the n'th symmetric power space $\mathrm{Sym}^n(\mathbb{C}^2)$. The dimension of this vector space is $n+1$. Hence there exists a homomorphism

$$\phi_{n+1} : Sl(2, \mathbb{C}) \longrightarrow Sl(n+1, \mathbb{C})$$

One can show that any $n+1$ dimensional irreducible representation of $Sl(2, \mathbb{C})$ is given by such ϕ_{n+1}.

We have the following equivalence ([Ram75, Prop. 2.1]),

(1) ϕ_{n+1} is irreducible.
(2) $\mathrm{Im}(\phi_{n+1}) \nsubseteq P$ $\forall$ parabolic subgroup P of $SL(W)$.

Lemma 5.5. *If E is a Γ-stable bundle on Y such that for a high degree Γ-invariant general curve C in Y $E|_c$ is Γ-stable and $M(E|_c) = SL(2)$ then $E|_C(\phi_{n+1})$ is a Γ-stable $SL(W)$-bundle on C by an extension of structure group via the principal homomorphism ϕ_{n+1}.*

Proof. Let $\rho : \Gamma_1 \longrightarrow SU(2)$ be the irreducible unitary representation such that $E|_c \simeq \tilde{C} \times_\rho \mathbb{C}^2$ (by propn 10 of [Ses69]) and $\mathcal{M}(E|_c) = SL(2)$. To show that $E|_c(\phi_{n+1})$ is Γ-stable $SL(W)$-bundle, we need to show that the representation

$$\eta = (\phi_{n+1} \circ \rho) : \Gamma_1 \longrightarrow K$$

is *irreducible*, where K is a maximal compact subgroup of $SL(W)$. I.e,

$$\mathfrak{z} := \{Y \in \mathfrak{sl}(W) | ad(x)Y = Y, \forall x \in \mathrm{Im}(\phi_{n+1}|_{SU(2)} \circ \rho\} = \{0\}$$

Let $Y \in \mathfrak{z}$, $\{x_n\}$ be an arbitary converging sequence in $\mathrm{Im}(\eta)$ such that $ad(x_n)Y = Y, \forall x_n$. Note that $\mathrm{Im}(\eta)$ is dense in $\mathrm{Im}(\phi_{n+1})$. By continuity of adjoint representaion and the fact that $\{x_n\}$ is arbitary sequence in $\mathrm{Im}(\eta)$, we see that $ad(x)(Y)$=Y $\forall x \in \mathrm{Im}(\phi_{n+1})$, hence $\mathfrak{z} = \{0\}$. □

We can now conclude the following:

Theorem 5.6. *The moduli space* $M_{SL(W)}(c)^s$ *of* Γ*-stable principal* $SL(W)$*-bundles on a smooth projective surface* X *is non–empty for* $c \gg 0$.

Proof. Let E be a $SL(2)$-stable bundle such that $\mathcal{M}(E|_C) = SL(2)$. By 5.5 we know that $E|_C(\phi_{n+1})$ is a Γ-stable $SL(W)$ bundle on C. Note that $E|_C(\phi_{n+1}) = E(\phi_{n+1})|_C$. We claim that $E(\phi_{n+1})$ is Γ-stable $SL(W)$ bundle on X. Suppose not, then there exists a Γ-parabolic reduction (P, σ) on a big Γ-invariant open subset U of X (codim$(X - U) \geq 2$) and a dominant character χ of P, such that the degree of the associated line bundle $E|_U(\chi) \geq 0$. Since C is a general curve and X is a surface C will sit inside U. Hence $\sigma|_C$ is a Γ-reduction to a parabolic P for $E|_C$. Now we note that (from functoriality of Chern classes)

$$\deg(E(\phi_{n+1})|_U(\chi)) \cdot \deg(C) = \deg(E(\phi_{n+1})|_C(\chi))$$

so it will contradict the stability of $E|_C$. The *largeness* of second Chern class of $E(\phi_{n+1})$ follows from *largeness* of second Chern class for E. □

Remark 5.7. The associated parabolic stable bundle is of type full flag and weights are given by monomials $\alpha_1^n, \alpha^{n-1}\alpha_2, \ldots, \alpha_2^n$ where α_1 and α_2's are the weights of the rank2 full flag parabolic stable bundle started with.

Acknowledgement

We are very grateful to Prof. V. Balaji and Prof. D. S. Nagaraj for constant encouragement and many fruiful discussions. We thank R. Parthasarathi for sharing his views and giving many helpfull remarks.

References

[BBN01] V. Balaji, I. Biswas and D. S. Nagaraj, Principal bundles over projective manifolds with parabolic structure over a divisor, *Tohuku Math. J.*, **53(2)** (2001) 193–227.

[BDP06a] V. Balaji, Arijit Dey and R. Parthasarathi, Parabolic bundles on algebraic surfaces 1-the Donaldson–Uhlenbeck compactification, *preprint*, math.AG/0601274 (2006).

[BDP06b] V. Balaji, Arijit Dey and R. Parthasarathi, Parabolic bundles on algebraic surfaces I - the Donaldson–Uhlenbeck compactification, *Preprint* (2006).

[Bis95] Indranil Biswas, Cohomology of parabolic line bundle, *Math. Res. Letter*, **2(6)** (1995) 783–790.

[Bis97a] Indranil Biswas, Chern class of parabolic bundle, *Math. Ann.*, **307** (1997) 511–529.

[Bis97b] Indranil Biswas, Parabolic bundles as orbifold bundles, *Duke Mathematical Journal*, **88(2)** (1997) 305–325.

[Gro57] A. Grothendieck, Sur quelques points d'algèbre homologiques, *Tohoku Math. J.*, **9(2)** (1957) 119–221.

[Kaw81] Y. Kawamata, Characterization of abelian varieties, *Composito Mathematica*, **43(2)** (1981) 253–276.

[KMM87] Y. Kawamata, K. Matsuda and K. Matsuki, Introduction to the minimal model problem in *algebraic geometry, sendai, Adv. Stud. Pure. Math*, **10** (1987) 283–360.

[MY92] M. Maruyama and K. Yokogawa, Moduli of parabolic stable sheaves, *Math. Ann.*, **293(1)** (1992) 77–99.

[Ram75] A. Ramanathan, Stable principal bundles on a compact riemann surface, *Math. Ann.*, **213** (1975) 129–152.

[Ser72] J. P. Serre, Properietes galoisiennes des points d'ordre fini des courbes elliptiques, *Inv. Math.* (1972) pages 259–331.

[Ses69] C. S. Seshadri, Moduli of π-vector bundles over an algebraic curve, *Questions on Algebraic Varieties*, C.I.M.E III Ciclo (1969).

Teichmüller Theory and Moduli Problem No. 10, 2010, pp. 167–178.

Limit Set of Quasiconformal Mapping Class Group on Asymptotic Teichmüller Space

Ege Fujikawa

Department of Mathematics, Chiba University, 1-33 Yayoi-cho, Inage, Chiba, 263-8522, Japan
e-mail: fujikawa@math.s.chiba-u.ac.jp

Abstract. In the infinite dimensional Teichmüller space for a Riemann surface of infinite type, the action of the quasiconformal mapping class group is not discontinuous, in general. In this paper, we first give a survey of our recent research for the discontinuous action. Then we investigate the dynamics of the action of the quasiconformal mapping class group on the asymptotic Teichmüller space by comparing that on the Teichmüller space.

Keywords. Riemann surface of infinite type, Teichmüller space, Teichmüller modular group.

2000 Subject Classification: 30F60, 37F30

1. Introduction

The Teichmüller space $T(R)$ of a Riemann surface R is the set of all Teichmüller equivalence classes of quasiconformal homeomorphisms of R, which has a complex Banach manifold structure. We consider a Riemann surface R of analytically infinite type. Then $T(R)$ is infinite dimensional. The quasiconformal mapping class group $\mathrm{MCG}(R)$ of R is the group of all homotopy equivalence classes of quasiconformal automorphisms of R, and every element of $\mathrm{MCG}(R)$ induces a biholomorphic automorphism of $T(R)$. Then we have the Teichmüller modular group $\mathrm{Mod}(R)$, which is coincident with the group of all biholomorphic automorphisms of $T(R)$. On the basis of the fact that the action of $\mathrm{MCG}(R)$ on $T(R)$ is not discontinuous in general, in a series of our papers [7,8] and [15], we have considered the dynamics of the action of $\mathrm{MCG}(R)$ and investigated several phenomena which appear only when the Teichmüller space is infinite dimensional. In Section 2, we give a survey of our recent research on the discontinuous action of $\mathrm{MCG}(R)$. In particular, we define a stationary subgroup of $\mathrm{MCG}(R)$ as a generalization of the mapping class group of a topologically finite Riemann surface. Then a subgroup of $\mathrm{MCG}(R)$ with stationary property acts on $T(R)$ discontinuously under the bounded geometry condition on a Riemann surface. Moreover, the pure mapping class group and the essentially trivial mapping class group have the stationary property.

The asymptotic Teichmüller space $AT(R)$ is the set of all asymptotic equivalence classes of quasiconformal homeomorphisms of R and this equivalence is defined similarly to Teichmüller equivalence by using asymptotically conformal homeomorphisms. This is a quotient space of $T(R)$ and is also endowed with the complex structure. The quasiconformal mapping class group $\mathrm{MCG}(R)$ also acts biholomorphically on $AT(R)$ and we have the asymptotic Teichmüller modular group $\mathrm{Mod}_A(R)$, which is a subgroup of the biholomorphic automorphism group of $AT(R)$. However, a non-trivial element of $\mathrm{MCG}(R)$ can induce a trivial element of $\mathrm{Mod}_A(R)$. Moreover $\mathrm{MCG}(R)$ can act trivially on $AT(R)$. In Section 3, we review several results on sufficient conditions for non-trivial actions of $\mathrm{MCG}(R)$ on $AT(R)$.

The region of discontinuity on $T(R)$ for a subgroup $G \subset \mathrm{MCG}(R)$ is the largest open subset where G acts discontinuously, and the limit set is the complement of the region of discontinuity. Similarly, we also define the region of discontinuity and the limit set on $AT(R)$. We would like to know a relationship between the limit set on $T(R)$ and that on $AT(R)$. In [9], we have obtained a Riemann surface R such that the limit set on $T(R)$ for $\mathrm{MCG}(R)$ is empty but the limit set on $AT(R)$ for $\mathrm{MCG}(R)$ is not empty. Namely the actions of the quasiconformal mapping class group on $T(R)$ and on $AT(R)$ are quite different. However, we explore the problem that all limit points on $T(R)$ are projected to the limit set on $AT(R)$. In Section 4, we give a partial solution of this problem by using the results in Sections 2 and 3. We already know that a point of the region of discontinuity on $T(R)$ can be projected to the limit set on $AT(R)$. However we also prove that, for a cyclic group generated by a conformal automorphism of infinite order, the region of discontinuity on $AT(R)$ is not empty.

2. Discontinuity of Teichmüller modular group

2.1 *Teichmüller modular group*

Throughout this paper, we assume that a Riemann surface R admits a hyperbolic structure and has non-abelian fundamental group. The *Teichmüller space* $T(R)$ of a Riemann surface R is the set of all equivalence classes $[f]$ of quasiconformal homeomorphisms f of R. Here we say that two quasiconformal homeomorphisms f_1 and f_2 of R are *equivalent* if there exists a conformal homeomorphism $h : f_1(R) \to f_2(R)$ such that $f_2^{-1} \circ h \circ f_1$ is homotopic to the identity. All homotopies are considered to be relative to the ideal boundary at infinity. A distance between two points $[f_1]$ and $[f_2]$ in $T(R)$ is defined by $d([f_1], [f_2]) = (1/2) \log K(f)$, where f is an extremal quasiconformal homeomorphism in the sense that its maximal dilatation $K(f)$ is minimal in the homotopy class of $f_2 \circ f_1^{-1}$. Then d is a complete distance on $T(R)$ which is called the Teichmüller distance.

A *quasiconformal mapping class* is the homotopy equivalence class $[g]$ of quasiconformal automorphisms g of a Riemann surface, and the *quasiconformal mapping class group* $\mathrm{MCG}(R)$ of R is the group of all quasiconformal mapping classes on R. If R is of topologically infinite type, then $\mathrm{MCG}(R)$ is, in general, not a countable group. Every element $[g] \in \mathrm{MCG}(R)$ induces a biholomorphic automorphism $[g]_*$ of $T(R)$ by $[f] \mapsto [f \circ g^{-1}]$, which is also isometric with respect to the Teichmüller distance. Let $\mathrm{Aut}(T(R))$ be the group of all biholomorphic automorphisms of $T(R)$. Then we have a homomorphism

$$\iota : \mathrm{MCG}(R) \to \mathrm{Aut}(T(R))$$

given by $[g] \mapsto [g]_*$, and we define the *Teichmüller modular group* of R by

$$\mathrm{Mod}(R) = \iota(\mathrm{MCG}(R)).$$

It is proved in [2] that the homomorphism ι is injective (faithful) for all Riemann surfaces R of non-exceptional type. See also [6] and [21] for other proofs. Here we say that a Riemann surface R is of *exceptional type* if R has finite hyperbolic area and satisfies $2g + n \leq 4$, where g is the genus of R and n is the number of punctures of R. The homomorphism ι is also surjective for every Riemann surface R of non-exceptional type, namely $\mathrm{Mod}(R) = \mathrm{Aut}(T(R))$. The proof is a combination of the results of [1] and [20]. See [10] for a survey of the proof.

2.2 *Pure mapping class group*

Definition 2.1. *We say that a subgroup $G \subset \mathrm{MCG}(R)$ acts at a point $p \in T(R)$* discontinuously *if the following equivalent conditions are satisfied:*

(a) *there exists a neighborhood U of p such that the number of elements $[g]_* \in \iota(G)$ satisfying $[g]_*(U) \cap U \neq \emptyset$ is finite.*

(b) *there exist no distinct elements $[g_n]_* \in \iota(G)$ such that $d([g_n]_*(p), p) \to 0$ as $n \to \infty$.*

(c) *the orbit $\iota(G)(p)$ is discrete and the stabilizer subgroup $\mathrm{Stab}_{\iota(G)}(p)$ is finite.*

Set

$$\Omega(G) = \{p \in T(R) | G \text{ acts at } p \text{ discontinuously}\}.$$

We call $\Omega(G)$ the *region of discontinuity* of G. By definition, the region of discontinuity is an open subset on $T(R)$.

For a Riemann surface R of analytically finite type, the quasiconformal mapping class group $\mathrm{MCG}(R)$ acts on $T(R)$ discontinuously, namely $\Omega(\mathrm{MCG}(R)) = T(R)$ (see [16, Section 8]). However, for a Riemann surface R of analytically infinite type, the action of $\mathrm{MCG}(R)$ is, in general, not discontinuous.

On the basis of this fact, Gardiner and Lakic [18] considered a special planar Riemann surface, the complement $\hat{\mathbb{C}} - \mathrm{C}$ of the standard middle-thirds Cantor set C in the unit interval as a subset of the complex sphere $\hat{\mathbb{C}}$. They defined the pure mapping class group $P(\hat{\mathbb{C}} - \mathrm{C})$ as the group of all quasiconformal mapping classes $[g] \in \mathrm{MCG}(\hat{\mathbb{C}} - \mathrm{C})$ such that g fixes all points of C, and proved the following.

Proposition 2.2. *The pure mapping class group $P(\hat{\mathbb{C}} - \mathrm{C})$ acts on the Teichmüller space $T(\hat{\mathbb{C}} - \mathrm{C})$ discontinuously.*

We extend Proposition 2.2 for general Riemann surfaces. First we define the pure mapping class group for all Riemann surfaces.

Definition 2.3. *The* pure mapping class group $P(R)$ *of a Riemann surface R is the group of all quasiconformal mapping classes* $[g] \in \mathrm{MCG}(R)$ *such that g fixes all non-cuspidal ends of R.*

Then we have the following, which is proved in [11].

Theorem 2.4. *Let R be a Riemann surface satisfying the bounded geometry condition and having more than two non-cuspidal ends. Then the pure mapping class group $P(R)$ acts on the Teichmüller space $T(R)$ discontinuously.*

For a definition of the bounded geometry condition, see the next section. An outline of a proof of Theorem 2.4 is given in Section 2.4.

2.3 *Bounded geometry*

Definition 2.5. *We say that a Riemann surface R satisfies the* bounded geometry condition *if it satisfies the following three conditions:*

(i) lower bound condition: *the injectivity radius at any point of R except cusp neighborhoods is uniformly bounded away from zero.*
(ii) upper bound condition: *there exists a subdomain R^* of R such that the injectivity radius at any point of R^* is uniformly bounded from above and that the simple closed curves in R^* carry the fundamental group of R.*
(iii) *R has no ideal boundary at infinity, namely the Fuchsian model of R is of the first kind.*

Moreover we say that a Riemann surface R satisfies the strong bounded geometry condition *if R satisfies conditions* (i), (iii) *and the following condition* (ii$'$).

(ii$'$) strong upper bound condition: *the injectivity radius at any point of R is uniformly bounded from above.*

The (strong) bounded geometry condition is invariant under quasiconformal deformations, and every non-universal normal cover of a Riemann surface of analytically finite type satisfies the bounded geometry condition (see [7, Proposition 3]). More generally, if a Riemann surface R has a uniform pants decomposition, then R satisfies the bounded geometry condition. Here we say that a Riemann surface R has a *uniform pants decomposition* if it is decomposed into pairs of pants with geodesic boundary possibly degenerating to a puncture such that the hyperbolic lengths of boundary geodesics of all pairs of pants in the decomposition are uniformly bounded from above and below. In particular, the complement of the Cantor set satisfies the bounded geometry condition. However we see that having a uniform pants decomposition does not imply the bounded geometry condition as the following proposition says.

Proposition 2.6. *There exists a Riemann surface R such that it satisfies the bounded geometry condition but it does not have a uniform pants decomposition.*

Proof. Let $\check{R}$ be a compact Riemann surface of genus 2, and $\dot{R} = \check{R} - \{\check{x}\}$ for a point $\check{x} \in \check{R}$. Let $\check{H}$ be a Fuchsian model of $\check{R}$ acting on the unit disc Δ, $F \subset \Delta$ a fundamental region of $\check{H}$, and $\pi : \Delta \to \check{R}$ the universal covering projection. We take a point $x \in F$ such that $\pi(x) = \check{x}$, and set $\dot{F} = \overline{F} - \{x\}$. Let $R = \bigcup_{h \in \check{H}} h(\dot{F})$, which is a normal covering surface of $\dot{R}$ embedded in Δ. Since $\dot{R}$ is of analytically finite type, we see that R satisfies the bounded geometry condition. On the other hand, R does not have a uniform pants decomposition. Indeed, for every pants decomposition of R, a sequence of boundary geodesics of pairs of pants converges to the boundary $\partial\Delta = \{z \in \mathbf{C} \mid |z| = 1\}$ in the Hausdorff topology. Since the hyperbolic length on R is greater than that on Δ, we conclude that the hyperbolic lengths of the boundary geodesics are not bounded from above. □

2.4 *Stationary subgroup*

In this section, we explain our approach to a proof of Theorem 2.4. More generally, we consider the action of a stationary subgroup of the quasiconformal mapping class group, which is a generalization of the mapping class group of a topologically finite Riemann surface.

Definition 2.7. *A subgroup G of* MCG(R) *is said to be* stationary *if there exists a compact subsurface W of R such that $g(W) \cap W \neq \emptyset$ for every representative g of every element of G.*

It is known that a sequence of normalized quasiconformal homeomorphisms whose maximal dilatations are uniformly bounded is sequentially compact in compact open topology. The stationary property of mapping classes corresponds to the normalization in this context and hence such a sequence of mapping classes also has the compactness property if they are uniformly bounded. By using this observation, we have the following.

Theorem 2.8. *Let R be a Riemann surface satisfying the bounded geometry condition. Then* (i) $\Omega(\mathrm{MCG}(R)) \neq \emptyset$, (ii) $\Omega(G) = T(R)$ *for every stationary subgroup G of* MCG(R).

See [7] and [8] for a proof of Theorem 2.8.

Remark. There exist Riemann surfaces R such that $\emptyset \neq \Omega(\mathrm{MCG}(R)) \subsetneqq T(R)$. A typical example is a non-universal normal covering surface of a Riemann surface of analytically finite type.

Remark. There exist a Riemann surface R and a subgroup G of MCG(R) such that G is non-stationary but $\Omega(G) = T(R)$. See [13, Proposition 3.1]. In the paper, we further constructed a Riemann surface R satisfying the bounded geometry condition such that MCG(R) is non-stationary but $\Omega(\mathrm{MCG}(R)) = T(R)$.

By Theorem 2.8 (ii), the following theorem proved in [11] completes a proof of Theorem 2.4.

Theorem 2.9. *If a Riemann surface R has more than two non-cuspidal ends, then the pure mapping class group $P(R)$ is stationary.*

We have another example of stationary subgroup.

Example 2.10. For a simple closed geodesic c on R, let $G_c(R)$ be the set of all elements $[g] \in \mathrm{MCG}(R)$ such that $g(c)$ is freely homotopic to c. Then $G_c(R)$ is stationary, and thus it acts on $T(R)$ discontinuously if R satisfies the bounded geometry condition. See [8] and [15].

2.5 *Essentially trivial mapping class group*

In this section, we define a subgroup of the pure mapping class group, which is called the essentially trivial mapping class group.

Definition 2.11. *A quasiconformal mapping class* $[g] \in \mathrm{MCG}(R)$ *is said to be* essentially trivial *if there exists a compact subsurface* V_g *of* R *such that, for each connected component* W *of* $R - V_g$ *that is not a cusp neighborhood, the restriction* $g|_W : W \to R$ *is homotopic to the inclusion map* $id|_W : W \hookrightarrow R$*. The* essentially trivial mapping class group $E(R)$ *is the group of all essentially trivial mapping classes.*

It is clear that $E(R)$ is a subgroup of $P(R)$. Thus Theorem 2.4 yields that $E(R)$ acts on $T(R)$ discontinuously if a Riemann surface R has the bounded geometry condition and has more than two non-cuspidal ends. Although Theorem 2.4 is not true for Riemann surfaces the number of whose non-cuspidal ends are at most two, the assumption on the number of ends can be removed for the discontinuous action of $E(R)$ as follows. For a proof, see [11].

Theorem 2.12. *Let* R *be an analytically infinite Riemann surface satisfying the bounded geometry condition. Then* $E(R)$ *acts on* $T(R)$ *discontinuously and freely.*

3. Asymptotic Teichmüller modular group on asymptotic Teichmüller space

3.1 *Asymptotic Teichmüller space*

In this section, we consider the asymptotic Teichmüller space of a Riemann surface R, which was introduced in [19] when R is the upper half-plane and in [2], [3] and [17] when R is an arbitrary hyperbolic Riemann surface. We say that a quasiconformal homeomorphism f on R is *asymptotically conformal* if for every $\epsilon > 0$, there exists a compact subset V of R such that the maximal dilatation $K(f|_{R-V})$ of the restriction of f to $R - V$ is less than $1 + \epsilon$. We say that two quasiconformal homeomorphisms f_1 and f_2 on R are *asymptotically equivalent* if there exists an asymptotically conformal homeomorphism $h : f_1(R) \to f_2(R)$ such that $f_2^{-1} \circ h \circ f_1$ is homotopic to the identity by a homotopy that keeps every point of the ideal boundary at infinity fixed throughout. The *asymptotic Teichmüller space* $AT(R)$ of a Riemann surface R is the set of all asymptotic equivalence classes $[[f]]$ of quasiconformal homeomorphisms f on R. The asymptotic Teichmüller space $AT(R)$ is of interest only when R is analytically infinite. Otherwise $AT(R)$ is trivial, that is, it consists of just one point. Conversely, if R is analytically infinite, then $AT(R)$ is not trivial. Since a conformal homeomorphism is asymptotically conformal, there is a natural projection

$\pi : T(R) \to AT(R)$ that maps each Teichmüller equivalence class $[f] \in T(R)$ to the asymptotic Teichmüller equivalence class $[[f]] \in AT(R)$. The asymptotic Teichmüller space $AT(R)$ has a complex manifold structure such that π is holomorphic. See also [4] and [5].

For a quasiconformal homeomorphism f of R, the *boundary dilatation* of f is defined by $H^*(f) = \inf K(f|_{R-V})$, where the infimum is taken over all compact subsets V of R. Furthermore, for a Teichmüller equivalence class $[f] \in T(R)$, the *boundary dilatation* of $[f]$ is defined by $H([f]) = \inf H^*(g)$, where the infimum is taken over all elements $g \in [f]$. A distance between two points $[[f_1]]$ and $[[f_2]]$ in $AT(R)$ is defined by $d_A([[f_1]], [[f_2]]) = (1/2) \log H([f_2 \circ f_1^{-1}])$, where $[f_2 \circ f_1^{-1}]$ is a Teichmüller equivalence class of $f_2 \circ f_1^{-1}$ in $T(f_1(R))$. Then d_A is a complete distance on $AT(R)$, which is called the asymptotic Teichmüller distance. For every point $[[f]] \in AT(R)$, there exists an asymptotically extremal element $f_0 \in [[f]]$ in the sense that $H([f]) = H^*(f_0)$.

3.2 *Non-trivial action of mapping class group on $AT(R)$*

Every element $[g] \in \mathrm{MCG}(R)$ induces a biholomorphic automorphism $[g]_{**}$ of $AT(R)$ by $[[f]] \mapsto [[f \circ g^{-1}]]$, which is also isometric with respect to d_A. See [4]. Let $\mathrm{Aut}(AT(R))$ be the group of all biholomorphic automorphisms of $AT(R)$. Then we have a homomorphism

$$\iota_A : \mathrm{MCG}(R) \to \mathrm{Aut}(AT(R))$$

given by $[g] \mapsto [g]_{**}$, and we define the *asymptotic Teichmüller modular group* of R (the geometric automorphism group of $AT(R)$) by

$$\mathrm{Mod}_A(R) = \iota_A(\mathrm{MCG}(R)).$$

It is different from the case of $\iota : \mathrm{MCG}(R) \to \mathrm{Aut}(T(R))$ that the homomorphism ι_A is not injective, namely $\mathrm{Ker}\,\iota_A \neq \{[id]\}$, unless R is either the unit disc or a once-punctured disc. Moreover there exists a Riemann surface R such that $\mathrm{MCG}(R) = \mathrm{Ker}\,\iota_A$ as follows.

Example 3.1. In [23], a Riemann surface R of analytically infinite type is constructed so that $\mathrm{MCG}(R) = \mathrm{Ker}\,\iota_A$. Note that this Riemann surface R does not satisfy the bounded geometry condition and $\mathrm{MCG}(R)$ consists of only countable number of elements.

On the basis of this fact, we give a sufficient condition for non-trivial action of $\mathrm{MCG}(R)$ in [9].

Proposition 3.2. *Let R be a Riemann surface of topologically infinite type. Suppose that R satisfies the upper bound condition. Then* $\mathrm{Ker}\,\iota_A \subsetneqq \mathrm{MCG}(R)$.

For a proof of Proposition 3.2, we have shown that there exists a quasiconformal automorphism of R that is not homotopic to any asymptotically conformal automorphism of R if R satisfies the upper bound condition. Then the base point $[[id]]$ of

$AT(R)$ is not a common fixed point of $\mathrm{Mod}_A(R)$ and we have the assertion. Since the upper bound condition is invariant under quasiconformal deformations, we can apply the same argument for all points $[[f]] \in AT(R)$ to prove that there exists a quasiconformal automorphism of $f(R)$ that is not homotopic to any asymptotically conformal automorphism of $f(R)$ and then $[[f]]$ is not a common fixed point of $\mathrm{Mod}_A(R)$. Thus we have the following extension of Proposition 3.2.

Theorem 3.3. *If a Riemann surface R has the upper bound condition, then* $\mathrm{Mod}_A(R)$ *does not have a common fixed point in $AT(R)$.*

3.3 *Elements of* $\mathrm{Ker}\,\iota_A$

We characterize the subgroup $\mathrm{Ker}\,\iota_A$. In [11], we prove the following.

Theorem 3.4. *The inclusion relation $E(R) \subset \mathrm{Ker}\,\iota_A \subset P(R)$ holds.*

On the other hand, the following theorem gives a condition for a quasiconformal mapping class which does not belong to $\mathrm{Ker}\,\iota_A$.

Theorem 3.5. *Let g be a quasiconformal automorphism of a Riemann surface R. Suppose there exists a constant $\delta > 1$ such that, for every compact subset E of R, there is a simple closed geodesic c on R outside of E satisfying either*

$$\frac{\ell(g(c))}{\ell(c)} \leq \frac{1}{\delta} \qquad \text{or} \qquad \frac{\ell(g(c))}{\ell(c)} \geq \delta.$$

Then g is not homotopic to any asymptotically conformal automorphism of R. In particular, $[g] \notin \mathrm{Ker}\,\iota_A$.

A proof was given in [9]. As an application of Theorem 3.5, we also proved the following.

Proposition 3.6. *Let R be a normal cover of a compact Riemann surface whose covering transformation group is a cyclic group $\langle g \rangle$ generated by a conformal automorphism g of R of infinite order. Then* $[g] \notin \mathrm{Ker}\,\iota_A$.

Furthermore, Proposition 3.6 is extended as follows in [14] and [24].

Theorem 3.7. *Let R be a Riemann surface and let g be a conformal automorphism of R of infinite order. Then* $[g] \notin \mathrm{Ker}\,\iota_A$.

4. Limit set and region of discontinuity of asymptotic Teichmüller modular group

For a subgroup $G \subset \mathrm{MCG}(R)$ and a point $p \in T(R)$, it is said that $q \in T(R)$ is a *limit point* of p for G if there exists a sequence $\{[g_n]_*\}_{n=1}^{\infty}$ of distinct elements of $\iota(G) \subset \mathrm{Mod}(R)$ such that $d([g_n]_*(p), q) \to 0$ as $n \to \infty$. The set of all limit points of p for G is denoted by $\Lambda(G, p)$, and the *limit set* for G is defined by $\Lambda(G) = \bigcup_{p \in T(R)} \Lambda(G, p)$. It is said that $p \in T(R)$ is a *recurrent point* for G if

$p \in \Lambda(G, p)$, and the set of all recurrent points for G is called the *recurrent set* for G and is denoted by $\mathrm{Rec}(G)$. It is evident from the definition that $\mathrm{Rec}(G) \subset \Lambda(G)$ and these sets are G-invariant. Moreover, we proved in [12, Proposition 2.2] that $\Lambda(G) = \mathrm{Rec}(G)$, which is a closed set. Then the complement $T(R) - \Omega(G)$ of the region of discontinuity $\Omega(G)$ is coincident with $\Lambda(G)$.

For a subgroup $G \subset \mathrm{MCG}(R)$, the region of discontinuity $\Omega_A(G)$ on the asymptotic Teichmüller space $AT(R)$ is defined as the set of all points $\hat{p} \in AT(R)$ where G acts discontinuously (cf. Definition 2.1), and the limit set $\Lambda_A(G)$ on $AT(R)$ is defined in a similar way as in the previous paragraph. Then the complement $AT(R) - \Omega_A(G)$ of the region of discontinuity $\Omega_A(G)$ is coincident with $\Lambda_A(G)$.

In this section, we investigate a relationship between the limit set on $T(R)$ and that on $AT(R)$. In [9], we have proved the following proposition, which implies that a point in the region of discontinuity on $T(R)$ can be mapped into the limit set on $AT(R)$ by the projection $\pi : T(R) \to AT(R)$.

Proposition 4.1. *There exists a Riemann surface R satisfying the bounded geometry condition such that $\Omega(\mathrm{MCG}(R)) = T(R)$ and $\Omega_A(\mathrm{MCG}(R)) \subsetneq AT(R)$.*

On the other hand, if a Riemann surface has a certain property, then the dynamics of the actions of the quasiconformal mapping class group on $T(R)$ and on $AT(R)$ are the same. The following theorem is also proved in [9].

Proposition 4.2. *Let R be a Riemann surface that does not satisfy the lower bound condition. Then $\Lambda(\mathrm{MCG}(R)) = T(R)$ and $\Lambda_A(\mathrm{MCG}(R)) = AT(R)$.*

We consider the problem whether $\pi(\Lambda(\mathrm{MCG}(R))) \subset \Lambda_A(\mathrm{MCG}(R))$ for all Riemann surfaces R. Proposition 4.2 is a special case of this problem. The following statement gives a partial solution of this problem.

Theorem 4.3. *Let g be a conformal automorphism of infinite order of a Riemann surface R. For the cyclic group $G = \langle [g] \rangle$ generated by $[g] \in \mathrm{MCG}(R)$, we have $\pi(\Lambda(G)) \subset \Lambda_A(G)$.*

Proof. We take a limit point $p \in \Lambda(G)$ arbitrarily. Then there exists a sequence $\{n_i\}$ of integers such that $d([g^{n_i}]_*(p), p) \to 0$ as $i \to \infty$. Then $d_A([g^{n_i}]_{**}(\hat{p}), \hat{p}) \to 0$ as $i \to \infty$ for the projection $\hat{p} = \pi(p)$. Since g is a conformal automorphism of infinite order, the element $[g]_{**} \in \mathrm{Mod}_A(R)$ is of infinite order. Indeed, if the order of $[g]_{**}$ is n, then $g^n \in \mathrm{Ker}\, \iota_A$. Since g is of infinite order, so is g^n. This contradicts Theorem 3.7. Hence we conclude that $\hat{p} \in \Lambda_A(G)$. □

We have another partial solution of the above problem.

Theorem 4.4. *Let R be a Riemann surface that has more than two non-cuspidal ends and satisfies the bounded geometry condition. Then the inclusion $\pi(\Lambda(\mathrm{MCG}(R))) \subset \Lambda_A(\mathrm{MCG}(R))$ holds.*

We will prove Theorem 4.4. First we note the following.

Lemma 4.5. *Let R be a Riemann surface that satisfies the bounded geometry condition and has more than two non-cuspidal ends. Then $\mathrm{Ker}\, \iota_A$ acts on $T(R)$ discontinuously.*

Proof. By Theorem 3.4, we have $\operatorname{Ker}\iota_A \subset P(R)$. Hence the assertion follows from Theorem 2.4. □

Proof of Theorem 4.4. We take a limit point $p \in \Lambda(\mathrm{MCG}(R))$ arbitrarily. Then there exists a sequence $[g_n]$ of distinct elements of $\mathrm{MCG}(R)$ such that $d([g_n]_*(p), p) \to 0$ as $n \to \infty$. Then $d_A([g_n]_{**}(\hat{p}), \hat{p}) \to 0$ for the projection $\hat{p} = \pi(p)$. We will show that $\{[g_n]_{**}\}_{n\in\mathbb{Z}} \subset \mathrm{Mod}_A(R)$ contains infinitely many elements. Then we conclude that $\hat{p} \in \Lambda_A(\mathrm{MCG}(R))$. Suppose to the contrary that $\{[g_n]_{**}\}_{n\in\mathbb{Z}}$ is a finite set $\{[g_1]_{**}, \ldots, [g_k]_{**}\}$ for some $k \geq 1$. Then there exists an integer i $(1 \leq i \leq k)$, say 1, such that $[g_n]_{**} = [g_1]_{**}$ for infinitely many n. Let $\gamma_n := g_n \circ g_1^{-1}$. Then $[\gamma_n] \in \operatorname{Ker}\iota_A$ and $d([\gamma_n]_*[g_1]_*(p)), p) = d([g_n]_*(p), p) \to 0$. This means that the point $p \in T(R)$ is a limit point for the subgroup $\operatorname{Ker}\iota_A$. This contradicts Lemma 4.5. Thus $\{[g_n]_{**}\}_{n\in\mathbb{Z}}$ contains infinitely many elements. □

In general, each inclusion in the relation in Theorem 3.4 is proper. However, in the forthcoming paper, we will prove that $E(R) = \operatorname{Ker}\iota_A$ if R satisfies the strong bounded geometry condition. Then Theorem 4.4 is true for all Riemann surfaces R satisfying the strong bounded geometry condition. Indeed, by Theorem 2.12, the essentially trivial mapping class group $E(R)$ acts on $T(R)$ discontinuously under the bounded geometry condition. Then so does $\operatorname{Ker}\iota_A$ if R satisfies the strong bounded geometry condition. By using this fact instead of Lemma 4.5 in the proof of Theorem 4.4, we have the assertion.

As we have seen in Proposition 4.1, the projection of a point in the region of discontinuity on $T(R)$ can be a limit point on $AT(R)$. However, under a certain condition, the region of discontinuity on $AT(R)$ is not empty as the following theorem says (cf. Theorem 2.8).

Theorem 4.6. *Let R be a Riemann surface that admits a conformal automorphism g of infinite order. Then $\Omega_A(\langle[g]\rangle) \neq \emptyset$.*

Remark. Under the assumption in Theorem 4.6, let $\hat{p}$ be a point in $\Omega_A(\langle[g]\rangle)$. Then by Theorem 4.3, all the preimages of $\hat{p}$ to $T(R)$ under π belong to $\Omega(\langle[g]\rangle)$.

We will prove Theorem 4.6. In the proof, we use the following estimate which is proved in [9]. Let $\ell(c)$ be the geodesic length for the free homotopy class of a simple closed curve c on a Riemann surface.

Lemma 4.7. *Let R be a Riemann surface and let c be a simple closed geodesic on R. Let E be a subset on R and $d = d_R(c, E)$ be the hyperbolic distance between c and E. If g is a K-quasiconformal homeomorphism of R onto another Riemann surface such that the restriction of g to $R - E$ is H-quasiconformal $(1 \leq H \leq K)$, then the inequality*

$$(1/\alpha)\cdot\ell(c) \leq \ell(g(c)) \leq \alpha\cdot\ell(c)$$

holds for a constant

$$\alpha = \alpha(K, H, d) = K - (K - H)\frac{2\arctan(\sinh d)}{\pi}.$$

We also use the following fundamental fact for a conformal automorphism of infinite order.

Lemma 4.8. *Let R be a Riemann surface that admits a conformal automorphism g of infinite order. Then there exist a simple closed geodesic c on R and a positive integer k such that the images of c under the cyclic group $\langle g^k \rangle$ generated by g^k are mutually disjoint.*

Proof. Take a simple closed geodesic c_0 on R arbitrarily. Since the group of all conformal automorphisms of R acts on R properly discontinuously, only finitely many of $\{g^n(c_0)\}_{n\in\mathbb{N}}$ have non-empty intersection with c_0. Let k be a positive integer that is greater than any power n of g^n satisfying $g^n(c_0) \cap c_0 \neq \emptyset$. Then the geodesics $\{g^{kn}(c_0)\}_{n\in\mathbb{N}}$ are mutually disjoint. Thus the simple closed geodesic $c := g^k(c_0)$ is a desired one. □

We begin to prove Theorem 4.6.

Proof of Theorem 4.6. We will deform the Riemann surface R by a quasiconformal homeomorphism and construct a point $\hat{p} \in AT(R)$ as follows. By Lemma 4.8, we can take a simple closed geodesic c on R such that the images of c under $\langle g^k \rangle$ are mutually disjoint. We rename γ for g^k and set $c_n = \gamma^n(c)$. Then c_n have the same hyperbolic length ℓ and the hyperbolic distance $d_R(x, c_n)$ between any point $x \in R$ and c_n tends to ∞ as $n \to \infty$. See the proof of [22, Proposition 1]. Consider a subsequence $\{c_{2^m}\}_{m\in\mathbb{N}}$. We can take a quasiconformal homeomorphism f of R and a positive constant ℓ' $(< \ell)$ such that $\ell(f(c_{2^m})) \leq \ell'$ for all m and that $\ell(f(c_n)) = \ell$ for $n \neq 2^m$. Let c'_n be the geodesic that is freely homotopic to $f(c_n)$.

We will show that the point $\hat{p} := [[f]] \in AT(R)$ belongs to $\Omega_A(\langle [g] \rangle)$. We have only to prove that $\hat{p} \in \Omega_A(\langle [\gamma] \rangle)$. Set $\gamma_p = f \circ \gamma \circ f^{-1}$. Suppose to the contrary that $\hat{p} \in \Lambda_A(\langle [\gamma] \rangle)$. Then there exists a sequence $\{n_i\}$ of integers such that $d_A([\gamma^{n_i}]_*(\hat{p}), \hat{p}) \to 0$ as $i \to \infty$. We take a constant H_0 such that $1 < H_0 < \ell/\ell'$ and may assume that the boundary dilatation of $\gamma_p^{n_i}$ satisfies $H(\gamma_p^{n_i}) < H_0$ for all sufficiently large i. Then for each i, there exists a compact subset E_i on $f(R)$ such that $K(\gamma_p^{n_i}|_{f(R)-E_i}) < H_0$. Let $\alpha = \alpha(K, H, d)$ be the constant in Lemma 4.7, which tends to H as $d \to \infty$. For each i, take a positive constant d_i such that $\alpha_i := \alpha(K(\gamma_p^{n_i}), H_0, d_i) < \ell/\ell'$. Then we take an integer m_i such that $d(E_i, c'_{2^{m_i}}) \geq d_i$ and that $2^{m_i} + n_i$ is not a power of 2. Then we have the inequality

$$\frac{\ell(\gamma_p^{n_i}(c'_{2^{m_i}}))}{\ell(c'_{2^{m_i}})} = \frac{\ell(c'_{2^{m_i}+n_i})}{\ell(c'_{2^{m_i}})} \geq \frac{\ell}{\ell'} > \alpha_i,$$

which contradicts Lemma 4.7. □

References

[1] C. J. Earle and F. P. Gardiner, Geometric isomorphisms between infinite dimensional Teichmüller spaces, *Trans. Amer. Math. Soc.*, **348** (1996) 1163–1190.

[2] C. J. Earle, F. P. Gardiner and N. Lakic, Teichmüller spaces with asymptotic conformal equivalence, I.H.E.S. (1995), preprint.

[3] C. J. Earle, F. P. Gardiner and N. Lakic, Asymptotic Teichmüller space, Part I: The complex structure, *Contemporary Math.*, **256** (2000) 17–38.

[4] C. J. Earle, F. P. Gardiner and N. Lakic, Asymptotic Teichmüller space, Part II: The metric structure, *Contemporary Math.*, **355** (2004) 187–219.

[5] C. J. Earle, V. Markovic and D. Saric, Barycentric extension and the Bers embedding for asymptotic Teichmüller space, *Contemporary Math.*, **311** (2002) 87–105.

[6] A. L. Epstein, Effectiveness of Teichmüller modular groups, In the tradition of Ahlfors and Bers, *Contemporary Math.*, **256** (2000) 69–74.

[7] E. Fujikawa, Limit sets and regions of discontinuity of Teichmüller modular groups, *Proc. Amer. Math. Soc.*, **132** (2004) 117–126.

[8] E. Fujikawa, Modular groups acting on infinite dimensional Teichmüller spaces, *Contemporary Math.*, **355** (2004) 239–253.

[9] E. Fujikawa, The action of geometric automorphisms of asymptotic Teichmüller spaces, *Michigan Math. J.*, **54** (2006) 269–282.

[10] E. Fujikawa, Another approach to the automorphism theorem for Teichmüller spaces, In the tradition of Ahlfors-Bers, IV, *Contemporary Math.*, **432** (2007) 39–44.

[11] E. Fujikawa, Pure mapping class group acting on Teichmüller space, *Conform. Geom. Dyn.*, **12** (2008) 227–239.

[12] E. Fujikawa and K. Matsuzaki, Recurrent and periodic points for isometries of L^∞ spaces, *Indiana Univ. Math. J.*, **55** (2006) 975–997.

[13] E. Fujikawa and K. Matsuzaki, Non-stationary and discontinuous quasiconformal mapping class groups, *Osaka J. Math.*, **44** (2007) 173–185.

[14] E. Fujikawa and K. Matsuzaki, Elliptic quasiconformal mapping classes acting on asymptotic Teichmüller spaces, Proceedings of the 15th International Conference on Finite or Infinite Dimensional Complex Analysis and Applications, Osaka Municipal Universities Press (2008) 169–173.

[15] E. Fujikawa, H. Shiga and M. Taniguchi, On the action of the mapping class group for Riemann surfaces of infinite type, *J. Math. Soc. Japan*, **56** (2004) 1069–1086.

[16] F. P. Gardiner, Teichmüller theory and quadratic differentials, Wiley-Interscience, New York (1987).

[17] F. P. Gardiner and N. Lakic, Quasiconformal Teichmüller Theory, Mathematical Surveys and Monographs, *Amer. Math. Soc.*, **76** (2000).

[18] F. P. Gardiner and N. Lakic, A vector field approach to mapping class actions, *Proc. London Math. Soc.*, **92** (2006) 403–427.

[19] F. P. Gardiner and D. P. Sullivan, Symmetric structure on a closed curve, *Amer. J. Math.*, **114** (1992) 683–736.

[20] V. Markovic, Biholomorphic maps between Teichmüller spaces, *Duke Math. J.*, **120** (2003) 405–431.

[21] K. Matsuzaki, Inclusion relations between the Bers embeddings of Teichmüller spaces, *Israel J. Math.*, **140** (2004) 113–124.

[22] K. Matsuzaki, A countable Teichmüller modular group, *Trans. Amer. Math. Soc.*, **357** (2005) 3119–3131.

[23] K. Matsuzaki, A quasiconformal mapping class group acting trivially on the asymptotic Teichmüller space, *Proc. Amer. Math. Soc.*, **135** (2007) 2573–2579.

[24] K. Matsuzaki, The action of elliptic modular transformations on asymptotic Teichmüller spaces, in this volume.

Teichmüller Theory and Moduli Problem No. 10, 2010, pp. 179–194.

Finite Earthquakes and the Associahedron

Frederick P. Gardiner and Jun Hu

Department of Mathematics, Brooklyn College, Brooklyn, NY 11210
e-mail: frederick.gardiner@gmail.com; jun@sci.brooklyn.cuny.edu

Abstract. Let K_{n+2} be the n-dimensional associahedron and T_n be the real analytic Teichmüller space of a set of $n+3$ variable labelled points cyclically arranged on the unit circle. By using earthquakes with finitely many leaves, we show there is a homeomorphism Φ between T_n and interior of K_{n+2}. Furthermore, by letting certain earthquake parameters approach ∞, we obtain a compactification $\bar{T}_n$ of T_n for which Φ naturally extends to a homeomorphism between $\bar{T}_n$ and K_{n+2}. The faces of $\bar{T}_n$ are realized as products of lower dimensional associahedra.

Introduction

The associahedron K_{n+2} is a complex of dimension n with a combinatorial structure determined by the different associations of $n+2$ letters. It consists of a large number of cubes of dimension n sewn together in a certain way. The exact number is the number of all of the different ways of associating a multiplication on $n+2$ ordered factors. Each cube is labelled by the maximal associations of the $n+2$ letters and the faces of the cubes are labelled by the partial associations. The cubes are sewn together along faces according to their labels. The associahedrons on three, four and five letters, K_3, K_4, and K_5, can be easily constructed as follows.

In the case $n = 1$ we have three factors a, b and c with two associations $a(bc)$ and $(ab)c$. K_3 is a union of two closed line segments labelled $a(bc)$ and $(ab)c$ which are joined together at a central point which is labelled abc. Clearly, K_3 can be embedded in $\mathbb{R}$. It is a one-dimensional complex.

In the case $n = 2$ we have 4 factors a, b, c and d with five associations,

$$((ab)c)d,\ (ab)(cd),\ a((bc)d),\ (a(bc))d,\ a(b(cd)).$$

The maximal and partial associations of $abcd$ are arranged in a partially ordered set in Figure 1. We view the diagram as a picture of the ordering and regard the association $abcd$ with no parentheses as its smallest element. Any association that can be obtained from another by adding matching pairs of parentheses is considered larger.

K_4 is a pentagon; its interior is labelled $abcd$, its five sides are labelled with the five partial associations of four letters, and its five vertices are labelled with the five maximal associations. Each vertex is adjacent to a side if the label on the vertex can be obtained by adding one pair of parentheses to the label on the adjoining side. We see that each edge of K_4 is a one-dimensional associahedron, the edges form a simple

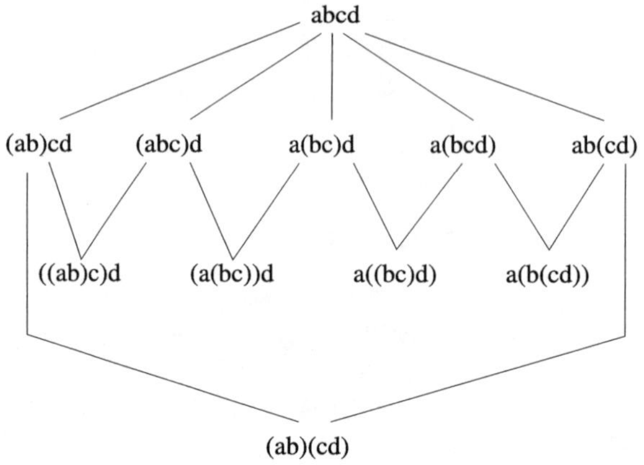

Figure 1. Partial ordering for the partial associations for K_4.

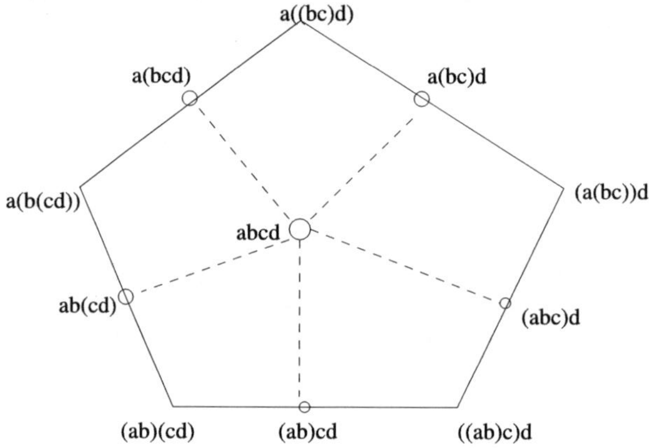

Figure 2. The associahedron K_4.

closed curve in $\mathbb{R}^2$ and K_4 is represented by the 2-dimensional complex shown in Figure 2.

For K_5 we have 5 factors a, b, c, d and e with 14 maximal associations:

$((ab)(cd))e,$	$a((bc)(de)),$		two stars
$a((b(cd))e),$	$((a(bc))d)e,$	$(ab)(c(de)),$	three left accordions
$a(b((cd)e)),$	$(a((bc)d))e,$	$a(b((cd)e)),$	three right accordions
$(((ab)c)d)e,$	$a(((bc)d)e),$	$(ab)((cd)e),$	six fans
$(a(bc))(de),$	$a(b(c(de))),$	$(a(b(cd)))e.$	

The explanation for why these associations are called stars, accordions and fans is given in the fourth paragraph of section 2 and in the caption to Figure 4. K_5 is represented by the three dimensional convex polyhedron shown in Figure 3. Its vertices are labelled by the fourteen maximal associations on five letters a, b, c, d and e listed above.

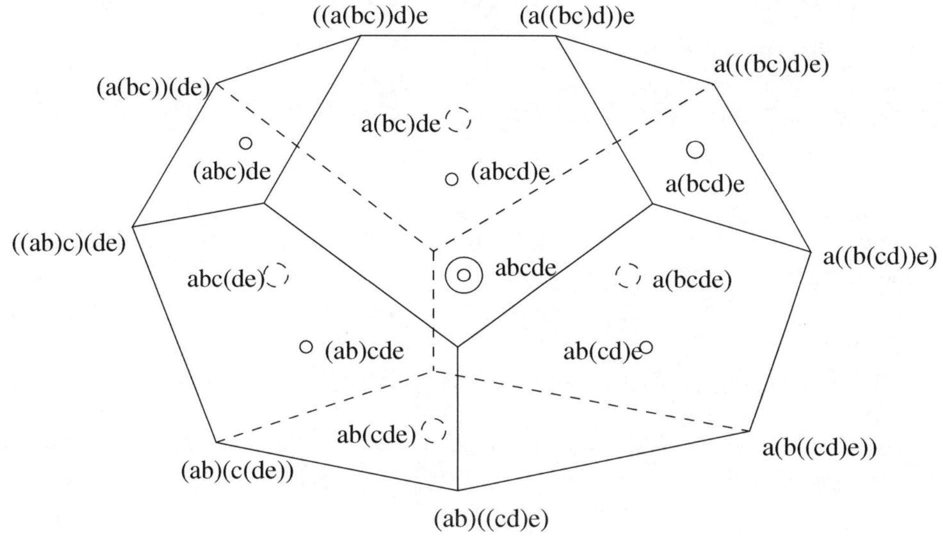

Figure 3. The associahedron K_5.

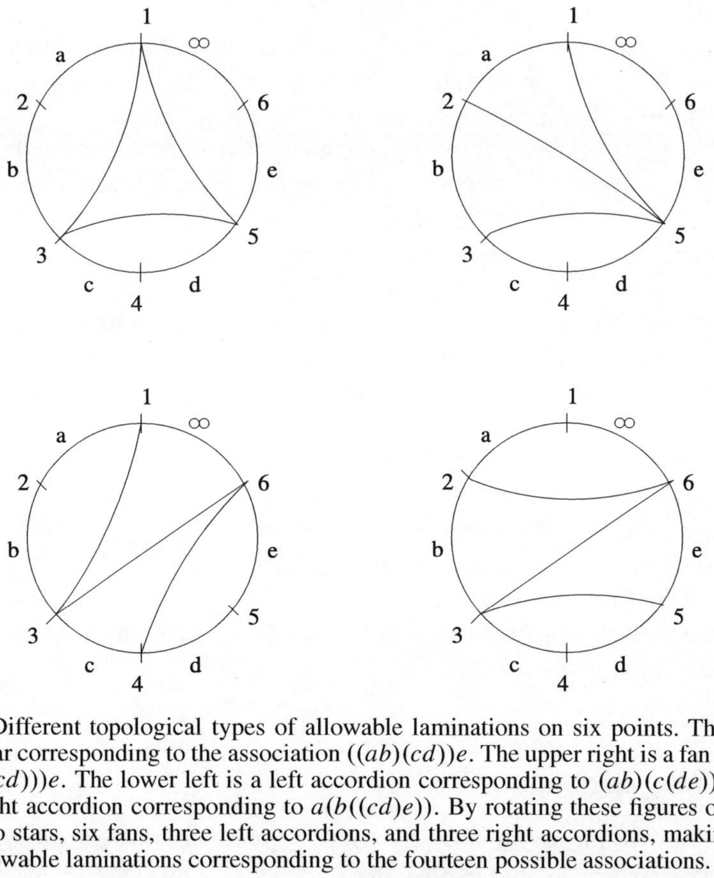

Figure 4. Different topological types of allowable laminations on six points. The upper left figure is a star corresponding to the association $((ab)(cd))e$. The upper right is a fan corresponding to $(a(b(cd)))e$. The lower left is a left accordion corresponding to $(ab)(c(de))$. The lower right is a right accordion corresponding to $a(b((cd)e))$. By rotating these figures one sees that there are two stars, six fans, three left accordions, and three right accordions, making a total of fourteen allowable laminations corresponding to the fourteen possible associations.

The number of faces of different dimension of K_5 breaks down as follows:

i) one three-dimensional cell,
ii) six pentagonal faces and three rectangular faces,
iii) twenty-one edges, and
iv) fourteen vertices.

In general, K_{n+2} is a simple convex polytope of dimension n (see [3]), whose faces are lower dimensional associahedra or products of lower dimensional associahedra. In the last section we review the algorithm for constructing K_{n+2}, $n \geq 2$ that appears in Devadoss [3]. It is obtained by truncating an n-dimensional simplex by codimension one hyperplanes.

Now we introduce the real analytic Teichmüller space T_n of dimension n. It is a certain set of equivalence classes of $n + 3$ variable points cyclically arranged on the unit circle. Two such labelled sets $S = \{p_1, \ldots, p_{n+3}\}$ and $S' = \{p'_1, \ldots, p'_{n+3}\}$ are considered equivalent if there exists an orientation preserving Möbius transformation A preserving the circle such that $A(p_j) = p'_j$ for $1 \leq j \leq n+3$. Another way to view T_n is as equivalence classes of orientation preserving homeomorphisms from the circle into the circle restricted to a fixed subset S of the circle consisting of $n+3$ points. Two such homeomorphisms h_0 and h_1 are equivalent if there is a Möbius transformation A such that $A \circ h_0(p) = h_1(p)$ for every p in S.

In this paper we show how the finite earthquake theorem ([8]) provides a homeomorphism Φ from T_n onto the interior of the associahedron K_{n+2}. By letting certain of the earthquake parameters approach ∞, a compactification $\bar{T}_n$ of T_n is realized in such a way that Φ extends to a homeomorphism between $\bar{T}_n$ and K_{n+2}. Thus, we obtain a natural cell structure for T_n without using quadratic differentials or Teichmüller's theorem. The relationship between the faces of K_{n+2} provides a combinatorial arrangement that shows how the pieces of T_n are sewn to form the interior of K_{n+2} and how the pieces of $\bar{T}_n$ form K_{n+2}.

There are interesting relationships between this cell structure and other important theorems from the Teichmüller theory of a finite set of variable points on a circle. In particular, Teichmüller's theorem, [15], [2], [6], the theorem on the existence of Jenkins-Strebel differentials with given heights, [14], [12], [9], and the infinitesimal earthquake theorem, [5]. These are topics that should be investigated further.

The theory of the associahedron was first developed by Stasheff [13] as a topic in topology and is also related to problems in homological algebra [11] and category theory [18]. In this paper we approach the subject from a different viewpoint, namely, the viewpoint of Teichmüller geometry.

We are grateful to Reza Chamanara, Bill Harvey, Dennis Sullivan and Noson Yanofsky for their help and encouragement.

1. The finite earthquake theorem

In this section we revisit a finite version of the well-known earthquake theorem of Thurston in [16], which tells us that any orientation-preserving homeomorphism of the unit circle is realized as the extension to the boundary of an earthquake map on the interior unit disk. An earthquake map on the unit disk is a piecewise Möbius transformation on domains separated by non-intersecting hyperbolic geodesics such that the comparisons of the Möbius transformations on different domains are hyperbolic, have their axes separating the domains, and shift in the same direction. The union of these non-intersecting lines is called the lamination for the earthquake and the amount of shifting is quantified by a transverse measure. Provided the amount of shifting is controlled, the restriction of the earthquake to the unit circle is a homeomorphism. In [16] Thurston gave a construction that shows that any homeomorphism of the circle is realized in this manner by an earthquake and that its measure and lamination are uniquely determined. Moreover, up to post composition by a Möbius transformation, the homeomorphism can be reconstructed from its lamination and measure.

In [8] Gardiner and Lakic gave a version of this theorem that applies to the case that the homeomorphism is replaced by a cyclic order preserving map h defined only on a finite subset S of the circle and mapping S bijectively to another subset $S' = h(S)$. In their formulation a lamination $\mathcal{L}$ is *allowable* if it consists of a finite number of non-intersecting geodesics joining points of S *and if it has the property that no line of* $\mathcal{L}$ *joins points adjacent in* S. Using their finite earthquake theorem and a limiting process, they deduce Thurston's theorem for arbitrary homeomorphisms, ([7] and [8]).

In this paper, consideration of only the allowable finite laminations is essential for two reasons:

i) The different topological types of allowable laminations for a finite set S with $n+3$ elements correspond in an obvious way to the different possible associations for the multiplication of $n+2$ letters.
ii) Without this property a finite earthquake is not uniquely determined up to post composition with a Möbius transformation by its lamination and measure.

To see how allowable finite laminations correspond to associations consider the pictures shown in Figure 4. Six points are marked on the unit circle and the intervals between them are labelled in counter-clockwise order with the symbols a, b, c, d, e, and ∞. Each line of the lamination divides the symbols into two sets. Because the finite lamination is allowable, each of these sets contains at least two symbols. To obtain the association corresponding to the lamination, we insert a pair of matching parentheses that encloses the set which does not include the symbol ∞. We insert such a pair of matching parentheses for every line of the lamination. In general, this procedure yields a one-to-one correspondence between the maximal allowable finite laminations on subsets of $n+3$ points and the full associations of $n+2$ letters.

A non-negative atomic measure σ associated to an allowable finite lamination $\mathcal{L}$ is a collection of non-negative numbers λ_j associated to lines l_j in $\mathcal{L}$. We write $\sigma(\{l_j\}) = \lambda_j$. Now we explain how the measure σ together with its lamination $\mathcal{L}$ induces a left earthquake that yields a one-to-one order-preserving map h from S onto another set $S' = h(S)$ of $n+3$ points in the unit circle. Note that $\mathcal{L}$ consists of at most $m \leq n$ lines. These lines cut the disk into $m+1$ pieces, which we call the strata of the lamination. We choose any line l_{j_0} of $\mathcal{L}$, stand on one side of that line and look across to the hyperbolic half-plane on the other side. We shift every point of that half plane to the left by an isometry with translation axis equal to the line l_{j_0} and translation length equal to λ_{j_0}. All the lines of the lamination lying in that half plane together with their endpoints on the circle are also shifted to the left. Having done this, denote by s_{j_0} the stratum adjacent to l_{j_0} in that shifted half plane. Now we move to each other geodesic boundary line l_{k_0} of s_{j_0} and repeat the same process on the other side of l_{k_0} as we have done along the line l_{j_0}. We repeat this process indefinitely until we have used every line of $\mathcal{L}$ lying in this half-plane. Then we turn around 180^0 and do the same thing in the half plane lying on the other side of l_{j_0}. In the end, we obtain a map E_{σ,j_0} which is discontinuous along the lines l_{j_0} but which extends to a homeomorphism of the boundary of the unit disc $\partial\mathbb{D}$. E_{σ,j_0} maps the finite set S one-to-one and onto a set S' and it depends on σ and the choice of the line l_{j_0} in $\mathcal{L}$. However, up to post composition by an orientation preserving isometry, it is independent of this choice. That is, if we had started at a different line l_{j_1} of $\mathcal{L}$ and followed the same procedure, then there would be an isometry $A_{0,1}$ of $\mathbb{D}$, such that

$$A_{0,1} \circ E_{\sigma,j_0} = E_{\sigma,j_1}.$$

Since up to post-composition by Möbius transformations the maps $E_{\sigma,j}$ do not depend on j, we denote the coset class $PSL(2,\mathbb{R}) \circ E_{\sigma,j}$ by E_σ.

Given this notation we can now state the finite earthquake theorem given by Gardiner and Lakic in [8].

Theorem 1 (The finite earthquake theorem). *Suppose*

$$S = \{p_1, \ldots, p_{n+3}\} \text{ and } S' = \{p'_1, \ldots, p'_{n+3}\}$$

are two finite subsets of the same cardinality, both arranged in counterclockwise cyclic order on the circle and assume $h(p_j) = p'_j$. *Then there exists a unique allowable lamination* $\mathcal{L}$ *for* S *and a unique measure* σ *supported on* $\mathcal{L}$ *such that, up to post composition by an orientation preserving isometry of* $\mathbb{D}$, *the left earthquake map* E_σ *maps* p_j *to* p'_j. *The measure* σ *and its corresponding lamination are uniquely determined by the locations of the points of* S *and* S'.

The key to the proof of this theorem is the following lemma.

Lemma 1 ([8]). *Let* h *be the map from* S *onto* S' *with* $h(p_i) = p'_i$, *where* $1 \le i \le n+3$. *There exists an orientation preserving Möbius transformation* A *preserving the unit disk such that the post composition* $A \circ h$ *of* h *by* A *fixes at least three points in* S *and moves all other points counter-clockwise.*

In [8] Gardiner and Lakic show inductively how to find the isometry A and the three fixed points of $A \circ h$. They also give a recurrence process that finds the lamination and the weights that are assigned to its lines. On his website, *http://comet.lehman.cuny.edu/lakic,* Lakic uses this recurrence to give a Maple program which finds the lamination and the measure.

In the following, we present another proof in which the isometry A is directly constructed as a composition of three isometries of different types, that is,

$$A = (\text{hyperbolic translation}) \circ (\text{parabolic translation}) \circ (\text{rotation}).$$

Proof. The proof is comprised of the following steps:

First, select a rotation A_1 around the Euclidean center of the unit disk such that $h_1 = A_1 \circ h$ has at least one fixed point p_j in S.

Second, find a parabolic translation A_2 fixing p_j and preserving the unit circle and such that $h_2 = A_2 \circ h_1$ has at least two fixed points p_k and p_j and all other points p_l are either fixed or moved counter-clockwise by h_2.

To see that this is possible, change coordinates so that the hyperbolic plane is realized by the upper half plane $\mathbb{H}$ and the point p_j is placed at ∞. Then all of the points p_k and $p'_k = h_1(p_k)$ are located on the real axis. Put b equal to the minimum value of $p'_k - p_k$ over all values of k not equal to j and let this minimum be realized by the value $p'_k - p_k$. Then by putting $A_2(z) = z - b$, $h_2 = A_2 \circ h_1$ fixes p_k and p_j and for every other point p_l, $p'_l = h_2(p_l) \geq p_l$.

Finally, observe that there exists a hyperbolic translation A_3 with translation axis joining p_k to p_j and such that $h_3 = A_3 \circ h_2$ has at least three fixed points p_j, p_k and p_l and all other points in S are either fixed or moved counter-clockwise by h_3.

To see that this is possible, change coordinates so that $p_j = \infty$ and $p_k = 0$. Assume there are points p_l on the negative real axis. Then $p_l \leq p'_l = h_2(p_l) < 0$ and we let λ_- be the largest possible value of p'_l/p_l among those values of l for which p_l is negative. Clearly, $0 < \lambda_- \leq 1$. By letting $A_3(z) = z/\lambda_-$, $h_3 = A_3 \circ h_2$ will fix p_j, p_k and p_l and, for every other point p_m in S on either the positive or negative real axis, $h_3(p_m) \geq p_m$.

Similarly, if there are points p_l on the positive real axis, then we let λ_+ be the smallest possible value of p'_l/p_l among those values of l for which p_l is positive. Clearly, $\lambda_+ \geq 1$. By taking A_3 of the same form $A_3(z) = z/\lambda_+$, we reach the same conclusion for $h_3 = A_3 \circ h_2$.

Note that if there are points p_l, $l \neq j, k$, of S on the both sides of the origin, one can use the points on either side to construct A_3. Thus, in this situation there are two candidates for A_3.

The desired isometry A is equal to $A_3 \circ A_2 \circ A_1$. □

To prove the finite earthquake theorem we start with this lemma. It guarantees there is an orientation preserving, non-Euclidean isometry A, such that $A \circ h$ fixes a finite subset of S consisting of three or more points and moves all other points of S to the left. Since there are at least three such points in S, at least two of them must be non-adjacent. Let l be the hyperbolic line that joins these two non-adjacent points and $\mathbb{H}_l$ the half-plane lying on one side of l that contains a third point of S. Since all points of S that lie on the boundary of $\mathbb{H}_l$ are moved to the left, there is one such point which is moved the least to the left. If B_l is a hyperbolic transformation with axis l that moves p_l to p'_l, then $B_l^{-1} \circ A \circ h$ still moves to the left all points of S that are on the boundary of the half-plane $\mathbb{H}_l$. Then we put weight $\sigma(\{l\})$ equal to the translation length of B_l, and repeat the same procedure for the restriction of $B_l^{-1} \circ A \circ h$ on the two half-planes on both sides of l.

To prove the uniqueness assume there are two finite earthquake representations $(E_\sigma, \mathcal{L})$ and $(E'_{\sigma'}, \mathcal{L}')$. We first show no line l of $\mathcal{L}$ can cross any line l' in $\mathcal{L}'$. If l in $\mathcal{L}$ crosses l' in $\mathcal{L}'$, let a and c be the endpoints of l and b and d be the endpoints of l', where a, b, c, d are in counter-clockwise order. Let B be the isometry that maps a, b, c to $h(a), h(b), h(c)$ respectively. Then $B^{-1} \circ h$ fixes a, b, c and moves d counter-clockwise. Now let B' be the hyperbolic isometry with translation axis connecting a to c and moving $B^{-1} \circ h(d)$ back to d. Then $B' \circ B^{-1} \circ h$ fixes a, c, d and moves b clockwise. Since l' belongs to $\mathcal{L}'$ and $B \circ (B')^{-1}$ is the isometry uniquely determined by the values of h at a, c and d, $B \circ (B')^{-1} \circ h$ must move b counter-clockwise. This is a contradiction. Thus no line in $\mathcal{L}$ can cross any line in $\mathcal{L}'$.

Since no line of $\mathcal{L}$ crosses a line of $\mathcal{L}'$, by adding lines to each of these laminations with weight zero, we may assume that $\mathcal{L} = \mathcal{L}.'$ Now we only need to prove the weight of each line l in $\mathcal{L}$ under E is equal to the weight of l under E'. If there is any line for $\mathcal{L}$ with a weight different from its weight for $\mathcal{L}'$, we may assume that line connects 0 to ∞. By following the step 3 in the proof of the previous lemma, the weight of this line is equal to $\ln \lambda_+ + \ln \frac{1}{\lambda_-}$. (If the negative or positive real axis contains no points, then the corresponding summand is zero.) According to the construction, this value is uniquely determined by h. Therefore, all lines have the same weights and the measured lamination is unique.

2. The combinatorial structure of T_n

In this section we use finite earthquakes to study the combinatorial structure of T_n. We first introduce an embedding of the associahedron K_{n+2} into a much higher dimensional vector space. Then we show T_n is homeomorphic to the interior of the embedded image of K_{n+2}.

Let P_{n+3} be a convex polygon in the Euclidean plane with $n + 3$ sides. Label the sides of the polygon in the counterclockwise direction with the symbols,

$$a_1, a_2, \ldots, a_{n+2}, \infty.$$

Consider the set Λ_n of all diagonals, that is, straight line segments joining non-adjacent vertices of P_{n+3}. Notice that Λ_n has N elements where

$$N = \binom{n+3}{2} - (n+3). \tag{1}$$

Let $I = [0, \infty)$, the positive real axis together with 0, and let $\bar{I} = [0, \infty]$, the compactification of I.

Definition. *We define A_{n+2} (resp. $\bar{A}_{n+2}$) to be the closed subset of the product space I^N (resp. $\bar{I}^N$) consisting of vectors $\upsilon = (\upsilon_1, \ldots, \upsilon_N)$ such that each $\upsilon_j \in I$ (resp. $\bar{I}^N$) and such that any two non-zero entries υ_i and υ_j of υ correspond to diagonals d_i and d_j of P_{n+3} that do not intersect.*

Corollary of the definition. *A_{n+2} and $\bar{A}_{n+2}$ are contractible Hausdorff spaces and $\bar{A}_{n+2}$ is compact.*

Proof. $\bar{A}_{n+2}$ is compact because it is a closed subset of the compact product $\bar{I}^N$. A_{n+2} and $\bar{A}_{n+2}$ are contractible because any contraction of I or of $\bar{I}$ can be extended to a contraction of A_{n+2} or $\bar{A}_{n+2}$ since the condition that υ be a vector in A_{n+2} or in $\bar{A}_{n+2}$ is preserved throughout the contraction. □

There is a partition of $\bar{A}_{n+2}$ into cells of dimensions $k \leq n$. By a finite *lamination* L of P_{n+3}, we shall mean a set of n or fewer, non-intersecting diagonals. Then we let $F(L)$ be the set of all vectors v in $\bar{A}_{n+2}$ for which the entries v_j corresponding to diagonals d_j in L are equal to ∞ and the other entries are not equal to infinity. Combinatorial formulas for the numbers of the laminations L consisting of fixed numbers of the lines in Λ_n are well-known, [17]. For example the number of L's of maximal size n is equal to the Catalan number

$$\frac{1}{n+2}\binom{2n+2}{n+1}. \tag{2}$$

For different laminations L of P_{n+3}, the sets $F(L)$ comprise the interiors of the faces of the associahedron of dimension n.

To introduce Teichmüller coordinates to A_{n+2}, we choose P_{n+3} to be the regular polygon circumscribed by the unit circle. Thus the set S of the vertices of P_{n+3} consists of $n+3$ equally spaced points on the unit circle. We also replace the diagonals by hyperbolic lines joining pairs of non-adjacent points of S. The Teichmüller space $T(S)$ consists of deformations of S to variable sets of $n+3$ cyclically ordered points also lying on the unit circle, factored by $PSL(2, \mathbb{R})$. By the finite earthquake theorem the points of $T(S)$ are parameterized by the vectors v in A_{n+2}. In fact, this parametrization introduces a continuous isomorphism between $T(S)$ (with the Teichmüller metric) and A_{n+2} (with the topology inherited from the product topology of I^N.)

The next theorem is justification for these statements. Before proceeding to it, we recall the definition of the Teichmüller's metric and the Teichmüller topology. The Teichmüller distance between two ordered $(n+3)$-tuples $p_1, \ldots, p_{n+3}$ and $p'_1, \ldots, p'_{n+3}$ is $\log K(h)$ where $K(h)$ is the smallest possible dilatation of a quasiconformal map h preserving the unit disk such that $h(p_j) = p'_j$, $1 \leq j \leq n+3$.

Lemma 2. *Let $p = (p_1, \ldots, p_{n+3})$ be an ordered set of $n+3$ points on the unit circle. Let U_ϵ be the set of variable points $p' = (p'_1, \ldots, p'_{n+3})$ with the property that there is a Möbius transformation A preserving the unit disk for which*

$$|p_j - A(p'_j)| < \epsilon \text{ for } 1 \leq j \leq n+3.$$

Then $U_\epsilon(p)$, $\epsilon > 0$, forms a neighborhood basis in the the Teichmüller topology at the point represented by p in T_n.

Proof. First assume p' is close to p in the Teichmüller topology, that is, for a given $\alpha > 0$, there is a quasiconformal map h preserving the unit disk such that $h(p_j) = p'_j$ and such that $K(h) < 1 + \alpha$. Then every extremal length problem for topological quadrilaterals in the unit disk with sides on the unit circle and vertices taken from the points of S must be K−quasi-preserved by h. Since these extremal lengths are strictly monotone functions of cross-ratios, the same is true about the cross ratios. That is, there exists a $\beta > 0$ depending on α and the spacing between the points of S, such that

$$\left|\log \frac{cr(a', b', c', d')}{cr(a, b, c, d)}\right| < \beta. \tag{3}$$

Here, a, b, c, d and a', b', c', d' are arbitrary quadruples taken from the sets S and S' with $h(a) = a', h(b) = b', h(c) = c', h(d) = d'$ and

$$cr(a, b, c, d) = \frac{b-a}{c-b} \cdot \frac{d-c}{d-a}.$$

We shall denote the quantity inside the absolute value on the left side of (3) by $cr_d^h(Q)$ and call it the cross ratio distortion of h on the quadruple $Q = \{a, b, c, d\}$. For an explanation of the functional relationship between cross ratio distortion and extremal length see [1] or [10].

To complete the proof we must show that if p' is in U_ϵ for small $\epsilon > 0$ then there is a quasiconformal map h preserving the unit disk and carrying p_j to p'_j such that $\log K(h) = \epsilon'$ for small positive ϵ' depending on ϵ and converging to 0 as ϵ converges to 0. Since $K(h) = K(A \circ h \circ B)$ for any Möbius transformations A and B, we may assume the points p_j and p'_j lie on the real axis. We may also assume $p_1 = p'_1 = \infty, p_2 = p'_2, p_3 = p'_3$ and $|p_j - p'_j| < \epsilon$ for $4 \leq j \leq n+3$. Now choose δ equal to the minimum distance $|p_{j+2} - p_{j+3}|$ for $1 \leq j \leq n$. For each such p_j construct a square R_j centered at p_j with side length equal to δ and with top and bottom parallel to the x-axis. Notice that the rectangles R_j are disjoint for different values of j. Also, the two diagonals of R_j intersect at p_j and cut R_j into four triangles. If $|p'_j - p_j| < \delta/2$ then p'_j lies in the interior of R_j and the piecewise affine map h that is the identity on the exterior of the union of the rectangles R_j and moves p_j to $p'_j = p_j + \epsilon_j$ is a quasiconformal. Moreover, it is easy to see that $K(h) \leq 1 + C\epsilon$, where $\epsilon = \max_j |\epsilon_j|$. □

Theorem 2. *The earthquake parametrization is a homeomorphism from T_n onto A_{n+2}.*

Proof. Let $S = \{p_1, \ldots, p_{n+3}\}$ be a fixed set of $n+3$ cyclically ordered, equally spaced points on the unit circle. A point in T_n is represented by another set of $n+3$ cyclically ordered points $S' = \{p'_1, \ldots, p'_{n+3}\}$. Two such sets S' and S'' represent the same point if and only if there is a Möbius transformation preserving the unit disc such that $A(p_j) = p'_j$ for $1 \leq j \leq n+3$.

The finite left earthquake parametrization yields a map $\Phi : A_{n+2} \to T_n$. By definition a vector $v \in A_{n+2}$ has a finite number of positive entries $v_1, \ldots, v_m$ with $m \leq n$ while all other entries are equal to zero. Each nonzero entry corresponds to a hyperbolic line l_j to which we assign the weight v_j. The totality of these lines comprise a finite lamination $\mathcal{L}$, and also by definition no two of these lines intersect and none of them join adjacent points of S. We have seen in the Introduction how the data consisting of these lines together with their weights produce a homeomorphism f of the unit circle, well-defined up to post composition by a Möbius transformation. We define $\Phi(v)$ to be the map f which takes S to $S' = f(S)$. By the finite earthquake theorem, Φ is a well-defined bijection from A_{n+2} onto T_n.

We claim that Φ is a homeomorphism. To show that Φ is continuous suppose that v is a vector in A_{n+2} and v' is another vector close to v. That means that every entry v'_j of v' is close to the corresponding entry of v_j of v in the sense that $|v'_j - v_j| < \epsilon$.

Reorder the entries of v so that $v_1, \ldots, v_m$ are positive and the rest of its entries are equal to zero. Note that by hypothesis the lines l_j corresponding to v_j for $1 \leq j \leq m$ form the lines of a finite lamination $\mathcal{L}$ and if $0 < \epsilon < \min_{1 \leq j \leq m} v_j$, then v'_j is also positive and, therefore, the lines of the lamination $\mathcal{L}'$ for $\Phi(v')$ include all of the lines of $\mathcal{L}$. There may be other lines of $\mathcal{L}'$, but since $|v'_k - v_k| < \epsilon$ for all k, the values of v'_k for $k > m$ must lie between 0 and ϵ. If we start the construction of the earthquake for v and for v' on the same stratum of $\mathcal{L}$, it is now clear that the two constructed maps h and h' will be nearly equal on S, and this shows that Φ is continuous.

We must show that Φ^{-1} is continuous, that is, that the earthquake parametrization maps T_n to A_{n+2} continuously. Let $[f]$ and $[g]$ be two points in T_n, and $(\sigma_f, \mathcal{L}_f)$ and $(\sigma_g, \mathcal{L}_g)$ be the corresponding earthquake measures and laminations for f and g. By Lemma 2 if $[f]$ and $[g]$ are close in the Teichmüller metric then the cross-ratio distortions of f and g on any quadruple of points contained in S are close. So we assume that the difference of cross-ratio distortions of f and g on every quadruple in S is less than ϵ.

Claim 1. If a geodesic l_f in $\mathcal{L}_f$ crosses a geodesic l_g in $\mathcal{L}_g$, then both the weights λ_f of l_f in σ_f and λ_g of l_g in σ_g are less than ϵ.

Denote the endpoints of l_f by a and c and the endpoints of l_g by b and d and label these endpoints so that the quadruple $Q = \{a, b, c, d\}$ is arranged counterclockwise on the unit circle. Let $cr_d^f(Q)$ and $cr_d^g(Q)$ be the cross-ratio distortions on the quadruple Q under f and g respectively. Under the earthquake map for f, the weights on all other geodesics in $\mathcal{L}_f$ may move b or d further to the left before the movement corresponding to λ_f. Therefore the cross-ratio distortion $cr_d^f(Q)$ is greater than or equal to the cross-ratio distortion of Q under the earthquake map E_l^f with only one geodesic l_l and weight λ_f. But that the cross-ratio distortion of Q under E_l^f is equal to λ_f. Therefore $cr_d^f(Q) \geq \lambda_f$. Similarly, $cr_d^g(\{b, c, d, a\}) \geq \lambda_g$ and so $cr_d^g(Q) = -cr_d^g(\{b, c, d, a\}) \leq -\lambda_g$. Therefore,

$$\epsilon > |cr_d^f(Q) - cr_d^g(Q)| \geq \lambda_f + \lambda_g.$$

Hence $\lambda_f < \epsilon$ and $\lambda_g < \epsilon$.

Claim 2. If a geodesic l_f is in $\mathcal{L}_f$ but not in $\mathcal{L}_g$ and $\mathcal{L}_g$ has no geodesic crossing l_f, then $\lambda_f < \epsilon$; vice versa, it is also true.

Denote the endpoints of l_f by a and c. Then in S, there must be a point b on one side of l_f and another point d on the other side of l_f such that the four points a, b, c, d lie in the Euclidean closure of a single stratum of $\mathcal{L}_g$. Now let Q be the quadruple consisting of a, b, c, d labelled counterclockwise on the unit circle (exchange the labels of b and d if necessary.) Then $cr_d^f(Q) \geq \lambda_f$ and $cr_d^g(Q) = 0$. Therefore

$$\epsilon > |cr_d^f(Q) - cr_d^g(Q)| \geq \lambda_f.$$

Claim 3. If a geodesic l is contained in both $\mathcal{L}_f$ and $\mathcal{L}_g$, then $|\lambda_f - \lambda_g| < \epsilon$.

The proof of this claim is similar to claim 2 except that now $cr_d^g(Q) = \lambda_g$. Thus

$$\epsilon > |cr_d^f(Q) - cr_d^g(Q)| \geq |\lambda_f - \lambda_g|.$$

Claims 1, 2 and 3 together imply that σ_f and σ_g have distance at most $2n\epsilon$ in the metric on A_{n+2}. Thus Φ^{-1} maps T_n to A_{n+2} continuously. □

3. The structures of the faces of the associahedron

In this section we study the Teichmüller structures of the interiors of the faces of the associahedron. More precisely, we show:

Theorem 3. *For every lamination L of P_{n+3}, $F(L)$ is isomorphic to a Teichmüller space or a geometric product of Teichmüller spaces. ($F(L)$ is defined after the definitions of A_{n+2} and $\overline{A_{n+2}}$ in section 2.) In particular, $F(\emptyset)$ is isomorphic to T_{n+3} and for any maximal lamination L consisting of n lines in Λ_n, $F(L)$ is a single point.*

Proof. Let U_j be the closure of j-th component of the complement of L in $\bar{\mathbb{D}}$. Each U_j contains a subset S_j of 3 or more points of S on its boundary, where $1 \leq j \leq m$. Let $P(j)$ be the Euclidean convex hull of S_j. Let $A(S, L, \lambda)$ be the set of those elements of A_{n+2} for which the coordinates of the diagonals in L are equal to the entries of a fixed vector $\lambda = (\lambda_1, \dots, \lambda_k)$ consisting of non-zero weights. Now let $A(S_j)$ be the set of those elements in A_{n+2} for which the coordinates of only those diagonals of $P(j)$ are allowed to be non-zero. Then $A(S, L, \lambda)$ is isomorphic to the direct sum of $A(S_j)$'s, that is,

$$A(S, L, \lambda) \cong A(S_1) \oplus A(S_2) \oplus \cdots \oplus A(S_m).$$

By using the finite earthquake theorem and by changing hyperbolic geodesics to the Euclidean segments connecting the same endpoints, one sees that the points of each $T(S_j)$ are parameterized by the vectors v in $A(S_j)$. Therefore, $A(S, L, \lambda)$ is isomorphic to the product space of $T(S_j)$'s, that is,

$$A(S, L, \lambda) \cong T(S_1) \times T(S_2) \times \cdots \times T(S_m).$$

Note that when L is empty, $A(S, L, \lambda) \cong T(S)$ and when L is maximal, that is, when it contains n lines, then $T(L, \lambda)$ reduces to a single point. By letting all of the weights $\lambda_1, \dots, \lambda_k$ simultaneously approach ∞ we obtain

$$F(L) \cong T(S_1) \times T(S_2) \times \cdots \times T(S_m).$$ □

4. The realization of $\bar{A}_{n+2}$ as a polytope

We choose an increasing homeomorphism from $\bar{I} = [0, \infty]$ onto $J = [0, 1]$. Then there is an induced homeomorphism H from $\bar{I}^N$ onto J^N. Since $\bar{A}_{n+2} \subset \bar{I}^N$, we can let $\bar{B}_{n+2} = H(\bar{A}_{n+2})$. We will show that there is a piecewise affine homeomorphism from $\bar{B}_{n+2}$ onto a closed bounded convex set in $\mathbb{R}^n$. More precisely, we obtain

Theorem 4. *There is a a piecewise affine homeomorphism h from $\bar{B}_{n+2}$ onto an n-dimensional simple convex polytope K_{n+2} (commonly called the associahedron of dimension n). The homeomorphism h carries $H \circ F(\emptyset)$ onto the interior of K_{n+2} and for any lamination L of P_{n+3}, it carries $H \circ F(L)$ onto the interior of a face of K_{n+2} of codimension k, where k is the number of the diagonals in L.*

Proof. To proceed with the proof we must first summarize the algorithm presented by Devadoss in [3] and [4] for constructing K_{n+2}. K_{n+2} is obtained by truncating an n-dimensional simplex by codimension-1 half-planes. We will see the above theorem follows as a natural consequence by cutting K_{n+2} into a union of topological cubes and labelling the faces of K_{n+2} with the finite subsets of non-intersecting diagonals of P_{n+3}. These subsets correspond to the full and partial associations of $n+2$ factors.

Consider the set of congruent regular polygons with $n+3$ sides labelled with the symbols $\{x_1, \dots, x_{n+2}, \infty\}$ in cyclic counterclockwise order and each with a different marked diagonal, that is, a line segment joining non-adjacent vertices. Denote the set of these marked polygons by $\mathcal{G}_n$. We will say that two elements G_1 and G_2 in $\mathcal{G}_n$ satisfy the *nonintersecting condition* if the superposition of G_1 onto G_2 by a congruence respecting the labeling of the sides does not have intersecting diagonals.

The diagonal of an element G of $\mathcal{G}_n$ divides the $(n+3)$-gon into two parts. The part that does not contain the side marked with ∞ is called the *free part* of G. Let $\mathcal{G}_n^i$ be the collection of the elements in $\mathcal{G}_n$ that contain i sides of P_{n+3} in their free parts. It is easy to see that the order of $\mathcal{G}_n^i$ is $n+3-i$, where $1 < i < n+2$. In particular, the order of $\mathcal{G}_n^2$ is $n+1$, which is the number of sides (codimension one faces) of an n-dimensional simplex W_n. Arbitrarily label each face of W_n by an element of $\mathcal{G}_n^2$.

Notice that the labels on some adjacent faces of W_n do not satisfy the nonintersecting condition. This is an obstruction to the simplex satisfying the associahedron condition. In order to overcome this obstruction, it is natural to truncate the vertices, bring in more faces and introduce new labels. In fact, this is how it is carried out in [3], and it is organized in an inductive way as follows.

Step I. Truncating two vertices.

Check all vertices (codimension n faces) of W_n and find the two vertices that satisfy the following condition. For each of these two there exists an element G in $\mathcal{G}_n^{n+3-2}$ such that G satisfies the nonintersecting condition with the labels of all faces adjacent to that vertex. Now truncate those two vertices off W_n by two codimension 1 half-planes and label the two new faces (two simplices of dimenson $n-1$) by the corresponding two elements in $\mathcal{G}_n^{n+3-2.}$ Then the labels of the new faces and their adjacent ones satisfy the nonintersecting condition. We continue to call the resulting simple convex polytope W_n. Figure 5 shows how K_4 is constructed.

Step II. Truncating some edges if $n \geq 3$.

Check the edges (codimension $n-1$ faces) of W_n and find those edges that satisfy the following condition. For each of them there exists an element G in $\mathcal{G}_n^{n+3-3}$ such that G satisfies the nonintersecting condition with the labels of all faces adjacent to that edge. Now truncate those edges W_n by codimension 1 half-planes and label

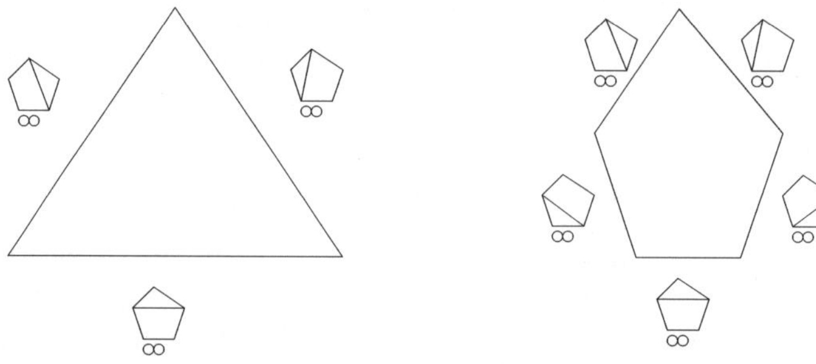

Figure 5. The construction of K_4 through truncation.

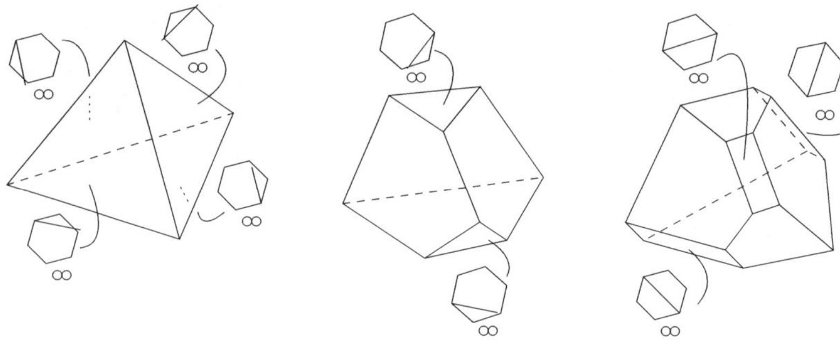

Figure 6. The construction of K_5 through truncation.

the new faces by the corresponding elements of $\mathcal{G}^{n+3-3}$. Then the labels of the new faces and their adjacent faces satisfy the nonintersecting condition. We continue to call the resulting simple convex polytope W_n. Figure 6 shows how K_5 is constructed.

Step III. If $n \geq 4$, inductively find the codimension i faces that the labels of their adjacent codimension 1 faces do not satisfy the nonintersecting condition, truncate them by codimension 1 half-planes, and finally label the new codimension 1 faces by the corresponding elements in $\mathcal{G}_n^{n+3-i}$ for $i = 4, 5, \ldots, n$. The resulting simple convex polytope is K_{n+2}.

In the course of truncating W_n to obtain K_{n+2}, the truncations of the faces of dimension i add $i+2$ new codimension 1 faces to the polytope and each of them is labelled by an element in $\mathcal{G}_n^{i+2}$, where $i = 0, 1, \ldots, n-2$. Therefore, the number of codimension 1 faces of K_{n+2} is equal to $\sum_{i=0}^{n-1} card(\mathcal{G}_n^{n+1-i})$.

Now we are ready to begin the proof of Theorem 4. Recall from the second paragraph of section 2 that Λ_n is the set of all possible diagonals of the polygon P_{n+3} with $n+3$ sides. Identify each diagonal in Λ_n with the element of $\mathcal{G}_n$ for which that diagonal is marked. Then the superposition of k non-intersecting elements of $\mathcal{G}_n$ serves as a label of a face of K_{n+2} of codimension k. For example, each maximal lamination L corresponds to the superposition of n elements of $\mathcal{G}_n$ with non-intersecting diagonals. The label for any vertex is the superposition of the labels of all of its adjacent faces. Now we view all vertices of K_{n+2} as the centers of the faces of dimension 0 and for each 1-dimensional face take any interior point to be its center. Then we inductively select the centers for the faces of K_{n+2} of higher and higher dimension. For each 2-dimensional face of K_{n+2}, we take as its center an interior point in the convex hull of the centers of the edges of that face; for each 3-dimensional face of K_{n+2}, we take as its center an interior point of the convex hull of the centers of all 2-dimensional faces of those 3-dimensional faces. We continue inductively until we obtain a center for K_{n+2} itself. In fact, we use the same regular $(n+3)$-gon with k nonintersecting diagonals as the label of a codimension k face and as a label of the the center of that face. See Figure 2 for the labels of the centers of K_4 by the associations on four factors and Figure 3 for the labels of the centers of some faces of K_5.

Given the label V of a vertex of K_{n+2}, denote by 2^V the collection of the labels of the faces of K_{n+2} whose diagonals form sub-collections of the diagonals of the label of V, and denote by L the corresponding maximal lamination for V. Let Δ be the convex hull of the set consisting of the centers of the faces with labels in 2^V and let Γ be the image under H of the closure of the cube generated by the vectors with non-negative coordinates for the diagonals in L and zero coordinates for the diagonals not in L.

Then there exists an affine homeomorphism from Γ onto Δ, which maps the vertices of Γ to the vertices of Δ respectively according to their labels and have linear extensions among the corresponding simplices. All these affine homeomorphisms are pasted together to provide a piecewise affine homeomorphism between $\bar{B}_{n+2}$ and K_{n+2}. Furthermore, this piecewise affine homeomorphism maps the faces of $\bar{B}_{n+2}$ onto the respective faces of K_{n+2}, according to their labels.

In general this construction realizes K_{n+2} as a union of cubes of dimension n, and there is one cube corresponding to each vertex V. For K_4 the cubes are the two-dimensional quadrilaterals shown in Figure 2. For K_5 there are fourteen such three-dimensional cubes. We leave it to the reader to try to depict one of these cubes starting with the drawing given in Figure 3.

This completes the proof of the theorem. □

References

[1] L. V. Ahlfors, Lectures on Quasiconformal Mapping, volume 10 of *Van Nostrand Studies*, Van Nostrand-Reinhold, Princeton, N. J. (1966).

[2] L. Bers, Quasiconformal mappings and Teichmüller's theorem, in Analytic Functions, Princeton University Press, Princeton, N. J. (1960) pages 89–119.

[3] S. Devadoss, Tessellations of moduli spaces and the mosaic operad, *Contemporary Mathematics of AMS*, **239** (1999) 91–114.

[4] S. Devadoss, Combinatorial equivalence of real moduli spaces, *Notices of the American Math. Soc.*, **51(6)** (2004) 620–626.

[5] F. P. Gardiner, Infinitesimal bending and twisting in one-dimensional dynamics, *Trans. Am. Math. Soc.*, **347(3)** (1995) 915–937.

[6] F. P. Gardiner and J. Hu, A short course on Teichmüller's theorem, In Proceedings of the Year on Teichmüller Theory and Moduli Problems, 2005–06, Harish Chandra Research Institute, Allahabad (2005) to appear.

[7] F. P. Gardiner, J. Hu, and N. Lakic. Earthquake curves, *Contemp. Math., AMS*, **311** (2002) 141–196.

[8] F. P. Gardiner and N. Lakic, Quasiconformal Teichmüller Theory, AMS, Providence, Rhode Island (2000).

[9] J. Hubbard and H. Masur, Quadratic differentials and foliatons. *Acta Math.*, **142** (1979) 221–274.

[10] O. Lehto and K. I. Virtanen, Quasiconformal Mapping in the Plane, Springer-Verlag, New York, Berlin (1965).

[11] S. MacLane, Natural associativity and commutativity, *Rice University Studies*, **49(4)** (1963) 28–46.

[12] A. Marden and K. Strebel, The heights theorem for quadratic differentials on Riemann surfaces, *Acta Math.*, **153** (1984) 153–211.

[13] J. Stasheff, Homotopy associativity of h-spaces i, *Trans. Amer. Math. Soc.*, **108** (1963) 275–292.

[14] K. Strebel, Quadratic Differentials, Springer-Verlag, Berlin & New York (1984).

[15] O. Teichmüller, Extremale quasikonforme Abbildungen und quadratische Differentiale, *Abh. Preuss. Akad.*, **22** (1939) 3–197.

[16] W. P. Thurston, Earthquakes in two-dimensional hyperbolic geometry, In Low-dimensional Topology and Kleinian groups, **112** pages 91–112. London Math. Soc. (1986).

[17] H. S. Wilf, Generatingfunctionology, Academic Press, San Diego, California, 2nd edition (1994).

[18] Noson Yanofsky, Obstructions to coherence: natural noncoherent associativity, *Journal of Pure and Applied Algebra*, **147** (2000) 175–213.

Teichmüller Theory and Moduli Problem No. 10, 2010, pp. 195–228.

A Short Course on Teichmüller's Theorem

F. P. Gardiner and Jun Hu

Department of Mathematics, Brooklyn College, Brooklyn, NY 11210
e-mail: fgardiner@gc.cuny.edu; jun@sci.brooklyn.cuny.edu

Abstract. We present a brief exposition of Teichmüller's theorem.

Introduction

An orientation preserving homeomorphism f from a Riemann surface X onto a Riemann surface Y is given. Teichmüller's problem is to find a quasiconformal homeomorhism in the homotopy class of f with minimal maximal dilatation, that is, to find a homeomorphism f_0 whose maximal dilatation $K(f_0)$ is as small as possible in its homotopy class.

Teichmüller's theorem states that the problem has a unique extremal solution provided that X is compact or compact except for a finite number of punctures, namely, a Riemann surface of finite analytic type. Moreover, except when f_0 is conformal, f_0 is equal to a stretch mapping along the horizontal trajectories of some uniquely determined holomorphic quadratic differential $\varphi(z)(dz)^2$, with $\iint_X |\varphi| dxdy = 1$, postcomposed by a conformal map. It turns out that even for arbitrary Riemann surfaces, whether or not they are of finite analytic type, this statement is generically true (see [20], [27]).

The goal of this course is to present a brief proof of the original Teichmüller theorem in a series of lectures and exercises on the following topics:

1. conformal maps and Riemann surfaces,
2. quasiconformal maps, dilatation and Beltrami coefficients,
3. extremal length,
4. the Beltrami equation,
5. the Reich-Strebel inequality and Teichmüller's uniqueness theorem,
6. the minimum norm principle,
7. the heights argument,
8. the Hamilton-Krushkal condition and Teichmüller's existence theorem,
9. trivial and infinitesimally trivial Beltrami differentials,
10. Teichmüller space and Teichmüller's metric,
11. infinitesimal Teichmüller's metric,
12. tangent vectors to Teichmüller spaces.

The course is organized as follows. The minimal necessary background and defining concepts in the statement of the theorem are introduced in the first four lectures.

The uniqueness part of Teichmüller's Theorem is presented in lectures 5, 6 and 7 and the existence part and the expression for extremal mapping in terms of a unique holomorphic quadratic differential are presented in lectures 8 and 9. Finally Teichmüller space, Teichmüller's metric, the infinitesimal form of Teichmüller's metric, and the structure of the tangent space to the Teichmüller space of an n-punctured Riemann sphere are explained in the lectures 10, 11 and 12.

There are many expositions of Teichmüller's theorem in the literature. However, our approach differs from others in that it highlights the minimum norm principle for measured foliations. [1,19,20,23,33,35,36]

These notes are meant merely to serve as a first introduction to the subject. There are many topics we have omitted which, if put together, make a unified subject. These include the equality of Kobayashi's and Teichmüller's metrics [37] [18], a result which shows that the complex structure of Teichmüller space determines its metric. There is also Royden's result that for compact surfaces of genus bigger than 2 the mapping class group is the full group of biholomorphic isomorphisms of $T(R)$. This result was extended by Earle and Kra to surfaces of finite analytic type [15] and [16]. Though still unknown for many infinite dimensional dynamical Teichmüller spaces, it has also been extended to Teichmüller spaces of Riemann surfaces of infinite type in varying degrees of generality by Earle, Gardiner [14] and Lakic [26] and by Markovic [17], [32].

One should add to these topics the theory of holomorphic motions, that is, the theorems of Mañé, Sad and Sullivan [30], Bers-Royden [11], Sullivan-Thurston [39] and Slodkowski [38]. There are also the beautiful results of Reich, Strebel [35] [36] and Hamilton [21] which relate infinitesimal extremality to global extremality and further result by Bozin, Lakic, Markovic and Mateljevic [13] that relates infinitesimal unique extremality to global unique extremality.

Unfortunately, it would take a whole book to explore all of these topics. We recommend [33], [23], [19] and [20].

For the most part, these notes were written while we attended the Harish Chandra Research Institute in Allahabad India. We wish to express our gratitude to all of the faculty and staff there for their hospitality and for making our visit so pleasurable. In addition, we thank the National Science Foundation of the USA and the Department of Science and Technology of India for supporting the "Year on Teichmüller Theory, 2005–06." We also owe special thanks to Krishnendu Gongopadhyay for his kind assistance in making the diagram that appears in Figure 1 of exercise 10.

1. Conformal maps and Riemann surfaces

A map $z \mapsto w = f(z) = u(z) + iv(z)$ from a domain U in $\mathbb{C}$ onto another domain $f(U)$ in $\mathbb{C}$ is conformal if it is a holomorphic homeomorphism. Thus we assume conformal maps are one-to-one and orientation preserving. In particular the first order approximation of such a map at a point $z_0 \in U$ has the form

$$f(z) = a_0 + a_1(z - z_0) + o(|z - z_0|),$$

with $a_1 \neq 0$. This means that in a neighborhood of z_0 the mapping f is approximated infinitesimally by the complex linear map $\Delta z \mapsto \Delta w = a_1 \Delta z$, where $a_1 = re^{i\theta}$, and $e^{i\theta}$ effects a rotation by angle θ and $r \neq 0$ effects a magnification by r. Such a linear map does not change shape. For example, the image of an equilateral triangle by $z \mapsto a_1 z$ is still an equilateral triangle.

Definition. *A Riemann surface is a connected Hausdorff topological space X together with a system of charts z_j mapping open sets U_j of X into the complex plane. The system of charts is assumed to cover X and on overlapping charts $z_j(U_i \cap U_j)$ the transition functions $z_i \circ (z_j)^{-1}$ are conformal.*

Any chart z is a holomorphic map from an open subset of X into $\mathbb{C}$ and therefore $dz = dx + i dy$ is a linear map from the tangent bundle to X into the tangent bundle to $\mathbb{C}$. From this viewpoint dx and dy are real-valued linear functionals on the tangent bundle to X, and thus they are elements of its cotangent bundle. Choose a basis $\left\{\frac{\partial}{\partial x}, \frac{\partial}{\partial y}\right\}$ dual to the basis $\{dx, dy\}$ in the sense that

$$\begin{pmatrix} \left(\frac{\partial}{\partial x}\right)(dx) & \left(\frac{\partial}{\partial x}\right)(dy) \\ \left(\frac{\partial}{\partial y}\right)(dx) & \left(\frac{\partial}{\partial y}\right)(dy) \end{pmatrix} = \begin{pmatrix} 1 & 0 \\ 0 & 1 \end{pmatrix}.$$

Exercise 1. By definition, the partial derivatives $\frac{\partial}{\partial z}$ and $\frac{\partial}{\partial \bar{z}}$ are the basis dual to $dz = dx + i dy$ and $d\bar{z} = dx - i dy$. Show that

$$\begin{cases} \frac{\partial}{\partial z} = \frac{1}{2}\left(\frac{\partial}{\partial x} - i\frac{\partial}{\partial y}\right) \\ \frac{\partial}{\partial \bar{z}} = \frac{1}{2}\left(\frac{\partial}{\partial x} + i\frac{\partial}{\partial y}\right). \end{cases}$$

Exercise 2. Show that if $f = u + iv$, then $\frac{\partial}{\partial \bar{z}} f = 0$ is equivalent to $u_x = v_y$ and $u_y = -v_x$.

Exercise 3. Using the notations $\frac{\partial}{\partial z} f = f_z$ and $\frac{\partial}{\partial \bar{z}} f = f_{\bar{z}}$, show that $u_x v_y - u_y v_x = |f_z|^2 - |f_{\bar{z}}|^2$.

Exercise 4. Show that $u_{xx} + u_{yy} = 4u_{z\bar{z}}$. By definition the Laplacian of u is $\Delta u = u_{xx} + u_{yy}$. A C^2 real-valued function $f(x, y)$ is called harmonic, subharmonic or superharmonic if $\Delta f = 0, \geq 0$, or ≤ 0, respectively. Show that the notions of harmonic, subharmonic and superharmonic are well-defined on a Riemann surface. That is, show that if z and w are two charts defined on overlapping domains in a Riemann surface X and if Δ_z is the Laplacian with respect to x and y with $z = x + iy$ and Δ_w is the Laplacian with respect to u and v with $w = u + iv$, then the Laplacian of f with respect to z or with respect to w are simultaneously, zero, positive or negative.

2. Quasiconformal maps, dilatation and Beltrami coefficients

If one thinks of a conformal structure as a shape, on an infinitesimal level conformal maps do not distort shape. Thus, in order to study maps that permit distortion of shape on an infinitesimal level, one must extend the discussion to maps that are not conformal. If at the same time one keeps the amount of distortion bounded, one is lead to the notion of a quasiconformal map.

We shall first consider what it means for a real linear map to be quasiconformal. By definition quasiconformal maps are locally approximable almost everywhere by orientation preserving real linear maps which that distort shape by a uniformly bounded amount. To begin our discussion we consider a real linear map given by

$$\begin{pmatrix} x \\ y \end{pmatrix} \mapsto \begin{pmatrix} u \\ v \end{pmatrix} = T \begin{pmatrix} x \\ y \end{pmatrix}$$

where

$$T = df = \begin{pmatrix} u_x & u_y \\ v_x & v_y \end{pmatrix} = \begin{pmatrix} a & b \\ c & d \end{pmatrix}.$$

Since we assume quasiconformal maps are orientation preserving, $ad - bc > 0$. The circle $|z|^2 = x^2 + y^2 = 1$ is mapped by T to the ellipse whose equation in the $w = u + iv-$plane is $|T^{-1}w|^2 = 1$. Expressed in terms of the inner product, this ellipse is the set of points satisfying

$$\left(T^{-1}\begin{pmatrix} u \\ v \end{pmatrix}, T^{-1}\begin{pmatrix} u \\ v \end{pmatrix}\right) = \left((T^{-1})^t T^{-1}\begin{pmatrix} u \\ v \end{pmatrix}, \begin{pmatrix} u \\ v \end{pmatrix}\right) = 1.$$

The distortion K of T is the eccentricity of this ellipse, that is, the length of its major axis divided by the length of its minor axis. This ratio is the same as the square root of the ratio of the larger eigenvalue λ_2 of $(T^{-1})^t T^{-1}$ to its smaller eigenvalue λ_1. Since $((T^{-1})^t T^{-1})^{-1} = TT^t$, this ratio is the same as the corresponding ratio for

$$TT^t = \begin{pmatrix} a^2 + b^2 & ac + bd \\ ac + bd & c^2 + d^2 \end{pmatrix}.$$

But $K = \left(\frac{\lambda_2}{\lambda_1}\right)^{1/2}$ implies $K + 1/K = \frac{\lambda_1 + \lambda_2}{\sqrt{\lambda_1 \lambda_2}}$. The numerator is the trace of TT^t and the denominator is the square root of the determinant of TT^t, which is the same as the determinant of T. Thus

$$K + 1/K = \frac{a^2 + b^2 + c^2 + d^2}{ad - bc}.$$

The complex distortion of a two-dimensional linear map is determined by the eccentricity K and by an angle and these two pieces of information are conveniently expressed in complex notation. Any real linear map from $\mathbb{C}$ to $\mathbb{C}$ has the complex form

$$w = T(z) = Az + B\bar{z},$$

with complex constants A and B. For orientation preserving T, the determinant is $|A|^2 - |B|^2 > 0$ and the formula can be written as

$$w = T(z) = A(z + \mu \bar{z}), \tag{1}$$

where $\mu = B/A$ and $|\mu| < 1$. In this form T is the stretch map $S(z) = z + \mu\bar{z}$ postcomposed by a multiplication, which is conformal and consists of a rotation through angle arg A and magnification by the factor $|A|$. Thus all of the distortion of shape caused by T is expressed by the complex number μ. Here μ is called the complex dilatation or the Beltrami coefficient. From it one can find angles of directions of maximal magnification and maximal shrinking, as well as the distortion. If we let $\mu = ke^{i\alpha}$ and $z = re^{i\theta}$, then

$$S(z) = re^{i\theta} + kre^{i\alpha-\theta} = re^{i\theta}(1 + ke^{i(\alpha-2\theta)})$$

and as θ moves from 0 to 2π, $S(z)$ sweeps out an ellipse whose maximum distance from the origin occurs when $\theta = \alpha/2$ or $\alpha/2 + \pi$ and whose minimum distance from the origin occurs when $\theta = \alpha/2 - \pi/2$ or $\alpha/2 + \pi/2$. The eccentricity of the ellipse is equal to $\frac{1+|\mu|}{1-|\mu|}$, we denote it by $K(T)$. Since it depends only on μ, we also denote it by $K(\mu)$.

After applying the stretch map, multiplication by A rotates the direction of maximal stretching by the angle arg A. Consider the composition $T_2 \circ T_1$ of two such real-linear maps $T_1 = A_1 \circ S_1$ and $T_2 = A_2 \circ S_2$. The image by T_1 of the direction of maximal stretching of S_1 may coincide with the direction of maximal stretching of S_2. Because the direction of maximal shrinking is orthogonal to the direction of maximal stretching, when the image by T_1 of the direction of maximal stretching of S_1 coincides with the direction of maximal stretching of S_2, the similar statement is true for the image of the direction of maximal shrinking. Thus these directions coincide precisely when $K(T_2 \circ T_1) = K(T_2)K(T_1)$ and otherwise $K(T_2 \circ T_1) < K(T_2)K(T_1)$.

Exercise 5. Let μ_1 and μ_2 be Beltrami coefficients of $T_1(z) = A_1 z + B_1\bar{z}$ and $T_2(z) = A_2 z + B_2\bar{z}$. Show that the Beltrami coefficient of $T_2 \circ (T_1)^{-1}$ is

$$\left(\frac{\mu_2 - \mu_1}{1 - \overline{\mu_1}\mu_2}\right)\frac{1}{\theta_1}, \tag{2}$$

where $\theta_1 = \overline{A_1}/A_1$.

Definition. *A C^1 orientation preserving homeomorphism $w = f(z)$ of a Riemann surface X onto another Riemann surface $f(X)$ is quasiconformal if*

$$K(f) = \sup_{z \in X} K_z(f) < \infty,$$

where $K_z(f) = \frac{1+k(z)}{1-k(z)}$ and $k(z) = |f_{\bar{z}}/f_z|$. If $K = K(f)$, then f is called K-quasiconformal.

Exercise 6. For a mapping f from a Riemann surface X to a Riemann surface Y, show that $K_z(f)$ and $K(f)$ are well-defined independently of choice of coordinate charts.

3. Extremal length

Let Γ be a family of curves on a Riemann surface. Every γ in Γ is assumed to be a countable union of open arcs or closed curves. The extremal length of Γ, $\Lambda(\Gamma)$, is a sort of average minimum length of the curves in Γ. It is an important quantity because it is invariant under conformal mappings and quasi-invariant under quasiconformal mappings, in the sense to be described precisely in Proposition 2. First we define the set of allowable metrics. A metric $\rho(z)|dz|$ is allowable if

- it is defined independently under choice of chart, that is, $\rho_1(z_1)|dz_1| = \rho_2(z_2)|dz_2|$, where ρ_1 and ρ_2 are representatives of ρ in terms of the charts z_1 and z_2.
- ρ is measurable and ≥ 0 everywhere, and
- $A(\rho) = \iint_X \rho^2 dxdy \neq 0$ or ∞.

For such an allowable ρ, define

$$L_\gamma = \int_\gamma \rho|dz|$$

if ρ is measurable along γ; otherwise define $L_\gamma = +\infty$. Let $L(\rho, \Gamma) = \inf L_\gamma(\rho)$, where the infimum is taken over all curves γ in Γ. The extremal length of the curve family Γ is

$$\Lambda(\Gamma) = \sup_\rho \frac{L(\rho, \Gamma)^2}{A(\rho)}, \tag{3}$$

where the supremum is taken over all allowable metrics ρ. Notice that the ratio in the surpremum is invariant if ρ is multiplied by a positive constant. Thus in attempting to evaluate $\Lambda(\Gamma)$ we may normalize in different ways. For example by putting $L(\rho)$ or $A(\rho)$ equal to 1, or by putting $L(\rho) = A(\rho)$. A metric ρ is said to be extremal if it realizes the supremum in (3).

Lemma 1. *Let X be a rectangle with width a and height b and Γ be the family of arcs in X that join the left side of X to the right side of X. Then $\Lambda(\Gamma) = a/b$ with extremal metric $\rho_0 = 1$.*

Proof. We show that $\rho_0 = 1$ is an extremal metric. Note that $L(\rho_0, \Gamma) = a$ and $A(\rho_0) = ab$. Therefore $\Lambda(\Gamma) \geq a^2/ab = a/b$. Let ρ be any competing metric and scale ρ so that $L(\rho, \Gamma) = a$. Then for every horizontal line segment γ crossing X, $a \leq \int_\gamma \rho(z)|dz| = \int_\gamma \rho(z)dx$, so $ab \leq \iint_X \rho(z)dxdy$. By Schwarz's inequality, $(ab)^2 \leq (\iint_X \rho^2 dxdy)(\iint_X dxdy)$. Hence $ab \leq A(\rho)$ and $a/b = a^2/ab \geq L(\rho)^2/A(\rho)$, which implies $\Lambda(\Gamma) \leq a/b$. Thus $\Lambda(\Gamma) = a/b$. □

Lemma 2. *Let $X = \{z : 1 < |z| < R\}$ and Γ be the family of closed curves in X homotopic in X to a core curve of X. By definition a core curve is any circle centered at the origin and lying in the interior of X. Then $\Lambda(\Gamma) = 2\pi/\log R$ with extremal metric $\rho_0(z) = 1/r$.*

Proof. We will show that $\rho_0 = 1/r$ is an extremal metric. Clearly $A(\rho_0) = 2\pi \log R$ and $L(\rho_0) = \int_0^{2\pi}(1/r)rd\theta = 2\pi$. Therefore,

$$\Lambda(\Gamma) \geq 2\pi/\log R.$$

For any allowable metric $\rho(z)|dz|$, $L(\rho) \leq \int_0^{2\pi} \rho(re^{i\theta})rd\theta$,

$$\frac{L(\rho)}{r} \leq \int_0^{2\pi} \rho(re^{i\theta})d\theta,$$

$$L(\rho)\log R \leq \iint_X \rho dr d\theta = \iint \frac{1}{\sqrt{r}}\sqrt{r}\rho dr d\theta.$$

So by Schwarz's inequality

$$(L(\rho)\log R)^2 \leq \iint \frac{1}{r} dr d\theta \iint \rho^2 r dr d\theta,$$

$$\frac{L(\rho)^2}{A(\rho)} \leq 2\pi / \log R.$$

Thus $\Lambda(\Gamma) \leq 2\pi/\log R$ and $\Lambda(\Gamma) = 2\pi/\log R$. □

Exercise 7. Let $X = \{z : 1 < |z| < R\}$ and Γ^t be the family of arcs in X that join $|z| = 1$ to $|z| = R$. Show that $\Lambda(\Gamma^t) = \log R/(2\pi)$ also with extremal metric $\rho_0(z) = 1/r$.

Now let $w = f(z)$ be a continuously differentiable and quasiconformal map from an annulus $X = \{z : 1 < |z| < R\}$ onto another annulus $f(X) = \{z : 1 < |z| < R'\}$ with $R' > R$ and assume μ is the Beltrami coefficient of f, that is,

$$f_{\bar{z}}(z) = \mu(z) f_z(z).$$

Then if Γ is the family of arcs in X joining its inner and outer boundary contours, $\Gamma^\mu = f(\Gamma)$ will be the family of arcs joining the inner and outer boundary contours of $f(A)$.

Note that $f_0 : z = re^{i\theta} \mapsto r^{K_0}e^{i\theta}$ maps $X = \{z : 1 < |z| < R\}$ onto $X' = \{\tilde{z} : 1 < |\tilde{z}| < R'\}$ provided that $R' = R^{K_0}$, and $\zeta = \log z$ and $\tilde{\zeta} = \log \tilde{z}$ are natural parameters for the quadratic differentials $(dz/z)^2$ and $(d\tilde{z}/\tilde{z})^2$ on X and X'. The mapping f_0 expressed in terms of ζ and $\tilde{\zeta}$ is given by $\zeta = \xi + i\eta \mapsto \tilde{\zeta} = K_0\xi + i\eta$. We let f be an arbitrary quasiconformal map from X onto X' also mapping the inner and outer contours of X onto the inner and outer contours of X'.

Proposition 1. *In the notation described above*

$$K_0 = \frac{\Lambda(\Gamma^\mu)}{\Lambda(\Gamma)} \leq \iint_X \frac{\left|1 + \mu\frac{\varphi}{|\varphi|}\right|^2}{1 - |\mu|^2} |\varphi| dx dy.$$

Moreover, if μ is the Beltrami coefficient of f, then $\|\mu\|_\infty > k_0$ unless

$$\mu = k_0|\varphi|/\varphi \ \ a.e.,$$

where $k_0 = \frac{K_0-1}{K_0+1}$ and $\varphi(z) = \frac{1}{(2\pi \log R)z^2}$.

Proof. Through the correspondence $z = \exp(\zeta)$, f determines a mapping from the rectangle $\{\zeta : 0 \le \xi \le \log R, 0 \le \eta \le 2\pi\}$ onto the annulus $\{z : 1 < |z| < R'\}$. Let γ be a radial arc in Γ joining the inner and outer boundary contours of X. Then

$$\log R' \le \int_{f(\gamma)} |dz| = \int_{\Gamma} |df| = \int_0^{\log R} |f_\zeta||1 + \tilde{\mu}|d\xi,$$

where $\tilde{\mu} = f_{\bar{\zeta}}/f_\zeta$. Integrating from $\eta = 0$ to $\eta = 2\pi$ we obtain

$$2\pi \log R' \le \iint |f_\zeta||1 + \tilde{\mu}|d\xi d\eta,$$

where the integral is over the rectangle $\{\zeta : 0 \le \xi \le \log R, 0 \le \eta \le 2\pi\}$. Introducing a factor of $\sqrt{1 - |\mu|^2}$ in both numerator and denominator and applying Schwarz's inequality together with the Jacobian change of variable formula (see exercise 3) yields

$$(2\pi \log R')^2 \le \iint |f_\zeta|^2(1 - |\tilde{\mu}|^2)d\xi d\eta \iint \frac{|1 + \tilde{\mu}|^2}{1 - |\tilde{\mu}|^2} d\xi d\eta.$$

Since the first integral on the right hand side is the area of the rectangle in the $\tilde{\zeta}$-plane, we get

$$2\pi \log R' \le \iint \frac{|1 + \tilde{\mu}|^2}{1 - |\tilde{\mu}|^2} d\xi d\eta.$$

Since $\zeta = \log z$, $\tilde{\mu} = \mu\bar{z}/z$, where $\mu = f_{\bar{z}}/f_z$. Noting that $\varphi(z) = \frac{1}{(2\pi \log R)z^2}$, the last inequality can be rewritten as

$$K_0 = \frac{\Lambda(\Gamma^\mu)}{\Lambda(\Gamma)} \le \iint_{1<|z|<R} \frac{\left|1 + \mu\frac{\varphi}{|\varphi|}\right|^2}{1 - |\mu|^2} |\varphi| dx dy. \tag{4}$$

Since

$$\frac{\left|1 + \mu\frac{\varphi}{|\varphi|}\right|^2}{1 - |\mu|^2} \le \frac{(1 + k(f))^2}{1 - k(f)^2} = \frac{1 + k(f)}{1 - k(f)}, \tag{5}$$

from (4) and (5) we obtain

$$\frac{1 + k_0}{1 - k_0} \le \frac{|1 + \mu\varphi/|\varphi||^2}{1 - |\mu|^2} \le \frac{1 + k(f)}{1 - k(f)}.$$

If $k(f) = k_0$, then these inequalities are equalities and hence $\mu\varphi/|\varphi| = k_0$, that is, $\mu = k_0|\varphi|/\varphi$ almost everywhere. □

Exercise 8. By the same technique as used in the previous theorem, except by using concentric circles centered at the origin in place of radial line segments, show that

$$\frac{\Lambda(\Gamma)}{\Lambda(\Gamma^\mu)} \le \iint_{1<|z|<R} \frac{|1 - \mu\varphi/|\varphi||^2}{1 - |\mu|^2} |\varphi| dx dy.$$

Conclude that for small complex numbers t,

$$\log \Lambda(\Gamma^{t\mu}) = \log \Lambda(\Gamma) + 2\mathrm{Re}\left(t \iint_{1<|z|<R} \mu\varphi dxdy\right) + o(t),$$

where $\Gamma^{t\mu} = f_t(\Gamma)$ and f_t satisfies the Beltrami equation $(f_t)_{\bar{z}} = t\mu(f_t)_z$ (see the next section).

Exercise 9. Consider a complex number μ in the unit disc and a number $M > 0$. Show that

$$\frac{|1-\mu e^{-i\theta}|^2}{1-|\mu|^2} = M$$

is the equation of a circle inside the unit disc and tangent to the unit circle with diameter $2M/(M+1)$ and tangent to the unit circle at the point $e^{i\theta}$.

Exercise 10. Consider the map f mapping the unit square onto a figure distorted by dyadic translations along subrectangles of the unit square whose vertical sides are located on the gaps of a standard middle-thirds Cantor set on the unit interval of the imaginary axis. On the largest subrectangle f is defined by the translation $z \mapsto z+1/2$ and on the two rectangles at the next two levels of the Cantor set f is defined by $z \mapsto z+1/4$ and $z \mapsto z+3/4$. The definition of f should be clear from the following picture:

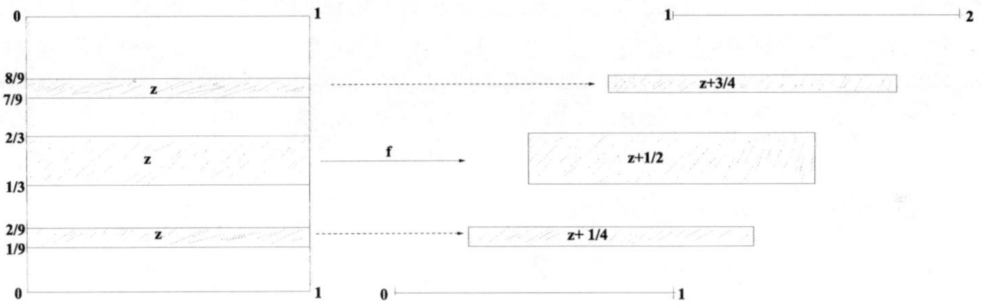

Figure 1.

In figure 1 the scaling on the y-axis is different from the scaling on the x-axis so the unit square appears as a rectangle. Show that f is a homeomorphism onto its image, and in fact, f is Hölder continuous with Hölder exponent $\alpha = \log 2/\log 3$. Furthermore, show that f is not quasiconformal by finding a sequence of curve families Γ_n such that there is no finite value K for which

$$\Lambda(f(\Gamma_n)) \leq K\Lambda(\Gamma_n).$$

Hint: Let Γ_1 be the curve family of arcs lying in the rectangle $\{0 \leq x \leq 1, 2/9 \leq y \leq 1/3\}$ with initial point at $(x, 2/9)$ and endpoint at $(x, 1/3)$. If β is such an arc, then $f(\beta)$ has initial point $(x, 2/9)$ and endpoint $(x+1/2, 1/3)$. Consider the analogous curve families Γ_n lying in the rectangles

$$\{0 \leq x \leq 1, 1/3 - 1/(3^{n+1}) \leq y \leq 1/3\}.$$

Proposition 2. *Suppose f is a C^1–quasiconformal mapping with dilatation K taking a Riemann surface X onto a Riemann surface X' and taking a curve family Γ onto a curve family Γ'. Then $K^{-1}\Lambda(\Gamma) \leq \Lambda(f(\Gamma)) \leq K\Lambda(\Gamma)$.*

Proof. Let w be a local parameter on X' and z be a local parameter on X and assume the mapping f takes z to w. Then for any allowable metric ρ on X,

$$\tilde{\rho}(w) = \left| \frac{\rho(z)}{|w_z| - |w_{\bar{z}}|} \right|$$

defines an allowable metric on X'. Then for $\gamma' = w(\gamma)$ and $w = u + iv = w(z)$, we have

$$\int_{\gamma'} \tilde{\rho}(w)|dw| = \int_\gamma \left| \frac{\rho(z)}{|w_z| - |w_{\bar{z}}|} \right| |w_z dz + w_{\bar{z}} d\bar{z}|$$

$$= \int_\gamma \left| \frac{\rho(z)}{|w_z| - |w_{\bar{z}}|} \right| |w_z + w_{\bar{z}} \frac{d\bar{z}}{dz}||dz| \geq \int_\gamma \rho(z)|dz|$$

and

$$\iint X' \tilde{\rho}(w)^2 du dv = \iint_X \rho(z)^2 \frac{|w_z| + |w_{\bar{z}}|}{|w_z| - |w_{\bar{z}}|} dx dy \leq K A(\rho).$$

This means that we can transport any metric ρ on X to a metric $\tilde{\rho}$ on X' in such a way that curves γ are transported to curves γ' that are at least as long, and such that the area of X' with respect to $\tilde{\rho}$ is no larger than K times the area of X with respect to ρ. Therefore, $\Lambda(\Gamma) \leq K\Lambda(f(\Gamma))$ and the reverse inequality follows by applying the same argument to the inverse mapping. □

4. The Beltrami equation

Suppose that Ω is a plane domain and $\mu(z)$ is a continuous complex valued function defined on Ω with $|\mu(z)| < 1$. Consider the equation

$$f_{\bar{z}}(z) = \mu(z) f_z(z). \tag{6}$$

(6) is called the Beltrami equation, a mapping f satisfying (6) is called a solution, and μ is called the Beltrami coefficient or the complex dilatation of f. We now suppose local homeomorphic solutions are known to exist in a neighborhood of every point z in Ω.

Exercise 11. Show that the set of such solutions define a Riemann surface structure on Ω. That is, show that transition functions are holomorphic.

We remark that for each t with $|t| < 1/\|\mu\|_\infty$, the construction of the previous exercise creates a one-parameter family of Riemann surface structures on X which yield the original structure when $t = 0$.

Now, we consider the following special case of (6). The Beltrami coefficient has the special form $\mu(z) = k|\varphi(z)|/\varphi(z)$ where $\varphi(z)$ is a holomorphic function defined on a plane domain Ω and $0 < k < 1$.

Exercise 12. Assume that $\varphi(z)$ is a holomorphic function defined on Ω. Show that (6) has a local homeomorphic solution when

$$\mu(z) = k\frac{|\varphi(z)|}{\varphi(z)}$$

in a neighborhood of every point where $\varphi(z) \neq 0$ and that μ defines a Riemann surface structure on Ω.

Hint: Let $\zeta(z) = \int \sqrt{\varphi(z)}dz$, in a neighborhood of any point z where $\varphi(z) \neq 0$. ζ is defined by taking an anti-derivative of a power series for $\sqrt{\varphi(z)}$. Note that ζ is defined ambiguously up to a plus or minus sign and an additive constant, but $\zeta = \xi + i\eta$ is a local homeomorphism. The function ζ is called a natural parameter with respect to the quadratic differential $\varphi(z)(dz)^2$. Then form

$$f(z) = \zeta + k\bar{\zeta} = \int \sqrt{\varphi(z)}dz + k\overline{\int \sqrt{\varphi(z)}dz}.$$

Show that $f_{\bar{\zeta}}/f_{\zeta} = k$, that $f_{\bar{z}}/f_z = k\frac{|\varphi|}{\varphi}$ and that f is K-quasiconformal with $K = \frac{1+k}{1-k}$.

Exercise 13. For a holomorphic quadratic differential $\varphi(z)(dz)^2$ defined on a Riemann surface X, the conditions $\varphi(z)(dz)^2 > 0$ and $\varphi(z)(dz)^2 < 0$ define parameterized curves on X called horizontal and vertical trajectories which are perpendicular. These trajectories are mapped to horizontal or vertical line segments in the $\zeta-$plane for any natural parameter ζ.

Exercise 14. In a neighborhood of $z = 0$ draw the critical horizontal trajectories of $z(dz)^2$, $z^2(dz)^2$, and $z^3(dz)^2$. In addition, draw regular horizontal trajectories which are near to $z = 0$. Tell how many prongs there are for the critical horizontal trajectories of $\varphi(z)(dz)^2 = z^n(dz)^2$ near to $z = 0$. Putting $\zeta = d\xi + id\eta = \int \sqrt{\varphi(z)}dz$, tell why it is not possible to find η as a single valued function of z in a neighborhood of $z = 0$ when n is an odd integer.

Ultimately you will learn that equation (6) admits global homeomorphic solutions on the Riemann sphere even if $\mu(z)$ is assumed only to be measurable. That is, if $\Omega = \mathbb{C} \cup \{\infty\}$ and $\mu(z)$ is measurable and $\|\mu\|_\infty < 1$, then there is a global homeomorphism f of $\mathbb{C} \cup \{\infty\}$ satisfying (6), where the partial derivatives with respect to z and $\bar{z}$ are taken in the generalized sense. (See [29], [3],[23]). Moreover, if the solution $f^\mu(z)$ is normalized to fix three points on the Riemann sphere, say 0, 1 and ∞, then it is unique and $w = f^\mu(z)$ is a holomorphic function of μ, ([12],[4]). This version of the existence theorem for equation (6) (with holomorphic dependence of the solution on the Beltrami coefficient) is sometimes called the *measurable Riemann mapping theorem* (denoted by MRMT). In this phrase the adjective measurable refers not to the theorem but to the coefficient in the Beltrami equation. Moreover, a central aspect of the theorem is not mentioned in its title, namely, the fact that dependence of the solution $w = f^\mu(z)$ on the coefficient μ is holomorphic.

Here we briefly outline the proof of MRMT with holomorphic dependence. It is based on properties of the following two singular integral operators:

$$P\mu(z) = -\frac{1}{\pi}\iint \mu(\zeta)\left\{\frac{1}{\zeta - z} - \frac{1}{\zeta}\right\} d\xi d\eta$$

and

$$T\mu(z) = -\frac{1}{\pi}\iint \mu(\zeta)\left\{\frac{1}{(\zeta - z)^2} - \frac{1}{\zeta^2}\right\} d\xi d\eta.$$

P is a smoothing operator in the sense that if $\mu(\zeta)$ has compact support and is in L^p with $p > 2$, then

$$|P\mu(z_1) - P\mu(z_2)| \leq C|z_1 - z_2|^{1-2/p},$$

and T preserves smoothness in the sense that for μ in L^p, $(p > 2)$ and with compact support

$$\|T\mu\|_p \leq C_p\|\mu\|_p,$$

with $C_p \to 1$ as p decreases to 2. Moreover, if partial derivatives are interpreted in the generalized sense, then

$$(P\mu(z))_{\bar{z}} = \mu(z) \tag{7}$$

and

$$(P\mu(z))_z = T\mu(z). \tag{8}$$

Then a (non-normalized) solution to (6) is given by

$$f^\mu(z) = z + P\mu(z) + P\mu T\mu(z) + P\mu T\mu T\mu(z) + \cdots. \tag{9}$$

We leave it to the reader to check that formal term-by-term differentiation of this infinite sum yields

$$\begin{aligned}(f^\mu(z))_{\bar{z}} &= \mu(z) + \mu T\mu(z) + \mu T\mu T\mu(z) + \cdots\\ &= \mu(z)\left(1 + T\mu(z) + T\mu T\mu(z) + T\mu T\mu T\mu(z) + \cdots\right),\end{aligned}$$

and

$$(f^\mu(z))_z = 1 + T\mu(z) + T\mu T\mu(z) + T\mu T\mu T\mu(z) + \cdots.$$

Here, we make a few comments about the meaning of (9) and the structure of the proof. We assume that $\|\mu\|_\infty = k < 1$ and that μ has compact support. In (9) we view μT as the operator given by first applying the linear map T and afterwards multiplying by μ. Thus, for example, the expression

$$\mu T\mu T\mu(z)$$

means first apply T to μ, then multiply the resulting L_p function by μ, then apply T again, and finally multiply by μ. Moreover, the norm of μT as an operator satisfies $\|\mu T\|_p \leq C_p k$. Since $k < 1$ and for p larger than 2 but sufficiently close to 2, C_p approaches 1, $C_p k$ is less than 1. Therefore, μT is an operator with norm less than 1, and $(I - \mu T)^{-1}$ is a bounded operator on L_p. So the solution (9) can be rewritten as a perturbation of the identity:

$$f^{\mu}(z) = z + P((I - \mu T)^{-1})(\mu).$$

Note that the final application of P smooths the L_p-function $(I - \mu T)^{-1}(\mu)$ and so the resulting solution is Hölder continuous with Hölder exponent $1 - 2/p$. Several more steps are involved in showing that the series yields a homeomorphism of the Riemann sphere for arbitrary μ in L_∞ with $\|\mu\|_\infty < 1$, and that the operators P and T satisfy properties (7) and (8) for generalized partial derivatives. For these steps we refer to [3] and [29].

We point out that if μ is replaced by $t\mu$ for a complex number t, then the solution $f^{t\mu}(z)$ is exhibited as a power series in t convergent for $|t| < 1/\|\mu\|_\infty$, its derivative at $t = 0$ is the first term of the series, namely, the vector field $V(z)\frac{\partial}{\partial z}$. When the solution is normalized so that $f^{t\mu}(z)$ fixes the points 0, 1 and ∞, then this vector field vanishes at those points and is given by the formula

$$\begin{aligned} V(z) &= -\frac{1}{\pi}\iint \mu(\zeta)\left\{\frac{1}{\zeta - z} - \frac{z}{\zeta - 1} + \frac{z-1}{\zeta}\right\} d\xi d\eta \\ &= -\frac{z(z-1)}{\pi}\iint \mu(\zeta)\left\{\frac{1}{\zeta(\zeta-1)(\zeta-z)}\right\} d\xi d\eta. \end{aligned} \tag{10}$$

5. The Reich-Strebel inequality and Teichmüller's uniqueness theorem

To generalize the example of the previous section, we assume $f_0 : X \to X^\mu$ is a mapping satisfying the Beltrami equation

$$f_{\bar{z}}(z) = \mu(z) f_z(z),$$

where $\mu = k_0|\varphi_0|/\varphi_0$ and φ_0 is a holomorphic quadratic differential on X with $\iint X|\varphi_0|dxdy = 1$. Such a Beltrami coefficient is said to be of Teichmüller form. We also assume that f is another mapping from X to X^μ which is homotopic to f_0.

For non-compact Riemann surfaces X the form of homotopy we require needs to be clarified. We require that there exists a map $f_t(z)$ continuous simultaneously in both variables such that it is equal to f_0 for $t = 0$ and equal to f for $t = 1$, and $f_0(p) = f_t(p) = f(p)$ for every t with $0 \le t \le 1$ and every boundary point p of X. For now we will assume that X is compact (and so has no boundary) or a punctured compact surface with only finitely many punctures. In the later case, the boundary of X consists of finitely many points.

Just as in the case of the annulus, we will obtain the following Reich-Strebel inequality

$$K_0 \le \iint_X \frac{\left|1 + \nu\frac{\varphi_0}{|\varphi_0|}\right|^2}{1 - |\nu|^2}|\varphi_0|dxdy, \tag{11}$$

where ν is the Beltrami coefficient of any competing map f. In this section, we first show how this inequality implies the next theorem.

Theorem 1 (Teichmüller's Uniqueness Theorem). *If a quasiconformal mapping f_0 from a Riemann surface X to another Riemann surface X' has Beltrami coefficient of Teichmüller form $k_0|\varphi_0|/\varphi_0$ for some holomorphic quadratic differential φ_0 with $\|\varphi_0\| = 1$ and $0 < k_0 < 1$, then any other quasiconformal mapping f^μ in the same homotopy class either has the same Beltrami coefficient or $\|\mu\|_\infty > k_0$. Moreover, if f_0 and f_1 are two mappings with Beltrami coefficients of Teichmüller form $k_0|\varphi_0|/\varphi_0$ and $k_1|\varphi_1|/\varphi_1$, then $k_0 = k_1$ and $\varphi_0 = \varphi_1$.*

Proof. Clearly,

$$\frac{\left|1+\mu\frac{\varphi_0}{|\varphi_0|}\right|^2}{1-|\mu|^2} \le \frac{(1+k(f))^2}{1-k(f)^2} = \frac{1+k(f)}{1-k(f)}. \tag{12}$$

By the Reich-Strebel inequality (11) and the above estimate, we obtain

$$K_0 \le \iint_X \frac{\left|1+\mu\frac{\varphi_0}{|\varphi_0|}\right|^2}{1-|\mu|^2}|\varphi_0|dxdy \le K(f^\mu), \tag{13}$$

that is,

$$\frac{1+k_0}{1-k_0} \le \iint_X \frac{\left|1+\mu\frac{\varphi_0}{|\varphi_0|}\right|^2}{1-|\mu|^2}|\varphi_0|dxdy \le \frac{1+k(f)}{1-k(f)}.$$

Therefore, if f^μ is also extremal in the same class, then $k(f) = k_0$ and hence these inequalities are equalities, which forces $\mu\varphi_0/|\varphi_0| = k_0$, that is, $\mu = k_0|\varphi_0|/\varphi_0$ almost everywhere. □

Before starting out the work of proving (11), we take note of two closely related inequalities.

Theorem 2. *Suppose f is a quasiconformal self map of X homotopic to the identity on X and φ is any holomorphic quadratic differential on X with*

$$\|\varphi\| = \iint_X |\varphi|dxdy = 1.$$

Then

$$1 \le \iint_X \frac{\left|1+\mu\frac{\varphi}{|\varphi|}\right|^2}{1-|\mu|^2}|\varphi|dxdy. \tag{14}$$

We note that (14) is the special case of (11) when $K_0 = 1$, and we postpone the proof until the end of section 7. For now we show only how Theorem 2 implies another inequality, which is a version of *The Main Inequality of Reich and Strebel* and which implies (11).

Theorem 3. *Let f and g be two quasiconformal maps from X to Y, with Beltrami differential coefficients μ_f and μ_g. Assume $g \circ f^{-1}$ is homotopic to the identity on Y,*

$w = u + iv = f(z)$, and let $\psi(w)(dw)^2$ be an integrable, holomorphic, quadratic differential on Y with $\iint_Y |\psi| du dv = 1$. Then

$$1 \leq \iint_Y \frac{\left|1 + \mu_g \frac{1}{\theta}\frac{\psi}{|\psi|}\right|^2}{1 - |\mu_g|^2} \frac{\left|1 - \mu_f \frac{1}{\theta}\frac{\psi}{|\psi|}\left(\frac{1+\mu_g \frac{\bar{\mu}_f}{\mu_f}\frac{\bar{\psi}}{|\psi|}}{1+\mu_g \frac{1}{\theta}\frac{\psi}{|\psi|}}\right)\right|^2}{1 - |\mu_f|^2} |\psi(w)| du dv. \tag{15}$$

where $\theta = \bar{p}/p$ and $p = f_z$.

Proof. Note that

$$\mu_{g \circ f^{-1}} = \left[\frac{\mu_g - \mu_f}{1 - \overline{\mu_f}\mu_g} \cdot \frac{1}{\theta}\right] \circ f^{-1}. \tag{16}$$

On using (14) for the surface Y and replacing μ by (16), for $\|\psi\| = 1$ we obtain

$$1 \leq \iint_Y \frac{\left|1 + \sigma \frac{\psi}{|\psi|}\right|^2}{1 - |\sigma|^2} |\psi(w)| du dv. \tag{17}$$

where

$$\sigma(w) = \left[\frac{\mu_g - \mu_f}{1 - \overline{\mu_f}\mu_g} \cdot \frac{1}{\theta}\right] \circ f^{-1}(w).$$

(17) simplifies to

$$1 \leq \iint_Y \frac{\left|(1 - \overline{\mu_f}\mu_g) + (\mu_g - \mu_f)\frac{1}{\theta}\frac{\psi}{|\psi|}\right|^2}{|1 - \overline{\mu_f}\mu_g|^2 - |\mu_g - \mu_f|^2} |\psi(w)| du dv. \tag{18}$$

The part inside the absolute value in the numerator of this integrand can be rewritten in the following way

$$(1 - \overline{\mu_f}\mu_g) + (\mu_g - \mu_f)\frac{1}{\theta}\frac{\psi}{|\psi|} = \left[1 + \mu_g \frac{1}{\theta}\frac{\psi}{|\psi|}\right]\left[1 - \mu_f \frac{1}{\theta}\frac{\psi}{|\psi|}\left(\frac{1 + \mu_g \frac{\overline{\mu_f}}{\mu_f}\frac{\bar{\psi}}{|\psi|}}{1 + \mu_g \frac{1}{\theta}\frac{\psi}{|\psi|}}\right)\right]$$

Therefore (18) can be rewritten as

$$1 \leq \iint_Y \frac{\left|1 + \mu_g \frac{1}{\theta}\frac{\psi}{|\psi|}\right|^2}{1 - |\mu_g|^2} \cdot \frac{\left|1 - \mu_f \frac{1}{\theta}\frac{\psi}{|\psi|}\left(\frac{1+\mu_g \frac{\overline{\mu_f}}{\mu_f}\frac{\overline{\psi}}{|\psi|}}{1+\mu_g \frac{1}{\theta}\frac{\psi}{|\psi|}}\right)\right|^2}{1 - |\mu_f|^2} \cdot |\psi(w)| du dv. \tag{19}$$

□

We now go to the proof of (11). We assume $w = f(z)$ is of Teichmüller form, with Beltrami coefficient $k\frac{|\varphi_0(z)|}{\varphi_0}$. This means that in terms of $\zeta = \int \sqrt{\varphi_0(z)} dz$, $f(\zeta) = \tilde{\zeta} = \tilde{\xi} + i\tilde{\eta} = K^{1/2}\xi + iK^{-1/2}\eta$ and the push-forward of φ_0 under f is a holomorphic quadratic differential on X' of the form $\psi_0(\tilde{\zeta})(d\tilde{\zeta})^2 = (d\tilde{\zeta})^2$ with norm also equal to 1. Note that in terms of the natural parameters ζ, and $\tilde{\zeta}$, $\theta \equiv 1$, $\psi \equiv 1$ and $\mu_f \equiv k$.

Applying (19) to this mapping f and to the quadratic differential $\psi_0(\tilde{\zeta})(d\tilde{\zeta})^2$, (19) becomes

$$1 \leq \iint_Y \frac{|1+\mu_g \circ f^{-1}(\tilde{\zeta})|^2}{1-|\mu_g \circ f^{-1}(\tilde{\zeta})|^2} \cdot \frac{(1-k)^2}{1-k^2} d\tilde{\xi} d\tilde{\eta}$$

$$= \frac{1}{K} \iint_X \frac{|1+\mu_g(\zeta)|^2}{1-|\mu_g(\zeta)|^2} d\xi d\eta = \frac{1}{K} \iint_X \frac{\left|1+\mu_g(z)\frac{\varphi_0(z)}{|\varphi_0(z)|}\right|^2}{1-|\mu_g|^2} |\varphi_0(z)| dxdy,$$

and then we obtain (11).

6. The minimum norm principle

Now suppose we are given two integrable quadratic differentials φ and ψ. Also suppose φ is holomorphic and ψ is locally L_1 with respect to $d\xi d\eta$, where $\zeta = \xi + i\eta$, the natural parameter measure for φ. For any vertical trajectory β of φ, we can form $h_\psi(\beta) = \int_\beta |\mathrm{Im}(\sqrt{\psi(z)}dz)|$, where the integral is unoriented and taken in the positive sense regardless of the orientation of β. In the unusual circumstance that β runs along a closed trajectory, by $h_\psi(\beta)$ we mean the arclength integral of $|\mathrm{Im}(\sqrt{\psi(z)}dz)|$ along the entire parametrization of β.

Theorem 4 (The Minimum Norm Principle). *Assume X is a Riemann surface of finite analytic type and φ is a holomorphic quadratic differential on X with $\|\varphi\| = \iint_X |\varphi| dxdy < \infty$. Let ψ be another locally integrable quadratic differential and assume there is a constant M such that for every non-critical vertical segment β for φ, one has $h_\varphi(\beta) \leq h_\psi(\beta) + M$. Then*

$$\|\varphi\| \leq \iint_X |\sqrt{\varphi}||\sqrt{\psi}| dxdy.$$

Proof. As usual, we let $\zeta = \xi + i\eta$ be a natural parameter for φ, that is, $(d\zeta)^2 = \varphi(z)(dz)^2$. Put

$$g(p) = \int_{\beta_p} |\mathrm{Im}(\sqrt{\psi(\zeta)}d\zeta)|,$$

where β_p is a vertical segment for φ with height b and midpoint p. Since the number of zeroes of φ on X is finite, the number of vertical trajectories of φ which lead into these zeroes is countable. Therefore, the set of regular vertical trajectories of φ is of full measure and g is defined on a set of full measure in X. We wish to establish the following inequality.

$$\iint_X g(p) d\xi d\eta \leq b \iint_X |\sqrt{\psi(\zeta)}| d\xi d\eta. \tag{20}$$

All that is required is a change of order of integration with respect to $d\eta$ along any regular vertical trajectory of φ. Since the transformation $\zeta = \xi + i\eta \mapsto \xi + i(\eta + t)$ has Jacobian identically equal to 1, this change of variable is justified so long as we can view this transformation as an almost everywhere defined self-mapping of X. However, since trajectories of a quadratic differential do not in general have globally defined orientation on X, it is not possible to distinguish between the maps $\zeta = \xi + i\eta \mapsto \xi + i(\eta + t)$ and $\zeta = \xi + i\eta \mapsto \xi - i(\eta + t)$. To avoid this problem, we pass to the double cover $\tilde{X}$ of X where $\sqrt{\varphi}$ can be globally defined, and where the norm of the lifted quadratic differential $|\tilde{\varphi}|$ on $\tilde{X}$ is exactly twice norm of φ on X. After these steps we end up with (20).

To go on with the proof, we now invoke the hypothesis that $b - M \leq \int_{\beta_p} |\mathrm{Im}(\sqrt{\psi} dz)|$, and so $b - M \leq g(p)$ for almost all p. From (20) this implies

$$(b - M) \iint_X d\xi d\eta \leq b \iint_X |\sqrt{\psi(\zeta)}| d\xi d\eta,$$

and dividing both sides by b and taking the limit as $b \to \infty$, we obtain

$$\iint_X d\xi d\eta \leq \iint_X |\sqrt{\psi(\zeta)}| d\xi d\eta,$$

where $\psi(\zeta)$ is the expression for the quadratic differential ψ in terms of the local parameter ζ. Since $d\zeta = \sqrt{\varphi(z)} dz$,

$$d\xi d\eta = \frac{1}{2}|d\zeta d\bar{\zeta}| = \frac{1}{2}|\varphi dz d\bar{z}| = |\varphi| dx dy,$$

and

$$\sqrt{\psi(\zeta)} = \frac{\sqrt{\psi(z)}}{\sqrt{\varphi(z)}},$$

by changing from the natural parameter ζ to the local parameter z, we obtain the desired inequality, namely,

$$\|\varphi\| \leq \iint_X |\sqrt{\varphi}||\sqrt{\psi}| dx dy. \qquad \square$$

7. The heights argument

Lemma 3. *Let φ be a holomorphic quadratic differential on X with $\|\varphi\| < \infty$. Let f be a quasiconformal self-mapping of X which is homotopic to the identity. Then there exists a constant M such that for every non-critical vertical segment β of φ, one has*

$$h_\varphi(\beta) \leq h_\varphi(f(\beta)) + M.$$

the constant M depends on φ and f but not on β.

Proof. Let $\bar{X}$ be the completion of X with the n punctures filled in. So $\bar{X}$ is compact with no punctures, and since f is quasiconformal, it extends to a quasiconformal map fixing the punctures. By hypothesis this extension is homotopic to the identity by a homotopy f_t which fixes the punctures pointwise. The infinitesimal form $|\varphi|^{1/2}|dz|$ determines a finite valued metric on X. To see that the distance from a point in X to a puncture is finite, one observes that φ has at most simple poles and so to find the length of a short arc ending at a puncture, one has to calculate an integral of the form $\int_0^a t^{-1/2}dt$, and this clearly converges.

Let f_t be the homotopy connecting f to the identity, so $f_0(p) = p$ and $f_1(p) = f(p)$. Let $l(p)$ be the infimum of the $\varphi-$lengths of all curves which go from p to $f(p)$ and which are homotopic with fixed endpoints to the curve $t \mapsto f_t(p)$. Clearly, $l(p)$ is a continuous function on the compact set $\bar{X}$. Let M_1 be the maximum of this function.

Let β be a noncritical vertical segment for φ with endpoints p and q. The segment β and the curve which consists of a curve homotopic to $t \mapsto f_t(p)$ followed by $f(\beta)$ and then followed by a curve homotopic to $t \mapsto f_{1-t}(q)$ is homotopic to β with fixed endpoints. From the following lemma it will follow that

$$h_\varphi(\beta) \leq h_\varphi(f_t(p)) + h_\varphi(f(\beta)) + h_\varphi(f_{1-t}(q)).$$

Since the first and third terms of this inequality are each bounded by M_1, the lemma follows with $M = 2M_1$. □

Lemma 4. *Let φ be a holomorphic quadratic differential on X. Suppose β is a segment of a vertical trajectory of φ and that γ is an arc in X with the same endpoints as β such that the arc β followed by the arc γ^{-1} forms a closed curve homotopic to a point. Then $h_\varphi(\beta) \leq h_\varphi(\gamma)$.*

Proof. It is sufficient to prove this inequality for lifts of the arcs β and γ in the universal cover $\tilde{X}$. If β followed by γ^{-1} forms a closed curve in X homotopic to a point, the lifts of these curves form a closed curve in $\tilde{X}$ contractible to a point in $\tilde{X}$. Because there can be only finitely many singularities of φ in a bounded simply connected domain containing this curve, β may be subdivided into a finite number of subintervals such that horizontal strips emanating from these subintervals must be crossed by γ. Since the height of the strip is constant, the height of β along each strips is less than or equal to the height of γ. By adding up over all of the strips, the lemma follows. □

Now our approach to obtain Theorem 2 is to apply the minimum norm principle of the previous section to the quadratic differential

$$\psi(z) = \varphi(f(z))f_z^2(z)\,(1 - \mu\varphi/|\varphi|)^2\,. \tag{21}$$

We leave it to the reader to check that this is a quadratic differential, and go on to show that for any vertical segment β relative to φ,

$$h_\psi(\beta) = h_\varphi(f(\beta)).$$

By definition,

$$h_\varphi(f(\beta)) = \int_{f(\beta)} |\mathrm{Im}(\sqrt{\varphi(f)}df)| = \int_\beta |\mathrm{Im}(\sqrt{\varphi(f(z))}f_z(1 + \mu(z)(d\bar{z}/dz))dz)|.$$

Note that for a natural parameter ζ, $d\zeta = \sqrt{\varphi(z)}dz$, and for a parametrization $\zeta = it$ of the vertical trajectory β, $\frac{\overline{d\zeta}}{d\zeta} = -1$. Therefore, $d\bar{z}/dz = -\varphi/|\varphi|$ along β. The final result is that $h_\varphi(f(\beta)) = h_\psi(\beta)$. Therefore, we obtain the following lemma.

Lemma 5. *There exists a constant M such that for all regular vertical trajectories β of φ,*

$$h_\varphi(\beta) \le h_\psi(\beta) + M.$$

Proof. Merely apply the previous lemma and the previously derived equality $h_\varphi(f(\beta)) = h_\psi$. □

Now we are ready to prove inequality (14) in Theorem 2. From the minimum norm principle,

$$\|\varphi\| \le \iint |\varphi(f(z))|^{1/2}|f_z| \quad |1 - \mu\varphi/|\varphi|| \quad |\varphi|^{1/2}dxdy.$$

Introducing a factor of $(1-|\mu|^2)^{1/2}$ into the numerator and denominator and applying Schwarz's inequality yields

$$\|\varphi\| \le \left(\iint |\varphi(f(z))||f_z|^2(1-|\mu|^2)dxdy\right)^{1/2}\left(\iint \frac{|1-\mu\varphi/|\varphi||^2}{1-|\mu|^2}|\varphi|dxdy\right)^{1/2}.$$

Since the first factor on the right hand side of this expression is simply equal to $\|\varphi\|^{1/2}$, we obtain

$$1 \le \iint \frac{|1-\mu\varphi/|\varphi||^2}{1-|\mu|^2}|\varphi|dxdy.$$

Replacing φ by $-\varphi$, we obtain (14).

8. The Hamilton-Krushkal condition and Teichmüller's existence theorem

Suppose $f : X \to Y$ is a quasiconformal map between Riemann surfaces and $f_{\bar z}(z)/f_z(z) = \mu(z)$ is its Beltrami differential coefficient. We shall say f is extremal in its homotopy class if no other mapping in the same class has smaller dilatation. If we let

$$K = \inf\{K(g) : \text{where } g \text{ is in the homotopy class of } f\},$$

then by definition there is a sequence of mappings g_n in the homotopy class of f with $K(g_n) < K + 1/n$. The sequence of the maps g_n is uniformly Hölder continuous, and therefore it is a normal family. Let g be a limit of a subsequence of $\{g_n\}_{n=1}^{\infty}$. Then g, is also Hölder continuous and $K(g) \le K$. Moreover, g is in the same homotopy class as that of f because homotopy equivalence is determined by the images of closed curves γ on X. That is to say, two maps g_0 and g_1 are homotopic if g_0 and g_1 take the same values at the punctures and if for every closed curve γ on X, $g_0(\gamma)$ is freely homotopic to $g_1(\gamma)$ in Y. Thus the limit g must be in the same homotopy class of f. We conclude that $K(g) = K$ and every homotopy class of a quasiconformal homeomorphism f has an extremal representative, which is the existence for Teichmüller's Theorem.

The more interesting and useful part of Teichmüller's Theorem is to find an explicit description of the unique extremal mapping, this is the interpretation through stretching and shrinking of the horizontal and vertical trajectories of a holomorphic quadratic differential. In other words, the Beltrami coefficient of the extremal mapping is of Teichmüller form. By pairing Beltrami differentials with holomorphic quadratic differential, Hamilton and Krushkal obtained the following useful condition.

Theorem 5 ([21][25] The Hamilton-Krushkal Condition). *If f^{μ} is extremal in its homotopy class, then*

$$k = (K-1)/(K+1) = \|\mu\|_{\infty} = \sup Re \iint_X \mu\varphi dxdy, \tag{22}$$

where the supremum is taken over all holomorphic quadratic differentials φ with $\|\varphi\| = \iint_X |\varphi| dxdy = 1$.

It is clear that for any μ,

$$\sup_{\|\varphi\|=1} \text{Re} \iint_X \mu\varphi dxdy \leq \|\mu\|_{\infty}.$$

We say that μ satisfies the Hamilton-Krushkal condition if the equality holds.

Proof. Start by assuming $k = \|\mu\|_{\infty} > k_0$, where k_0 is the supremum in (22). By the Hahn-Banach and the Riesz representation theorems, there exists an L_{∞}-complex valued function ν such that

$$\iint_X \mu\varphi dxdy = \iint_X \nu\varphi dxdy$$

such that $\|\nu\|_{\infty} = k_0$, the supremum in (22). Hence $\mu - \nu$ is infinitesimally trivial, and we shall show in the next section that this implies there exists a smooth curve σ_t of Beltrami coefficients such that

$$\|\sigma_t - t(\mu - \nu)\|_{\infty} = O(t^2),$$

and such that for each t, f^{σ_t} is a self-map of X homotopic to the identity. For brevity let $\sigma_t = \sigma$ and form $f^{\tau} = f^{\mu} \circ (f^{\sigma})^{-1}$. Clearly, f^{τ} is in the same homotopy class as μ, and we will show that, for sufficiently small $t > 0$, $\|\tau\|_{\infty} < \|\mu\|_{\infty}$, which contradicts with the assumption that μ is extremal in its class and completes the proof of the theorem. Note that

$$\tau(f^{\sigma}(z)) = \frac{\mu - \sigma}{1 - \bar{\sigma}\mu} \cdot \frac{1}{\theta}, \tag{23}$$

where $\theta = \bar{p}/p$ and $p = f_z^{\sigma}$. This implies

$$|\tau \circ f^{\sigma}|^2 = \frac{|\mu|^2 - 2\,\text{Re}\,\mu\bar{\sigma} + |\sigma|^2}{1 - 2\,\text{Re}\,\mu\bar{\sigma} + |\mu\sigma|^2},$$

which gives

$$|\tau \circ f^\sigma| = |\mu| - \frac{1-|\mu|^2}{|\mu|} \operatorname{Re}(\mu\bar{\sigma}) + O(t^2).$$

Replacing σ by $t(\mu - \nu)$, we obtain

$$|\tau \circ f^\sigma| = |\mu| - t\frac{1-|\mu|^2}{|\mu|} \operatorname{Re}(|\mu|^2 - \mu\bar{\nu}) + O(t^2). \tag{24}$$

Since $k_0 = \|\nu\|_\infty < k = \|\mu\|_\infty$, by putting

$$S_1 = \{z \in X : |\mu(z)| \leq (k+k_0)/2\}, \quad \text{and}$$
$$S_2 = \{z \in X : (k+k_0)/2 < |\mu(z)| \leq k\},$$

we obtain $S_1 \cup S_2 = X$ and (23) implies there exist $\delta_1 > 0$ and $c_1 > 0$ such that for $0 < t < \delta_1$,

$$|\tau \circ f^\sigma(z)| \leq k - c_1 t \quad \text{for } z \text{ in } S_1.$$

For z in S_2 the coefficient of t in (24) is bounded below by

$$\frac{1-k^2}{k} \cdot \left[\left(\frac{k+k_0}{2}\right)^2 - k_0 k\right] = \frac{1-k^2}{k} \cdot \left(\frac{k-k_0}{2}\right)^2 > 0.$$

Therefore, (24) implies there exist $\delta_2 > 0$ and $c_2 > 0$ such that for $0 < t < \delta_2$,

$$|\tau \circ f^\sigma| \leq k - c_2 t \text{ for } z \text{ in } S_2.$$

Putting these two statements for S_1 and for S_2 together, we find that $\|\tau\|_\infty < k$ for sufficiently small $t > 0$, and this proves the theorem. □

Theorem 6. *Suppose X is a Riemann surface of finite analytic type and assume that it is given the finite dimensionality of the linear space holomorphic quadratic differentials on X. Let f be a quasiconformal map from X to another Riemann surface Y. Then in the homotopy class of f there exists an extremal representative f_0 homotopic to f and its Beltrami coefficient is of the form $k_0|\varphi_0|/\varphi_0$. The constant k_0 is uniquely determined by f and if $k_0 > 0$, then φ is uniquely determined up to positive multiple.*

Proof. By the remark preceding Theorem 5 we know that in any homotopy class of quasiconformal mappings from one surface to another there exists an extremal representative. We also know from Theorem 5 that the extremality of f in its Teichmüller class forces the linear condition

$$\|\mu\|_\infty = \sup \operatorname{Re} \iint_X \mu\varphi dxdy,$$

where the supremum is taken over all holomorphic quadratic differentials with $\iint |\varphi| dxdy = 1$. By the finite dimensionality of the space of integrable quadratic differentials on X, there is a quadratic differential φ_0 with $\|\varphi_0\| = 1$ such that

$$\sup \operatorname{Re} \iint_X \mu\varphi dxdy = \operatorname{Re} \iint_X \mu\varphi_0 dxdy,$$

and then

$$\iint [\|\mu\|_\infty |\varphi_0| - \mathrm{Re}(\mu\varphi_0)] = 0. \tag{25}$$

Since

$$|\mu\varphi_0| \le \|\mu\|_\infty |\varphi_0|,$$

the equality (25) forces that $\mu = \|\mu\|_\infty \frac{|\varphi_0|}{\varphi_0}$ almost everywhere. □

9. Trivial and infinitesimally trivial Beltrami differentials

Let $M(X)$ be the space of Beltrami coefficients μ with $\|\mu\|_\infty < 1$. Since $M(X)$ is the open unit ball in a complex Banach space, it is also a complex manifold. The natural map Φ from $M(X)$ onto $T(X)$ takes a Beltrami coefficient μ onto its Teichmüller equivalence class $\Phi(\mu)$, and the complex structure on $T(X)$ is the one which makes Φ holomorphic.

The preimage under Φ of the Teichmüller class of the identity corresponds to those mappings f from X to X' for which there is a conformal map c from X' to X such that $c \circ f$ is a quasiconformal self-mapping of X homotopic to the identity. We denote by $M_0(X)$ the subset in $M(X)$ consisting of of Beltrami coefficients of such mappings. It is called the space of trivial Beltrami coefficients. By Theorem 2 every Beltrami coefficient in $M_0(X)$ satisfies inequality (14). Note that by changing φ to $-\varphi$, the numerator in the integrand of the integral in the right hand side of (14) (at the end of Section 7) can be replaced by

$$\left|1 - \mu\frac{\varphi}{|\varphi|}\right|^2,$$

and multiplying this by $|\varphi|$, we obtain

$$\left|1 - \mu\frac{\varphi}{|\varphi|}\right|^2 |\varphi| = |\varphi|(1 - |\mu|^2) - 2\mathrm{Re}\ \mu\varphi + 2|\mu|^2|\varphi|.$$

Since $\iint_X |\varphi| dxdy = 1$, from (14) we obtain

$$\mathrm{Re} \iint_X \frac{\mu\varphi}{1 - |\mu|^2} dxdy \le \iint_X \frac{|\mu|^2|\varphi|}{1 - |\mu|^2} dxdy. \tag{26}$$

Since this form of the inequality is unchanged if φ is multiplied by a positive number, in (26) we only need to impose the condition that $\|\varphi\| = \iint_X |\varphi| dxdy < \infty$. By reversing these steps, it is also clear that (26) implies (14).

Now suppose μ_t is a smooth curve of Beltrami coefficients passing through the zero Beltrami coefficient at time $t = 0$ with tangent vector ν, that is, suppose

$$\left\|\frac{\mu_t - t\nu}{t}\right\|_\infty \to 0 \tag{27}$$

as $t \to 0$. By letting t be a real parameter and a purely imaginary parameter respectively and calculating the first order variation in inequality (26), we obtain that if ν is a tangent vector to a μ_t lying in $M_0(X)$, then

$$\iint_X \nu\varphi dxdy = 0 \tag{28}$$

for all holomorphic quadratic differentials φ with $\|\varphi\| < \infty$.

Definition. *A Beltrami differential ν is called infinitesimally trivial if it satisfies (28) for all integrable holomorphic quadratic differentials φ.*

We have already seen that the burden of proof of the Hamilton-Krushkal condition in Theorem 5 is to show the converse of (27), that is, we need the following theorem.

Theorem 7. *If a Beltrami differential ν is infinitesimally trivial, then there exists a curve of trivial Beltrami coefficients μ_t passing through the origin at time $t = 0$ with the property that*

$$\left\| \frac{\mu_t - t\nu}{t} \right\|_\infty \to 0$$

as $t \to 0$.

Proof. There are many references for this well-known theorem, [1,3,6,9,19,20,33,34]. We first give a proof in the case where X is the Riemann sphere with n points removed. After treating this case, we outline the proof for any surface of genus g with n punctures, provided that $3g - 3 + n > 0$. □

The case X has genus zero

Note that in the genus zero case, the Möbius transformations are triply-transitive, and so any two such surfaces are conformal if $n \leq 3$, and in that case the Teichmüller space of X consists of just one point, $M_0(X) = M(X)$, and there is nothing to prove. If n is 4 or more, by the action of a Möbius transformation we can move three of the points to 0, 1 and ∞, and label the remaining $n - 3$ points by $p_1, \ldots, p_{n-3}$. Thus $X = \mathbb{C}\backslash\{0, 1, p_1, \ldots, p_{n-3}\}$, and an arbitrary integrable, holomorphic quadratic differential on X has the form

$$\varphi(z)(dz)^2 = \frac{a_0 + a_1 z + \cdots + a_{n-4}z^{n-4}}{z(z-1)(z-p_1)\cdots(z-p_{n-3})}(dz)^2, \tag{29}$$

where $a_0, \ldots, a_{n-4}$ are arbitrary complex numbers. This quadratic differential has at most simple poles at n points, namely, 0, 1, $p_1, \ldots, p_{n-3}$ and at ∞. To see the order of the pole at ∞ we need to express $\varphi(z)(dz)^2$ in terms of the local parameter $w = 1/z$ which vanishes at $z = \infty$. Using the equation $\varphi^z(z)(dz)^2 = \varphi^w(w)(dw)^2$, we obtain

$$\varphi^w(w) = \frac{1}{w^4} \cdot \frac{a_0 + a_1(1/w) + \cdots + a_n(1/w)^{n-4}}{1/w((1/w)-1)((1/w)-p_1)\cdots((1/w)-p_{n-3})}.$$

Multiplying both the numerator and denominator by w^{n-4}, we see that

$$\varphi^{w}(w) = \frac{1}{w} \cdot \frac{a_0 w^{n-4} + a_1 w^{n-5} + \cdots + a_n}{1(1-w)(1-wp_1)\cdots(1-wp_{n-3})},$$

which has a simple pole at $w = 0$ unless $a_n = 0$, in which case it is holomorphic at $w = 0$. Since $\varphi(z)(dz)^2$ has at most a simple zero at any one of these n points, it is integrable in a neighborhood of each of these points and therefore integrable over $\mathbb{C} \cup \{\infty\}$.

Consider a family of disjoint discs D_j centered at p_j, $1 \le j \le n-3$, with positive radii, and let $\Psi : M(X) \to (\mathbb{C}\backslash\{0,1\})^{n-3}$ be the map which assigns to μ in $M(X)$ the vector $(f^{\mu}(p_1), \ldots, f^{\mu}(p_{n-3}))$, where f^{μ} is the quasiconformal self-mapping of $\mathbb{C} \cup \{\infty\}$ normalized to fix 0, 1 and ∞. Because of the theory of the Beltrami equation [4], each coordinate $f^{\mu}(p_j)$ depends continuously and holomorphically on μ. In particular, there exists a number $\delta > 0$, such that if $\|\mu\|_{\infty} < \delta$, then $f^{\mu}(p_j)$ is in D_j for each j. The map Ψ is also locally surjective, holomorphic and $\Psi^{-1}(p_1, \ldots, p_{n-3}) \cap \{\mu \in M(X) \text{ with } \|\mu\|_{\infty} < \delta\} = \{\mu \in M_0(X) \text{ with } \|\mu\|_{\infty} < \delta\}$. Therefore, $M_0(X)$ has a manifold structure at $\mu = 0$, and a vector ν is a tangent vector to $M_0(X)$ at $\mu = 0$ if $D\Psi_0(\nu) = (0, \ldots, 0)$. But from the theory of the Beltrami equation

$$f^{t\nu}(p_j) = p_j - t\frac{1}{\pi}\iint_{\mathbb{C}} \frac{p_j(p_j-1)}{z(z-1)(z-p_j)}\nu(z)\,dxdy + O(t^2).$$

Since we are assuming that ν is infinitesimally trivial and since for each j, with $1 \le j \le n-3$, $\frac{p_j(p_j-1)}{z(z-1)(z-p_j)}(dz)^2$ is an integrable holomorphic quadratic differential on X, we conclude that $D\Psi_0(\nu) = (0, \ldots, 0)$. Therefore, there is a curve μ_t in $M_0(X)$ tangent to ν.

The case X

We briefly explain this proof in several steps. For more complete explanations we refer to any of the following references: [1], [5], [7], [19], [22], [23], [24].

Step 1 (Uniformization and Poincaré series). The first step is to realize the Riemann surface X as the upper half plane $\mathbb{H}$ factored by a Fuchsian group Γ, that is, a discrete group of Möbius transformations preserving $\mathbb{H}$. Then the Riemann surface X is conformal to the factor space $\mathbb{H}/\Gamma$ and integrable quadratic differentials φ on X can be realized as automorphic functions φ in $\mathbb{H}$ satisfying the following properties:

1. $\varphi(z)$ is holomorphic in $\mathbb{H}$,
2. φ is automorphic in the sense that

$$\varphi(A(z))A'(z)^2 = \varphi(z) \tag{30}$$

for all A in Γ,

3. $\varphi(z)$ is integrable over X in the sense that

$$\iint_{X/\Gamma} |\varphi(z)|dxdy = \iint_{\omega} |\varphi(z)|dxdy < \infty,$$

where ω is a fundamental domain for Γ in $\mathbb{H}$.

Now we begin with the elementary identity

$$\iint_{\mathbb{H}} \frac{d\xi d\eta}{|\zeta - z|^4} = \frac{\pi}{4} y^{-2}, \tag{31}$$

where $\zeta = \xi + i\eta$ and $z = x + iy$ are in the upper and lower half planes, respectively. The integral in (31) is invariant under the translation $\zeta \mapsto \zeta - x$, and so is equal to

$$\iint_{\mathbb{H}} \frac{d\xi d\eta}{|\zeta - iy|^4}.$$

Make the substitution $w = \frac{\zeta + iy}{\zeta - iy},$ and observe that $\zeta \mapsto w$ is a conformal mapping carrying $\mathbb{H}$ onto the unit disc $|w| < 1$. Moreover, if $w = u + iv$, then since

$$\frac{dw}{d\zeta} = \frac{-2iy}{(\zeta - iy)^2},$$

$$\iint_{\mathbb{H}} \frac{4y^2 d\xi d\eta}{|\zeta - iy|^4} = \iint_{|w|<1} dudv = \pi,$$

and (31) follows.

From (31) we see that if $\varphi_z(\zeta) = \sum_{A \in \Gamma} \frac{A'(\zeta)^2}{(A(\zeta) - z)^4}$, then $\varphi_z(\zeta)$ is given by an absolutely convergent series because

$$\|\varphi_z\| = \iint_{X/\Gamma} \left| \sum_{A \in \Gamma} \frac{A'(\zeta)^2}{(A(\zeta) - z)^4} \right| d\xi d\eta$$

$$\leq \iint_{\omega} \sum_{A \in \Gamma} \left| \frac{A'(\zeta)^2}{(A(\zeta) - z)^4} \right| d\xi d\eta = \iint_{\mathbb{H}} \frac{d\xi d\eta}{|\zeta - z|^4}.$$

Since the hypothesis of Theorem 7 says that $\iint_X \nu \varphi dxdy = 0$ for every integrable holomorphic quadratic differential, on replacing φ by φ_z we conclude that

$$0 = \iint_{\omega} \nu \varphi_z d\xi d\eta = \iint_{\mathbb{H}} \frac{\nu(\zeta)}{(\zeta - z)^4} d\xi d\eta, \tag{32}$$

for every z in the lower half plane.

Step 2 (The Bers' embedding). . We may identify a Beltrami coefficient μ on X with a measurable complex valued function μ defined on $\mathbb{H}$ satisfying $\|\mu\|_\infty < 1$, and

$$\mu(A(\zeta)) \frac{\overline{A'(\zeta)}}{A'(\zeta)} = \mu(\zeta). \tag{33}$$

Such a coefficient represents a point $[\mu]$ in $T(X)$ and by putting μ identically equal to zero in the lower half plane, we may associate to μ the unique quasiconformal homeomorpism of $\mathbb{C}$ that solves the Beltrami equation

$$f^{\mu}_{\bar{z}}(z) = \mu(z) f^{\mu}_{z}(z)$$

and fixes the points 0, 1 and ∞. This homeomorphism has the following properties.

1. It is holomorphic in the lower half plane.
2. It yields by conjugation an isomorphism χ of Γ into a discrete subgroup of $PSL(2, \mathbb{C})$ by the equation
$$f^{\mu}(A(z)) = \chi(A)(f^{\mu}(z)).$$
3. Since Γ is the covering group of a Riemann surface of finite analytic type its limit set is the entire real axis and the values of f^{μ} on the real axis normalized to fix 0, 1 and ∞ determine and are determined by the Teichmüller equivalence class of μ.
4. Since f^{μ} applied to the lower half plane is a simply connected region and a Jordan domain, $f^{\mu}(z)$ realizes the Riemann mapping from the lower half plane to its image and this Riemann mapping determines and is determined by the Teichmüller equivalence class of μ.

Because of these properties, it is natural to form the Schwarzian derivative $\{f, z\}$ of $f = f^{\mu}$ for values for z in the lower half plane. The Schwarzian derivative of a function f is defined by

$$S_f(z) = N_f(z)' - \frac{1}{2} N_f(z)^2 = 6 \lim_{w \to z} \frac{\partial^2}{\partial w \partial z} \log \frac{f(w) - f(z)}{w - z}, \tag{34}$$

where

$$N_f(z) = \frac{f''(z)}{f'(z)}.$$

The map Φ taking the Teichmüller equivalence class of μ in $M(X)$ to the Schwarzian derivative of f^{μ} in $B(X)$ is called the Bers' embedding. Φ maps an equivariant Beltrami coefficient, that is, a Beltrami coefficient μ satisfying $\|\mu\|_{\infty} < 1$ and (33), to a holomorphic quadratic differential φ satisfying (30) and

$$\|y^2 \varphi(z)\|_{\infty} \leq 3/2. \tag{35}$$

In fact, Φ is a one-to-one holomorphic mapping from $T(X)$ onto an open set in the Banach space $B(X)$ of functions $\varphi(z)$ holomorphic in the lower half plane satisfying (30) and

$$\|y^2 \varphi(z)\|_{\infty} < \infty.$$

For this result we cite [3]. Although we do not need this result in full generality, the first step in proving it is the next step of the argument we are currently following.

Step 3 (A section for Φ). If a map Φ from $T(X)$ onto B is given and $\Phi(0) = 0$, s is called a local section of Φ at 0 if s maps an open neighborhood U of 0 in B onto an open neighborhood of 0 in $T(X)$ and $\Phi \circ s = I$, the identity on U.

Lemma 6 (Ahlfors-Weill). *The map $s : \varphi \mapsto -2y^2\varphi(\bar{z})$ provides a section for $\Phi : M(X) \to B$. That is, if $\|2y^2\varphi(z)\|_\infty < 1$, then $\Phi \circ s(\varphi) = \varphi$, or what is the same,*

$$\{f^{2y^2\varphi(\bar{z})}, z\} = \varphi(z).$$

Proof. If φ is given and you want to solve for f the equation

$$\{f, z\} = \varphi(z),$$

you can follow the following procedure. You solve for two linearly independent solutions η_1 and η_2 of the equation $\eta'' = -\frac{1}{2}\varphi\eta$ normalized by $\eta_1'\eta_2 - \eta_2'\eta_1 \equiv 1$ and then form

$$\hat{f}(z) = \begin{cases} \frac{\eta_1(\bar{z})+(z-\bar{z})\eta_1'(\bar{z})}{\eta_2(\bar{z})+(z-\bar{z})\eta_2'(\bar{z})} & \text{for } z \text{ in the upper half plane,} \\ \frac{\eta_1(z)}{\eta_2(z)} & \text{for } z \text{ in the lower half plane.} \end{cases} \tag{36}$$

In the case that $\|y^2\varphi(z)\|_\infty < 1$, $\hat{f}$ is a quasiconformal self mapping of the entire plane, holomorphic in the lower half plane and for z in the lower half plane $\{f, z\} = \varphi(z)$. All of the details of this calculation are given in [3] and in [19], pages 97–102. □

Step 4 (Construction of μ_t in Theorem 7). Let ν be a Beltrami coefficient satisfying (32). We need to calculate $\dot{\Phi} = \lim_{t\to 0}(1/t)\Phi(t\nu) = 0$, and in order to do this we view $\Phi(t\nu)$ as the composition of the maps $t\nu \mapsto f^{t\nu}$ and the map $f^{t\nu} \mapsto \{f^{t\nu}, z\}$. From the power series solution to the Beltrami equation the derivative of the first map is given by (10) with μ replaced by ν. Note that if $f^t(z) = z + t\dot{f} + o(t)$, then since $\{f, z\} = \frac{f'''}{f'} - (3/2)\left(\frac{f''}{f'}\right)^2$, $\lim_{t\to 0}(1/t)\{f^t, z\} = \dot{f}'''$, where

$$\dot{f}(z) = -\frac{1}{\pi}\iint_{\mathbb{C}} \nu(\zeta)\left\{\frac{1}{\zeta - z} - \frac{z}{\zeta - 1} + \frac{z-1}{\zeta}\right\} d\xi d\eta.$$

Since ν is identically equal to zero in the lower half plane,

$$\dot{\Phi}(t\nu)|_{t=0} = \dot{f}''' = \frac{6}{\pi}\iint_{\mathbb{H}} \frac{\nu(\zeta)}{(\zeta - z)^4} d\xi d\eta$$

and by (32) this is equal to zero for all z in the lower half plane.

Since $\Phi(t\nu) = \varphi^t = t\varphi_1 + t^2\varphi_2 + t^3\varphi_3 + \cdots$ is a convergent power series in the Banach space B, we see from the previous paragraph and formula (32) that $\varphi_1 \equiv 0$. This implies that the holomorphic curve $\nu_t = -2y^2\varphi^t(\bar{z})$ has vanishing first order term, that is, $\|\nu_t(z)\|_\infty = \|2y^2\varphi^t(\bar{z})\|_\infty \le Ct^2$.

Now form $f^{\mu_t} = (f^{\nu_t})^{-1} \circ f^{t\nu}$. This composition is well-defined because each of the mappings $(f^{\nu_t})^{-1}$ and $f^{t\nu}$ are normalized self-mappings of the complex plane. If we let $\tilde{\nu}$ be the Beltrami coefficient of $(f^{\nu_t})^{-1}$ and $\hat{\nu} = \tilde{\nu}\theta$, where $\theta = \bar{p}/p$ and $p = f_z^{t\nu}$, then $\|\hat{\nu}\|_\infty = \|\nu_t\|_\infty \le Ct^2$. But from the composition formula for Beltrami coefficients, we have

$$\mu_t = \frac{t\nu + \hat{\nu}}{1 + \overline{t\nu}\hat{\nu}} = t\nu + \hat{\nu}\left(\frac{1 - |t\nu|^2}{1 + \overline{t\nu}\hat{\nu}}\right) = t\nu + O(t^2).$$

But since $t\nu$ and ν_t have the same image under Φ, they are equivalent Beltrami coefficients and so μ_t is in $M_0(X)$. □

10. Teichmüller space and Teichmüller's metric

For any Riemann surface X, Teichmüller's space $T(X)$ consists of the quasiconformal maps f from X to variable Riemann surfaces $f(X)$ factored by an equivalence relation. Two such maps f_0 and f_1 are considered to be equivalent if there is a conformal map c from $f_0(X)$ to $f_1(X)$ and a homotopy g_t such that $g_0 = c \circ f_0$, $g_1 = f_1$, $g_t(z)$ is a jointly continuous map from X to $c \circ f_0(X)$, and $g_t(p) = c \circ f_0(p) = f_1(p)$ for every p in the ideal boundary of X.

The reader may not know what we mean by the ideal boundary of X, and if so we refer to any of the various books on Teichmüller theory in the bibliography. Since these notes attempt to treat only the case that X is of finite analytic type, X has no ideal boundary if X is compact or its ideal boundary consists of finitely many punctures.

For a given map f from X to a variable Riemann surface Y, we denote its equivalence class by $[f, Y]$.

Definition. *The Teichmüller metric $d : T(X) \times T(X) \to \mathbb{R}_+ \cup \{0\}$ is defined by*

$$d([f_0, Y_0], [f_1, Y_1]) = \inf \frac{1}{2} \log K(g),$$

where the infimum is taken over all g in the equivalence class of $f_1 \circ (f_0)^{-1}$.

Exercise 15. Show that d provides a complete metric on $T(X)$.

Hint: The triangle inequality follows from properties of the infimum and the fact that dilatation satisfies the inequality $K(f \circ g) \leq K(f)K(g)$. Symmetry follows from the fact that $K(g) = K(g^{-1})$. It is obvious that $d \geq 0$ because by definition $K \geq 1$. Finally, you need to show that if $d([f_0, Y_0], [f_1, Y_1]) = 0$, then $[f_0, Y_0]$ and $[f_1, Y_1]$ are the same equivalence class. That is, you must show that there is a conformal map c from $f_0(Y_0)$ to $f_1(Y_1)$ such that $c \circ f_0$ is homotopic to f_1 through a homotopy that fixes boundary points. The completeness of d follows from properties of quasiconformal mappings.

Exercise 16. Show that

$$d([f_0, Y_0], [f_1, Y_1]) = \inf \frac{1}{2} \log \frac{1 + \left\| \frac{\mu_1 - \mu_0}{1 - \overline{\mu_0}\mu_1} \right\|_\infty}{1 - \left\| \frac{\mu_1 - \mu_0}{1 - \overline{\mu_0}\mu_1} \right\|_\infty},$$

where the infimum is taken over Beltrami coefficients of all possible maps in the equivalence classes $[f_0, Y_0]$ and $[f_1, Y_1]$.

Theorem 8. *The Teichmüller space of a Riemann surface of finite analytic type of genus g with n punctures is homeomorphic to a cell of dimension $3g - 3 + n$ provided that this number is positive.*

Proof. From Teichmüller's theorem we know that each equivalence class $[f, Y]$ except the equivalence class of the identity is represented by a Beltrami coefficient of the form $k|\varphi|/\varphi$, where k and φ are uniquely determined with $0 < k < 1$ and $\|\varphi\| = 1$. We let $Q(X)$ be the Banach space of integrable holomorphic quadratic differentials on X and consider the map Ψ from the open unit ball of $Q(X)$ into $T(X)$ given by

$$\varphi \mapsto [f^{k|\varphi|/\varphi}, f^{k|\varphi|/\varphi}(X)],$$

where $k = \|\varphi\|$. This map is injective and surjective by Teichmüller's uniqueness and existence theorems. To show that it is continuous consider a sequence φ_n in the open unit ball of $Q(X)$ converging in norm to an element φ of this ball. If we put $k_n = \|\varphi_n\|$ and $k = \|\varphi\|$, then k_n converges to k and $0 \le k < 1$. If $k = 0$, then the Beltrami coefficients $k_n|\varphi_n|/\varphi_n$ converge uniformly to 0, and the Teichmüller distance from $[f^{k_n|\varphi_n|/\varphi_n}, Y_n]$ to $[identity, X]$ is no greater than $(1/2)\log\frac{1+k_n}{1-k_n}$. This shows that Ψ is continuous at 0. If $k > 0$, then the Beltrami coefficients $\mu_n = k_n|\varphi_n|/\varphi_n$ converge to $\mu = k|\varphi|/\varphi$ in the bounded pointwise sense. From the theory of quasiconformal mapping, this is enough to guarantee that the maps f^{μ_n} converge normally to the map f^{μ}, and because $T(X)$ is locally compact, this is enough to guarantee that $[f^{\mu_n}, Y_n]$ converges to $[f^{\mu}, Y]$ in the Teichmüller metric.

Finally, we must show that Ψ applied to a compact set is compact. A compact set C in the open ball B is closed and bounded away from the unit sphere. If $\varphi_n \in C$ converges to $\varphi \in C$, then there exists $r < 1$ such that $\|\varphi_n\| \le r < 1$. Thus $K(f_n) \le \frac{1+r}{1-r}$, where f_n is a normalized family of quasiconformal mappings with Beltrami coefficients of the form

$$\|\varphi_n\|\frac{|\varphi|}{\varphi}.$$

Thus $\Psi(C)$ forms a bounded set in $T(X)$. By the same argument as given in the previous paragraph, this set is also closed. □

11. Infinitesimal Teichmüller's metric

Let X be a Riemann surface of finite analytic type, f be a quasiconformal map from X to another Riemann surface Y, and μ be the Beltrami coefficient of f. Let K_0 be the extremal dilatation of a mapping in the class $[f, Y]$. We have already seen that the main inequality gives the following upper bound for K_0.

$$K_0 \le \sup \iint_X \frac{\left|1 + \mu\frac{\varphi}{|\varphi|}\right|^2}{1 - |\mu|^2}|\varphi|dxdy, \tag{37}$$

where the supremum is taken over holomorphic quadratic differentials φ with $\|\varphi\| = 1$. Moreover, if $K_0 > 1$ and if f^{μ} is extremal in its class, then the supremum is realized by a unique quadratic differential φ and the Beltrami coefficient of any extremal mapping f_0 in the class of f is equal to $k_0|\varphi|/\varphi$.

Exercise 17. Show that the main inequality also gives a lower bound for K_0, namely,

$$\frac{1}{K_0} \leq \iint_X \frac{\left|1 - \mu \frac{\varphi}{|\varphi|}\right|^2}{1 - |\mu|^2} |\varphi| dx dy, \tag{38}$$

for every holomorphic quadratic φ with $\|\varphi\| = 1$.

Exercise 18. Replace μ by $t\mu$ in (37) and in (38) and show that for any Beltrami coefficient μ,

$$d([id, X], [f^{t\mu}, f^{t\mu}(X)]) = t\left(\sup\left|\operatorname{Re} \iint_X \mu\varphi dx dy\right|\right) + O(t^2), \tag{39}$$

where the supremum is taken over all holomorphic quadratic differentials φ with $\|\varphi\| = 1$.

The coefficient of t in the expression on the right hand side of (39) is called the infinitesimal form of Teichmüller's metric at the base point. It transports to every point $[f^\mu, f^\mu(X)]$ of Teichmüller space with tangent vector ν at that point and is given by the global form expressed by the following formula:

$$F([\mu], \nu) = \sup\left|\operatorname{Re} \iint_X \varphi(w)\left[\frac{\nu}{1 - |\mu|^2} \cdot \frac{1}{\theta}\right] du dv\right|, \tag{40}$$

where $w = f^\mu(z)$, $\theta = \bar{p}/p$, $p = \frac{\partial}{\partial z} f^\mu$, $w = u + iv$, and the supremum is taken over all holomorphic quadratic differentials on the Riemann surface $f^\mu(X)$ with norm equal to 1.

It is not difficult to see from these formulas that Teichmüller's metric is equal to the integral of its infinitesimal form. That is, the Teichmüller distance between any two points is equal to the infimum of the arc length integrals with respect to F in (40), where the infimum is taken over all differentiable curves joining the two points, and any curve of the form $\gamma(t) = t\nu$, $0 \leq t \leq 1$, will realize this infimum.

Instead of pursuing this calculation we prefer to go on to an interpretation of the Teichmüller infinitesimal norm of a tangent vector to Teichmüller space in the case that the Riemann surface X is the Riemann sphere $\mathbb{C} \cup \{\infty\}$ with $n \geq 4$ points removed.

12. Tangent vectors to Teichmüller space

In this section we study the infinitesimal Teichmüller norm on vector fields tangent to Teichmüller space. The most direct way to illustrate the theory is to consider only the case when X is a Riemann surface equal to the Riemann sphere $\mathbb{C} \cup \{\infty\}$ minus $n \geq 4$ points. For simplicity we temporarily assume that the n points are in the finite plane, $\mathbb{C}$, and that they are the finite set $E = \{a_k\}_{k=1}^n$. In that case the quadratic differentials can be identified with holomorphic functions of z of the form

$$\varphi(z) = \sum_{k=1}^{n} \frac{\rho_k}{z - a_k},$$

where the residues ρ_k satisfy the equations

$$\begin{aligned}\sum_{k=1}^{n} \rho_k &= 0,\\ \sum_{k=1}^{n} \rho_k a_k &= 0,\\ \sum_{k=1}^{n} \rho_k a_k^2 &= 0.\end{aligned} \tag{41}$$

Exercise 19. Let $w = 1/z$ be a local parameter at $z = \infty$. Verify that the assumption that $\varphi^w(w)\left(\frac{dw}{dz}\right)^2 = \varphi(z)$ is a holomorphic quadratic differential at $w = 0$ forces the side conditions (41). Also verify that

$$\varphi(z) = \frac{p(z)}{\prod_{k=1}^{n}(z - a_k)},$$

where $p(z)$ is a polynomial of degree less than or equal to $n - 4$.

Given a Beltrami differential μ, its infinitesimal Teichmüller norm is the smallest possible value of $\|\nu\|_\infty$ where ν is a Beltrami differential with the property that

$$\iint_X \nu\varphi dxdy = \iint_X \mu\varphi dxdy$$

for all φ in $Q(X)$. By the Hahn-Banach and Riesz representation theorems such a ν exists.

Definition. *We denote this number $\|\nu\|_\infty$ by $\|\mu\|_T$ and call it the Teichmüller infinitesimal norm of μ. It is the norm of μ considered as a linear functional on the Banach space $Q(X)$.*

If we let $L_\infty(X)$ be the vector space of measurable Beltrami differentials $\mu(z)\frac{\overline{dz}}{dz}$ defined on X, then clearly

$$(\varphi, \mu) = \iint_X \mu\varphi dxdy$$

defines a pairing. On putting $N = \{\mu : (\varphi, \mu) = 0 \text{ for all } \varphi \in Q(X)\}$, this pairing establishes an isomorphism between the $Q(X)^*$, the dual space to $Q(X)$, and $L_\infty(X)/N$.

Another way to construct $Q(X)^*$ is to introduce the vector space Z of vectors v_j assigned to the points a_j of E factored by an equivalence relation. Two vectors $V = (v_1, \ldots, v_n)$ and $W = (w_1, \ldots, w_n)$ are considered equivalent if there is a quadratic polynomial $p(z) = az^2 + bz + c$ such that $v_j - w_j = p(a_j)$ for every j.

Exercise 20. Use (41) to show that the pairing

$$(V, \varphi) = \sum_{j=1}^{n} v_j \rho_j,$$

where $V = (v_1, \dots, v_n)$ and $\varphi(z) = \sum_{j=1}^{n} \frac{\rho_j}{z-a_j}$, is well-defined and non-degenerate, and establishes an isomorphism between Z and $Q(X)^*$.

Now we set up an isomorphism between $L_\infty(X)/N$ and Z which satisfies the identity

$$-\frac{\pi}{2}(V, \varphi) = \iint_X \mu\varphi dxdy. \tag{42}$$

The formula for V in terms of μ is

$$V(z) = -\frac{(z-a_1)(z-a_2)(z-a_3)}{\pi} \iint \frac{\mu(\zeta)}{(\zeta-a_1)(\zeta-a_2)(\zeta-a_3)(\zeta-z)} d\xi d\eta,$$

and the important properties are that

1. V is continuous and has generalized first partial derivatives,
2. $V_{\bar{z}}(z) = \mu$ and
3. $V(z) = O(|z|^2)$ as $z \to \infty$.

In order to do the following exercise you do not need this formula; all you need are these three properties.

Exercise 21. Prove equation (42) by using the identity

$$d(V(z)\varphi(z)dz) = V_{\bar{z}}(z)\varphi(z)\overline{dz}dz$$

and Stokes' theorem in the region consisting of the complement of n very small discs of radius ϵ centered at the points of E.

The point of this exercise is that the linear functional determined by μ in $L_\infty(X)/N$ depends precisely on the values of the vector field V at the points of E, up to the addition of a quadratic polynomial. This is reflected in the infinitesimal form of Teichmüller's metric. According to (39) the infinitesimal displacement in Teichmüller's metric caused by a tangent vector μ is

$$\sup_{\|\varphi\|=1} \left| \iint_X \mu\varphi dxdy \right| = \sup_{\|\varphi\|=1} \left| \frac{\pi}{2} \sum_{j=1}^{n} \rho_j v_j \right|.$$

In this way the velocity of the moving Riemann surface in $T(X)$ is expressed in terms of the velocities of the points in E and the quadratic differentials of norm one on $X = \mathbb{C} \cup \{\infty\} \backslash E$.

References

[1] W. Abikoff, The Real Analytic Theory of Teichmüller space, volume 820 of Springer Lecture Notes in Mathematics, Springer-Verlag (1980).
[2] L. V. Ahlfors, On quasiconformal mappings, *J. Anal. Math.*, **4** (1954) 1–58.
[3] L. V. Ahlfors, Lectures on Quasiconformal Mapping, volume 38 of University Lecture Series, *Amer. Math. Soc.* (2006).

[4] L. V. Ahlfors and L. Bers, Riemann's mapping theorem for variable metrics, *Annals of Math.*, **72** (1961) 385–404.

[5] L. V. Ahlfors and L. Sario, Riemann Surfaces, Princeton University Press, Princeton (1960).

[6] L. Bers, Quasiconformal mappings and Teichmüller's theorem, in Analytic Functions, Princeton University Press, Princeton, N. J. (1960) pages 89–119.

[7] L. Bers, Automorphic forms and general Teichmüller spaces, In *Proceedings of the Conference on Complex Analysis (Minneapolis 1964)*, pages 109–113, Berlin (1965) Springer.

[8] L. Bers, A non-standard integral equation with applications to quasiconformal mapping, *Acta Math.*, **116** (1966) 113–134.

[9] L. Bers, Extremal quasiconformal mappings, *Annals of Math. Studies*, **66** (1971) 27–52.

[10] L. Bers, Fibre spaces over Teichmüller spaces, *Acta Math.*, **130** (1973) 89–126.

[11] L. Bers and H. Royden, Holomorphic families of injections, *Acta Math.*, **157** (1986) 259–286.

[12] B. Bojarski, Generalized solutions of a system of differential equations of first order and elliptic type with discontinuous coefficients, *Math. Sbornik*, **85** (1957) 451–503.

[13] V. Bozin, N. Lakic, M. Markovic and M. Mateljevic, Unique extremality, *J. d'Analyse Math.*, **75** (1998) 299–337.

[14] C. J. Earle and F. P. Gardiner, Geometric isomorphisms between infinite dimensional Teichmüller spaces, *Trans. Am. Math. Soc.*, **348(3)** (1996) 1163–1190.

[15] C. J. Earle and I. Kra, On holomorphic mappings between Teichmüller spaces, In L. V. Ahlfors et al., editor, *Contributions to Analysis*, pages 107–124, New York (1974) Academic Press.

[16] C. J. Earle and I. Kra, On isometries between Teichmüller spaces, *Duke Math. J.*, **41** (1974) 583–591.

[17] C. J. Earle and V. Markovic, Isometries between the spaces of L^1 holomorphic quadratic differentials on Riemann surfaces of finite type, *Duke Math. J.*, **120(2)** (2003) 433–440.

[18] F. P. Gardiner, Approximation of infinite dimensional Teichmüller spaces, *Trans. Amer. Math. Soc.*, **282(1)** (1984) 367–383.

[19] F. P. Gardiner, Teichmüller Theory and Quadratic Differentials, John Wiley & Sons, New York (1987)

[20] F. P. Gardiner and N. Lakic, Quasiconformal Teichmüller Theory, AMS, Providence, Rhode Island (2000).

[21] R. S. Hamilton, Extremal quasiconformal mappings with prescribed boundary values, *Trans. Amer. Math. Soc.*, **138** (1969) 399–406.

[22] W. J. Harvey, Discrete Groups and Automorphic Functions, Academic Press (1977).

[23] Y. Imayoshi and M. Taniguchi, An Introduction to Teichmüller Spaces, Springer-Verlag, Tokyo (1992).

[24] I. Kra, Automorphic Forms and Kleinian Groups, Benjamin, Reading, Mass. (1972).

[25] S. L. Krushkal, Extremal quasiconformal mappings, *Sib. Math. J.*, **10** (1969) 411–418.

[26] N. Lakic, An isometry theorem for quadratic differentials on Riemann surfaces of finite genus, *Trans. Amer. Math. Soc.*, **349** (1997) 2951–2967.

[27] N. Lakic, Strebel points, *Contemporary Math.*, **211** (1997) 417–431.

[28] O. Lehto, Univalent Functions and Teichmüller Spaces, Springer-Verlag, New York (1987).

[29] O. Lehto and K. I. Virtanen, Quasiconformal Mapping in the Plane, Springer-Verlag, New York, Berlin (1965).

[30] R. Màñé, P. Sad and D. Sullivan, On the dynamics of rational maps, *Ann. Èc. Norm. Sup.*, **96** (1983) 193–217.

[31] A. Marden and K. Strebel, The heights theorem for quadratic differentials on Riemann surfaces, *Acta Math.*, **153** (1984) 153–211.

[32] V. Markovic, Biholomorphic maps between Teichmueller spaces, *Duke Math. J.*, **120(2)** (2003) 405–431.

[33] S. Nag, The Complex Analytic Theory of Teichmüller Spaces, Canadian Mathematical Society, Wiley-Interscience, New York (1987).

[34] E. Reich, On the decomposition of a class of plane quasiconformal mappings, *Comment. Math. Helv.*, **53** (1978) 15–27.

[35] E. Reich and K. Strebel, Teichmüller mappings which keep the boundary pointwise fixed, *Ann. Math. Studies*, **66** (1971) 365–367.

[36] E. Reich and K. Strebel, Extremal quasiconformal mappings with given boundary values, Contributions to Analysis (1974) pages 375–392.

[37] H. Royden, Automorphisms and isometries of Teichmüller space, In Advances in the Theory of Riemann Surfaces (1971) pages 369–384.

[38] Z. Slodkowski, Holomorphic motions and polynomial hulls, *Proc. Amer. Math. Soc.*, **111** (1991) 347–355.

[39] D. P. Sullivan and W. P. Thurston, Extending holomorphic motions, *Acta Math.*, **157** (1986) 243–257.

[40] O. Teichmüller, Untersuchungen über konforme und quasikonforme Abbildungen, *Deutsche Math.*, **3** (1938) 621–678.

[41] O. Teichmüller, Eine Verschärfung des Dreikreisesatzes, *Deutsche Math.*, **4** (1939) 16–22.

[42] O. Teichmüller, Extremale quasikonforme Abbildungen und quadratische Differentiale, *Abh. Preuss. Akad.*, **22** (1939) 3–197.

[43] O. Teichmüller, Über extremalprobleme der konformen Geometrie, *Deutsche Math.*, **6** (1941) 50–77.

Teichmüller Theory and Moduli Problem No. 10, 2010, pp. 229–246.

Prime Order Automorphisms of Riemann Surfaces

Jane Gilman

Mathematics Department, Rutgers University, Newark, NJ 07102
e-mail: gilman@andromeda.rutgers.edu

Abstract. Recently there has been renewed interest in the mapping-class group of a compact surface of genus $g \geq 2$ and also in its finite order elements. A finite order element of the mapping-class group will be a conformal automorphism on some Riemann surface of genus g. Here we give the details of the proof that there is an *adapted basis* for any conformal automorphism of prime order on a surface of genus g and extend the original result to apply to fixed point free automorphisms. An adapted basis is one that reflects the action of the automorphism in the optimal manner described below. The proof uses the Schreier-Reidemeister rewriting process. We find some new consequences of the existence of an adapted basis. We also construct an explicit example of such a basis and compute its intersection matrix.

1. Introduction

For a conformal automorphism of a compact Riemann surface the notion of an *adapted* homology basis was developed as part of a proof that the Riemann space (also known as the Moduli Space) of a punctured surface had the structure of a quasi-projective variety [6,8]. An adapted basis is one which reflects the action of the conformal automorphism in an optimal way. Such an action would be reflected in the structure of the period matrix of the surface in a useful manner. More recently Rodriguez, Riera, Gonzalez and others have used the notion of a basis adapted to a group of automorphisms to obtain information about abelian varieties and especially, the Prym variety. (See [23], [14] and the references given there.)

In this paper we survey earlier results about the matrix representation of a prime order automorphism with respect to an adapted basis and the corresponding intersection matrix for such a basis. The proof of the existence of an adapted basis uses the Schreier-Reidemeister rewriting process. Here we give full details of the application of the Schreier-Reidemeister rewriting process used in [10] to construct the adapted basis. We have been told that the application in [10] was too sketchy for some readers to follow. We find some new consequences of the existence of an adapted basis and extend the result to the case of a fixed point free automorphism. We also construct an explicit example of such a basis and compute its intersection matrix.

The paper is organized as follows. In section 2 we fix notation and terminology and we review the conjugacy invariants for an element of the mapping-class group of prime order and basic facts about homology. Section 3 introduces the notion of

This work was partially supported by grants from the NSA and the Rutgers Research Council and by the Yale University Mathematics Department.

an adapted homology basis, section 4 discusses the existence of such bases and section 5 the intersection numbers of elements in an adapted basis. Section 6 fixes some matrix notation. In section 7 the Schreier-Reidemeister rewriting process is explained and the calculation is carried out in detail (section 7.2). The case for a fixed point free automorphism is carried out in section 8 and some corollaries are drawn in section 10.

2. Preliminaries

2.1 *Notation and terminology*

We let h be a conformal automorphism of a compact Riemann surface S of genus $g \geq 2$. Then h will have a finite number, t, of fixed points. We let S_0 be the quotient of S under the action of the cyclic group generated by h so that $S_0 = S/\langle h \rangle$ and let g_0 be its genus. If h is of prime order p with $p \geq 2$, then the Riemann-Hurwitz relation shows that $2g = 2pg_0 + (p-1)(t-2)$. If $p = 2$, of course, this implies that t will be even.

2.2 *Equivalent languages*

We emphasize that h can be thought of in a number of equivalent ways using different terminology. For a compact Riemann surface of genus $g \geq 2$, homotopy classes of homeomorphisms of surfaces are the same as isotopy classes. Therefore, h can be thought of as a representative of a homotopy class or an isotopy class. Further, every isotopy class of finite order contains an element of finite order so that h can be thought of as a homeomorphism of finite order. For every finite order homeomorphism of a surface there is a Riemann surface on which its action is conformal. A conformal homeomorphism of finite order up to homotopy is finite. We use the language of conformal maps, but observe that all of our results can be formulated using these other classes of homeomorphisms.

We remind the reader that the Mapping-class group of a compact surface of genus g is also known as the Teichmüller Modular group or the Modular group, for short. We write $MCG(S)$ or $MCG(S_g)$ for the mapping class group of the surface S using the g when we need to emphasize that S is a compact surface of genus g. The Torelli Modular group or the Torelli group for short is homeomorphisms of S modulo those that induce the identity on homology and the homology of a surface is the abelianized homotopy. There is surjective map π from the mapping-class group onto $Sp(2g, \mathbb{Z})$ that assigns to a homeomorphism the matrix of its action on a canonical homology basis (see 2.4).

Since h can be thought of as a finite representative of a finite order mapping-class, we will always treat it as finite. For ease of exposition we use the language of conformal maps and do not distinguish between a homeomorphism that is of finite order or that is of finite order up to homotopy or isotopy, a finite order representative for the homotopy class, a conformal representative for the class or the class itself. That is, between the topological map, its homotopy class or a finite order representative or a conformal representative.

For ease of exposition in what follows we first assume that $t > 0$. We treat the case $t = 0$ separately in section 8.

2.3 *Conjugacy invariants for prime order mapping classes or conformal automorphisms*

In determining the conjugacy class of a prime order element of the mapping class group, we may first assume that the element h is actually conformal and of finite order. If h has t fixed points $p_1, \ldots, p_t$ then in a neighborhood of p_i it is a (counterclockwise) rotation through an angle $s_i \frac{2\pi}{p}$ for some s_i with $0 < s_i < p$. Nieslen showed topologically that the conjugacy class of h is completely determined by the rotation angles (see page 53 of [22]). The numbers s_i are called *the rotation numbers*. The set of rotation numbers determine the conjugacy class.

One can also consider *the complementary rotation numbers* n_i where $n_i \cdot s_i \equiv 1(p)$. Here $\equiv \ (p)$ denotes equivalence modulo p and we let $\mathbb{Z}_p$ denote the integers modulo p.

Alternately we can uniformize S and S_0 by a pair of Fuchsian groups F and F_0 so that $S = U/F$, $S_0 = U/F_0$ and the action of the conformal map h is isomorphic to the action of F_0/F on S. In this setting, h is determined by a surjective homomorphism $\phi : F_0 \to \mathbb{Z}_p$. If F is the kernel of ϕ, then F_0/F acts in a natural way on $S = U/F$. The homomorphism ϕ will take t elliptic elements of F_0 onto non-trivial elements n_i of $\mathbb{Z}_p$ and these elements will satisfy $\Sigma_{i=1}^{t} n_i \equiv 0(p)$. One uses Harvey's list [15] of automorphisms of F_0 to assure that the conjugacy class of h depends only upon the number of n_i taking on a possible value and not the order in which the n_i appear in any list of complementary rotation numbers. This is the proof given in [9] where it is also shown that every such t-tuple of integers determines a conformal map. We will discuss this further in section 7.2.

Since the order in which the complementary rotation numbers are listed does not matter, we can instead count m_j, the number of n_i equal to j. Then we have $\Sigma_{i=1}^{p-1} i \cdot m_i \equiv 0 \ (p)$ and the conjugacy class is also determined by the $(p-1)$-tuple, $(m_1, \ldots, m_{p-1})$.

2.4 *Homology*

We recall the following facts about Riemann surfaces.

The homology group of a compact Riemann surface of genus g is the abelianized homotopy. Therefore, a homology basis for S will contain $2g$ homologously independent curves. Every surface has a *canonical homology basis*, a set of $2g$ simple closed curves, $a_1, \ldots, a_g; b_1, \ldots, b_g$ with the property that for all i and j *the intersection numbers* satisfy $a_i \times a_j = 0$, $b_i \times b_j = 0$ and $a_i \times b_j = \delta_{ij} = -b_j \times a_i$ where δ_{ij} is the Kronecker delta.

3. Adapted homology bases

Roughly speaking a homology basis for S is *adapted to h* if it reflects the action of h in a simple manner: for each curve γ in the basis either all of the images of γ under powers of h are also in the basis or the basis contains all but one of the images of γ under powers of h and the omitted curve is homologous to the negative of the sum of the images of γ under the other powers of h.

To be more precise

Definition 3.1. *A homology basis for S is adapted to h if for each γ_0 in the basis there is a curve γ with $\gamma_0 = h^k(\gamma)$ for some integer k and either*

(1) $\gamma, h(\gamma), \ldots h^{p-1}(\gamma)$ *are all in the basis, or*
(2) $\gamma, h(\gamma), \ldots h^{p-2}(\gamma)$ *are all in the basis and* $h^{p-1}(\gamma) \approx^h -(h(\gamma)+h^2(\gamma)+\cdots+h^{p-2}(\gamma))$. *Here $\approx^h$ denotes is homologous to.*

If $\mathcal{A}$ is an adapted homology basis, we define the adapted matrix *to be the matrix of h with respect to the adapted basis and denote it by $M_{\mathcal{A}}(h)$. We shorten $M_{\mathcal{A}}(h)$ to $M_{\mathcal{A}}$ when h is clearly understood.*

4. Existence of adapted Homology Bases

It is known that

Theorem 4.1 (Gilman 1977 [10]). *There is a homology basis adapted to h. In particular, if $g \geq 2$, $t \geq 2$, g_0 are as above, then the adapted basis has $2p \times g_0$ elements of type (1) above and $(p-1)(t-2)$ elements of type (2).*

and thus it follows that

Corollary 4.2. [10] *The adapted matrix $M_{\mathcal{A}}(h)$ will be composed of diagonal blocks, $2g_0$ of which are $p \times p$ permutation matrices with 1's along the super diagonal and 1 in the leftmost entry of the last row and t of which are $(p-1) \times (p-1)$ matrices with 1's along the super diagonal and all entries in the last row -1.*

A proof of theorem 4.1 is given in section 7.2.

Remark 4.3. We adopt the following convention. When we pass from homotopy to homology, we use the same notation for the homology class of the curve as for the curve or its homotopy class, but write $\approx^h$ instead of $=$. It will be clear from the context which we mean.

5. Intersection Matrix for an Adapted Homology Basis

So far information about $M_{\mathcal{A}}(h)$ seems to depend only on t and not upon the $(p-1)$-tuple $(m_1, \ldots, m_{p-1})$ (or equivalently, upon the set of integers $\{n_1, \ldots, n_t\}$) which determines the conjugacy class of h in the mapping-class group. However, while the $2pg_0$ curves can be extended to a canonical homology basis for h, the rest of the basis cannot and its intersection matrix, $I_{\mathcal{A}}$ depends upon these integers.

In [12] the intersection matrix for the adapted basis was computed.

Remark 5.1. For any one reading the original paper [12] note that m and n the letters denoting the rotation numbers and the complementary rotation numbers have been interchanged. To avoid confusion, we use u and v in this section when presenting result from [12].

The adapted basis consisted of the curves of type (1):

$$\{A_w, B_w, w = 1, \ldots, g_0\} \cup \{h^j(A_w), h^j(B_w), j = 1, \ldots, p-1\}$$

and (some of) the curves of type (2):

$$X_{i,v_i}, h^j(X_{i,v_i}),\ i = 1, \ldots, (p-1),\ j = 1, \ldots, p-2,\ v_i = 1, \ldots, u_i.$$

A lexicographical order is placed on X_{i,v_i} so that $(r, v_r) < (s, v_s)$ if and only if $r < s$ or $r = s$ and $v_r < v_s$. The $t-2$ curves X_{s,v_s} with the largest subscript pairs are to be included in the homology basis. Let $\hat{s}$ be the smallest integer s such that $u_s \neq 0$ and let $\hat{q}$ be chosen so that $\hat{q}\hat{s} \equiv 1(p)$. For any integer v let $[v]$ denote the least non-negative residue of $\hat{q}v$ modulo p. Thus the integer $[v]$ satisfies $0 \leq [v] \leq p-1$ and $\hat{s} \cdot [v] \equiv v \bmod(\mathrm{p})$.

Theorem 5.2 (Gilman-Patterson, 1981 [12]). *If $(u_1, \ldots, u_{p-1})$ determines the conjugacy class of h in the mapping-class group, then the surface S has a homology basis consisting of:*

(1) $h^j(A_w), h^j(B_w)$ *where* $1 \leq w \leq g_0, 0 \leq j \leq p-1$.
(2) $h^k(X_{s,v_s})$ *where* $0 \leq k \leq p-2$ *and for all pairs* (s, v_s) *with* $1 \leq s \leq p-1, 1 \leq v_s \leq u_s$ *except that the two smallest pairs are omitted.*
The intersection numbers for the elements of the adapted basis are given by

(a) $h^j(A_w) \times h^j(B_w) = 1$
(b) *If* $(r, v_r) < (s, v_s)$, *then*

$$h^0(X_{r,v_r}) \times h^k(X_{s,v_s}) = \begin{cases} 1 & \text{if } [k] < [r] \leq [k+s] \\ -1 & \text{if } [k+s] < [r] \leq [k] \end{cases}$$

$$h^0(X_{s,v_s}) \times h^k(X_{s,v_s}) = \begin{cases} 1 & \text{if } [k] \leq [s] < [k+s] \\ -1 & \text{if } [k+s] < [s] < [k] \end{cases}$$

(3) *All other intersection numbers are* 0 *except for those following from the above by applying the identities below to arbitrary homology classes C and D.*
$C \times D = -D \times C$
$h^j(C) \times h^k(D) = h^0(C) \times h^{k-j}(D)$, *($k-j$ reduced modulo p).*

Proof. For details we refer the reader to [12]. Basically, the proof of this theorem comes from a careful interpretation of the isomorphism between covering groups, fundamental groups, and defining subgroups of coverings and their relation to words corresponding to closed curves on the quotient surface that lift to closed curves. □

6. Matrix forms

We can write the results of theorems 4.1 and 5.2 and corollary 4.2 in an explicit matrix form. To do so we fix notation for some matrices. We will use the various explicit forms in subsequent sections.

We let $M_{\tilde{A}}$ denote the matrix of the action of h on an adapted basis and $I_{\tilde{A}}$ be the corresponding intersection matrix. Further, we let $M_{h_{CAN}}$ be the matrix of the action of h on a canonical homology basis. The corresponding intersection matrix is denoted by $I_{h_{CAN}}$. If we let I_k denote the $k \times k$ identity matrix, then $I_{h_{CAN}}$ is (conjugate to) the $2g \times 2g$ matrix $= \begin{pmatrix} 0 & I_g \\ -I_g & 0 \end{pmatrix}$.

However, we prefer to replace it by the following block matrix where $q = \frac{(p-1)(t-2)}{2}$

$$I_{h_{CAN}} = \begin{pmatrix} 0 & I_{pg_0} & 0 & 0 \\ -I_{pg_0} & 0 & 0 & 0 \\ 0 & 0 & 0 & I_q \\ 0 & 0 & -I_q & 0 \end{pmatrix}.$$

We denote the $p \times p$ permutation matrix by

$$M_{p\times p} = \begin{pmatrix} 0 & 1 & 0 & 0 & \dots & 0 & 0 \\ 0 & 0 & 1 & 0 & \dots & 0 & 0 \\ 0 & 0 & 0 & 1 & \dots & 0 & 0 \\ & & & & \ddots & & \\ 0 & 0 & 0 & 0 & \dots & 1 & 0 \\ 0 & 0 & 0 & 0 & \dots & 0 & 1 \\ 1 & 0 & 0 & 0 & \dots & 0 & 0 \end{pmatrix};$$

the $(p-1) \times (p-1)$ non-permutation matrix of corollary 4.2 by

$$N_{(p-1)\times(p-1)} = \begin{pmatrix} 0 & 1 & 0 & 0 & \dots & 0 & 0 \\ 0 & 0 & 1 & 0 & \dots & 0 & 0 \\ 0 & 0 & 0 & 1 & \dots & 0 & 0 \\ & & & & \ddots & & \\ 0 & 0 & 0 & 0 & \dots & 1 & 0 \\ 0 & 0 & 0 & 0 & \dots & 0 & 1 \\ -1 & -1 & -1 & -1 & \dots & -1 & -1 \end{pmatrix}.$$

Thus we have the $2g_0p \times 2g_0p$ block matrix

$$M_{A_{2g_0,p\times p}} = \begin{pmatrix} M_{p\times p} & 0 & 0 \dots & 0 & 0 \\ 0 & M_{p\times p} & 0 \dots & 0 & 0 \\ & & \ddots & & \\ 0 & 0 & 0 \dots & M_{p\times p} & 0 \\ 0 & 0 & 0 \dots & 0 & M_{p\times p} \end{pmatrix}$$

and the $(t-2)(p-1) \times (t-1)(p-1)$ block matrix

$$N_{\mathcal{A}_{(t-2),(p-1)\times(p-1)}} = \begin{pmatrix} N_{(p-1)\times(p-1)} & 0 & 0 \ldots & 0 & 0 \\ 0 & N_{(p-1)\times(p-1)} & 0 \ldots & 0 & 0 \\ & & \ddots & & \\ 0 & 0 & 0 \ldots & N_{(p-1)\times(p-1)} & 0 \\ 0 & 0 & 0 \ldots & 0 & N_{(p-1)\times(p-1)} \end{pmatrix}$$

so that the $2g \times 2g$ matrix $M_{\mathcal{A}}$ breaks into blocks and can be written as

$$M_{\mathcal{A}} = \begin{pmatrix} M_{\mathcal{A}_{2g_0,p\times p}} & 0 \\ 0 & N_{\mathcal{A}_{(t-2),(p-1)\times(p-1)}} \end{pmatrix}$$

where the blocks are of appropriate size. The basis can be rearranged so that $2g \times 2g$ matrix $M_{\tilde{\mathcal{A}}}$ corresponding to the rearranged basis breaks into blocks

$$M_{\tilde{\mathcal{A}}} = \begin{pmatrix} M_{\mathcal{A}_{g_0,p\times p}} & 0 & 0 \\ 0 & M_{\mathcal{A}_{g_0,p\times p}} & 0 \\ 0 & 0 & N_{\mathcal{A}_{(t-2),(p-1)\times(p-1)}} \end{pmatrix}.$$

Here the submatrix

$$\begin{pmatrix} M_{\mathcal{A}_{g_0,p\times p}} & 0 \\ 0 & M_{\mathcal{A}_{g_0,p\times p}} \end{pmatrix}$$

is a symplectic matrix.

Remark 6.1. We note that if $p = 2$, $M_{p\times p}$ reduces to $\begin{pmatrix} 0 & 1 \\ 1 & 0 \end{pmatrix}$ and $N_{(p-1)\times(p-1)}$ to the 1×1 matrix -1.

We obtain the corollary

Corollary 6.2. *Let S be a compact Riemann surface of genus g and assume that S has a conformal automorphism h of prime order $p \geq 2$. Assume that h has t fixed points where $t \geq 2$. Let S_0 be the quotient surface $S_0 = S/\langle h \rangle$ where $\langle h \rangle$ denotes the cyclic group generated by h and let g_0 be the genus of S_0 so that $2g = 2pg_0 + (t-2)(p-1)$.*

There is a homology basis on which the action of h is given by the $2g \times 2g$ matrix $M_{\tilde{\mathcal{A}}}$.

The matrix $M_{\tilde{\mathcal{A}}}$ contains a $2g_0p \times 2g_0p$ symplectic submatrix, but is not a symplectic matrix except in the special case $t = 2$.

The point here is that while two automorphisms with the same number of fixed points will have the *same* matrix representation with respect to an adapted basis, the intersection matrices will not be the same and, therefore, the corresponding two matrix representations in the symplectic group will not be conjugate.

There is an algorithm to replace $M_{\tilde{\mathcal{A}}}$ by the symplectic matrix $M_{h_{CAN}}$ by replacing the submatrix $N_{(t-2),(p-1)\times(p-1)}$ by a symplectic matrix of the same size (see [11]). We will call this matrix $N_{symp\tilde{\mathcal{A}}}$ and give an example in section 9.

We note that $I_{\tilde{A}}$ is of the form

$$\begin{pmatrix} 0 & I_{pg_0} & 0 & 0 \\ -I_{pg_0} & 0 & 0 & 0 \\ 0 & 0 & B_1 & B_2 \\ 0 & 0 & B_3 & B_4 \end{pmatrix}$$

where the blocks B_i are of the appropriate dimension and we let B denote the matrix

$$\begin{pmatrix} B_1 & B_2 \\ B_3 & B_4 \end{pmatrix}.$$

7. Schreier-Reidemeister rewriting

If we begin with an arbitrary finitely presented group G_0 and a subgroup G, the Schreier-Reidemeister rewriting process tells one how to obtain a presentation for G_0 from the presentation for G. In our case the larger group G_0 will correspond to the group uniformizing S_0 and the subgroup G corresponds to the group uniformizing S.

In particular, one chooses a special set of coset representatives for G modulo G_0, called Schreier representatives, and uses these to find a set of generators for G. These generators are labeled by the original generators of the group and the coset representative.

7.1 *The relation between the action of the homeomorphism and the surface kernel subgroup*

We may assume that $S_0 = U/F_0$ where F_0 is the Fuchsian group with presentation

$$\langle a_1, \ldots, a_{g_0}, b_1, \ldots, b_{g_0}, x_1, \ldots x_t \mid x_1 \cdots x_t (\Pi_{i=1}^{g}[a_i, b_i]) = 1; x_i^p = 1 \rangle. \qquad (1)$$

We summarize the result of [9] repeating some facts about the n_i and m_j. We let $\phi : F_0 \to \mathbb{Z}_p$ be given by

$$\phi(a_i) = \phi(b_i) = 0 \quad \forall i = 1, \ldots, g_0 \text{ and } \phi(x_j) = n_j \neq 0 \quad \forall j = 1, \ldots, t.$$

The n_i satisfy $\Sigma_{i=1}^{t} n_i \equiv 0 \ (p)$. If $F = Ker\ \phi$, then $S = U/F$. Moreover, F_0/F acts on S with quotient S_0. Conjugation by x_1 acts on F and if h is the induced conformal map on S, $\langle h \rangle$ is isomorphic to the action induced by this conjugation and the conjugacy class of h in the mapping-class group is determined by the set of n_i. Replacing h by a conjugate we may assume that $0 < n_i \leq n_j < p$ if $i < j$.

Since m_i is the number of j such that $\phi(x_j) = i$, we have $\Sigma_{i=1}^{p-1} i m_i \equiv 0 \ (p)$ and the conjugacy class of h is also completely determined by $(m_1, \ldots, m_{p-1})$.

When we need to emphasize the relation of h to ϕ, we write h_ϕ to mean the automorphism determined by conjugation by x_1. The conjugacy class of h^2, would then be determined by the homomorphism ψ with $\psi(x_j) \equiv 2\phi(x_j) \ (p)$ or by conjugation by x_1^2.

We note that in [9] results are written in greatest possible generality so that S_0 has punctures, some of which are fixed by the automorphism and others of which are not. Here, we use the results of [9] for compact surfaces. F_0 is sometimes called surface kernel and ϕ the surface kernel homomorphism [13,15,24].

Any other map from F_0 onto $\mathbb{Z}_p$ with the same $(m_1, \ldots, m_{p-1})$ and with $\phi(x_j) \neq 0 \ \forall j$ will yield an automorphism conjugate to h.

7.2 *The rewriting*

We want to apply the rewriting process to words in the generators of this presentation for F_0 to obtain a presentation for F. We choose coset representatives for $H = \langle h \rangle$ as $x_1, x_1^2, x_1^3, \ldots, x_1^p$ and observe that these are a *Schreier* system. (see page 93 of [21]). That is, every initial segment of a representative is again a representative.

The Schreier right coset function assigns to a word W in the generators of F_0, its coset representative $\overline{W}$ and $\overline{W} = x_1^q$ if $\phi(W) = \phi(x_1^q)$.

If $\mathfrak{a}$ is a generator of F_0, set $S_{K,\,\mathfrak{a}} = K\mathfrak{a}\overline{K\mathfrak{a}}^{-1}$. The rewriting process τ assigns to a word that is in the kernel of the map ϕ, a word written in the specific generators, $S_{K,\,\mathfrak{a}}$ for F. Namely, if $\mathfrak{a}_w$, $w = 1, \ldots, r$ are generators for F_0 and

$$U = \mathfrak{a}_{v_1}^{\epsilon_1} \mathfrak{a}_{v_2}^{\epsilon_2} \cdots \mathfrak{a}_{v_r}^{\epsilon_r} \quad (\epsilon_i = \pm 1),$$

defines an element of F, then (corollary 2.7.2 page 90 of [21])

$$\tau(U) = S_{K_1,\,\mathfrak{a}_{v_1}}^{\epsilon_1} S_{K_2,\,\mathfrak{a}_{v_2}}^{\epsilon_2} \cdots S_{K_r,\,\mathfrak{a}_{v_r}}^{\epsilon_r}$$

where K_j is the representative of the initial segment of U preceding $\mathfrak{a}_{v_j}$ if $\epsilon_j = 1$ and K_j is the coset representative of U up to and including $\mathfrak{a}_{v_j}^{-1}$ if $\epsilon_j = -1$.

In our case each $\mathfrak{a}_v$ stands for some generator of F_0, that is one of the a_i or b_i or x_j. We apply theorem 2.8 of [21] to see

Theorem 7.1. *Let F_0 have the presentation given by equation (1). Then F has presentation*

$$\langle S_{K,\,a_i},\ S_{K,\,b_i},\ i = 1, \ldots, g_0;\ S_{K,x_j},\ j = 1, \ldots t \,| \tag{2}$$

$$\tau(K \cdot x_1 \cdots x_t(\Pi_{i=1}^{g_0}[a_i, b_i]) \cdot K^{-1}) = 1,\ \tau(K x_j^p K^{-1}) = 1\rangle. \tag{3}$$

Proof. Let K run over a complete set of coset representatives for $\phi : F \to H$. Then F_0 has generators

$$S_{K,\,a_i}, S_{K,\,b_i} \quad i = 1, \ldots, g_0$$

$$S_{K,\,x_j}, \quad j = 1, \ldots t$$

and relations

$$\tau(K(x_1 \cdots x_t \Pi_{i=1}^{g_0}[a_i, b_i])K^{-1}) = 1 \tag{4}$$

$$\tau(K x_j^p K^{-1}) = 1 \tag{5}$$

□

We want to simplify this presentation and eliminate generators and relations so that there is a single defining relation for the subgroup. We first assume that $\phi(x_1) = h$. We note that if we can find a homology basis adapted to h, we can easily find a homology basis adapted to any power of h and, therefore, this assumption will not be significant.

Definition 7.2. *A relation is evenly worded if for each generator A that occurs in the relation A^{-1} also occurs. The generators A and B occurring in a relation are linked if the relation is of the form $W_0AW_1BW_2A^{-1}W_3B^{-1}W_4$ where the $W_i, i = 0, \ldots, 4$ are words in the generators not involving $A^{\pm 1}$ or $B^{\pm 1}$. The relation is fully linked if each generator A occurring in the relation is linked to a unique distinct generator.*

We will show:

Theorem 7.3. *Let F_0 have the presentation given by equation (1). Then F has presentation*

$$\langle h^j(A_i), h^j(B_i), i = 1, \ldots, g_0, j = 0, \ldots, p-1 : h^j(X_i),$$
$$i = 3, \ldots, t, j = 0, \ldots, p-2 | \hat{\hat{R}} = 1\rangle.$$

The relation $\hat{\hat{R}}$ is the single defining relation for the group F. The single defining relation $\hat{\hat{R}}$ is evenly worded and fully linked.

and

Corollary 7.4. *The homology basis obtained by abelianizing the basis in theorem 7.3 gives a homology basis adapted to h.*

We note that an explicit formula for $\hat{\hat{R}}$ is given in [11].

Proof. If we let $\phi(x_1) = h$ and $\phi(K) = \phi(x_1)^r$, then we have $S_{K,X_j} = x_1^{r\phi(x_1)} \cdot x_j \cdot \overline{K \cdot x_j}^{-1}$. Thus if $X_j = x_j \cdot \overline{x_j}^{-1}$, then $S_{K,x_j} = h^r(X_j)$.

We begin by rewriting the generators and relations using this notation.

First we find $\tau(\overbrace{x_1x_1\cdots x_1}^{p\text{–factors}}) = 1$. Setting $X_1 = S_{x_1^p,x_1}$ since $\overline{1} = x_1^p$, we have

$$\tau(x_1^p) = X_1 \cdot h(X_1) \cdot h^2(X_1) \cdots h^{p-2}(X_1)h^{p-1}(X_1) = 1. \tag{6}$$

Similarly, if $\phi(x_j) = n_j$, and we set $X_j = S_{\overline{1},x_j}$, then if $\overline{K} = x_1^s$, then we can write $S_{K,x_j} = h^{s \cdot n_j}(X_j)$.

This tells us that

$$\tau(x_j^p) = X_j \cdot h^{n_j}(X_j) \cdot h^{2n_j}(X_j) \cdots h^{(p-2)n_j}(X_j)h^{(p-1)n_j}(X_j) = 1. \tag{7}$$

In particular, we will make special use of this when $j = 2$

$$\tau(x_2^p) = X_2 \cdot h^{n_2}(X_j) \cdot h^{2n_2}(X_2) \cdots h^{(p-2)n_2}(X_2)h^{(p-1)n_2}(X_2) = 1. \tag{8}$$

Note that in deriving all equations we are free to make use of the fact (see [21]) that

$$S_{M,\,x_1} \approx 1 \forall \text{ Schreier representatives } M \tag{9}$$

where $\approx$ denotes *freely equal to*. This eliminates the p generators, $S_{x_1^j, x_1}, j = 1, \ldots, p$.

Recall that a word is freely equal to another word if one word can be obtained from the other by inserting and deleting expressions of the form $x \cdot x^{-1}$. This means in particular that these words represent the same element in every group.

Now equation (7) is a relation in the fundamental group. We remind the reader that for a compact Riemann surface, homology is abelianized homotopy so that when abelianized, it reduces to

$$h^{p-1}(X_j) \approx^h - X_j - h(X_j) - \cdots - h^{p-2}(X_j) \tag{10}$$

where $\approx^h$ denotes *is homologous to*.

We also note that the $\tau(Kx_j^p K^{-1}) = 1$ do not give us any additional relations for $\bar{K} \neq 1$, but merely a conjugate relation already implied by $\tau(x_j^p) = 1$.

Next we set $A_i = S_{\bar{1},\, a_i}$ and $h^r(A_i) = S_{x_1^r,\, a_i}$, $B_i = S_{1,\, b_i}$ and $h^r(B_i) = S_{x_1^r,\, b_i}$, then the relation $R = 1$, where $R = x_1 \cdots x_t(\Pi_{i=1}^{g_0}[a_i, b_i])$, yields $\tau(R) = 1$ and we obtain

$$X_1 h^{n_1}(X_2) h^{n_1 \cdot n_2}(X_2) \cdots h^{n_1 \cdot n_2 \cdots n_{t-2}}(X_{t-1}) \cdot h^{n_1 \cdot n_2 \cdots n_{t-1}}(X_t)(\Pi_{i=1}^{g_0}[A_i, B_i]) = 1 \tag{11}$$

and using the fact that $X_1 \approx 1$, we have

$$h^{n_1}(X_2) h^{n_1 \cdot n_2}(X_3) \cdots h^{n_1 \cdot n_2 \cdots n_{t-2}}(X_{t-1}) \cdot h^{n_1 \cdot n_2 \cdots n_{t-1}}(X_t)(\Pi_{i=1}^{g_0}[A_i, B_i] = 1 \tag{12}$$

Similarly, we obtain the relations $\tau(KRK^{-1}) = 1$. We can solve equation (12) for $(h(X_2))^{-n_1}$ to obtain

$$(h(X_2))^{-n_1} = h^{n_1 \cdot n_2}(X_3) \cdots h^{n_1 \cdot n_2 \cdots n_{t-2}}(X_{t-1}) \cdot h^{n_1 \cdot n_2 \cdots n_{t-1}}(X_t)(\Pi_{i=1}^{g_0}[A_i, B_i]) \tag{13}$$

Now each of the relations $\tau(KRK^{-1})$ allows us to solve for $h^q(X_2)$ for some q and for each K we obtain a different q. Therefore, we can substitute equation (13) and its images under powers of h into equation (8) (i.e. the relation $\tau(x_2^p) = 1$). We thus eliminate all of the generators of the form S_{M, x_2} for each coset representative M and all of the relations $\tau(KRK^{-1})$. We obtain one new relation $\hat{R}$ from equation (8). This relation involves $h^w(X_j)$ for every $w = 0, 1, \ldots, p-2$ and every $j = 3, \ldots, t$. We also note that for each q the sequence $\Pi_{i=1}^{g_0}[h^q(A_i), h^q(B_i)]$ occurs in $\hat{R}$. Using equation (7) to replace the generator $h^{p-1}(X_j)$ by a word in the $h^{-v}(X_j)$ with $v = 0, \ldots, p-1$ we eliminate those relations and we obtain a single defining relation $\hat{\hat{R}}$ involving each of the following generators below and their inverses exactly once. The generators are

$$h^d(X_j) \quad j = 3, \ldots, t \text{ and } d = 0, 1, \ldots, p-2$$

and

$$h^d(A_i), h^d(B_i) \quad i = 1, \ldots, g_0 \text{ and } d = 0, 1, \ldots, p-1$$

We obtain $2g_0p+(t-2)(p-1)$ generators and a single defining relation. It is fairly straight forward to check that the relation has the last two properties of the theorem.

□

We recapitulate. The idea of this proof is that for each K, we can solve $\tau(KRK^{-1}) = 1$ for an appropriate image S_{K',x_2} where K' depends upon K. The appropriate image of S_{K',x_2} is placed on the left of the equation, and we then substitute the right hand side of the solution into the relation $\tau(x_2^p) = 1$. This yields one relation $\hat{\mathcal{R}}$ which involves each A_i, B_i generator and their inverses and all of their images under powers of h and each S_{K,x_j} and the images under powers of h but no inverses. We still have the finite order relations $\tau(x_j^p)$, $j = 3, \ldots, t$. The $\tau(Kx_j^pK^{-1}) = 1$ are merely permutations of the relation $\tau(x_j^p) = 1$ so we can eliminate all but one of these. For each $j = 3, \ldots, t$, $\tau(x_j^p)$ can be solved for $h^{p-1}(S_{\overline{1},x_j})$. It will be a word in the inverses of all of the other S_{K,x_j}. We substitute these into $\hat{\mathcal{R}}$ and obtain a single defining relation $\hat{\hat{R}}$ in which every generator and its inverse occurs exactly once.

8. The Case $t = 0$

If the number of fixed points is zero, we can still find an adapted basis. The calculations are slightly different. We have $2g = 2p(g_0 - 1) + 2$. The presentation given by (1) for the group F_0 becomes

$$\langle a_1, \ldots, a_{g_0}, b_1, \ldots, b_{g_0} | \ (\Pi_{i=1}^{g_0}[a_i, b_i]) = 1; \rangle. \tag{14}$$

Again, replacing h by a conjugate if necessary, the map $\phi : F_0 \to \mathbb{Z}_p$ can be taken to be

$$\phi(a_i) = \phi(b_i) = 0 \quad \forall i = 2, \ldots, g_0 \text{ and } \phi(a_1) = 1 \text{ and } \phi(b_1) = 0$$

Using the rewriting with coset representatives $1, a_1, \ldots, a_1^{p-1}$, note that $S_{a_1^k,a_1} \approx 1$ for $k = 0, \ldots, p-2$. Let $A = S_{a_1^{p-1},a_1}$. Then h acts on $Ker\ \phi$ via conjugation by a_1. We have $h(A) = A$

We let $\{h^k(A_j), h^k(B_j), j = 2, \ldots, g_0, k = 0, \ldots, p-1\}$ be as in the proof of Theorem 7.3 and let $B = S_{1,b_1}$.

We let $P = \Pi_{i=2}^{g_0}[A_i, B_i]$.

Then we can compute that

$$\tau(R) = 1 \implies h(B)B^{-1}T = 1$$

$$\tau(a_1^kRa_1^{-k}) = 1 \implies h^k(B)(h^{k-1}(B))^{-1}h^k(P) = 1 \forall k = 1, \ldots p-2$$

$$\text{and } \tau(a_1^{p-1}Ra_1^{-(p-1)}) = 1 \implies ABA^{-1}(h^{p-1}(B))^{-1}h^{p-1}(P) = 1 \tag{15}$$

We eliminate generators using (15) and let $\alpha = A$ and $\beta = h^{p-1}(B)$ to obtain generators

$$\{\alpha, \beta\} \cup \{h^k(A_j), h^k(B_j), j = 2, \ldots, g_0, k = 0, \ldots, p-1\}$$

and the single defining relation $\beta\alpha\beta^{-1} = h^{p-1}(P)\alpha\Pi_{i=k}^{p-2}h^k(P)$. Further we calculate that $h(\alpha) \approx^h \alpha$ and $h(\beta) \approx^h \beta$. Note that P is a product of commutators. Thus the matrix representation for h on $S = U/F$ where $F = \text{Ker}\phi$ is given by $2(g_0 - 1)$ permutation matrices $M_{p\times p}$ and one two by two identity matrix.

Thc basis is a canonical homology basis. We have $\alpha \times \beta = 1$. Further α and β are disjoint from any other lifted curves in the basis. Each lift of a generator other than a_1 or b_1 is a simple closed curve and is disjoint from all other curves except for the corresponding $h^k(B_j)$. That is, $h^k(A_i) \times h^r(B_j) = \delta_{ij} \cdot \delta_{kr}$ for all integers $i, j \in \{2, \ldots, p-1\}$ and $k, r \in \{0, \ldots, p-1\}$. We replace α or β by an appropriate conjugate if necessary. We note that the curves α and β are by default of type (1) in definition 3.1

We have obtained the following version of Theorem 5.2 when $t = 0$

Theorem 8.1. *If $t = 0$, then the surface S has a canonical homology basis consisting of:*

(1) $h^j(A_w), h^j(B_w)$ *where* $2 \leq w \leq g_0, 0 \leq j \leq p-1$.
(2) α, β *where* $h^k(\alpha) \approx^h \alpha, h^k(\beta) \approx^h \beta, 0 \leq k \leq p$
(3) *The intersection numbers for the elements of the adapted basis are given by*
   (a) $h^j(A_w) \times h^j(B_w) = 1$
   (b) $\alpha \times \beta = 1$
   (c) *All other intersection numbers are* 0 *except for those that following from the above by applying the identities below to arbitrary homology classes C and D.*
   $C \times D = -D \times C$
   $h^j(C) \times h^k(D) = h^0(C) \times h^{k-j}(D)$, *($k - j$ reduced modulo p.)*

9. Example, $p = 3, t = 5, (1, 1, 2, 1, 1)$

In this section we work out the specific example with $p = 3$ and $t = 5$. Assume $\phi(x_1) = 1, \phi(x_2) = 1, \phi(x_3) = 2, \phi(x_4) = 1, \phi(x_5) = 1$ so that $(n_1, \ldots, n_5) = (1, 1, 2, 1, 1)$ and $(m_1, m_2) = (4, 1)$.

First replacing h by a conjugate, we may assume that $\phi(x_1) = 1, \phi(x_2) = 1, \phi(x_3) = 1, \phi(x_4) = 1, \phi(x_5) = 2$. We choose as coset representatives x_1, x_1^2 and x_1^3.

For any g_0, we have generators

$$h^q(A_i), h^q(B_i), q = 0, \ldots, p-1 = 2, i = 1 \ldots, g_0.$$

and

$$S_{\overline{x_1^r}, x_j}, \quad r = 1, 2, 3, j = 2, 3, 4.$$

We also have by equation (9)

$$S_{\overline{x_1^r},x_1} \approx 1, r = 1, 2, 3.$$

and, therefore, these generators and the relation $\tau(x_1^3)$ drop out of the set of generators and relations to be considered.

We have

$$\tau(x_j^3) = S_{\overline{1},x_j} \cdot S_{\overline{x_j},x_j} \cdot S_{\overline{x_j^2},x_j} \tag{16}$$

Set $S_{\overline{1},x_1} = Y_1$. Then $h(Y_1) = S_{x_1,x_1}$ and $h^2(X_j) = S_{x_1^2,x_j}$;

Set $S_{\overline{x_1},x_2} = Y_2$ Then $h(Y_2) = S_{x_1^2,x_2}$ and $h^2(Y_2) = S_{x_1^3,x_2}$;

Similarly, set $S_{\overline{x_1x_2},x_3} = Y_3$. Then $h(Y_3) = S_{x_1^3,x_3}$ and $h^2(Y_3) = S_{x_1,x_3}$; If $S_{\overline{x_1x_2x_3},x_4} = Y_4$ Then $h(Y_4) = S_{x_1,x_4}$ and $h^2(Y_4) = S_{x_1^2,x_4}$; and finally if $S_{\overline{x_1x_2x_3x_4},x_5} = Y_5$, then $h(Y_5) = S_{x_1^2,x_5}$ and $h^2(Y_5) = S_{x_1^3,x_j}$;

Use this notation and use $\tau(x_j^3) = 1$ to see that

$$\begin{aligned}
h^2(Y_2) \cdot Y_2 \cdot h(Y_2) &= 1 \\
h(Y_3) \cdot h^2(Y_3) \cdot Y_3 &= 1 \\
Y_4 \cdot h(Y_4) \cdot h^2(Y_4) &= 1 \\
h^2(Y_5) \cdot h(Y_5) \cdot Y_5 &= 1
\end{aligned} \tag{17}$$

We compute

$$\tau(R) = S_{\overline{1},x_1} \cdot S_{\overline{x_1},x_2} \cdot S_{\overline{x_1x_2},x_3} \cdot S_{\overline{x_1x_2x_3},x_4} \cdot S_{\overline{x_1x_2x_3x_4},x_5} \cdot (\Pi_{i=1}^{g_0}[A_i, B_i]) = 1. \tag{18}$$

Using $S_{\overline{x_1^r},x_1} \approx 1, r = 1, 2, 3$ and solving for $(Y_2)^{-1}$ in equation (18), we have

$$(S_{x_1,x_2})^{-1} = Y_2^{-1} = Y_3 \cdot Y_4 \cdot Y_5 \cdot (\Pi_{i=1}^{g_0}[A_i, B_i]) = 1. \tag{19}$$

and

$$(h(Y_2))^{-1} = h(Y_3) \cdot h(Y_4) \cdot h(Y_5) \cdot (\Pi_{i=1}^{g_0}[h(A_i), h(B_i)]) = 1. \tag{20}$$

$$(h^2(Y_2))^{-1} = h^2(Y_3) \cdot h^2(Y_4) \cdot h^2(Y_5) \cdot (\Pi_{i=1}^{g_0}[h^2(A_i), h^2(B_i)]) = 1. \tag{21}$$

Using $h^2(Y_2)Y_2h(Y_2) = 1$ and letting $P = (\Pi_{i=1}^{g_0}[A_i, B_i])$, we have

$$h(Y_3) \cdot h(Y_4) \cdot h(Y_5) \cdot h(P) \cdot Y_3 \cdot Y_4 \cdot Y_5 \cdot P) \cdot h^2(Y_3) \cdot h^2(Y_4) \cdot h^2(Y_5) \cdot h^2(P) = 1 \tag{22}$$

We use equation (17) to replace the $h^2(Y_j)$ and obtain the relation

$$\begin{aligned}
h(Y_3) \cdot h(Y_4) \cdot h(Y_5) \cdot h(P) \cdot Y_3 \cdot Y_4 \cdot Y_5 \cdot P) \cdot (h(Y_3))^{-1} \cdot (Y_3)^{-1} \\
\cdot (h(Y_4))^{-1} \cdot (Y_4)^{-1} \cdot (Y_5)^{-1} \cdot (h(Y_5))^{-1} \cdot h^2(P) = 1.
\end{aligned} \tag{23}$$

Equation (23) is the relation $\hat{\hat{R}} = 1$.

We use the notation of section 6, in particular the definition of the matrix B given at the end of that section. We can compute from the formulas for intersection numbers in Theorem 5.2 that the relevant part of the intersection matrix, B, is the 6×6 submatrix that gives the intersection matrix for the curves in the basis given in the order

$$X_{1,3}, h(X_{1,3}), X_{1,4}, h(X_{1,4}), X_{2,1}, h(X_{2,1})$$

is

$$B = I_{\hat{\hat{R}}} = \begin{pmatrix} 0 & 1 & 1 & 0 & 1 & -1 \\ -1 & 0 & -1 & 1 & 0 & 1 \\ -1 & 1 & 0 & 1 & 1 & -1 \\ 0 & -1 & -1 & 0 & 0 & 1 \\ -1 & 0 & -1 & 0 & 0 & 0 \\ 1 & -1 & 1 & -1 & 0 & 0 \end{pmatrix}.$$

That is, the matrix $I_{\tilde{A}}$ breaks up into blocks $\begin{pmatrix} 0 & I_{pg_0} & 0 \\ -I_{pg_0} & 0 & 0 \\ 0 & 0 & I_{\hat{\hat{R}}} \end{pmatrix}$

We now rearrange the relation. To simply the notation we let $a = Y_3$, $b = Y_4$ and $c = Y_5$. So that the relation becomes

$$h(a)\cdot h(b)\cdot h(c)\cdot h(P)\cdot a\cdot b\cdot c\cdot P\cdot (h(a))^{-1}a^{-1}(h(b))^{-1}\cdot b^{-1}\cdot c^{-1}\cdot (h(c))^{-1}\cdot h^2(P) = 1. \tag{24}$$

We can also make the simplifying assumption, replacing the elements that occur in P, $h(P)$ and $h^2(P)$ by conjugates, that we are merely working with the symbol

$$h(a)\cdot h(b)\cdot h(c)\cdot a\cdot b\cdot c\cdot (h(a))^{-1}a^{-1}(h(b))^{-1}\cdot b^{-1}\cdot c^{-1}\cdot (h(c))^{-1} = 1. \tag{25}$$

We replace generators and relations using the algorithm of [11] as follows:

Let $M = h(a)\cdot W_1\cdot h(b)W_2\cdot (h(a))^{-1}$ where $W_1 = \emptyset$, $W_2 = h(c)\cdot a\cdot b\cdot c$. Set $W_3 = a^{-1}$ and $W_4 = b^{-1}\cdot c^{-1}\cdot (h(c))^{-1}$.

Let $N = W_3 W_2 (h(a))^{-1}$. Then

$$\begin{aligned} &h(a)\cdot h(b)\cdot h(c)\cdot a\cdot b\cdot c\cdot (h(a))^{-1}a^{-1}(h(b))^{-1}\cdot b^{-1}\cdot c^{-1}\cdot (h(c))^{-1} \\ &= [M,N]W_3W_2W_1W_4 \\ &= [M,N]\cdot a^{-1}\cdot h(c)abcb^{-1}\cdot c^{-1}\cdot (h(c))^{-1} = 1. \end{aligned} \tag{26}$$

At this point one can proceed by inspection and let $[b,c]^*$ denote the conjugate of $[b,c]$ by $a^{-1}\cdot h(c)\cdot a$ to obtain

$$[M,N]\cdot [b,c]^*\cdot [a^{-1}, h(c)] = 1.$$

However, to follow the algorithm carefully, we would set $\tilde{M} = a^{-1}\cdot h(c)\cdot a$ and $\tilde{N} = b\cdot c\cdot b^{-1}\cdot c^{-1}\cdot a$.

Then equation (26) becomes

$$[M,N]\cdot [\tilde{M},\tilde{N}]\cdot [b,c] = 1.$$

Thus the canonical homology basis is given by

$$\{h^j(A_i), h^j(B_i)\}, i = 1, \ldots, g_0, j = 0, \ldots, p-1 \cup \{M, N, \tilde{M}, \tilde{N}, b, c\}.$$

The reordered basis $\{M, \tilde{M}, b, N, \tilde{N}, c\}$ has intersection matrix

$$\begin{pmatrix} 0 & 0 & 0 & 1 & 0 & 0 \\ 0 & 0 & 0 & 0 & 1 & 0 \\ 0 & 0 & 0 & 0 & 0 & 1 \\ -1 & 0 & 0 & 0 & 0 & 0 \\ 0 & -1 & 0 & 0 & 0 & 0 \\ 0 & 0 & -1 & 0 & 0 & 0 \end{pmatrix}.$$

We can compute the action of h on these last six elements of the homology basis. First we note that $h(b) \approx^h M - c - b - a - h(c)$ and $h(a) \approx^h -N + h(c) + b + c$. Therefore,

$$c \mapsto h(c) \approx^h \tilde{M}$$

$$h(c) \mapsto -c - h(c) \approx^h -c - \tilde{M}$$

$$a \mapsto h(a) \approx^h -N + h(c) + b + c \approx^h -N + \tilde{M} + b + c$$

$$b \mapsto h(b) \approx^h M - c - b - a - h(c) \approx^h M - c - b - \tilde{N} - \tilde{M}$$

$$M \mapsto h(M) \approx^h -\tilde{M} - N, \quad \text{and}$$

$$N \mapsto h(N) \approx^h -c + M - N.$$

Thus the matrix of the action of h with respect to the ordered basis $M, \tilde{M}, b, N, \tilde{N}, c$ is the submatrix we have been seeking. Namely,

$$N_{symp\tilde{A}} = \begin{pmatrix} 0 & 1 & 0 & -1 & 0 & 0 \\ 0 & -1 & 0 & 1 & 0 & -1 \\ 1 & -1 & -1 & -0 & -1 & -1 \\ 1 & 0 & 0 & -1 & 0 & -1 \\ 0 & 1 & 1 & -1 & 0 & -1 \\ 0 & 1 & 0 & 0 & 0 & 0 \end{pmatrix}$$

One can verify that this 6×6 matrix really is a submatrix of a symplectic matrix, as it should be.

10. Remarks

Recall (section 2.2) that there is a surjective map $\pi : MCG(S_g) \to SP(2g, \mathbb{Z})$. It is well known that the restriction of π to elements (mapping-classes) of finite order is an isomorphism. It is shown in [10] that Theorem 4.1 implies a stronger result than this which we note for completeness.

Corollary 10.1 ([10]). *If h is a conformal automorphism of S of genus $g \geq 2$ and if there are two pairs of curves C_1, D_1 and C_2, D_2 with $C_i \times C_j = D_i \times D_j = 0$ for $i = 1, 2$ and $j = 1, 2$ and $C_i \times D_j = \delta_{ij}$ where δ_{ij} is the Kronecker delta, and $h(C_i) \approx^h C_i$ and $h(D_i) \approx^h D_i$ for $i = 1, 2$, then h is the identity.*

Proof. If h is of prime order, simply write each of the four curves as a sum of the curves in the adapted homology basis, apply h and equate coefficients. If h is not of prime order, apply this to every power that is of prime order to see that each must be the identity. □

We obtain immediately,

Corollary 10.2. *If two mapping-classes of finite order have the same action on homology, then they are equal. Equivalently, the restriction of π to elements of finite order is an isomorphism.*

The idea of an adapted homology basis predates Thurston's notation of a reducible mapping-class. However, it is clear that the notions are related. A homeomorphism h is *reducible* if h fixes a *a partition* on the surface, that is, a set of disjoint simple closed curves on the surface. A mapping-class is reducible if it contains a reducible representative.

Corollary 10.3. *If h represents a mapping-class of prime order, $p \geq 2$, and $g_0 \neq 0$, then h is a reducible mapping-class.*

Proof. Replace h by a conformal representative if necessary. An element of type (1) (definition 3.1) in an adapted basis for h taken along with its images gives a set of closed curves on the surface, fixed by the homeomorphism of the surface. Once we have shown that these are simple closed curves (as we do in theorem 5.2), is clear that the element h is reducible, that is, h fixes a *a partition*, a set of disjoint simple closed curves on the surface. □

11. Acknowledgements

The author thanks Yair Minsky and the Yale Mathematics Department for their hospitality and support while some of this work was carried out.

We give a long, but by no means exhaustive bibliography.

References

[1] P. Buser and G. Peter, Courtois, *Finite parts of the spectrum of a Riemann surface. Math. Ann.* **287** (1990), no. 3, 523–530. MR 92c:58146.

[2] P. Buser and M. Seppälä, Computing on Rimeann Surfaces, *Topology and Teichmüller Spaces*, World Scientific, (1996) 5–30.

[3] P. Buser and M. Seppälä, Symmetric pants decompositions of Riemann surfaces, *Duke Math. J.* **67** (1992), no. 1, 39–55.

[4] P. Buser and R. Silhol, Geodesics, periods, and equations of real hyperelliptic curves, *Duke Math. J.* **108** (2001), no. 2, 211–250.

[5] M. D. E. Conder and R. S. Kulkarni, Infinite families of automorphism groups of Riemann surfaces in Discrete groups and geometry (Birmingham, 1991), 47–56, *London Math. Soc. Lecture Note Ser.*, **173** (1992) Cambridge Univ. Press, Cambridge.
[6] J. Gilman, Relative Modular Groups in Teichmüller Spaces, PhD thesis, Columbia University, (1971).
[7] J. Gilman, Compact Riemann Surfaces with Conformal Involutions, *Proc. Amer. Math. Soc.*, **37** (1973) 105–107.
[8] J. Gilman, On the Moduli of Compact Riemann Surfaces with a Finite Number of Punctures Discontinuous Groups and Riemann Surfaces, *Annals of Math. Studies*, **79** (1974) 181–205.
[9] J. Gilman, On Conjugacy Classes in the Teichmüller Modular Group, *Mich. Math. J.*, **23** (1976) 53–63.
[10] J. Gilman, A Matrix Representation for Automorphisms of Riemann Surfaces, Lin. Alg. and its Applications, **17** (1977) 139–147.
[11] J. Gilman, Canonical Symplectic Representations for Prime Order Conjugacy Classes of the Mapping Class Croup, *J. Algebra*, **318** (2007) 430–455.
[12] J. Gilman and D. Patterson, Intersection Matrices for Adapted Bases, Proc. 1978 Stony Brook Conference, *Annals of Math. Studies* **97** (1981) 149–166.
[13] J. Gilman, Structures of Elliptic Irreducible Subroups of the Mapping-class Group, *Proc. London Math Soc.* **47(3)** (1983) 27–42.
[14] V. Gonzalez-Aguilera and R. E. Rodriguez, On principally polarized abelian varieties and Riemann surfaces associated to prisms and pyramids in Lipa's legacy, *Contemp. Math.*, **211** AMS (1997) 269–284,
[15] W. J. Harvey, On branch loci in Teichmüller space, *Trans. Amer. Math. Soc.* **153** (1971) 387–399.
[16] R. S. Kulkarni, Riemann surfaces admitting large automorphism groups (1995) 63–79, *Contemp. Math., 201, Amer. Math. Soc.*, Providence, RI.
[17] R. S. Kulkarni, Infinite families of surface symmetries, *Israel J. Math.* **76** (1991) no. 3, 337–343.
[18] R. S. Kulkarni, A note on Wiman and Accola-Maclachlan surfaces, *Ann. Acad. Sci. Fenn. Ser. A I Math.*, **16(1)** (1991) 83–94.
[19] R. S. Kulkarni and C. Maclachlan, Cyclic p-groups of symmetries of surfaces, *Glasgow Math. J.* **33(2)** (1991) 213–221.
[20] A. M. MacBeath, Action of automorphisms of a compact Riemann surface on the First Homology Group, *Bull. Lon. Math Soc.*, **5** (1973) 103–108.
[21] Magnus, Karrass and Solitar, Combinbatorial Group Theory, *J. Wiley*, (1966) 253–274.
[22] J. Nielsen, Die Structur periodsiher Transformationen von Flachen, *D. K. Dan. Vidensk. Selsk. Math-fys. Medd* **XV** (1937) 1–77.
[23] G. Riera, Gonzalo and R. E. Rodrguez, Riemann surfaces and abelian varieties with an automorphism of prime order, *Duke Math. J.*, **69** (1993) no. 1, 199–217. MR 93k:14059.
[24] D. Singerman, Finitely maximal Fuchsian groups, *J. London Math. Soc.* **(2)6** (1972) 29–38.
[25] G. Springer, Introduction to Riemann Surfaces Addison-Wesley, (1957).
[26] A. Weaver, Hyperelliptic surfaces and their moduli, *Geom. Dedicata* **103** (2004) 69–87.

Teichmüller Theory and Moduli Problem No. 10, 2010, pp. 247–280.

On Complex Curves and Complex Surfaces Defined Over Number Fields

Ernesto Girondo[1] and Gabino González-Diez[2]*

[1]Departamento de Matemáticas, Universidad Autónoma de Madrid, 28049 Madrid, Spain
and
Instituto de Ciencias Matemáticas, CSIC-UAM-UC3M-UCM
[2]Departamento de Matemáticas, IMAFF, Consejo Superior de Investigaciones Científicas, 28006 Madrid, Spain
e-mail: ernesto.girondo@uam.es; gabino.gonzalez@uam.es

Abstract. Belyi's theorem states that a compact Riemann surface C can be defined over a number field if and only if there is on it a meromorphic function with three critical values. Such functions (resp. Riemann surfaces) are called *Belyi functions* (resp. *Belyi surfaces*). Alternatively Belyi surfaces can be characterized as those which contain a proper Zariski open subset uniformised by a *torsion free* subgroup of the classical modular group $\mathbb{PSL}_2(\mathbb{Z})$. In the first part of this survey article we discuss this result and the companion theory of Grothendieck's *dessins d'enfants*. In the second part we establish a result analogous to Belyi's theorem in complex dimension two. It turns out that the role of Belyi functions is now played by (*composed*) *Lefschetz pencils* with three critical values while the analogous to torsion free subgroups of the modular group will be certain extensions of them acting on a *Bergman domain* of $\mathbb{C}^2$. These groups were first introduced by Bers and Griffiths.

1. Introduction

This article is an extended version of the talk given by the second author at the International Workshop on *Teichmüller Theory and Moduli Problems* held at the Harish-Chandra Research Institute (HRI), Allahabad (India), on January 2006. It is intended to be of expository nature and for this reason proofs are merely outlined or explained via examples.

The paper is divided in two parts. The first part starts with Belyi's theorem, which states that a compact Riemann surface (= complex algebraic curve) C can be defined over a number field if and only if there is a meromorphic function $f : C \to \mathbb{P}^1$, called a *Belyi function*, with only three critical values, say $0, 1, \infty$ ([4]). From the uniformization point of view, this is readily seen to be equivalent to saying that C contains a finite set Σ such that $C \setminus \Sigma$ can be uniformized by a torsion free subgroup of the classical modular group $\mathbb{PSL}_2(\mathbb{Z})$.

Belyi's theorem has attracted much attention ever since Grothendieck noticed in his *Esquisse d'un Programme* ([14]) that it implies amazing interrelations between algebraic curves defined over number fields and a certain class of graphs embedded in a topological surface, which he named *dessins d'enfants*.

*Both autors were partially supported by the MCyT research project MTM2006-01859

The proof of the "only if" part of this theorem results from a surprisingly simple construction of Belyi (in Grothendieck's words "*jamais sans doute un résultat profond et déroutant ne fut démontré en si peu de lignes!*", [14]). The construction is illustrated in an explicit example in Section 4.

The "if" part can be deduced from an old criterion of rationality due to Weil ([30]). The proof we present here is based on a criterion of our own which, although less powerful than Weil's, we find easier to handle. And, in fact, it will be used in a different setting in the second part of the article.

The absolute Galois group $\mathrm{Gal}(\bar{\mathbb{Q}})$, i.e. the group of field automorphisms of the algebraic closure of $\mathbb{Q}$, is a basic object of interest in algebraic number theory. The astonishing fact about Belyi's theorem is that it permits to define an action of $\mathrm{Gal}(\bar{\mathbb{Q}})$ on these merely topological, or combinatorial, objects called dessins. The hope to gain understanding of this still mysterious group via this action seems to have been the main reason that led Grothendieck to the introducion of the theory of dessins. In the first part of the paper this action is described in detail in several explicit examples.

In the second part the goal is to establish a result analogous to Belyi's theorem in dimension 2, that is for complex surfaces. It will turn out that in this case the role of Belyi functions is going to be played by (composed) Lefschetz pencils with three critical values (*Lefschetz functions*). As for the point of view of uniformization the analogous result goes as follows. It is known that any complex projective surface S possesses a closed subvariety Σ such that the Zariski open set $S\backslash\Sigma$ can be uniformized in the form $S\backslash\Sigma = \mathbb{B}/G$ where $\mathbb{B}$ is a Bergman domain in $\mathbb{C}^2$ and G is a group of biholomorphic transformations of $\mathbb{B}$ which, roughly speaking, is an extension of a Fuchsian group Γ of type $(0, r)$ by a Kleinian group K (Bers-Griffiths Uniformization, [5,13]). Now, except for an exceptional case, the complex surfaces that can be defined over a number field are those for which Γ can be chosen to be a torsion free subgroup of $\mathbb{P}\mathrm{SL}_2(\mathbb{Z})$.

The proof we outline here uses, mainly, tools that belong to Teichmüller theory.

Part I: Complex Curves

2. Riemann surfaces defined over $\bar{\mathbb{Q}}$

Probably, the feature that makes the theory of (compact) Riemann Surfaces so intensively atractive is the equivalence between the following three classes of objects:

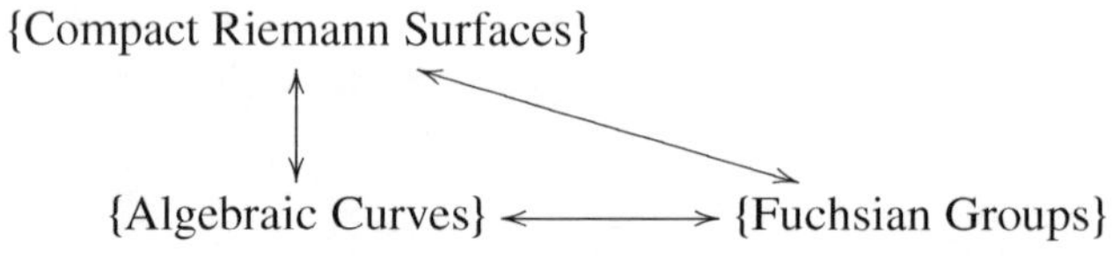

According to this equivalence a given Riemann surface C may be described as $C_F = \{F(x, y) = 0\}$ for an irreducible polynomial $F \in \mathbb{C}[x, y]$, but also as $C \simeq \mathbb{H}/\Gamma$ where $\mathbb{H}$ is the upper half plane and $\Gamma \subset \mathbb{P}\mathrm{SL}_2(\mathbb{R})$ is a Fuchsian group.

A natural problem from the point of view of arithmetic is to determine which Riemann surfaces are definable over the field of algebraic numbers $\bar{\mathbb{Q}}$, i.e. which Riemann surfaces correspond to curves C_F with $F \in \bar{\mathbb{Q}}[x, y]$.

One difficulty comes from the fact that different algebraic curves may correspond to the same Riemann surface. The curve $y^2 = x^3 - \pi$, for example, is not directly defined over $\bar{\mathbb{Q}}$, but it is isomorphic to $\{y^2 = x^3 - 1\}$ via the isomorphism

$$\{y^2 = x^3 - \pi\} \longrightarrow \{y^2 = x^3 - 1\}$$

$$(x, y) \longmapsto (x/\sqrt[3]{\pi}, y/\sqrt{\pi})$$

Recall that the moduli space $\mathcal{M}_g$, whose points are in one to one correspondence with isomorphy classes of compact Riemann surfaces of genus g, can be obtained as a quotient

$$\mathcal{M}_g = \frac{\mathcal{T}_g}{\mathrm{Mod}_g}$$

where $\mathcal{T}_g$ is the Teichmüller space of genus g, of complex dimension $(3g - 3)$, and Mod_g is the mapping class group. The mapping class group is known to act properly and discontinuously on $\mathcal{T}_g$ as a group of biholomorphic transformations. Therefore $\mathcal{M}_g$ is an analytic space of complex dimension $3g - 3$ (see [24]).

Now the subset representing Riemann surfaces definable over $\bar{\mathbb{Q}}$ is clearly countable. Hence most of the Riemann surfaces are not arithmetic.

In the following result we summarize several characterizations of the arithmeticity of a Riemann surface.

Theorem 1. *The following conditions are equivalent:*

1. *C is defined over $\bar{\mathbb{Q}}$.*
2. *There exists a covering $f : C \to \widehat{\mathbb{C}}$ ramified over $\{0, 1, \infty\}$, where $\widehat{\mathbb{C}} := \mathbb{C} \cup \{\infty\} = \mathbb{P}^1$ is the Riemann sphere.*
3. *C can be described as $C = \mathbb{H}/K$, where K is a finite index subgroup of a Fuchsian triangle group.*
4. *C is isomorphic to the compactification of $\mathbb{H}/\Gamma$, where Γ is a finite index subgroup of the modular group $\mathbb{P}\mathrm{SL}_2(\mathbb{Z})$.*

The equivalence between 1 and 2 is nowadays known as *Belyi's theorem*, since Belyi provided an algorithm showing $1 \Rightarrow 2$ (the implication $2 \Rightarrow 1$ was already known by that time). This is the reason why Riemann surfaces defined over $\bar{\mathbb{Q}}$ or, equivalently, over a number field, are often called *Belyi surfaces*. Similarly, the corresponding coverings $f : C \to \widehat{\mathbb{C}}$ are usually referred to as *Belyi functions*.

Part of the importance of Belyi surfaces comes from the fact that $\mathrm{Gal}(\bar{\mathbb{Q}})$, the *absolute Galois group*, acts on the set of *Belyi pairs*. A Belyi pair (C, f) consists of a Belyi surface with a Belyi function defined on it. Two such pairs (C_1, f_1) and (C_2, f_2) are considered equivalent when they are so as ramified coverings, that is, when there exists an isomorphism of Riemann surfaces $F : C_1 \to C_2$ such that the diagram

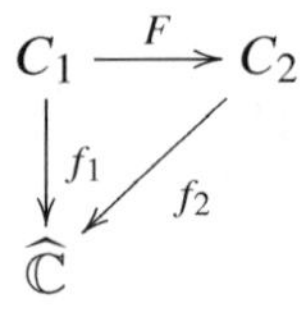

commutes.

2.1 *The action of* **Gal**($\bar{\mathbb{Q}}$) *on Belyi pairs*

It is well known that in the dictionary

$$\{\text{Compact Riemann Surfaces}\} \longleftrightarrow \{\text{Algebraic Curves}\}$$
$$C \longleftrightarrow C_F$$

meromorphic functions correspond to rational functions, that is

$$\{\text{Meromorphic Functions } f \text{ in } C\} \longleftrightarrow \{\text{Rational Functions } R \text{ in } C_F\}$$

Therefore a given Belyi pair (C, f) may also be represented as (C_F, R), where C_F is the algebraic curve corresponding to C and R is the rational function corresponding to f. According to Theorem 1 the polynomial $F(x, y)$ can be chosen so as to have coefficients in a number field. Furthermore, the following stronger version of Belyi's theorem holds.

Theorem 2. *Every Belyi pair (C, f) can be represented by a pair (C_F, R) in which both the algebraic curve $F(x, y)$ and the rational function $R(x, y)$ have coefficients in $\bar{\mathbb{Q}}$.*

Now let $\sigma \in \mathrm{Gal}(\bar{\mathbb{Q}})$. Given a polynomial $P(x, y) = \sum_{i,j} a_{i,j} x^i y^j \in \bar{\mathbb{Q}}[x, y]$, we denote by P^σ the polynomial obtained after applying σ to the coefficients, that is $P^\sigma(x, y) = \sum_{i,j} \sigma(a_{i,j}) x^i y^j$.

This action on the coefficients of polynomials induces an action of $\mathrm{Gal}(\bar{\mathbb{Q}})$ on Belyi pairs. If (C, f) is represented as (C_F, R), then we put

$$\sigma(C, f) = (C_{F^\sigma}, R^\sigma).$$

3. Grothendieck's dessins

Belyi functions can be understood in a surprisingly simple combinatorial way, by means of certain graphs which Grothendieck named dessins d'enfants (child's drawings).

Definition 3. *A* dessin d'enfant *is a pair $(X, \mathcal{D})$ where X is an oriented compact topological surface, and $\mathcal{D} \subset X$ is a graph such that:*

1. *$\mathcal{D}$ is connected.*
2. *$\mathcal{D}$ is bicolored, i.e. the vertices have been given either white or black color and two vertices connected by an edge have always distinct colors.*
3. *$X \backslash \mathcal{D}$ is the union of finitely many topological discs, which we call* faces *of $\mathcal{D}$.*

The genus *of $(X, \mathcal{D})$ is simply the genus of the topological surface X.*

Two dessins $(X_1, \mathcal{D}_1)$ and $(X_2, \mathcal{D}_2)$ are considered equivalent if there is an orientation-preserving homeomorphism from X_1 to X_2 whose restriction to $\mathcal{D}_1$ induces a homeomorphism of the colored graphs $\mathcal{D}_1$ and $\mathcal{D}_2$.

There is a very convenient way of encoding all the information describing a dessin by means of two permutations of its edges. This is done as follows.

Suppose $\mathcal{D}$ has N edges, and label them with integer numbers from 1 to N. Consider one of the white vertices, and look at the edges incident to it by turning around the vertex in positive sense, according to the orientation of the surface. This procedure gives a cycle, and the same strategy applied to every white vertex produces a permutation $\sigma_0 \in \mathbb{S}_N$, namely the product of all these cycles. The same construction applied to the black vertices defines a second permutation σ_1.

Note that $\sigma_1 \cdot \sigma_0$ sends the edge labeled k to another edge that belongs to one of the two faces meeting at edge k (see Figure 1). In particular, every cycle of $\sigma_1 \cdot \sigma_0$ corresponds to a face of $\mathcal{D}$, the length of the cycle being half the number of edges of the corresponding face (an edge incident at both sides with the same face counts twice). We thus see that the dessin can be reconstructed from the cycle structure of σ_0, σ_1, and $\sigma_1 \cdot \sigma_0$.

Let us describe how σ_0 and σ_1 determine the dessin in a concrete example.

Suppose $\sigma_0 = (1,2)(3,4)(5,6)$ and $\sigma_1 = (4,1,5)(3,2,6)$. The corresponding dessin $\mathcal{D}$ has six edges, three white vertices of degree two (since σ_0 has three cycles of length two), and two black vertices of degree three. Moreover $\sigma_1 \cdot \sigma_0 = (1,6,4,2,5,3)$, hence $\mathcal{D}$ has only one face. As for the genus of the surface where $\mathcal{D}$ is embedded, the Euler-Poincaré formula gives

$$2 - 2g = \text{faces} - \text{edges} + \text{vertices} = 1 - 6 + 5 = 0,$$

thus $\mathcal{D}$ is embedded in a topological torus $\mathbb{T}$ (see Figure 2, where $\mathbb{T}$ is also depicted as a rectangle whose opposite sides are identified).

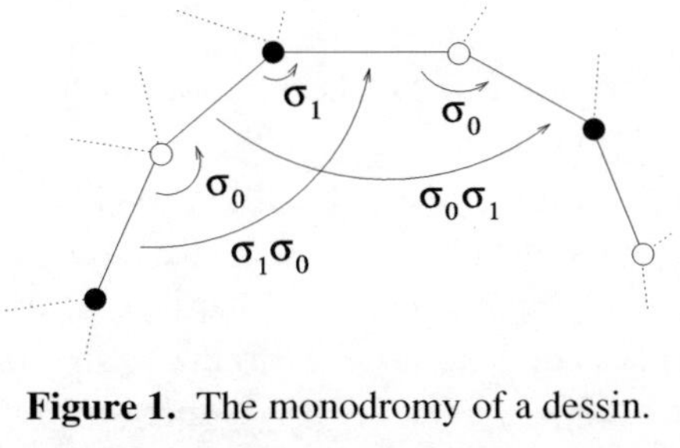

Figure 1. The monodromy of a dessin.

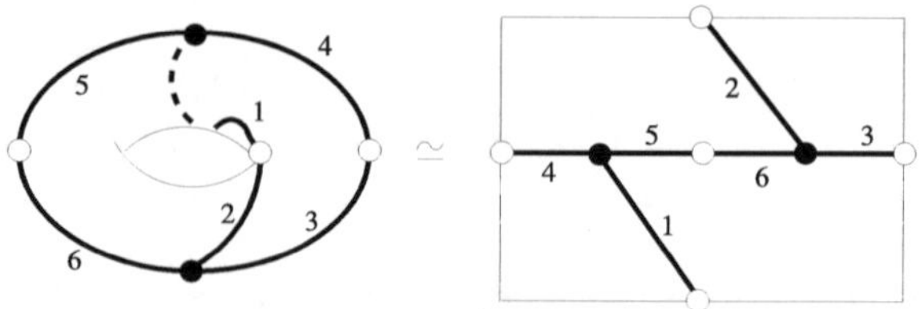

Figure 2. A dessin in a topological torus.

This construction entitles us to speak of the *permutation representation* of $\mathcal{D}$ and to write

$$\mathcal{D} = (\sigma_0, \sigma_1)$$

or, equivalently, $\mathcal{D} = (\sigma_0, \sigma_1, \sigma_\infty)$, with $\sigma_\infty = (\sigma_0\sigma_1)^{-1}$. Conjugate representations will be regarded as equivalent.

The crucial point in the theory of dessins is the fact first noticed by Grothendieck that a dessin defines in a very precise way a Belyi pair. This is a most striking fact. In more precise terms it means that out of the combinatorics of the inclusion of the graph $\mathcal{D}$ in the topological surface X, a mere collection of purely topological data, one is able to endow X with a Riemann surface structure $X_{\mathcal{D}}$ and a meromorphic function $f : X_{\mathcal{D}} \to \mathbb{P}^1$ which, in fact, are defined over a number field.

Later we will show that the correspondence can be reversed, from Belyi pairs to dessins. But let us first describe the construction of the Belyi pair associated to the dessin in our particular example. It will transpire that the construction works for general dessins.

3.1 *The Belyi pair associated to a dessin*

Mark the center (meaning a point in the interior) of each of the faces (in our example, just one) of $\mathbb{T}\backslash\mathcal{D}$ with the symbol $\times$, and draw a simple path from it to each vertex (in our example, 2 black and 3 white). This way we get a triangulation $\mathcal{T}(\mathcal{D})$, i.e. a subdivision of the faces into triangles whose three vertices are of the three distinct types: $\circ$, $\bullet$ and $\times$. Now colour the triangles white or black, according to whether or not the circuit $\circ \to \bullet \to \times \to \circ$ occurs clockwise (see Figure 3).

Choose a white triangle T_1, and a homeomorphism $f_1 : T_1 \to \bar{\mathbb{H}}^+ := \mathbb{H}\cup(\mathbb{R}\cup\{\infty\})$ such that

$$f_1 : \begin{cases} \partial T_1 & \to & \mathbb{R} \cup \{\infty\} \\ \circ & \to & 0 \\ \bullet & \to & 1 \\ \times & \to & \infty \end{cases} \tag{1}$$

Then take a triangle T_2 adjoint to T_1 (therefore a black triangle), and map it to the lower half-plane via a homeomorphism $f_2 : T_2 \to \bar{\mathbb{H}}^-$, which coincides with f_1 on their common edge and behaves on ∂T_2 according to (1). Going on with this process we end up with a continuous mapping $f_{\mathcal{D}} : \mathbb{T} = (\cup T_i) \to \widehat{\mathbb{C}}$, that is even a covering mapping if we puncture the points $\circ$,$\bullet$ and $\times$ (that is, the vertices of the triangles).

There is on $\mathbb{T}$ a unique Riemann surface structure $\mathbb{T}_{\mathcal{D}}$ for which $f_{\mathcal{D}} : \mathbb{T}_{\mathcal{D}} \to \widehat{\mathbb{C}}$ becomes a holomorphic function, in fact a Belyi function all whose ramification points

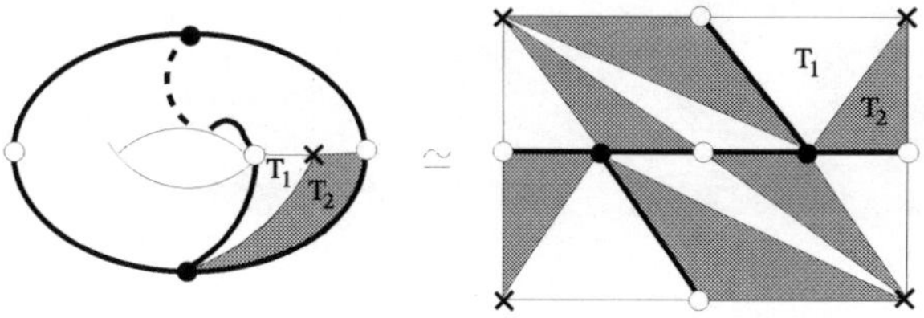

Figure 3. Triangulation associated to a dessin.

are vertices of some triangle. The Belyi pair we obtain does not depend on the various choices made along the construction, such as the division of the faces or the different homeomorphisms. As we shall see in Section 3.3 it is completely determined by the representation $\mathcal{D} = (\sigma_0, \sigma_1, \sigma_\infty)$.

3.2 *The dessin associated to a Belyi function*

The description of the correspondence in the reverse direction is quite simple. Given a Belyi function $f : C \to \widehat{\mathbb{C}}$, we merely take $\mathcal{D}_f = f^{-1}([0, 1])$, the preimage of the real interval $[0, 1]$. Then $\mathcal{D}_f$ is a dessin d'enfant, and the following properties hold.

Proposition 4. *Let $\mathcal{D}_f = f^{-1}([0, 1])$. Then*

- *The degree of f gives the number of edges of $\mathcal{D}_f$.*
- *The inverse images $f^{-1}([0, 1])$, $f^{-1}([-\infty, 0])$ and $f^{-1}([1, \infty])$ consist of a union of segments that together form the triangulation $\mathcal{T}(\mathcal{D}_f)$.*
- *The set of white (resp. black) vertices of $\mathcal{D}_f$ is precisely $f^{-1}(0)$ (resp $f^{-1}(1)$). The degree of a vertex equals the ramification index of f at that vertex. Similarly, the centers of faces can be chosen to be the preimages of infinity, and the ramification index at each of them is half the number of edges of the corresponding face.*
- *The inverse images $f^{-1}(\bar{\mathbb{H}}^+)$ and $f^{-1}(\bar{\mathbb{H}}^-)$ are respectively the black and white triangles of the triangulation $\mathcal{T}(\mathcal{D}_f)$.*

Thus, we have a clear description of the correspondence

{Dessins d'enfants}	$\longleftrightarrow$	{Belyi pairs}
$\mathcal{D} \subset X$	$\longrightarrow$	$(X_{\mathcal{D}}, f_{\mathcal{D}})$
$\mathcal{D}_f = f^{-1}([0, 1])$	$\longleftarrow$	(C, f)

3.3 *The monodromy of a dessin*

3.3.1 *Monodromy of a covering of the Riemann sphere* Recall that the monodromy of a degree d covering $f : C \longrightarrow \mathbb{P}^1$ ramified over $\{q_1, \ldots, q_r\} \subset \mathbb{P}^1$, is a group homomorphism

$$\begin{aligned} \mathrm{Mon} : \pi_1(\mathbb{P}^1 \backslash \{q_i\}, p) &\longrightarrow \mathrm{Bij}\{f^{-1}(p)\} = \mathbb{S}_d \\ \gamma &\longmapsto \mathrm{Mon}(\gamma) \end{aligned}$$

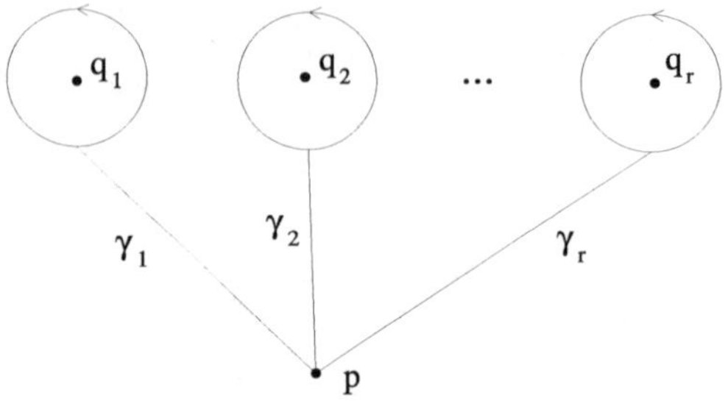

Figure 4. Generators of $\pi_1(\mathbb{P}^1\backslash\{q_i\})$.

where $p \in \mathbb{P}^1\backslash\{q_i\}$ is the base point with fibre $f^{-1}(p) = \{x_1, \dots, x_d\}$ and $\mathrm{Mon}(\gamma)(x_i)$ is the end point of the path in C obtained by lifting the loop γ with initial point $x_i \in C$.

It is well known that once the branching values $\{q_i\}$ are fixed, the homomorphism Mon characterizes the covering. We will sometimes represent the monodromy as

$$\mathrm{Mon} \equiv (\sigma_1, \dots, \sigma_r) \tag{2}$$

where $\sigma_i = \mathrm{Mon}(\gamma_i)$ and γ_i is a loop surrounding q_i as in Figure 4. This is because the loops γ_i generate $\pi_1(\mathbb{P}^1\backslash\{q_i\})$ subject to the single relation $\prod \gamma_i = 1$. In particular, in the above representation we must also have $\prod \sigma_i = 1$.

The connectedness of C implies that the group generated by the permutations σ_i is a transitive subgroup of $\mathbb{S}_d$, known as the *monodromy group* of f. Conversely, it is also well known that any r-tuple of elements of $\mathbb{S}_d$, $(\sigma_1, \dots, \sigma_r)$, which generates a transitive subgroup of $\mathbb{S}_d$ and satisfies the relation $\prod \sigma_i = 1$, arises as the monodromy of a degree d ramified covering of $\mathbb{P}^1$.

3.3.2 *Monodromy of a dessin* It is interesting to observe that in this language the permutation representation of a dessin $\mathcal{D}$ is nothing but the monodromy of $f_{\mathcal{D}}$. This can be easily seen by choosing $p = 1/2$ as base point of the fundamental group of $\mathbb{P}^1\backslash\{0, 1, \infty\}$. Then σ_0 and σ_1 are simply the images of loops around 0 and 1 (as in Figure 4) under the monodromy homomorphism of $f_{\mathcal{D}}$.

This is the reason why the permutation representation of a dessin is often called the *monodromy* of a dessin. Similarly the group generated by σ_0 and σ_1 will be called the *monodromy group* of $\mathcal{D}$.

It follows that there is a correspondence between the following three classes of objects:

$$\begin{array}{ccc}
\left\{\begin{array}{c}\text{Belyi pairs } (C, f)\\ \text{with } \deg(f) = n\end{array}\right\} & \xrightarrow{f^{-1}([0,1])} & \left\{\begin{array}{c}\text{Dessins d'enfants}\\ \text{with } n \text{ edges}\end{array}\right\} \\
\Big\downarrow \text{Mon} & \swarrow \text{Perm.rep.} & \\
\left\{\begin{array}{c}\text{Ordered pairs of}\\ \text{permutations generating}\\ \text{transitive subgroups of } \mathbb{S}_n\end{array}\right\} & &
\end{array}$$

It is clear from what has gone above that these are bijective correspondences.

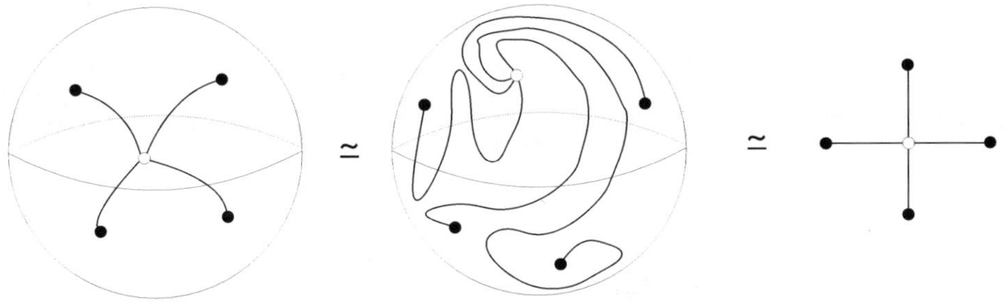

Figure 5. The dessin on the sphere corresponding to $\beta(z) = z^4$.

4. Some examples

We shall now show some explicit examples of Belyi pairs and the corresponding dessins.

The first examples will be of genus 0, the simplest case since a half of the Belyi pair, namely the Riemann surface, is already known (the Riemann sphere $\widehat{\mathbb{C}}$).

Example 5. Consider $\mathcal{D} = (\sigma_0, \sigma_1)$, where $\sigma_0 = (1, 2, 3, 4), \sigma_1 = \mathrm{Id} \in \mathbb{S}_4$. It can be easily checked that it corresponds to the dessin in Figure 5, where the first two pictures represent two possible concrete aspects of $\mathcal{D}$ (remember the equivalence up to homeomorphisms). It is in fact more convenient to depict $\mathcal{D}$ as in the right hand side of Figure 5, where the sphere has been mapped to the plane by stereographic projection.

We would like to describe explicitly the Belyi pair $(\mathbb{S}^2_{\mathcal{D}}, f_{\mathcal{D}})$ associated to this dessin. Since the only Riemann surface of genus zero is $\widehat{\mathbb{C}}$, we must have a commutative diagram of the form

$$\begin{array}{ccc} \mathbb{S}^2_{\mathcal{D}} & \xrightarrow{\ g\ } & \widehat{\mathbb{C}} \\ \Big\downarrow f_{\mathcal{D}} & \swarrow \beta & \\ \widehat{\mathbb{C}} & & \end{array}$$

where g is an isomorphism of Riemann surfaces and β is a rational function. Thus, our goal is to find the rational function β such that $\beta^{-1}([0, 1]) = g(\mathcal{D})$.

The following information about β follows from Proposition 4.

- β has degree 4.
- $\beta^{-1}(\infty)$ has only one point.
- $\beta^{-1}(0)$ has only one point, whose ramification index is four.
- $\beta^{-1}(1)$ has four non-ramified points.

Note that by composing g with a Möbius transformation we can assume that the center of the face of $\mathcal{D}$, the white vertex, and a chosen black vertex correspond via g to ∞, 0 and 1. It follows that the Belyi function corresponding to the dessin in Figure 5 is $\beta(z) = z^4$.

In the same way, the dessin in the sphere given by a star with n arms (monodromy $\sigma_0 = (1, \ldots, n)$, $\sigma_1 = \mathrm{Id}$) has $\beta(z) = z^n$ as associated Belyi function.

Example 6. ([22]) Let us consider a slightly more interesting case: the dessin in the left hand side of Figure 6.

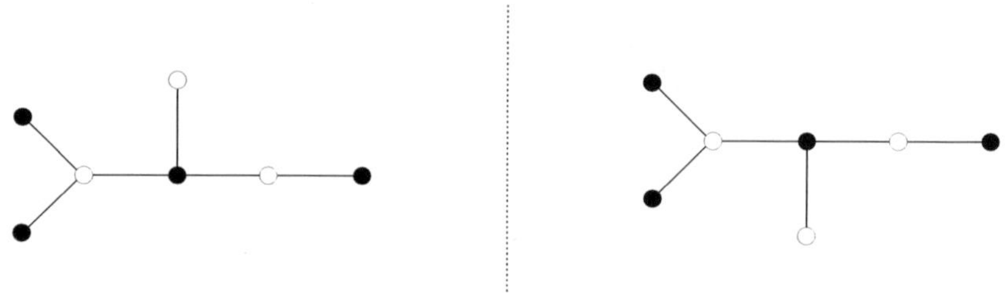

Figure 6. Two conjugate trees.

Choose the white vertices of degrees 3 and 2 to be $z = 0$ and $z = 1$ respectively, and let $z = a$ be the third white vertex. Then, according to Proposition 4

$$\beta(z) = z^3(z-1)^2(z-a),$$

and the cuestion is whether or not it is possible to determine the value of a.

Let us compute the derivative of β:

$$\begin{aligned}\beta'(z) &= 3z^2(z-1)^2(z-a) + 2z^3(z-1)(z-a) + z^3(z-1)^2 \\ &= z^2(z-1)(6z^2 + (-5a-4)z + 3a)\end{aligned}$$

Now, again by Proposition 4, β must have a ramification point of order 3 (different from 0 or 1), therefore that point must occur as double zero of β'. Then the discriminant of the polynomial $P(z) = 6z^2 + (-5a-4)z + 3a$ must vanish, that is

$$25a^2 - 32a + 16 = 0,$$

hence $a = \frac{4}{25}(4+3i)$ or $a = \frac{4}{25}(4-3i)$.

If we draw the preimage of the interval $[0, 1]$ by the two Belyi functions corresponding to the two possible choices of a, we find the two graphs of Figure 6. Because these two values of a are Galois-conjugate of each other, the two dessins will be said to be *conjugate dessins* (see Section 5).

See [22] for more examples of this kind.

Example 7. Let us study in detail an example of a dessin in a genus 1 topological surface. Consider again the dessin $\mathcal{D}$ in Figure 2, whose monodromy is $\sigma_0 = (1,6)(2,4)(3,5)$ and $\sigma_1 = (1,2,3)(4,5,6)$.

Since the construction of the Belyi pair depends only on the topology, we can replace each of the triangles in Figure 3 by an Euclidean triangle of angles $\frac{\pi}{2}$, $\frac{\pi}{3}$ and $\frac{\pi}{6}$ at the vertices $\circ$, $\bullet$ and $\times$ respectively. This way we obtain a quadrangle with identified opposite sides (see the left part of Figure 7). We can now cut the torus along the dessin, and paste the pieces together to form a regular hexagon with opposite sides identified (right part of Figure 7). We think of the hexagon as centered at the origin of the Euclidean plane, and $\mathcal{D}$ can be seen now at the boundary.

With this highly symmetric representation of $\mathcal{D}$ at our disposal, we can be more precise about the construction, explained in Section 3, of the associated function $f_{\mathcal{D}}$. Recall that to define the function $f_{\mathcal{D}}$, one begins with a homeomorphism $f_1 : T_1 \to \bar{\mathbb{H}}^+$. Let T_1 be as in Figure 7. We choose a homeomorphism f_1, and extend it to $f_2 : T_2 \to \bar{\mathbb{H}}^-$ by $f_2(z) = f_1(\bar{z})$, where T_2 is also shown in Figure 7.

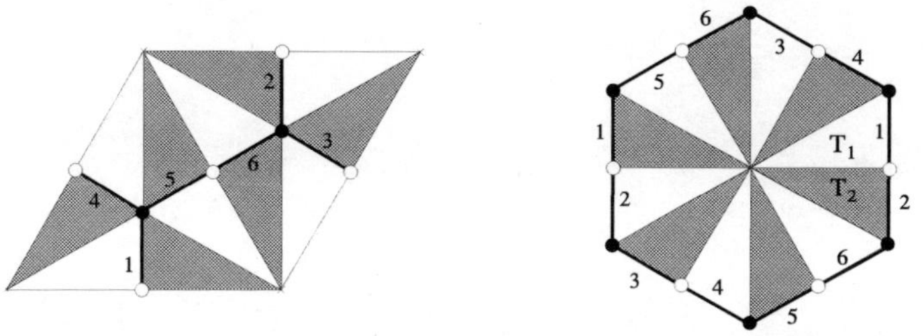

Figure 7. Further topological representations of the dessin in Figure 2.

Then we can define $f_{\mathcal{D}}$ on each triangle as follows: $f_{\mathcal{D}} = f_1$ in T_1, $f_{\mathcal{D}} = f_2$ in T_2, and $f_{\mathcal{D}} = f_j \circ R^k_{\pi/3}$ in the remaining triangles, where $R_{\pi/3}$ is the rotation through angle $\pi/3$ around the origin, $j = 1$ or 2 for the white and black triangles respectively, and k is the necessary exponent for the composition to make sense.

Now consider the mapping $\gamma : \mathbb{T}_{\mathcal{D}} \to \mathbb{T}_{\mathcal{D}}$ induced by $R_{\pi/3}$ in the Belyi surface $\mathbb{T}_{\mathcal{D}}$. It is clear that γ is a homeomorphism. Moreover, since $f_{\mathcal{D}} \circ \gamma = f_{\mathcal{D}}$, it follows that γ is an (order 6) automorphism of $\mathbb{T}_{\mathcal{D}}$. Furthermore γ has at least one fixed point, namely the center of our hexagon. Now, it is well known, by the classical theory of elliptic curves, that there is a unique torus with such an automorphism, namely

$$\begin{aligned} \{y^2 = x^3 - 1\} &\longrightarrow \{y^2 = x^3 - 1\} \\ (x, y) &\longmapsto (\xi_3 x, -y) \end{aligned}$$

where $\xi = \exp 2\pi i/3$. This ensures that $\mathbb{T}_{\mathcal{D}}$ corresponds to the algebraic curve $y^2 = x^3 - 1$. Accordingly, $f_{\mathcal{D}}$ can be described by $(x, y) \mapsto x^3 - 1$.

All the previous examples show in concrete cases the correspondence from dessins to Belyi pairs. Let us focus now in the reverse construction, from Belyi pairs to dessins, by studying in detail an example in genus 1.

Example 8. Let C be the Riemann surface $\{y^2 = x(x-1)(x-\sqrt{2})\} \cup \{\infty\}$, and construct a Belyi function f on it by means of the following composition of maps

$$\begin{array}{l} \{y^2 = x(x-1)(x-\sqrt{2})\} \cup \{\infty\} \\ \quad \downarrow {\scriptstyle (x,y)\mapsto x} \\ \quad \widehat{\mathbb{C}} \\ \quad \downarrow {\scriptstyle t\mapsto t^2-2} \text{ (minimal polinomial of } \sqrt{2}) \\ \quad \widehat{\mathbb{C}} \\ \quad \downarrow {\scriptstyle u\mapsto -1/u} \\ \quad \widehat{\mathbb{C}} \\ \quad \downarrow {\scriptstyle z\mapsto 4z(1-z)} \text{ (Belyi polynomial } P_{1/2}, \text{ see Section 6.1)} \\ \quad \widehat{\mathbb{C}} \end{array}$$

where the curved arrow f goes from the curve directly to the last $\widehat{\mathbb{C}}$.

that is

$$f(x,y) = \frac{-4(x^2-1)}{(x^2-2)^2}.$$

We can track back the ramification indices over the three ramification values 0, 1, and ∞ from the bottom to the top in the previous diagram, and we obtain

$$\begin{array}{ccc}
\left\{\begin{array}{c} \infty^{(4)} \\ (1,0)^{(2)} \\ (-1,\sqrt{-2-\sqrt{8}})^{(1)} \\ (-1,-\sqrt{-2-\sqrt{8}})^{(1)} \end{array}\right\} & \{(0,0)^{(8)}\} & \left\{\begin{array}{c} (\sqrt{2},0)^{(4)} \\ (-\sqrt{2},\sqrt{-4-\sqrt{32}})^{(2)} \\ (-\sqrt{2},-\sqrt{-4-\sqrt{32}})^{(2)} \end{array}\right\} \\
\downarrow & \downarrow & \downarrow \\
\{\infty^{(2)}, 1^{(1)}, -1^{(1)}\} & \{0^{(4)}\} & \{\sqrt{2}^{(2)}, -\sqrt{2}^{(2)}\} \\
\downarrow & \downarrow & \downarrow \\
\{\infty^{(1)}, -1^{(1)}\} & \{-2^{(2)}\} & \{0^{(2)}\} \\
\downarrow & \downarrow & \downarrow \\
\{0^{(1)}, 1^{(1)}\} & \{1/2^{(2)}\} & \{\infty^{(2)}\} \\
\downarrow & \downarrow & \downarrow \\
\{0\} & \{1\} & \{\infty\}
\end{array}$$

where the superindices denote branching indices.

Now we find the preimage of the real interval [0, 1] by f, performing inverse images step by step in the composition of maps that defines f. The result is shown in Figure 8, where the algebraic curve C is constructed by gluing two copies of the complex plane along cuts from 0 to 1 and from $\sqrt{2}$ to ∞.

5. The action of Gal($\bar{\mathbb{Q}}$) on dessins

We already described how the absolute Galois group acts on Belyi pairs. Now, having established the equivalence between Belyi pairs and dessins, we can make the Galois elements $\sigma \in \mathrm{Gal}(\bar{\mathbb{Q}})$ act on the dessins themselves, by means of the rule

$$\begin{array}{ccc}
\mathcal{D} & \longrightarrow & \sigma(\mathcal{D}) \\
\updownarrow & & \updownarrow \\
(X_{\mathcal{D}}, f_{\mathcal{D}}) & \overset{\sigma}{\longleftrightarrow} & (X^{\sigma}_{\mathcal{D}}, f^{\sigma}_{\mathcal{D}})
\end{array}$$

Let us have a second look at our examples in Section 4.

In the genus 0 case, Galois conjugation affects only the Belyi function and not the Riemann surface (there is only one Riemann surface of genus 0). As the trees in Example 5 come from polynomials defined over the rationals, all of them remain

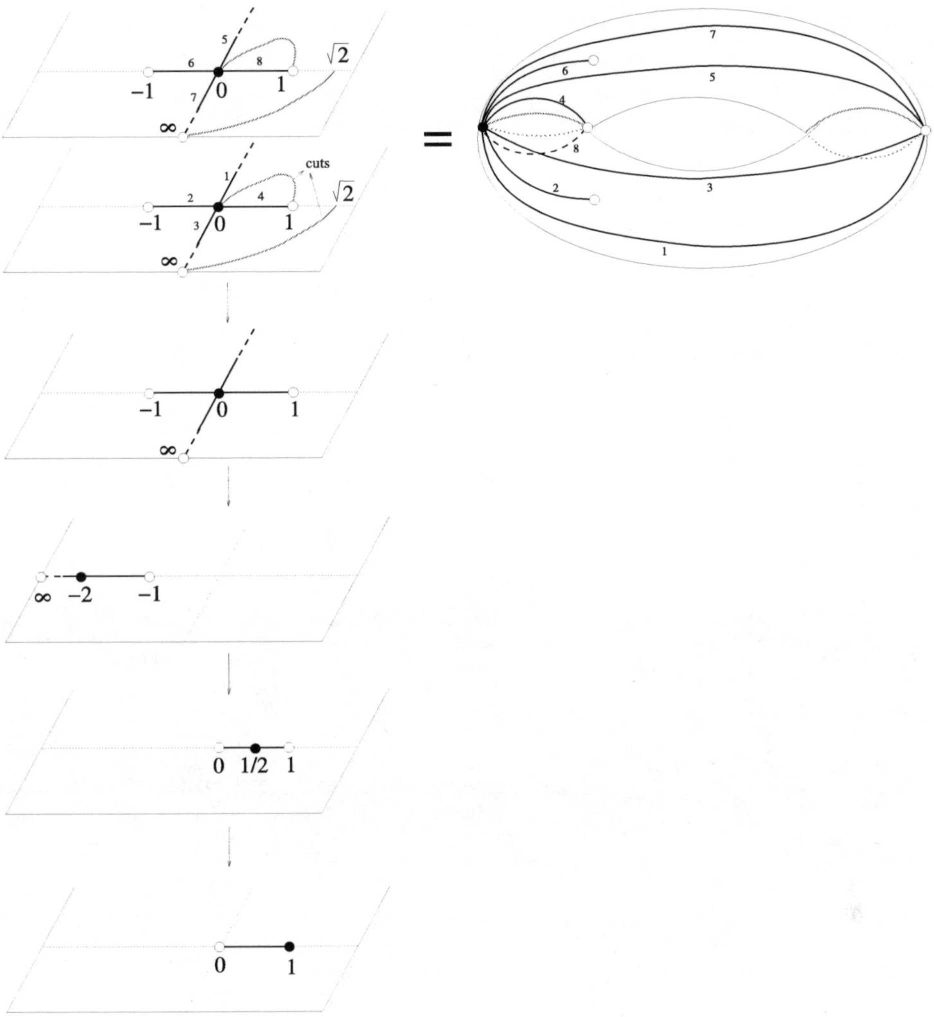

Figure 8. $\mathcal{D} = f^{-1}([0, 1])$; f as in Example 8.

fixed by the whole action of $\text{Gal}(\bar{\mathbb{Q}})$. In other words, we have $\sigma(\mathcal{D}) = \mathcal{D}$ for every $\sigma \in \text{Gal}(\bar{\mathbb{Q}})$.

Something different happens in Example 6. Every σ such that $\sigma(i) = -i$ transforms the parameter a into $\bar{a}$, hence sends the tree in the left hand side of Figure 6 to the tree in the right hand side of the same Figure and viceversa. On the contrary, both trees remain invariant by Galois elements fixing i. It can be seen that these two trees form a *complete orbit* of the action of $\text{Gal}(\bar{\mathbb{Q}})$ on dessins. This is done by computing certain invariants of the action of $\text{Gal}(\bar{\mathbb{Q}})$ and showing that no other graph has the same invariants (see Proposition 10).

The action can be more involved in genus 1 since the conjugation affects also the Riemann surface (nevertheless the conjugate Riemann surface will still have genus 1, see Proposition 10). But again this is not the case of the dessin $\mathcal{D}$ in Example 7. As both $X_{\mathcal{D}}$ and $f_{\mathcal{D}}$ are defined over the rationals, every Galois element fixes $\mathcal{D}$.

Example 8 is much more interesting. Let $\sigma \in \mathrm{Gal}(\bar{\mathbb{Q}})$ such that $\sigma(\sqrt{2}) = -\sqrt{2}$. Consider the conjugate Belyi pair (C^σ, f^σ). As f was defined over $\mathbb{Q}$, f^σ is given by the same formula as f, namely

$$f^\sigma(x, y) = \frac{4(x^2 - 1)}{x^4},$$

but now f^σ must be viewed as a function on the conjugate Riemann surface $C^\sigma = \{y^2 = x(x-1)(x+\sqrt{2})\} \cup \{\infty\}$, which is not isomorphic to C since $J(\sqrt{2}) \neq J(-\sqrt{2})$ where $J(\lambda) = \frac{\lambda^2-\lambda+1}{\lambda^2(\lambda-1)^2}$ is the classical J-invariant that classifies elliptic curves. To depict $\sigma(\mathcal{D})$ we can follow the same strategy as the one we used for $\mathcal{D}$, that is, we follow the track of the ramification through the four maps in which f^σ has been decomposed. We obtain

$$\begin{array}{ccc}
\left\{\begin{array}{c} \infty^{(4)} \\ (1,0)^{(2)} \\ (-1, \sqrt{-2+\sqrt{8}})^{(1)} \\ (-1, -\sqrt{-2+\sqrt{8}})^{(1)} \end{array}\right\} & \{(0,0)^{(8)}\} & \left\{\begin{array}{c} (-\sqrt{2}, 0)^{(4)} \\ (\sqrt{2}, \sqrt{-4+\sqrt{32}})^{(2)} \\ (\sqrt{2}, -\sqrt{-4+\sqrt{32}})^{(2)} \end{array}\right\} \\
\downarrow & \downarrow & \downarrow \\
\{\infty^{(2)}, 1^{(1)}, -1^{(1)}\} & \{0^{(4)}\} & \{\sqrt{2}^{(2)}, -\sqrt{2}^{(2)}\} \\
\downarrow & \downarrow & \downarrow \\
\{\infty^{(1)}, -1^{(1)}\} & \{-2^{(2)}\} & \{0^{(2)}\} \\
\downarrow & \downarrow & \downarrow \\
\{0^{(1)}, 1^{(1)}\} & \{1/2^{(2)}\} & \{\infty^{(2)}\} \\
\downarrow & \downarrow & \downarrow \\
\{0\} & \{1\} & \{\infty\}
\end{array}$$

where the superindices denote branching indices. Now compute step by step the preimage of the real interval $[0, 1]$. See Figure 9, where the final picture of $\sigma(\mathcal{D})$ is drawn on the topological torus. For another example of this kind, see [32].

Remark 9. An alternative way to obtain the branching table corresponding to (C^σ, f^σ) is to observe that $f(p) = 0$ (resp. 1 or ∞) if and only if $f^\sigma(p^\sigma) = \sigma(0) = 0$ (resp. $\sigma(1) = 1$ or $\sigma(\infty) = \infty$) and that the branching orders are also preserved (although this is a little more difficult). Therefore the branching table corresponding to (C^σ, f^σ) can be obtained by simply transforming the branching table for (C, f) under σ.

5.1 *Some invariants*

Trying to get some insight of the group $\mathrm{Gal}(\bar{\mathbb{Q}})$ by studying its action on dessins is probably the main goal that led Grothendieck to the introduction of the theory of dessins. One would like therefore to have a good collection of invariants at hand.

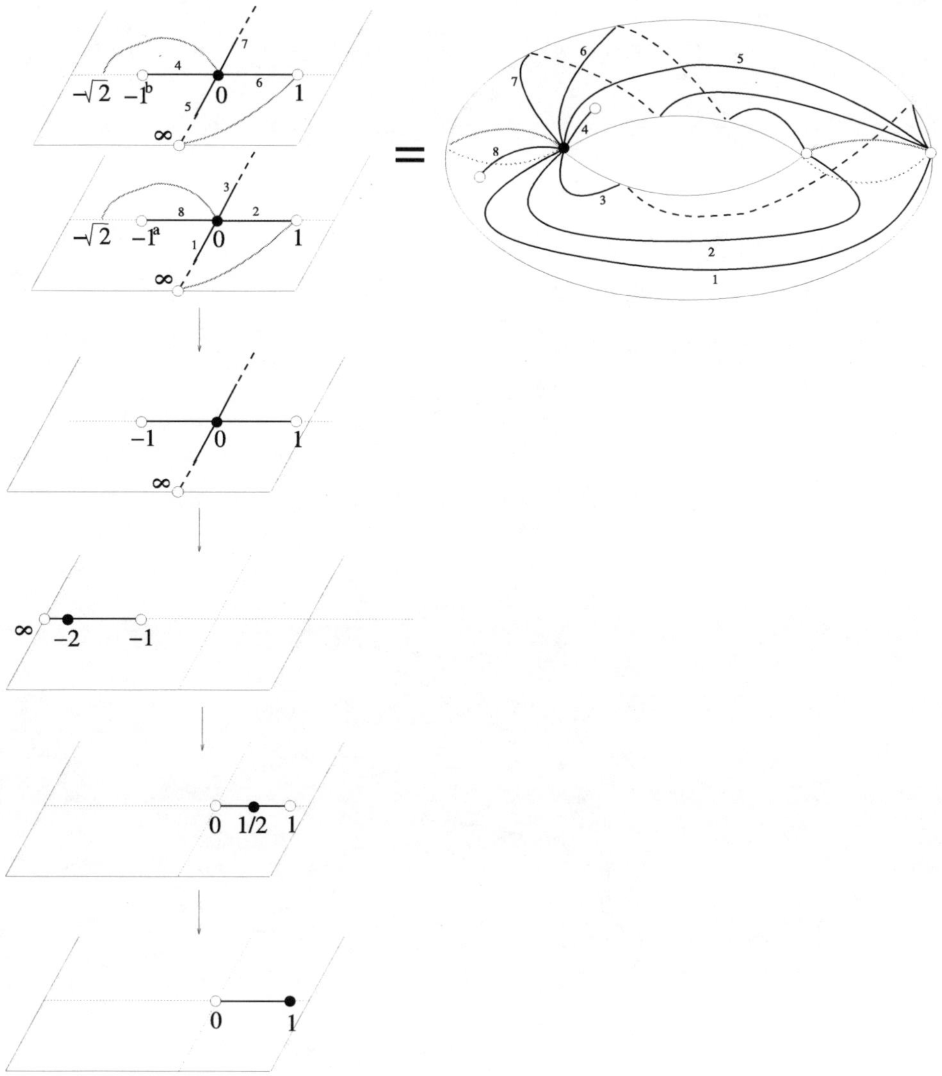

Figure 9. Description of $\sigma(\mathcal{D}) = (f^\sigma)^{-1}([0, 1])$; f as in Example 8.

Proposition 10. *Let $\mathcal{D}$ be a dessin. The following properties of $\mathcal{D}$ remain invariant under the action of the absolute Galois group.*

- *The number of white vertices, black vertices, edges and faces.*
- *The degrees of the white vertices, the degrees of the black vertices, and the degrees of the faces.*
- *The genus.*
- *The monodromy group.*

Those in the previous list are the most simple invariants, but some others have also been studied. Anyway, there does not exist a *complete* list, in the sense that there exist explicit examples of non-conjugate dessins for which all known invariants can be computed and happen to agree.

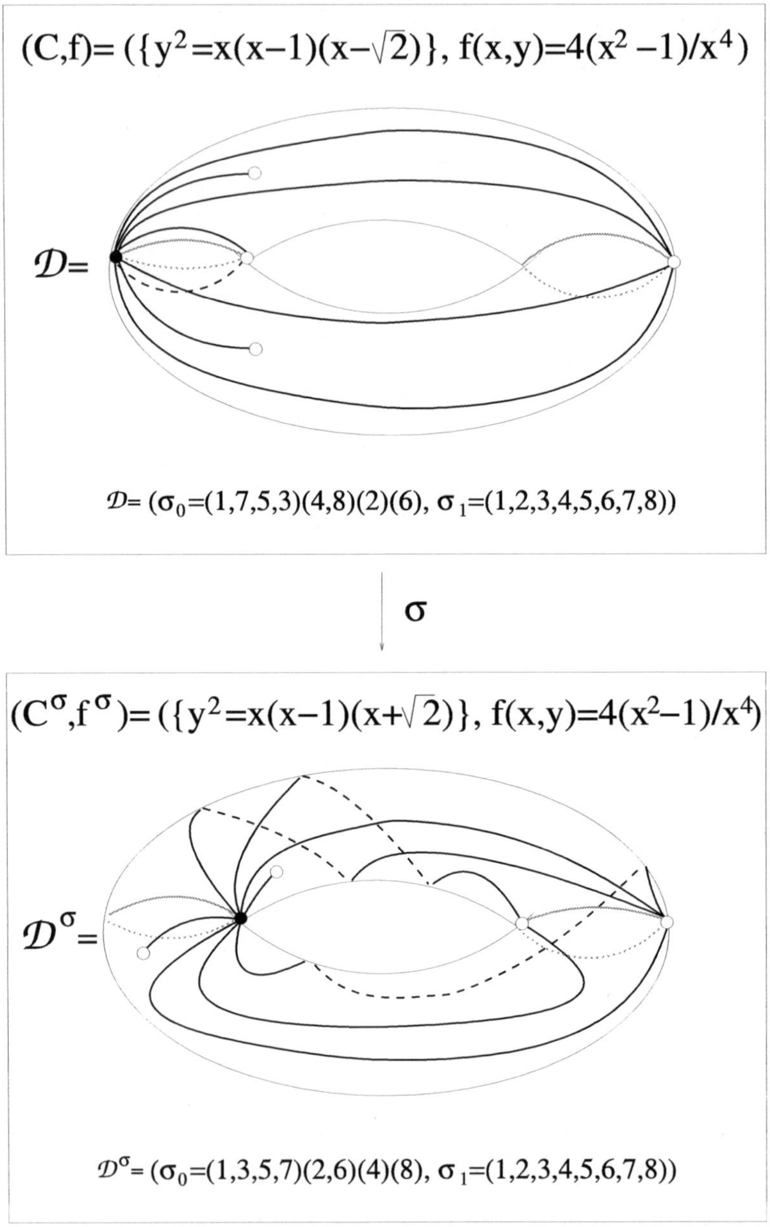

Figure 10. The accion of $\sigma(\sqrt{2}) = -\sqrt{2}$ on the dessin in Figure 8.

We warn the reader that Proposition 10 is not completely trivial. Note that the conjugation by an element of the absolute Galois group is a highly discontinuous operation, hence it is not clear why any of the objects in the list should be preserved. The way to show it is by means of the correspondence between dessins and morphisms $f : C \to \mathbb{P}^1$, on which $\mathrm{Gal}(\bar{\mathbb{Q}})$ acts nicely. Indeed it is not hard to see (Remark 9) that this action preserves degree, ramification, Hurwitz formula, hence genus, etc.

5.2 *Faithfulness of the action of* $\mathrm{Gal}(\bar{\mathbb{Q}})$ *on dessins*

In order to obtain information about the absolute Galois group by making it act on dessins, it is natural to ask whether this action is faithful. In fact it is.

Theorem 11. *The restriction of the action of* $\mathrm{Gal}(\bar{\mathbb{Q}})$ *to dessins of genus g is faithful for every g.*

For the case of genus zero, one can be even more restrictive, as the action is faithful even on trees. This was first shown by Lenstra (see Schneps article in [27]).

In genus 1 the result is quite obvious and follows from the classical theory of elliptic curves. Given $\sigma \in \mathrm{Gal}(\bar{\mathbb{Q}})$ and an algebraic number j such that $\sigma(j) \neq j$, let C_λ : $y^2 = x(x-1)(x-\lambda)$ be the elliptic curve whose Jacobi invariant $J(\lambda) = \frac{\lambda^2-\lambda+1}{\lambda^2(\lambda-1)^2}$ is precisely j. Then $C_\lambda^\sigma = C_{\sigma(\lambda)}$ has Jacobi invariant $\sigma(j)$ and hence it is not isomorphic to C_λ.

For genus g greater than one, it can be shown that the action is already faithful when restricted to hyperelliptic Riemann surfaces of the form

$$\{y^2 = (x-1)(x-2)\cdots(x-(2g+1))(x-(a+n))\},$$

where $a \in \bar{\mathbb{Q}}$ (see [8]).

6. The proof of Belyi's theorem

We want to prove the following statement:

$$\boxed{C \text{ is defined over } \bar{\mathbb{Q}} \Longleftrightarrow \text{There exists } f : C \to \widehat{\mathbb{C}} \text{ ramified over } \{0, 1, \infty\}}$$

6.1 *The* only if *part [4]*

The proof of the implication $\Rightarrow$, which is the real contribution of Belyi, is based on the clever algorithm illustrated in Example 8. Concerning the ramification values, it is rather obvious that

$$\{\text{Ram. values of } f \circ g\} = f(\{\text{Ram. values of } g\}) \cup \{\text{Ram. values of } f\},$$

and this is the reason why a suitable composition of maps may reduce the number of ramification values. Belyi's algorithm consists of the following steps:

1. Find a morphism to $\mathbb{P}^1$ that ramifies only above algebraic points. This is possible because C is defined over $\bar{\mathbb{Q}}$.
2. Compose with a rational function to push all the ramification values inside $\mathbb{Q} \cup \{\infty\}$. The minimal polynomial $m(x)$ of the ramification values of the first step is used at this point. Note that the new branch points created in this process will be roots of $m'(x)$. Since $\deg(m'(x)) < \deg(m(x))$ it is clear that after repeating this argument finitely many times one is left with only rational critical values.

3. The next step is to compose with a polynomial that sends one of the rational ramification values, $\lambda = \frac{m}{m+n}$, inside the set $\{0, 1, \infty\}$. This is precisely Belyi's polynomial P_λ, defined as follows.

$$P_\lambda(x) = \frac{(m+n)^{m+n}}{m^m n^n} x^m (1-x)^n$$

   P_λ enjoys the following properties.
   - P_λ ramifies only at $x = 0, 1, \infty$ and λ.
   - $P_\lambda(0) = 0$, $P_\lambda(1) = 0$, $P_\lambda(\infty) = \infty$ and $P_\lambda(\lambda) = 1$.
   - P_λ sends the remaining rational branching values to rational values.
4. Finally, compose with Belyi's polynomial to the remaining branching values, one by one.

Belyi's argument allows us to give a slightly more general, but equivalent, version of Belyi's theorem.

Theorem 12. *C is defined over a number field if and only if there exists a covering $f : C \longrightarrow \mathbb{P}^1$ ramified over algebraic values.*

6.2 *A criterion for a variety to be defined over* $\bar{\mathbb{Q}}$

The *if* part in Belyi's theorem was previously known, as a consequence of some results of Weil [30]. Instead of Weil's criterion we will apply a criterion given in [9] which will be also used in the second part of the paper (for other proofs, see [20,31]).

Let $X \subset \mathbb{P}^n(\mathbb{C})$ be an irreducible projective variety of arbitrary dimension. Extending the terminology we have employed in the one dimensional case, we shall say that *X is defined over a field $K \subset \mathbb{C}$* if there is a finite collection of homogeneous polynomials with coefficients in K

$$\left\{P_\alpha(X_0, \ldots, X_n) = \sum a_\nu X_0^{\nu_0} \ldots X_n^{\nu_n}\right\}_\alpha$$

whose zero set $Z(P_\alpha)$ is X. We shall say that *X can be defined over K* if it is isomorphic to a variety defined over K.

Likewise, we shall say that a morphism $f : X \to Y$ between irreducible varieties $X \subset \mathbb{P}^n(\mathbb{C})$ and $Y \subset \mathbb{P}^r(\mathbb{C})$ *is defined over K* if X and Y are defined over K and there is an open cover $\{U_j\}$ of X such that $f_{|U_j} \equiv (F_{j,0}, \ldots, F_{j,r})$ for some homogeneous polynomials $F_{j,k} = F_{j,k}(X_0, \ldots, X_n)$ with coefficients in K ([28]). We will say that *$f : X \to Y$ can be defined over K* if it is equivalent to a morphism $f_0 : X \to Y$ defined over K. Here f and f_0 being equivalent means that there are automorphisms $h_1 : X \simeq X$ and $h_2 : Y \simeq Y$ such that the following diagram commutes

$$\begin{array}{ccc} X & \xrightarrow{f} & Y \\ h_1 \downarrow & & \downarrow h_2 \\ X & \xrightarrow{f_0} & Y \end{array}$$

We are interested in the question of whether a given variety X can be defined over a *number field*, or equivalently, over $\bar{\mathbb{Q}}$, the field of algebraic numbers. Let $\mathrm{Gal}(\mathbb{C}) =$

Gal($\mathbb{C}/\mathbb{Q}$) denote the group of all field automorphisms of $\mathbb{C}$. The action of Gal($\mathbb{C}$) on complex varieties and morphisms will be denoted in the same way as we did in the first part of the paper.

In [9] the following criterion was established.

Criterion 13. ([9]) The following conditions relative to an irreducible variety $X \subset \mathbb{P}^n(\mathbb{C})$ (resp. a morphism $f : X \to Y$ between irreducible projective varieties defined over a number field) are equivalent

i) X (resp. $f : X \to Y$) can be defined over a number field.
ii) The family $\{X^\sigma\}_{\sigma \in Gal(\mathbb{C})}$ (resp. $\{f^\sigma : X^\sigma \to Y^\sigma\}_{\sigma \in Gal(\mathbb{C})}$) contains only finitely many isomorphism classes of complex projective varieties (resp. of morphisms).

The i) $\Rightarrow$ ii) part of this criterion is clear. By elementary Galois Theory an algebraic number gets transformed by Gal($\mathbb{C}$) into only finitely many complex numbers. The proof of the part ii) $\Rightarrow$ i) is more difficult but, at the same time, rather elementary, in the sense that it only uses trascendental field theory.

Example 14. For the curve $y^2 = x^3 - \pi$ considered in Section 2 we have

$$\mathrm{Gal}(\mathbb{C})(y^2 = x^3 - \pi) = \{y^2 = x^3 - e\}_{e \text{ transcendental}}$$

This family contains only one isomorphism class, since for any pair of curves in the family we have the following isomorphism

$$\begin{array}{ccc} y^2 = x^3 - \pi & \simeq & y^2 = x^3 - e \\ (x, y) & \to & (x\sqrt[3]{e/\pi},\, y\sqrt{e/\pi}) \end{array}$$

6.3 *The* if *part*

We are now in position to prove the implication $\Leftarrow$) in Belyi's theorem (as stated in Theorem 12), as well as Theorem 2.

Let $f : C \to \mathbb{P}^1$ be a surjective morphism with branch values $y_1, \ldots, y_r \in \mathbb{P}^1(\mathbb{Q}) = \mathbb{Q} \cup \{\infty\}$. For any $\sigma \in$ Gal($\mathbb{C}$) the morphism $f^\sigma : C^\sigma \to \mathbb{P}^1$ has same degree as $f : C \to \mathbb{P}^1$. By hypothesis, the family $\{(y_1^\sigma, \ldots, y_r^\sigma)\}_{\sigma \in \mathrm{Gal}(\mathbb{C})}$ contains only finitely many distinct r-tuples. Therefore there are finitely many possible monodromy homomorphisms

$$\pi_1 : \mathbb{P}^1 \backslash \{y_i^\sigma\} \longrightarrow \mathbb{S}_d$$

and hence (Section 3.3) finitely many nonisomorphic covering curves C^σ and covering maps f^σ. Now, apply Criterion 13.

Part II: Belyi's theorem for complex surfaces

Let now S be a *complex surface*, that is a compact holomorphic manifold of complex dimension 2. Naturally, we shall say that S can be defined over a number field if it is biholomorphic to a projective surface defined over $\bar{\mathbb{Q}}$. Note that a complex surface need not be algebraic, that is, need not be isomorphic to a projective surface.

It is natural to ask whether a result similar to Belyi's theorem, say in the form given in Theorem 12, holds for complex surfaces. To be more explicit, we can ask the following

Question 1. Is it true that S can be defined over $\bar{\mathbb{Q}}$ if and only if there is a meromorphic function $f \in \mathcal{M}(S)$ such that its critical values lie all in $P^1(\bar{\mathbb{Q}}) = \bar{\mathbb{Q}} \cup \{\infty\}$?

7. Minimal surfaces

The answer to Question 1, in the precise form it has been formulated, cannot be affirmative. To see this, consider first a surface S defined over $\bar{\mathbb{Q}}$ and a meromorphic function $f \in \mathcal{M}(S)$ whose critical values $\mathrm{crit}(f) = \{q_i\}$ lie in $\mathbb{P}^1(\bar{\mathbb{Q}})$. Then, choose $q \in \mathbb{P}^1(\bar{\mathbb{Q}})$, $q \neq q_i$, and blow up S at one (or several) points $P \in f^{-1}(q)$ (Figure 11). This gives us a new surface $\pi^P : S^P \to S$ and a new meromorphic function $f \circ \pi^P \in \mathcal{M}(S_P)$ whose critical value set $\{q_i, q\}$ is again contained in $\mathbb{P}^1(\bar{\mathbb{Q}})$. Now, one cannot expect S^P to be defined over a number field for an arbitrary choice of *transcental points* $P \in f^{-1}(q)$. This leads us to introduce the concept of minimal surfaces.

Recall that S is termed *minimal* when it does not contain genus zero Riemann surfaces with self-intersection -1. These are called *exceptional* or (-1)-*curves*. Exceptional curves arise when a surface is blown up. By this we mean that given a

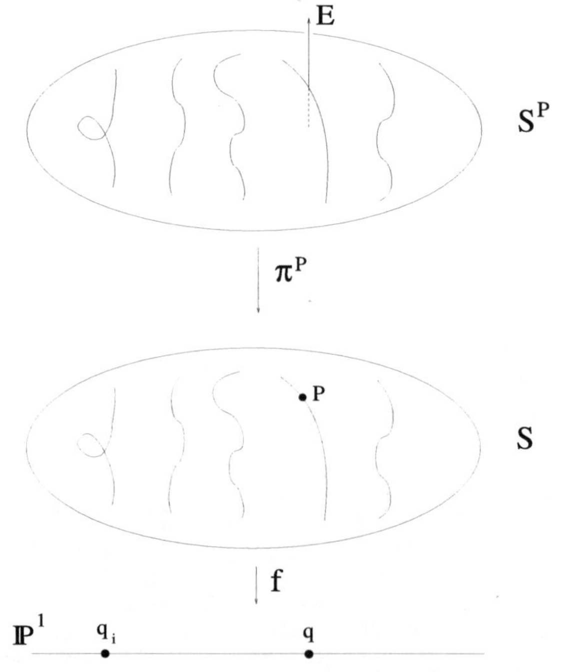

Figure 11. Blowing up S at the point P.

complex surface S and an exceptional curve E in it, there is a complex surface S_E and a holomorphic mapping $\pi_E : S \to S_E$, called *blowing down* or *contraction map*, so that E maps to a point $P \in S_E$ in such a way that π_E equals π^P, the blow up of S_E at the point P. For an arbitrary complex surface S, a minimal surface $S_{\min}$ can be obtained by a sequence of contractions

$$S = S_0 \to S_1 \to \cdots \to S_n = S_{\min}$$

A minimal surface obtained in this way is called *a minimal model of* S ([3,2]).

Suppose that S is defined over $\bar{\mathbb{Q}}$ and let $E \subset S$ be an exceptional curve. Then, not surprisingly, E is also defined over $\bar{\mathbb{Q}}$ ([9]). Now, since blowing down an exceptional curve is an operation in Algebraic Geometry that works over any algebraic closed field, there can be no doubt that the contracted surface S_E and the point $P = \pi(E) \in S_E$ are both defined over $\bar{\mathbb{Q}}$. As the same argument can be applied at every next step, we see that if a complex surface S is defined over a number field then so must be any minimal model of S. Thus, we have

Proposition 15. *A complex surface S can be defined over a number field if and only if it can be obtained out of a minimal surface $S_{\min}$ defined over $\bar{\mathbb{Q}}$ by a finite sequence of blow-ups centered at points also defined over $\bar{\mathbb{Q}}$.*

Therefore in our search for a Belyi criterion for complex surfaces we can restrict ourselves to minimal ones.

8. Bertini-Lefschetz theory of pencils

Recall that by classical results of Bertini, given a projective surface $S \subset \mathbb{P}^n$ there is a *pencil of hyperplanes* $\{H_\lambda = \lambda_0 H_0 + \lambda_1 H_1\}_\lambda$ with $\lambda = (\lambda_0, \lambda_1) \in \mathbb{P}^1$, so that the hyperplane sections $S_\lambda = S \cap H_\lambda$ are *generically* non singular connected algebraic curves and $S = \cup_{\lambda \in \mathbb{P}^1} S_\lambda$. Moreover, the rule

$$f(x) = \lambda \in \mathbb{P}^1 \text{ if and only if } x \in S_\lambda$$

defines a *meromorphic function* $f \in \mathcal{M}(S)$ with nonempty *base locus* $B = \cap_{\lambda \in \mathbb{P}^1} S_\lambda = Z(H_0, H_1) \cap S$. By results of Lefschetz (see [21]), this meromorphic function $f : S \dashrightarrow \mathbb{P}^1$ can be chosen so as to satisfy the following requirements.

i) $f : S \backslash B \to \mathbb{P}^1$ is a holomorphic submersion outside a finite set of critical points $\mathrm{Crit}(f) = \{x_1, \ldots, x_d\}$, no two of them in a same fibre, which therefore correspond bijectively to the critical values $\mathrm{crit}(f) = \{q_1 = f(x_1), \ldots, q_d = f(x_d)\}$
ii) at each critical point x_i, f is locally of the form $(z_1, z_2) \to z_1^2 + z_2^2$, and
iii) at each base point b_k, f is locally of the form $(z_1, z_2) \to z_1/z_2$

Definition 16. *A Lefschetz pencil (L.P.) on a complex surface S is a meromorphic function $f \in \mathcal{M}(S)$ which has a nonempty base locus $B = \{b_1, \ldots, b_r\}$ and satisfies conditions i) to iii) above.*

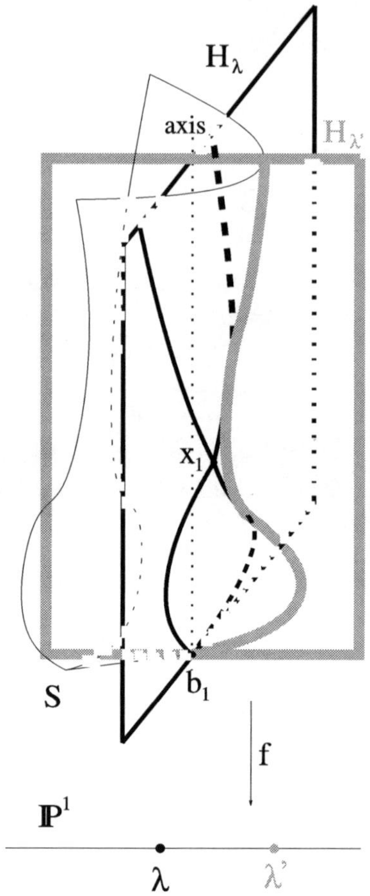

Figure 12. Bertini-Lefschetz theorem.

We observe that $f : S \dashrightarrow \mathbb{P}^1$ is not well defined at the points in B, that is what the dashed arrow is meant to emphasize. However, associated to a L.P. there are two well defined holomorphic maps which will permit us to regard our complex surfaces either as

1. A family of Riemann surfaces of type (g, r) or
2. A family of compact Riemann surfaces of genus g with r sections.

Let us describe these two maps.

1. $f : S \backslash B \longrightarrow \mathbb{P}^1$.

   Away from the critical values $\lambda = q_i$ the fibres $f^{-1}(\lambda)$ are Riemann surfaces of finite type (g, r), that is compact Riemann surfaces of genus g with r punctures. The closure of each of the fibres in S is obtained by including the points $\{b_1, \ldots, b_r\}$.
2. $\widetilde{f} : \widetilde{S} \to \mathbb{P}^1$

   This is the *Lefschetz fibration* associated to the L.P. Here $\widetilde{S}$ is the surface obtained by blowing up S at the points in $B = \{b_1, \ldots, b_r\}$. Recall that, roughly,

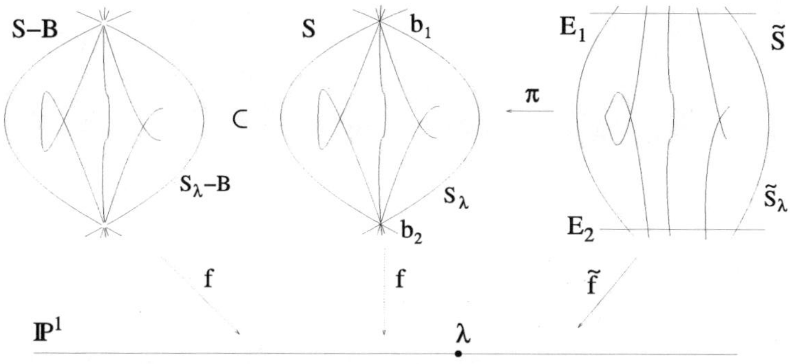

Figure 13. The two maps associated to a Lefschetz pencil.

what the blowing up operation $\pi : \widetilde{S} \to S$ does is to replace each point $b_i \in B$ by a projective line $\mathbb{P}^1_{b_i}$ consisting of points b_i^λ, one for each Riemann surface S_λ so that one may coherently define $\widetilde{f}(b_I^\lambda) = \lambda$. As for the fibres of $\widetilde{f}$, the map π induces an isomorphism between $\widetilde{S}_\lambda = \widetilde{f}^{-1}(\lambda)$ and $f^{-1}(\lambda) \cup B$. It follows that the fibration $\widetilde{f} : \widetilde{S} \to \mathbb{P}^1$ comes equipped with r sections, one for each base point.

It is known that all nonsingular fibres $f^{-1}(\lambda)$ (resp. $\widetilde{f}^{-1}(\lambda)$) are connected Riemann surfaces of a constant genus g, which is usually referred to as the *genus of the pencil* (see e.g. [12], 8.1). Similarly the pair (g, r) will be called the type of the pencil. Accordingly a pencil will be said *of hyperbolic type* if $2g - 2 + r > 0$.

Having established the fact that algebraic complex surfaces admit Lefschetz pencils, the next fact to be observed is that, in the other direction, the results of Kodaira on complex surfaces (see [2]) show that the existence of a L.P. on an arbitrary complex surface forces that surface to be algebraic. Since for a complex surface S to be defined over $\bar{\mathbb{Q}}$ it has to be first of all algebraic, it is clear at this stage that Question 1 should be modified as follows.

Question 2. Is it true that S can be defined over $\bar{\mathbb{Q}}$ if and only if there is a L.P. $f \in M(S)$ such that $\mathrm{crit}(f) \subset \mathbb{P}^1(\bar{\mathbb{Q}})$?

It turns out that except for a very restricted class of complex surfaces the answer to this second question is going to be affirmative.

9. Monodromy of Lefschetz pencils

As in the case of morphisms between Riemann surfaces, Lefschetz pencils have also a monodromy attached to them.

9.1 *Definition of the monodromy homomorphism*

In analogy with Section 3.3 the monodromy of a L.P. $f : S \dashrightarrow \mathbb{P}^1$, with $\mathrm{crit}(f) = \{q_i\}$ is going to be a group homomorphism Mon from $\pi_1(\mathbb{P}^1 \backslash \{q_i\})$ to the group of bijections of a nonsingular fibre $\mathrm{Bij}(\widetilde{f}^{-1}(q))$ (or $\mathrm{Bij}(f^{-1}(q))$, depending on which

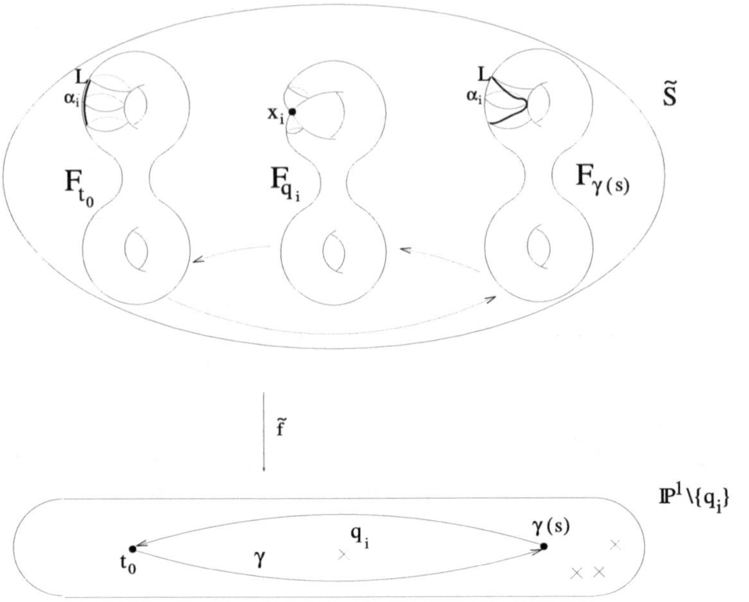

Figure 14. The monodromy of a Lefschetz pencil. As we walk along γ, the curve α_i rotates by 360 degrees.

of the two families of Riemann surfaces associated to our Lefschetz pencil we are working with).

Let us discuss the construction of this monodromy homomorphism for the family $\tilde{f} : \tilde{S} \to \mathbb{P}^1\backslash\{q_i\}$ of Riemann surfaces of genus g, the other case being similar. We choose first a base point $t_0 \in \mathbb{P}^1\backslash\{q_i\}$ with fibre F_{t_0}. Since from the topological point of view this family of Riemann surfaces is locally trivial, it is clear that for any given loop $\gamma : I = [0,1] \to \mathbb{P}^1\backslash\{q_i\}$ with base point $\gamma(0) = \gamma(1) = t_0$, one can define a continuous family of orientation preserving homeomorphisms $\{\varphi_s : F_{t_0} \to F_{\gamma(s)}\}_{s\in I}$ such that φ_0 is the identity map. For $t = 1$ we obtain a bijection $\varphi_\gamma : F_{t_0} \to F_{t_0}$ that defines our group element $\varphi_\gamma \in \mathrm{Bij}(F_{t_0})$.

Moreover, by definition, φ_γ is not a mere bijection of F_{t_0}. It is actually an orientation preserving homeomorphism of F_{t_0}, thereby giving rise to a mapping class $[\varphi_\gamma] \in \mathrm{Mod}_g$. Therefore we have a well defined group homomorphism

$$\begin{aligned} \mathrm{Mon} : \pi_1(\mathbb{P}^1\backslash\{q_i\}) &\longrightarrow \mathrm{Mod}_g \\ \gamma &\longmapsto [\varphi_\gamma] \end{aligned}$$

which will be referred to as the *monodromy of the Lefschetz pencil* ([7,12,6]). This same terminology will be used to refer to the companion homomorphism

$$\pi_1(\mathbb{P}^1\backslash\{q_i\}) \longmapsto \mathrm{Mod}_{g,r}$$

arising in a parallel way from the family of Riemann surfaces of type (g, r) given by $f : S\backslash B \mapsto \mathbb{P}^1\backslash\{q_i\}$.

It is well known that Mod_g is generated by *Dehn twists* along nontrivial loops of our reference Riemann surface F_{t_0} and in fact the homomorphism Mon can be described in terms of these generators in a very explicit manner.

To do that we associate to each critical value $q_i, i = 1, \dots, d$, the following objects:

1. A simple loop γ_i in $\mathbb{P}^1 \backslash \{q_i, t_0\}$ as in Figure 4.
2. A curve $\alpha_i \subset F_{t_0}$ which shrinks to the critical point x_i as we *retract* F_{t_0} to the singular fibre F_{q_i} (α_i is the so called *vanishing cycle* corresponding to the node x_i).
3. The (right) Dehn twist D_{α_i} along the curve α_i. This is a homeomorphism of F_{t_0} which rotates 360 degrees the curve α_i, and keeps fixed the complement of a small tubular neighborhood of α_i.

It can be seen ([7]) that $\mathrm{Mon}(\gamma_i) = D_{\alpha_i}$. Of course, the identity $\prod \gamma_i = 1$ implies that $\prod D_{\alpha_i} = 1$.

We will sometimes write

$$\mathrm{Mon} = (D_{\alpha_1}, \dots, D_{\alpha_d})$$

which is analogous to the representation (2) in Section 3.3.

From the point of view of Teichmüller theory the monodromy map can be interpreted in a very interesting way. As usual, let us denote by $\mathcal{M}_{g,r}$ the moduli space of Riemann surfaces of type (g, r). This is a complex analytic space whose points correspond bijectively to isomorphic classes of Riemann surfaces of type (g, r). It is a fundamental fact in analytic Teichmüller theory (see [24,5]) that $\mathcal{M}_{g,r}$ can be obtained as the quotient of Teichmüller space $\mathcal{T}_{g,r}$ by the group of all its biholomorphic automorphisms which turns out to be $\mathrm{Mod}_{g,r}$. In other words we have $\mathcal{M}_{g,r} = \mathcal{T}_{g,r}/\mathrm{Mod}_{g,r}$. Of course when $r = 0$ we simply write $\mathcal{T}_{g,0} = \mathcal{T}_g$, $\mathrm{Mod}_{g,0} = \mathrm{Mod}_g$ and $\mathcal{M}_{g,0} = \mathcal{M}_g$. By the very nature of moduli space there is a holomorphic map $\phi : \mathbb{P}^1 \backslash \{q_i\} \to \mathcal{M}_{g,r}$, *the classifying map*, defined by sending an arbitrary point $q \in \mathbb{P}^1 \backslash \{q_i\}$ to the point $\phi(q) \in \mathcal{M}_{g,r}$ representing the Riemann surface F_q. Let now Γ be the Fuchsian group uniformizing $\mathbb{P}^1 \backslash \{q_i\}$ so that $\mathbb{P}^1 \backslash \{q_i\} \simeq \mathbb{H}/\Gamma$ and $\Gamma \simeq \pi_1(\mathbb{P}^1 \backslash \{q_i\})$. Then the map $\phi : \mathbb{P}^1 \backslash \{q_i\} = \mathbb{H}/\Gamma \to \mathcal{M}_{g,r} = \mathcal{T}_{g,r}/\mathrm{Mod}_{g,r}$ lifts to a holomorphic map $\widetilde{\phi} : \mathbb{H} \to \mathcal{T}_{g,r}$ and the monodromy morphism Mon is nothing but the homomorphism

$$\mathrm{Mon} = \phi_* : \Gamma \to \mathrm{Mod}_{g,r}$$

defined by the property

$$\widetilde{\phi} \circ \gamma(z) = \phi_*(\gamma)(\widetilde{\phi}(z)), \text{ for any } z \in \mathbb{H}. \tag{3}$$

Of course the choice of a different lift $\varphi \circ \widetilde{\phi}$, for some $\varphi \in \mathrm{Mod}_{g,r}$, would result on a conjugate homomorphism $(\varphi \circ \widetilde{\phi})_*(\gamma) = \varphi \circ \phi_*(\gamma) \circ \varphi^{-1}$.

9.2 *The monodromy characterizes the pencil*

As in the case of mappings between Riemann surfaces, Lefschetz pencils too are characterized by their monodromies. This is a very interesting result due to Imayoshi and Shiga. To make it precise we need some terminology.

A L.P. of type (g, r), $f : S \dashrightarrow \mathbb{P}^1$, is said to be *locally trivial* if the corresponding classifying map $\phi : \mathbb{P}^1 \backslash \{q_i\} \to \mathcal{M}_{g,r}$ is constant; that is, if the fibres are all isomorphic to each other.

Theorem 17 ([18]). *A non locally constant L.P. of hyperbolic type (g, r) is determined by its monodromy morphism* $\mathrm{Mon} : \pi_1(\mathbb{P}^1 \backslash \{q_i\}) \to \mathrm{Mod}_{g,r}$ *up to finitely many choices.*

The main idea of the proof is as follows. By construction, the classifying map ϕ determines the family of Riemann surfaces (perhaps up to a finite number of choices if the generic fibre has nontrivial automorphisms). Therefore, with the preceding notation, all one has to show is that $\mathrm{Mon} = \phi_*$ determines the map $\widetilde{\phi}$, hence the map ϕ.

The strategy to do this is rather standard. The map $\widetilde{\phi}$ is determined by its limit values at the points $\xi \in \partial\mathbb{H}$, so one has to show that these limit values are determined by ϕ_*. For that, one writes $\xi = \lim \gamma_n(z)$ for a suitable sequence of elements $\gamma_n \in \Gamma$ and, in view of equation (3), one sets $\widetilde{\phi}(\xi) = \lim \phi_*(\gamma_n)(\widetilde{\phi}(z))$.

9.3 *Finiteness (Arakelov's theorem)*

In the case of ramified coverings of $\mathbb{P}^1$, dealt with in the first part of the article, the finiteness of the number of monodromies

$$\mathrm{Mon} : \pi_1(\mathbb{P}^1 \backslash \{q_i\}) \to \mathbb{S}_d$$

was an obvious consequence of the finiteness of the symmetric group $\mathbb{S}_d$.

On the contrary, in the case of Lefschetz pencils the target group $\mathrm{Mod}_{g,r}$ is an infinite group and consequently there are infinititly many homomorphisms from $\pi_1(\mathbb{P}^1 \backslash \{q_i\}$ to $\mathrm{Mod}_{g,r}$; one for each r-tuple $(D_1, \ldots, D_d)$ of Dehn twists (or, for that matter, arbitrary elements of $\mathrm{Mod}_{g,r}$) satisfying $\prod D_i = 1$. However, only finitely many of them may arise as monodromy maps of L.P. with $\mathrm{crit}(f) = \{q_i\}$.

Theorem 18. *Let $2g - 2 + r > 0$. Then, up to conjugation, only finitely many group homomorphisms*

$$\chi : \pi_1(\mathbb{P}^1 \backslash \{q_i\}) \to \mathrm{Mod}_{g,r}$$

arise as the monodromy map of a L.P. $f : S \dashrightarrow \mathbb{P}^1$ with $\mathrm{crit}(f) = \{q_i\}$.

By Section 9.1 the above theorem is equivalent to the statement that there are only finitely many non constant holomorphic maps $\phi : \mathbb{P}^1 \backslash \{q_i\} \to \mathcal{M}_{g,r}$. This is actually a very deep result. What lies behind it is nothing less that Arakelov's theorem [1], which states the finiteness of families of compact Riemann surfaces over a a Riemann surface of finite type. This result is known in the literature by several other names: Parshin-Arakelov theorem, Mordell Conjecture for the function field case, Geometric Shafarevich Conjecture, etc. The particular presentation above comes from a proof of a more general version of Arakelov's theorem, which includes also the case of families of Riemann surfaces with punctures, due to Imayoshi and Shiga [18] (see also the nice account given by Macmullen in [23]). The special feature about this proof is that it is carried out entirely within the framework of Teichmüller theory. And in fact, most of the fundamental facts of the analytic theory of Teichmüller spaces are involved in the proof. To mention some: the boundedness of $\mathcal{T}_{g,r}$ inside $\mathbb{C}^{3g-3+r}$, pants decompositions of hyperbolic surfaces, Mumford's compactness theorem, the equality of Teichmüller and Kobayashi distances on $\mathcal{T}_{g,r}$, the fact that the group of isometries of $\mathcal{T}_{g,r}$ coincides with the mapping class group $\text{Mod}_{g,r}$, etc.

10. Belyi's theorem for complex surfaces

In order to formulate our characterization of complex surfaces definable over $\bar{\mathbb{Q}}$ we need one more definition.

Definition 19. *By a Lefschetz function we shall refer to a meromorphic function $h \in \mathcal{M}(S)$ obtained as composition of a Lefschetz pencil $f : S \dashrightarrow \mathbb{P}^1$ with a rational function $\beta : \mathbb{P}^1 \to \mathbb{P}^1$.*

Theorem 20 ([10]). *Let S be a minimal complex surface S which is not a nonrational ruled surface. The following statements are equivalent.*

a) *S can be defined over a number field.*
b) *S admits a Lefschetz pencil $f \in \mathcal{M}(S)$ with critical values $q_1, \ldots, q_r$ in $\mathbb{P}^1(\bar{\mathbb{Q}})$.*
c) *S admits a Lefschetz function $h \in \mathcal{M}(S)$ with three critical values, say* $0, 1, \infty$.

For a nonrational ruled surface $p : S \to C$ the result is still true if in b) and c) we make the additional requirement that the curve C and the projection of the critical points of f on it are also defined over a number field.

We recall that S is called *rational* (resp. *ruled)* if it is bimeromorphically equivalent to $\mathbb{P}^2$ (*resp.* $C \times \mathbb{P}^1$, for some Riemann surface C). A ruled surface is rational if and only if $C \equiv \mathbb{P}^1$.

In view of Proposition 15, the above result can be reformulated as follows

Theorem 21 ([10]). *A complex surface S can be defined over a number field if and only if it is either a minimal surface admitting a Lefschetz function $h = \beta \circ f$ with only three critical values, or it is obtained from one such surface by a finite sequence of blow-ups centered at points with coordinates in $\bar{\mathbb{Q}}$. In the case of nonrational ruled surfaces $p : S \to C$, we must in addition require that the curve C and the image on it of the critical points of the Lefschetz pencil f are defined over a number field too.*

10.1 *The proof of Theorem 20*

b $\Leftrightarrow$ *c*)

What this equivalence means is that if there exists a L.P. $f_1 \in \mathcal{M}(S)$ and a rational function $\beta : \mathbb{P}^1 \to \mathbb{P}^1$ such that $\beta \circ f_1$ has only three critical values then there is a L.P. $f \in \mathcal{M}(S)$ such that $\text{crit}(f) \subset \mathbb{P}^1(\bar{\mathbb{Q}})$. This can be seen by suitable application of the classical Belyi theorem.

$a \Rightarrow b$)

This part is easy. As we have recalled in Section 8, Bertini's theory provides S with a Lefschetz pencil structure. Moreover if S is defined over $\bar{\mathbb{Q}}$ and we take our pencil of hyperplanes $\{H_\lambda\}_\lambda$ defined also over $\bar{\mathbb{Q}}$, then the corresponding Lefschetz pencil will have critical points x_i and critical values q_i defined over $\bar{\mathbb{Q}}$ too.

$b \Rightarrow a$)

As it has been mentioned before, the existence of a L.P. $f \in \mathcal{M}(S)$ ensures that S is algebraic. What remains to be shown is that if, moreover, $\text{crit}(f) = \{q_i\} \subset \mathbb{P}^1(\bar{\mathbb{Q}})$ then S can be defined over a number field. As in the Riemann surface case this can be achieved in three steps.

Step 1. A non locally constant family of punctured Riemann surfaces of hyperbolic type $f : S \backslash B \to \mathbb{P}^1$ is determined by its monodromy homomorphism $\text{Mon} : \pi_1(\mathbb{P}^1 \backslash \{q_i\}) \to \text{Mod}_{g,r}$ up to finitely many choices.

Step 2. There are only finitely many monodromy homomorphisms

$$\chi : \pi_1(\mathbb{P}^1 \backslash \{q_i\}) \to \text{Mod}_{g,r}$$

arising from a Lefschetz pencil on a minimal complex surface with pre-assigned critical values $\{q_i\}$.

Step 3. Apply Criterion 13.

Steps 1 and 2 correspond to Theorems 17 and 18 respectively. As for Step 3, one has only to observe that the condition $\text{crit}(f) \subset \mathbb{P}^1(\bar{\mathbb{Q}})$ obviously implies that the family of critical value sets $\text{crit}(f^\sigma) = \{q_i^\sigma\}$ obtained by letting the group $\text{Gal}(\mathbb{C})$ act on the pair (S, f) contains only finitely many different sets.

In the above discussion two cases have been excluded, namely families of Riemann surfaces of type $(0, r)$, $r \leq 3$ and locally constant families.

In the first instance S is a rational surface, thus S is either $\mathbb{P}^2$ or one of the countably many ruled surfaces F_n of Hirzebruch (see [2] or [3]) which are known to be defined over $\bar{\mathbb{Q}}$. The second case accounts for the ruled surfaces whose exceptional behavior is contemplated in Theorem 20.

This behaviour is what Figure 15 is intended to illustrate. There, S is a ruled surface obtained by first blowing up one or several points $P = P_i = (c_i, q_i) \in C \times \mathbb{P}^1$ and then blowing down the line of the ruling on which P lies, which the blowing up map π^P has turned into an exceptional line. This process is known as an elementary transformation ([2]). In Figure 15 we use the same notation L for this line in $C \times \mathbb{P}^1$ and for its strict transform in $\widetilde{C \times \mathbb{P}^1}$. The composition $f = p_2 \circ \pi^P \circ \pi_L^{-1}$ defines a L.P. on S with $\text{crit}(f) = \{q_i\}$. If the points P_i are chosen so that $q_i \in \bar{\mathbb{Q}}$, then this L.P. satisfies the requirements in Theorem 20. But if the points $c_i \in C$ are trascendental there is no reason to think that S should be definable over a number field, even if C is.

We close this section by mentioning related work done independently by I. Ronkine [26]. Also with related aim (but different point of view) is the article by [25] by Paranjape.

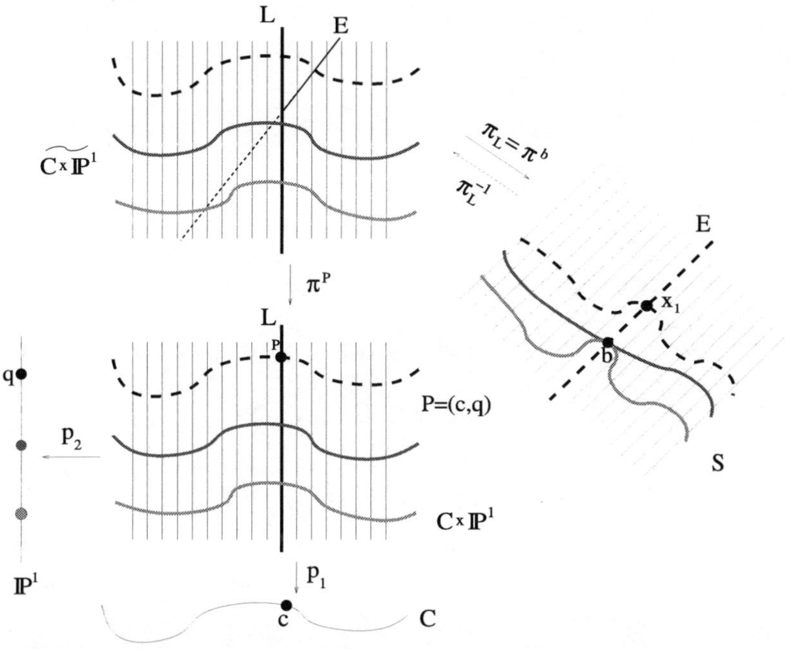

Figure 15. The Lefschetz pencil $f = p_2 \circ \pi^P \circ \pi_L^{-1}$ on the ruled surface S.

11. Belyi's theorem via Griffiths uniformization

According to Theorem 1, a compact Riemann surface C is a Belyi surface if and only if there is a finite set $\Sigma \subset C$ such that $C \setminus \Sigma$ is isomorphic to a quotient of the form $\mathbb{H}/\Gamma$ where Γ is a torsion free finite index subgroup of $\mathbb{P}\mathrm{SL}_2(\mathbb{Z})$.

This characterization may be regarded as a manifestation of how the arithmetic nature of an algebraic curve is reflected in that of its uniformizing group. This phenomenon can be also paralleled in the 2-dimensional case. Now the role of Fuchsian uniformization of algebraic curves is going to be played by Griffiths uniformization of algebraic surfaces.

11.1 *Uniformization of certain Zariski open sets of an algebraic surface*

A domain $\mathbb{B}$ in $\mathbb{C}^2$ is called a *Bergman domain* if it is the set of pairs (t, z) such that $t \in \mathbb{H}$, and $z \in D_t$ where D_t is a bounded Jordan domain whose boundary curve admits a parametric representation

$$z = W(t, e^{i\theta}),\ 0 \leq \theta \leq 2\pi$$

W being, for each fixed θ, a holomorphic function of t (see [5]). We can thus write $\mathbb{B} = \cup_{t \in \mathbb{H}} D_t$. It is clear that by choosing an isomorphism between the upper half plane and the unit disc $\mathbb{D}$ we can obtain an equivalent domain $\mathbb{B}_1$ in which $\mathbb{H}$ is replaced by $\mathbb{D}$. We shall say that $\mathbb{B}$ is bounded if $\mathbb{B}_1$ is.

By a *Bers transformation* of a Bergman domain $\mathbb{B} \subset \mathbb{H} \times \mathbb{C}$ we shall mean a holomorphic isomorphism $g(t, z) = (\widehat{g}(t), g_t(z))$ where $\widehat{g} \in \mathbb{P}\mathrm{SL}_2(\mathbb{R})$ is a real

Möbius transformation and $g_t : D_t \to D_{\widehat{g}(t)}$ is a biholomorphic map. If G is a group of Bers transformations acting freely on $\mathbb{B}$, there is an obvious short exact sequence

$$1 \hookrightarrow K \to G \xrightarrow{\rho} \Gamma \to 1$$

where the epimorphism ρ is defined by $\rho(g) = \widehat{g}$. We see that while Γ acts on $\mathbb{H}$, the group K acts freely on each simply connected region D_t as a Kleinian group K_t whose quotient space D_t/K_t is a Riemann surface. We will say that G is a *Bers-Griffiths extension* of Γ if for each $t \in \mathbb{H}$ the Riemann surface D_t/K_t is of finite hyperbolic type (p, r). In that case K itself will be said to be of type (p, r).

It is clear that in this situation G gives rise to a holomorphic fibration $f : \mathbb{B}/G \to \mathbb{H}/\Gamma$ whose fibre over a coset $[t] \in \mathbb{H}/\Gamma$, $t \in \mathbb{H}$, is the Riemann surface D_t/K_t. We shall be interested in the case in which Γ is a Fuchsian group of finite type.

Definition 22. *We shall say that a 2-dimensional complex manifold U admits a Griffiths uniformization if its holomorphic universal cover $\widetilde{U} \to U$ is isomorphic to a bounded Bergman domain of $\mathbb{C}^2$ and its covering group G is a Bers-Griffiths extension of some Fuchsian group of finite type.*

Griffiths uniformization theorem for algebraic surfaces states that every algebraic surface contains a Zariski open set which admits a Griffiths uniformization ([13], see also [5]).

11.2 *Characterization of complex surfaces defined over $\bar{\mathbb{Q}}$ via Griffiths uniformization*

Theorem 23 ([10]). *A minimal non ruled surface $S \subset \mathbb{P}^n(\mathbb{C})$ can be defined over a number field if and only if it contains a Zariski open set U admitting a Griffiths uniformization such that the uniformizing group G is a Bers-Griffiths extension of a genus zero finite index torsion free subgroup of $\mathbb{P}\mathrm{SL}_2(\mathbb{Z})$.*

The proof goes as follows. Let $f : S \dashrightarrow \mathbb{P}^1$ be a Lefschetz pencil with base locus $B = \{b_1, \ldots, b_r\}$, critical points $\{x_1, \ldots, x_d\}$ and critical values $\{q_1, \ldots, q_d\}$ in $\mathbb{P}^1(\bar{\mathbb{Q}})$. Then, by restriction to the regular values, one obtains a a family of *r-punctured* nonsingular Riemann surfaces $f : U \to \mathbb{P}^1 \backslash \{q_i\}$, where $U \subset S$ is the Zariski open set $U = (S \backslash B) \backslash f^{-1}\{q_i\}$. Moreover the genus p of the fibres must be strictly positive, for otherwise the surface would be ruled.

Let $\Gamma \subset \mathbb{P}\mathrm{SL}_2(\mathbb{R})$ be the Fuchsian group uniformizing $\mathbb{P}^1 \backslash \{q_i\}$. Then the results in [13] imply that U admits a Griffiths uniformization such that the covering transformations group G is a Bers-Griffiths extension of Γ by a group K of type (p, r). We would like to show that $\Gamma \subset \mathbb{P}\mathrm{SL}_2(\mathbb{Z})$. But, the values q_i being algebraic numbers, this is a straightforward consequence of classical Belyi theory.

Conversely, let us assume that our surface S contains a Zariski open set $U \subset S$ admitting a Griffiths uniformization $U \simeq \widetilde{U}/G$, where G is a Bers-Griffiths extension of a finite index torsion free subgroup $\Gamma \subset \mathbb{P}\mathrm{SL}_2(\mathbb{Z})$ of genus zero by a group of type

(p, r) with $2p - 2 + r > 0$. Then, as noted above, the action of G induces a holomorphic fibration of Riemann surfaces of finite hyperbolic type $f : U \to \mathbb{P}^1 \backslash \{q_i\}$ where $\mathbb{P}^1 \backslash \{q_i\}$ is isomorphic to $\mathbb{H}/\Gamma$, hence, by the classical Belyi theorem, we can assume $\{q_i\} \subset \mathbb{P}^1(\bar{\mathbb{Q}})$. At this point one would like to bring the proof to an end by extending f to a L.P. on S whose singular fibres occur at the points q_i. However, this does not seem to be entirely straightforward. So, instead, using first compactification techniques of Imayoshi ([16], see also [15] and [17]) and then resolution of singularities, one extends f to a morphism

$$Y \xrightarrow{\pi} \widehat{U} \xrightarrow{f} \mathbb{P}^1$$

where $\widehat{U}$ is a compact normal surface bimeromorphic to S and $\pi : Y \to \widehat{U}$ is a resolution of singularities of $\widehat{U}$ (see e.g. [2] III.6.1). We see that Y is a surface bimeromorphic to $\widehat{U}$, and hence to the algebraic surface S. Therefore Y is a projective surface birationally equivalent to S. In fact, being non ruled, S is the unique minimal model of Y. Moreover, by the GAGA principle, the map $f_1 = f \circ \pi : Y \to \mathbb{P}^1$ is a regular map. In particular the action of the Galois group Gal($\mathbb{C}$) on $f_1 : Y \to \mathbb{P}^1$ is well defined. Thus, we are again in position to apply Criterion 13 which together with a suitable version of Arakelov's Theorem allows one to conclude the desired result.

Remark 24. It should be noted that Theorem 23 does not hold for arbitrary minimal ruled surfaces $p : S \to C$, even if the base curve C is defined over $\bar{\mathbb{Q}}$. This can be seen as follows. Any such surface contains a Zariski open set $V \simeq (C \backslash \Sigma) \times \mathbb{P}^1$, with Σ a finite subset of C, therefore, it surely contains a smaller one of the form $U \simeq (C \backslash \Sigma) \times (\mathbb{P}^1 \backslash \{0, 1, \infty\})$. The latter admits a Griffiths uniformization with uniformizing group $G \simeq \Gamma(2) \times K$, where K is the Fuchsian group uniformizing $(C \backslash \Sigma)$, so that $C \backslash \Sigma = \mathbb{D}/K$, and the action of G is the obvious product action on $\mathbb{H} \times \mathbb{D}$. But, on the other hand, the moduli space of minimal ruled surfaces over a given curve C of genus g is known to depend on $3g - 3$ complex parameters (see [3]), thus, for mere cardinality reasons, not all of them can be defined over $\bar{\mathbb{Q}}$.

11.3 *Dessins on 4-manifolds?*

Having established for complex surfaces a result analogous to Belyi's theorem for complex curves, one wonders how far the analogy goes. In particular, since we spent a good deal of the first part of this paper discussing the notion of dessins d'enfants and the action of Gal($\bar{\mathbb{Q}}$) on them, it is natural to ask if this theory can also be extended to complex dimension two. Let us make more precise what we mean by that.

1. First of all, given a compact oriented 4-manifold M, one would like to single out a distinguished class of suitable topological decompositions of M (that would play the role of dessins) which in a natural way correspond to pairs (M, f) where f is a differentiable map of a certain type from M to $\mathbb{S}^2$.

This first requirement is satisfactorily solved by the so called topological Lefschetz pencils.

Definition 25. *A topological Lefschetz pencil (T.L.P.)* $f : M \dashrightarrow \mathbb{S}^2$ *on a four-manifold comprises an oriented four-manifold* M *and a surjective map* f *defined on the complement of a finite set* $B = \{b_i, i = 1, \ldots, r\}$ *to the sphere with finitely many critical points* $\{x_j, j = 1, \ldots, d\}$*, all in distinct fibres, such that*

i) *Near each* x_j *there are local complex coordinates with respect to which the map* f *takes the form* $f(z_1, z_2) = z_1^2 + z_2^2$.
ii) *Near each* b_i *there are complex coordinates with respect to which the map* f *takes the form* $f(z_1, z_2) = z_1/z_2$.

From the topological point of view L.P and T.L.P. are of course indistinguishable.

Using the function $|f|^2$ as a Morse function on M (or rather the four-manifold $\widetilde{M}$ obtained by blowing up the points $\{b_i\}$), Kas ([19]) gave a handlebody decomposition $\mathcal{D}$ of M on which the elementary pieces are of the form $\Sigma_g \times D$, where D is the topological disc and Σ_g is the topological surface homeomorphic to the generic fibre. We can call this type of handlebody decomposition of a four-manifold M a *Kas handlebody decomposition.*

It is known ([19,6,12]) that there is an essentially bijective correspondence between Kas handlebody decompositions $\mathcal{D}$, T.L.P. (M, f) and monodromy representations Mon $= (D_{\alpha_1}, \ldots, D_{\alpha_d})$ as defined in Section 9.1. Thus, it looks natural to reserve the term *four-dimensional dessin* to any of the following equivalent objects

$$\mathcal{D} \equiv (M, f) \equiv (D_{\alpha_1}, \ldots, D_{\alpha_d}).$$

2. Once we have decided what a possible candidate for the concept of four-dimensional dessins could be, one needs to be able to associate to each dessin $\mathcal{D}$ a holomorphic structure on M for which f becomes a holomorphic L.P. This will make of M a projective variety on which the action of Gal($\bar{\mathbb{Q}}$) would make sense.

This is the point where difficulties emerge for not all T.L.P. can be endowed with a complex structure. One way to see this is that, because of fundamental work by Donaldson [6] and Gompf [11], one knows that T.L.P. are in correspondence with symplectic four-manifolds. But, on the other hand, as it was first observed by Thurston ([29]), not all symplectic four-manifolds are Kähler manifolds.

Thus, it seems that in order to make sense of four-dimensional dessins one needs first to understand which monodromies $\mathcal{D} \equiv (D_{\alpha_1}, \ldots, D_{\alpha_d})$ arise as monodromies of (complex) L.P.

Acknowledgement

The second author would like to express his gratitude to the H.R.I. and its director R. Kulkarni for his hospitality during his visit.

References

[1] S. Ju Arakelov, Families of algebraic curves with fixed degeneracies, *Izv. Akad. Nauk SSSR Ser. Mat.*, **35** (1971) 1269–1293.

[2] W. Barth, K. Hulek, C. Peters and A. Van de Ven, Compact complex surfaces. Ergebnisse der Mathematik und ihrer Grenzgebiete (3), **4**, Springer-Verlag, Berlin, Heildelberg, (2004).

[3] A. Beauville, Surfaces algebriques complexes, *Astérisque*, **54** (1978).

[4] G. Belyi, On Galois extensions of a maximal cyclotomic field, *Math, USSR Izv.*, **14** (1980) no. 2, 247–256.

[5] L. Bers, Uniformization, moduli, and Kleinian groups, *Bull. London Math. Soc.*, **4** (1972) 257–300.

[6] S. K. Donaldson, Lefschetz fibrations in symplectic geometry, Proceedings of the International Congress of Mathematicians, Vol. II (Berlin, 1998). Doc. Math. 1998, Extra Vol. II, 309–314.

[7] C. J. Earle and P. L. Sipe, Families of Riemann surfaces over the punctured disk, *Pacific J. Math.*, **150** (1991) no. 1, 79–96.

[8] E. Girondo and G. González-Diez, A note on the action of the absolute Galois group on dessins, *Bull. London Math. Soc.*, **39** (2007) no. 5, 721–723.

[9] G. González-Diez, Variations on Belyi's theorem, *Quart. J. Math.*, **57** (2006) 339–354.

[10] G. González-Diez, Belyi's theorem for complex surfaces, *Amer. J. Math.*, **130** (2008) no. 1, 59–74.

[11] R. E. Gompf, The topology of symplectic manifolds, *Turkish J. Math.*, **25** (2001) no. 1, 43–59.

[12] R. E. Gompf and A. I. Stipsicz, 4-Manifolds and Kirby Calculus. G.S.M. Vol. 20, AMS.

[13] P. A. Griffiths, Complex-analytic properties of certain Zariski open sets on algebraic varieties, *Ann. of Math.*, **94** (1971) no. 2, 21–51.

[14] A. Grothendieck, Esquisse d'un Programme. In Geometric Galois actions. Around Grothendieck's Esquisse d'un programme, *London. Math. Soc.*, L.N.Series 242. (L. Schneps and P. Lochak, editors).

[15] Y. Imayoshi, Holomorphic families of Riemann surfaces and Teichmüller spaces, Riemann surfaces and related topics. Proceedings of the 1978 Stony Brook Conference, pp. 277–300, *Ann. of Math. Stud.*, **97**, Princeton Univ. Press (1981).

[16] Y. Imayoshi, Holomorphic families of Riemann surfaces and Teichmüller spaces II, *Tôhoku Math. J.*, (2) **31** (1979) no. 4, 469–489.

[17] Y. Imayoshi, Holomorphic families of Riemann surfaces and Teichmüller spaces III, Bimeromorphic embedding of algebraic surfaces into projective spaces by automorphic forms, *Tôhoku Math. J., (2)* **33** (1981) no. 2, 227–247.

[18] Y. Imayoshi and H. Shiga, A finiteness theorem for holomorphic families of Riemann surfaces, In Holomorphic Functions and Moduli II, MSRI. **28** (1998) 207–219.

[19] A. Kas, On the handlebody decomposition associated to a Lefschetz fibration, *Pacific J. Math.*, **89** (1980) no. 1, 89–104.

[20] B. Köck, Belyi's theorem revisited, *Beiträge Algebra Geom.*, **45** (2004) no. 1, 253–265.

[21] K. Lamotke, The topology of complex projective varieties after S. Lefschetz, *Topology*, **20** (1981) no. 1, 15–51.

[22] S. K. Lando and A. K. Zvonkin, Graphs on surfaces and their applications, Encyclopaedia of Mathematical Sciences, **141** (II), Berlin, Springer (2004).

[23] C. T. McMullen, From dynamics on surfaces to rational points on curves, *Bull. Amer. Math. Soc.*, (N.S.) **37** (2000) no. 2, 119–140.

[24] S. Nag, The complex analytic theory of Teichmüller spaces, Wiley-Interscience Publication (1998).

[25] K. H. Paranjape, A geometric characterization of arithmetic varieties, *Proc. Indian Acad. Sci. (Math. Sci.)*, **112** (2002) no. 3, 383–391.

[26] I. Ronkine, Eine höherdimensionale Variante des Satzes von Belyi. Diplomarbeit, Rheinische Friedrich-Wilhelms-Universität Bonn (2003).

[27] L. Schneps (editor), The Grothendieck theory of Dessins d'Enfants, *London. Math. Soc, L.N. Series*, **200** Cambridge University Press (1994).

[28] I. R. Shafarevich, Basic Algebraic Geometry, Springer Verlag (1977).

[29] W. P. Thurston, Some simple examples of symplectic manifolds, *Proc. Amer. Math. Soc.*, **55** (1976) no. 2, 467–468.

[30] A. Weil, The field of definition of a variety, *Amer. J. Math*, **78** (1956) 509–524.

[31] J. Wolfart, The "obvious" part of Belyi's Theorem and Riemann surfaces with many automorphisms. In Geometric Galois actions 1, *London Math. Soc. Lecture Note Ser.*, 242, Cambridge University Press, Cambridge (1997), 97–112. (L. Schneps and P. Lochak, editors).

[32] J. Wolfart, ABC for polynomials, dessins d'enfants, and uniformization – a survey. In Proceedings der ELAZ-Konferenz 2004. Steiner Verlag, Stuttgart (2006). (W. Schwarz, J. Steuding, editors), 313–345.

Teichmüller Theory and Moduli Problem No. 10, 2010, pp. 281–371.

Moduli Spaces for Principal Bundles in Large Characteristic

Tomás L. Gómez[1], Adrian Langer[2], Alexander H. W. Schmitt[3] and Ignacio Sols[4]

[1]IMAFF, Consejo Superior de Investigaciones Científicas, Serrano 113bis, 28006 Madrid, Spain; and Facultad de Matemáticas, Universidad Complutense de Madrid, 28040 Madrid, Spain
[2]Institute of Mathematics, Warsaw University, Ul. Banacha 2, PL-02-097 Warszawa, Poland, and Institute of Mathematics, Polish Academy of Sciences, Ul. Śniadeckich 8, P.O. Box 21, PL-00-956 Warszawa, Poland
[3]Freie Universität Berlin, Institut für Mathematik, Arnimallee 3, D-14195 Berlin, Germany
[4]Departamento de Algebra, Facultad de Ciencias Matemáticas, Pza. de las Ciencias, 3, Universidad Complutense de Madrid, E-28040 Madrid, Spain
e-mail: tomas.gomez@mat.csic.es; alan@mimuw.edu.pl; alexander.schmitt@fu-berlin.de; isols@mat.ucm.es

Abstract. We introduce a unified formalism which allows to construct and compactify moduli spaces of stable principal bundles over fields of characteristic zero or large positive characteristic. The compactification will depend on the choice of a representation of the structure group whose kernel lies in the center.

Keywords. Singular principal bundle, moduli space, Geometric Invariant Theory, positive characteristic, semistable reduction.

1. Introduction

Let G be an arbitrary reductive group and X a smooth projective scheme over a field k of characteristic p. In this article, we address the problem of constructing and compactifying the moduli space of stable principal G-bundles on X. This article complements [6] as well as the papers [8], [27], and [28], dealing with the characteristic zero theory. The reader is referred to the introduction to [6] for a survey on the history and the motivation of the subject. Here, we briefly mention the novelties of the current paper.

- We refine the method of [27] and [28] in order to cover non-semisimple reductive groups, too. In particular, the Gieseker-Maruyama moduli spaces are recovered. We emphasize that this involves non-trivial and very subtle modifications of the techniques of those papers.
- We develop the formalism for constructing the moduli space via certain non-faithful representations, such as the adjoint one. It turns out that it works over curves in any characteristic and over arbitrary projective base manifolds in characteristic zero. Together with the results of [6] and [11], this brings

Ramanathan's results to positive characteristic. When X is a curve, G is semi-simple, and p is large enough, the projective moduli space was constructed in [2] by different methods.

- We include a new and hopefully simpler approach to obtain the semistable reduction theorem in characteristic zero and large positive characteristic.
- Last but not least, we fill a gap in the literature by supplying the construction of the moduli space of decorated sheaves over projective base manifolds over fields of arbitrary characteristic.

As far as characteristic zero is concerned, this paper gives the state of the art account on the subject and provides a unified formulation of the results of the formerly competing works [8] on the one hand and [27] and [28] on the other hand. In addition, it clarifies to what extent these techniques are suited for attacking the problem in positive characteristic.

We now come to the more detailed presentation of the contents of our work. In order to define the notion of stability and construct the moduli space we need to choose a representation. If X is a curve, the result is independent of the representation, but if $\dim(X) > 1$ both the notions of stability and the moduli space will in general depend on this choice. We distinguish two cases: faithful representations and faithless representations with kernel in the center.

1.1 *Faithful representations*

Let G be an arbitrary reductive group which is not necessarily connected[1]. Fix a faithful representation $\rho\colon G \longrightarrow \mathrm{GL}(V)$, such that $\rho(G) \subseteq \mathrm{SL}(V)$.[2] If $p > 0$, we assume that the group G and the representation ρ are both defined over the integers. Suppose the G-module V decomposes as a direct sum $V_1 \oplus \cdots \oplus V_t$ of G-modules. Then, $\rho(G) \subseteq (\mathrm{GL}(V_1) \times \cdots \times \mathrm{GL}(V_t)) \cap \mathrm{SL}(V)$. Assume that the radical $\mathscr{R}(G)$ of G maps to the center of $\mathrm{GL}(V_1) \times \cdots \times \mathrm{GL}(V_t)$. (This may be achieved as follows: The radical $\mathscr{R}(G)$ is a torus ([3], 11.21, Proposition, p. 158), so that $\rho_{|\mathscr{R}(G)}$ may be diagonalized, i.e., $V = V_1 \oplus \cdots \oplus V_t$ as $\mathscr{R}(G)$-module. The V_i are non-trivial eigenspaces for different characters of $\mathscr{R}(G)$. Since $\mathscr{R}(G)$ lies in the center of G, the above decomposition of V is indeed a decomposition of V as a G-module.) We split the representation, so that we may construct principal G-bundles from principal $(\mathrm{GL}(V_1) \times \cdots \times \mathrm{GL}(V_t))$-bundles, i.e., tuples of vector bundles, by reducing the structure group. This means that we use the representation $\rho'\colon G \longrightarrow \mathrm{GL}(V_1) \times \cdots \times \mathrm{GL}(V_t)$ rather than ρ. The reason behind this is that ρ' maps the radical of G to the center of $\mathrm{GL}(V_1) \times \cdots \times \mathrm{GL}(V_t)$. This condition grants that we may prevent one parameter subgroups of the radical from contributing to the semistability concept. Recall that, in characteristic zero, the concept of Ramanathan semistability behaves well under extension of the structure group via $G \longrightarrow G'$, if and only if the radical of G is mapped to the radical of G' [23].

---

[1] Observe that Ramanathan needs the assumption of connectedness!

[2] For non-connected groups, it suffices to require that the image of G^0 lies in $\mathrm{SL}(V)$. We will not elaborate on this, but the reader may have a look at [28], Section 6.1.

In the sequel, we will use the following abbreviations: $\underline{V} := (V_1, \dots, V_t)$; $\underline{\mathscr{A}} = (\mathscr{A}_1, \dots, \mathscr{A}_t)$ stands for a tuple of torsion free coherent sheaves, such that $\mathrm{rk}(\mathscr{A}_i) = \dim_k(V_i)$, $i = 1, \dots, t$, and $\det(\mathscr{A}) \cong \mathscr{O}_X$, $\mathscr{A} := \mathscr{A}_1 \oplus \cdots \oplus \mathscr{A}_t$;

$$\mathscr{H}(\underline{\mathscr{A}}, \underline{V}) := \mathscr{H}om(\mathscr{A}_1, V_1^\vee \otimes \mathscr{O}_X) \times_X \cdots \times_X \mathscr{H}om(\mathscr{A}_t, V_t^\vee \otimes \mathscr{O}_X)$$

$$:= \mathscr{S}pec(\mathscr{S}ym^\star(\mathscr{A}_1 \otimes V_1 \oplus \cdots \oplus \mathscr{A}_t \otimes V_t))$$

$$\mathscr{I}(\underline{\mathscr{A}}, \underline{V}) := \mathscr{I}som(V_1 \otimes \mathscr{O}_U, \mathscr{A}_{1|U}^\vee) \times_U \cdots \times_U \mathscr{I}som(V_t \otimes \mathscr{O}_U, \mathscr{A}_{t|U}^\vee),$$

U being the maximal open subset where all the $\mathscr{A}_i$ are locally free. We will look at pairs $(\underline{\mathscr{A}}, \tau)$ with $\underline{\mathscr{A}}$ as above and $\tau\colon \mathscr{S}ym^\star(\mathscr{A}_1 \otimes V_1 \oplus \cdots \oplus \mathscr{A}_t \otimes V_t)^G \longrightarrow \mathscr{O}_X$ a homomorphism of $\mathscr{O}_X$-algebras which is non-trivial in the sense that the induced section $\sigma\colon X \longrightarrow \mathscr{S}pec(\mathscr{S}ym^\star(\mathscr{A}_1 \otimes V_1 \oplus \cdots \oplus \mathscr{A}_t \otimes V_t)^G)$ be not the zero section. Such a pair is called a *pseudo G-bundle*.

Definition. *A singular principal G-bundle*[3] *is a pseudo G-bundle, such that* $\sigma_{|U}(U) \subset \mathscr{I}(\underline{\mathscr{A}}, \underline{V})/G$.

In the case of a singular principal G-bundle $(\underline{\mathscr{A}}, \tau)$, we get a principal G-bundle $\mathscr{P}(\underline{\mathscr{A}}, \tau)$ over U, defined by means of base change:

$$\begin{array}{ccc} \mathscr{P}(\underline{\mathscr{A}}, \tau) & \longrightarrow & \mathscr{I}(\underline{\mathscr{A}}, \underline{V}) \\ \big\downarrow & & \big\downarrow \\ U & \xrightarrow{\sigma_{|U}} & \mathscr{I}(\underline{\mathscr{A}}, \underline{V})/G. \end{array}$$

Semistability. We now define the notion of semistability for a singular principal G-bundle $(\underline{\mathscr{A}}, \tau)$. For this, let $\lambda\colon \mathbb{G}_m(k) \longrightarrow G$ be a one parameter subgroup of G. This yields a parabolic subgroup $Q_G(\lambda)$ (see (6) below) and a weighted filtration $(\underline{V}_\bullet(\lambda), \underline{\alpha}_\bullet(\lambda))$ of the tuple $\underline{V}$, i.e., for each i, $(V_{i\bullet}(\lambda_i), \alpha_{i\bullet}(\lambda_i))$ is the weighted flag in V_i associated with $\lambda_i\colon \mathbb{G}_m(k) \longrightarrow \mathrm{GL}(V_i)$ (see Section 2.3). Here, we have used the convention of writing a one parameter subgroup λ of $\mathrm{GL}(V_1) \times \cdots \times \mathrm{GL}(V_t)$ in the form $\lambda = (\lambda_1, \dots, \lambda_t)$ with $\lambda_i\colon \mathbb{G}_m(k) \longrightarrow \mathrm{GL}(V_i)$, $i = 1, \dots, t$. Then, a *reduction of* $(\underline{\mathscr{A}}, \tau)$ *to* λ is a section $\beta\colon U' \longrightarrow \mathscr{P}(\underline{\mathscr{A}}, \tau)_{|U'}/Q_G(\lambda)$ over an open subset $U' \subseteq U$ with $\mathrm{codim}_X(X \setminus U') \geq 2$. This defines a weighted filtration $(\underline{\mathscr{A}}_\bullet(\beta), \underline{\alpha}_\bullet(\beta))$ of $\underline{\mathscr{A}}$: Here, $\underline{\alpha}_\bullet(\beta) = (\alpha_{i\bullet}(\beta), i = 1, \dots, t)$ with $\alpha_{i\bullet}(\beta) = (\alpha_{i,s_i}, \dots, \alpha_{i,1})$, if $\underline{\alpha}_\bullet(\lambda_i) = (\alpha_{i,1}, \dots, \alpha_{i,s_i})$, $i = 1, \dots, t$, and the filtration $\mathscr{A}_{i\bullet}(\beta)\colon \{0\} \subsetneq \mathscr{A}_{i,1} \subsetneq \cdots \subsetneq \mathscr{A}_{i,s_i} \subsetneq \mathscr{A}_i$, $i = 1, \dots, t$, is obtained as follows. The section

$$\begin{aligned} \beta_i\colon U' \xrightarrow{\beta} \mathscr{P}(\mathscr{A}, \tau)_{|U'}/Q_G(\lambda) &\hookrightarrow \mathscr{I}(\underline{\mathscr{A}}, \underline{V})_{|U'}/Q_{\mathrm{GL}(V_1)\times\cdots\times\mathrm{GL}(V_t)}(\lambda) \\ &= (\mathscr{I}som(V_1 \otimes \mathscr{O}_{U'}, \mathscr{A}_{1|U'}^\vee)/Q_{\mathrm{GL}(V_1)}(\lambda_1)) \times_{U'} \cdots \\ &\quad \times_{U'} (\mathscr{I}som(V_t \otimes \mathscr{O}_{U'}, \mathscr{A}_{t|U'}^\vee)/Q_{\mathrm{GL}(V_t)}(\lambda_t)) \\ &\xrightarrow{\pi_i} \mathscr{I}som(V_i \otimes \mathscr{O}_{U'}, \mathscr{A}_{i|U'}^\vee)/Q_{\mathrm{GL}(V_i)}(\lambda_i) \end{aligned}$$

[3]Here, we deviate from the original terminology in [26] and [28], because we now know that we need only the "honest" objects in our compactification.

yields a filtration

$$\{0\} \subsetneq \mathscr{A}'_{i,1} \subsetneq \cdots \subsetneq \mathscr{A}'_{i,s_i} \subsetneq \mathscr{A}^{\vee}_{i|U'}$$

of $\mathscr{A}^{\vee}_{i|U'}$ by subbundles with $\mathrm{rk}(\mathscr{A}'_{i,j}) = \dim_k(V_{i,j})$, $j = 1, \ldots, s_i$. This is because $Q_{\mathrm{GL}(V_i)}(\lambda_i)$ is the $\mathrm{GL}(V_i)$-stabilizer of the flag $V_{i\bullet}(\lambda_i)$ and, thus, $\mathscr{I}som(V_i \otimes \mathscr{O}_{U'}, \mathscr{A}^{\vee}_{i|U'})/Q_{\mathrm{GL}(V_i)}(\lambda_i) \longrightarrow U'$ is the bundle of flags in the fibers of $\mathscr{A}^{\vee}_{i|U'}$ having the same dimensions as the members of the flag $V_{i\bullet}(\lambda_i)$. We define $\mathscr{A}''_{i,j} := \ker(\mathscr{A}_{i|U'} \longrightarrow \mathscr{A}'^{\vee}_{s_i+1-j})$, $j = 1, \ldots, s_i$, so that we obtain a filtration

$$\{0\} \subsetneq \mathscr{A}''_{i,1} \subsetneq \cdots \subsetneq \mathscr{A}''_{i,s_i} \subsetneq \mathscr{A}_{i|U'}$$

of $\mathscr{A}_{i|U'}$ by subbundles. Let $\iota\colon U' \longrightarrow X$ be the inclusion and define $\mathscr{A}_{i,j}$ as the saturation of $\mathscr{A}_i \cap \iota_\star(\mathscr{A}''_{i,j})$, $j = 1, \ldots, s_i$. This is the filtration we denote by $\mathscr{A}_{i\bullet}(\beta)$. It is worth noting that, if $\lambda' = g \cdot \lambda \cdot g^{-1}$ for some $g \in G$, then any reduction to λ may also be interpreted as a reduction to λ'.

For every weighted filtration $(\underline{\mathscr{A}}_\bullet, \underline{\alpha}_\bullet)$ of $\underline{\mathscr{A}}$, we set

$$M(\underline{\mathscr{A}}_\bullet, \underline{\alpha}_\bullet) := \sum_{i=1}^{t} \sum_{j=1}^{s_i} \alpha_{i,j} (\mathrm{rk}\, \mathscr{A}_{i,j} \cdot P(\mathscr{A}_i) - \mathrm{rk}\, \mathscr{A}_i \cdot P(\mathscr{A}_{i,j})).$$

Definition. *A singular principal G-bundle* $(\underline{\mathscr{A}}, \tau)$ *is (semi)stable, if for every one parameter subgroup* $\lambda\colon \mathbb{G}_m(k) \longrightarrow \mathscr{D}(G) = [G, G]$ *and every reduction* β *of* $(\underline{\mathscr{A}}, \tau)$ *to* λ*, we have*

$$M(\underline{\mathscr{A}}, \tau; \beta) := M(\underline{\mathscr{A}}_\bullet(\beta), \underline{\alpha}_\bullet(\beta))(\succeq)0\,.$$

(The standard notation "$(\succeq)$" means that "$\succeq$" is used in the definition of semistability, and "$\succ$" in the definition of stability [14], and we use the lexicographic ordering of polynomials.)

Finally, there is a notion of S-equivalence which will be explained in Section 4.2. We have the implications

$$\begin{aligned} \mathscr{P}(\underline{\mathscr{A}}, \tau) \text{ is Ramanathan-stable} &\Longrightarrow (\underline{\mathscr{A}}, \tau) \text{ is stable} \\ &\Longrightarrow (\underline{\mathscr{A}}, \tau) \text{ is semistable} \\ &\Longrightarrow \mathscr{P}(\underline{\mathscr{A}}, \tau) \text{ is Ramanathan-semistable.} \end{aligned}$$

More precisely, in our language, Ramanathan's notion of (semi)stability becomes

$$L(\underline{\mathscr{A}}, \tau; \beta) := \sum_{i=1}^{t} \sum_{j=1}^{s_i} \alpha_{i,j} (\mathrm{rk}\, \mathscr{A}_{i,j} \cdot \deg(\mathscr{A}_i) - \mathrm{rk}\, \mathscr{A}_i \cdot \deg(\mathscr{A}_{i,j}))(\ge)0 \qquad (1)$$

for every one parameter subgroup $\lambda\colon \mathbb{G}_m(k) \longrightarrow \mathscr{D}(G)$ and every reduction β of $(\mathscr{A}, \tau)$ to λ. Here, deg stands for the degree with respect to the chosen polarization.

Remark. It is easy to check from the definition that the condition of semistability has to be checked only for the indivisible one parameter subgroups which define maximal parabolic subgroups.

For a fixed tuple $\underline{P} = (P_1, \dots, P_t)$ of Hilbert polynomials, we define the moduli functors

$$\underline{\mathrm{M}}(\rho)^{(\mathrm{s})\mathrm{s}}_{\underline{P}} \colon \underline{\mathrm{Sch}}_k \longrightarrow \underline{\mathrm{Set}}$$

$$S \longmapsto \left\{ \begin{array}{l} \text{Isomorphy classes of families of} \\ \text{(semi)stable singular} \\ \text{principal } G\text{-bundles } (\underline{\mathscr{A}}, \tau), \\ \text{such that } P(\mathscr{A}_i) = P_i,\, i = 1, \dots, t \end{array} \right\}.$$

First Main Theorem. *Let ρ be a faithful representation of G satisfying the conditions at the beginning of this section. If $p \gg 0$ (depending on ρ) or $p = 0$, then the coarse moduli space for the functors $\underline{\mathrm{M}}(\rho)^{(\mathrm{s})\mathrm{s}}_{\underline{P}}$ exists as a projective scheme $\mathrm{M}(\rho)^{\mathrm{ss}}_{\underline{P}}$.*

(The exact meaning of "coarse moduli space" will be explained in Section 2.1.)

Example. Let us mention some examples which will be explained in more detail in Section 6.

i) If G is semisimple, then the radical $\mathscr{R}(G)$ of G is trivial, so that $t = 1$, and we recover the exact statement of the main theorem of [28], but now also in large characteristic.
ii) If $G = \mathrm{GL}(V)$, we take the representation $\rho\colon \mathrm{GL}(V) \longrightarrow \mathrm{SL}(V \oplus k)$, $m \longmapsto (m, \det(m)^{-1})$. Giving a singular principal G-bundle is then equivalent to giving a triple $(\mathscr{A}, \mathscr{L}, \varepsilon)$ where $\mathscr{A}$ is a torsion free sheaf of rank $\dim_k(V)$, $\mathscr{L}$ is a line bundle, and $\varepsilon\colon \mathscr{L}^\vee \longrightarrow \det(\mathscr{A})$ is an isomorphism. Giving the isomorphy class of this triple is equivalent to giving only the isomorphy class of the torsion free sheaf $\mathscr{A}$. The notion of (semi)stability turns out to be the Gieseker (semi)stability of $\mathscr{A}$, i.e., that
$$\frac{P(\mathscr{A}')}{\mathrm{rk}\,\mathscr{A}'} (\preceq) \frac{P(\mathscr{A})}{\mathrm{rk}\,\mathscr{A}},$$
for all proper saturated subsheaves $\mathscr{A}'$, and our moduli space becomes the Gieseker-Maruyama moduli space of (semi)stable torsion free sheaves.
iii) If $\mathrm{Char}(k) \neq 2$, and G is one of the classical groups $\mathrm{O}_r(k)$, $\mathrm{SO}_r(k)$, and $\mathrm{Sp}_r(k)$, we recover, for the natural choices of ρ, the moduli of semistable orthogonal sheaves, special orthogonal sheaves, and symplectic sheaves, respectively, constructed in [7]. (We publish in this form the main result of [7] as a particular case of our general construction). Let G be for instance $\mathrm{Sp}_r(k)$. To give a singular principal G-bundle on X, for the fundamental representation $\rho\colon \mathrm{Sp}_r(k) \hookrightarrow \mathrm{SL}_r(k)$, is equivalent to give a symplectic sheaf $(\mathscr{A}, \varphi_{\mathscr{A}})$ of rank r, i.e., a torsion free sheaf $\mathscr{A}$ of this rank, decorated with a skew-symmetric homomorphism $\varphi_{\mathscr{A}}\colon \mathscr{A} \otimes \mathscr{A} \longrightarrow \mathscr{O}_X$, such that $\mathscr{A} \longrightarrow \mathscr{A}^\vee$ is a monomorphism. For any subsheaf $\mathscr{A}' \subseteq \mathscr{A}$, the *orthogonal complement* $\mathscr{A}'^\perp \subseteq \mathscr{A}$ is defined as the kernel of $\mathscr{A} \longrightarrow \mathscr{A}^\vee \longrightarrow \mathscr{A}'^\vee$. Our notion of (semi)stability then translates into the condition that, for any isotropic subsheaf $\{0\} \subseteq \mathscr{A}' \subseteq \mathscr{A}'^\perp \subseteq \mathscr{A}$, one has
$$P(\mathscr{A}') + P(\mathscr{A}'^\perp) (\preceq) P(\mathscr{A}).$$

For $\rho\colon \mathrm{O}_r(k) \longrightarrow \mathrm{SL}_{r+1}(k)$, one obtains the analogous definition and moduli of semistable orthogonal sheaves; and for $\rho\colon \mathrm{SO}_r(k) \longrightarrow \mathrm{SL}_r(k)$ one obtains the same definition of semistability, and corresponding moduli, but for special orthogonal sheaves, i.e., for orthogonal sheaves $(\mathscr{A}, \varphi_{\mathscr{A}})$ together with an isomorphism $\psi_{\mathscr{A}}\colon \det \mathscr{A} \cong \mathscr{O}_X$, such that $\psi^2_{\mathscr{A}} = \det \varphi_{\mathscr{A}}$.

iv) For the realizations ρ of simple groups as groups of automorphisms of a non-associative algebra $(V, \varphi\colon V \otimes V \longrightarrow V)$—such as the algebra of octonions for G_2—, singular principal G-bundles are φ-algebra sheaves $(\mathscr{A}, \varphi_{\mathscr{A}})$ of rank $\dim_k(V)$, i.e., torsion free sheaves $\mathscr{A}$ of this rank, decorated with a homomorphism $\varphi_{\mathscr{A}}\colon \mathscr{A} \otimes \mathscr{A} \longrightarrow \mathscr{A}^{\vee\vee}$, such that, at any point $x \in X$ where $\mathscr{A}$ is locally free, the algebra structure $\varphi_{\mathscr{A}}(x)$ is isomorphic to φ. Its (semi)stability translates to

$$\sum_{i\in\mathbb{Z}} (\mathrm{rk}(\mathscr{A}_i)P(\mathscr{A}) - \mathrm{rk}(\mathscr{A})P(\mathscr{A}_i))(\succeq)0,$$

for any balanced algebra filtration $\mathscr{A}_\bullet \subset \mathscr{A}$ (i.e., $\mathbb{Z}$-indexed filtration with $\sum i(\mathrm{rk}\,\mathscr{A}_i - \mathrm{rk}\,\mathscr{A}_{i-1}) = 0$ and $\varphi_{\mathscr{A}}(\mathscr{A}_i, \mathscr{A}_j) \subseteq \mathscr{A}_{i+j}$). We thus obtain projective moduli spaces for these objects.

v) If G is of adjoint type over a field of characteristic zero, Example iv) applies in particular to the adjoint representation. Since G is of finite index in the automorphism group of the Lie algebra structure, the notion of (semi)stability does not change, as shown in Section 6.

1.2 *Faithless representations*

In this case, instead of using singular principal G-bundles, we have to consider principal G-sheaves that we will now define. Let us fix a representation $\rho\colon G \longrightarrow \mathrm{GL}(V)$, such that the image is contained in $\mathrm{SL}(V)$ and the kernel group scheme of ρ is contained in the center group scheme $\mathscr{Z}(G)$. We can factor ρ through a faithful representation $\rho'\colon G' \longrightarrow \mathrm{GL}(V)$ of $G' = G/\ker(\rho)$. We assume for simplicity that G' is semisimple. In this paper, we use the construction of moduli spaces of principal G-bundles for ρ' to construct the moduli space of principal G-bundles for ρ. Let us recall that a *rational principal G-bundle* on the polarized manifold $(X, \mathscr{O}_X(1))$ is a principal G-bundle defined over an open subset $U \subseteq X$ with $\mathrm{codim}_X(X \setminus U) \geq 2$. Let $U_{\mathscr{A}}$ denote the maximal open subset where the torsion free sheaf $\mathscr{A}$ is locally free.

Definition. *A principal ρ-sheaf is a triple $\mathfrak{P} = (\mathscr{P}, \mathscr{A}, \psi)$, where $\mathscr{A}$ is a torsion free sheaf on X, $\mathscr{P}$ is a principal G-bundle on $U_{\mathscr{A}}$, and $\psi\colon \mathscr{P}(V^\vee) \longrightarrow \mathscr{A}_{|U_{\mathscr{A}}}$ is an isomorphism.*

Note that, if ρ is faithful, giving a principal ρ-sheaf is equivalent to giving a singular principal G-bundle. The *degree of* $\mathfrak{P}$ is the homomorphism $d_{\mathfrak{P}}\colon X^\star(G) \longrightarrow \mathbb{Z}$, such that, for any character χ, $d_{\mathfrak{P}}(\chi)$ is the degree of the unique line bundle extending the line bundle $\mathscr{P}(\chi)$ from $U_{\mathscr{A}}$ to X. We say that a principal ρ-sheaf $\mathfrak{P}$ is *(semi)stable*, if the corresponding singular principal G'-bundle is (semi)stable. Let us fix a polynomial

P and a homomorphism $d\colon X^\star(G) \longrightarrow \mathbb{Z}$. We define the moduli functor of principal ρ-sheaves of degree d with Hilbert polynomial P

$$\underline{\mathrm{M}}(\rho)^{\mathrm{(s)s}}_{P,d}\colon \underline{\mathrm{Sch}}_k \longrightarrow \underline{\mathrm{Set}}$$

by

$$S \longmapsto \left\{ \begin{array}{l} \text{Isomorphy classes of families of} \\ \text{(semi)stable principal } \rho\text{-sheaves} \\ \text{of degree } d \text{ with Hilbert polynomial } P \end{array} \right\}.$$

Second Main Theorem. *Assume either that X is a curve or that* $\mathrm{Char}(k) = 0$. *Then, the coarse moduli space for the functors $\underline{\mathrm{M}}(\rho)^{\mathrm{(s)s}}_{\underline{P},d}$ exists as a quasi-projective scheme $\mathrm{M}(\rho)^{\mathrm{ss}}_{\underline{P},d}$. It is projective under the assumptions in* [11] *or* [6].

For higher dimensional varieties in positive characteristic, the existence of the moduli space depends on a certain conjecture on relative purity of the second flat cohomology group of finite diagonalizable group schemes (see Conjecture 5.2.5). This conjecture is motivated by some well known results about relative purity of étale cohomology for smooth pairs (see, e.g., [1]). In the special case that ρ is the adjoint representation, our result generalizes the results of [8] and [25]. A more precise comparison to the results and techniques of [8] and [25] can be found at the end of Section 5.5.

Example. vi) This theorem allows to enlarge Example iv), for $\dim(X) = 1$ or $\mathrm{Char}(k) = 0$, to the case of an arbitrary reductive group G, because the kernel of its adjoint representation is the center. We thus recover, as principal ρ-sheaves, the objects that were called principal G-sheaves in [8] with the same condition of (semi)stability and projective moduli.

Let $\mathscr{P}$ be a rational principal G-bundle on the polarized manifold $(X, \mathscr{O}_X(1))$. Then, $\mathscr{P}$ is called *slope stable*, if for any big open subset $U' \subseteq U$ and any reduction $\sigma\colon U' \longrightarrow (\mathscr{P}_{|U'})/P$ of $\mathscr{P}_{|U'}$ to a parabolic subgroup $P \subset G$, the degree of the pull back to U' of the line bundle associated with any non-trivial anti-dominant character of P is positive. The objects occurring in the definition of the coarse moduli scheme generalize stable rational principal G-bundles. More precisely, for any faithful representation $G \longrightarrow \mathrm{GL}(V)$ with image in $\mathrm{SL}(V)$, a slope stable principal G-bundle corresponds to a stable singular principal G-bundle. Therefore, the moduli spaces of singular principal G-bundles for different faithful representations can be thought of as different compactifications of one (quasi-projective) moduli space of slope stable rational principal G-bundles. Similarly, for any representation $G \longrightarrow \mathrm{GL}(V)$ whose kernel is central, a slope stable principal G-bundle corresponds to a stable principal ρ-sheaf. As above, the moduli spaces of principal ρ-sheaves can also be thought of as different compactifications of the moduli space of slope stable rational principal G-bundles.

Notation

We work over an algebraically closed field k of characteristic $p \geq 0$. A *scheme* will be a locally noetherian scheme over k. For a vector bundle $\mathscr{E}$ over a scheme X, we set $\mathbb{P}(\mathscr{E}) := \mathrm{Proj}(\mathscr{S}ym^{\star}(\mathscr{E}))$, i.e., $\mathbb{P}(\mathscr{E})$ is the projective bundle of hyperplanes in the fibers of $\mathscr{E}$. An open subset $U \subseteq X$ is said to be *big*, if $\mathrm{codim}_X(X \setminus U) \geq 2$. The degree $\deg(\mathscr{E})$ and the Hilbert polynomial $P(\mathscr{E})$ of a torsion free coherent $\mathscr{O}_X$-module $\mathscr{E}$ are taken with respect to the fixed polarization $\mathscr{O}_X(1)$. We set $[x]_+ := \max\{0, x\}$, $x \in \mathbb{R}$.

2. Preliminaries

In this section, we collect different results which will be needed throughout the construction of the moduli space for singular principal G-bundles for a reductive group G via a faithful representation $G \longrightarrow \mathrm{SL}(V)$. Some of the concepts have already been mentioned in the introduction.

2.1 *Functors, universal spaces, and quotients*

In this paper, we will deal with moduli functors of the form

$$\underline{\mathrm{M}}^{(\mathrm{s})\mathrm{s}} \colon \underline{\mathrm{Sch}}_k \longrightarrow \underline{\mathrm{Set}}$$

$$S \longmapsto \left\{ \begin{array}{l} \text{Isomorphy classes of families} \\ \text{of (semi)stable objects} \end{array} \right\}.$$

In each case, we define S-equivalence on the set of isomorphy classes of semistable objects (e.g., semistable sheaves or principal G-bundles with fixed numerical data) which, restricted to stable objects, reduces to isomorphism. Assuming we have the moduli functors and S-equivalence, we introduce the following convenient terminology: A *coarse moduli scheme* for the functors $\underline{\mathrm{M}}^{(\mathrm{s})\mathrm{s}}$ consists of a scheme M^{ss}, an open subscheme $\mathrm{M}^{\mathrm{s}} \subseteq \mathrm{M}^{\mathrm{ss}}$, and natural transformations of functors

$$\theta^{(\mathrm{s})\mathrm{s}} \colon \underline{\mathrm{M}}^{(\mathrm{s})\mathrm{s}} \longrightarrow h_{\mathrm{M}^{(\mathrm{s})\mathrm{s}}}$$

with the following properties:

1. The space $\mathrm{M}^{(\mathrm{s})\mathrm{s}}$ corepresents $\underline{\mathrm{M}}^{(\mathrm{s})\mathrm{s}}$ with respect to $\alpha = \theta^{(\mathrm{s})\mathrm{s}}$. It does so uniformly, if $\mathrm{Char}(k) > 0$, and universally, if $\mathrm{Char}(k) = 0$. (See [14], Definition 2.2.1. Observe that "uniformly" refers to the base change property for *flat* morphisms ϕ in that definition.)
2. The map $\theta^{\mathrm{s}}(k) \colon \underline{\mathrm{M}}^{\mathrm{s}}(k) \longrightarrow \mathrm{M}^{\mathrm{s}}(k)$ is a bijection between the set of isomorphy classes of stable objects and the closed points of M^{s}.
3. The map $\theta^{\mathrm{ss}}(k) \colon \underline{\mathrm{M}}^{\mathrm{ss}}(k) \longrightarrow \mathrm{M}^{\mathrm{ss}}(k)$ induces a bijection between the set of S-equivalence classes of semistable objects and the closed points of M^{ss}.

The difference between positive and zero characteristic in the above definition comes from our use of Geometric Invariant Theory, as GIT quotients in positive characteristic are not necessarily universal categorical. For $G = \mathrm{GL}(V)$, one can in fact show that, in positive characteristic, the moduli space of stable sheaves universally corepresents the moduli functor (see [18], Theorem 0.2). This follows from the fact that stable sheaves are simple. However, even in characteristic zero, the sheaves corresponding to stable principal G-bundles on a curve are no longer simple (see [24], Remark 4.1), so this proof fails in general.

Let $\rho\colon G \longrightarrow \mathrm{GL}(V)$ be a representation of G, such that the kernel of ρ is a finite group subscheme of $\mathscr{Z}(G)$. Let $\mathscr{S}$ be a set of isomorphy classes of principal ρ-sheaves on X. Let $\widetilde{\mathbf{F}}^{\mathscr{S}}$ be the sheaf associated with the presheaf $\mathbf{F}^{\mathscr{S}}\colon \underline{\mathrm{Sch}_k} \longrightarrow \underline{\mathrm{Set}}$, sending a k-scheme T to the set of isomorphy classes of families of principal ρ-sheaves in $\mathscr{S}$ parameterized by T (and defined by pull back on morphisms). For simplicity, we consider from now on sheaves in the Zariski topology which is enough for constructing the moduli space of principal ρ-sheaves. However, everything (with the appropriate changes in the definitions) works, if we replace the Zariski topology by the étale topology.

Let H be a k-group scheme acting on a k-scheme M. Let $\mathbf{M/H}\colon \underline{\mathrm{Sch}_k} \longrightarrow \underline{\mathrm{Set}}$ be the sheaf on $\underline{\mathrm{Sch}_{k,\mathrm{Zar}}}$ associated with the presheaf sending T to the quotient $\mathrm{Hom}(T, M)/\mathrm{Hom}(T, H)$. We say that M is a *universal space with group H for the set $\mathscr{S}$*, if there is an isomorphism of sheaves $\widetilde{\mathbf{F}}^{\mathscr{S}} \longrightarrow \mathbf{M/H}$. Let H be a k-group scheme acting on a k-scheme M. Let $\mathfrak{P}_M$ be a family of principal ρ-sheaves parameterized by M. Assume there is a lifting of the action of H to $\mathfrak{P}_M$, i.e., there is an isomorphism

$$\Lambda\colon \bar{m}^\star\mathfrak{P}_M \xrightarrow{\cong} \bar{p}_M^\star\mathfrak{P}_M.$$

We say that $\mathfrak{P}_M$ is a *universal family with group H* for the set $\mathscr{S}$, if the following conditions are satisfied:

1. *Local universal property:* Given a family $\mathfrak{P}_S$ of principal ρ-sheaves in $\mathscr{S}$ parameterized by S and a closed point $s \in S$, there is an open neighborhood $i\colon S_0 \hookrightarrow S$ of s and a morphism $t\colon S_0 \longrightarrow M$ such that $\bar{\imath}^\star\mathfrak{P}_S \cong \bar{t}^\star\mathfrak{P}_M$.
2. *Gluing property:* Given two morphisms $t_1, t_2\colon S \longrightarrow M$ and an isomorphism $\beta\colon \bar{t}_2^\star\mathfrak{P}_M \longrightarrow \bar{t}_1^\star\mathfrak{P}_M$, there is a unique morphism $h\colon S \longrightarrow H$, such that $t_2 = h[t_1]$ and $\overline{(h\, t_1)}^\star\Lambda = \beta$.

Obviously, if $\mathfrak{P}_M$ is a universal family with group H for $\mathscr{S}$, then M is a universal space with group H for the set $\mathscr{S}$. Assume that M is a universal space with group H for the set $\mathscr{S}$. Then, M admits a universal family with group H, if and only if the quotient stack $[M/H]$ is isomorphic to the moduli stack of principal ρ-sheaves in $\mathscr{S}$ (both considered in the Zariski topology).

2.2 *A little representation theory*

Let $\kappa\colon \mathrm{GL}_r(k) \longrightarrow \mathrm{GL}(V)$ be a representation. It is called a *polynomial representation*, if it extends to a (multiplicative) map $\bar{\kappa}\colon M_r(k) \longrightarrow \mathrm{End}(V)$. We say that κ is *homogeneous of degree $a \in \mathbb{Z}$*, if

$$\kappa(z \cdot \mathbb{E}_r) = z^a \cdot \mathrm{id}_V, \quad \forall z \in \mathbb{G}_m(k).$$

Let $P(r,a)$ be the abelian category of homogeneous polynomial representations of $\mathrm{GL}_r(k)$ of degree a. It comes with the duality functor

$$\begin{aligned} ^\star\colon P(r,a) &\longrightarrow P(r,a) \\ \kappa &\longmapsto (\kappa \circ \mathrm{id}^\vee_{\mathrm{GL}_r(k)})^\vee. \end{aligned}$$

Here, $.^\vee$ stands for the corresponding contragredient representation. An example for a representation in $P(r,a)$ is the *a-th divided power* $(\mathrm{Sym}^a(\mathrm{id}_{\mathrm{GL}_r(k)}))^\star$, i.e., the representation of $\mathrm{GL}_r(k)$ on

$$D^a(W) := (\mathrm{Sym}^a(W^\vee))^\vee, \quad W := k^r.$$

Note that $(W^{\otimes a})^\star = W^{\otimes a}$. More generally, we look, for $u, \upsilon > 0$, at the $\mathrm{GL}_r(k)$-module

$$\mathbb{D}^{u,\upsilon}(W) := \bigoplus_{\substack{(u_1,\dots,u_\upsilon):\\ u_i\geq 0, \sum_{i=1}^{\upsilon} u_i = u}} (D^{u_1}(W)\otimes\cdots\otimes D^{u_\upsilon}(W)). \tag{2}$$

Lemma 2.2.1. *Let $\kappa\colon \mathrm{GL}_r(k) \longrightarrow \mathrm{GL}(U)$ be a homogeneous polynomial representation of degree u. Then, there exists an integer $\upsilon > 0$, such that U is a quotient of the $\mathrm{GL}(U)$-module $\mathbb{D}^{u,\upsilon}(W)$. If κ is homogeneous, but not polynomial, then it is a quotient of $\mathbb{D}^{u,\upsilon}(W)\otimes(\bigwedge^r W)^{\otimes -w}$ for some $w > 0$.*

Proof. The proof of Proposition 5.3 in [17] shows that any representation $\kappa'\colon \mathrm{GL}_r(k) \longrightarrow \mathrm{GL}(U')$ in $P(r,u)$ is, for suitable $\upsilon > 0$, a sub-representation of the representation of $\mathrm{GL}_r(k)$ on the vector space

$$\bigoplus_{\substack{(u_1,\dots,u_\upsilon):\\ u_i\geq 0, \sum_{i=1}^{\upsilon} u_i = u}} (\mathrm{Sym}^{u_1}(W)\otimes\cdots\otimes \mathrm{Sym}^{u_\upsilon}(W)).$$

Applying this result to the dual $\kappa^\star\colon \mathrm{GL}_r(k) \longrightarrow \mathrm{GL}(U^\star)$ of κ proves the first assertion.

The second assertion follows from the obvious fact that $U\otimes(\bigwedge^r W)^{\otimes w}$ will be a polynomial representation for large w. □

Remark 2.2.2. As is apparent from the construction in [17], the above result also holds over the ring of integers.

Corollary 2.2.3. *Let $\kappa\colon \mathrm{GL}_r(k) \longrightarrow \mathrm{GL}(V)$ be a homogeneous polynomial representation of degree a, and suppose that the characteristic of the base field is zero or bigger than a. Then, there exists $b > 0$, such that V is a quotient of the $\mathrm{GL}_r(k)$-module $W_{a,b} := (W^{\otimes a})^{\oplus b}$, $W := k^r$.*

Proof. If $p > a$, then one has $D^a(W) \cong \mathrm{Sym}^a(W)$. This fact obviously implies the claim. □

2.3 *Background from GIT*

We recall some notation and results from Geometric Invariant Theory. Let G be a reductive group over the field k and $\kappa\colon G \longrightarrow \mathrm{GL}(W)$ a representation on the finite dimensional k-vector space W. This yields the action

$$\begin{aligned}\alpha\colon G \times W &\longrightarrow W\\ (g, w) &\longmapsto \kappa(g)(w).\end{aligned}$$

Recall that a one parameter subgroup is a homomorphism

$$\lambda\colon \mathbb{G}_m(k) \longrightarrow G.$$

Such a one parameter subgroup defines a decomposition

$$W = \bigoplus_{\gamma \in \mathbb{Z}} W^{\gamma}$$

with

$$W^{\gamma} = \{\, w \in W | \kappa(\lambda(z))(w) = z^{\gamma} \cdot w,\ \forall z \in \mathbb{G}_m(k)\}, \quad \gamma \in \mathbb{Z}.$$

Let $\gamma_1 < \cdots < \gamma_{s+1}$ be the integers with $W^{\gamma} \neq \{0\}$ and $\gamma_\bullet(\lambda) := (\gamma_1, \ldots, \gamma_{s+1})$. We define the flag

$$\begin{aligned}W_\bullet(\lambda) : \{0\} \subsetneq W_1 := W^{\gamma_1} \subsetneq W_2 := W^{\gamma_1} \oplus W^{\gamma_2} \subsetneq \cdots\\ \subsetneq W_s := W^{\gamma_1} \oplus \cdots \oplus W^{\gamma_s} \subsetneq W_{s+1} := W\end{aligned}$$

and the tuple $\alpha_\bullet(\lambda) := (\alpha_1, \ldots, \alpha_s)$ of positive rational numbers with

$$\alpha_i := \frac{\gamma_{i+1} - \gamma_i}{\dim_k(W)}, \quad i = 1, \ldots, s.$$

In general, we will refer to $(W_\bullet(\lambda), \gamma_\bullet(\lambda))$ as the *weighted flag of* λ. If λ is a one parameter subgroup of the special linear group $\mathrm{SL}(W)$, we will refer to $(W_\bullet(\lambda), \alpha_\bullet(\lambda))$ as the *weighted flag of* λ. (This abuse of notation is justified by the fact that $\gamma_\bullet(\lambda)$ can be computed from $\alpha_\bullet(\lambda)$, in this case.) For a point $w \in W \setminus \{0\}$, we define

$$\mu_\kappa(\lambda, w) := \max\{\, \gamma_i \mid w \text{ has a non-trivial component in } W_{\gamma_i}, i = 1, \ldots, s+1\}.$$

Note that, for $G = \mathrm{GL}(V_1) \times \cdots \times \mathrm{GL}(V_t)$ and $\lambda^j = (\lambda^j_1, \ldots, \lambda^j_t)\colon \mathbb{G}_m(k) \longrightarrow G$, $j = 1, 2$,

$$\begin{aligned}&\mu_\kappa(\lambda^1, w) = \mu_\kappa(\lambda^2, w),\\ \text{if} \quad &(V_{i\bullet}(\lambda^1_i), \gamma_\bullet(\lambda^1_i)) = (V_{i\bullet}(\lambda^2_i), \gamma_\bullet(\lambda^2_i)), \quad \text{for } i = 1, \ldots, t.\end{aligned} \tag{3}$$

Similarly, for $G = \mathrm{SL}(V_1) \times \cdots \times \mathrm{SL}(V_t)$ and $\lambda^j = (\lambda^j_1, \ldots, \lambda^j_t)\colon \mathbb{G}_m(k) \longrightarrow G$, $j = 1, 2$,

$$\begin{aligned}&\mu_\kappa(\lambda^1, w) = \mu_\kappa(\lambda^2, w),\\ \text{if} \quad &(V_{i\bullet}(\lambda^1_i), \alpha_\bullet(\lambda^1_i)) = (V_{i\bullet}(\lambda^2_i), \alpha_\bullet(\lambda^2_i)), \quad \text{for } i = 1, \ldots, t.\end{aligned} \tag{4}$$

(See [22], Proposition 2.7, Chapter 2. Note that we take the weighted flags in the V_i, $i = 1, \ldots, t$, and not in W.)

Suppose we are given a projective scheme X, a G-action $\bar{\sigma} : G \times X \longrightarrow X$, and a linearization $\sigma : G \times \mathscr{L} \longrightarrow \mathscr{L}$ of this action in the line bundle $\mathscr{L}$. For a point $x \in X$ and a one parameter subgroup λ, we get the point

$$x_\infty(\lambda) := \lim_{z \to \infty} \bar{\sigma}(\lambda(z), x).$$

This point is fixed under the action $\mathbb{G}_m(k) \times X \longrightarrow X$, $(z, x) \longmapsto \bar{\sigma}(\lambda(z), x)$. Therefore, $\mathbb{G}_m(k)$ acts on the fiber $\mathscr{L}\langle x_\infty(\lambda)\rangle$. This action is of the form $l \longmapsto z^\gamma \cdot l$, $z \in \mathbb{G}_m(k)$, $l \in \mathscr{L}\langle x_\infty(\lambda)\rangle$, and we set

$$\mu_\sigma(\lambda, x) := -\gamma.$$

For a representation κ of G as above, we get the action

$$\begin{aligned} \bar{\sigma} : G \times \mathbb{P}(W^\vee) &\longrightarrow \mathbb{P}(W^\vee) \\ (g, [w]) &\longmapsto [\kappa(g)(w)] \end{aligned}$$

together with an induced linearization σ in $\mathscr{O}_{\mathbb{P}(W^\vee)}(1)$. One checks that

$$\mu_\kappa(\lambda, w) = \mu_\sigma(\lambda, [w]), \quad \forall w \in W \backslash \{0\}, \lambda : \mathbb{G}_m(k) \longrightarrow G. \tag{5}$$

Finally, we recall that a one parameter subgroup $\lambda : \mathbb{G}_m(k) \longrightarrow G$ gives the parabolic subgroup

$$Q_G(\lambda) := \{ g \in G | \lim_{z \to \infty} \lambda(z) \cdot g \cdot \lambda(z)^{-1} \text{ exists in } G \}. \tag{6}$$

The unipotent radical of $Q_G(\lambda)$ is the subgroup

$$\mathscr{R}_u(Q_G(\lambda)) := \{ g \in G | \lim_{z \to \infty} \lambda(z) \cdot g \cdot \lambda(z)^{-1} = e \}.$$

Remark 2.3.1. In the book [34], one defines the parabolic subgroup

$$P_G(\lambda) := \{ g \in G | \lim_{z \to 0} \lambda(z) \cdot g \cdot \lambda(z)^{-1} \text{ exists in } G \},$$

i.e.,

$$P_G(\lambda) = Q_G(-\lambda).$$

Therefore, every parabolic subgroup of Q is of the shape $Q_G(\lambda)$ for an appropriate one parameter subgroup λ of G. We have chosen a different convention, because it is compatible with our GIT notation.

Some specific quotient problems. Again, we are led to consider the case when the base manifold is just a point. Here, the situation is, however, more involved due to the presence of a non-trivial center. This will amount to the fact that we have to look also at certain torus quotients of the spaces that we initially define.

Let G be a reductive linear algebraic group defined over k. We fix a representation $\rho\colon G \longrightarrow \mathrm{GL}(V)$ on the finite dimensional k-vector space V with $\rho(G) \subseteq (\mathrm{GL}(V_1)\times \cdots \times \mathrm{GL}(V_t)) \cap \mathrm{SL}(V)$. Set $r_i := \dim_k(V_i)$, $\rho_i\colon G \longrightarrow \mathrm{GL}(V_i)$, $i = 1, \ldots, t$.

We look at the representation

$$R\colon (\mathrm{GL}_{r_1}(k) \times \cdots \times \mathrm{GL}_{r_t}(k)) \times G \longrightarrow \mathrm{GL}\left(\bigoplus_{i=1}^{t} \mathrm{Hom}(k^{r_i}, V_i^{\vee})\right)$$

$$(g_1, \ldots, g_t, g) \longmapsto \left((h_1, \ldots, h_t) \in \bigoplus_{i=1}^{t} \mathrm{Hom}(k^{r_i}, V_i^{\vee}) \longmapsto (h_1', \ldots, h_t')\right)$$

with $h_i'(w)(v) = h_i(g_i^{-1} \cdot w)(\rho_i(g^{-1})(v))$, $w \in k^{r_i}, v \in V_i, i = 1, \ldots, t$.

The representation R provides actions of $(\mathrm{GL}_{r_1}(k) \times \cdots \times \mathrm{GL}_{r_t}(k)) \times G$ on

$$\bigoplus_{i=1}^{t}(V_i \otimes k^{r_i})^{\vee} = \bigoplus_{i=1}^{t} \mathrm{Hom}(k^{r_i}, V_i^{\vee}) \quad \text{and} \quad \mathbb{P}\left(\bigoplus_{i=1}^{t} \mathrm{Hom}(k^{r_i}, V_i^{\vee})^{\vee}\right)$$

and induces $(\mathrm{GL}_{r_1}(k) \times \cdots \times \mathrm{GL}_{r_t}(k))$-actions on the categorical quotients

$$\mathbb{H} := \left(\bigoplus_{i=1}^{t}(V_i \otimes k^{r_i})^{\vee}\right) /\!\!/ G \quad \text{and} \quad \bar{\mathbb{H}} := \mathbb{P}\left(\bigoplus_{i=1}^{t} \mathrm{Hom}(k^{r_i}, V_i^{\vee})^{\vee}\right) /\!\!/ G$$

$$\overset{[6],\ 2.1.7}{=} (\mathbb{H}\setminus\{0\}) /\!\!/ \mathbb{G}_m(k).$$

Here, we have used the embedding $\mathbb{G}_m(k) \hookrightarrow \mathrm{GL}_{r_1}(k) \times \cdots \times \mathrm{GL}_{r_t}(k)$, $z \longmapsto (z\cdot\mathbb{E}_{r_1}, \ldots, z\cdot\mathbb{E}_{r_t})$. The coordinate algebra of $\mathbb{H}$ is $\mathrm{Sym}^{\star}(k^{r_1} \otimes_k V_1 \oplus \cdots \oplus k^{r_t} \otimes_k V_t)^G$. For $s > 0$, we set

$$\mathbb{W}_s := \bigoplus_{i=1}^{s} \mathbb{U}_i, \quad \mathbb{U}_i := (\mathrm{Sym}^i(k^{r_1} \otimes_k V_1 \oplus \cdots \oplus k^{r_t} \otimes_k V_t)^G)^{\vee}, \quad i \geq 0.$$

If s is so large that $\bigoplus_{i=0}^{s} \mathrm{Sym}^i(k^{r_1} \otimes_k V_1 \oplus \cdots \oplus k^{r_t} \otimes_k V_t)^G$ contains a set of generators for the algebra $\mathrm{Sym}^{\star}(k^{r_1} \otimes_k V_1 \oplus \cdots \oplus k^{r_t} \otimes_k V_t)^G$, then we have a $(\mathrm{GL}_{r_1}(k) \times \cdots \times \mathrm{GL}_{r_t}(k))$-equivariant surjection of algebras

$$\mathrm{Sym}^{\star}(\mathbb{W}_s^{\vee}) \longrightarrow \mathrm{Sym}^{\star}(k^{r_1} \otimes_k V_1 \oplus \cdots \oplus k^{r_t} \otimes_k V_t)^G,$$

and, thus, a $(\mathrm{GL}_{r_1}(k) \times \cdots \times \mathrm{GL}_{r_t}(k))$-equivariant embedding

$$\iota_s\colon \mathbb{H} \hookrightarrow \mathbb{W}_s.$$

Set $\widetilde{G} := (\mathrm{GL}_{r_1}(k) \times \cdots \times \mathrm{GL}_{r_t}(k)) \cap \mathrm{SL}_r(k)$, $r := r_1 + \cdots + r_t$, and

$$\mathbb{I} := (\mathrm{Isom}(k^{r_1}, V_1^{\vee}) \times \cdots \times \mathrm{Isom}(k^{r_t}, V_t^{\vee}))/G \quad (\cong (\mathrm{GL}_{r_1}(k) \times \cdots \times \mathrm{GL}_{r_t}(k))/G).$$

Lemma 2.3.2.

i) *Every point $\iota_s(i)$, $i \in \mathbb{I}$, is $\widetilde{G}$-polystable.*
ii) *A point $\iota_s(h)$, $h \in \mathbb{H}\setminus\mathbb{I}$, is not $\widetilde{G}$-semistable.*

Proof. Ad i). We choose bases for $V_i^{\vee}$, $i = 1, \ldots, t$. This provides us with the function

$$\mathfrak{d}\colon \bigoplus_{i=1}^{t} \operatorname{Hom}(k^{r_i}, V_i^{\vee}) \longrightarrow k$$

$$\underline{f} = (f_1, \ldots, f_t) \longmapsto \det(f_1 \oplus \cdots \oplus f_t)$$

which is invariant under the action of $\widetilde{G} \times G$. Therefore, it descends to a (non constant) $\widetilde{G}$-invariant function on $\mathbb{H}$, called again $\mathfrak{d}$. For any $i \in \mathbb{I}$, we clearly have $\mathfrak{d}(i) \neq 0$, so that $\iota_s(i)$ is $\widetilde{G}$-semistable ([22], Lemma A.1.2). Furthermore, for any $\underline{f} \in \operatorname{Isom}(k^{r_1}, V_1^{\vee}) \times \cdots \times \operatorname{Isom}(k^{r_t}, V_t^{\vee})$, the $(\widetilde{G} \times G)$-orbit of $\underline{f}$ is just the level set $\mathfrak{d}^{-1}(z)$ for an appropriate $z \in \mathbb{G}_m(k)$. In particular, it is closed. The image of this orbit is the $\widetilde{G}$-orbit of $i := [\underline{f}]$ in $\mathbb{H}$ which is, therefore, closed. Since ι_s is a closed, $\widetilde{G}$-equivariant embedding, the orbit of $\iota_s(i)$ is closed, too.

Ad ii). It is easy to see from the construction that the ring of $\widetilde{G}$-invariant functions on $\mathbb{H}$ is generated by $\mathfrak{d}$. This makes the asserted property evident. □

An important consequence is the following.

Proposition 2.3.3. *Fix bases for $V_1, \ldots, V_t$ in order to obtain an isomorphism*

$$\operatorname{GL}(V_1) \times \cdots \times \operatorname{GL}(V_t) \cong \operatorname{GL}_{r_1}(k) \times \cdots \times \operatorname{GL}_{r_t}(k)$$

and a $(\operatorname{GL}_{r_1}(k) \times \cdots \times \operatorname{GL}_{r_t}(k))$*-equivariant isomorphism*

$$\phi\colon (\operatorname{GL}_{r_1}(k) \times \cdots \times \operatorname{GL}_{r_t}(k))/G \longrightarrow (\operatorname{Isom}(k^{r_1}, V_1^{\vee}) \times \cdots \times \operatorname{Isom}(k^{r_t}, V_t^{\vee}))/G.$$

Suppose that $x = \iota_s(i)$ for some $i = \phi(g) \in \mathbb{I}$. Then, for a one parameter subgroup $\lambda\colon \mathbb{G}_m(k) \longrightarrow \widetilde{G}$, the following conditions are equivalent:

i) $\mu_{\kappa_s}(\lambda, x) = 0$, *$\kappa_s$ being the representation of $\widetilde{G}$ on $\mathbb{W}_s$.*
ii) *There is a one parameter subgroup $\lambda'\colon \mathbb{G}_m(k) \longrightarrow g \cdot G \cdot g^{-1}$ with*

$$(W_{i\bullet}(\lambda_i), \gamma_\bullet(\lambda_i)) = (W_{i\bullet}(\lambda'_i), \gamma_\bullet(\lambda'_i)), \quad W_i := \mathbb{C}^{r_i},\ i = 1, \ldots, t.$$

Proof. The proof is a straightforward generalization of the proof for the case $t = 1$, i.e., Proposition 2.1.4 in [6]. □

In the next step, we have to consider some quotients of the varieties just defined. Let $T' \subset \operatorname{GL}_{r_1}(k) \times \cdots \times \operatorname{GL}_{r_t}(k)$ be the $(t-1)$-dimensional torus that is given as the image of the homomorphism

$$\mathbb{G}_m(k)^{\times(t-1)} \longrightarrow \operatorname{GL}_{r_1}(k) \times \cdots \times \operatorname{GL}_{r_t}(k)$$

$$(z_1, \ldots, z_{t-1}) \longmapsto (z_1 \cdot \mathbb{E}_{r_1}, \ldots, z_{t-1} \cdot \mathbb{E}_{r_{t-1}}, (z_1 \cdot \cdots \cdot z_{t-1})^{-1} \cdot \mathbb{E}_{r_t}),$$

and $T \subset \operatorname{GL}_{r_1}(k) \times \cdots \times \operatorname{GL}_{r_t}(k)$ the t-dimensional torus that is the image of the homomorphism

$$\mathbb{G}_m(k)^{\times t} \longrightarrow \operatorname{GL}_{r_1}(k) \times \cdots \times \operatorname{GL}_{r_t}(k)$$

$$(z_1, \ldots, z_t) \longmapsto (z_1 \cdot \mathbb{E}_{r_1}, \ldots, z_t \cdot \mathbb{E}_{r_t}).$$

We want to describe the varieties

$$\widetilde{\mathbb{H}} := \mathbb{H}/\!\!/ T' \quad \text{and} \quad \widetilde{\widetilde{\mathbb{H}}} := \mathbb{P}\left(\bigoplus_{i=1}^{t} \mathrm{Hom}(k^{r_i}, V_i^{\vee})^{\vee}\right) /\!\!/ (T' \times G) \overset{[6],\, 2.1.7}{=} (\mathbb{H}\setminus\{0\})/\!\!/ T.$$

Note that

$$\mathbb{S} := (k^{r_1} \otimes_k V_1) \otimes_k \cdots \otimes_k (k^{r_t} \otimes_k V_t)$$

is a direct summand of $\mathrm{Sym}^t(k^{r_1} \otimes_k V_1 \oplus \cdots \oplus k^{r_t} \otimes_k V_t)$, and, thus, $\mathrm{Sym}^\star(\mathbb{S})^G$ is a subalgebra of $\mathrm{Sym}^\star(k^{r_1} \otimes_k V_1 \oplus \cdots \oplus k^{r_t} \otimes_k V_t)^G$. We easily check that $\mathrm{Sym}^\star(\mathbb{S})^G$ is the coordinate algebra of $\widetilde{\mathbb{H}}$. Likewise, we define $\widetilde{\mathbb{U}}_i$ as the image of $\mathbb{U}_i$ under the quotient map

$$(\mathrm{Sym}^\star(k^{r_1} \otimes_k V_1 \oplus \cdots \oplus k^{r_t} \otimes_k V_t)^G)^\vee \longrightarrow (\mathrm{Sym}^\star(\mathbb{S})^G)^\vee,$$

and declare

$$\widetilde{\mathbb{W}}_s := \bigoplus_{i=1}^{s} \widetilde{\mathbb{U}}_i, \quad i \geq 0.$$

We find the commutative diagram

$$\begin{array}{ccc}
\mathrm{Sym}^\star(\widetilde{\mathbb{W}}_s^\vee) & \longrightarrow & \mathrm{Sym}^\star(\mathbb{S})^G \\
\big\downarrow & & \big\downarrow \\
\mathrm{Sym}^\star(\mathbb{W}_s^\vee) & \twoheadrightarrow & \mathrm{Sym}^\star(k^{r_1} \otimes_k V_1 \oplus \cdots \oplus k^{r_t} \otimes_k V_t)^G
\end{array}$$

which induces

$$\begin{array}{ccc}
\mathbb{H} & \overset{\iota_s}{\hookrightarrow} & \mathbb{W}_s \\
\big\downarrow & & \big\downarrow \\
\widetilde{\mathbb{H}} & \overset{\widetilde{\iota}_s}{\longrightarrow} & \widetilde{\mathbb{W}}_s.
\end{array}$$

Again, for $s \gg 0, \widetilde{\iota}_s$ will be a closed embedding. Define $S := \mathrm{SL}_{r_1}(k) \times \cdots \times \mathrm{SL}_{r_t}(k)$. Note that we have surjective homomorphisms

$$S \times T' \longrightarrow (\mathrm{GL}_{r_1}(k) \times \cdots \times \mathrm{GL}_{r_t}(k)) \cap \mathrm{SL}_r(k),$$
$$S \times T \longrightarrow \mathrm{GL}_{r_1}(k) \times \cdots \times \mathrm{GL}_{r_t}(k).$$

The actions of $S \times T'$ on $\mathbb{H}$ and $\mathbb{W}_s$ induce actions of S on $\widetilde{\mathbb{H}}$ and $\widetilde{\mathbb{W}}_s$, and $\widetilde{\iota}_s$ is S-equivariant.

Set $\widetilde{\mathbb{I}} := (\mathrm{Isom}(k^{r_1}, V_1^\vee) \times \cdots \times \mathrm{Isom}(k^{r_t}, V_t^\vee))/(T' \times G)$ ($\cong (\mathrm{GL}_{r_1}(k) \times \cdots \times \mathrm{GL}_{r_t}(k))/(T' \times G)$). This is a dense open subset of $\widetilde{\mathbb{H}}$. The semistability of points $\widetilde{\iota}_s(h)$, $h \in \widetilde{\mathbb{H}}$, with respect to the action of the group S is described by the following result.

Lemma 2.3.4.

i) *Every point* $\widetilde{\iota}_s(i)$, $i \in \widetilde{\mathbb{I}}$, *is S-polystable.*
ii) *A point* $\widetilde{\iota}_s(h)$, $h \in \widetilde{\mathbb{H}}\backslash\widetilde{\mathbb{I}}$, *is not S-semistable.*

Proof. The proof is completely analogous to the one of Lemma 2.3.2 □

A key result is now the following.

Proposition 2.3.5. *Fix bases for* $V_1, \ldots, V_t$ *in order to obtain a* $(\mathrm{GL}_{r_1}(k) \times \cdots \times \mathrm{GL}_{r_t}(k))$*-equivariant isomorphism*

$$\widetilde{\phi}\colon (\mathrm{GL}_{r_1}(k) \times \cdots \times \mathrm{GL}_{r_t}(k))/(T' \times G) \\ \longrightarrow (\mathrm{Isom}(k^{r_1}, V_1^\vee) \times \cdots \times \mathrm{Isom}(k^{r_t}, V_t^\vee))/(T' \times G).$$

Suppose that $x = \widetilde{\iota}_s(i)$ *for some* $i = \widetilde{\phi}(g) \in \widetilde{\mathbb{I}}$. *Then, for a one parameter subgroup* $\lambda\colon \mathbb{G}_m(k) \longrightarrow S$, *the following conditions are equivalent:*

i) $\mu_{\widetilde{\kappa}_s}(\lambda, x) = 0$, $\widetilde{\kappa}_s$ *being the representation of S on* $\widetilde{\mathbb{W}}_s$.
ii) *There is a one parameter subgroup* $\lambda'\colon \mathbb{G}_m(k) \longrightarrow g \cdot [G, G] \cdot g^{-1}$ *with*

$$(W_{i\bullet}(\lambda_i), \alpha_\bullet(\lambda_i)) = (W_{i\bullet}(\lambda_i'), \alpha_\bullet(\lambda_i')), \quad W_i := \mathbb{C}^{r_i},\ i = 1, \ldots, t.$$

Proof. The strategy is the same as in the proof of Proposition 2.1.4 in [6]. A little extra difficulty is that we wish to deal with one parameter subgroups of the derived group $[G, G]$.

We may clearly assume $g = (\mathbb{E}_{r_1}, \ldots, \mathbb{E}_{r_t})$. We first show "ii)$\Longrightarrow$i)". Since $[G, G]$ is contained in the $(\mathrm{GL}_{r_1}(k) \times \cdots \times \mathrm{GL}_{r_t}(k))$-stabilizer of x, we have $\mu_{\widetilde{\kappa}_s}(\lambda', x) = 0$ for any one parameter subgroup $\lambda'\colon \mathbb{G}_m(k) \longrightarrow [G, G]$. Now, (4) implies the claim.

We turn to the implication "i)$\Longrightarrow$ii)". By Lemma 2.3.4, i), there exists an element $g' \in S$, such that

$$x' := \lim_{z\to\infty} \lambda(z) \cdot x = \phi(g').$$

By Proposition 2.1.2 in [6], we may choose

$$g' \in \mathscr{R}_u(Q_{\mathrm{GL}_{r_1}(k)\times\cdots\times\mathrm{GL}_{r_t}(k)}(\lambda)) = \mathscr{R}_u(Q_{\mathrm{SL}_{r_1}(k)\times\cdots\times\mathrm{SL}_{r_t}(k)}(\lambda)) \\ = \mathscr{R}_u(Q_{\mathrm{SL}_{r_1}(k)}(\lambda_1)) \times \cdots \times \mathscr{R}_u(Q_{\mathrm{SL}_{r_t}(k)}(\lambda_t)) \subset S.$$

In particular, the element g' fixes the flag $W_{i\bullet}(\lambda_i)$, $i = 1, \ldots, t$. Since λ fixes x', it lies in $(g' \cdot \widetilde{(T' \times G)} \cdot g'^{-1}) \cap S$. Here, $\widetilde{(T' \times G)}$ is the image of $T' \times G$ in $\mathrm{GL}_{r_1}(k) \times \cdots \times \mathrm{GL}_{r_t}(k)$. Setting $\lambda' := g'^{-1} \cdot \lambda \cdot g'$, we obviously have $(W_{i\bullet}(\lambda_i), \alpha_\bullet(\lambda_i)) = (W_{i\bullet}(\lambda_i'), \alpha_\bullet(\lambda_i'))$, $i = 1, \ldots, t$, and λ' is a one parameter subgroup of $\widetilde{(T' \times G)} \cap S = ((\mathscr{Z}(G) \cap S) \cdot [G, G])^0$. Since $\mathscr{Z}(G) \cap S$ is a finite group, λ' must be a one parameter subgroup of $[G, G]$. □

Next, we look at the categorical quotient

$$\bar{\mathbb{H}} = \mathrm{Proj}(\mathrm{Sym}^\star(k^{r_1} \otimes_k V_1 \oplus \cdots \oplus k^{r_t} \otimes_k V_t)^G).$$

For any positive integer d, we define

$$\mathrm{Sym}^{(d)}(k^{r_1} \otimes_k V_1 \oplus \cdots \oplus k^{r_t} \otimes_k V_t)^G := \bigoplus_{i=0}^{\infty} \mathrm{Sym}^{id}(k^{r_1} \otimes_k V_1 \oplus \cdots \oplus k^{r_t} \otimes_k V_t)^G.$$

Then,

$$\begin{aligned} &\mathrm{Proj}(\mathrm{Sym}^{\star}(k^{r_1} \otimes_k V_1 \oplus \cdots \oplus k^{r_t} \otimes_k V_t)^G) \\ &\cong \mathrm{Proj}(\mathrm{Sym}^{(d)}(k^{r_1} \otimes_k V_1 \oplus \cdots \oplus k^{r_t} \otimes_k V_t)^G). \end{aligned}$$

We can choose s, such that

a) $\mathrm{Sym}^{\star}(k^{r_1} \otimes_k V_1 \oplus \cdots \oplus k^{r_t} \otimes_k V_t)^G$ is generated by elements in degree $\leq s$.
b) $\mathrm{Sym}^{(s!)}(k^{r_1} \otimes_k V_1 \oplus \cdots \oplus k^{r_t} \otimes_k V_t)^G$ is generated by elements in degree 1, i.e., by the elements in the vector space $\mathbb{V}_s := \mathrm{Sym}^{s!}(k^{r_1} \otimes_k V_1 \oplus \cdots \oplus k^{r_t} \otimes_k V_t)^G$.

Obviously, there is a natural surjection

$$\mathrm{Sym}^{\star}(\mathbb{V}_s) \longrightarrow \mathrm{Sym}^{(s!)}(k^{r_1} \otimes_k V_1 \oplus \cdots \oplus k^{r_t} \otimes_k V_t)^G.$$

This defines a closed and $(\mathrm{GL}_{r_1}(k) \times \cdots \times \mathrm{GL}_{r_s}(k))$-equivariant embedding

$$\bar{\iota}_s \colon \bar{\mathbb{H}} \hookrightarrow \mathbb{P}(\mathbb{V}_s).$$

We also define

$$\mathcal{O}_{\bar{\mathbb{H}}}(s!) := \bar{\iota}_s^{\star} \mathcal{O}_{\mathbb{P}(\mathbb{V}_s)}(1).$$

Note that

$$\mathcal{O}_{\bar{\mathbb{H}}}((s+1)!) = \mathcal{O}_{\bar{\mathbb{H}}}(s!)^{\otimes (s+1)}. \tag{7}$$

Lemma 2.3.6. *For a positive integer s, such that* a) *and* b) *as above are satisfied, a G-semistable point $\underline{f} \in \bigoplus_{i=1}^t \mathrm{Hom}(k^{r_i}, V_i^{\vee})$, and a one parameter subgroup $\lambda \colon \mathbb{G}_m(k) \longrightarrow \widetilde{G}$, one has*

$$\mu_{\kappa_s}(\lambda, h) > (= / <)\, 0 \iff \mu_{\sigma_s}(\lambda, \bar{h}) > (= / <)\, 0.$$

Here, $h := \iota_s([\underline{f}])$, $\bar{h} := \bar{\iota}_s([\underline{f}])$, and σ_s is the linearization of the $\widetilde{G}$-action on $\bar{\mathbb{H}}$ in $\mathcal{O}_{\bar{\mathbb{H}}}(s!)$.

Proof. We introduce the vector space

$$\begin{aligned} \widehat{\mathbb{V}}_s := \bigoplus (&\mathrm{Sym}^{d_1}((k^{r_1} \otimes V_1 \otimes \cdots \otimes k^{r_t} \otimes V_t)^G) \otimes \cdots \\ &\otimes \mathrm{Sym}^{d_s}(\mathrm{Sym}^s(k^{r_1} \otimes V_1 \otimes \cdots \otimes k^{r_t} \otimes V_t)^G)), \end{aligned} \tag{8}$$

the sum running over all tuples $(d_1, \ldots, d_s)$ with $d_i \geq 0$, $i = 1, \ldots, s$, and $\sum i d_i = s!$. There is the canonical surjection $\widehat{\mathbb{V}}_s \longrightarrow \mathbb{V}_s$, and we obtain the following commutative diagram

$$\begin{array}{ccccc}
\left(\bigoplus_{i=1}^{t} \mathrm{Hom}(k^{r_i}, V_i^{\vee})\right)/\!\!/G & & \xhookrightarrow{\quad \iota_s \quad} & & \mathbb{W}_s \setminus \{0\} \\
\Big\downarrow {\scriptstyle \mathbb{G}_m(k)\text{-quotient}} & & & & \Big\downarrow {\scriptstyle \alpha} \\
\mathbb{P}\left(\left(\bigoplus_{i=1}^{t} \mathrm{Hom}(k^{r_i}, V_i^{\vee})\right)^{\vee}\right)/\!\!/G & \xhookrightarrow{\ \bar{\iota}_s\ } & \mathbb{P}(\mathbb{V}_s) & \xhookrightarrow{\ \hat{\iota}_s\ } & \mathbb{P}(\widehat{\mathbb{V}}_s).
\end{array}$$

The morphism α factorizes over the quotient with respect to the $\mathbb{G}_m(k)$-action on $\mathbb{W}_s$ that is given on $\mathbb{U}_i$ by scalar multiplication with z^{-i}, $i = 1, \ldots, s$, $z \in \mathbb{G}_m(k)$. The morphism α can be explicitly described: An element $(l_1, \ldots, l_s) \in \mathbb{W}_s$ with

$$l_i \colon \mathrm{Sym}^i(k^{r_1} \otimes V_1 \otimes \cdots \otimes k^{r_t} \otimes V_t)^G \longrightarrow k, \quad i = 1, \ldots, s,$$

is mapped to the class

$$\left[\bigoplus_{\substack{\underline{d}=(d_1,\ldots,d_s):\\ d_i \geq 0, \sum i d_i = s!}} l_{\underline{d}}\right] : \widehat{\mathbb{V}}_s \longrightarrow k$$

where the component $l_{\underline{d}}$ is given as

$$u_1 \otimes \cdots \otimes u_s \longmapsto (l_1(u_1^1) \cdot \ldots \cdot l_1(u_1^{d_1})) \cdot \ldots \cdot (l_s(u_s^1) \cdot \ldots \cdot l_s(u_s^{d_s})), \quad u_i = [u_i^1 \otimes \cdots \otimes u_i^{d_i}].$$

With this description, one easily sees

$$\mu_{\kappa_s}(\lambda, h) > (= / <)\, 0 \quad \Longleftrightarrow \quad \mu_{\widehat{\sigma}_s}(\lambda, \alpha(h)) > (= / <)\, 0$$

for all $\lambda \colon \mathbb{G}_m(k) \longrightarrow \widetilde{G}$ and all $h \in \mathbb{W}_s \setminus \{0\}$. Here, $\widehat{\sigma}_s$ is the linearization of the $\widetilde{G}$-action on $\mathbb{P}(\widehat{\mathbb{V}}_s)$ in $\mathcal{O}_{\mathbb{P}(\widehat{\mathbb{V}}_s)}(1)$. The fact that $\mu_{\sigma_s}(\lambda, v) = \mu_{\widehat{\sigma}_s}(\lambda, \hat{\iota}_s(v))$, for $v \in \mathbb{P}(\mathbb{V}_s)$, $\lambda \colon \mathbb{G}_m(k) \longrightarrow \widetilde{G}$, together with the above diagram, implies the claim. □

Similarly to the above, we have

$$\widetilde{\widetilde{\mathbb{H}}} = \mathrm{Proj}(\mathrm{Sym}^{\star}(\mathbb{S})^G).$$

We set $\mathrm{Sym}^{\langle s \rangle}(\mathbb{S}) := \mathrm{Sym}^{\star}(\mathbb{S}) \cap \mathrm{Sym}^s(k^{r_1} \otimes_k V_1 \oplus \cdots \oplus k^{r_t} \otimes_k V_t)$, $s \geq 0$. Choosing s large enough, we may assume that

a) $\mathrm{Sym}^{\star}(\mathbb{S})^G$ is generated by the elements that have degree $\leq s$ as elements of $\mathrm{Sym}^{\star}(k^{r_1} \otimes_k V_1 \oplus \cdots \oplus k^{r_t} \otimes_k V_t)^G$.
b) $\mathrm{Sym}^{\star}(\mathbb{S})^G \cap \mathrm{Sym}^{(s!)}(k^{r_1} \otimes_k V_1 \oplus \cdots \oplus k^{r_t} \otimes_k V_t)^G$ is generated by the elements in the vector space $\widetilde{\mathbb{V}}_s := \mathrm{Sym}^{\langle s! \rangle}(\mathbb{S})^G$.

We have the commutative diagram

$$\begin{array}{ccc}
\mathrm{Sym}^{\star}(\widetilde{\mathbb{V}}_s) & \longrightarrow\!\!\!\!\!\rightarrow & \mathrm{Sym}^{\star}(\mathbb{S})^G \\
\Big\downarrow & & \Big\downarrow \\
\mathrm{Sym}^{\star}(\mathbb{V}_s) & \longrightarrow\!\!\!\!\!\rightarrow & \mathrm{Sym}^{\star}(k^{r_1} \otimes_k V_1 \oplus \cdots \oplus k^{r_t} \otimes_k V_t)^G.
\end{array}$$

This gives rise to the commutative diagram

$$\begin{array}{ccc} \bar{\mathbb{H}} & \overset{\bar{\iota}_s}{\hookrightarrow} & \mathbb{P}(\mathbb{V}_s) \\ \downarrow & & \downarrow \\ \widetilde{\bar{\mathbb{H}}} & \overset{\widetilde{\bar{\iota}}_s}{\hookrightarrow} & \mathbb{P}(\widetilde{\mathbb{V}}_s). \end{array} \tag{9}$$

For the line bundle

$$\mathscr{O}_{\widetilde{\bar{\mathbb{H}}}}(s!) := \widetilde{\bar{\iota}}_s^{\star} \mathscr{O}_{\mathbb{P}(\widetilde{\mathbb{V}}_s)}(1),$$

(7) and Diagram (9) imply

$$\mathscr{O}_{\widetilde{\bar{\mathbb{H}}}}((s+1)!) = \mathscr{O}_{\widetilde{\bar{\mathbb{H}}}}(s!)^{\otimes(s+1)}. \tag{10}$$

Lemma 2.3.7. *Let s be a positive integer, such that* c) *and* d) *as above are satisfied, and let $\underline{f} \in \bigoplus_{i=1}^{t} \operatorname{Hom}(k^{r_i}, V_i^{\vee})$ be a $(T' \times G)$-semistable point. Set $h := \widetilde{\iota}_s([\underline{f}])$ and $\bar{h} := \widetilde{\bar{\iota}}_s([\underline{f}])$. Then, for any one parameter subgroup $\lambda\colon \mathbb{G}_m(k) \longrightarrow S$, we have*

$$\mu_{\widetilde{\kappa}_s}(\lambda, h) > (= / <)\, 0 \quad \Longleftrightarrow \quad \mu_{\widetilde{\sigma}_s}(\lambda, \bar{h}) > (= / <)\, 0.$$

Here, $\widetilde{\sigma}_s$ is the linearization of the S-action on $\widetilde{\bar{\mathbb{H}}}$ in $\mathscr{O}_{\widetilde{\bar{\mathbb{H}}}}(s!)$. In particular, $\bar{h}$ is S-semistable, if and only if $\underline{f} \in \mathsf{X}_{i=1}^{t} \operatorname{Isom}(k^{r_i}, V_i^{\vee})$.

Proof. The proof runs as the one of Lemma 2.3.6. □

Computation of some weights. The representation $\widehat{\mathbb{V}}_s$ of $\mathrm{GL}_r(k) = \mathrm{GL}(W)$ is polynomial. Hence, according to Corollary 2.2.3, there is a surjective homomorphism of $\mathrm{GL}(r)$-modules

$$\pi : W_{a,b} = (W^{\otimes a})^{\oplus b} \longrightarrow \widehat{\mathbb{V}}_s, \quad a := s!. \tag{11}$$

Suppose we have a decomposition $W = W_1 \oplus \cdots \oplus W_t$ with $\dim_k(W_i) = r_i$, $i = 1, \ldots, t$. This yields the embedding

$$\begin{aligned} \mathbb{G}_m(k)^{\times(t-1)} &\longrightarrow \mathrm{GL}(W_1) \times \cdots \times \mathrm{GL}(W_t) \\ (z_1, \ldots, z_{t-1}) &\longmapsto (z_1 \cdot \mathrm{id}_{W_1}, \ldots, z_{t-1} \cdot \mathrm{id}_{W_{t-1}}, (z_1 \cdot \cdots \cdot z_{t-1})^{-1} \cdot \mathrm{id}_{W_t}). \end{aligned}$$

Let T' be the image of this homomorphism. We want to study the quotient $\mathbb{P}(W_{a,b})/\!\!/T'$. For a suitable positive integer d, we have a closed embedding

$$\eta_d : \mathbb{P}(W_{a,b})/\!\!/T' \hookrightarrow \mathbb{P}(\mathrm{Sym}^d(W_{a,b})^{T'}).$$

Lemma 2.3.8. *For every $d > 0$, the vector space $\mathrm{Sym}^d(W_{a,b})^{T'}$ is a quotient of*

$$\widetilde{W}^d_{a,b} := ((W_1 \otimes \cdots \otimes W_t)^{\otimes a(d)})^{\oplus b(d)}$$

where $a(d) = ad/t$.

Proof. Write

$$W_{a,b} = \bigoplus_{\substack{\underline{a}=(a_1,\ldots,a_t):a_i\geq 0,\ \sum a_i = a \\ \beta\in\{1,\ldots,b\}}} W_{\underline{a},\beta}$$

where $W_{\underline{a}} := W_1^{\otimes a_1} \otimes \cdots \otimes W_t^{\otimes a_t}$, and $W_{\underline{a},\beta}$ is $W_{\underline{a}}$ viewed as a subspace of the β-th copy of $W^{\otimes a}$ in $W_{a,b}$. Fix a bijection

$$\kappa\colon \{1,\ldots,u\} \longrightarrow \{\underline{a} = (a_1,\ldots,a_t) | a_i \geq 0,\ \sum a_i = a\} \times \{1,\ldots,b\}$$

and let

$$\bar{\kappa}\colon \{1,\ldots,u\} \longrightarrow \{\underline{a} = (a_1,\ldots,a_t) | a_i \geq 0,\ \sum a_i = a\}$$

be the composition of κ with the projection onto the first factor. Then,

$$\mathrm{Sym}^d(W_{a,b})^{T'} = \bigoplus \mathrm{Sym}^{d_1}(W_{\kappa(1)}) \otimes \cdots \otimes \mathrm{Sym}^{d_u}(W_{\kappa(u)}).$$

This sum runs over all $(d_1,\ldots,d_u)$ that are subject to the constraints

$$d_1 + \cdots + d_u = d \quad \text{and} \quad a(d) := \sum_{i=1}^{u} d_i \cdot a_j^i = \sum_{i=1}^{u} d_i \cdot a_k^i \quad \forall j,k \in \{1,\ldots,t\}.$$

Here, we have written $\bar{\kappa}(i) = (a_1^i,\ldots,a_t^i)$, $i = 1,\ldots,u$. Obviously, $a(d) = ad/t$, and $\mathrm{Sym}^{d_1}(W_{\kappa(1)}) \otimes \cdots \otimes \mathrm{Sym}^{d_u}(W_{\kappa(u)})$ is a quotient of $(W_1 \otimes \cdots \otimes W_t)^{\otimes a(d)}$. Thus, $\mathrm{Sym}^d(W_{a,b})^{T'}$ is a quotient of $((W_1 \otimes \cdots \otimes W_t)^{\otimes a(d)})^{\oplus b(d)}$ for a suitable positive integer $b(d)$. □

Note that we have the commutative diagram of S-equivariant morphisms

$$\begin{array}{ccccc} \mathbb{P}(\mathbb{V}_s) & \hookrightarrow & \mathbb{P}(\widehat{\mathbb{V}}_s) & \hookrightarrow & \mathbb{P}(W_{a,b}) \\ \downarrow & & & & \downarrow \\ \mathbb{P}(\widetilde{\mathbb{V}}_s) & & \xrightarrow{\quad\theta_d\quad} & & \mathbb{P}(\widetilde{W}^d_{a,b}). \end{array} \tag{12}$$

Denote by $\widetilde{\sigma}^d_{a,b}$ the canonical linearization of the S-action on $\mathbb{P}(\widetilde{W}^d_{a,b})$ in $\mathcal{O}_{\mathbb{P}(\widetilde{W}^d_{a,b})}(1)$. For any point $[l] \in \mathbb{P}(\widetilde{\mathbb{V}}_s)$, we find

$$\mu_{\widetilde{\sigma}_s}(\lambda,[l]) = \frac{1}{d} \cdot \mu_{\widetilde{\sigma}^d_{a,b}}(\lambda, \vartheta_d[l]) \stackrel{(5)}{=} \mu_{\widetilde{\kappa}^d_{a,b}}(\lambda,[\widetilde{l}]), \quad [\widetilde{l}] = \theta_d[l]. \tag{13}$$

Here, $\widetilde{\kappa}^d_{a,b}$ is the representation of S on $\widetilde{W}^d_{a,b}$.

Let us write, for non-negative integers a and b, $\widetilde{\kappa}_{a,b}$ for the representation of $\mathrm{GL}_{r_1}(k) \times \cdots \times \mathrm{GL}_{r_t}(k) \cong \mathrm{GL}(W_1) \times \cdots \times \mathrm{GL}(W_t)$ on

$$\widetilde{W}_{a,b} := ((W_1 \otimes \cdots \otimes W_t)^{\otimes a})^{\oplus b}.$$

For a one parameter subgroup $\lambda\colon \mathbb{G}_m(k) \longrightarrow S$ and a map

$$\widetilde{l}\colon ((W_1 \otimes \cdots \otimes W_t)^{\otimes a})^{\oplus b} \longrightarrow k,$$

we have the formula

$$\mu_{\widetilde{\kappa}_{a,b}}(\lambda, [\widetilde{l}]) = -\min\{\gamma^1_{j^1_1} + \cdots + \gamma^1_{j^1_a} + \cdots + \gamma^t_{j^t_1} + \cdots + \gamma^t_{j^t_a} \mid (j^i_1, \ldots, j^i_a) \in \{1, \ldots, s_i + 1\}^{\times a}, i = 1, \ldots, t : \widetilde{l}_{|(W_{1,j^1_1} \otimes \cdots \otimes W_{1,j^1_a} \otimes \cdots \otimes W_{t,j^t_1} \otimes \cdots \otimes W_{t,j^t_a})^{\oplus b}} \not\equiv 0\}. \tag{14}$$

Here, $(\gamma^i_1, \ldots, \gamma^i_{s_i+1})$ and $W_{i\bullet}(\lambda_i) : \{0\} \subsetneq W_{i,1} \subsetneq \cdots \subsetneq W_{i,s_i} \subsetneq W_i$ are the data associated with the one parameter subgroup $\lambda_i : \mathbb{G}_m(k) \longrightarrow \mathrm{SL}_{r_i}(k)$, $i = 1, \ldots, t$, $\lambda = (\lambda_1, \ldots, \lambda_t)$, and the standard action of S on $W = k^r = W_1 \oplus \cdots \oplus W_t$.

We will also need another formula for determining the μ-function. For this, let $\lambda = (\lambda_1, \ldots, \lambda_t)$ be a one parameter subgroup of S. Choose bases $\underline{w}^i = (w^i_1, \ldots, w^i_{r_i})$ for W_i which consist of eigenvectors for λ_i, $i = 1, \ldots, t$. For $\underline{e} = (e^1, \ldots, e^t) \in I := \{1, \ldots, r_1\} \times \cdots \times \{1, \ldots, r_t\}$, we set

$$w_{\underline{e}} := w^1_{e^1} \otimes \cdots \otimes w^t_{e^t} \in W_1 \otimes \cdots \otimes W_t,$$

for $\underline{\iota} = (\underline{e}_1, \ldots, \underline{e}_a) \in I^{\times a}$, we define

$$w_{\underline{\iota}} := w_{\underline{e}_1} \otimes \cdots \otimes w_{\underline{e}_a} \in (W_1 \otimes \cdots \otimes W_t)^{\otimes a},$$

and, finally, for $k \in \{1, \ldots, b\}$ and $\underline{\iota} \in I^{\times a}$,

$$w^k_{\underline{\iota}} = (0, \ldots, 0, w_{\underline{\iota}}, 0, \ldots, 0) \in ((W_1 \otimes \cdots \otimes W_t)^{\otimes a})^{\oplus b},$$

$w_{\underline{\iota}}$ occupying the k-th slot. Then, as k varies over $\{1, \ldots, b\}$ and $\underline{\iota}$ over $I^{\times a}$, the $w^k_{\underline{\iota}}$ form a basis for $((W_1 \otimes \cdots \otimes W_t)^{\otimes a})^{\oplus b}$. Let $w^{k\vee}_{\underline{\iota}}$ be the dual basis for $((W_1 \otimes \cdots \otimes W_t)^{\otimes a})^{\oplus b\vee}$. Suppose $[\widetilde{l}] \in \mathbb{P}(\widetilde{W}_{a,b})$, $\widetilde{l} = \sum a^k_{\underline{\iota}} w^{k\vee}_{\underline{\iota}}$. Then, there exist $k_0 \in \{1, \ldots, b\}$ and $\underline{\iota}_0 \in I^{\times a}$ with

$$\mu_{\widetilde{\kappa}_{a,b}}(\lambda, [\widetilde{l}]) = \mu_{\widetilde{\kappa}_{a,b}}(\lambda, [w^{k_0\vee}_{\underline{\iota}_0}]).$$

Finally, if $\lambda = (\lambda_1, \ldots, \lambda_t)$ with $\lambda_i = \lambda(\underline{w}^i, \gamma^{(j_i)}_{r_i})$, $i = 1, \ldots, t$, (see [27], Section 1.3), then

$$\mu_{\widetilde{\kappa}_{a,b}}(\lambda, [w^{k_0\vee}_{\underline{\iota}_0}]) = \sum_{i=1}^{t} (\nu_i r_i - a j_i), \tag{15}$$

with

$$\nu_i = \#\{e^i_f \leq j_i \mid \underline{\iota}_0 = (\underline{e}_1, \ldots, \underline{e}_a),\ \underline{e}_f = (e^1_f, \ldots, e^t_f),\ f = 1, \ldots, a\},$$
$$i = 1, \ldots, t.$$

2.4 *G-linearized sheaves*

Assume that $\rho : G \longrightarrow \mathrm{GL}(V)$ is any representation. Let B be a scheme and $\mathscr{A}$ a coherent $\mathscr{O}_B$-module. Equip B with the trivial G-action. We obtain the G-linearized sheaf $\mathscr{A} \otimes V$. It follows easily from the universal property of the symmetric algebra ([10], Section (9.4.1)) that $\mathscr{S}ym^\star(\mathscr{A} \otimes V)$ inherits a G-linearization. Note that the algebra $\mathscr{S}ym^\star(\mathscr{A} \otimes V)$ is naturally graded and that the G-linearization preserves this grading. Let $\mathscr{S}ym^\star(\mathscr{A} \otimes V)^G$ be the sub-algebra of G-invariant elements in $\mathscr{S}ym^\star(\mathscr{A} \otimes V)$. The G-linearization provides a π-invariant action of G on

$$\mathscr{H}om(\mathscr{A}, V^\vee \otimes \mathscr{O}_B) := \mathscr{S}pec(\mathscr{S}ym^\star(\mathscr{A} \otimes V)),$$

$\pi\colon \mathscr{S}ym^\star(\mathscr{A} \otimes V) \longrightarrow B$ being the natural projection. Then, the categorical quotient of the scheme $\mathscr{S}pec(\mathscr{S}ym^\star(\mathscr{A} \otimes V))$ by the G-action is given through

$$\mathscr{S}pec(\mathscr{S}ym^\star(\mathscr{A} \otimes V))/\!\!/G = \mathscr{S}pec(\mathscr{S}ym^\star(\mathscr{A} \otimes V)^G) \xrightarrow{\bar{\pi}} B.$$

In characteristic zero, the construction commutes with base change. In positive characteristic, we have to be more careful. Let $f\colon A \longrightarrow B$ be a morphism of schemes. The natural isomorphism

$$f^\star(\mathscr{S}ym^\star(\mathscr{A} \otimes V)) \longrightarrow \mathscr{S}ym^\star(f^\star(\mathscr{A}) \otimes V)$$

of G-linearized sheaves gives rise to the homomorphism

$$\mathrm{bc}(f)\colon f^\star(\mathscr{S}ym^\star(\mathscr{A} \otimes V)^G) \longrightarrow \mathscr{S}ym^\star(f^\star(\mathscr{A}) \otimes V)^G.$$

Now, assume that we are in the setting of Section 1.1. For every i, we have the decomposition

$$\mathscr{S}ym^i(\mathscr{A} \otimes V) = \bigoplus_{i_1+\cdots+i_t=i} (\mathscr{S}ym^{i_1}(\mathscr{A}_1 \otimes V_1) \otimes \cdots \otimes \mathscr{S}ym^{i_t}(\mathscr{A}_t \otimes V_t)) \quad (16)$$

which is preserved by the base change isomorphism and the G-linearization. Thus, there is also the base change homomorphism

$$\mathrm{bc}(f)\colon f^\star(\mathscr{S}ym^\star(\mathscr{A}_1 \otimes V_1 \oplus \cdots \oplus \mathscr{A}_t \otimes V_t)^G) \longrightarrow$$
$$\mathscr{S}ym^\star(f^\star(\mathscr{A}_1) \otimes V_1 \oplus \cdots \oplus f^\star(\mathscr{A}_t) \otimes V_t)^G.$$

Lemma 2.4.1. *If $\mathscr{A}_i$ is locally free, $i = 1, \ldots, t$, then* $\mathrm{bc}(f)$ *is an isomorphism.*

Proof. This is an immediate consequence of Lemma 2.3.1 in [6]. □

Likewise, Lemma 2.3.2 in [6] implies:

Lemma 2.4.2. *Let $\psi_i\colon \mathscr{A}_i' \longrightarrow \mathscr{A}_i$, $i = 1, \ldots, t$, be surjective homomorphisms of locally free $\mathscr{O}_B$-modules. Then, the induced homomorphism*

$$\mathscr{S}ym^\star(\mathscr{A}_1' \otimes V_1 \oplus \cdots \oplus \mathscr{A}_t' \otimes V_t)^G \longrightarrow \mathscr{S}ym^\star(\mathscr{A}_1 \otimes V_1 \oplus \cdots \oplus \mathscr{A}_t \otimes V_t)^G$$

of $\mathscr{O}_B$-algebras is surjective, too.

2.5 *Semistable rational principal G-bundles*

We now review the formalism introduced by Ramanathan and compare it with our setup. A *rational principal G-bundle on X* is a pair $(U, \mathscr{P})$ which consists of a big open subset $U \subseteq X$ and a principal G-bundle $\mathscr{P}$ on U. Such a rational principal G-bundle is said to be *(semi)stable*, if for every open subset $U' \subseteq U$ which is big in X, every parabolic subgroup P of G, every reduction $\beta\colon U' \longrightarrow \mathscr{P}_{|U'}/P$ of the structure group of $\mathscr{P}$ to P over U', and every anti-dominant character χ on P, we have

$$\deg(\mathscr{L}(\beta, \chi))(\geq)0.$$

Note that the anti-dominant character χ and the P-bundle $\mathscr{P}_{|U'} \longrightarrow \mathscr{P}_{|U'}/P$ define a line bundle on $\mathscr{P}_{|U'}/P$. Its pullback to U' via β is the line bundle $\mathscr{L}(\beta, \chi)$. Since U' is big in X, it makes sense to speak about the degree of $\mathscr{L}(\beta, \chi)$.

We fix a pair (B, T) which consists of a Borel subgroup $B \subset G$ and a maximal torus $T \subset B$. If P and P' are conjugate in G, a reduction β of a principal G-bundle to P may equally be interpreted as a reduction to P'. Thus, it suffices to consider parabolic subgroups of the type $P_G(\lambda)$, $\lambda\colon \mathbb{G}_m(k) \longrightarrow T$ a one parameter subgroup, which contain B. Here, we use the convention

$$P_G(\lambda) := Q_G(-\lambda) = \{\, g \in G \mid \lim_{z \to 0} \lambda(z) \cdot g \cdot \lambda(z)^{-1} \text{ exists in } G\}.$$

Let $X_\star(T)$ and $X^\star(T)$ be the free $\mathbb{Z}$-modules of one parameter subgroups and characters of T, respectively. We have the canonical pairing $\langle \cdot, \cdot \rangle\colon X_\star(T) \times X^\star(T) \longrightarrow \mathbb{Z}$. Set $X_{\star,\mathbb{K}}(T) := X_\star(T) \otimes_{\mathbb{Z}} \mathbb{K}$ and $X^\star_{\mathbb{K}}(T) := X^\star(T) \otimes_{\mathbb{Z}} \mathbb{K}$, and let $\langle \cdot, \cdot \rangle_{\mathbb{K}}\colon X_{\star,\mathbb{K}}(T) \times X^\star_{\mathbb{K}}(T) \longrightarrow \mathbb{K}$ be the $\mathbb{K}$-bilinear extension of $\langle \cdot, \cdot \rangle$, $\mathbb{K} = \mathbb{Q}, \mathbb{R}$. Let T_1 be the radical of G, and T_2 a maximal torus of the derived group $\mathscr{D}(G)$, such that $T_1 \times T_2$ surjects onto T. This surjection induces isomorphisms

$$X_{\star,\mathbb{K}}(T_1) \oplus X_{\star,\mathbb{K}}(T_2) \cong X_{\star,\mathbb{K}}(T)$$

$$\text{and} \quad X^\star_{\mathbb{K}}(T) \cong X^\star_{\mathbb{K}}(T_1) \oplus X^\star_{\mathbb{K}}(T_2), \quad \mathbb{K} = \mathbb{Q}, \mathbb{R}.$$

These isomorphisms are compatible with the products $\langle \cdot, \cdot \rangle_{\mathbb{K}}$ and $\langle \cdot, \cdot \rangle^{\sim}_{\mathbb{K}}\colon X_{\star,\mathbb{K}}(T_1 \times T_2) \times X^\star_{\mathbb{K}}(T_1 \times T_2) \longrightarrow \mathbb{K}$, the latter one being the direct sum of the products of T_1 and T_2. We see that $\langle \cdot, \cdot \rangle_{\mathbb{K}}$ is trivial on $X_{\star,\mathbb{K}}(T_1) \times X^\star_{\mathbb{K}}(T_2)$ and $X_{\star,\mathbb{K}}(T_2) \times X^\star_{\mathbb{K}}(T_1)$, $\mathbb{K} = \mathbb{Q}, \mathbb{R}$. Finally, suppose $(\cdot, \cdot)^\star\colon X^\star_{\mathbb{R}}(T) \times X^\star_{\mathbb{R}}(T) \longrightarrow \mathbb{R}$ is a scalar product which is invariant under the Weyl group $W(T) = \mathscr{N}(T)/T$. This also yields the product $(\cdot, \cdot)_\star\colon X_{\star,\mathbb{R}}(T) \times X_{\star,\mathbb{R}}(T) \longrightarrow \mathbb{R}$. We assume that $(\cdot, \cdot)^\star$ is defined over $\mathbb{Q}$ and that $X^\star_{\mathbb{R}}(T_1)$ and $X^\star_{\mathbb{R}}(T_2)$ are orthogonal with respect to $(\cdot, \cdot)^\star$. Then, we get $(\cdot, \cdot)^\star_2\colon X^\star_{\mathbb{R}}(T_2) \times X^\star_{\mathbb{R}}(T_2) \longrightarrow \mathbb{R}$.

The datum (B, T) defines the set of positive roots R and the set $R^\vee$ of coroots (see [34]). Let $\mathscr{C} \subset X_{\star,\mathbb{R}}(T_2)$ be the cone spanned by the elements of $R^\vee$ and $\mathscr{D} \subset X^\star_{\mathbb{R}}(T_2)$ the dual cone of $\mathscr{C}$ with respect to $\langle \cdot, \cdot \rangle^2_{\mathbb{R}}$, the restriction of $\langle \cdot, \cdot \rangle_{\mathbb{R}}$ to $X_{\star,\mathbb{R}}(T_2) \times X^\star_{\mathbb{R}}(T_2)$. Equivalently, the cone $\mathscr{D}$ may be characterized as being the dual cone of the cone spanned by the elements in R with respect to $(\cdot, \cdot)^\star_2$. Indeed, one has

$\langle \cdot, \alpha^\vee \rangle_{\mathbb{R}} = 2(\cdot, \alpha)^\star/(\alpha, \alpha)^\star$, $\alpha \in R$. Now, a character $\chi \in X^\star(T)$ is called *dominant*, if it lies in $\mathscr{D}$, and *anti-dominant*, if $-\chi$ lies in $\mathscr{D}$. A character χ of a parabolic subgroup containing B is called *(anti-)dominant*, if its restriction to T is (anti-)dominant.

Remark 2.5.1. Note that a dominant character is, by definition, trivial on the center of G.

In the definition of semistability, we may clearly assume that $(\chi, \alpha)^\star > 0$ for every $\alpha \in R$ with $\langle \lambda, \alpha \rangle > 0$, if $P = P_G(\lambda)$. Otherwise, we may choose λ', such that $(\chi, \alpha)^\star > 0$, if and only if $\langle \lambda', \alpha \rangle > 0$. Then, $P_G(\lambda')$ is a parabolic subgroup which contains $P_G(\lambda)$, χ is induced by a character χ' on $P_G(\lambda')$, the reduction β defines the reduction $\beta'\colon \mathscr{P}_{|U'}/P_G(\lambda) \longrightarrow \mathscr{P}_{|U'}/P_G(\lambda')$, and $\mathscr{L}(\beta, \chi) = \mathscr{L}(\beta', \chi')$. Every one parameter subgroup λ of T defines a character χ_λ of T, such that $\langle \lambda, \chi \rangle = (\chi_\lambda, \chi)^\star$ for all $\chi \in X^\star(T)$. If λ lands in T_2, then $\chi_\lambda \in X^\star_{\mathbb{Q}}(T_2)$. Finally, observe that the cone $\mathscr{C}'$ of one parameter subgroups λ of T_2, such that $B \subseteq P_G(\lambda)$, is dual to the cone spanned by the roots. Thus,

$$\forall \lambda \in X_\star(T_2): \quad B \subseteq P_G(\lambda) \quad \Longleftrightarrow \quad \chi_\lambda \in \mathscr{D}.$$

If one of those conditions is verified, then $(P_G(\lambda), \chi_\lambda)$ consists of a parabolic subgroup containing B and a dominant character on it. Similarly, if $Q_G(\lambda)$ contains B, then χ_λ is an anti-dominant character on $Q_G(\lambda)$. Our discussion shows:

Lemma 2.5.2. *A rational principal G-bundle $(U, \mathscr{P})$ is (semi)stable, if and only if for every open subset $U' \subseteq U$ which is big in X, every one parameter subgroup $\lambda \in \mathscr{C}' \cap X_{\star,\mathbb{Q}}(T_2)$, and every reduction $\beta\colon U' \longrightarrow \mathscr{P}_{|U'}/Q_G(\lambda)$ of the structure group of $\mathscr{P}$ to $Q_G(\lambda)$ over U', we have*

$$\deg(\mathscr{L}(\beta, \chi_\lambda))(\geq)0.$$

For any one parameter subgroup λ of G, we may find a pair (B', T') consisting of a Borel subgroup B' of G and a maximal torus $T' \subset B'$, such that $\lambda \in X_\star(T')$ and $B' \subseteq Q_G(\lambda)$. Then, there exists an element $g \in G$, such that $(g \cdot B \cdot g^{-1}, g \cdot T \cdot g^{-1}) = (B', T')$, and we obtain

$$\begin{aligned}(\cdot, \cdot)^{\star\prime}\colon X^\star_{\mathbb{R}}(T') \times X^\star_{\mathbb{R}}(T') &\longrightarrow \mathbb{R} \\ (\chi, \chi') &\longmapsto (\chi(g^{-1} \cdot\cdot\cdot g), \chi'(g^{-1} \cdot\cdot\cdot g))^\star.\end{aligned}$$

Since $(\cdot, \cdot)^\star$ is invariant under $W(T)$, the product $(\cdot, \cdot)^{\star\prime}$ does not depend on g. In particular, we obtain a character χ_λ on $Q_G(\lambda)$. This character does not depend on (B', T') (see [28], (2.31)). We conclude:

Lemma 2.5.3. *A rational principal G-bundle $(U, \mathscr{P})$ is (semi)stable, if and only if for every open subset $U' \subseteq U$ which is big in X, every one parameter subgroup $\lambda\colon \mathbb{G}_m(k) \longrightarrow [G, G]$, and every reduction $\beta\colon U' \longrightarrow \mathscr{P}_{|U'}/Q_G(\lambda)$ of the structure group of $\mathscr{P}$ to $Q_G(\lambda)$ over U', we have*

$$\deg(\mathscr{L}(\beta, \chi_\lambda))(\geq)0.$$

Suppose that $\rho\colon G \longrightarrow \mathrm{GL}(V)$ is a faithful representation with $\rho(G) \subseteq \widetilde{G} \cap \mathrm{SL}(V)$, $\widetilde{G} := \mathrm{GL}(V_1) \times \cdots \times \mathrm{GL}(V_t)$. We may assume that T maps to the maximal torus $\widetilde{T} \subset \widetilde{G}$, consisting of the diagonal matrices. The character group $X^\star(\widetilde{T})$ is freely generated by the characters $e_i : \mathrm{diag}(\lambda_1, \ldots, \lambda_n) \longmapsto \lambda_i$, $i = 1, \ldots, n$. We define

$$(\cdot,\cdot)^\star_{\widetilde{T}} \colon X^\star_{\mathbb{R}}(\widetilde{T}) \times X^\star_{\mathbb{R}}(\widetilde{T}) \longrightarrow \mathbb{R}$$

$$\left(\sum_{i=1}^n x_i \cdot e_i, \sum_{i=1}^n y_i \cdot e_i\right) \longmapsto \sum_{i=1}^n x_i y_i.$$

The scalar product $(\cdot,\cdot)^\star_{\widetilde{T}}$ is clearly defined over $\mathbb{Q}$ and invariant under the Weyl group $W(\widetilde{T})$. We have $\mathscr{D}(\widetilde{G}) = \mathrm{SL}(V_1)\times\cdots\times\mathrm{SL}(V_t)$ and $\widetilde{T}_1 := \mathscr{R}(\widetilde{G}) = \{ (z_1\cdot \mathrm{id}_{V_1}, \ldots, z_t\cdot \mathrm{id}_{V_t}) \mid z_i \in \mathbb{G}_m(k),\ i = 1, \ldots, t\}$, and one easily checks that $X^\star_{\mathbb{R}}(\widetilde{T}_1) \perp X^\star_{\mathbb{R}}(\widetilde{T}_2)$ with respect to $(\cdot,\cdot)^\star_{\widetilde{T}}$, $\widetilde{T}_2 := \widetilde{T} \cap (\mathrm{SL}(V_1) \times \cdots \times \mathrm{SL}(V_t))$. The product $(\cdot,\cdot)^\star_{\widetilde{T}}$ therefore restricts to a scalar product $(\cdot,\cdot)^\star$ on $X^\star_{\mathbb{R}}(T)$ with the properties we asked for. We find a nice formula for $\deg(\mathscr{L}(\beta, \chi_\lambda))$: Indeed, if $(U, \mathscr{P})$ is a rational principal G-bundle, and if $\underline{\mathscr{E}} = (\mathscr{E}_1, \ldots, \mathscr{E}_t)$ is the tuple of vector bundles on U associated with $\mathscr{P}$ by means of ρ, then we have, for every one parameter subgroup $\lambda\colon \mathbb{G}_m(k) \longrightarrow [G, G]$, $\lambda = (\lambda_1, \ldots, \lambda_t)$ as one parameter subgroup of $\mathrm{SL}(V_1) \times \cdots \times \mathrm{SL}(V_t)$,

$$\iota\colon \mathscr{P}/Q_G(\lambda) \hookrightarrow \left(\mathop{\times}_{i=1}^{t} \mathscr{I}som(V_i, \mathscr{E}_i)\right) / Q_{\widetilde{G}}(\lambda).$$

As usual, we obtain a weighted filtration $(\underline{V}_\bullet(\lambda), \underline{\alpha}_\bullet(\lambda))$ of $\underline{V} = (V_1, \ldots, V_t)$, that is, a weighted filtration $((V_{i\bullet}(\lambda_i), \alpha_{i\bullet}(\lambda_i))$ of V_i for each $i = 1, \ldots, t$, and, for every reduction $\beta\colon U' \longrightarrow \mathscr{P}_{|U'}/Q_G(\lambda)$ over a big open subset $U' \subseteq U$, the reduction $\iota \circ \beta$ corresponds to filtrations

$$\mathscr{E}_{i\bullet}(\beta) : \{0\} \subsetneq \mathscr{E}_{i,1} \subsetneq \cdots \subsetneq \mathscr{E}_{i,s_i} \subsetneq \mathscr{E}_{i|U'}, \quad i = 1, \ldots, t,$$

by subbundles with $\mathrm{rk}(\mathscr{E}_{i,j}) = \dim_k(V_{i,j})$, $j = 1, \ldots, s_i$, $i = 1, \ldots, t$. With the weighted filtration $(\underline{\mathscr{E}}_\bullet(\beta), \underline{\alpha}_\bullet(\lambda))$, of $\underline{\mathscr{E}} = (\mathscr{E}_{1|U'}, \ldots, \mathscr{E}_{t|U'})$, we find

$$\begin{aligned}\deg(\mathscr{L}(\beta, \chi_\lambda)) &= L(\underline{\mathscr{E}}_\bullet(\beta), \underline{\alpha}_\bullet(\lambda)) \\ &= \sum_{i=1}^t \sum_{j=1}^{s_i} \alpha_{i,j}(\mathrm{rk}(\mathscr{E}_{i,j}) \cdot \deg(\mathscr{E}_i) - \mathrm{rk}(\mathscr{E}_i) \cdot \deg(\mathscr{E}_{i,j})).\end{aligned}$$

This equality is proved along the lines of Example 2.15 in [29]. Thus, we conclude:

Lemma 2.5.4. *A rational principal G-bundle $(U, \mathscr{P})$ is (semi)stable, if and only if, for every open subset $U' \subseteq U$ which is big in X, every one parameter subgroup $\lambda\colon \mathbb{G}_m(k) \longrightarrow [G, G]$, and every reduction $\beta\colon U' \longrightarrow \mathscr{P}_{|U'}/Q_G(\lambda)$ of the structure group of $\mathscr{P}$ to $Q_G(\lambda)$ over U', we have*

$$L(\underline{\mathscr{E}}_\bullet(\beta), \underline{\alpha}_\bullet(\lambda))(\geq)0.$$

Remark 2.5.5. If $(U, \mathscr{P})$ is given as a singular principal G-bundle $(\underline{\mathscr{A}}, \tau)$, then, in the notation of the introduction, we have

$$\mathscr{A}_{i|U} = \mathscr{E}_i^\vee,\ \mathscr{A}_{i,j|U} = \ker(\mathscr{E}_i^\vee \longrightarrow \mathscr{E}_{i,s_i+1-j}^\vee) \quad \text{and}$$

$$\alpha_{i,j;\beta} = \alpha_{i,s_i+1-j}, \quad j = 1, \dots, s_i,\ i = 1, \dots, t.$$

Then, one readily verifies

$$L(\underline{\mathscr{A}}, \tau; \beta) = L(\underline{\mathscr{E}}_\bullet(\beta), \underline{\alpha}_\bullet(\beta)).$$

This proves that the definition of slope semistability (1) given in the introduction is the original definition of Ramanathan. We have arrived at our notion of semistability, by replacing degrees with Hilbert polynomials. Thus, our semistability concept is the "Gieseker version" of Ramanathan semistability.

3. The theory of decorated tuples of sheaves

A theory of decorated tuples of sheaves has already been considered in [30]. (The assumption on the characteristic of the ground field is now obsolete, by [18] and [19].) However, the scope in [30] is very general and makes some restrictions on the stability parameters necessary. These restrictions have to be overcome in our situation in order to find the appropriate semistability concept for singular principal bundles. In this chapter, we will therefore present the necessary material in a self-contained manner. In particular, the results of [7] will be covered.

3.1 *Basic notions*

Suppose we are given a decomposition $W := k^r = W_1 \oplus \cdots \oplus W_t$ with $\dim_k(W_i) =: r_i$, $i = 1, \dots, t$. We fix integers $a > 0$, $b > 0$, and c and a line bundle $\mathscr{L}$ on X. A *tuple of sheaves with a decoration of type* $(\mathrm{I} : a, b, c; \mathscr{L})$ is a tuple $(\underline{\mathscr{E}}, \phi)$ which consists of a tuple $\underline{\mathscr{E}} = (\mathscr{E}_1, \dots, \mathscr{E}_t)$ of torsion free sheaves $\mathscr{E}_i$ of rank r_i on X, $i = 1, \dots, t$, and a non-trivial homomorphism

$$\phi\colon (\mathscr{E}_1^{\otimes a} \otimes \cdots \otimes \mathscr{E}_t^{\otimes a})^{\oplus b} \longrightarrow \det(\mathscr{E})^{\otimes c} \otimes \mathscr{L}, \quad \mathscr{E} := \mathscr{E}_1 \oplus \cdots \oplus \mathscr{E}_t.$$

A *homomorphism* between two tuples of sheaves $(\underline{\mathscr{E}}^1, \phi_1)$ and $(\underline{\mathscr{E}}^2, \phi_2)$ with a decoration of type $(\mathrm{I} : a, b, c; \mathscr{L})$ is a tuple $(\underline{\psi}, z)$ where $\psi_i\colon \mathscr{E}_i^1 \longrightarrow \mathscr{E}_i^2$ are homomorphisms for all i, $z \in k$, and

$$(\psi_1^{\otimes a} \otimes \cdots \otimes \psi_t^{\otimes a})^{\oplus b} \circ \phi_2 = (\det(\psi)^{\otimes c} \otimes (z \cdot \mathrm{id}_{\mathscr{L}})) \circ \phi_1,$$

$$\psi := \psi_1 \oplus \cdots \oplus \psi_t\colon \mathscr{E}^1 \longrightarrow \mathscr{E}^2.$$

A *weighted filtration* $(\underline{\mathscr{E}}_\bullet, \underline{\alpha}_\bullet)$ *of the tuple* $\underline{\mathscr{E}}$ consists of filtrations

$$\{0\} \subsetneq \mathscr{E}_{i,1} \subsetneq \cdots \subsetneq \mathscr{E}_{i,s_i} \subsetneq \mathscr{E}_{i,s_i+1} = \mathscr{E}_i \tag{17}$$

of $\mathscr{E}_i$ by saturated subsheaves and of positive rational numbers $\alpha_{i,j}$, $i = 1, \ldots, t$, $j = 1, \ldots, s_i$. The *trivial weighted filtration* occurs for $s_i = 0$. The *standard weight vectors* are

$$\gamma_{r_i}^{(j)} := (\underbrace{j - r_i, \ldots, j - r_i}_{j\times}, \underbrace{j, \ldots, j}_{(r_i - j)\times}), \quad j = 1, \ldots, r_i - 1. \tag{18}$$

Given a weighted filtration $(\underline{\mathscr{E}}_\bullet, \underline{\alpha}_\bullet)$ of the tuple $\underline{\mathscr{E}}$, we obtain the *associated weight vectors for each i*

$$(\underbrace{\gamma_1^i, \ldots, \gamma_1^i}_{r_{i,1}\times}, \underbrace{\gamma_2^i, \ldots, \gamma_2^i}_{(r_{i,2}-r_{i,1})\times}, \ldots, \underbrace{\gamma_{s_i+1}^i, \ldots, \gamma_{s_i+1}^i}_{(r_i - r_{i,s_i})\times}) := \sum_{j=1}^{s_i} \alpha_{i,j} \cdot \gamma_{r_i}^{(\operatorname{rk} \mathscr{E}_{i,j})}.$$

For $s_i = 0$, we get (0).

Suppose we are given a weighted filtration $(\underline{\mathscr{E}}_\bullet, \underline{\alpha}_\bullet)$ of $\underline{\mathscr{E}}$. We then set

$$M(\underline{\mathscr{E}}_\bullet, \underline{\alpha}_\bullet) := \sum_{i=1}^{t} \sum_{j=1}^{s_i} \alpha_{i,j} \cdot (\operatorname{rk}(\mathscr{E}_{i,j}) \cdot P(\mathscr{E}_i) - \operatorname{rk}(\mathscr{E}_i) \cdot P(\mathscr{E}_{i,j})),$$

$$L(\underline{\mathscr{E}}_\bullet, \underline{\alpha}_\bullet) := \sum_{i=1}^{t} \sum_{j=1}^{s_i} \alpha_{i,j} \cdot (\operatorname{rk}(\mathscr{E}_{i,j}) \cdot \deg(\mathscr{E}_i) - \operatorname{rk}(\mathscr{E}_i) \cdot \deg(\mathscr{E}_{i,j})).$$

(A trivial weighted filtration $(\{0\} \subsetneq \mathscr{E}_i, ())$ contributes zero to either sum.) Let $\mathscr{J}$ be the set of functions

$$J : [1, \ldots, t] \times [1, \ldots, a] \longrightarrow \mathbb{N}$$
$$(i, u) \longmapsto J(i, u) =: j_u^i \tag{19}$$

with $1 \leq j_u^i \leq s_i + 1$. For a tuple of sheaves $(\underline{\mathscr{E}}, \phi)$ with a decoration of type $(\mathrm{I}: a, b, c; \mathscr{L})$ and a filtration $(\underline{\mathscr{E}}_\bullet, \underline{\alpha}_\bullet)$, we also declare

$$\mu(\underline{\mathscr{E}}_\bullet, \underline{\alpha}_\bullet; \phi) := -\min_{J \in \mathscr{J}} \left\{ \sum_{i=1}^{t} (\gamma_{j_1^i}^i + \cdots + \gamma_{j_a^i}^i) | \phi_{|\left(\bigotimes_{i=1}^t \mathscr{E}_{i,j_1^i} \otimes \cdots \otimes \mathscr{E}_{i,j_a^i}\right)^{\oplus b}} \not\equiv 0 \right\}. \tag{20}$$

Fix a positive polynomial $\delta \in \mathbb{Q}[x]$ of degree at most $\dim(X) - 1$. Now, we say that a tuple of sheaves $(\underline{\mathscr{E}}, \phi)$ with a decoration of type $(\mathrm{I}: a, b, c; \mathscr{L})$ is *δ-(semi) stable*, if the inequality

$$M(\underline{\mathscr{E}}_\bullet, \underline{\alpha}_\bullet) + \delta \cdot \mu(\underline{\mathscr{E}}_\bullet, \underline{\alpha}_\bullet; \phi)(\succeq)0 \tag{21}$$

holds for all non-trivial filtrations $(\underline{\mathscr{E}}_\bullet, \underline{\alpha}_\bullet)$ of $\underline{\mathscr{E}}$. Here, a weighted filtration is *non-trivial*, if there exists an index $i_0 \in \{1, \ldots, t\}$ with $s_{i_0} > 0$. If $\bar{\delta}$ is a positive rational number, we call a tuple of sheaves $(\underline{\mathscr{E}}, \phi)$ with a decoration of type $(\mathrm{I}: a, b, c; \mathscr{L})$ *$\bar{\delta}$-slope (semi)stable*, if the inequality

$$L(\underline{\mathscr{E}}_\bullet, \underline{\alpha}_\bullet) + \bar{\delta} \cdot \mu(\underline{\mathscr{E}}_\bullet, \underline{\alpha}_\bullet; \phi)(\succeq)0$$

holds for all non-trivial filtrations $(\underline{\mathscr{E}}_\bullet, \underline{\alpha}_\bullet)$ of $\underline{\mathscr{E}}$.

A *family of decorated tuples of type* I *parameterized by the scheme* T is the datum $(\underline{\mathscr{E}}_T, \varphi_T, \mathscr{N})$, consisting of a tuple of torsion free sheaves $\underline{\mathscr{E}}_T = (\mathscr{E}_{T,1}, \ldots, \mathscr{E}_{T,t})$ on $X \times T$, flat over T, a line bundle $\mathscr{N}$ on T and a homomorphism

$$\varphi_T : \left(\bigotimes_{i=1}^{t} \mathscr{E}_{T,i}{}^{\otimes a} \right)^{\oplus b} \longrightarrow \det(\mathscr{E}_T)^{\otimes c} \otimes \pi_X^\star \mathscr{L} \otimes \pi_T^\star \mathscr{N} \,, \tag{22}$$

where, as usual, $\mathscr{E}_T = \mathscr{E}_{T,1} \oplus \cdots \oplus \mathscr{E}_{T,t}$. We say it is a *family of δ-(semi) stable decorated sheaves*, if the restriction to $X \times \{t\}$ is a δ-(semi) stable decorated tuple of sheaves, for every closed point $t \in T$. If $\mathscr{N}$ is not specified, we take $\mathscr{N} = \mathscr{O}_T$.

A *homomorphism between two families* $(\underline{\mathscr{E}}_T, \varphi_T, \mathscr{N})$ *and* $(\underline{\mathscr{E}}'_T, \varphi'_T, \mathscr{N}')$ is a family of homomorphisms $(f_i \colon \mathscr{E}_{T,i} \longrightarrow \mathscr{E}'_{T,i}, \alpha \colon \mathscr{N} \longrightarrow \mathscr{N}')$, such that the diagram

$$\begin{array}{ccc}
(\bigotimes_{i=1}^t \mathscr{E}_{T,i}{}^{\otimes a})^{\oplus b} & \xrightarrow{(\bigotimes_{i=1}^t f_i^{\otimes a})^{\oplus b}} & (\bigotimes_{i=1}^t \mathscr{E}'_{T,i}{}^{\otimes a})^{\oplus b} \\
\downarrow \varphi_T & & \downarrow \varphi'_T \\
\det(\mathscr{E}_T)^{\otimes c} \otimes \pi_X^\star \mathscr{L} \otimes \pi_T^\star \mathscr{N} & \xrightarrow{(\det f)^{\otimes c} \otimes \mathrm{id}_{\pi_X^\star \mathscr{L}} \otimes \pi_T^\star \alpha} & \det(\mathscr{E}'_T)^{\otimes c} \otimes \pi_X^\star \mathscr{L} \otimes \pi_T^\star \mathscr{N}'
\end{array}$$

commutes. Let

$$\underline{\mathrm{M}}^{\delta\text{-(s)s}} := \underline{\mathrm{M}}(\mathrm{I} : a, b, c; \mathscr{L})_{\underline{P}}^{\delta\text{-(s)s}}$$

be the contravariant functor from the category $\underline{\mathrm{Sch}}_k$ of k-schemes, locally of finite type, to the category $\underline{\mathrm{Sets}}$ of sets that sends a scheme T to the set of isomorphy classes of families of δ-(semi)stable decorated sheaves of type $(\mathrm{I} : a, b, c; \mathscr{L})$ with Hilbert polynomials $\underline{P} = (P_1, \ldots, P_t)$ parameterized by T, and sends a morphism $T' \longrightarrow T$ to the map defined by pullback.

A *tuple of sheaves with a decoration of type* $(\mathrm{II} : a, b, c; \mathscr{L})$ is a tuple $(\underline{\mathscr{E}}, \phi)$ which consists of a tuple $\underline{\mathscr{E}} = (\mathscr{E}_1, \ldots, \mathscr{E}_t)$ of torsion free sheaves $\mathscr{E}_i$ of rank r_i on X, $i = 1, \ldots, t$, and a non-trivial homomorphism

$$\phi \colon (\mathscr{E}^{\otimes a})^{\oplus b} \longrightarrow \det(\mathscr{E})^{\otimes c} \otimes \mathscr{L}.$$

The notion of a *homomorphism* of such decorated tuples is defined as before.

Remark 3.1.1. Since the tensor powers of $\mathscr{E}$ decompose as

$$\mathscr{E}^{\otimes a} = \bigoplus_{\substack{\underline{a}=(a_1,\ldots,a_t): \\ a_i \geq 0,\ a_1+\cdots+a_t=a}} \mathscr{E}_1^{\otimes a_1} \otimes \cdots \otimes \mathscr{E}_t^{\otimes a_t},$$

the decoration ϕ is composed of decorations

$$\phi_{\underline{a}} \colon (\mathscr{E}_1^{\otimes a_1} \otimes \cdots \otimes \mathscr{E}_t^{\otimes a_t})^{\oplus b} \longrightarrow \det(\mathscr{E})^{\otimes c} \otimes \mathscr{L}, \quad \underline{a} = (a_1, \ldots, a_t),\ a_1 + \cdots + a_t = a.$$

Next, we choose d, $a(d)$, $b(d)$, and $c(d)$ as in Lemma 2.3.8. Then, we may associate with every tuple of sheaves $(\underline{\mathscr{E}}, \phi)$ with a decoration of type $(\mathrm{II} : a, b, c; \mathscr{L})$ a tuple

$(\underline{\mathscr{E}}, \widetilde{\phi})$ with a decoration of type $(\mathrm{I} : a(d), b(d), c(d); \mathscr{L}^{\otimes d})$ as follows: Let $U \subseteq X$ be the maximal open subset where $\mathscr{E}_{i|U}$ is locally free for $i = 1, \ldots, t$. Then, the construction outlined in Lemma 2.3.8 can be used to associate with $\phi_{|U}$ a homomorphism

$$\widehat{\phi}\colon ((\mathscr{E}_{1|U} \otimes \cdots \otimes \mathscr{E}_{t|U})^{\otimes a(d)})^{\oplus b(d)} \longrightarrow \det(\mathscr{E}_{|U})^{\otimes c(d)} \otimes \mathscr{L}^{\otimes d}_{|U}.$$

Together with the inclusion $\imath\colon U \longrightarrow X$, this yields

$$\begin{aligned}\widetilde{\phi}\colon & ((\mathscr{E}_1 \otimes \cdots \otimes \mathscr{E}_t)^{\otimes a(d)})^{\oplus b(d)} \\ & \longrightarrow \imath_\star((\mathscr{E}_{1|U} \otimes \cdots \otimes \mathscr{E}_{t|U})^{\otimes a(d)})^{\oplus b(d)} \xrightarrow{\imath_\star(\widehat{\phi})} \\ & \xrightarrow{\imath_\star(\widehat{\phi})} \imath_\star(\det(\mathscr{E}_{|U})^{\otimes c(d)} \otimes \mathscr{L}^{\otimes d}_{|U}) \\ & = \det(\mathscr{E})^{\otimes c(d)} \otimes \mathscr{L}^{\otimes d}.\end{aligned}$$

Again, we fix the stability parameter, i.e., a positive polynomial $\delta = (\bar{\delta}/(n-1)!) \cdot x^{n-1} + \cdots \in \mathbb{Q}[x]$ of degree at most $n - 1 = \dim(X) - 1$, and say that a tuple of sheaves $(\underline{\mathscr{E}}, \phi)$ with a decoration of type $(\mathrm{II} : a, b, c; \mathscr{L})$ is *δ-(semi)stable*, if a) $\widetilde{\phi} \not\equiv 0$ and b) the tuple of sheaves $(\underline{\mathscr{E}}, \widetilde{\phi})$ with a decoration of type $(\mathrm{I} : a(d), b(d), c(d); \mathscr{L}^{\otimes d})$ is $((1/d) \cdot \delta)$-(semi)stable. If $\bar{\delta}$ is a positive rational number, a tuple of sheaves $(\underline{\mathscr{E}}, \phi)$ with a decoration of type $(\mathrm{II} : a, b, c; \mathscr{L})$ is said to be *$\bar{\delta}$-slope (semi)stable*, if a) $\widetilde{\phi} \not\equiv 0$ and b) the tuple of sheaves $(\underline{\mathscr{E}}, \widetilde{\phi})$ with a decoration of type $(\mathrm{I} : a(d), b(d), c(d); \mathscr{L}^{\otimes d})$ is $((1/d) \cdot \bar{\delta})$-slope (semi)stable. Note that, for $\delta = (\bar{\delta}/(n-1)!) \cdot x^{n-1} + \cdots$, we have

$$(\underline{\mathscr{E}}, \phi) \text{ is } \delta\text{-semistable} \quad \Longrightarrow \quad (\underline{\mathscr{E}}, \phi) \text{ is } \bar{\delta}\text{-slope semistable.} \tag{23}$$

Remark 3.1.2.

i) The notions of δ-(semi)stability and $\bar{\delta}$-(semi)stability are independent of the concrete choice of the number d.
ii) Using Corollary 2.3.2 in [6], the assignment $\phi \longmapsto \widetilde{\phi}$ can be promoted to give a natural transformation from the moduli functor of δ-(semi)stable tuples of sheaves with a decoration of type $(\mathrm{II} : a, b, c; \mathscr{L})$ (and fixed Hilbert polynomials) into the moduli functor for δ-(semi)stable tuples of sheaves with a decoration of type $(\mathrm{I} : a(d), b(d), c(d); \mathscr{L}^{\otimes d})$.

3.2 *Global boundedness*

We need some preparations. Denote by $F^k\colon X \longrightarrow X$ the k-th power of the Frobenius morphism, $k \geq 0$, if $\mathrm{Char}(k) > 0$. Let us set

$$L_X := \left[\lim_{k \longrightarrow \infty} \frac{\mu_{\max}((F^k)^\star \Omega_X)}{p^{k+1}}\right]_+,$$

if $\mathrm{Char}(k) = p$, and $L_X := 0$, if $\mathrm{Char}(k) = 0$. Note that, if A is a nef divisor, such that $T_X(A)$ is globally generated, then, in the case of positive characteristic, we have $L_X \leq AH^{n-1}/p$ [18].

Lemma 3.2.1. *If $\mathscr{E}_1$ and $\mathscr{E}_2$ are two $\mathscr{O}_X$-modules with torsion in codimension at least 2 of rank r_1 and r_2, respectively, we have*

$$\mu_{\max}(\mathscr{E}_1 \otimes \mathscr{E}_2) \leq \mu_{\max}(\mathscr{E}_1) + \mu_{\max}(\mathscr{E}_2) + (r_1 + r_2 - 2) \cdot L_X$$

and

$$\mu_{\min}(\mathscr{E}_1 \otimes \mathscr{E}_2) \geq \mu_{\min}(\mathscr{E}_1) + \mu_{\min}(\mathscr{E}_2) - (r_1 + r_2 - 2) \cdot L_X.$$

Proof. Let us sketch the proof. By [18], Theorem 2.7, we may choose k so large that $F^{k\star}\mathscr{E}_1$ and $F^{k\star}\mathscr{E}_2$ have strong Harder-Narasimhan filtrations. A result of Ramanan-Ramanathan asserts that the tensor product of two strongly slope semistable sheaves[4] is again strongly slope semistable (see [23], Theorem 3.23, [19], Theorem A.3). This easily implies

$$\mu_{\max}(F^{k\star}(\mathscr{E}_1 \otimes \mathscr{E}_2)) = \mu_{\max}(F^{k\star}\mathscr{E}_1) + \mu_{\max}(F^{k\star}\mathscr{E}_2).$$

Altogether, we see that

$$\begin{aligned}\mu_{\max}(\mathscr{E}_1 \otimes \mathscr{E}_2) &\leq \frac{1}{p^k} \cdot \mu_{\max}(F^{k\star}(\mathscr{E}_1 \otimes \mathscr{E}_2)) = \frac{1}{p^k}(\mu_{\max}(F^{k\star}\mathscr{E}_1) + \mu_{\max}(F^{k\star}\mathscr{E}_2)) \\ &\leq \mu_{\max}(\mathscr{E}_1) + (r_1 - 1) \cdot L_X + \mu_{\max}(\mathscr{E}_2) + (r_2 - 1) \cdot L_X.\end{aligned}$$

Here, the first inequality is obvious and the last one results from [18], Corollary 6.2. □

Theorem 3.2.2. *Fix Hilbert polynomials $\underline{P} = (P_1, \ldots, P_t)$, a, b, c, $\mathscr{L}$, and $\mathrm{J} \in \{\mathrm{I}, \mathrm{II}\}$. Then, the set of isomorphy classes of torsion free sheaves $\mathscr{F}$, such that there exist a tuple of sheaves $(\underline{\mathscr{E}}, \phi)$ with a decoration of type $(\mathrm{J} : a, b, c; \mathscr{L})$, such that $P(\mathscr{E}_i) = P_i$, $i = 1, \ldots, t$, a positive rational number $\bar{\delta}$, such that $(\underline{\mathscr{E}}, \phi)$ is $\bar{\delta}$-slope semistable, and an index $i_0 \in \{1, \ldots, t\}$ with $\mathscr{F} = \mathscr{E}_{i_0}$ is bounded.*

Proof. We discuss the case $\mathrm{J} = \mathrm{II}$, the other case resulting from a trivial modification of the arguments. Suppose $(\underline{\mathscr{E}}, \phi)$ is a tuple of torsion free $\mathscr{O}_X$-modules with a decoration of type $(\mathrm{II} : a, b, c; \mathscr{L})$, such that $P(\mathscr{E}_i) = P_i$, $i = 1, \ldots, t$. Let $\bar{\delta} > 0$ be a rational number, such that $(\underline{\mathscr{E}}, \phi)$ is $\bar{\delta}$-slope semistable. If all the $\mathscr{E}_i$ are slope semistable, there is nothing to do. Otherwise, we let $\widehat{I} \subseteq I := \{1, \ldots, t\}$ be the set of indexes $\hat{\imath}$, such that $\mathscr{E}_{\hat{\imath}}$ is not slope semistable. For $\hat{\imath} \in \widehat{I}$, let

$$\mathscr{E}_{\hat{\imath}\bullet} : \{0\} = \mathscr{E}_{\hat{\imath},0} \subsetneq \mathscr{E}_{\hat{\imath},1} \subsetneq \mathscr{E}_{\hat{\imath},2} \subsetneq \cdots \subsetneq \mathscr{E}_{\hat{\imath},h_{\hat{\imath}}} \subsetneq \mathscr{E}_{\hat{\imath},h_{\hat{\imath}}+1} = \mathscr{E}_{\hat{\imath}}$$

be the slope Harder-Narasimhan filtration of $\mathscr{E}_{\hat{\imath}}$. This means that $\mathscr{E}_{\hat{\imath},j}$ is a saturated subsheaf of $\mathscr{E}_{\hat{\imath}}$, such that $\mathscr{E}_{\hat{\imath}}^j = \mathscr{E}_{\hat{\imath},j}/\mathscr{E}_{\hat{\imath},j-1}$ is slope semistable, $j = 1, \ldots, h_{\hat{\imath}} + 1$, and

$$\mu(\mathscr{E}_{\hat{\imath}}^1) > \mu(\mathscr{E}_{\hat{\imath}}^2) > \cdots > \mu(\mathscr{E}_{\hat{\imath}}^{h_{\hat{\imath}}+1}).$$

[4] An $\mathscr{O}_X$-module with torsion in codimension two is called *strongly slope semistable*, if all of its Frobenius pull backs are slope semistable.

We use $r_{\hat{\imath},j} := \operatorname{rk} \mathscr{E}_{\hat{\imath},j}$, $r_{\hat{\imath}}^j := \operatorname{rk} \mathscr{E}_{\hat{\imath}}^j$, and $\mu_{\hat{\imath}}^j := \mu(\mathscr{E}_{\hat{\imath}}^j)$, $j = 1, \ldots, h_{\hat{\imath}} + 1$, $\hat{\imath} \in \widehat{I}$. Define

$$C(\mathscr{E}_{\hat{\imath}\bullet}) = \left\{ \underline{\gamma}_{\hat{\imath}} = (\gamma_{\hat{\imath},1}, \ldots, \gamma_{\hat{\imath},h_{\hat{\imath}}+1}) \in \mathbb{R}^{h_{\hat{\imath}}+1} | \gamma_{\hat{\imath},1} \le \gamma_{\hat{\imath},2} \le \cdots \le \gamma_{\hat{\imath},h_{\hat{\imath}}+1}, \right.$$

$$\left. \sum_{j=1}^{h_{\hat{\imath}}+1} \gamma_{\hat{\imath},j} \cdot r_{\hat{\imath}}^j = 0 \right\},$$

and

$$C(\underline{\mathscr{E}}) := \mathop{\mathsf{X}}_{\hat{\imath} \in \widehat{I}} C(\mathscr{E}_{\hat{\imath}\bullet}).$$

We equip $\mathbb{R}^h$, $h := \sum_{\hat{\imath} \in \widehat{I}} (h_{\hat{\imath}} + 1)$, with the maximum norm $\|\cdot\|$. Note that $C(\underline{\mathscr{E}})$ is a cone in $\mathbb{R}^h$. For all $\underline{v} = (\underline{\gamma}_{\hat{\imath}}, \hat{\imath} \in \widehat{I}) \in C(\underline{\mathscr{E}}) \backslash \{0\}$, we have

$$\sum_{\hat{\imath} \in \widehat{I}} \sum_{j=1}^{h_{\hat{\imath}}} \frac{\gamma_{\hat{\imath},j+1} - \gamma_{\hat{\imath},j}}{r_{\hat{\imath}}} (r_{\hat{\imath}} \cdot \deg(\mathscr{E}_{\hat{\imath},j}) - r_{\hat{\imath},j} \cdot \deg(\mathscr{E}_{\hat{\imath}})) < 0,$$

so that the $\bar{\delta}$-slope semistability of $(\underline{\mathscr{E}}, \phi)$ implies

$$f(\underline{v}) := \mu(\underline{\mathscr{E}}_\bullet(\underline{v}), \underline{\alpha}_\bullet(\underline{v}); \phi) > 0.$$

Here, we set $(\mathscr{E}_{i\bullet}(\underline{v}), \alpha_{i\bullet}(\underline{v})) := (\{0\} \subsetneq \mathscr{E}_i, ())$, for $i \notin \widehat{I}$, $\mathscr{E}_{\hat{\imath}\bullet}(\underline{v}) := \mathscr{E}_{\hat{\imath}\bullet}$, and

$$\alpha_{\hat{\imath}\bullet}(\underline{v}) := \left(\frac{\gamma_{\hat{\imath},2} - \gamma_{\hat{\imath},1}}{r_{\hat{\imath}}}, \ldots, \frac{\gamma_{\hat{\imath},h_{\hat{\imath}}+1} - \gamma_{\hat{\imath},h_{\hat{\imath}}}}{r_{\hat{\imath}}} \right), \quad \hat{\imath} \in \widehat{I}.$$

Consider the set

$$K := C(\underline{\mathscr{E}}) \cap \{\underline{v} \in \mathbb{R}^h \mid \|\underline{v}\| = 1\}.$$

Obviously K is a compact set and f is piecewise linear, hence continuous, so that f attains its positive infimum on K. We claim that we can find a lower bound $C_0 > 0$ for this infimum which depends only on the input data. This will be clear once we show that, a priori, there are only finitely many possibilities for the function f. To this end, one checks that f depends only on the $r_{\hat{\imath}}^j$, $j = 1, \ldots, h_{\hat{\imath}} + 1$, $i \in \widehat{I}$, and the tuples $(j_1^i, \ldots, j_{a(d)}^i)$, $j_k^i \in \{1, \ldots, h_i + 1\}$, $k = 1, \ldots, a(d)$, $i = 1, \ldots, t$, $h_i := 0$ for $i \in I \backslash \widehat{I}$, with

$$\widetilde{\phi}_{|(\mathscr{E}_{1,j_1^1} \otimes \cdots \otimes \mathscr{E}_{1,j_{a(d)}^1} \otimes \cdots \otimes \mathscr{E}_{t,j_1^t} \otimes \cdots \otimes \mathscr{E}_{t,j_{a(d)}^t})^{\oplus b(d)}} \not\equiv 0.$$

These data live in a finite set which is defined in terms of the input data $r_1, \ldots, r_t$, and $a(d)$.

Now, take tuples $(j_1^i, \ldots, j_{a(d)}^i)$, $j_k^i \in \{1, \ldots, h_i + 1\}$, $k = 1, \ldots, a(d)$, $i = 1, \ldots, t$, $h_i := 0$ for $i \in I \backslash \widehat{I}$, with

$$\widetilde{\phi}_{|(\mathscr{E}_{1,j_1^1} \otimes \cdots \otimes \mathscr{E}_{1,j_{a(d)}^1} \otimes \cdots \otimes \mathscr{E}_{t,j_1^t} \otimes \cdots \otimes \mathscr{E}_{t,j_{a(d)}^t})^{\oplus b(d)}} \not\equiv 0.$$

Then,

$$\mu_{\min}(\mathcal{E}_{1,j_1^1}\otimes\cdots\otimes\mathcal{E}_{1,j_{a(d)}^1}\otimes\cdots\otimes\mathcal{E}_{t,j_1^t}\otimes\cdots\otimes\mathcal{E}_{t,j_{a(d)}^t})\le c(d)\cdot\deg(\mathcal{E})+d\cdot\deg(\mathcal{L}).$$

Note that $\mu_{\min}(\mathcal{E}_{\hat{\imath},j})=\mu_{\hat{\imath}}^{j}$, $j=1,\dots,h_{\hat{\imath}}+1$, $\hat{\imath}\in\widehat{I}$, and $\mu_{\min}(\mathcal{E}_i)=\mu(\mathcal{E}_i)$, $i\in I\backslash\widehat{I}$. Lemma 3.2.1 thus shows

$$a(d)\sum_{i\in I\backslash\widehat{I}}\mu(\mathcal{E}_i)+\sum_{\hat{\imath}\in\widehat{I}}(\mu_{\hat{\imath}}^{j_1^{\hat{\imath}}}+\cdots+\mu_{\hat{\imath}}^{j_{a(d)}^{\hat{\imath}}})\le C_1$$

$$:=a(d)\sum_{i=1}^{t}(r_i-1)L_X+c(d)\deg(\mathcal{E})+d\deg(\mathcal{L}).$$

With

$$C_2:=\max\left\{-a(d)\cdot\sum_{i\in J}\mu(\mathcal{E}_i)|J\subseteq I\right\},$$

we may write this as

$$\sum_{\hat{\imath}\in\widehat{I}}(\mu_{\hat{\imath}}^{j_1^{\hat{\imath}}}+\cdots+\mu_{\hat{\imath}}^{j_{a(d)}^{\hat{\imath}}})\le C:=C_1+C_2. \tag{24}$$

Take the point $\underline{v}=(\underline{\gamma}_{\hat{\imath}},\hat{\imath}\in\widehat{I})\in\mathbb{R}^h$ with

$$\underline{\gamma}_{\hat{\imath}}:=(\mu(\mathcal{E}_{\hat{\imath}})-\mu_{\hat{\imath}}^{1},\dots,\mu(\mathcal{E}_{\hat{\imath}})-\mu_{\hat{\imath}}^{h_{\hat{\imath}}+1}),\quad \hat{\imath}\in\widehat{I}.$$

By construction, $\underline{v}\in C(\underline{\mathcal{E}})\backslash\{0\}$ and

$$f(\underline{v})=\mu(\underline{\mathcal{E}}_{\bullet}(\underline{v}),\underline{\alpha}_{\bullet}(\underline{v});\phi)\le C+a(d)\cdot\sum_{\hat{\imath}\in\widehat{I}}\mu(\mathcal{E}_{\hat{\imath}})\le C'$$

with

$$C':=C+\max\left\{a(d)\cdot\sum_{i\in J}\mu(\mathcal{E}_i)|J\subseteq I\right\}.$$

But f is linear on each ray, so that

$$f(\underline{v})=\|\underline{v}\|\cdot f\left(\frac{\underline{v}}{\|\underline{v}\|}\right)\ge C_0\cdot\|\underline{v}\|.$$

Now, this shows that there is an index $i_0\in\widehat{I}$ with either

$$\mu_{i_0}^{1}-\mu(\mathcal{E}_{i_0})=\|\underline{v}\|\le C'':=\frac{C'}{C_0}\quad\text{and}\quad\mu_{\hat{\imath}}^{1}-\mu(\mathcal{E}_{\hat{\imath}})\le\mu_{i_0}^{1}-\mu(\mathcal{E}_{i_0}),\quad\forall\hat{\imath}\in\widehat{I},$$

or

$$\mu(\mathcal{E}_{i_0})-\mu_{i_0}^{h_{i_0}+1}=\|\underline{v}\|\le C''\quad\text{and}\quad\mu(\mathcal{E}_{\hat{\imath}})-\mu_{\hat{\imath}}^{h_{\hat{\imath}}+1}\le\mu(\mathcal{E}_{i_0})-\mu_{i_0}^{h_{i_0}+1},\quad\forall\hat{\imath}\in\widehat{I},$$

i.e., for all $\hat{\imath} \in \widehat{I}$, either

$$\mu_{\max}(\mathscr{E}_{\hat{\imath}}) \le \mu_{\max}(\mathscr{E}_{i_0}) - \mu(\mathscr{E}_{i_0}) + \mu(\mathscr{E}_{\hat{\imath}}) \le C''' := C'' + \max\{i \in I \mid \mu(\mathscr{E}_i)\}$$
$$\text{or} \quad \mu_{\min}(\mathscr{E}_{\hat{\imath}}) \ge \mu_{\min}(\mathscr{E}_{i_0}) - \mu(\mathscr{E}_{i_0}) + \mu(\mathscr{E}_{\hat{\imath}}) \ge C'''' := -C'' + \min\{i \in I \mid \mu(\mathscr{E}_i)\}.$$

The theorem now follows from [18], Theorem 4.2. □

Remark 3.2.3. Call a tuple $(\underline{\mathscr{E}}, \phi)$ of torsion free sheaves with a decoration of type $(\mathrm{I} : a, b, c; \mathscr{L})$ *pre-semistable*, if it satisfies the condition $L(\underline{\mathscr{E}}_\bullet, \underline{\alpha}_\bullet) \ge 0$ for every weighted filtration $(\underline{\mathscr{E}}_\bullet, \underline{\alpha}_\bullet)$ of $\underline{\mathscr{E}}$ with $\mu(\underline{\mathscr{E}}_\bullet, \underline{\alpha}_\bullet; \phi) \le 0$. The proof of Theorem 3.2.2 actually shows that the set of isomorphy classes of torsion free sheaves $\mathscr{F}$, such that there exist a pre-semistable tuple $(\underline{\mathscr{E}}, \phi)$ of torsion free sheaves with a decoration of type $(\mathrm{I} : a, b, c; \mathscr{L})$ with $P(\mathscr{E}_i) = P_i$, $i = 1, \ldots, t$, and an index $i \in \{1, \ldots, t\}$ with $\mathscr{F} \cong \mathscr{E}_i$, is bounded.

3.3 *S-equivalence*

First, suppose we are given a δ-semistable tuple of sheaves $(\underline{\mathscr{E}}, \phi)$ with a decoration of type $(\mathrm{I} : a, b, c; \mathscr{L})$ and a non-trivial weighted filtration $(\underline{\mathscr{E}}_\bullet, \underline{\alpha}_\bullet)$ with

$$M(\underline{\mathscr{E}}_\bullet, \underline{\alpha}_\bullet) + \delta \cdot \mu(\underline{\mathscr{E}}_\bullet, \underline{\alpha}_\bullet; \phi) = 0.$$

We want to define the *associated admissible deformation* $\mathrm{df}_{(\underline{\mathscr{E}}_\bullet, \underline{\alpha}_\bullet)}(\underline{\mathscr{E}}, \phi) = (\underline{\mathscr{E}}_{\mathrm{df}}, \phi_{\mathrm{df}})$. Of course, $\underline{\mathscr{E}}_{\mathrm{df}} = (\mathscr{E}_{i,\mathrm{df}}, i = 1, \ldots, t)$ with $\mathscr{E}_{i,\mathrm{df}} := \bigoplus_{j=0}^{s_i} \mathscr{E}_{i,j+1}/\mathscr{E}_{i,j}$, $i = 1, \ldots, t$. Let U be the maximal (big!) open subset where all the $\mathscr{E}_{i,\mathrm{df}}$ are locally free. We may choose a one parameter subgroup

$$\lambda = (\lambda_1, \ldots, \lambda_t) \colon \mathbb{G}_m(k) \longrightarrow \mathrm{SL}_{r_1}(k) \times \cdots \times \mathrm{SL}_{r_t}(k),$$

such that the weighted flags $(\underline{W}_\bullet(\underline{\lambda}), \underline{\alpha}_\bullet(\underline{\lambda}))$ in $\underline{W} = (W_i := k^{r_i}, i = 1, \ldots, t)$ satisfy

- $W_{i\bullet}(\lambda_i) : \{0\} \subsetneq W_{i,1} \subsetneq \cdots \subsetneq W_{i,s_i} \subsetneq k^{r_i}$ with $\dim_k(W_{i,j}) = \mathrm{rk}(\mathscr{E}_{i,j})$, $j = 1, \ldots, s_i$, $i = 1, \ldots, t$;
- $\alpha_{i\bullet}(\lambda_i) = \alpha_{i\bullet}$, $i = 1, \ldots, t$.

Then, the tuple of filtrations $(\mathscr{E}_{i\bullet}, i = 1, \ldots, t)$ corresponds to a reduction of the structure group of the $(\mathrm{GL}_{r_1}(k) \times \cdots \times \mathrm{GL}_{r_t}(k))$-bundle $\mathsf{X}_{i=1}^t \mathscr{I}som(\mathscr{O}_X^{\oplus r_i}, \mathscr{E}_i)$ to $Q(\lambda) := Q_{\mathrm{GL}_{r_1}(k) \times \cdots \times \mathrm{GL}_{r_t}(k)}(\lambda)$. On the other hand, λ defines a decomposition

$$\widetilde{W}_{a,b,c} = U_{\gamma_1} \oplus \cdots \oplus U_{\gamma_{s+1}}, \quad \gamma_1 < \cdots < \gamma_{s+1}.$$

Now, observe that $Q(\lambda)$ fixes the flag

$$\begin{aligned}\{0\} := U_0 \subsetneq U_1 := U_{\gamma_1} \subsetneq U_2 := (U_{\gamma_1} \oplus U_{\gamma_2}) \subsetneq \cdots \subsetneq U_s \\ := (U_1 \oplus \cdots \oplus U_{\gamma_s}) \subsetneq \widetilde{W}_{a,b,c}.\end{aligned} \tag{25}$$

Thus, we obtain a $Q(\lambda)$-module structure on

$$\bigoplus_{k=0}^{s} U_{k+1}/U_k \cong \widetilde{W}_{a,b,c}. \tag{26}$$

Next, we write $Q(\lambda) = \mathscr{R}_u(Q(\lambda)) \rtimes L(\lambda)$ where

$$\begin{aligned} L(\lambda) \cong \mathrm{GL}(W_{1,1}/W_{1,0}) \times \cdots \times \mathrm{GL}(k^{r_1}/W_{1,s_1}) \\ \times \cdots \times \mathrm{GL}(W_{t,1}/W_{t,0}) \times \cdots \times \mathrm{GL}(k^{r_t}/W_{t,s_t}) \end{aligned}$$

is the centralizer of λ. Note that (26) is an isomorphism of $L(\lambda)$-modules. The process of passing from $\underline{\mathscr{E}}$ to $\underline{\mathscr{E}}_{\mathrm{df}}$ corresponds to first reducing the structure group to $Q(\lambda)$, then extending it to $L(\lambda)$ via $Q(\lambda) \longrightarrow Q(\lambda)/\mathscr{R}_u(Q(\lambda)) \cong L(\lambda)$, and then extending it to $\mathrm{GL}_{r_1}(k) \times \cdots \times \mathrm{GL}_{r_t}(k)$ via the inclusion $L(\lambda) \subset \mathrm{GL}_{r_1}(k) \times \cdots \times \mathrm{GL}_{r_t}(k)$. By (25), $((\mathscr{E}_{1|U} \otimes \cdots \otimes \mathscr{E}_{t|U})^{\otimes a})^{\oplus b} \otimes (\det(\mathscr{E}_{|U}))^{\otimes -c}$ has a filtration

$$\{0\} \subsetneq \mathscr{U}_1 \subsetneq \mathscr{U}_2 \subsetneq \cdots \subsetneq \mathscr{U}_s \subsetneq ((\mathscr{E}_{1|U} \otimes \cdots \otimes \mathscr{E}_{t|U})^{\otimes a})^{\oplus b} \otimes (\det(\mathscr{E}_{|U}))^{\otimes -c},$$

and, by (26), we have a canonical isomorphism

$$((\mathscr{E}_{1,\mathrm{df}|U} \otimes \cdots \otimes \mathscr{E}_{t,\mathrm{df}|U})^{\otimes a})^{\oplus b} \otimes (\det(\mathscr{E}_{\mathrm{df}|U}))^{\otimes -c} \cong \bigoplus_{k=0}^{s} \mathscr{U}_{k+1}/\mathscr{U}_k.$$

Now, the restriction ϕ_1 of $\phi_{|U}$ to $\mathscr{U}_1$ is non-trivial, and thus we may define $\widehat{\phi}_{\mathrm{df}}$ as ϕ_1 on $\mathscr{U}_1$ and as zero on the other components. Then, we finally obtain

$$\begin{aligned} \phi_{\mathrm{df}}\colon ((\mathscr{E}_1 \otimes \cdots \otimes \mathscr{E}_t)^{\otimes a})^{\oplus b} &\longrightarrow \iota_\star((\mathscr{E}_{1,\mathrm{df}|U} \otimes \cdots \otimes \mathscr{E}_{t,\mathrm{df}|U})^{\otimes a})^{\otimes b}) \\ &\xrightarrow{\iota_\star(\widehat{\phi}_{\mathrm{df}})} \iota_\star(\det(\mathscr{E}_{|U})^{\otimes c} \otimes \mathscr{L}_{|U}) = \det(\mathscr{E})^{\otimes c} \otimes \mathscr{L}, \end{aligned}$$

$\iota\colon U \longrightarrow X$ being the inclusion. The equivalence relation *S-equivalence* is now generated by

$$(\underline{\mathscr{E}}, \phi) \sim_{\mathrm{S}} \mathrm{df}_{(\underline{\mathscr{E}}_\bullet, \underline{\alpha}_\bullet)}(\underline{\mathscr{E}}, \phi)$$

for every non-trivial weighted filtration $(\underline{\mathscr{E}}_\bullet, \underline{\alpha}_\bullet)$ with

$$M(\underline{\mathscr{E}}_\bullet, \underline{\alpha}_\bullet) + \delta \cdot \mu(\underline{\mathscr{E}}_\bullet, \underline{\alpha}_\bullet; \phi) = 0.$$

Remark 3.3.1.

i) Another way of looking at S-equivalence is the following: With the notation as above, we may choose an open subset $U \subseteq X$ (no longer big), such that we have isomorphisms $\psi_i\colon \mathscr{E}_{i|U} \cong k^{r_i} \otimes \mathscr{O}_U$ with $\psi(\mathscr{E}_{i,j}) = W_{i,j} \otimes \mathscr{O}_U$ for $j = 1, \ldots, s_i$, $i = 1, \ldots, t$. For such trivializations, we obtain, from $\phi_{|U}$, the morphism

$$\sigma\colon U \longrightarrow \mathbb{P}(((\mathscr{E}_{1|U} \otimes \cdots \otimes \mathscr{E}_{t|U})^{\otimes a})^{\oplus b}) \overset{\text{“}\psi\text{”}}{\cong} \mathbb{P}(\widetilde{W}_{a,b,c}) \times U \longrightarrow \mathbb{P}(\widetilde{W}_{a,b,c}).$$

For the morphism $\sigma_{\mathrm{gr}}\colon U \longrightarrow \mathbb{P}(\widetilde{W}_{a,b,c})$ associated with $\phi_{\mathrm{df}|U}$, we find the relationship

$$\sigma_{\mathrm{df}}(x) = \lim_{z\to\infty} \lambda(z) \cdot \sigma(x), \quad x \in U. \tag{27}$$

ii) Let us call a δ-semistable tuple of sheaves $(\underline{\mathscr{E}}, \phi)$ with a decoration of type $(\mathrm{I} : a, b, c; \mathscr{L})$ *δ-polystable*, if $(\underline{\mathscr{E}}, \phi)$ is isomorphic to $\mathrm{df}_{(\underline{\mathscr{E}}_\bullet, \underline{\alpha}_\bullet)}(\underline{\mathscr{E}}, \phi)$, for every non-trivial weighted filtration $(\underline{\mathscr{E}}_\bullet, \underline{\alpha}_\bullet)$ with

$$M(\underline{\mathscr{E}}_\bullet, \underline{\alpha}_\bullet) + \delta \cdot \mu(\underline{\mathscr{E}}_\bullet, \underline{\alpha}_\bullet; \phi) = 0.$$

Then, the GIT construction implies the following properties of S-equivalence.

Lemma.

i) *Two δ-polystable decorated tuples of sheaves $(\underline{\mathscr{E}}_1, \phi_1)$ and $(\underline{\mathscr{E}}_2, \phi_2)$ are S-equivalent, if and only if they are isomorphic,*
ii) *For any δ-semistable tuple of sheaves $(\underline{\mathscr{E}}, \phi)$ with a decoration of type* $(\mathrm{I} : a, b, c; \mathscr{L})$*, there exists a δ-polystable decorated tuple* $\mathrm{gr}(\underline{\mathscr{E}}, \phi)$ *which is S-equivalent to $(\underline{\mathscr{E}}, \phi)$. The decorated tuple* $\mathrm{gr}(\underline{\mathscr{E}}, \phi)$ *is unique up to isomorphy.*

Next, we call two δ-semistable tuples of sheaves $(\underline{\mathscr{E}}^1, \phi_1)$ and $(\underline{\mathscr{E}}^2, \phi_2)$ with a decoration of type $(\mathrm{II} : a, b, c; \mathscr{L})$ *S-equivalent*, if the associated tuples $(\underline{\mathscr{E}}^1, \widetilde{\phi}_1)$ and $(\underline{\mathscr{E}}^2, \widetilde{\phi}_2)$ with a decoration of type $(\mathrm{I} : a(d), b(d), c(d); \mathscr{L}^{\otimes d})$ are S-equivalent.

Remark 3.3.2. Let $(\underline{\mathscr{E}}, \phi)$ be a tuple of sheaves with a decoration of type $(\mathrm{II} : a, b, c; \mathscr{L})$ and $(\underline{\mathscr{E}}_\bullet, \underline{\alpha}_\bullet)$ a non-trivial weighted filtration. For every tuple $\underline{a} = (a_1, \ldots, a_t)$ with $a_1 + \cdots + a_t = a$ and $\phi_{\underline{a}} \not\equiv 0$, we may declare

$$\mu(\underline{\mathscr{E}}_\bullet, \underline{\alpha}_\bullet; \phi_{\underline{a}})$$

by the obvious modification of the original definition. Thus, we also have

$$\widetilde{\mu}(\underline{\mathscr{E}}_\bullet, \underline{\alpha}_\bullet; \phi) := \max\{\mu(\underline{\mathscr{E}}_\bullet, \underline{\alpha}_\bullet; \phi_{\underline{a}}) | \phi_{\underline{a}} \not\equiv 0\}. \tag{28}$$

One has the following relations among the different quantities

$$\widetilde{\mu}(\underline{\mathscr{E}}_\bullet, \underline{\alpha}_\bullet; \phi) = \frac{1}{d} \cdot \widetilde{\mu}(\underline{\mathscr{E}}_\bullet, \underline{\alpha}_\bullet; \mathrm{Sym}^d(\phi)) \geq \frac{1}{d} \cdot \mu(\underline{\mathscr{E}}_\bullet, \underline{\alpha}_\bullet; \widetilde{\phi}). \tag{29}$$

Suppose we are given a non-trivial weighted filtration $(\underline{\mathscr{E}}_\bullet, \underline{\alpha}_\bullet)$ for which equality holds in (29) and for which

$$M(\underline{\mathscr{E}}_\bullet, \underline{\alpha}_\bullet) + \delta \cdot \widetilde{\mu}(\underline{\mathscr{E}}_\bullet, \underline{\alpha}_\bullet; \phi) = 0.$$

By a procedure similar to the one before, we may define $\mathrm{df}_{(\underline{\mathscr{E}}_\bullet, \underline{\alpha}_\bullet)}(\underline{\mathscr{E}}, \phi) = (\underline{\mathscr{E}}_{\mathrm{df}}, \phi_{\mathrm{df}})$. One checks

$$\mathrm{df}_{(\underline{\mathscr{E}}_\bullet, \underline{\alpha}_\bullet)}(\underline{\mathscr{E}}, \widetilde{\phi}) = (\underline{\mathscr{E}}_{\mathrm{df}}, \widetilde{\phi}_{\mathrm{df}}). \tag{30}$$

Both the definition of δ-semistability and S-equivalence invoke the assignment $\phi \longmapsto \widetilde{\phi}$. In particular, we have the equivalence relation

$(\underline{\mathscr{E}}_1, \phi_1) \sim_A (\underline{\mathscr{E}}_2, \phi_2)$, if and only if the associated tuples $(\underline{\mathscr{E}}_1, \widetilde{\phi}_1)$ and $(\underline{\mathscr{E}}_2, \widetilde{\phi}_2)$ with a decoration of type $(\mathrm{I} : a(d), b(d), c(d); \mathscr{L}^{\otimes d})$ are isomorphic.

We would like to understand these procedures in the framework of GIT. First, we define the algebraic group

$$\widetilde{T} := \{(z_1, \ldots, z_t) \in \mathbb{G}_m(k)^{\times t} \mid z_1 \cdot \cdots \cdot z_t = 1\}.$$

A one parameter subgroup $\lambda\colon \mathbb{G}_m(k) \longrightarrow \widetilde{T}$ is specified by integers $\gamma_1,\ldots, \gamma_t$ with $\gamma_1 + \cdots + \gamma_t = 0$:

$$\lambda(z) = (z^{\gamma_1}, \ldots, z^{\gamma_t}), \quad z \in \mathbb{G}_m(k).$$

Recall the notation of Remark 3.1.1. For a tuple of sheaves $(\underline{\mathscr{E}}, \phi)$ with a decoration of type $(\mathrm{II} : a, b, c; \mathscr{L})$ and a one parameter subgroup λ of $\widetilde{T}$, we set

$$\mu(\lambda, \phi) := -\min\{a_1\gamma_1 + \cdots + a_t\gamma_t | \phi_{(a_1,\ldots,a_t)} \not\equiv 0\}.$$

Lemma 3.3.3. *We have $\widetilde{\phi} \not\equiv 0$, if and only if $\mu(\lambda, \phi) \geq 0$ holds for every one parameter subgroup $\lambda\colon \mathbb{G}_m(k) \longrightarrow \widetilde{T}$.*

Proof. This follows from easy GIT considerations at the generic point of X. (See proof of Proposition 3.3.4.) □

Next, suppose we are given a decorated tuple $(\underline{\mathscr{E}}, \phi)$ with $\widetilde{\phi} \not\equiv 0$ and a one parameter subgroup λ of $\widetilde{T}$, such that $\mu(\lambda, \phi) = 0$. This yields the *associated admissible deformation* $\mathrm{df}_\lambda(\underline{\mathscr{E}}, \phi) = (\underline{\mathscr{E}}_{\mathrm{df}}, \underline{\phi}_{\mathrm{df}})$ with $\underline{\mathscr{E}}_{\mathrm{df}} = \mathscr{E}$ and

$$\phi_{\mathrm{df},(a_1,\ldots,a_t)} := \begin{cases} \phi_{(a_1,\ldots,a_t)}, & \text{if } a_1\gamma_1 + \cdots + a_t\gamma_t = 0 \\ 0, & \text{else} \end{cases}.$$

Let "$\sim_B$" be the equivalence relation that is generated by $(\underline{\mathscr{E}}, \phi) \sim_B \mathrm{df}_\lambda(\underline{\mathscr{E}}, \phi)$ for any one parameter subgroup λ of $\widetilde{T}$ with $\mu(\lambda, \phi) = 0$. Now suppose we are given a family $\mathfrak{S}$ of isomorphy classes of δ-semistable tuples of sheaves with a decoration of type $(\mathrm{II} : a, b, c; \mathscr{L})$, such that both ϕ and $\widetilde{\phi}$ are surjective over a big open subset of X for every decorated tuple $(\underline{\mathscr{E}}, \phi) \in \mathfrak{S}$ and such that every admissible deformation $\mathrm{df}_\lambda(\underline{\mathscr{E}}, \phi)$ of an element $(\underline{\mathscr{E}}, \phi) \in \mathfrak{S}$ still lies in $\mathfrak{S}$.

Proposition 3.3.4. *Two tuples of sheaves $(\underline{\mathscr{E}}_1, \phi_1)$ and $(\underline{\mathscr{E}}_2, \phi_2)$ in $\mathfrak{S}$ are equivalent with respect to the equivalence relation "$\sim_A$", if and only if they are equivalent with respect to the equivalence relation "$\sim_B$".*

Proof. To begin with, we have to introduce a bunch of notation. Let η be the generic point of X, $K := k(X)$ the function field, and $\bar{K}$ an algebraic closure of K. Fix K-vector spaces W_i with $\dim_K(W_i) = \mathrm{rk}(\mathscr{E}_i^j)$, $i = 1, \ldots, t$, $j = 1, 2$. Set $W := W_1 \oplus \cdots \oplus W_t$, $\bar{W}_i := \bar{W} \otimes_K \bar{K}$, and $\bar{W} := \bar{W}_1 \oplus \cdots \oplus \bar{W}_t$. Given $f \in W^\vee_{a,b,c}$, set $\bar{f} := f \otimes \mathrm{id}_{\bar{K}}$. In general, for an object O defined over K, we let $\bar{O}$ be the corresponding object defined over $\bar{K}$.

Standard GIT in $\mathbb{P}(\bar{W}_{a,b,c})$ (compare Section 2.3, "Computation of some weights") shows the following.

Lemma.

i) *A point $[\bar{f}] \in \mathbb{P}(\bar{W}_{a,b,c})$ is semistable with respect to the action of $\bar{T}$, if and only if*

$$\mu(\lambda, \bar{f}) = -\min\{a_1\gamma_1 + \cdots + a_t\gamma_t \mid \bar{f}_{(a_1,\ldots,a_t)} \not\equiv 0\} \geq 0,$$

for every set of integers $\gamma_1, \ldots, \gamma_t$ with $\gamma_1 + \cdots + \gamma_t = 0$, and $\lambda\colon \mathbb{G}_m(\bar{K}) \longrightarrow \bar{T}$, $z \longmapsto (z^{\gamma_1}, \ldots, z^{\gamma_t})$, $\bar{T} := \{ (z_1, \ldots, z_t) \in \mathbb{G}_m(\bar{K})^{\times t} \mid z_1 \cdot \cdots \cdot z_t = 1 \}$.

ii) *Let "$\sim_C$" be the equivalence relation that identifies two points in the semistable locus of $\mathbb{P}(\bar{W}_{a,b,c})$, if their orbit closures intersect inside the semistable locus of $\mathbb{P}(\bar{W}_{a,b,c})$. Then, by the Hilbert-Mumford-criterion, "$\sim_C$" is the equivalence relation generated by $[\bar{f}] \sim [\bar{f}_{\mathrm{df}}]$ for every one parameter subgroup λ of $\bar{T}$ with $\mu(\lambda, \bar{f}) = 0$. Here,*

$$\bar{f}_{\mathrm{df},(a_1,\ldots,a_t)} := \begin{cases} \bar{f}_{(a_1,\ldots,a_t)}, & \text{if } a_1\gamma_1 + \cdots + a_t\gamma_t = 0 \\ 0, & \text{else} \end{cases}.$$

On the other hand, using the embedding $\mathbb{P}(\bar{W}_{a,b,c})/\!\!/\bar{T} \hookrightarrow \mathbb{P}(\tilde{\bar{W}}_{a(d),b(d),c(d)})$ for $d \gg 0$, one sees that "$\sim_C$" agrees with the equivalence relation

$$[\bar{f}_1] \sim [\bar{f}_2] \quad \Longleftrightarrow \quad [\tilde{\bar{f}}_1] = [\tilde{\bar{f}}_2].$$

We call of point $[f] \in \mathbb{P}(W_{a,b,c})$ *semistable*, if

$$-\min\{a_1\gamma_1 + \cdots + a_t\gamma_t \,|\, f_{(a_1,\ldots,a_t)} \not\equiv 0\} \geq 0,$$

holds for every set of integers $\gamma_1, \ldots, \gamma_t$ with $\gamma_1 + \cdots + \gamma_t = 0$. An immediate consequence is the following.

Corollary. *Let "$\sim_A$" be the equivalence relation on the semistable K-valued points in the projective space $\mathbb{P}(W_{a,b,c})$ given through*

$$[f_1] \sim_A [f_2] \quad :\Longleftrightarrow \quad [\tilde{f}_1] = [\tilde{f}_2],$$

and "$\sim_B$" the one that is generated by $[f] \sim_B [f_{\mathrm{df}}]$ for every set of integers $\gamma_1, \ldots, \gamma_t$ with $\gamma_1 + \cdots + \gamma_t = 0$ and $\min\{a_1\gamma_1 + \cdots + a_t\gamma_t \,|\, f_{(a_1,\ldots,a_t)} \not\equiv 0\} = 0$. *Again,*

$$f_{\mathrm{df},(a_1,\ldots,a_t)} := \begin{cases} f_{(a_1,\ldots,a_t)}, & \text{if } a_1\gamma_1 + \cdots + a_t\gamma_t = 0 \\ 0, & \text{else} \end{cases}.$$

Then, "$\sim_A$" and "$\sim_B$" are equal as equivalence relations.

Now, we may prove the statement of the proposition. First, suppose $(\underline{\mathscr{E}}^1, \phi_1) \sim_B (\underline{\mathscr{E}}^2, \phi_2)$. This means that $(\underline{\mathscr{E}}^1, \tilde{\phi}_1) \sim_B (\underline{\mathscr{E}}^2, \tilde{\phi}_2)$ are isomorphic. In particular, there are isomorphisms $\psi_i \colon \mathscr{E}_i^1 \longrightarrow \mathscr{E}_i^2$, $i = 1, \ldots, t$, so that we may assume $\underline{\mathscr{E}}^1 = \underline{\mathscr{E}}^2 =: \underline{\mathscr{E}}$. Choose trivializations $\mathscr{E}_{i,\eta} \cong W_i$, $i = 1, \ldots, t$. Let $f_{1,2} \in W^{\vee}_{a,b,c}$ be the points defined by these trivializations and $\phi_{1,2}$. For any set of integers $\gamma_1, \ldots, \gamma_t$ with $\gamma_1 + \cdots + \gamma_t = 0$ and the corresponding one parameter subgroup λ of $\tilde{T}$, one checks

$$\mu(\lambda, \phi_{1,2}) = -\min\{a_1\gamma_1 + \cdots + a_t\gamma_t \,|\, f_{1,2(a_1,\ldots,a_t)} \not\equiv 0\}.$$

We see that f_1 and f_2 are semistable and $f_1 \sim_B f_2$ in the sense of the corollary. Note that for a one parameter subgroup λ of $\tilde{T}$ defined by $\gamma_1, \ldots, \gamma_t$, we have, for the corresponding admissible deformations of ϕ_1 and f_1, respectively,

$$\phi_{1,\mathrm{df};\eta} = f_{1,\mathrm{df}}.$$

Now, if we perform all the admissible deformations on $(\underline{\mathscr{E}}, \phi_1)$ which carry $[f_1]$ into $[f_2]$, then we arrive at a decorated tuple of sheaves $(\underline{\mathscr{E}}, \phi_2')$, such that $(\underline{\mathscr{E}}, \phi_1) \sim_A (\underline{\mathscr{E}}, \phi_2')$ and $[f_2'] = [f_2] \in \mathbb{P}(W_{a,b,c})$. By our assumption, we may find a big open subset $U \subseteq X$, such that

$$\phi'_{2|U}, \phi_{2|U} \colon ((\mathscr{E}_{|U}^{\otimes a})^{\oplus b}) \otimes \det(\mathscr{E}_{|U})^{\otimes -c} \longrightarrow \mathscr{L}_{|U}$$

are surjective homomorphisms, so that they define morphisms

$$\sigma', \sigma \colon U \longrightarrow \mathbb{P}(((\mathscr{E}_{|U}^{\otimes a})^{\oplus b}) \otimes \det(\mathscr{E}_{|U})^{\otimes -c}).$$

We know that σ' and σ agree at the generic point, whence they are equal. Since U is big, the universal property of projective bundles implies that $\phi'_{2|U} = z \cdot \phi_{2|U}$ for some $z \in \mathbb{G}_m(k)$. Thus, we obviously have $\phi_2 = z \cdot \phi_2'$. Finally, one immediately checks that $(\underline{\mathscr{E}}, \phi_2')$ is isomorphic to $(\underline{\mathscr{E}}, z \cdot \phi_2') = (\underline{\mathscr{E}}, \phi_2)$.

Conversely, suppose that $(\underline{\mathscr{E}}^1, \phi_1) \sim_A (\underline{\mathscr{E}}^2, \phi_2)$. By the Corollary, $\widetilde{\phi}_1$ and $\widetilde{\phi}_2$ define the same point in $\mathbb{P}(\widetilde{W}_{a,b,c})$. An argument similar to the one given above shows that $(\underline{\mathscr{E}}^1, \widetilde{\phi}_1)$ and $(\underline{\mathscr{E}}^2, \widetilde{\phi}_2)$ are isomorphic. □

We call a tuple of sheaves $(\underline{\mathscr{E}}, \phi)$ with a decoration of type $(\mathrm{II} : a, b, c; \mathscr{L})$ *A-polystable*, if $(\underline{\mathscr{E}}, \phi)$ is isomorphic to $\mathrm{df}_\lambda(\underline{\mathscr{E}}, \phi)$ for every one parameter subgroup λ of $\widetilde{T}$ with $\mu(\lambda, \phi) = 0$. Finally, a tuple of sheaves $(\underline{\mathscr{E}}, \phi)$ with a decoration of type $(\mathrm{II} : a, b, c; \mathscr{L})$ is called *δ-polystable*, if it is a) δ-semistable, b) A-polystable, and c) $(\underline{\mathscr{E}}, \widetilde{\phi})$ is $((1/d) \cdot \delta)$-polystable.

3.4 *Moduli spaces for decorated tuples (type I)*

In this section, we will generalize the results of [27] and [7]. The following moduli spaces are the basis of the subsequent constructions.

Theorem 3.4.1. *Fix polynomials* $\underline{P} = (P_1, \ldots, P_t)$ *of degree* $n = \dim(X)$, *integers* a, b, c, *and a line bundle* $\mathscr{L}$ *on* X. *Let* δ *be a polynomial of degree at most* $n - 1$. *There exists a projective coarse moduli space* $\mathrm{M}(\mathrm{I} : a, b, c; \mathscr{L})_{\underline{P}}^{\delta\text{-ss}}$ *for the functors* $\underline{\mathrm{M}}(\mathrm{I} : a, b, c; \mathscr{L})_{\underline{P}}^{\delta\text{-(s)s}}$.

To prove this theorem, we start by recalling some boundedness results. After some technical lemmas, we reformulate the stability condition for decorated sheaves in terms of the dimensions of some spaces of sections (Theorem 3.4.6). Then, we construct a quasi-projective parameter space $\mathfrak{Z}$, with a polarization which depends on δ, for decorated sheaves together with a choice of bases of the vector spaces $H^0(\mathscr{E}_i(m))$, $i = 1, \ldots, t$. To get rid of this choice, we take a GIT quotient. As usual, the crucial step is to show that a point of $\mathfrak{Z}$ is GIT-(semi)stable, if and only if it corresponds to a δ-(semi)stable decorated tuple of sheaves. This is done in Theorem 3.4.18 (where the alternative definition of δ-(semi)stability proved in Theorem 3.4.6 is used). Finally, a standard argument shows that the GIT quotient corepresents the functor of families of decorated tuples. The moduli space obtained in this way is projective, because there is a natural compactification $\overline{\mathfrak{Z}}$ of $\mathfrak{Z}$, such that all the points in $\overline{\mathfrak{Z}} \setminus \mathfrak{Z}$ are GIT-unstable, and then the GIT quotient of $\mathfrak{Z}$ is equal to the GIT quotient of $\overline{\mathfrak{Z}}$, and this is projective, because $\overline{\mathfrak{Z}}$ is so.

The following theorem was proved for characteristic 0 in [20], and for arbitrary characteristic in [18].

Theorem 3.4.2. *The family of torsion free sheaves $\mathscr{E}$ with fixed Hilbert polynomial P, such that $\mu_{\max}(\mathscr{E}) \leq C$, for a fixed constant C, is bounded.*

Lemma 3.4.3. *Let X be a smooth projective scheme with a polarization $\mathscr{O}_X(1)$, $d := \deg \mathscr{O}_X(1)$, and let $r > 0$ be an integer. Then, there exists a constant B with the following property: for every torsion free sheaf $\mathscr{E}$ with $0 < \operatorname{rk}(\mathscr{E}) \leq r$, we have*

$$h^0(\mathscr{E}) \leq \frac{1}{d^{n-1}n!}((\operatorname{rk}(\mathscr{E}) - 1)([\mu_{\max}(\mathscr{E}) + B]_+)^n + ([\mu_{\min}(\mathscr{E}) + B]_+)^n).$$

Proof. For characteristic 0, this was proved by Simpson (see [33], Corollary 1.7, and [13], Lemma 2.2). For arbitrary characteristic, the estimate

$$\frac{h^0(\mathscr{E})}{\operatorname{rk}\mathscr{E}} \leq d\binom{\frac{\mu(\mathscr{E})}{d} + f(r) + n}{n} \leq \frac{1}{d^{n-1}n!}(\mu(\mathscr{E}) + B)^n$$

holds for a slope semistable torsion free sheaf $\mathscr{E}$. The first inequality was proved in [19] with $d = \deg \mathscr{O}_X(1)$ and $f(r)$ an explicitly given function. The second one is obtained by setting $B = (f(r) + n)d$. If $\mathscr{E}$ is any torsion free sheaf, let $\mathscr{E}^i$ be its Harder-Narasimhan factors. Using $h^0(\mathscr{E}) \leq \sum h^0(\mathscr{E}^i)$, the result follows. □

Lemma 3.4.4. *There is an integer A_1 (depending only on the Hilbert polynomials $P(\mathscr{E}_i)$, $i = 1, \ldots, t$, the integers a, b, c, and the line bundle $\mathscr{L}$), such that it is enough to check the stability condition* (21) *for weighted filtrations with* ***integers*** *$\alpha_{i,j} \leq A_1$ for all i, j.*

Proof. In the definition of stability, if we allow some (but not all) of the integers $\alpha_{i,j}$ to be zero, then we can assume that the filtration gives a full flag on the general fiber, i.e., $s_i + 1 = \operatorname{rk}\mathscr{E}_i =: r_i$ and $\operatorname{rk}\mathscr{E}_{i,j} = j$. We take this point of view only during this proof.

Let $\mathscr{J}$ be as in (19), but with $s_i = r_i - 1$. Note that (20) is a piecewise linear function of $\gamma \in C$, where $C \subset \bigoplus_i \mathbb{Z}^{r_i}$ is the cone defined by $\gamma_1^i \leq \ldots \leq \gamma_{r_i}^i$ and $\gamma_1^i + \ldots + \gamma_{r_i}^i = 0$ for all i. This is because it is defined as the minimum among a finite set of linear functions, namely the functions $\sum_{i=1}^t(\gamma_{j_1^i}^i + \cdots + \gamma_{j_a^i}^i)$ for $J \in \mathscr{J}$. There is a decomposition of $C = \bigcup_{I \in \mathscr{I}} C_I$ into a finite number of subcones C_I, such that the function (20) is linear on each cone C_I.

Checking (semi)stability with a vector γ is clearly equivalent to checking it with a multiple of that vector. Choose one vector on each one-dimensional edge of each cone C_I, and call this set of vectors S. Taking a multiple if necessary, we may assume that the coordinates of all vectors in S are divisible by r_i for all i. Therefore, all vectors in S come from a collection of integer weights $\alpha_{i,j} \geq 0$, $i = 1, \ldots, t$, $j = 1, \ldots, r_i - 1$, given by the formula $\gamma^i = \sum_{j=1}^{r_i-1} \alpha_{i,j}\gamma_{r_i}^{(j)}$ (the fact that the coordinates

γ^i are divisible by r_i implies that the numbers $\alpha_{i,j}$ are integers). Hence, to obtain the finite set S of vectors, it is enough to consider a finite set of values for $\alpha_{i,j}$, and hence there is a maximum value A_1.

Finally, it is easy to see that it is enough to check (21) for the weights associated with the vectors in S. Indeed, first note that since the first term in (21) is linear on C, then it is also linear on each C_I. Then the left hand side of (21) is linear on each C_I, and hence it is enough to check it on all the edges of all the cones C_I. □

Lemma 3.4.5. *Let $(\underline{\mathscr{E}_\bullet}, \underline{\alpha_\bullet})$ be a weighted filtration as above, and let $\mathbb{T}'$ be a subset of $\mathbb{T} = \{(i,j) \mid i = 1, \ldots, t,\ j = 1, \ldots, s_i\}$. Let $(\underline{\mathscr{E}'_\bullet}, \underline{\alpha'_\bullet})$ be the subfiltration obtained by considering only those subsheaves $\mathscr{E}_{i,j}$ and weights $\alpha_{i,j}$ for which $(i,j) \in \mathbb{T}'$. Then*

$$\mu(\underline{\mathscr{E}_\bullet}, \underline{\alpha_\bullet}; \varphi) \ \geq \ \mu(\underline{\mathscr{E}'_\bullet}, \underline{\alpha'_\bullet}; \varphi) - a \cdot \sum_{(i,j)\in\mathbb{T}\setminus\mathbb{T}'} \alpha_{i,j} r_{i,j}.$$

Proof. Let $\mathscr{J}$ be as in (19). Given $J \in \mathscr{J}$, we have

$$(\gamma^i_{j^i_1} + \cdots + \gamma^i_{j^i_a}) = \sum_{j=1}^{s_i} \alpha_{i,j}(a\, r_{i,j} - \nu_{i,j}(J)\, r_i),$$

where $\nu_{i,j}(J) = \#\{u \mid 1 \leq u \leq a,\ j^i_u \leq j\}$. If $J \in \mathscr{J}$ gives the minimum in (20), we set $\varepsilon_{i,j}(\underline{\mathscr{E}_\bullet}, \underline{\alpha_\bullet}; \varphi) := \nu_{i,j}(J)$ (or just $\varepsilon_{i,j}(\underline{\mathscr{E}_\bullet})$, if the rest of the data is clear from the context). Then,

$$\mu(\underline{\mathscr{E}_\bullet}, \underline{\alpha_\bullet}; \varphi) = -\sum_{i=1}^{t}\sum_{j=1}^{s_i} \alpha_{i,j}(a\, r_{i,j} - \varepsilon_{i,j}(\underline{\mathscr{E}_\bullet})\, r_i). \tag{31}$$

We index the filtration $(\underline{\mathscr{E}'_\bullet}, \underline{\alpha'_\bullet})$ with $\mathbb{T}'$. Let $J' \in \mathscr{J}$ be a multi-index giving the minimum in (20) for the filtration $(\underline{\mathscr{E}'_\bullet}, \underline{\alpha'_\bullet})$. In particular, we have $\varphi_{|(\bigotimes_{i=1}^t \mathscr{E}_{i,j'^i_1}\otimes\cdots\otimes\mathscr{E}_{i,j'^i_a})^{\oplus b}}$ $\neq 0$. Then

$$\begin{aligned}
-\mu(\underline{\mathscr{E}_\bullet}, \underline{\alpha_\bullet}; \varphi) &= \min_{J\in\mathscr{J}} \left\{ \sum_{i=1}^{t} (\gamma^i_{j^i_1} + \cdots + \gamma^i_{j^i_a}) | \varphi_{|(\otimes_{i=1}^t \mathscr{E}_{i,j^i_1}\otimes\cdots\otimes\mathscr{E}_{i,j^i_a})^{\oplus b}} \neq 0 \right\} \\
&\leq \sum_{i=1}^{t} (\gamma^i_{j'^i_1} + \cdots + \gamma^i_{j'^i_a}) \\
&= \sum_{i=1}^{t}\sum_{j=1}^{s_i} \alpha_{i,j}(a\, r_{i,j} - \nu_{i,j}(J')\, r_i) \\
&= \sum_{i=1}^{t}\sum_{j=1}^{s_i} \alpha_{i,j}(a\, r_{i,j} - \varepsilon_{i,j}(\underline{\mathscr{E}'_\bullet})\, r_i)
\end{aligned}$$

$$= \sum_{(i,j)\in\mathbb{T}'} \alpha_{i,j}(a\, r_{i,j} - \varepsilon_{i,j}(\underline{\mathscr{E}}'_\bullet)\, r_i)$$
$$+ \sum_{(i,j)\in\mathbb{T}\setminus\mathbb{T}'} \alpha_{i,j}(a\, r_{i,j} - \varepsilon_{i,j}(\underline{\mathscr{E}}'_\bullet)\, r_i)$$
$$\leq -\mu(\underline{\mathscr{E}}'_\bullet, \underline{\alpha}'_\bullet; \varphi) + a \cdot \sum_{(i,j)\in\mathbb{T}\setminus\mathbb{T}'} \alpha_{i,j} r_{i,j}\, .$$

This is the claimed inequality. □

Theorem 3.4.6. *There is an integer N_0, such that, if $m \geq N_0$, the following properties of decorated tuples $(\underline{\mathscr{E}}, \varphi)$, with $\mathscr{E}_i$ torsion free and $P(\mathscr{E}_i) = P_i$ for all i, are equivalent:*

1. *$(\underline{\mathscr{E}}, \varphi)$ is (semi)stable.*
2. *$P_i(m) \leq h^0(\mathscr{E}_i(m))$ and for every weighted filtration $(\underline{\mathscr{E}}_\bullet, \underline{\alpha}_\bullet)$ as in* (17)

$$\left(\sum_{i=1}^{t}\sum_{j=1}^{s_i} \alpha_{i,j}(r_{i,j} P_i(m) - r_i\, h^0(\mathscr{E}_{i,j}(m)))\right) + \delta(m)\cdot \mu(\underline{\mathscr{E}}_\bullet, \underline{\alpha}_\bullet; \varphi)\ (\geq)\ 0.$$

3. *For every weighted filtration $(\underline{\mathscr{E}}_\bullet, \underline{\alpha}_\bullet)$ as in* (17)

$$\left(\sum_{i=1}^{t}\sum_{j=1}^{s_i} \alpha_{i,j}(r_i h^0(\widehat{\mathscr{E}}_{i,j}(m)) - \widehat{r}_{i,j} P_i(m))\right) + \delta(m)\cdot \mu(\underline{\mathscr{E}}_\bullet, \underline{\alpha}_\bullet; \varphi)\ (\geq)\ 0.$$

Here, $\widehat{\mathscr{E}}_{i,j} = \mathscr{E}_i/\mathscr{E}_{i,j}$ and $\widehat{r}_{i,j} = \operatorname{rk}\widehat{\mathscr{E}}_{i,j}$.

Furthermore, for any decorated tuple $(\underline{\mathscr{E}}, \varphi)$ satisfying these conditions, $\mathscr{E}_i$ is m-regular for all i.

Recall that a sheaf $\mathscr{E}$ is called *m-regular*, if $h^u(\mathscr{E}(m-u)) = 0$ for $u > 0$. If $\mathscr{E}$ is m-regular, then $\mathscr{E}(m)$ is generated by global sections, and it is m'-regular for any $m' > m$. The set of decorated tuples $(\underline{\mathscr{E}}, \varphi)$, with $\mathscr{E}_i$ torsion free and $P(\mathscr{E}_i) = P_i$, $i = 1, \ldots, t$, satisfying the weak version of Conditions 1.-3. will be called $\mathscr{S}^{s}$, $\mathscr{S}'_m$, and $\mathscr{S}''_m$. Before we prove this theorem we need some lemmas.

Lemma 3.4.7. *There are integers N_1, C, such that, if $(\underline{\mathscr{E}}, \varphi)$ belongs to $\mathscr{S} = \mathscr{S}^{s} \cup \bigcup_{m\geq N_1} \mathscr{S}''_m$, then, for all saturated weighted filtrations, the following holds for all i and j:*

$$\deg(\mathscr{E}_{i,j}) - r_{i,j}\mu_{i,a} \leq C, \tag{32}$$

(where $\mu_{i,a} = (d_i - a\bar{\delta})/r_i)$ and either $-C \leq \deg(\mathscr{E}_{i,j}) - r_{i,j}\mu_{i,a}$ or

1. *$r_i\, h^0(\mathscr{E}_{i,j}(m)) < r_{i,j}(P_i(m) - a\delta(m))$, if $(\underline{\mathscr{E}}, \varphi) \in \mathscr{S}^{s}$ and $m \geq N_1$.*
2. *$\widehat{r}_{i,j}(P_i - a\,\delta) \prec r_i(P(\widehat{\mathscr{E}}_{i,j}) - a\,\delta)$, if $(\underline{\mathscr{E}}, \varphi) \in \bigcup_{m\geq N_1} \mathscr{S}''_m$.*

Proof. Let B be as in Lemma 3.4.3. Choose C large enough, such that $C > a\bar{\delta}$ and the leading coefficient of $G_i - (P_i - a\delta)/r_i$ is negative. Here,

$$G_i(m) = \frac{1}{d^{n-1}n!}\left(\left(1 - \frac{1}{r_i}\right)(\mu_{i,a} + a\bar{\delta} + md + B)^n + \frac{1}{r_i}\left(\mu_{i,a} - \frac{1}{r_i}C + md + B\right)^n\right). \tag{33}$$

Choose N_1 large enough, such that, for $m \geq N_1$ and for all i,

$$\delta(m) \geq 0, \tag{34}$$

$$\mu_{i,a} - \frac{C}{r_i} + md + B > 0, \tag{35}$$

$$G_i(m) - \frac{P_i(m) - a\delta(m)}{r_i} < 0. \tag{36}$$

Since the filtration is assumed to be saturated and since $\mathscr{E}_i$ is torsion free, we have $0 < r_{i,j} < r_i$.

Case 1. Suppose $(\underline{\mathscr{E}}, \varphi) \in \mathscr{S}^{s}$. For each i, j, consider the one step filtration $\{0\} \subsetneq \mathscr{E}_{i,j} \subsetneq \mathscr{E}_i$. The leading coefficient of the δ-semistability condition applied to this filtration, together with $C > a\bar{\delta}$, implies (32).

Let $\mathscr{E}_{i,j,\max} \subset \mathscr{E}_{i,j}$ be the term in the Harder-Narasimhan filtration with maximal slope. Therefore, the same argument applied to $\mathscr{E}_{i,j,\max}$ gives

$$\mu_{\max}(\mathscr{E}_{i,j}) = \mu(\mathscr{E}_{i,j,\max}) < \mu_{i,a} + a\,\bar{\delta}. \tag{37}$$

Now assume that the first alternative does not hold, i.e.,

$$-C > \deg(\mathscr{E}_{i,j}) - r_{i,j}\mu_{i,a}.$$

This gives

$$\mu_{\min}(\mathscr{E}_{i,j}) \leq \mu(\mathscr{E}_{i,j}) < \mu_{i,a} - \frac{C}{r_i}. \tag{38}$$

Combining Lemma 3.4.3 with (35) - (38), we have

$$r_i\, h^0(\mathscr{E}_{i,j}(m)) < r_{i,j}(P_i(m) - a\,\delta(m)).$$

Case 2. Suppose $(\underline{\mathscr{E}}, \varphi) \in \mathscr{S}''_m$ for some $m \geq N_1$. For each i, j, consider the quotient $\widehat{\mathscr{E}}_{i,j} = \mathscr{E}_i/\mathscr{E}_{i,j}$. Let $\widehat{\mathscr{E}}_{i,j,\min}$ be the last factor of the Harder-Narasimhan filtration of $\widehat{\mathscr{E}}_{i,j}$ (i.e., $\mu(\widehat{\mathscr{E}}_{i,j,\min}) = \mu_{\min}(\widehat{\mathscr{E}}_{i,j})$). Let $\mathscr{E}'$ be defined by the exact sequence

$$0 \longrightarrow \mathscr{E}' \longrightarrow \mathscr{E}_{i,j} \longrightarrow \widehat{\mathscr{E}}_{i,j,\min} \longrightarrow 0\,,$$

and consider the one step filtration $\{0\} \subsetneq \mathscr{E}' \subsetneq \mathscr{E}_i$. Equations (33) and (35) imply that $0 < G_i(m)$. Then, a short calculation, using (36), the fact that $(\underline{\mathscr{E}}, \varphi) \in \mathscr{S}''_m$, (34), and Lemma 3.4.3, shows

$$G_i(m) < \frac{h^0(\widehat{\mathscr{E}}_{i,j,\min}(m))}{\mathrm{rk}(\widehat{\mathscr{E}}_{i,j,\min})} \leq \frac{1}{d^{n-1}n!}(\mu_{\min}(\widehat{\mathscr{E}}_{i,j}) + md + B)^n.$$

It can be seen that, if this inequality of polynomials holds for some $m \geq N_1$, then it holds for all larger values of m. Hence, choosing m large enough and looking at the coefficients, we have

$$\mu_{\min}(\widehat{\mathscr{E}}_{i,j}) \geq \mu_{i,a} + \left(1 - \frac{1}{r_i}\right) a\,\bar{\delta} - \frac{C}{r_i^2}.$$

A short calculation, using this, $\mu_{\min}(\widehat{\mathscr{E}}_{i,j}) \leq \mu(\widehat{\mathscr{E}}_{i,j})$, and $0 < \mathrm{rk}(\widehat{\mathscr{E}}_{i,j}) < \mathrm{rk}(\mathscr{E}_i)$ (whence $\mathrm{rk}(\mathscr{E}_i) > 1$), yields (32).

Now assume that the first alternative does not hold, i.e.,

$$-C > \deg(\mathscr{E}_{i,j}) - r_{i,j}\mu_{i,a}.$$

It follows that $\widehat{r}_{i,j}\mu_{i,a} < \deg(\widehat{\mathscr{E}}_{i,j}) - a\,\bar{\delta}$, and hence

$$\widehat{r}_{i,j}(P_i - a\delta) \prec r_i(P(\widehat{\mathscr{E}}_{i,j}) - a\delta).$$

This concludes the proof. □

Lemma 3.4.8. *The set $\mathscr{S} = \mathscr{S}^{s} \cup \bigcup_{m \geq N_1} \mathscr{S}''_m$ is bounded.*

Proof. Let $(\underline{\mathscr{E}}, \varphi) \in \mathscr{S}$. Let $\mathscr{E}'$ be a subsheaf of $\mathscr{E}_i$ for some i, and $\overline{\mathscr{E}}'$ the saturated subsheaf of $\mathscr{E}_i$ generated by $\mathscr{E}'$. Using Lemma 3.4.7, we infer

$$\frac{\deg(\mathscr{E}')}{\mathrm{rk}(\mathscr{E}')} \leq \frac{\deg(\overline{\mathscr{E}}')}{\mathrm{rk}(\overline{\mathscr{E}}')} \leq \mu_{i,a} + \frac{C}{\mathrm{rk}(\overline{\mathscr{E}}')} \leq \mu_{i,a} + C.$$

Then, by Theorem 3.4.2, the set $\mathscr{S}$ is bounded. □

Lemma 3.4.9. *Let $\mathscr{S}_0$ be the set of sheaves $\mathscr{E}'$, such that $\mathscr{E}'$ is a saturated subsheaf of $\mathscr{E}_i$ for some $(\underline{\mathscr{E}}, \varphi) \in \mathscr{S}$ and $i \in \{1, \dots, t\}$, and furthermore*

$$|\deg(\mathscr{E}') - r'\mu_{i,a}| \leq C.$$

Then, $\mathscr{S}_0$ is bounded.

Proof. Let $\mathscr{E}' \in \mathscr{S}_0$. The sheaf $\mathscr{E}'' = \mathscr{E}_i/\mathscr{E}'$ is torsion free, and $|\deg(\mathscr{E}'')|$ is uniformly bounded, because the set $\mathscr{S}$ is bounded and

$$|\deg(\mathscr{E}'')| \leq |\deg(\mathscr{E}_i)| + |\deg(\mathscr{E}')| \leq \max_{\underline{\mathscr{E}} \in \mathscr{S}} |\deg(\mathscr{E}_i)| + C + r|\mu_{i,a}|.$$

A lemma of Grothendieck ([9], Lemma 2.5) implies that the set of torsion free quotients $\mathscr{E}''$ of sheaves in a bounded set with $|\deg(\mathscr{E}'')| \leq C''$ for some fixed constant C'' is bounded, too. Therefore, $\mathscr{S}_0$ is also bounded. □

Lemma 3.4.10. *There is an integer N_2, such that for every weighted filtration $(\underline{\mathscr{E}}_\bullet, \underline{\alpha}_\bullet)$ as in* (17) *with $\mathscr{E}_{i,j} \in \mathscr{S}_0$, the inequality of polynomials*

$$\left(\sum_{i=1}^{t}\sum_{j=1}^{s_i} \alpha_{i,j}(r_{i,j}P_i - r_i P(\mathscr{E}_{i,j}))\right) + \delta \cdot \mu(\underline{\mathscr{E}}_\bullet, \underline{\alpha}_\bullet; \varphi) \;(\succeq)\; 0$$

holds, if and only if it holds for a particular value of $m \geq N_2$.

Proof. Since $\mathscr{S}_0$ is bounded, the set that consists of the polynomials δ, P_0, $r'P_0$, and $P(\mathscr{E}')$ for $\mathscr{E}' \in \mathscr{S}_0$ is finite. On the other hand, thanks to Lemma 3.4.4, we only need to consider a finite number of values for $\alpha_{i,j}$. Hence, the result follows. □

Proof of Theorem 3.4.6. Let $N_0 > \max\{N_1, N_2\}$ and such that all sheaves in $\mathscr{S}$ and $\mathscr{S}_0$ are N_0-regular, and $\bigotimes_{i=1}^{t} \mathscr{E}_{i,j_1^i} \otimes \cdots \otimes \mathscr{E}_{i,j_a^i}$ is $(at N_0)$-regular for all $J \in \mathscr{J}$ (see (19) for the definition of $\mathscr{J}$) and $\mathscr{E}_{i,j}$ in $\mathscr{S}_0$.
2. ⇒ 3. Let $(\underline{\mathscr{E}}, \varphi) \in \mathscr{S}'_m$. Consider a weighted filtration $(\underline{\mathscr{E}}_\bullet, \underline{\alpha}_\bullet)$. Then

$$\left(\sum_{i=1}^{t}\sum_{j=1}^{s_i} \alpha_{i,j}(r_i h^0(\widehat{\mathscr{E}}_{i,j}(m)) - \widehat{r}_{i,j} P_i(m))\right) + \delta(m)\cdot\mu(\underline{\mathscr{E}}_\bullet, \underline{\alpha}_\bullet; \varphi)$$

$$\geq \left(\sum_{i=1}^{t}\sum_{j=1}^{s_i} \alpha_{i,j}(r_{i,j} P_i(m) - r_i h^0(\mathscr{E}_{i,j}(m)))\right) + \delta(m)\cdot\mu(\underline{\mathscr{E}}_\bullet, \underline{\alpha}_\bullet; \varphi)(\geq)0.$$

1. ⇒ 2. Let $(\underline{\mathscr{E}}, \varphi) \in \mathscr{S}^s$ and consider a saturated weighted filtration as in (17). Since, for all i, $\mathscr{E}_i$ is N_0-regular, $P_i(m) = h^0(\mathscr{E}_i(m))$. If $\mathscr{E}_{i,j} \in \mathscr{S}_0$, then $P(\mathscr{E}_{i,j})(m) = h^0(\mathscr{E}_{i,j}(m))$. If $\mathscr{E}_{i,j} \notin \mathscr{S}_0$, then the second alternative of Lemma 3.4.7 holds, and then

$$r_i\, h^0(\mathscr{E}_{i,j}(m)) \;<\; r_{i,j}(P_i(m) - a\,\delta(m)).$$

Let $\mathbb{T}' \subset \mathbb{T} = \{(i,j) \mid i = 1,\ldots,t,\ j = 1,\ldots,s_i\}$ be the subset of those (i,j) for which $\mathscr{E}_{i,j} \in \mathscr{S}_0$. Let $(\underline{\mathscr{E}}'_\bullet, \underline{\alpha}'_\bullet)$ be the corresponding subfiltration. Lemma 3.4.5 and a short calculation show that

$$\left(\sum_{i=1}^{t}\sum_{j=1}^{s_i} \alpha_{i,j}(r_{i,j} P_i(m) - r_i\, h^0(\mathscr{E}_{i,j}(m)))\right) + \delta(m)\cdot\mu(\underline{\mathscr{E}}_\bullet, \underline{\alpha}_\bullet; \varphi)$$

$$\geq \left(\sum_{(i,j)\in\mathbb{T}'} \alpha_{i,j}(r_{i,j} P_i(m) - r_i\, P(\mathscr{E}_i)(m))\right) + \delta(m)\cdot\mu(\underline{\mathscr{E}}'_\bullet, \underline{\alpha}'_\bullet; \varphi)$$

$$+ \left(\sum_{(i,j)\in\mathbb{T}\setminus\mathbb{T}'} \alpha_{i,j}(r_{i,j} P_i(m) - r_i\, h^0(\mathscr{E}_{i,j}(m)) + a r_{i,j}\delta(m))\right)$$

$$\geq \left(\sum_{(i,j)\in\mathbb{T}'} \alpha_{i,j}(r_{i,j} P_i(m) - r_i\, P(\mathscr{E}_{i,j})(m))\right) + \delta(m)\cdot\mu(\underline{\mathscr{E}}'_\bullet, \underline{\alpha}'_\bullet; \varphi)(\geq)0. \quad (39)$$

The condition that $\mathscr{E}_{i,j}$ is saturated can be dropped, because $h^0(\mathscr{E}_{i,j}(m)) \leq h^0(\overline{\mathscr{E}}_{i,j}(m))$, and we have $\mu(\underline{\mathscr{E}}_\bullet, \underline{\alpha}_\bullet; \varphi) = \mu(\overline{\underline{\mathscr{E}}}_\bullet, \underline{\alpha}_\bullet; \varphi)$. Here, $\overline{\mathscr{E}}_{i,j}$ is the saturated subsheaf generated by $\mathscr{E}_{i,j}$ in $\mathscr{E}_i$.
3. ⇒ 1. Let $(\underline{\mathscr{E}}, \varphi) \in \mathscr{S}''_m$ and consider a saturated weighted filtration $(\underline{\mathscr{E}}_\bullet, \underline{\alpha}_\bullet)$. Since, for all i, $\mathscr{E}_i$ is N_0-regular, $P_i(m) = h^0(\mathscr{E}_i(m))$. If $\mathscr{E}_{i,j} \in \mathscr{S}_0$, then

$P(\mathscr{E}_{i,j})(m) = h^0(\mathscr{E}_{i,j}(m))$. Hence, Hypothesis 3 applied to the subfiltration $(\underline{\mathscr{E}}'_\bullet, \underline{\alpha}'_\bullet)$ obtained by the terms with $\mathscr{E}_{i,j} \in \mathscr{S}_0$ implies

$$\left(\sum_{(i,j):\mathscr{E}_{i,j}\in\mathscr{S}_0} \alpha_{i,j}(r_i P(\widehat{\mathscr{E}}_{i,j})(m) - \widehat{r}_{i,j} P_i(m)) \right) + \delta(m) \cdot \mu(\underline{\mathscr{E}}'_\bullet, \underline{\alpha}'_\bullet; \varphi) \ (\geq) \ 0 .$$

This is equivalent to

$$\left(\sum_{(i,j):\mathscr{E}_{i,j}\in\mathscr{S}_0} \alpha_{i,j}(r_{i,j} P_i(m) - r_i P(\mathscr{E}_{i,j})(m)) \right) + \delta(m) \cdot \mu(\underline{\mathscr{E}}'_\bullet, \underline{\alpha}'_\bullet; \varphi) \ (\geq) \ 0 ,$$

and, by Lemma 3.4.10, this is, in turn, equivalent to

$$\left(\sum_{(i,j):\mathscr{E}_{i,j}\in\mathscr{S}_0} \alpha_{i,j}(r_{i,j} P_i - r_i P(\mathscr{E}_{i,j})) \right) + \delta \cdot \mu(\underline{\mathscr{E}}'_\bullet, \underline{\alpha}'_\bullet; \varphi) \ (\succeq) \ 0. \tag{40}$$

If $\mathscr{E}_{i,j} \notin \mathscr{S}_0$, then the second alternative of Lemma 3.4.7 holds, and then

$$r_{i,j} P_i - r_i P(\mathscr{E}_{i,j}) - a\, r_{i,j}\, \delta \succ 0 . \tag{41}$$

Using Lemma 3.4.5, (40), and (41), we deduce

$$\left(\sum_{i=1}^{t} \sum_{j=1}^{s_i} \alpha_{i,j}(r_{i,j} P_i - r_i P(\mathscr{E}_{i,j})) \right) + \delta \cdot \mu(\underline{\mathscr{E}}_\bullet, \underline{\alpha}_\bullet; \varphi) \ (\succeq) \ 0 .$$

Again, we can drop the condition that the filtration is saturated, and this finishes the proof of Theorem 3.4.6 □

Corollary 3.4.11. *Let $(\underline{\mathscr{E}}, \varphi)$ be δ-semistable, $m \geq N_0$, and assume that there is a weighted filtration $(\underline{\mathscr{E}}_\bullet, \underline{\alpha}_\bullet)$ with*

$$\left(\sum_{i=1}^{t} \sum_{j=1}^{s_i} \alpha_{i,j}(r_{i,j} P_i(m) - r_i h^0(\mathscr{E}_{i,j}(m))) \right) + \delta(m) \cdot \mu(\underline{\mathscr{E}}_\bullet, \underline{\alpha}_\bullet; \varphi) \ = \ 0 .$$

Then, $\mathscr{E}_{i,j} \in \mathscr{S}_0$ and $h^0(\mathscr{E}_{i,j}(m)) = P(\mathscr{E}_{i,j})(m)$, for all i, j.

Proof. By the proof of the part "1.⇒2." of Theorem 3.4.6, if we have this equality, then all inequalities in (39) are equalities, hence $\mathbb{T}' = \mathbb{T}$, $\mathscr{E}_{i,j} \in \mathscr{S}_0$, for all i, j, and the result follows. □

Note that, in Theorem 3.4.6, we are assuming that $\mathscr{E}_i$ is torsion free for all i. To handle the general case, we will use the following lemma

Lemma 3.4.12. *Let $(\underline{\mathscr{E}}, \varphi)$ be a decorated tuple. Assume there is a family $(\underline{\mathscr{E}}_C, \varphi_C)$ parameterized by a smooth curve C, such that $(\underline{\mathscr{E}}_0, \varphi_0) = (\underline{\mathscr{E}}, \varphi)$ for a point $0 \in C$*

and $(\mathscr{E}_i)_t$ is torsion free for $t \neq 0$. Then, there exist a decorated tuple of sheaves $(\underline{\mathscr{F}}, \psi)$ and a homomorphism

$$(\underline{\mathscr{E}}, \varphi) \stackrel{\beta}{\longrightarrow} (\underline{\mathscr{F}}, \psi),$$

such that $\mathscr{F}_i$ is torsion free and $P(\mathscr{E}_i) = P(\mathscr{F}_i)$, for all i, and the sequence

$$0 \longrightarrow T(\mathscr{E}_i) \longrightarrow \mathscr{E}_i \stackrel{\beta_i}{\longrightarrow} \mathscr{F}_i,$$

with $T(\mathscr{E}_i)$ the torsion subsheaf of $\mathscr{E}_i$, $i = 1, \ldots, t$, is exact.

Proof. The family is given by a tuple $(\underline{\mathscr{E}}_C, \varphi_C, \mathscr{N})$ as in (22). Shrinking C, we can assume that $\mathscr{N}$ is trivial. Let $U = (X \times C) \setminus \mathrm{Supp}(T(\underline{\mathscr{E}}_0))$ (by the *support of* $T(\underline{\mathscr{E}}_0)$, we mean the union of the supports of $T((\mathscr{E}_i)_0)$ for all i). Let $\mathscr{F}_{iC} = j_\star(\mathscr{E}_{iC|U})$. Since it has no C-torsion, $\mathscr{F}_{iC}$ is flat over C. The natural map $\widetilde{\beta}\colon \underline{\mathscr{E}}_C \longrightarrow \underline{\mathscr{F}}_C$ is an isomorphism on U. Hence, we have a homomorphism $\psi_U := \varphi_{C|U}$ on U, and this extends to a homomorphism ψ_C on $X \times C$. Finally, define $(\underline{\mathscr{F}}, \psi) = (\underline{\mathscr{F}}_0, \psi_0)$, and let β be the homomorphism induced by $\widetilde{\beta}$. □

The GIT construction. As before, we have fixed polynomials P_i of degree $n = \dim(X)$, $i = 1, \ldots, t$, integers a, b and c, and a line bundle $\mathscr{L}$. Let d_i be the degree of a torsion free sheaf whose Hilbert polynomial is P_i, $i = 1, \ldots, t$.

Let $m \geq N_0$ be large enough, such that, for all $u > 0$, all line bundles $\mathscr{M}$ of degree $d_1 + \cdots + d_t$, and all $m > N$, we have $h^u(\mathscr{M}^{\otimes c} \otimes \mathscr{L}(atm)) = 0$ for all $u > 0$ and $\mathscr{M}^{\otimes c} \otimes \mathscr{L}(atm)$ is generated by global sections. Let U_i be a vector space of dimension $h_i = P_i(m)$, $i = 1, \ldots, t$. The choice of m implies that, if $(\underline{\mathscr{E}}, \varphi)$ is δ-semistable, then $\mathscr{E}_i(m)$ is generated by global sections for all i, and $h^u(\mathscr{E}_i(m)) = 0$ for $u > 0$.

Let $(\underline{g}, \underline{\mathscr{E}}, \varphi)$ be a tuple where $(\underline{\mathscr{E}}, \varphi)$ is a δ-semistable decorated tuple of sheaves and $\underline{g} = (g_1, \ldots, g_t)$ is a tuple of isomorphisms $g_i\colon U_i \longrightarrow H^0(\mathscr{E}_i(m))$, $i = 1, \ldots, t$. This induces quotients

$$q_i\colon U_i \otimes \mathscr{O}_X(-m) \longrightarrow \mathscr{E}_i,\ i = 1, \ldots, t, \tag{42}$$

and, hence, a point $\underline{q} = (q_1, \ldots, q_t)$ in $\overline{\mathfrak{Q}} = \overline{\mathfrak{Q}}_1 \times \cdots \times \overline{\mathfrak{Q}}_t$, the product of the projective quot schemes $\overline{\mathfrak{Q}}_i$ of quotients of $U_i \otimes \mathscr{O}_X(-m)$ with Hilbert polynomial P_i. The scheme $\overline{\mathfrak{Q}}$ has universal quotients $q_{\overline{\mathfrak{Q}},i}\colon U_i \otimes \pi_X^\star \mathscr{O}_X(-m) \longrightarrow \mathfrak{E}_{\overline{\mathfrak{Q}},i}$. Let $\mathfrak{E}_{\overline{\mathfrak{Q}}} = \mathfrak{E}_{\overline{\mathfrak{Q}},1} \oplus \cdots \oplus \mathfrak{E}_{\overline{\mathfrak{Q}},t}$. The tuple $(\underline{g}, \underline{\mathscr{E}}, \varphi)$ also induces a linear map

$$\begin{aligned}\Phi\colon \left(\bigotimes_{i=1}^t U_i^{\otimes a}\right)^{\oplus b} &\longrightarrow H^0\left(\left(\bigotimes_{i=1}^t \mathscr{E}_i(m)^{\otimes a}\right)^{\oplus b}\right)\\ &\longrightarrow H^0(\det(\mathscr{E})^{\otimes c} \otimes \mathscr{L}(atm)),\end{aligned} \tag{43}$$

and, hence, a point $(\underline{q}, [\Phi])$ in the projective bundle $\bar{\mathfrak{H}}$ over $\overline{\mathfrak{Q}}$ defined as

$$\bar{\mathfrak{H}} := \mathbb{P}\left(\left(\bigotimes_{i=1}^t U_i^{\otimes a}\right)^{\oplus b} \otimes \left(\pi_{\overline{\mathfrak{Q}}\star}(\det(\mathfrak{E}_{\overline{\mathfrak{Q}}})^{\otimes c} \otimes \pi_X^\star \mathscr{L}(atm))\right)^\vee\right) \stackrel{\pi}{\longrightarrow} \bar{\mathfrak{Q}}.$$

The schemes $\bar{\mathfrak{Q}}$ and $\bar{\mathfrak{H}}$ carry natural actions by $\mathrm{SL}(U_1) \times \cdots \times \mathrm{SL}(U_t)$, and π is equivariant.

A point $(\underline{q}, [\Phi]) \in \overline{\mathfrak{H}}$ associated with a tuple $(\underline{g}, \underline{\mathscr{E}}, \varphi)$ has the property that the homomorphism Φ in (43), composed with the evaluation map, factors as

$$\begin{array}{ccc} \left(\bigotimes_{i=1}^{t} U_i^{\otimes a}\right)^{\oplus b} \otimes \mathscr{O}_X & \xrightarrow{\left(\bigotimes_{i=1}^{t} q_i^{\otimes a}\right)^{\oplus b}} & \left(\mathscr{E}(m)^{\otimes a}\right)^{\oplus b} \\ \downarrow \Phi & & \\ H^0(\det(\mathscr{E})^{\otimes c} \otimes \mathscr{L}(atm)) \otimes \mathscr{O}_X & & \downarrow \varphi \\ \downarrow \mathrm{ev} & & \\ \det(\mathscr{E})^{\otimes c} \otimes \mathscr{L}(atm). & \swarrow & \end{array} \tag{44}$$

Consider the relative version of the homomorphisms in (44), i.e., the following commutative diagram on $X \times \bar{\mathfrak{H}}$:

$$\begin{array}{ccccccccc} 0 & \longrightarrow & \mathscr{K} & \longrightarrow & \left(\bigotimes_{i=1}^{t} U_i^{\otimes a}\right)^{\oplus b} \otimes \mathscr{O}_{X \times \overline{\mathfrak{H}}} & \longrightarrow & (\mathrm{id}_X \times \pi)^{\star}\left(\bigotimes_{i=1}^{t} \mathfrak{E}_{\overline{\mathfrak{Q}},i}(m)^{\otimes a}\right)^{\oplus b} & \longrightarrow & 0 \\ & & & \searrow^{f} & \downarrow \Phi_{\overline{\mathfrak{H}}} & & & & \\ & & & & \mathscr{A} := (\mathrm{id}_X \times \pi)^{\star} \det(\mathfrak{E}_{\overline{\mathfrak{Q}}})^{\otimes c} \otimes \pi_X^{\star} \mathscr{L}(atm). & & & & \end{array} \tag{45}$$

Here, $\Phi_{\bar{\mathfrak{H}}}$ is the relative version of the composition $\mathrm{ev} \circ \Phi$ in Diagram (44).

The points $(\underline{q}, [\Phi])$ where the restriction $\Phi_{\overline{\mathfrak{H}}|X \times \{(q,[\Phi])\}}$ factors through $\left(\bigotimes_{i=1}^{t} \mathscr{E}_i(m)^{\otimes a}\right)^{\oplus b}$ (as in (44)) are the points where the restriction $f_{|X \times \{(q,[\Phi])\}}$ is identically zero. We will need the following lemma (see [26], Proposition 2.1, or [8], Lemma 0.9).

Lemma 3.4.13. *Let Y be a scheme, and let $f \colon \mathscr{G} \longrightarrow \mathscr{F}$ be a homomorphism of coherent sheaves on $X \times Y$. Assume that $\mathscr{F}$ is flat over Y. Then, there is a unique closed subscheme $\iota \colon Z \hookrightarrow Y$ satisfying the following universal property: Given a Cartesian diagram*

$$\begin{array}{ccc} X \times S & \xrightarrow{\bar{h}} & X \times Y \\ \downarrow p_S & & \downarrow p \\ S & \xrightarrow{h} & Y, \end{array}$$

$\bar{h}^{\star} f = 0$, if and only if h factors through Z.

Let $\overline{\mathfrak{Z}}' \hookrightarrow \bar{\mathfrak{H}}$ be the scheme given by this lemma for $Y = \overline{\mathfrak{H}}$ and the homomorphism $f \colon \mathscr{K} \longrightarrow \mathscr{A}$, let $\mathfrak{Z} \subset \overline{\mathfrak{Z}}'$ be the open subscheme of points, such that the corresponding sheaves $\mathscr{E}_i$ are torsion free and the induced homomorphisms $H^0(q_i(m)) \colon U_i \longrightarrow H^0(\mathscr{E}_i(m))$ are isomorphisms. Let $\overline{\mathfrak{Z}}$ be the closure of $\mathfrak{Z}$ in $\overline{\mathfrak{Z}}'$, and let $\iota \colon \overline{\mathfrak{Z}} \hookrightarrow \mathfrak{H}$ be the inclusion. Then $\bar{\iota}^{\star} f = 0$, and there is the following commutative diagram on $X \times \overline{\mathfrak{Z}}$:

$$\begin{array}{ccccccc}
\bar{\iota}^\star \mathscr{K} & \longrightarrow & (\bigotimes_{i=1}^t U_i^{\otimes a})^{\oplus b} \otimes \mathscr{O}_{X\times\overline{\mathfrak{Z}}} & \longrightarrow & (\bar{\iota}^\star \bigotimes_{i=1}^t (\mathrm{id}_X \times \pi)^\star \mathscr{E}_{\overline{\mathfrak{Q}},i}(m)^{\otimes a})^{\oplus b} & \longrightarrow & 0 \\
& \searrow^{\bar{\iota}^\star f} & \downarrow{\scriptstyle \bar{\iota}^\star \widetilde{\Phi}} & \swarrow_{\widetilde{\varphi}} & & & \\
& & \bar{\iota}^\star \mathscr{A} & & & &
\end{array}$$

. (46)

Hence, there is a tautological family of based decorated tuples of sheaves parameterized by $\mathfrak{Z}$ with

$$\varphi_{\overline{\mathfrak{Z}}}\colon \left(\bigotimes_{i=1}^{t} \mathfrak{E}_{\overline{\mathfrak{Z}},i}{}^{\otimes a}\right)^{\oplus b} \longrightarrow \det(\mathfrak{E}_{\overline{\mathfrak{Z}}})^{\otimes c} \otimes \pi_X^\star \mathscr{L} \otimes \pi_{\overline{\mathfrak{Z}}}^\star \mathscr{N}, \tag{47}$$

where, as usual, $\mathfrak{E}_{\overline{\mathfrak{Z}}} = \mathfrak{E}_{\overline{\mathfrak{Z}},1} \oplus \cdots \oplus \mathfrak{E}_{\overline{\mathfrak{Z}},t}$.

Given a point $(\underline{q}, [\Phi])$ in $\overline{\mathfrak{Z}}$, using the tautological family (47), we can recover the tuple $(\underline{q}, \underline{\mathscr{E}}, \varphi)$ up to isomorphism, and, if $H^0(q_i(m))\colon U_i \longrightarrow H^0(\mathscr{E}_i(m))$ is an isomorphism, $i = 1, \ldots, t$, then we recover the tuple $(\underline{g}, \underline{\mathscr{E}}, \varphi)$ (where $g_i = H^0(q_i(m))$) up to isomorphism, i.e., if $(\underline{g}', \underline{\mathscr{E}}', \varphi')$ is another tuple corresponding to the same point, then there exists an isomorphism $(\underline{f}, \alpha)$ between $(\underline{\mathscr{E}}, \varphi)$ and $(\underline{\mathscr{E}}', \varphi')$ given by $f_i\colon \mathscr{E}_i \longrightarrow \mathscr{E}_i'$ with $H^0(f_i(m)) \circ q_i = q_i'$. To get rid of the choice of g_i, we have to take the quotient by $\mathrm{GL}(U_1) \times \cdots \times \mathrm{GL}(U_t)$. But, if $\lambda_i \in \mathbb{G}_m(k)$, $i = 1, \ldots, t$, $((g_1, \ldots, g_t), \underline{\mathscr{E}}, \varphi)$ and $((\lambda_1 g_1, \ldots, \lambda_t g_t), \underline{\mathscr{E}}, \varphi)$ correspond to the same point, and hence it is enough to divide by the action of

$$\widetilde{S} = \mathrm{SL}(U_1) \times \cdots \times \mathrm{SL}(U_t).$$

In fact, the moduli space will be the GIT quotient of $\overline{\mathfrak{Z}}$ by this group. To apply GIT, we need a linearization of the action in an ample line bundle of $\overline{\mathfrak{Z}}$. We will obtain this now, using Simpson's ideas. Let $l > m$ be an integer, and $W = H^0(\mathscr{O}_X(l-m))$. The quotient q_i induces homomorphisms:

$$q_i\colon U_i \otimes \mathscr{O}_X(l-m) \longrightarrow \mathscr{E}_i(l)$$

$$q_i'\colon U_i \otimes W \longrightarrow H^0(\mathscr{E}_i(l))$$

$$q_i''\colon \bigwedge^{P_i(l)} (U_i \otimes W) \longrightarrow \bigwedge^{P_i(l)} H^0(\mathscr{E}_i(l)) \cong k.$$

If l is large enough, these homomorphisms are surjective, and give Grothendieck's embedding

$$\overline{\mathfrak{Q}}_i \longrightarrow \mathbb{P}\Big(\bigwedge^{P_i(l)} (U_i \otimes W)\Big),\ i = 1, \ldots, t,$$

and, hence, a very ample line bundle $\mathscr{O}_{\overline{\mathfrak{Q}}_i}(1)$ on $\overline{\mathfrak{Q}}_i$ (depending on m and l).

Fix a Poincaré bundle $\mathscr{P}$ on $J \times X$, where $J = \mathrm{Pic}^{d_1+\cdots+d_t}(X)$. The classifying morphism $\overline{\mathfrak{Q}} \longrightarrow J$ produces a morphism

$$\overline{\mathfrak{H}} \longrightarrow P = \mathbb{P}\left(\left(\bigotimes_{i=1}^{t} U_i^{\otimes a}\right)^{\oplus b} \otimes (\pi_{J\star}(\mathscr{P}^{\otimes c} \otimes \pi_X^\star \mathscr{L}(atm)))^\vee\right). \tag{48}$$

Note that $\pi_{J\star}(\mathscr{P}^{\otimes c} \otimes \pi_X^\star \mathscr{L}(atm))$ is locally free, because m has been chosen sufficiently large. Thus, P is a projective bundle over J. Replacing $\mathscr{P}$ with another Poincaré bundle defined by tensorizing with the pullback of the dual of a sufficiently ample line bundle on J, we can assume that $\mathscr{O}_P(1)$ is very ample (this line bundle depends on m). Using the previous morphism, we obtain the finite (and injective) morphism

$$\overline{\mathfrak{Z}} \longrightarrow \overline{\mathfrak{Q}}_1 \times \cdots \times \overline{\mathfrak{Q}}_t \times_J P$$

and, hence, the $\mathrm{SL}(U_1) \times \cdots \times \mathrm{SL}(U_t)$-linearized polarization

$$\mathscr{O}_{\overline{\mathfrak{Z}}}(\varepsilon_1, \ldots, \varepsilon_t; e) := \pi^\star_{\overline{\mathfrak{Q}}_1} \mathscr{O}_{\overline{\mathfrak{Q}}_1}(\varepsilon_1) \otimes \cdots \otimes \pi^\star_{\overline{\mathfrak{Q}}_t} \mathscr{O}_{\overline{\mathfrak{Q}}_t}(\varepsilon_t) \otimes \pi_P^\star \mathscr{O}_P(e). \tag{49}$$

We have denoted by $\pi_{\overline{\mathfrak{Q}}_i}$ and π_P the projections from $\overline{\mathfrak{Z}}$ to $\overline{\mathfrak{Q}}_i$ and P, respectively. Furthermore, we suppose that $\underline{\varepsilon} = (\varepsilon_1, \ldots, \varepsilon_t)$ and e are integers with

$$\frac{e}{\varepsilon_i} = \frac{P_i(l)\delta(m) - \delta(l)P_i(m)}{P_i(m) - a\delta(m)}, \qquad i = 1, \ldots, t. \tag{50}$$

In Proposition 3.4.16, we will identify the GIT-(semi)stable points in $\overline{\mathfrak{Z}}$, using the Hilbert-Mumford criterion. In Theorem 3.4.18, we relate filtrations of tuples of sheaves with filtrations of the tuple of vector spaces $\underline{U} = (U_1, \ldots, U_t)$ to prove that GIT-(semi)stable points of $\overline{\mathfrak{Z}}$ coincide with the points associated with δ-(semi)stable points. In particular, the GIT-semistable points of $\overline{\mathfrak{Z}}$ lie in $\mathfrak{Z}$. If $\widetilde{S}$ acts on a projective scheme Y with a given linearization, we have the quantity $\mu(\lambda, y)$ (see Section 2.3).

A *weighted filtration* $(\underline{U}_\bullet, \underline{\alpha}_\bullet)$ *of the tuple of vector spaces* $\underline{U} = (U_1, \ldots, U_t)$ consists of filtrations of vector spaces

$$\{0\} \subsetneq U_{i,1} \subsetneq U_{i,2} \subsetneq \cdots \subsetneq U_{i,s_i} \subsetneq U_{i,s_i+1} = U_i \tag{51}$$

and positive integers $\alpha_{i,j} > 0$, $i = 1, \ldots, t$, $j = 1, \ldots, s_i$. It can happen that $s_i = 0$ for some i. If $s_i = 1$ (one step filtration of $\mathscr{E}_i$) and $\alpha_{i,1}$ is not specified, it is understood that its value is 1.

Consider the vector defined as $\Gamma^i = \sum_{j=1}^{s_i} \alpha_{i,j} \Gamma^{(\dim_k(U_{i,j}))}_{\dim_k(U_i)}$ with

$$\Gamma^{(s)}_u = (\underbrace{s-u, \ldots, s-u}_{s\times}, \underbrace{s, \ldots, s}_{(u-s)\times}), \qquad 1 \le s \le u-1. \tag{52}$$

The vector $\Gamma^i \in \mathbb{Z}^{\dim_k(U_i)}$ has $s_i + 1$ different entries which we denote $\Gamma^i_1 < \cdots < \Gamma^i_{s_i+1}$. In other words, Γ^i_j is the $\dim_k(U_{i,j})$-th component of Γ^i. If $s_i = 0$, we set $\Gamma^i_1 = 0$.

Let $\mathscr{J}$ be the set of functions defined in (19). Define

$$\mu(\underline{U}_\bullet, \underline{\alpha}_\bullet; \Phi) = -\min_{J \in \mathscr{J}} \left\{ \sum_{i=1}^{t} (\Gamma^i_{j^i_1} + \cdots + \Gamma^i_{j^i_a}) | \Phi_{|(\bigotimes_{i=1}^t U_{i,j^i_1} \otimes \cdots \otimes U_{i,j^i_a})^{\oplus b}} \not\equiv 0 \right\}. \tag{53}$$

Recall (Lemma 3.4.5) that $\nu_{i,j}(J) = \#\{u \mid 1 \le u \le a,\ j^i_u \le j\}$. If $J \in \mathscr{J}$ is a multi-index giving the minimum in (53), we will set $\varepsilon_{i,j}(\underline{U}_\bullet, \underline{\alpha}_\bullet; \Phi) := \nu_i(J)$ (or just $\varepsilon_{i,j}(\underline{U}_\bullet)$, if the rest of the data is clear from the context). Then, we have, as in (31),

$$\mu(\underline{U}_\bullet, \underline{\alpha}_\bullet; \Phi) = -\sum_{i=1}^{t}\sum_{j=1}^{s_i} \alpha_{i,j}\big(a \dim_k(U_{i,j}) - \varepsilon_{i,j}(\underline{U}_\bullet)\dim_k(U_i)\big). \tag{54}$$

Given a subspace $U' \subset U$ of a vector space U and a quotient $q\colon U \otimes \mathscr{O}_X(-m) \longrightarrow \mathscr{E}$, we define the subsheaf $\mathscr{E}_{U'}$ of $\mathscr{E}$ as the image of the restriction of q to U':

$$\begin{array}{ccc} U \otimes \mathscr{O}_X(-m) & \longrightarrow\!\!\!\!\!\rightarrow & \mathscr{E} \\ \uparrow & & \uparrow \\ U' \otimes \mathscr{O}_X(-m) & \longrightarrow\!\!\!\!\!\rightarrow & \mathscr{E}_{U'}. \end{array}$$

In particular, $\mathscr{E}_{U'}(m)$ is generated by global sections.

On the other hand, if the quotient $q\colon U \otimes \mathscr{O}_X(-m) \longrightarrow \mathscr{E}$ induces an injection $U \hookrightarrow H^0(\mathscr{E}(m))$ (we will later show that all quotients coming from GIT-semistable points of $\mathfrak{Z}$ satisfy this property), and if $\mathscr{E}' \subset \mathscr{E}$ is a subsheaf, we define

$$U_{\mathscr{E}'} = U \cap H^0(\mathscr{E}'(m)).$$

The following two lemmas are easy to check.

Lemma 3.4.14. *Given a point $(\underline{q}, [\Phi]) \in \overline{\mathfrak{Z}}$, such that, for all i, $H^0(q_i(m))$ is injective, and a weighted filtration $(\underline{\mathscr{E}}_\bullet, \underline{\alpha}_\bullet)$ of $\underline{\mathscr{E}}$, we have:*

1. $\mathscr{E}_{iU_{i\mathscr{E}_{i,j}}} \subset \mathscr{E}_{i,j}$.
2. *For all $J \in \mathscr{J}$, if $\varphi|_{(\bigotimes_{i=1}^t \mathscr{E}_{i,j^i_1}\otimes\cdots\otimes\mathscr{E}_{i,j^i_a})^{\oplus b}} \equiv 0$, then $\Phi|_{(\bigotimes_{i=1}^t U_{i\mathscr{E}_{i,j^i_1}}\otimes\cdots\otimes U_{i\mathscr{E}_{i,j^i_a}})^{\oplus b}}$ $\equiv 0$.*
3. $-\sum_{i=1}^t\sum_{j=1}^{s_i}\alpha_{i,j}\,\varepsilon_{i,j}(\underline{\mathscr{E}}_\bullet, \underline{\alpha}_\bullet; \varphi) \le -\sum_{i=1}^t\sum_{j=1}^{s_i}\alpha_{i,j}\,\varepsilon_{i,j}(\underline{U_{\mathscr{E}_\bullet}}, \underline{\alpha}_\bullet; \Phi)$.

Furthermore, if, for all i, q_i induces an isomorphism $U_i \cong H^0(\mathscr{E}_i(m))$, all $\mathscr{E}_{i,j}$ are m-regular and, for all $J \in \mathscr{J}$, $\bigotimes_{i=1}^t \mathscr{E}_{i,j^i_1} \otimes \cdots \otimes \mathscr{E}_{i,j^i_a}$ is (atm)-regular, then 1. becomes an equality, 2. becomes "if and only if" and 3. an equality.

Lemma 3.4.15. *Given a point $(\underline{q}, [\Phi]) \in \overline{\mathfrak{Z}}$, such that q_i induces an injection $U_i \hookrightarrow H^0(\mathscr{E}_i(m))$, and a weighted filtration $(\underline{U}_\bullet, \underline{\alpha}_\bullet)$ of $\underline{U}$, we have:*

1. $U_i \subset U_{i\mathscr{E}_{iU_{i,j}}}$.
2. *For all $J \in \mathscr{J}$, $\varphi|_{(\bigotimes_{i=1}^t \mathscr{E}_{iU_{i,j^i_1}}\otimes\cdots\otimes\mathscr{E}_{iU_{i,j^i_a}})^{\oplus b}} \equiv 0$, if and only if $\Phi|_{(\bigotimes_{i=1}^t U_{i,j^i_1}\otimes\cdots\otimes U_{i,j^i_a})^{\oplus b}}$ $\equiv 0$.*
3. $-\sum_{i=1}^t\sum_{j=1}^{s_i}\alpha_{i,j}\,\varepsilon_{i,j}(\underline{\mathscr{E}_{U_\bullet}}, \underline{\alpha}_\bullet; \varphi) = -\sum_{i=1}^t\sum_{j=1}^{s_i}\alpha_{i,j}\,\varepsilon_{i,j}(\underline{U}_\bullet, \underline{\alpha}_\bullet; \Phi)$.

Proposition 3.4.16. *For sufficiently large l, the point $(\underline{q}, [\Phi])$ in $\overline{\mathfrak{Z}}$ is GIT-(semi)stable with respect to $\mathcal{O}_{\overline{\mathfrak{Z}}}(\underline{\varepsilon}, e)$, if and only if for every weighted filtration $(\underline{U}_\bullet, \underline{\alpha}_\bullet)$ of $\underline{U}$*

$$\sum_{i=1}^{t}\left(\varepsilon_i \sum_{j=1}^{s_i} \alpha_{i,j}(\dim_k(U_i)\, P(\mathscr{E}_{iU_{i,j}})(l) - \dim_k(U_{i,j}) P_i(l))\right) + e\,\mu(\underline{U}_\bullet, \underline{\alpha}_\bullet; \Phi)\ (\succeq)\ 0. \tag{55}$$

Furthermore, there is an integer A_2 (depending only on m, P_i, $i = 1, \dots, t$, a, b, c, and $\mathscr{L}$), such that it is enough to consider weighted filtrations with $\alpha_{i,j} \le A_2$.

Proof. Given m, the sheaves $\mathscr{E}_{iU'}$ for $U' \subset U_i$ form a bounded family. So, if l is large enough, we will have

$$\dim_k q_i'(U' \otimes W) = h^0(\mathscr{E}_{iU'}(l)) = P(\mathscr{E}_{iU'})(l)$$

for all subspaces $U' \subset U_i$. By the Hilbert-Mumford criterion, a point is GIT-(semi)stable, if and only if, for all one-parameter subgroups λ of $\widetilde{S}$,

$$\mu(\lambda, (\underline{q}, [\Phi])) = \sum_{i=1}^{t}(\varepsilon_i \mu(\lambda, q_i)) + e\,\mu(\lambda, [\Phi])\ (\succeq)\ 0.$$

A one-parameter subgroup of $\widetilde{S}$ is specified by bases $\{e_{i,1}, \dots, e_{i,h_i}\}$ of U_i and vectors $\Gamma^i \in \mathbb{Z}^{h_i}$ with non-decreasing coordinates, $i = 1, \dots, t$. This defines a weighted filtration $(\underline{U}_\bullet, \underline{\alpha}_\bullet)$ of $\underline{U}$ as follows: Let $\Gamma^i_1 < \dots < \Gamma^i_{t+1}$ be the different values of Γ^i, let $U_{i,j}$ be the vector space generated by all $e_{i,u}$ which are eigenvectors of the action of λ on U_i with eigenvalue $\le \Gamma^i_j$, and let $\alpha_{i,j} = (\Gamma^i_{j+1} - \Gamma^i_j)/h_i$. Set $\mathscr{J}_i = \{1, \dots, s_i + 1\}^{\times P_i(l)}$. By [33] or [13], we have

$$\begin{aligned}\mu(\lambda, q_i) &= -\min_{(j_1,\dots,j_{P_i(l)})\in\mathscr{J}_i}\{(\Gamma^i_{j_1} + \dots + \Gamma^i_{j_{P(l)}}) | q''_{i|(U_{i,j_1}\otimes W)\wedge\dots\wedge(U_{i,j_{P(l)}}\otimes W)} \not\equiv 0\}\\ &= \sum_{j=1}^{s_i} \alpha_{i,j}(\dim_k(U_i)\, P(\mathscr{E}_{iU_{i,j}})(l) - \dim_k(U_{i,j})\, P_i(l))\end{aligned}$$

and, by direct computation,

$$\begin{aligned}\mu(\lambda, [\Phi]) &= -\min_{J\in\mathscr{J}}\left\{\sum_{i=1}^{t}(\Gamma^i_{j^i_1} + \dots + \Gamma^i_{j^i_a}) | \Phi_{|(\otimes_{i=1}^t U_{i,j^i_1}\otimes\dots\otimes U_{i,j^i_a})^{\oplus b}} \not\equiv 0\right\}\\ &= \mu(\underline{U}_\bullet, \underline{\alpha}_\bullet; \Phi).\end{aligned}$$

The last statement follows from an argument similar to the proof of Lemma 3.4.4. □

Proposition 3.4.17. *The point $(\underline{q}, [\Phi]) \in \overline{\mathfrak{Z}}$ is GIT-(semi)stable, if and only if, for all weighted filtrations $(\underline{\mathscr{E}}_\bullet, \underline{\alpha}_\bullet)$ of $\underline{\mathscr{E}}$,*

$$\begin{aligned}\sum_{i=1}^{t}\sum_{j=1}^{s_i} \alpha_{i,j}((P(\mathscr{E}_{i,j}) - \varepsilon_{i,j}(\underline{\mathscr{E}}_\bullet)\delta)(\dim_k(U_i) - a\delta(m))\\ - (\dim_k(U_{i\mathscr{E}_{i,j}}) - \varepsilon_{i,j}(\underline{\mathscr{E}}_\bullet)\delta(m))(P_i - a\delta))(\succeq)0.\end{aligned} \tag{56}$$

Furthermore, if $(\underline{q}, [\Phi])$ is GIT-semistable, then the induced map $f_{q_i}\colon U_i \longrightarrow H^0(\mathcal{E}_i(m))$ is injective.

Proof. First, we prove that, if $(\underline{q}, [\Phi])$ is GIT-semistable, then the induced linear map f_{q_i} is injective. Let U' be its kernel and consider the filtration $\{0\} \subseteq U' \subsetneq U_i$. We have $\mathcal{E}_{iU'} = \{0\}$ and $\mu(\{0\} \subseteq U' \subsetneq U_i; \Phi) = a \dim_k(U')$. Applying Proposition 3.4.16, we have

$$-\varepsilon_i \dim_k(U')P_i(l) - ea \dim_k(U') \geq 0,$$

and hence $U' = \{0\}$.

Using (50) and (54), the inequality of Proposition 3.4.16 becomes

$$\sum_{i=1}^{t}\sum_{j=1}^{s_i} \alpha_{i,j}((P(\mathcal{E}_{iU_{i,j}})(l) - \varepsilon_{i,j}(\underline{U}_{\bullet})\delta(l))(\dim_k(U_i) - a\delta(m)) - (\dim_k(U_{i,j}) - \varepsilon_{i,j}(\underline{U}_{\bullet})\,\delta(m))(P_i(l) - a\delta(l)))(\geq)0. \tag{57}$$

An argument similar to Lemma 3.4.10 (using A_2 instead of A_1) shows that we can take l large enough (depending only on m, a, b, c, P_i, $i = 1, \ldots, t$, $\mathcal{L}$, and δ), such that this inequality holds for l, if and only if it holds as an inequality of polynomials.

Now, assume that $(\underline{q}, [\Phi])$ is GIT-(semi)stable. Take a weighted filtration $(\underline{\mathcal{E}}_{\bullet}, \underline{\alpha}_{\bullet})$ of $\underline{\mathcal{E}}$. Then, Lemma 3.4.14 and (57), applied to the associated weighted filtration $(\underline{U}_{\underline{\mathcal{E}}_{\bullet}}, \underline{\alpha}_{\bullet})$ of $\underline{U}$, give (56). On the other hand, assume that (56) holds. Take a weighted filtration $(\underline{U}_{\bullet}, \underline{\alpha}_{\bullet})$ of $\underline{U}$. Then, Lemma 3.4.15 and (56), applied to the associated weighted filtration $(\underline{\mathcal{E}}_{\underline{U}_{\bullet}}, \underline{\alpha}_{\bullet})$ of $\underline{\mathcal{E}}$, give (57), and it follows that $(\underline{q}, [\Phi])$ is GIT-(semi)stable. □

Theorem 3.4.18. *Assume $m > N$. For l sufficiently large, a point $(\underline{q}, [\Phi])$ in $\overline{\mathfrak{Z}}$ is GIT-(semi)stable, if and only if the corresponding decorated tuple of sheaves $(\underline{\mathcal{E}}, \varphi)$ is δ-(semi)stable and the linear map $f_{q_i}\colon U_i \longrightarrow H^0(\mathcal{E}_i(m))$ induced by q_i is an isomorphism for all i.*

Proof. We prove this in two steps:

Step 1. $(\underline{q}, [\Phi])$ is GIT-semistable $\Longrightarrow$ $(\underline{\mathcal{E}}, \varphi)$ is δ-semistable and $H^0(q_i(n))$ is an isomorphism, for all i.

Taking the leading coefficient in (56) gives

$$\sum_{i=1}^{t}\sum_{j=1}^{s_i} \alpha_{i,j}(r_{i,j}(\dim_k(U_i) - a\delta(m)) - (\dim_k(U_{i\mathcal{E}_{i,j}}) - \varepsilon_{i,j}(\underline{\mathcal{E}}_{\bullet})\delta(m))r_i) \geq 0.$$

Note that, even if $(\underline{q}, [\Phi])$ is GIT-stable, we only get the weak inequality here. This implies

$$\left(\sum_{i=1}^{t}\sum_{j=1}^{s_i} \alpha_{i,j}(r_i h^0(\widehat{\mathcal{E}}_{i,j}(m)) - \widehat{r}_{i,j} P_i(m))\right) + \mu(\underline{\mathcal{E}}_{\bullet}, \underline{\alpha}_{\bullet}; \varphi)\delta(m) \geq 0. \tag{58}$$

To be able to apply Theorem 3.4.6, we still need to show that $\mathcal{E}_i$ is torsion free for all i. By Lemma 3.4.12, there exists a decorated tuple of sheaves $(\underline{\mathcal{F}}, \psi)$, with $\mathcal{F}_i$ torsion free for all i, such that $P(\mathcal{E}_i) = P(\mathcal{F}_i)$ and the sequence

$$\{0\} \longrightarrow T(\mathcal{E}_i) \longrightarrow \mathcal{E}_i \xrightarrow{\beta_i} \mathcal{F}_i$$

is exact for all i. Consider a weighted filtration $(\underline{\mathcal{F}}_\bullet, \underline{\alpha}_\bullet)$ of $\underline{\mathcal{F}}$. Let $\widehat{\mathcal{F}}_{i,j} = \mathcal{F}_i/\mathcal{F}_{i,j}$, and let $\widehat{\mathcal{E}}_{i,j}$ be the image of $\mathcal{E}_i$ in $\widehat{\mathcal{F}}_{i,j}$. Let $\mathcal{E}_{i,j}$ be the kernel of $\mathcal{E}_i \longrightarrow \widehat{\mathcal{E}}_{i,j}$. Then $\mathrm{rk}(\mathcal{F}_{i,j}) = \mathrm{rk}(\mathcal{E}_{i,j}) = r_{i,j}$, $h^0(\widehat{\mathcal{F}}_{i,j}(m)) \geq h^0(\widehat{\mathcal{E}}_{i,j}(m))$, and $\mu(\underline{\mathcal{F}}_\bullet, \underline{\alpha}_\bullet; \psi) = \mu(\underline{\mathcal{E}}_\bullet, \underline{\alpha}_\bullet; \varphi)$. Using this and applying (58) to $\mathcal{E}_{i,j}$, we get

$$\left(\sum_{i=1}^{t}\sum_{j=1}^{s_i} \alpha_{i,j}(r_i h^0(\widehat{\mathcal{F}}_{i,j}(m)) - \widehat{r}_{i,j} P_i(m))\right) + \delta(m)\cdot \mu(\underline{\mathcal{F}}_\bullet, \underline{\alpha}_\bullet; \varphi) \geq 0.$$

Hence, Theorem 3.4.6 implies that $(\underline{\mathcal{F}}, \psi)$ is δ-semistable.

Next, we will show that $T(\underline{\mathcal{E}}) = \{0\}$, and hence, since $P(\mathcal{E}_i) = P(\mathcal{F}_i)$, for all i, we will conclude that $(\underline{\mathcal{E}}, \varphi)$ is isomorphic to $(\underline{\mathcal{F}}, \psi)$. Define $\underline{\mathcal{E}}''$ to be the image of $\underline{\mathcal{E}}$ in $\underline{\mathcal{F}}$. Then,

$$P_i(m) - a\delta(m) = h^0(\mathcal{F}_i(m)) - a\delta(m) \geq h^0(\mathcal{E}_i''(m)) - a\delta(m) \geq P_i(m) - a\delta(m),$$

where the last inequality follows from (58) applied to the one step filtration $\{0\} \subseteq T(\mathcal{E}_i) \subsetneq \mathcal{E}_i$. Hence, equality holds at all places and $h^0(\mathcal{F}_i(m)) = h^0(\mathcal{E}_i''(m))$. Since $\mathcal{F}_i$ is globally generated, $\mathcal{F}_i = \mathcal{E}_i''$, and hence $T(\mathcal{E}_i) = \{0\}$ for all i.

Finally, we have seen that f_{q_i} is injective. Since $(\underline{\mathcal{E}}, \varphi)$ is δ-semistable and $\dim_k(U_i) = h^0(\mathcal{E}_i(m))$, we see that f_{q_i} is an isomorphism.

Step 2. $(\underline{\mathcal{E}}, \varphi)$ *is* δ*-stable or strictly* δ*-semistable and* $H^0(q_i(m))$ *is an isomorphism* $\Longrightarrow$ $(\underline{q}, [\Phi])$ *is GIT-stable or strictly GIT-semistable, respectively.*

Since f_{q_i} is an isomorphism, we have $U_{i\mathcal{E}'} = H^0(\mathcal{E}'(m))$, for any subsheaf $\mathcal{E}' \subset \mathcal{E}_i$. Then, Theorem 3.4.6 implies that for all weighted filtrations

$$\left(\sum_{i=1}^{t}\sum_{i=1}^{s_i} \alpha_{i,j}(r_{i,j} P_i(m) - r_i \dim_k(U_{i\mathcal{E}_{i,j}}))\right) + \delta(m)\cdot \mu(\underline{\mathcal{E}}_\bullet, \underline{\alpha}_\bullet; \varphi)\ (\geq)\ 0. \quad (59)$$

If the inequality is strict, then

$$\sum_{i=1}^{t}\sum_{j=1}^{s_i} \alpha_{i,j}((P(\mathcal{E}_{i,j}) - \varepsilon_{i,j}(\underline{\mathcal{E}}_\bullet)\delta)(\dim_k(U_i) - a\delta(m))$$
$$- (\dim_k(U_{i\mathcal{E}_{i,j}}) - \varepsilon_{i,j}(\underline{\mathcal{E}}_\bullet)\delta(m))(P_i - a\delta)) \succ 0.$$

If $(\underline{\mathcal{E}}, \varphi)$ is strictly δ-semistable, then, by Theorem 3.4.6, there is a filtration giving equality in (59). Corollary 3.4.11 implies that $h^0(\mathcal{E}_{i,j}(m)) = P(\mathcal{E}_{i,j})(m)$, and, by Lemma 3.4.10,

$$\left(\sum_{i=1}^{t}\sum_{j=1}^{s_i} \alpha_{i,j}(r_{i,j} P_i - r_i P(\mathcal{E}_{i,j}))\right) + \delta\cdot \mu(\underline{\mathcal{E}}_\bullet, \underline{\alpha}_\bullet; \varphi) = 0.$$

A short calculation using this and (59) gives

$$\sum_{i=1}^{t}\sum_{j=1}^{s_i}\alpha_{i,j}((P(\mathscr{E}_{i,j})-\varepsilon_{i,j}(\underline{\mathscr{E}_\bullet})\delta)(\dim_k(U_i)-a\delta(m))$$
$$-(\dim_k(U_{i\mathscr{E}_{i,j}})-\varepsilon_{i,j}(\underline{\mathscr{E}_\bullet})\delta(m))(P_i-a\delta)) = 0\,.$$

So, we finish by using Proposition 3.4.17. □

Proof of Theorem 3.4.1. The main ingredient of the proof is Theorem 3.4.18, showing that GIT-(semi)stable points correspond to δ-(semi)stable decorated sheaves.

Recall the definition of $\overline{\mathfrak{Z}}$ and its polarization in (49). Let $\mathrm{M}^{\delta\text{-ss}} := \mathrm{M}(\mathrm{I} : a, b, c;\mathscr{L})^{\delta\text{-ss}}_{\underline{P}}$ and $\mathrm{M}^{\delta\text{-s}} := \mathrm{M}(\mathrm{I} : a, b, c;\mathscr{L})^{\delta\text{-s}}_{\underline{P}}$ be the GIT quotients of $\overline{\mathfrak{Z}}$ and $\overline{\mathfrak{Z}}^{\mathrm{s}}$, respectively, by $\widetilde{S}$. Since $\overline{\mathfrak{Z}}$ is projective, $\mathrm{M}^{\delta\text{-ss}}$ is also projective. GIT gives that $\mathrm{M}^{\delta\text{-s}}$ is an open subset of the projective scheme $\mathrm{M}^{\delta\text{-ss}}$. The restriction $\overline{\mathfrak{Z}}^{\mathrm{s}} \longrightarrow \mathrm{M}^{\delta\text{-s}}$ to the stable part is a geometric quotient, i.e., the fibers are $\widetilde{S}$-orbits, and hence the points of $\mathrm{M}^{\delta\text{-s}}$ correspond to isomorphism classes of δ-stable decorated sheaves.

It only remains to show that $\mathrm{M}^{\delta\text{-ss}}$ corepresents the functor $\underline{\mathrm{M}}^{\delta\text{-ss}}$. This is done using standard arguments. Let $(\underline{\mathscr{E}}_T, \varphi_T, \mathscr{N})$ be a family of δ-semistable decorated sheaves (cf. (22)) parameterized by a scheme T. Then, $\mathscr{U}_i := \pi_{T\star}(\mathscr{E}_{iT}\otimes\pi_X^\star\mathscr{O}_X(m))$ is locally free on T for all i. The family $\bigoplus_i \mathscr{E}_{iT}$ gives a map $\Delta\colon T \longrightarrow \mathrm{Pic}^d(X)$, sending $\tau \in T$ to $\det(\mathscr{E}_\tau)$. Cover T with small open sets T_u. For each u, we can find an isomorphism

$$\beta_{T_u}\colon \det(\mathscr{E}_{T_u}) \longrightarrow \overline{\Delta_u}^\star\mathscr{P}$$

(where $\mathscr{P}$ is the Poincaré bundle in the definition of P in (48)), and trivializations

$$g_{iT_u}\colon U_i\otimes\mathscr{O}_{T_u} \longrightarrow \mathscr{U}_{i|T_u}, \quad i = 1,\ldots,t.$$

Using this trivialization, we obtain families of quotients

$$q_{iT_u}\colon U_i\otimes\pi_X^\star\mathscr{O}_X(-m) \longrightarrow \mathscr{E}_{iT_u}, \quad i = 1,\ldots,t,$$

parameterized by T_u, giving a map $T_u \longrightarrow \overline{\mathfrak{Q}}_i$ for all i. Using the quotients q_{iT_u} and the isomorphism β_{T_u}, we have another family of quotients parameterized by T_u

$$\left(\bigotimes_{i=1}^{t} U_i^{\otimes a}\right)^{\oplus b}\otimes(\pi_{T_u\star}(\overline{\Delta_u}^\star\mathscr{P}^{\otimes c}\otimes\pi_X^\star\mathscr{L}(atm)))^\vee \longrightarrow \mathscr{N}.$$

Then, using the universal property of $\overline{\mathfrak{H}}$, we obtain a morphism to $\overline{\mathfrak{H}}$, and, by Lemma 3.4.13, this morphism factors through $\overline{\mathfrak{Z}}'$. Since a δ-semistable decorated tuple of sheaves gives a GIT-semistable point (Theorem 3.4.18), the image is in $\mathfrak{Z}^{\mathrm{ss}}\subset\mathfrak{Z}\subset\overline{\mathfrak{Z}}$. Composing with the geometric quotient to $\mathrm{M}^{\delta\text{-ss}}$, we obtain maps

$$\hat{f}_u\colon T_u \xrightarrow{f_u} \mathfrak{Z}^{\mathrm{ss}} \longrightarrow \mathrm{M}^{\delta\text{-ss}}.$$

The morphism $\hat{f}_u$ is independent of the choice of the isomorphism β_{T_u}: A different choice of isomorphism g_{iT_u} will change f_u to $h_u \cdot f_u$, where $h_u: T_u \longrightarrow \mathrm{GL}(U_1) \times \cdots \times \mathrm{GL}(U_t)$. Then, the morphisms $\hat{f}_u$ glue to give a morphism

$$\hat{f}: T \longrightarrow \mathrm{M}^{\delta\text{-ss}}.$$

Hence, we have a natural transformation

$$\underline{\mathrm{M}}^{\delta\text{-ss}} \longrightarrow h_{\mathrm{M}^{\delta\text{-ss}}}.$$

Recall that there is a tautological family (47) of decorated sheaves parameterized by $\overline{\mathfrak{Z}}$. By restriction to $\mathfrak{Z}^{\mathrm{ss}}$, we obtain a tautological family of δ-semistable decorated sheaves parameterized by $\mathfrak{Z}^{\mathrm{ss}}$. If $\underline{\mathrm{M}}^{\delta\text{-ss}} \longrightarrow h_Y$ is another natural transformation, the tautological family defines a map $\mathfrak{Z}^{\mathrm{ss}} \longrightarrow Y$. This map factors through the quotient $\underline{\mathrm{M}}^{\delta\text{-ss}}$, and it is easy to see that this proves that $\mathrm{M}^{\delta\text{-ss}}$ corepresents the functor $\underline{\mathrm{M}}^{\delta\text{-ss}}$. Therefore, Theorem 3.4.1 has been established. □

3.5 *Asymptotic semistability*

Let $\mathfrak{Z}$ be the parameter space defined in Section 3.4. For every polynomial δ as above, let U_δ be the set of points $z \in \mathfrak{Z}$ for which $(\underline{\mathscr{E}}_z, \phi_z)$ is a δ-semistable tuple of type I. This is an open subscheme of $\mathfrak{Z}$. This follows from the GIT construction of the moduli space given in Section 3.4. We also define

$$U_{\preceq\delta} := \bigcup_{0\prec\delta'\preceq\delta} U_{\delta'} \quad \text{and} \quad U^{\mathrm{ss}} := \bigcup_{\delta\succ 0} U_\delta.$$

These are clearly open subschemes of $\mathfrak{Z}$, and $U_{\preceq\delta_1} \subseteq U_{\preceq\delta_2}$, if $\delta_1 \preceq \delta_2$. Since U^{ss} is a quasi-projective scheme, there exists a polynomial δ_0 with

$$U_{\preceq\delta_0} = U^{\mathrm{ss}}. \tag{60}$$

Proposition 3.5.1. *Suppose we are given polynomials $\delta_0 \preceq \delta_1 \preceq \delta_2$, then*

$$U_{\delta_1} \supseteq U_{\delta_2}.$$

Proof. Let $z \in \mathfrak{Z}$ be a point, such that $(\underline{\mathscr{E}}_z, \phi_z)$ is δ_2-semistable but not δ_1-semistable. Then, there exists a non-trivial filtration $(\underline{\mathscr{E}}_\bullet, \underline{\alpha}_\bullet)$ of $\underline{\mathscr{E}}_z$ with

$$M(\underline{\mathscr{E}}_\bullet, \underline{\alpha}_\bullet) + \delta_2 \cdot \mu(\underline{\mathscr{E}}_\bullet, \underline{\alpha}_\bullet; \phi) \succeq 0$$
$$\text{and} \quad M(\underline{\mathscr{E}}_\bullet, \underline{\alpha}_\bullet) + \delta_1 \cdot \mu(\underline{\mathscr{E}}_\bullet, \underline{\alpha}_\bullet; \phi) \prec 0.$$

But then, we must have $M(\underline{\mathscr{E}}_\bullet, \underline{\alpha}_\bullet) \prec 0$ and $\mu(\underline{\mathscr{E}}_\bullet, \underline{\alpha}_\bullet; \phi) > 0$. For any $\delta \preceq \delta_1$ (in particular, for any $\delta \preceq \delta_0$), we find

$$M(\underline{\mathscr{E}}_\bullet, \underline{\alpha}_\bullet) + \delta \cdot \mu(\underline{\mathscr{E}}_\bullet, \underline{\alpha}_\bullet; \phi) \preceq M(\underline{\mathscr{E}}_\bullet, \underline{\alpha}_\bullet) + \delta_1 \cdot \mu(\underline{\mathscr{E}}_\bullet, \underline{\alpha}_\bullet; \phi) \prec 0.$$

This tells us $z \in U_{\preceq\delta_2} \subseteq U^{\mathrm{ss}}$ but $z \notin U_{\preceq\delta_0}$, a contradiction to (60). □

We form

$$U_\infty := \bigcap_{\delta \succeq \delta_0} U_\delta.$$

It will be our task to show that i) U_∞ is open and ii) to characterize the points in U_∞ by an intrinsic semistability condition.

Call a sheaf $(\underline{\mathscr{E}}, \phi)$ with a decoration of type $(\mathrm{I} : a, b, c; \mathscr{L})$ *asymptotically (semi)stable*, if one has a) $\widetilde{\phi} \not\equiv 0$, b) $\mu(\underline{\mathscr{E}}_\bullet, \underline{\alpha}_\bullet; \phi) \geq 0$ for any non-trivial filtration $(\underline{\mathscr{E}}_\bullet, \underline{\alpha}_\bullet)$ of $\underline{\mathscr{E}}$ and c) $M(\underline{\mathscr{E}}_\bullet, \underline{\alpha}_\bullet)(\succeq)0$ for any non-trivial filtration $(\underline{\mathscr{E}}_\bullet, \underline{\alpha}_\bullet)$ of $\underline{\mathscr{E}}$ with $\mu(\underline{\mathscr{E}}_\bullet, \underline{\alpha}_\bullet; \phi) = 0$. Let U_{as} be the set of points $z \in \mathfrak{Z}$, such that $(\underline{\mathscr{E}}_z, \phi_z)$ is asymptotically semistable.

Proposition 3.5.2. *Fix the Hilbert polynomial P. Then, there is a polynomial $\delta_\star$, such that every asymptotically (semi)stable sheaf $(\underline{\mathscr{E}}, \phi)$ with a decoration of type* $(\mathrm{I} : a, b, c; \mathscr{L})$ *and $P(\mathscr{E}) = P$ is δ-(semi)stable for all polynomials $\delta \succeq \delta_\star$.*

Proof. First, note that asymptotically semistable decorated sheaves are pre-semistable. Thus, by Remark 3.2.3, i), the set of isomorphy classes of torsion free sheaves $\mathscr{E}_i$ with Hilbert polynomial P_i, such that $\mathscr{E}_i$ is a summand in an asymptotically semistable tuple of sheaves $(\underline{\mathscr{E}}, \phi)$ with a decoration of type $(\mathrm{I} : a, b, c; \mathscr{L})$, is bounded for $i = 1, \ldots, t$. Let $\mathfrak{S}_i$ be the corresponding set, $i = 1, \ldots, t$.

As in Lemma 3.4.4 (see also [27] and [29]), one can see that there are finite sets $\mathscr{T}_i$ (which are independent of δ), such that the condition of δ-(semi)stability has to be verified only for non-trivial weighted filtrations $(\underline{\mathscr{E}}_\bullet, \underline{\alpha}_\bullet)$ of $\underline{\mathscr{E}}$, such that $(\mathrm{rk}\,\mathscr{E}_{i1}, \ldots, \mathrm{rk}\,\mathscr{E}_{is}, \alpha_{i\bullet}) \in \mathscr{T}_i$, $i = 1, \ldots, t$. The weight vectors associated with the elements in $\mathscr{T}_i$, $i = 1, \ldots, t$, are integral. This grants that $\mu(\underline{\mathscr{E}}_\bullet, \underline{\alpha}_\bullet; \phi)$ is an integer for such weighted filtrations. Now, there is a negative polynomial ε_0, depending on $\underline{P}$, $\max\{\, \mu_{\max}(\mathscr{E}_i) \mid [\mathscr{E}_i] \in \mathfrak{S}_i \,\}$, and $\mathscr{T}_i$, $i = 1, \ldots, t$, such that

$$M(\underline{\mathscr{E}}_\bullet, \underline{\alpha}_\bullet) \succ \varepsilon_0$$

holds for all non-trivial weighted filtrations $(\underline{\mathscr{E}}_\bullet, \underline{\alpha}_\bullet)$ of a tuple $\underline{\mathscr{E}}$ of torsion free sheaves with $[\mathscr{E}_i] \in \mathfrak{S}_i$ and $(\mathrm{rk}\,\mathscr{E}_{i1}, \ldots, \mathrm{rk}\,\mathscr{E}_{is}, \alpha_{i\bullet}) \in \mathscr{T}_i$, $i = 1, \ldots, t$. Suppose $\delta \succeq -\varepsilon_0$. Let $(\underline{\mathscr{E}}, \phi)$ be an asymptotically semistable tuple of sheaves with a decoration of type $(\mathrm{I} : a, b, c; \mathscr{L})$ and let $(\underline{\mathscr{E}}_\bullet, \underline{\alpha}_\bullet)$ be a weighted filtration of $\underline{\mathscr{E}}$ with $(\mathrm{rk}\,\mathscr{E}_{i1}, \ldots, \mathrm{rk}\,\mathscr{E}_{is}, \alpha_{i\bullet}) \in \mathscr{T}_i$, $i = 1, \ldots, t$. If $\mu(\underline{\mathscr{E}}_\bullet, \underline{\alpha}_\bullet; \phi) = 0$, then $M(\underline{\mathscr{E}}_\bullet, \underline{\alpha}_\bullet) + \delta \cdot \mu(\underline{\mathscr{E}}_\bullet, \underline{\alpha}_\bullet; \phi)(\succeq)0$, by asymptotic (semi)stability. On the other hand, if $\mu(\underline{\mathscr{E}}_\bullet, \underline{\alpha}_\bullet; \phi) > 0$ (and, thus, ≥ 1, because it is an integer), we have

$$M(\underline{\mathscr{E}}_\bullet, \underline{\alpha}_\bullet) + \delta \cdot \mu(\underline{\mathscr{E}}_\bullet, \underline{\alpha}_\bullet; \phi) \succ \varepsilon_0 + \delta \succeq 0.$$

This establishes the δ-(semi)stability of $(\underline{\mathscr{E}}, \phi)$. □

The next statement is a converse to the above result.

Proposition 3.5.3. *There is a polynomial $\delta_{\star\star}$, such that every sheaf $(\underline{\mathscr{E}}, \phi)$ with a decoration of type* $(\mathrm{I} : a, b, c; \mathscr{L})$ *and Hilbert polynomial $\underline{P}$ which is δ-semistable for some polynomial $\delta \succeq \delta_{\star\star}$ is also asymptotically semistable.*

See [6], proof of Corollary 4.2.2, p. 1212f.

Corollary 3.5.4. *There is a polynomial $\delta_{\star\star\star}$, such that for any two polynomials $\delta_2 \succeq \delta_1 \succeq \delta_{\star\star\star}$, $i = 1, 2$, one has*

$$U_{\delta_1} = U_{\delta_2}.$$

In particular, we have

$$U_\infty = U_{\mathrm{as}} = U_{\delta_{\star\star\star}}.$$

Proof. We choose $\delta_{\star\star\star} := \max\{\delta_0, \delta_\star, \delta_{\star\star}\}$. By Proposition 3.5.2 and 3.5.3, we have

$$U_{\delta_1} \subseteq U_{\mathrm{as}} \subseteq U_{\delta_2}.$$

By Proposition 3.5.1, we also have the converse inclusion, whence $U_{\delta_1} = U_{\delta_2}$. □

3.6 *Moduli spaces for decorated tuples (type II)*

For tuples of sheaves with a decoration of type $(\mathrm{II} : a, b, c; \mathscr{L})$, we can derive only a weaker result. We will first construct a parameter space, and then the moduli space.

Fix the input data a, b, c, $\mathscr{L}$, and the Hilbert polynomials $\underline{P} = (P_1, \ldots, P_t)$. We now sketch the construction of a parameter space for "all" semistable tuples with a decoration of type $(\mathrm{II} : a, b, c; \mathscr{L})$. First, we choose a natural number n, set $p_i := P_i(n)$, $i = 1, \ldots, t$, and fix complex vector spaces U_i of dimension p_i, $i = 1, \ldots, t$. Then, we have the quasi-projective quot scheme $\mathfrak{Q}_i$ that parameterizes quotients $q_i : U_i \otimes \mathscr{O}_X(-n) \longrightarrow \mathscr{E}_i$ where $\mathscr{E}_i$ is a torsion free sheaf with Hilbert polynomial P_i, such that $H^j(\mathscr{E}_i(n)) = \{0\}$, $j > 0$, and $H^0(q_i(n))$ is an isomorphism, $i = 1, \ldots, t$.

By Theorem 3.2.2, we may find an n_0, such that for every $n \geq n_0$, every positive polynomial δ of degree at most $\dim(X) - 1$, and every δ-semistable tuple $(\underline{\mathscr{E}} = (\mathscr{E}_1, \ldots, \mathscr{E}_t), \phi)$ with a decoration of type $(\mathrm{II} : a, b, c; \mathscr{L})$, such that $P(\mathscr{E}_i) = P_i$, $i = 1, \ldots, t$, there exist points of the form $[q_i : U_i \otimes \mathscr{O}_X(-n) \longrightarrow \mathscr{E}_i] \in \mathfrak{Q}_i$, $i = 1, \ldots, t$.

Set $\mathfrak{Q} := \times_{i=1}^t \mathfrak{Q}_i$. Over $X \times \mathfrak{Q}$, there are the universal quotients

$$\mathfrak{q}_{\mathfrak{Q},i} : U_i \otimes \pi_X^\star \mathscr{O}_X(-n) \longrightarrow \mathfrak{E}_{\mathfrak{Q},i}.$$

Next, define

$$\mathfrak{H} := \mathbb{P}((U^{\otimes a})^{\oplus b} \otimes \pi_{\mathfrak{Q}\star}(\det(\mathfrak{E}_{\mathfrak{Q}})^{\otimes c} \otimes \pi_X^\star \mathscr{L}(an))^\vee),$$

$U := U_1 \oplus \cdots \oplus U_t$, $\mathfrak{E}_{\mathfrak{Q}} := \mathfrak{E}_{\mathfrak{Q},1} \oplus \cdots \oplus \mathfrak{E}_{\mathfrak{Q},t}$. There is the natural projection $\pi : \mathfrak{H} \longrightarrow \mathfrak{Q}$. Let

$$\mathfrak{q}_{\mathfrak{H},i} : U_i \otimes \pi_X^\star \mathscr{O}_X(-n) \longrightarrow \mathfrak{E}_{\mathfrak{H},i}$$

be the pullback to $X \times \mathfrak{H}$ of $\mathfrak{q}_{\mathfrak{Q},i}$ via $\mathrm{id}_X \times \pi$, $i = 1, \ldots, t$, and $\mathfrak{E}_{\mathfrak{H}} := \mathfrak{E}_{\mathfrak{H},1} \oplus \cdots \oplus \mathfrak{E}_{\mathfrak{H},t}$. Then, there is the universal homomorphism

$$f_{\mathfrak{H}} \colon ((U \otimes \pi_X^\star \mathscr{O}_X(-n))^{\otimes a})^{\oplus b} \longrightarrow \det(\mathfrak{E}_{\mathfrak{H}})^{\otimes c} \otimes \pi_X^\star \mathscr{L} \otimes \mathscr{O}_{\mathfrak{H}}(1).$$

We define the closed subscheme $\mathfrak{Z} \hookrightarrow \mathfrak{H}$ by the condition that $f_{\mathfrak{H}|X \times \mathfrak{Z}}$ factorizes over $(\mathfrak{E}^{\otimes a}_{\mathfrak{H}|X \times \mathfrak{Z}})^{\oplus b}$ (see Lemma 3.4.13). Then, there is the universal family $(\underline{\mathfrak{E}}_{\mathfrak{Z}} = (\mathfrak{E}_{\mathfrak{Z},1}, \ldots, \mathfrak{E}_{\mathfrak{Z},t}), \phi_{\mathfrak{Z}})$ where $\mathfrak{E}_{\mathfrak{Z},i}$ is a $\mathfrak{Z}$-flat family of torsion free sheaves with fixed Hilbert polynomial P_i on $X \times \mathfrak{Z}$, $i = 1, \ldots, t$. For every positive polynomial δ of degree at most $\dim(X) - 1$ and every δ-semistable tuple of sheaves with a decoration of type $(\mathrm{II} : a, b, c; \mathscr{L})$ with $P(\mathscr{E}_i) = P_i$, $i = 1, \ldots, t$, there is a point $z \in \mathfrak{Z}$, such that $(\underline{\mathscr{E}}, \phi)$ is isomorphic to $(\underline{\mathscr{E}}_z, \phi_z) := (\underline{\mathfrak{E}}_{\mathfrak{Z}|X \times \{z\}}, \phi_{\mathfrak{Z}|X \times \{z\}})$.

On $\mathfrak{H}$ and $\mathfrak{Z}$, we have actions of the group

$$\widetilde{G} := \mathrm{GL}(U_1) \times \cdots \times \mathrm{GL}(U_t).$$

Define

$$T_1 \cong \mathbb{G}_m(k) := \{ (z \cdot \mathrm{id}_{U_1}, \ldots, z \cdot \mathrm{id}_{U_t}) \in \widetilde{G} \mid z \in \mathbb{G}_m(k) \}$$

$$T_2 \cong \mathbb{G}_m(k)^{\times (t-1)} := \{ (z_1 \cdot \mathrm{id}_{U_1}, \ldots, z_t \cdot \mathrm{id}_{U_t}) \in \widetilde{G} \mid z_i \in \mathbb{G}_m(k), \; i = 1, \ldots, t : z_1 \cdot \cdots \cdot z_t = 1 \}.$$

Then, we have a natural surjective homomorphism

$$T_1 \times T_2 \times \widetilde{S} \longrightarrow \widetilde{G}.$$

We may therefore establish the statements for the action of the group $T_1 \times T_2 \times \widetilde{S}$. In order to verify the existence of the good quotient, we may take the quotients in several steps, according to Example 2.1.6, ii), in [6]. Since T_1 acts trivially, we are left with the action of the group $T_2 \times \widetilde{S}$. Since T_2 is linearly reductive, the quotient $\mathfrak{Z}/\!\!/T_2$ is a closed subscheme of $\mathfrak{H}/\!\!/T_2$, and it is easy to see that two points z_1 and $z_2 \in \mathfrak{Z}$ lie in the same fiber of $\pi_1 \colon \mathfrak{Z}^0 \longrightarrow \mathfrak{Z}/\!\!/T_2$, if and only if $(\underline{\mathscr{E}}_{z_1}, \phi_{z_1}) \sim_A (\underline{\mathscr{E}}_{z_2}, \phi_{z_2})$ (see Section 3.3). Here, $\mathfrak{Z}^0$ is the open subscheme of T_2-semistable points z. By Lemma 3.3.3, it agrees with the open subset of points z, such $\widetilde{\phi}_z \not\equiv 0$. For $d \gg 0$, we associate with the universal family $(\underline{\mathfrak{E}}_{\mathfrak{Z}}, \phi_{\mathfrak{Z}})$ of tuples with a decoration of type $(\mathrm{II} : a, b, c; \mathscr{L})$ a family $(\underline{\mathfrak{E}}_{\mathfrak{Z}}, \widetilde{\phi}_{\mathfrak{Z}})$ of tuples with a decoration of type $(\mathrm{I} : a(d), b(d), c(d); \mathscr{L}^{\otimes d})$. This defines a morphism $\mathfrak{Z}^0 \longrightarrow \mathfrak{Z}_{\mathrm{I}} := \mathfrak{Z}_{\mathrm{I}}(a(d), b(d), c(d); \mathscr{L}^{\otimes d})$ and descends to a projective and $\widetilde{S}$-equivariant morphism

$$\pi_2 \colon \mathfrak{Z}/\!\!/T_2 \longrightarrow \mathfrak{Z}_{\mathrm{I}}.$$

Now,

$$\widetilde{U}_\delta := \pi_1^{-1}(\pi_2^{-1}(U_{(1/d) \cdot \delta}))$$

is the set of points $z \in \mathfrak{Z}$, such that $(\underline{\mathscr{E}}_z, \phi_z)$ is δ-semistable. By Lemma 2.1.7 in [6], we would be done, if we knew that π_2 is injective. Hence, we get the following weaker result.

Lemma 3.6.1. *Let* $\widetilde{\mathfrak{Z}} \subseteq \mathfrak{Z}$ *be a closed subscheme, and define* $\mathfrak{S}$ *as the class of decorated tuples* $(\underline{\mathscr{E}}, \phi)$ *which are isomorphic to a decorated tuple of the form* $(\underline{\mathscr{E}}_z, \phi_z)$ *for a point* $z \in \widetilde{\mathfrak{Z}} \cap \widetilde{U}_\delta$. *Assume that* $\mathfrak{S}$ *meets the requirements stated before Proposition* 3.3.4, *Then, the good quotient* $\widetilde{\mathfrak{Z}}/\!\!/\widetilde{G}$ *exists as a projective scheme. Its closed points correspond to the S-equivalence classes of decorated tuples from the class* $\mathfrak{S}$.

Proof. Since T_2 is linearly reductive, the good quotient $\widetilde{\mathfrak{Z}}/\!\!/T_2$ exists as a closed subscheme of $\mathfrak{Z}/\!\!/T_2$. Proposition 3.3.4 grants that the restriction of π_2 to $\widetilde{\mathfrak{Z}}/\!\!/T_2$ is injective. Therefore, $\pi_{2|\widetilde{\mathfrak{Z}}/\!\!/T_2}$ is finite, because it is also proper. Thus, Lemma 2.1.7 in [6] yields the existence and projectivity of the quotient $\widetilde{\mathfrak{Z}}/\!\!/\widetilde{G}$. By the GIT construction of the moduli space for $((1/d)\cdot\delta)$-semistable tuples with a decoration of type $(\mathrm{II} : \underline{a}, b, c; \mathscr{L})$, two points z_1 and z_2 in $\widetilde{\mathfrak{Z}}^0$ give the same point in $\widetilde{\mathfrak{Z}}/\!\!/\widetilde{G}$, if and only if $(\underline{\mathscr{E}}_{z_1}, \widetilde{\phi}_{z_1})$ and $(\underline{\mathscr{E}}_{z_2}, \widetilde{\phi}_{z_2})$ are S-equivalent. Thus, the statement about the closed points is also true. □

To conclude, let us stress again why we need the above, a little strange looking result. As in the case of a semisimple group, we would like to solve the moduli problem for singular principal bundles by recurrence to the theory of decorated tuples of sheaves. The decorated tuples of sheaves we obtain by this method are naturally of type II. Unfortunately, the general semistability concept for tuples of sheaves with a decoration of type II involves also one parameter subgroups which do not lie in the derived group (see [30]). On the other hand, the semistability concept for tuples with a decoration of type I depends only on one parameter subgroups of the derived group. Therefore, the natural idea is to relate the semistability concept of tuples of type II to the one of tuples of type I. However, this is possible only under some additional technical assumptions as the above lemma shows. Luckily, these technical assumptions are satisfied by the objects to which we will apply the theory.

4. Construction of the moduli space for reductive groups via faithful representations

In this section, we will prove the fundamental existence theorem on moduli spaces of semistable singular principal G-bundles, based on the choice of a faithful representation of the structure group.

4.1 *Singular principal bundles*

In the introduction, we have already given the formal definition for pseudo G-bundles and singular principal G-bundles. Here, we would like to motivate this definition and give some basic properties. Recall that we fix a faithful representation $\rho\colon G \longrightarrow \mathrm{GL}(V)$, such that the image is contained in $(\mathrm{GL}(V_1) \times \cdots \times \mathrm{GL}(V_t)) \cap \mathrm{SL}(V)$ and such that the radical $\mathscr{R}(G)$ of G maps to the center of $\mathrm{GL}(V_1) \times \cdots \times \mathrm{GL}(V_t)$.

Assume that we are given a principal G-bundle $\mathscr{P}$ on an algebraic variety Y. By means of ρ, we find the associated principal $(\mathrm{GL}(V_1) \times \cdots \times \mathrm{GL}(V_t))$-bundle $\rho_\star(\mathscr{P})$. The principal bundle $\rho_\star(\mathscr{P})$ identifies with a tuple $\underline{\mathscr{P}} = (\mathscr{P}_1, \dots, \mathscr{P}_t)$ where $\mathscr{P}_i$ is a principal $\mathrm{GL}(V_i)$-bundle, $i = 1, \dots, t$. The principal $\mathrm{GL}(V_i)$-bundle $\mathscr{P}_i$ may be written as the frame bundle of a vector bundle $\mathscr{E}_i$, $i = 1, \dots, t$. In other words, we find a tuple $\underline{\mathscr{E}} = (\mathscr{E}_1, \dots, \mathscr{E}_t)$ of vector bundles, such that

$$\mathscr{P}_i = \mathscr{I}som(V_i \otimes \mathscr{O}_X, \mathscr{E}_i), \quad i = 1, \dots, t.$$

Of course, we cannot reconstruct $\mathscr{P}$ from the tuple $\underline{\mathscr{E}}$, but we also have the diagram

$$\begin{array}{ccc} \mathscr{P} & \xrightarrow{G\text{-equivariant}} & \mathscr{I}(\underline{\mathscr{E}}^\vee, \underline{V}) \\ & \searrow \quad \swarrow & \\ & Y & \end{array}.$$

Here, $\underline{V} = (V_1, \dots, V_t)$, $\underline{\mathscr{E}}^\vee = (\mathscr{E}_1^\vee, \dots, \mathscr{E}_t^\vee)$ (duals are introduced in order to be conformal with the notation in the introduction), and

$$\mathscr{I}(\underline{\mathscr{E}}^\vee, \underline{V}) := \mathscr{I}som(V_1 \otimes \mathscr{O}_Y, \mathscr{E}_1) \times_Y \cdots \times_Y \mathscr{I}som(V_t \otimes \mathscr{O}_Y, \mathscr{E}_t).$$

If we take the G-quotients in the above diagram, we get a section $\sigma \colon Y \longrightarrow \mathscr{I}(\underline{\mathscr{E}}^\vee, \underline{V})/G$ as an additional datum. In fact, the pair $(\underline{\mathscr{E}}, \sigma)$ allows to reconstruct $\mathscr{P}$. More precisely, given a pair $(\underline{\mathscr{E}}, \sigma)$ with $\underline{\mathscr{E}} = (\mathscr{E}_1, \dots, \mathscr{E}_t)$ a tuple of vector bundles of the correct ranks and $\sigma \colon Y \longrightarrow \mathscr{I}(\underline{\mathscr{E}}^\vee, \underline{V})/G$ a section, we form the Cartesian diagram

$$\begin{array}{ccc} \mathscr{P} & \longrightarrow & \mathscr{I}(\underline{\mathscr{E}}^\vee, \underline{V}) \\ \big\downarrow & & \big\downarrow {\scriptstyle G\text{-bundle}} \\ Y & \xrightarrow{\sigma} & \mathscr{I}(\underline{\mathscr{E}}^\vee, \underline{V})/G \end{array}$$

in order to construct the principal G-bundle $\mathscr{P}$. Observe that there is a natural notion of isomorphy on the set of pairs $(\underline{\mathscr{E}}, \sigma)$. It is easy to check that the assignment $\mathscr{P} \longmapsto (\underline{\mathscr{E}}, \sigma)$ outlined above yields a bijection between the set of isomorphy classes of principal G-bundles on Y and the set of isomorphy classes of pairs $(\underline{\mathscr{E}}, \sigma)$. Note that a similar construction was used by Ramanathan [25] to get from principal bundles to vector bundles with a section in an associated object. Of course, today we know much more about the theory of such pairs than at the time when Ramanathan was preparing his work (compare Section 6.3). However, we will need slightly more general objects before we can invoke the machinery of "decorated tuples of sheaves". For this, we form

$$\begin{aligned} \mathscr{H}(\underline{\mathscr{E}}^\vee, \underline{V}) &:= \mathscr{H}om(V_1 \otimes \mathscr{O}_Y, \mathscr{E}_1) \times_Y \cdots \times_Y \mathscr{H}om(V_t \otimes \mathscr{O}_Y, \mathscr{E}_t) \\ &= \mathscr{H}om(\mathscr{E}_1^\vee, V_1^\vee \otimes \mathscr{O}_X) \times_Y \cdots \times_Y \mathscr{H}om(\mathscr{E}_t^\vee, V_t^\vee \otimes \mathscr{O}_Y) \\ &= \mathscr{S}pec(\mathscr{S}ym^\star(\mathscr{E}_1^\vee \otimes V_1 \oplus \cdots \oplus \mathscr{E}_t^\vee \otimes V_t)). \end{aligned}$$

Then, we have the open embedding

$$\mathscr{I}(\underline{\mathscr{E}}^{\vee}, \underline{V}) \subset \mathscr{H}(\underline{\mathscr{E}}^{\vee}, \underline{V}).$$

Next, we use that the image of our faithful representation takes its image in the special linear group $\mathrm{SL}(V)$. This condition grants that the above inclusion descends to the quotient (see [26], page 1189), i.e., that there is an open embedding

$$\mathscr{I}(\underline{\mathscr{E}}^{\vee}, \underline{V})/G \subset \mathscr{H}(\underline{\mathscr{E}}^{\vee}, \underline{V})/\!\!/G.$$

The space $\mathscr{H}(\underline{\mathscr{E}}^{\vee}, \underline{V})/\!\!/G$ is affine over Y. It is the relative spectrum of the sheaf

$$\mathscr{S}ym^{\star}(\mathscr{E}_1^{\vee} \otimes V_1 \oplus \cdots \oplus \mathscr{E}_t^{\vee} \otimes V_t)^G$$

of $\mathscr{O}_Y$-algebras. Thus, given a pair $(\underline{\mathscr{E}}, \sigma)$ as above, the section

$$\sigma : Y \longrightarrow \mathscr{I}(\underline{\mathscr{E}}^{\vee}, \underline{V})/G \subset \mathscr{H}(\underline{\mathscr{E}}^{\vee}, \underline{V})/\!\!/G$$

corresponds to a homomorphism

$$\tau : \mathscr{S}ym^{\star}(\mathscr{E}_1^{\vee} \otimes V_1 \oplus \cdots \oplus \mathscr{E}_t^{\vee} \otimes V_t)^G \longrightarrow \mathscr{O}_Y$$

of $\mathscr{O}_Y$-algebras. Since dualizing is not a well-behaved operation for torsion-free sheaves, we replace the vector bundle $\mathscr{E}_i^{\vee}$ in the above picture by a torsion-free sheaf $\mathscr{A}_i$, $i = 1, \ldots, t$.

Suppose Y is a smooth algebraic variety. Then, a *pseudo G-bundle* is a pair $(\underline{\mathscr{A}}, \tau)$, consisting of a tuple $\underline{\mathscr{A}} = (\mathscr{A}_1, \ldots, \mathscr{A}_t)$ of torsion free sheaves, such that $\mathrm{rk}(\mathscr{A}_i) = \dim_k(V_i)$, $i = 1, \ldots, t$, and $\det(\mathscr{A}) \cong \mathscr{O}_X$, $\mathscr{A} := \mathscr{A}_1 \oplus \cdots \oplus \mathscr{A}_t$, and a non-trivial homomorphism

$$\tau : \mathscr{S}ym^{\star}(\mathscr{A}_1 \otimes V_1 \oplus \cdots \oplus \mathscr{A}_t \otimes V_t)^G \longrightarrow \mathscr{O}_Y$$

of $\mathscr{O}_Y$-algebras. Here, non-trivial means that τ is not just the projection onto the degree zero component. This is equivalent to the condition that the induced section

$$\sigma : Y \longrightarrow \mathscr{H}(\underline{\mathscr{A}}, \underline{V})$$

is not the zero section. The question is to what extent a pseudo G-bundle may be interpreted as a degenerate version of a principal G-bundle.

In order to answer this question, we return to the projective base manifold $Y = X$. Let

$$\pi : \mathscr{H}(\underline{\mathscr{A}}, \underline{V}) \longrightarrow X \quad \text{and} \quad \bar{\pi} : \mathscr{H}(\underline{\mathscr{A}}, \underline{V})/\!\!/G \longrightarrow X$$

be the usual affine schemes over X. By the universal property of $\mathscr{H}(\underline{\mathscr{A}}, \underline{V})$, there is the tautological homomorphism

$$h\colon \pi^\star(\mathscr{A}) \longrightarrow V^\vee \otimes \mathscr{O}_{\mathscr{H}(\underline{\mathscr{A}},\underline{V})}.$$

Its determinant is a G-invariant homomorphism between trivial sheaves, so that it descends to a homomorphism

$$\Delta\colon \bar{\pi}^\star(\mathscr{A}) \longrightarrow \bigwedge^{\dim_k(V)} V^\vee \otimes \mathscr{O}_{\mathscr{H}(\underline{\mathscr{A}},\underline{V})/\!\!/G}.$$

Then, by means of pullback via σ, this yields

$$\mathfrak{d} := \sigma^\star(\Delta)\colon \det(\mathscr{A}) \longrightarrow \bigwedge^{\dim_k(V)} V^\vee \otimes \mathscr{O}_X.$$

Note that $\mathfrak{d}$ is either an isomorphism or identically zero. In other words, we proved the following:

Lemma 4.1.1. *Let $(\underline{\mathscr{A}}, \tau)$ be a pseudo G-bundle, let $U_{\mathscr{A}}$ be the maximal open subset where $\mathscr{A} = \mathscr{A}_1 \oplus \cdots \oplus \mathscr{A}_t$ is locally free and*

$$\sigma : U_{\mathscr{A}} \longrightarrow \mathscr{H}(\underline{\mathscr{A}}_{|U_{\mathscr{A}}}, \underline{V})/\!\!/G$$

the section defined by τ. Then, either

$$\sigma(U_{\mathscr{A}}) \subset \mathscr{I}(\underline{\mathscr{A}}_{|U_{\mathscr{A}}}, \underline{V})/G$$

or

$$\sigma(U_{\mathscr{A}}) \subset (\mathscr{H}(\underline{\mathscr{A}}_{|U_{\mathscr{A}}}, \underline{V})/\!\!/G)\backslash(\mathscr{I}(\underline{\mathscr{A}}_{|U_{\mathscr{A}}}, \underline{V})/G).$$

In the former case, we call $(\underline{\mathscr{A}}, \tau)$ a *singular principal G-bundle*, and we may then form the base change diagram

$$\begin{array}{ccc}
\mathscr{P}(\underline{\mathscr{A}}, \tau) & \longrightarrow & \mathscr{I}(\underline{\mathscr{A}}_{|U_{\mathscr{A}}}, \underline{V}) \\
\big\downarrow & & G\text{-}\big\downarrow\text{bundle} \\
U_{\mathscr{A}} & \xrightarrow{\sigma_{|U_{\mathscr{A}}}} & \mathscr{I}(\underline{\mathscr{A}}_{|U_{\mathscr{A}}}, \underline{V})/G.
\end{array}$$

Note that $(U_{\mathscr{A}}, \mathscr{P}(\underline{\mathscr{A}}, \tau))$ is a rational principal G-bundle in the sense of Ramanathan.

In the second case, we have a very degenerate object. For example, if G is the symplectic group $\mathrm{Sp}_r(k)$ with its "standard" representation on k^r, then the datum of a pseudo G-bundle is equivalent to the datum of a torsion free sheaf $\mathscr{A}$ and a non-zero anti-symmetric bilinear form $\beta\colon \mathscr{A} \otimes \mathscr{A} \longrightarrow \mathscr{O}_X$ (cf. [7], [26]). The singular principal G-bundles correspond to those objects where the induced homomorphism $\phi\colon \mathscr{A} \longrightarrow \mathscr{A}^\vee$ is an isomorphism over $U_{\mathscr{A}}$ while the degenerate objects are those where ϕ has everywhere rank less than r.

The semistable reduction theorem for semistable singular principal G-bundles is now equivalent to the fact that the semistability concept forces a semistable pseudo G-bundle to be a singular principal G-bundle.

4.2 *S-equivalence*

The semistability concept for singular principal G-bundles has been defined in the introduction. Here, we give the definition of the notion of S-equivalence on the set of semistable singular principal G-bundles.

Let $(\underline{\mathscr{A}}, \tau)$ be a semistable singular principal G-bundle, let $\lambda\colon G \longrightarrow \mathrm{GL}(V)$ be a one parameter subgroup and β a reduction of $(\underline{\mathscr{A}}, \tau)$ to λ defined over the big open subset U', such that

$$M(\underline{\mathscr{A}}, \tau; \beta) = 0.$$

We want to define the *associated admissible deformation* $\mathrm{df}_\beta(\underline{\mathscr{A}}, \tau) = (\underline{\mathscr{A}}_{\mathrm{df}}, \tau_{\mathrm{df}})$. Again, we define

$$\underline{\mathscr{A}}_{\mathrm{df}} := (\mathscr{A}_{1,\mathrm{df}}, \dots, \mathscr{A}_{t,\mathrm{df}}) \quad \text{with} \quad \mathscr{A}_{i,\mathrm{df}} := \bigoplus_{j=1}^{s_i+1} \mathscr{A}_{i,j}/\mathscr{A}_{i,j-1}, \qquad i = 1, \dots, t.$$

Since $\mathrm{Sym}^\star(k^{r_1} \otimes V_1 \oplus \cdots \oplus k^{r_t} \otimes V_t)^G$ is a locally finite module, λ yields weights $0 = \gamma_1 < \gamma_2 < \cdots$, a decomposition into (infinitely many) weight spaces, and a filtration

$$\{0\} =: U_0 \subsetneq U_1 \subsetneq \cdots \subsetneq \mathrm{Sym}^\star(k^{r_1} \otimes V_1 \oplus \cdots \oplus k^{r_t} \otimes V_t)^G.$$

On the level of sheaves, we find the corresponding filtration

$$\{0\} = \mathscr{U}_0 \subsetneq \mathscr{U}_1 \subsetneq \cdots \subsetneq \mathscr{S}ym^\star(\mathscr{A}_{1|U'} \otimes V_1 \oplus \cdots \oplus \mathscr{A}_{1|U'} \otimes V_t)^G.$$

If the index i corresponds to the weight γ, j to γ', and k to $\gamma + \gamma'$, we have

$$\mathscr{U}_i \cdot \mathscr{U}_j \subseteq \mathscr{U}_k. \tag{61}$$

In particular, $\mathscr{U}_1$ is a subalgebra. By (61), we have the isomorphism

$$\mathscr{S}ym^\star(\mathscr{A}_{1,\mathrm{df}|U'} \otimes V_1 \oplus \cdots \oplus \mathscr{A}_{t,\mathrm{df}|U'} \otimes V_t)^G \cong \bigoplus_{i\geq 0} \mathscr{U}_{i+1}/\mathscr{U}_i$$

of $\mathscr{O}_{U'}$-algebras. Thus, we may define

$$\widetilde{\tau}\colon \mathscr{S}ym^\star(\mathscr{A}_{1,\mathrm{df}|U'} \otimes V_1 \oplus \cdots \oplus \mathscr{A}_{t,\mathrm{df}|U'} \otimes V_t)^G \longrightarrow \mathscr{O}_{U'}$$

as the projection onto the sub-algebra $\mathscr{U}_1$ followed by the restriction of $\tau_{|U'}$ to $\mathscr{U}_1$. Then, $\widetilde{\tau}$ extends to

$$\tau_{\mathrm{df}}\colon \mathscr{S}ym^\star(\mathscr{A}_{1,\mathrm{df}} \otimes V_1 \oplus \cdots \oplus \mathscr{A}_{t,\mathrm{df}} \otimes V_t)^G \longrightarrow \mathscr{O}_X.$$

We say that the semistable singular principal G-bundle $(\underline{\mathscr{A}}, \tau)$ is *polystable*, if it is isomorphic to $\mathrm{df}_\beta(\underline{\mathscr{A}}, \tau)$ for every one parameter subgroup λ and every reduction β of $(\underline{\mathscr{A}}, \tau)$ to λ, such that

$$M(\underline{\mathscr{A}}, \tau; \beta) = 0.$$

The GIT construction of the moduli spaces implies the following fact.

Lemma 4.2.1. *For every semistable singular principal G-bundle $(\underline{\mathscr{A}}, \tau)$, there exists a polystable singular principal G-bundle* $\mathrm{gr}(\underline{\mathscr{A}}, \tau)$ *which is an admissible deformation of $(\underline{\mathscr{A}}, \tau)$. The singular principal G-bundle* $\mathrm{gr}(\underline{\mathscr{A}}, \tau)$ *is uniquely determined up to isomorphy.*

In fact, any sequence of (non-trivial) admissible deformations will "end" after finitely many steps in $\mathrm{gr}(\underline{\mathscr{A}}, \tau)$. We call two semistable singular principal G-bundles $(\underline{\mathscr{A}}, \tau)$ and $(\underline{\mathscr{A}}', \tau')$ *S-equivalent*, if $\mathrm{gr}(\underline{\mathscr{A}}, \tau)$ and $\mathrm{gr}(\underline{\mathscr{A}}', \tau')$ are isomorphic.

Remark 4.2.2. Without reference to the GIT construction, we might explain "S-equivalence" as the equivalence relation generated by "$(\underline{\mathscr{A}}, \tau) \sim_S \mathrm{df}_\beta(\underline{\mathscr{A}}, \tau)$". This viewpoint is useful for some proofs.

The above discussion has been carried out in a way which makes it clear that S-equivalence among semistable singular principal G-bundles and S-equivalence among associated decorated sheaves are the same thing (see Corollary 4.4.3). However, one can formulate the whole concept more nicely in the language of principal bundles. So, let $(\underline{\mathscr{A}}, \tau)$ be a semistable singular principal G-bundle, λ a one parameter subgroup, and β a reduction to λ which is defined over the open subset, say, U', such that

$$M(\underline{\mathscr{A}}, \tau; \beta) = 0.$$

This reduction gives rise to a principal $Q_G(\lambda)$-bundle $\mathscr{Q}$ over U', such that $\mathscr{P}(\underline{\mathscr{A}}, \tau)_{|U'}$ is obtained from $\mathscr{Q}$ by extension of the structure group via $Q_G(\lambda) \subset G$. We construct the new principal G-bundle $\mathscr{P}'$ over U' from $\mathscr{Q}$ by extending the structure group via $Q_G(\lambda) \longrightarrow L_G(\lambda) \subset G$. Since the diagram

$$\begin{array}{ccccc}
Q_G(\lambda) & \twoheadrightarrow & L_G(\lambda) & \hookrightarrow & G \\
\downarrow & & \downarrow & & \downarrow \\
Q_{\mathrm{GL}(V_1)\times\cdots\times\mathrm{GL}(V_t)}(\lambda) & \twoheadrightarrow & L_{\mathrm{GL}(V_1)\times\cdots\times\mathrm{GL}(V_t)}(\lambda) & \hookrightarrow & \mathrm{GL}(V_1)\times\cdots\times\mathrm{GL}(V_t)
\end{array}$$

is commutative, it is clear that the bundle $\rho_\star(\mathscr{P}')$ corresponds to the tuple $\underline{\mathscr{A}}^\vee_{\mathrm{df}|U'} = (\mathscr{A}^\vee_{\mathrm{df},1|U'}, \ldots, \mathscr{A}^\vee_{\mathrm{df},t|U'})$. As explained before, the bundle $\mathscr{P}' \subset \mathscr{I}(\underline{\mathscr{A}}_{|U'}, \underline{V})$ is described by a homomorphism

$$\widetilde{\tau}\colon \mathscr{S}ym^\star(\mathscr{A}_{\mathrm{df},1|U'} \otimes V_1 \oplus \cdots \oplus \mathscr{A}_{\mathrm{df},t|U'} \otimes V_t)^G \longrightarrow \mathscr{O}_{U'}$$

which extends to

$$\tau_{\mathrm{df}}\colon \mathscr{S}ym^\star(\mathscr{A}_{\mathrm{df},1} \otimes V_1 \oplus \cdots \oplus \mathscr{A}_{\mathrm{df},t} \otimes V_t)^G \longrightarrow \mathscr{O}_X.$$

4.3 *Associated decorations of type I and type II*

In Section 2.3, "Some specific quotient problems", we discussed the theory when the base manifold X is just a point. We will now use these results to treat the theory of singular principal G-bundles within the framework of decorated tuples of sheaves. In characteristic zero, this works very well whereas we need to be a bit more careful in positive characteristic. Therefore, we will first outline the construction in characteristic zero and then speak about the necessary modifications in positive characteristic.

Associated decorations in characteristic zero. Let $(\underline{\mathscr{A}}, \tau)$ be a pseudo G-bundle, and let U be the maximal open subset where $\mathscr{A} = \mathscr{A}_1 \oplus \cdots \oplus \mathscr{A}_t$ is locally free, $\underline{\mathscr{A}} = (\mathscr{A}_1, \ldots, \mathscr{A}_t)$. In (8), we introduced the vector space $\widehat{\mathbb{V}}_s$. The tuple $\underline{\mathscr{A}}_{|U} = (\mathscr{A}_{1|U}, \ldots, \mathscr{A}_{t|U})$ and the vector space $\widehat{\mathbb{V}}_s$ yield the vector bundle $\widehat{\mathscr{V}_s}$ on U and the surjection

$$\sigma : \mathscr{S}ym^\star(\widehat{\mathscr{V}_s}) \longrightarrow \mathscr{S}ym^{(s!)}(\mathscr{A}_{1|U} \otimes V_1 \oplus \cdots \oplus \mathscr{A}_{t|U} \otimes V_t)^G$$

of $\mathscr{O}_U$-algebras. Let $\widetilde{\tau}_s$ be the restriction of $\tau_{|U}$ to the subalgebra

$$\mathscr{S}ym^{(s!)}(\mathscr{A}_{1|U} \otimes V_1 \oplus \cdots \oplus \mathscr{A}_{t|U} \otimes V_t)^G.$$

Lemma 4.3.1. *Let $(\underline{\mathscr{A}}, \tau_1)$ and $(\underline{\mathscr{A}}, \tau_2)$ be two pseudo G-bundles, such that $\widetilde{\tau}_{1,s} = \widetilde{\tau}_{2,s}$. Then, there exists a root of unity $\zeta \in k$, such that the isomorphisms $\zeta \cdot \mathrm{id}_{\mathscr{A}_i}$, $i = 1, \ldots, t$, induce an isomorphism between $(\underline{\mathscr{A}}, \tau_1)$ and $(\underline{\mathscr{A}}, \tau_2)$.*

Proof. The proof is a straightforward generalization of [6], Lemma 5.1.1. □

Obviously, $\tau_{|U}$ is determined by $\sigma \circ \tau_{|U}$, and $\sigma \circ \tau_{|U}$ is determined by its restriction

$$\phi' : \widehat{\mathscr{V}_s} \longrightarrow \mathscr{O}_U$$

to $\widehat{\mathscr{V}_s}$. Since the $\mathrm{GL}(r)$-module $\widehat{\mathbb{V}}_s$ is polynomial, we have seen in Corollary 2.2.3 that there exist non-negative integers a, b, such that it is a quotient module of $(k^{r\otimes a})^{\oplus b}$. In our situation, this means that there is a surjection

$$\pi : (\mathscr{A}_{|U}^{\otimes a})^{\oplus b} \longrightarrow \mathscr{V}_s.$$

Altogether, we see that τ gives rise to a homomorphism

$$\phi'' : (\mathscr{A}_{|U}^{\otimes a})^{\oplus b} \longrightarrow \mathscr{O}_U$$

and, consequently, to a tuple of sheaves $(\underline{\mathscr{A}}, \phi)$ with a decoration of type $(\mathrm{II} : a, b, 0; \mathscr{O}_X)$:

$$\phi : (\mathscr{A}^{\otimes a})^{\oplus b} \longrightarrow \iota_\star((\mathscr{A}_{|U}^{\otimes a})^{\oplus b}) \xrightarrow{\iota_\star(\phi'')} \iota_\star \mathscr{O}_U = \mathscr{O}_X.$$

As usual, $\iota : U \subseteq X$ is the inclusion morphism. Corollary 2.3.2 in [6] implies that the same construction can be used to assign to a family $(\underline{\mathscr{A}}_S, \tau_S)$ of pseudo G-bundles parameterized by the scheme S a family $(\underline{\mathscr{A}}_S, \tau_S)$ of tuples of sheaves with a decoration of type $(\mathrm{II} : a, b, 0; \mathscr{O}_X)$ over $X \times S$.

Lemma 4.3.2. *The assignment $(\underline{\mathscr{A}}, \tau) \longmapsto (\underline{\mathscr{A}}, \phi)$ gives rise to an injection of the set of isomorphy classes of pseudo G-bundles into the set of isomorphy classes of tuples of sheaves with a decoration of type* $(\mathrm{II} : a, b, 0; \mathscr{O}_X)$.

Proof. This is an immediate consequence of Lemma 4.3.1. □

Associated decorations in positive characteristic. We will restrict ourselves to the case $t = 1$, the generalization to $t > 1$ being obvious. The reductive group G may be assumed to be defined over the ring of integers $\mathbb{Z}$. Since the coordinate algebra $\mathbb{Z}[G]$ is a free $\mathbb{Z}$-module, the usual arguments ([3], 1.10) show that there are a free $\mathbb{Z}$-module $W_{\mathbb{Z}}$ of finite rank and a faithful representation $\rho'\colon G \longrightarrow \mathrm{GL}(W_{\mathbb{Z}})$ which is defined over $\mathbb{Z}$. Since the determinant representation and its inverse are defined over $\mathbb{Z}$, we also find the representation

$$\rho := \rho' \oplus (\det \circ \rho')^{-1} \colon G \longrightarrow \mathrm{SL}(V_{\mathbb{Z}}), \quad V_{\mathbb{Z}} := W_{\mathbb{Z}} \oplus \mathbb{Z},$$

over the integers. By Seshadri's generalization of GIT relative to base varieties which are defined over Nagata rings [32], the algebra

$$\mathrm{Sym}^{\star}(V_{\mathbb{Z}} \otimes \mathbb{Z}^r)^G$$

is a finitely generated $\mathbb{Z}$-algebra, and we have the good quotients

$$\pi : \mathrm{Hom}(V_{\mathbb{Z}}, \mathbb{Z}^r) \longrightarrow \mathrm{Hom}(V_{\mathbb{Z}}, \mathbb{Z}^r)/\!\!/G := \mathrm{Spec}(\mathrm{Sym}^{\star}(V_{\mathbb{Z}} \otimes \mathbb{Z}^r)^G) \longrightarrow \mathrm{Spec}(\mathbb{Z})$$

and

$$\begin{aligned}\bar{\pi} : \mathbb{P}(\mathrm{Hom}(V_{\mathbb{Z}}, \mathbb{Z}^r)^{\vee}) \dashrightarrow\ & \mathbb{P}(\mathrm{Hom}(V_{\mathbb{Z}}, \mathbb{Z}^r)^{\vee})/\!\!/G \\ & := \mathrm{Proj}(\mathrm{Sym}^{\star}(V_{\mathbb{Z}} \otimes \mathbb{Z}^r)^G) \longrightarrow \mathrm{Spec}(\mathbb{Z}).\end{aligned}$$

The quotient

$$\pi^0\colon \mathrm{Isom}(V_{\mathbb{Z}}, \mathbb{Z}^r) \longrightarrow \mathrm{Isom}(V_{\mathbb{Z}}, \mathbb{Z}^r)/G$$

is a principal G-bundle and thus a universal categorical quotient. However, the quotients π and $\bar{\pi}$ are not necessarily universal categorical quotients. This fact accounts for the slight modifications which we do have to make. The good quotient parameterizes orbits of geometric points with respect to the equivalence relation that two points map to the same point in the quotient, if and only if the closures of their orbits do intersect. This implies the following.

Lemma 4.3.3. *Let* $Z \hookrightarrow \mathrm{Spec}(\mathbb{Z})$ *be a closed subscheme. Then, the canonical morphisms*

$$(\mathrm{Hom}(V_{\mathbb{Z}}, \mathbb{Z}^r) \times_{\mathrm{Spec}(\mathbb{Z})} Z)/\!\!/G \longrightarrow (\mathrm{Hom}(V_{\mathbb{Z}}, \mathbb{Z}^r)/\!\!/G) \times_{\mathrm{Spec}(\mathbb{Z})} Z$$

and

$$(\mathbb{P}(\mathrm{Hom}(V_{\mathbb{Z}}, \mathbb{Z}^r)^{\vee}) \times_{\mathrm{Spec}(\mathbb{Z})} Z)/\!\!/G \longrightarrow (\mathbb{P}(\mathrm{Hom}(V_{\mathbb{Z}}, \mathbb{Z}^r)^{\vee})/\!\!/G) \times_{\mathrm{Spec}(\mathbb{Z})} Z$$

are bijective on geometric points.

Let us write

$$(\mathrm{Hom}(V_{\mathbb{Z}}, \mathbb{Z}^r) \times_{\mathrm{Spec}(\mathbb{Z})} Z)\widetilde{/\!\!/}G := (\mathrm{Hom}(V_{\mathbb{Z}}, \mathbb{Z}^r)/\!\!/G) \times_{\mathrm{Spec}(\mathbb{Z})} Z$$

and

$$(\mathbb{P}(\mathrm{Hom}(V_{\mathbb{Z}}, \mathbb{Z}^r)^\vee) \times_{\mathrm{Spec}(\mathbb{Z})} Z)\widetilde{/\!\!/}G := (\mathbb{P}(\mathrm{Hom}(V_{\mathbb{Z}}, \mathbb{Z}^r)^\vee)/\!\!/G) \times_{\mathrm{Spec}(\mathbb{Z})} Z.$$

Next, assume that $\mathscr{A}$ is a locally free sheaf on the scheme U which is of finite type over $\mathrm{Spec}(k)$, k our usual ground field. Then, we may easily construct the geometric quotient

$$\begin{aligned}\widetilde{\mathscr{H}} &:= \mathscr{H}om(\mathscr{A}, V^\vee \otimes \mathscr{O}_U)\widetilde{/\!\!/}G \\ &:= (\mathscr{I}som(k^r \otimes \mathscr{O}_U, \mathscr{A}) \times_{\mathrm{Spec}(k)} \mathrm{Hom}(V, k^r)\widetilde{/\!\!/}G)/\mathrm{GL}_r(k),\end{aligned}$$

using local trivializations. The construction of $\mathscr{H}om(\mathscr{A}, V^\vee \otimes \mathscr{O}_U)\widetilde{/\!\!/}G$ clearly commutes with base change. Moreover, we have the natural morphism

$$\mathscr{H}om(\mathscr{A}, V^\vee \otimes \mathscr{O}_U)/\!\!/G \longrightarrow \mathscr{H}om(\mathscr{A}, V^\vee \otimes \mathscr{O}_U)\widetilde{/\!\!/}G$$

which is bijective, by Lemma 4.3.3. This construction has an algebraic counterpart. Define $\widetilde{\pi}\colon \widetilde{\mathscr{H}} \longrightarrow U$ as the projection map, and let

$$\widetilde{\mathscr{S}ym}^\star(\mathscr{A} \otimes V)^G$$

be the sheaf $\widetilde{\pi}_\star(\mathscr{O}_{\widetilde{\mathscr{H}}})$. Then, we obtain the homomorphism

$$\mathrm{ps}(\mathscr{A})\colon \widetilde{\mathscr{S}ym}^\star(\mathscr{A} \otimes V)^G \longrightarrow \mathscr{S}ym^\star(\mathscr{A} \otimes V)^G$$

that induces the bijective map $\mathscr{H}om(\mathscr{A}, V^\vee \otimes \mathscr{O}_U)/\!\!/G \longrightarrow \widetilde{\mathscr{H}}$.

Suppose that the algebra $\mathrm{Sym}^\star(V_{\mathbb{Z}} \otimes \mathbb{Z}^r)^G$ is generated in degrees $\le s$. By Lemma 2.2.1, we may write

$$\bigoplus_{\substack{(d_1,\ldots,d_s):\\ d_i\ge 0,\ \sum id_i = s!}} \mathrm{Sym}^{d_1}((V_{\mathbb{Z}} \otimes \mathbb{Z}^r)^G) \otimes \cdots \otimes \mathrm{Sym}^{d_s}(\mathrm{Sym}^i(V_{\mathbb{Z}} \otimes \mathbb{Z}^r)^G)$$

as the quotient of $\mathbb{D}^{s!,\upsilon}(\mathbb{Z}^r)$, for an appropriate integer $\upsilon > 0$.

Assume that $p > s!$. Then, $\mathbb{D}^{s!,\upsilon}(\mathbb{Z}^r) \otimes_{\mathbb{Z}} k$ is a quotient of $((k^r)^{\otimes s!})^{\oplus b}$, for an appropriate integer $b > 0$. Let S be a scheme of finite type over k, $(\mathscr{A}_S, \tau_S)$ a family of pseudo G-bundles parameterized by S, and $U \subset X \times S$ the maximal open subset where $\mathscr{A}$ is locally free. Then, we obtain the homomorphism

$$\begin{aligned}&\mathscr{S}ym^\star((\mathscr{A}_{S|U}^{\otimes s!})^{\oplus b}) \\ &\twoheadrightarrow \bigoplus_{\substack{(d_1,\ldots,d_s):\\ d_i\ge 0,\ \sum id_i = s!}} \mathscr{S}ym^{d_1}((V \otimes \mathscr{A}_{S|U})^G) \otimes \cdots \otimes \mathscr{S}ym^{d_s}(\widetilde{\mathscr{S}ym}^i(V \otimes \mathscr{A}_{S|U})^G) \\ &\overset{\text{“ps}(\mathscr{A})\text{”}}{\to} \bigoplus_{\substack{(d_1,\ldots,d_s):\\ d_i\ge 0,\ \sum id_i = s!}} \mathscr{S}ym^{d_1}((V \otimes \mathscr{A}_{S|U})^G) \otimes \cdots \otimes \mathscr{S}ym^{d_s}(\mathscr{S}ym^i(V \otimes \mathscr{A}_{S|U})^G) \\ &\overset{\text{“}\tau\text{”}}{\to} \mathscr{O}_U.\end{aligned}$$

As before, we may therefore associate with $(\mathscr{A}_S, \tau_S)$ a family $(\mathscr{A}_S, \phi_S)$ of decorated sheaves. As remarked above, this assignment is compatible with base change and gives consequently rise to a natural transformation between the respective functors. Note that the injectivity statements in Lemma 4.3.1 and 4.3.2 remain valid, thanks to Lemma 4.3.3.

We conclude this paragraph with the following result.

Lemma 4.3.4. *Let G be a reductive algebraic group, X and Y projective schemes equipped with a G-action, and $\pi : X \longrightarrow Y$ a finite and G-equivariant morphism. Suppose $\mathscr{L}$ is a G-linearized ample line bundle on Y. Then, for any point $x \in X$ and any one parameter subgroup $\lambda : \mathbb{G}_m(k) \longrightarrow G$, one has*

$$\mu_{\pi^\star(\mathscr{L})}(\lambda, x) = \mu_{\mathscr{L}}(\lambda, \pi(x)).$$

Proof. This is Lemma 2.1 in [28] and also holds in positive characteristic: Simply replace the G-module splitting by a splitting of the induced $\mathbb{G}_m(k)$-module. □

We will apply this lemma to the finite morphism

$$\mathbb{P}(\mathrm{Hom}(V, k^r)^\vee)/\!\!/G \longrightarrow \mathbb{P}(\mathrm{Hom}(V, k^r)^\vee)\widetilde{/\!\!/}G.$$

Semistable pseudo G-bundles. We fix an integer d as in Lemma 2.3.8. For every positive polynomial δ of degree at most $\dim(X) - 1$ with rational coefficients, we have the notion of δ-(semi)stability for tuples of sheaves with a decoration of type $(\mathrm{II} : a, b, c; \mathscr{L})$.

Now, set $\widetilde{\delta} := s! \cdot \delta$. A pseudo G-bundle $(\underline{\mathscr{A}}, \tau)$ is said to be *$\widetilde{\delta}$-(semi)stable*, if the associated tuple of sheaves $(\underline{\mathscr{A}}, \phi)$ with a decoration of type $(\mathrm{II} : a, b, c; \mathscr{O}_X)$ is δ-(semi)stable.

Remark 4.3.5.

i) If $(\underline{\mathscr{A}}, \tau)$ is a singular principal G-bundle, then Diagram (12) shows that $\widetilde{\phi}$ is non-zero and surjective over the maximal open subset where $\mathscr{A} = \mathscr{A}_1 \oplus \cdots \oplus \mathscr{A}_t$ is locally free, i.e., $\widetilde{\phi}$ is surjective over a big open subset.
ii) By (10), the notion of (semi)stability for pseudo G-bundles depends only on δ and not on s.

Let $(\underline{\mathscr{A}}, \tau)$ be a pseudo G-bundle, and let $(\underline{\mathscr{A}}, \phi)$ be its associated tuple of sheaves with a decoration of type $(\mathrm{II} : a, b, c; \mathscr{O}_X)$. We assume $\widetilde{\phi} \not\equiv 0$ (e.g., $(\underline{\mathscr{A}}, \tau)$ might be a singular principal G-bundle, by Remark 4.3.5) and let $(\underline{\mathscr{A}}, \widetilde{\phi})$ be the associated tuple of sheaves with a decoration of type $(\mathrm{I} : a(d), b(d), c(d); \mathscr{O}_X)$. For every non-trivial weighted filtration $(\underline{\mathscr{A}}_\bullet, \underline{\alpha}_\bullet)$ of $\mathscr{A} = (\mathscr{A}_1, \ldots, \mathscr{A}_t)$, we have defined the quantities

$$\widetilde{\mu}(\underline{\mathscr{A}}_\bullet, \underline{\alpha}_\bullet; \phi) \text{ (see (28)) } \quad \text{and} \quad \mu(\underline{\mathscr{A}}_\bullet, \underline{\alpha}_\bullet; \widetilde{\phi}) \text{ (see (20)).}$$

Lemma 4.3.6. *Let $(\underline{\mathscr{A}}, \tau)$ be a singular principal G-bundle and $(\underline{\mathscr{A}}_\bullet, \underline{\alpha}_\bullet)$ a non-trivial weighted filtration of $\mathscr{A}$. Then, the following conditions are equivalent:*

1. $\widetilde{\mu}(\underline{\mathscr{A}}_\bullet, \underline{\alpha}_\bullet; \phi) = 0.$
2. $\mu(\underline{\mathscr{A}}_\bullet, \underline{\alpha}_\bullet; \widetilde{\phi}) = 0.$

3. *There is a reduction β of $(\underline{\mathscr{A}}; \tau)$ to a one parameter subgroup λ of $[G, G]$ with*

$$(\underline{\mathscr{A}}_\bullet, \underline{\alpha}_\bullet) = (\underline{\mathscr{A}}_\bullet(\beta), \underline{\alpha}_\bullet(\beta)).$$

Proof. We start with "1. $\Rightarrow$ 3." and assume for simplicity $t = 1$, the general case being analogous. Let $\lambda' \colon \mathbb{G}_m(k) \longrightarrow \mathrm{SL}_r(k) = \mathrm{SL}(V^\vee)$ be a one parameter subgroup. By composing with the representation $\mathrm{SL}(V^\vee) \longrightarrow \mathrm{SL}(V)$, we get a one parameter subgroup of $\mathrm{SL}(V)$ which we call λ' again, for simplicity. We may assume that the associated weighted flag

$$(V_\bullet(\lambda') : \{0\} \subsetneq V_1 \subsetneq \cdots \subsetneq V_{s'} \subsetneq V, \alpha_\bullet(\lambda'))$$

satisfies $s' = s$, $\dim_k(V_i) = \mathrm{rk}\, \mathscr{A}_i'$, $\mathscr{A}_i' = \ker(\mathscr{A}^\vee \longrightarrow \mathscr{A}_{s+1-i}^\vee)$, $i = 1, \dots, s$, and $\alpha_\bullet(\lambda') = (\alpha_s, \dots, \alpha_1)$, if $\alpha_\bullet = (\alpha_1, \dots, \alpha_s)$. Then, the weighted filtration $(\mathscr{A}_\bullet, \alpha_\bullet)$ is associated with a reduction β' of the $\mathrm{GL}(V)$-principal bundle $\mathscr{I}som(V, \mathscr{A}^\vee_{|U'})$ to λ' with U' the maximal open subset where $\mathscr{A}$ is locally free and all the $\mathscr{A}_i'$ are subbundles. We may choose an open subset $\widetilde{U} \subseteq X$, such that there is a trivialization $\psi \colon \mathscr{A}^\vee_{|\widetilde{U}} \longrightarrow V \otimes \mathscr{O}_{\widetilde{U}}$ with $\psi(\mathscr{A}_i') = V_i \otimes \mathscr{O}_{\widetilde{U}}$, $\mathscr{A}_i' := \ker(\mathscr{A}^\vee_{|\widetilde{U}} \longrightarrow \mathscr{A}^\vee_{s+1-i|\widetilde{U}})$, $i = 1, \dots, s$. By definition of $\widetilde{\mu}(\mathscr{A}_\bullet, \alpha_\bullet; \tau)$, (13) - (14), Proposition 2.3.3, and Lemma 4.3.4, we see that there is a one parameter subgroup $\lambda \colon \mathbb{G}_m(k) \longrightarrow G$, such that

$$(V_\bullet(\lambda), \alpha_\bullet(\lambda)) = (V_\bullet(\lambda'), \alpha_\bullet(\lambda')).$$

With the principal bundles $\mathscr{P}(\mathscr{A}, \tau)$ and $\mathscr{I}som(V, \mathscr{A}^\vee_{|U})$, we may associate group schemes $\mathscr{G} \subset \mathscr{G}\mathscr{L}(V)$ over U. Now, $\mathscr{G}\mathscr{L}(V)$ acts on $\mathscr{I}som(V, \mathscr{A}^\vee_{|U})/\mathscr{Q}_{\mathrm{GL}(V)}(\lambda)$, and the stabilizer of the section $\beta' \colon U' \longrightarrow \mathscr{I}som(V, \mathscr{A}^\vee_{|U})/Q_{\mathrm{GL}(V)}(\lambda)$ is a parabolic subgroup $\mathscr{Q} \subset \mathscr{G}\mathscr{L}(V)_{|U'}$, such that

$$\mathscr{G}\mathscr{L}(V)_{|U'}/\mathscr{Q} = \mathscr{I}som(V, \mathscr{A}^\vee_{|U'})/Q_{\mathrm{GL}(V)}(\lambda).$$

The intersection $\mathscr{Q}_\mathscr{G} := \mathscr{Q} \cap \mathscr{G}_{|U'}$ is a parabolic subgroup with $\mathscr{G}_{|U'}/\mathscr{Q}_\mathscr{G} = \mathscr{P}(\mathscr{A}, \tau)_{|U'}/Q_Q(\lambda)$. (This can be seen as follows: Let C be the set of conjugacy classes of parabolic subgroups of G. There is a scheme $\mathscr{P}ar_p(\mathscr{G}_{|U'}/U')$ over U', such that giving a parabolic subgroup $\mathscr{Q}_\mathscr{G}$ of $\mathscr{G}_{|U'}$ whose fibers belong to $p \in C$ is the same as giving a section $U' \longrightarrow \mathscr{P}ar(\mathscr{G}_{|U'}/U')$ ([4], p. 443ff). It is easy to see that $\mathscr{P}ar_p(\mathscr{G}_{|U'}/U') \cong \mathscr{P}(\mathscr{A}, \tau)_{|U'}/Q_p$, Q_p being a representative for p (compare [23], p. 281). Finally, $\mathscr{G}_{|U'}/\mathscr{Q}_\mathscr{G} \cong \mathscr{P}ar_p(\mathscr{G}_{|U'}/U')$ ([4], Corollaire 3.6, page 445).) Therefore, we have the commutative diagram

$$\begin{array}{ccc} \mathscr{Q}_\mathscr{G} & \hookrightarrow & \mathscr{Q} \\ \downarrow & & \downarrow \\ \mathscr{G}_{|U'} & \hookrightarrow & \mathscr{G}\mathscr{L}(V)_{|U'}. \end{array}$$

Taking $\mathscr{Q}_{\mathscr{G}}$-quotients in the left hand column and $\mathscr{Q}$-quotients in the right hand column yields the commutative diagram

$$\begin{array}{ccc} U' & = & U' \\ \beta\downarrow & & \downarrow\beta' \\ \mathscr{P}(\mathscr{A},\tau)_{|U'}/Q_Q(\lambda) & \hookrightarrow & \mathscr{I}som(V,\mathscr{A}^{\vee}_{|U})/Q_{\mathrm{GL}(V)}(\lambda) \end{array}$$

and settles the claim.

The proof of "2. $\Rightarrow$ 3." is analogous to the previous proof, using Proposition 2.3.5 instead of 2.3.3, and the implications "3. $\Rightarrow$ 1." and "3. $\Rightarrow$ 2." are easy exercises. □

4.4 *The semistable reduction theorem*

Theorem 4.4.1. *Assume that* Char$(k) = 0$ *or that* ρ *is defined over* $\mathbb{Z}$ *and* $p \gg 0$. *Fix a tuple* $\underline{P} = (P_1, \dots, P_t)$ *of Hilbert polynomials. Then, there exists a positive polynomial* $\widetilde{\delta}_0$ *of degree* $\dim(X) - 1$, *such that for every polynomial* $\widetilde{\delta} \succ \widetilde{\delta}_0$, *the following properties hold true:*

a) *A singular principal* G*-bundle* $(\underline{\mathscr{A}}, \tau)$ *with* $P(\mathscr{A}_i) = P_i$, $i = 1, \dots, t$, *is (semi)stable as defined in the introduction, if and only if it is* $\widetilde{\delta}$*-(semi)stable.*
b) *If* $(\underline{\mathscr{A}}, \tau)$ *is a* $\widetilde{\delta}$*-semistable pseudo* G*-bundle with* $P(\mathscr{A}_i) = P_i$, $i = 1, \dots, t$, *then it is a singular principal* G*-bundle.*

Proof. Ad a). The first assertion follows immediately from Lemma 4.3.6 and the results from Section "Global Boundedness".

Ad b). We assume $t = 1$, the easy generalization to $t > 0$ being left to the reader. We will begin with the case that the base variety is a point. Let $f\colon V \longrightarrow (k^r)^{\vee}$ be a homomorphism which is neither a G-nullform nor an isomorphism. The assignment $(g, h) \longmapsto h \circ g^{-1}$, $g \in G$, $h\colon V \longrightarrow V$, yields a G-module structure on $\mathrm{Hom}(V, V)$. The linear map

$$\begin{aligned} \mathrm{Hom}(V, V) &\longrightarrow \mathrm{Hom}(V, (k^r)^{\vee}) \\ h &\longmapsto f \circ h \end{aligned}$$

is then a homomorphism of G-modules which maps id_V to f. Therefore, we obtain the commutative diagram

$$\begin{array}{ccccc} \mathrm{Sym}(V \otimes k^r)^G & \hookrightarrow & \mathrm{Sym}(V \otimes k^r) & \xrightarrow{\mathrm{ev}_f} & k \\ \overline{f}^{\star}\downarrow & & \downarrow f^{\star} & & \| \\ \mathrm{Sym}(V \otimes V^{\vee})^G & \hookrightarrow & \mathrm{Sym}(V \otimes V^{\vee}) & \xrightarrow{\mathrm{ev}_{\mathrm{id}_V}} & k \end{array}.$$

Note that the point $\bar{f} \in \mathrm{Hom}(V, (k^r)^{\vee})/\!\!/G$ that corresponds to the homomorphism $\mathrm{Sym}(V \otimes k^r)^G \longrightarrow k$ is the image of f under the quotient map. We introduce the representation

$$\widetilde{\mathbb{V}}(k^r) := \bigoplus_{\substack{(d_1,\dots,d_s):\\ d_i\ge 0,\ \sum i d_i = s!}} (\mathrm{Sym}^{d_1}((V\otimes k^r)^G)\otimes\cdots\otimes \mathrm{Sym}^{d_s}(\widetilde{\mathrm{Sym}}^s(V\otimes k^r)^G)),$$

$\widetilde{\mathrm{Sym}}^i(V\otimes k^r)^G := (\mathrm{Sym}^i(V_{\mathbb{Z}}\otimes\mathbb{Z}^r)^G)\otimes_{\mathbb{Z}} k$, $i>0$, of $\mathrm{GL}_r(k)$. Recall that we have introduced above $\mathrm{Hom}(V, W^\vee)\widetilde{/\!\!/}G$ for every k-vector space W of rank r and that we have defined $\widetilde{\mathrm{Sym}}^\star(V\otimes W)^G$ as its coordinate algebra. Hence, we may define the $\mathrm{GL}(V^\vee)$-module $\mathbb{V}(V^\vee)$ by analogy. Note that the construction is functorial, i.e., a linear map $W\longrightarrow W'$ yields a natural morphism $\mathrm{Hom}(V, W'^\vee)\widetilde{/\!\!/}G\longrightarrow \mathrm{Hom}(V, W^\vee)\widetilde{/\!\!/}G$. Using these conventions, we arrive at the commutative diagram

$$\begin{array}{ccc} \widetilde{\mathbb{V}}(k^r) & \longrightarrow & k \\ {\scriptstyle ``f"}\downarrow & & \| \\ \widetilde{\mathbb{V}}(V^\vee) & \longrightarrow & k \end{array}.$$

According to Corollary 2.2.3, we may find a surjection

$$\psi\colon ((k^r)^{\otimes s!})^{\oplus b}\longrightarrow \widetilde{\mathbb{V}}(k^r)$$

of $\mathrm{GL}_r(k)$-modules.

Claim. There is a surjection $\psi_V\colon ((V^\vee)^{\otimes s!})^{\oplus b}\longrightarrow \widetilde{\mathbb{V}}(V^\vee)$ of $\mathrm{GL}(V^\vee)$-modules, such that the diagram

$$\begin{array}{ccc} ((k^r)^{\otimes s!})^{\oplus b} & \xrightarrow{\ \psi\ } & \widetilde{\mathbb{V}}(k^r) \\ {\scriptstyle ``f"}\downarrow & & \downarrow{\scriptstyle ``f"} \\ ((V^\vee)^{\otimes s!})^{\oplus b} & \xrightarrow{\ \psi_V\ } & \widetilde{\mathbb{V}}(V^\vee) \end{array}$$

commutes.

We now continue, assuming this claim. Note that we have the commutative diagram

$$\begin{array}{ccc} ((k^r)^{\otimes s!})^{\oplus b} & \xrightarrow{\ \phi\ } & k \\ {\scriptstyle ``f"}\downarrow & & \| \\ ((V^\vee)^{\otimes s!})^{\oplus b} & \longrightarrow & k \end{array}. \qquad (62)$$

The class $[\phi]$ is the image of $[\bar f]$ under the embedding $\mathbb{P}(\mathrm{Hom}(V,(k^r)^\vee))/\!\!/G\hookrightarrow \mathbb{P}(((k^r)^{\otimes s!})^{\oplus b})$. Next, ϕ gives rise to the "transposed" maps

$$\phi_i^T : k^r \longrightarrow (((k^r)^\vee)^{\otimes(s!-1)})^{\oplus b},\quad i=1,\dots,a.$$

We set

$$K := \bigcap_{i=1}^{a}\ker(\phi_i^T).$$

Since ϕ is non-trivial, K is a proper subspace. On the other hand, $K \neq \{0\}$, by (62). Finally, one checks that $\mu(\lambda, [\phi]) < 0$, for every one parameter subgroup $\lambda\colon \mathbb{G}_m(k) \longrightarrow \mathrm{SL}_r(k)$ with weighted flag $(\{0\} \subsetneq K \subsetneq k^r, (1))$.

Now, let $(\mathscr{A}, \tau)$ be a pseudo G-bundle with associated decorated sheaf $(\mathscr{A}, \phi)$,

$$\phi\colon (\mathscr{A}^{\otimes s!})^{\oplus b} \longrightarrow \mathscr{O}_X.$$

As before, we obtain the transposed maps $\phi_i^T\colon \mathscr{A} \longrightarrow ((\mathscr{A}^\vee)^{\otimes(s!-1)})^{\oplus b}$, $i = 1, \ldots, a$. Then, $\mathscr{K} := \bigcap_{i=1}^a \ker(\phi_i^T)$ is a saturated subsheaf. The above arguments show that $\mu(\{0\} \subsetneq \mathscr{K} \subsetneq k^r, (1); \phi) < 0$. By Proposition 3.5.3, $(\mathscr{A}, \tau)$ is not asymptotically semistable.

Finally, we verify the claim. We choose an isomorphism $e_0\colon k^r \longrightarrow V^\vee$ which induces also an isomorphism $\mathrm{GL}_r(k) \longrightarrow \mathrm{GL}(V^\vee)$. Then, we obviously can find a surjection $\psi_V\colon ((V^\vee)^{\otimes s!})^{\oplus b} \longrightarrow \widetilde{\mathbb{V}}(V^\vee)$ of $\mathrm{GL}(V^\vee)$-modules, such that the diagram

$$\begin{array}{ccc}
((k^r)^{\otimes s!})^{\oplus b} & \xrightarrow{\psi} & \widetilde{\mathbb{V}}(k^r) \\
{\scriptstyle \text{“}e_0\text{”}}\big\downarrow \cong & & \cong \big\downarrow{\scriptstyle \text{“}e_0\text{”}} \\
((V^\vee)^{\otimes s!})^{\oplus b} & \xrightarrow{\psi_V} & \widetilde{\mathbb{V}}(V^\vee)
\end{array}$$

commutes. Next, the diagram

$$\begin{array}{ccc}
((k^r)^{\otimes s!})^{\oplus b} & \xrightarrow{\psi} & \widetilde{\mathbb{V}}(k^r) \\
{\scriptstyle \text{“}e\text{”}}\big\downarrow & & \big\downarrow{\scriptstyle \text{“}e\text{”}} \\
((k^r)^{\otimes s!})^{\oplus b} & \xrightarrow{\psi} & \widetilde{\mathbb{V}}(k^r)
\end{array}$$

commutes for every isomorphism $e\colon k^r \longrightarrow k^r$, because ψ is a homomorphism of $\mathrm{GL}_r(k)$-modules. But then, it also commutes for every endomorphism $e\colon k^r \longrightarrow k^r$, by continuity. Since there exists an endomorphism $e\colon k^r \longrightarrow k^r$ with $f = e_0 \circ e$, we are done. □

Consequences for S-equivalence. Recall from Section 3.3 that we have introduced the equivalence relation "$\sim_B$" on the set of tuples of sheaves $(\underline{\mathscr{E}}, \phi)$ with a decoration of type $(\mathrm{II} : a, b, c; \mathscr{O}_X)$. By Remark 4.3.5, i), this induces an equivalence relation on the set of singular principal G-bundles. For simplicity, we denote this equivalence relation again by "$\sim_B$".

Lemma 4.4.2. *The equivalence relation "$\sim_B$" on the set of singular principal G-bundles agrees with "isomorphy".*

Proof. Let $(\underline{\mathscr{A}}, \tau)$ be a singular principal G-bundle and $(\underline{\mathscr{A}}, \phi)$ the associated tuple of sheaves with a decoration of type $(\mathrm{II} : a, b, c; \mathscr{O}_X)$, and let $\lambda\colon \mathbb{G}_m(k) \longrightarrow \widetilde{T}$ be a one parameter subgroup with $\mu(\lambda, \phi) = 0$. The top line of Diagram (12), Lemma 2.3.6, and Proposition 2.3.3 imply that

$$\mathrm{df}_\lambda(\underline{\mathscr{A}}, \phi) = (\underline{\mathscr{A}}, \phi). \tag{63}$$

In fact, in Proposition 2.3.3, we have $\lambda' = \lambda$, because λ corresponds to a one parameter subgroup of the center of $\widetilde{G}$. Our assertion (63) results from the fact that $\mathscr{Z}(G)$ acts trivially on $(\mathrm{GL}_{r_1}(k) \times \cdots \times \mathrm{GL}_{r_t}(k))/G$. Now, (63) and Lemma 4.3.2 imply the claim. □

Corollary 4.4.3. *Fix a tuple* $\underline{P} = (P_1, \ldots, P_t)$ *of Hilbert polynomials and a polynomial* $\widetilde{\delta} \succ \widetilde{\delta}_0$ *of degree* $\dim(X) - 1$ *with rational coefficients (see Theorem* 4.4.1*).*

i) *The assignment* $(\underline{\mathscr{A}}, \tau) \longmapsto (\underline{\mathscr{A}}, \widetilde{\phi})$ *induces an injection of the set of isomorphy classes of (semi)stable singular principal G-bundles with Hilbert polynomials* $\underline{P}$ *into the set of isomorphy classes of* δ*-(semi)stable tuples of sheaves with a decoration of type* $(\mathrm{I} : a, b, c;\, \mathscr{O}_X)$ *and Hilbert polynomials* $\underline{P}$*,* $\delta := \widetilde{\delta}/s!$*.*
ii) *Two semistable singular principal G-bundles* $(\underline{\mathscr{A}}, \tau)$ *and* $(\underline{\mathscr{A}}', \tau')$ *with Hilbert polynomials* $\underline{P}$ *are S-equivalent (in the sense of Section 4.2), if and only if the associated decorated tuples of sheaves* $(\underline{\mathscr{A}}, \phi)$ *and* $(\underline{\mathscr{A}}', \phi')$ *are S-equivalent (in the sense of Section* 3.3*).*

Proof. Ad i). We define

$$\mathfrak{S} := \left\{ \begin{array}{l} \delta\text{-semistable tuples of sheaves } (\underline{\mathscr{A}}, \phi) \text{ with a decoration of type} \\ (\mathrm{II} : a, b, c;\, \mathscr{O}_X) \text{ which are associated with semistable} \\ \text{singular principal } G\text{-bundles } (\underline{\mathscr{A}}, \tau) \text{ with } P(\mathscr{A}_i) = P_i,\, i = 1, \ldots, t \end{array} \right\}.$$

Lemma 4.4.2 implies that $\mathfrak{S}$ contains every admissible deformation $\mathrm{df}_\lambda(\underline{\mathscr{A}}, \phi)$ of an element $(\underline{\mathscr{A}}, \phi)$ from $\mathfrak{S}$. Therefore, Proposition 3.3.4, Lemma 4.4.2, and Lemma 4.3.2 immediately yield assertion i).

Ad ii). For this assertion, we use Remark 3.3.2. Let $(\underline{\mathscr{A}}, \widetilde{\phi})$ and $(\underline{\mathscr{A}}', \widetilde{\phi}')$ be the associated tuples of sheaves with a decoration of type $(\mathrm{I} : a(d), b(d), c(d);\, \mathscr{O}_X)$. First, assume that $(\underline{\mathscr{A}}, \tau)$ and $(\underline{\mathscr{A}}', \tau')$ are S-equivalent. We may assume that $(\underline{\mathscr{A}}', \tau')$ is an admissible deformation of $(\underline{\mathscr{A}}, \tau)$. Thus, there are a one parameter subgroup $\lambda \colon \mathbb{G}_m(k) \longrightarrow [G, G]$ and a reduction β of $(\underline{\mathscr{A}}, \tau)$ to λ with

$$M(\underline{\mathscr{A}}_\bullet(\beta), \underline{\alpha}_\bullet(\beta)) = 0,$$

such that $(\underline{\mathscr{A}}', \tau') = \mathrm{df}_\beta(\underline{\mathscr{A}}, \tau)$. Lemma 4.3.6 implies

$$\widetilde{\mu}(\underline{\mathscr{A}}_\bullet(\beta), \underline{\alpha}_\bullet(\beta); \phi) = 0 \quad \text{and} \quad \mu(\underline{\mathscr{A}}_\bullet(\beta), \underline{\alpha}_\bullet(\beta); \widetilde{\phi}) = 0. \tag{64}$$

By definition,

$$(\underline{\mathscr{A}}', \phi') = \mathrm{df}_{(\underline{\mathscr{A}}_\bullet(\beta), \underline{\alpha}_\bullet(\beta))}(\underline{\mathscr{A}}, \phi).$$

Then, (64) and (30) imply

$$(\underline{\mathscr{A}}', \widetilde{\phi}') = \mathrm{df}_{(\underline{\mathscr{A}}_\bullet(\beta), \underline{\alpha}_\bullet(\beta))}(\underline{\mathscr{A}}, \widetilde{\phi}),$$

as desired.

Now, we assume that $(\underline{\mathscr{A}}, \phi)$ and $(\underline{\mathscr{A}}', \phi')$ are S-equivalent. By definition, this means that $(\underline{\mathscr{A}}, \widetilde{\phi})$ and $(\underline{\mathscr{A}}', \widetilde{\phi}')$ are S-equivalent. We may assume that

$$(\underline{\mathscr{A}}', \widetilde{\phi}') = \mathrm{df}_{(\underline{\mathscr{A}}_\bullet, \underline{\alpha}_\bullet)}(\underline{\mathscr{A}}, \widetilde{\phi})$$

for a non-trivial weighted filtration $(\underline{\mathscr{A}}_\bullet, \underline{\alpha}_\bullet)$ with $\mu(\underline{\mathscr{A}}_\bullet, \underline{\alpha}_\bullet; \widetilde{\phi}) = 0$. By Lemma 4.3.6,

$$(\underline{\mathscr{A}}_\bullet, \underline{\alpha}_\bullet) = (\underline{\mathscr{A}}_\bullet(\beta), \underline{\alpha}_\bullet(\beta))$$

for some reduction β of $(\underline{\mathscr{A}}, \tau)$ to a one parameter subgroup λ of $[G, G]$ with $M(\underline{\mathscr{A}}_\bullet(\beta), \underline{\alpha}_\bullet(\beta)) = 0$. By a reasoning similar to the one before, we see that $(\underline{\mathscr{A}}', \tau')$ and $\mathrm{df}_\beta(\underline{\mathscr{A}}, \tau)$ yield tuples of sheaves with a decoration of type $(\mathrm{I} : a(d), b(d), c(d); \mathscr{O}_X)$ which are both isomorphic to $(\underline{\mathscr{A}}', \phi')$. By Part i), $(\underline{\mathscr{A}}', \tau')$ must be isomorphic to $\mathrm{df}_\beta(\underline{\mathscr{A}}, \tau)$, and we are done. □

4.5 *Construction of the moduli space*

We fix the tuple $\underline{P} = (P_1, \ldots, P_t)$ of Hilbert polynomials, a natural number m, set $p_i := P_i(m)$, $i = 1, \ldots, t$, and fix complex vector spaces U_i of dimension p_i, $i = 1, \ldots, t$. Then, we have the quasi-projective quot scheme $\mathfrak{Q}_i$ that parameterizes quotients $q_i : U_i \otimes \mathscr{O}_X(-m) \longrightarrow \mathscr{A}_i$ where $\mathscr{A}_i$ is a torsion free sheaf with Hilbert polynomial P_i, such that $H^j(\mathscr{A}_i(m)) = \{0\}$, $j > 0$, and $H^0(q_i(m))$ is an isomorphism, $i = 1, \ldots, t$.

By Theorem 3.2.2 and Theorem 4.4.1, we can choose m_0 in such a way that, for every $m \geq m_0$, every positive polynomial δ of degree at most $\dim(X) - 1$, and every semistable singular principal G-bundle $(\underline{\mathscr{A}} = (\mathscr{A}_1, \ldots, \mathscr{A}_t), \tau)$ with $P(\mathscr{A}_i) = P_i$, $i = 1, \ldots, t$, there exist points of the form $[q_i : U_i \otimes \mathscr{O}_X(-m) \longrightarrow \mathscr{A}_i] \in \mathfrak{Q}_i$, $i = 1, \ldots, t$.

Set $\mathfrak{Q} := \times_{i=1}^t \mathfrak{Q}_i$. Over $X \times \mathfrak{Q}$, there are the universal quotients

$$\mathfrak{q}_{\mathfrak{Q},i} : U_i \otimes \pi_X^\star \mathscr{O}_X(-m) \longrightarrow \mathfrak{A}_{\mathfrak{Q},i}, \quad i = 1, \ldots, t.$$

There is a scheme

$$\mathfrak{Y} \longrightarrow \mathfrak{Q}$$

which carries a universal family $(\underline{\mathscr{A}}_{\mathfrak{Y}}, \tau_{\mathfrak{Y}})$ and which is equipped with a group action

$$\Gamma : \underbrace{(\mathrm{GL}(U_1) \times \cdots \times \mathrm{GL}(U_t))}_{=:\widetilde{G}} \times \mathfrak{Y} \longrightarrow \mathfrak{Y},$$

such that the following results hold true:

Proposition 4.5.1 (Local universal property). *Let S be a scheme and $(\underline{\mathscr{A}}_S, \tau_S)$ a family of $\widetilde{\delta}$-semistable pseudo G-bundles with Hilbert polynomials $\underline{P}$ parameterized by S. Then, there exist a covering of S by open subschemes S_i, $i \in I$, and morphisms $\beta_i : S_i \longrightarrow \mathfrak{Y}$, $i \in I$, such that the family $(\underline{\mathscr{A}}_{S|S_i}, \tau_{S|S_i})$ is isomorphic to the pullback of the universal family on $X \times \mathfrak{Y}$ by $\beta_i \times \mathrm{id}_X$ for all $i \in I$.*

Proposition 4.5.2 (Gluing property). *Let S be a scheme and $\beta_i : S \longrightarrow \mathfrak{Y}$, $i = 1, 2$, two morphisms, such that the pullback of the universal family via $\beta_1 \times \mathrm{id}_X$ is isomorphic to its pullback via $\beta_2 \times \mathrm{id}_X$. Then, there is a morphism $\Xi : S \longrightarrow \widetilde{G}$, such that β_2 equals the morphism*

$$S \stackrel{\Xi \times \beta_1}{\longrightarrow} \widetilde{G} \times \mathfrak{Y} \stackrel{\Gamma}{\longrightarrow} \mathfrak{Y}.$$

The construction is given for $t = 1$ in [26], §5.1, and [6], §5.3. Using the results of Section 2.4, the generalization to $t > 1$ is straightforward whence omitted.

Remark 4.5.3. The universal family is equipped with a $\widetilde{G}$-linearization. If one fixes, in the above proposition, an isomorphism between its pullbacks via $\beta_1 \times \mathrm{id}_X$ and $\beta_2 \times \mathrm{id}_X$, then there is a unique morphism $\Xi\colon S \longrightarrow \widetilde{G}$ that satisfies the stated properties and, in addition, that the given isomorphism is induced by pullback via $(\Xi \times \beta_1 \times \mathrm{id}_X)$ from the linearization of $(\underline{\mathscr{A}}_{\mathfrak{Y}}, \tau_{\mathfrak{Y}})$. This fact simply expresses that the moduli stack for semistable singular principal G-bundles will be the quotient stack of an appropriate open subscheme of the parameter space $\mathfrak{Y}$.

As before, we set

$$T_1 \cong \mathbb{G}_m(k) := \{(z \cdot \mathrm{id}_{U_1}, \ldots, z \cdot \mathrm{id}_{U_t}) \in \widetilde{G} \mid z \in \mathbb{G}_m(k)\}$$

$$T_2 \cong \mathbb{G}_m(k)^{\times(t-1)} := \{(z_1 \cdot \mathrm{id}_{U_1}, \ldots, z_t \cdot \mathrm{id}_{U_t}) \in \widetilde{G} | z_i \in \mathbb{G}_m(k),$$
$$i = 1, \ldots, t : z_1 \cdot \cdots \cdot z_t = 1\}.$$

Then, we have the natural surjective homomorphism

$$T_1 \times T_2 \times (\mathrm{SL}(U_1) \times \cdots \times \mathrm{SL}(U_t)) \longrightarrow (\mathrm{GL}(U_1) \times \cdots \times \mathrm{GL}(U_t)).$$

Therefore, we may construct the categorical quotient in several steps.

We also fix non-negative integers a, b, c, and a stability parameter δ, such that we have a natural transformation of the functor of (semi)stable singular principal G-bundles with Hilbert polynomials $\underline{P}$ into the functor of δ-(semi)stable tuples of sheaves with a decoration of type $(\mathrm{II} : a, b, c; \mathscr{O}_X)$ and Hilbert polynomials $\underline{P}$. In Section 3.4, we have already constructed a parameter space $\mathfrak{Z}_{\mathrm{II}}$ for these objects. We may assume that m is so large that all the constructions of moduli spaces for decorated tuples of sheaves go through with this m.

The universal family $(\underline{\mathscr{A}}_{\mathfrak{Y}}, \tau_{\mathfrak{Y}})$ provides us with a $T_1 \times T_2 \times (\mathrm{SL}(U_1) \times \cdots \times \mathrm{SL}(U_t))$-equivariant rational map

$$\phi\colon \mathfrak{Y} \dashrightarrow \mathfrak{Z}_{\mathrm{II}}$$

which is defined exactly in those points $y \in \mathfrak{Y}$ which correspond to pseudo G-bundles, i.e., for which $\tau_{\mathfrak{Y}|X\times\{y\}}$ is non-trivial. This map is T_1-invariant and, consequently, descends to a morphism

$$\bar{\phi}\colon \mathfrak{Y}/\!\!/T_1 \longrightarrow \mathfrak{Z}_{\mathrm{II}}.$$

Since $\mathfrak{Y}/\!\!/T_1$ is projective over $\mathfrak{Q}$, the morphism $\bar{\phi}$ is proper. By Lemma 4.3.2, it is also injective, so that it is finite. Let $\widetilde{\mathfrak{Z}}$ be the scheme theoretic image of $\bar{\phi}$.

Remark 4.5.4.

i) Note that Theorem 4.4.1 implies that the preimage of $\widetilde{U}_\delta$ under the map

$$\mathfrak{Y} \dashrightarrow \mathfrak{Y}/\!\!/T_1 \longrightarrow \mathfrak{Z}_{\mathrm{II}}$$

is exactly the set of semistable singular principal G-bundles.

ii) By i) and Lemma 4.4.2, the closed subscheme $\widetilde{\mathfrak{Z}}$ fulfills the assumption of Lemma 3.6.1. Therefore, the good quotient

$$\widetilde{\mathfrak{Z}}/\!\!/(T_2 \times (\mathrm{SL}(U_1) \times \cdots \times \mathrm{SL}(U_t)))$$

exists as a projective scheme.

Remark 4.5.4, ii), and Lemma 2.1.7 in [6] imply that

$$\begin{aligned}\mathrm{M}(\rho)^{\mathrm{ss}}_{\underline{P}} &:= (\mathfrak{Y}/\!\!/T_1)/\!\!/(T_2 \times (\mathrm{SL}(U_1) \times \cdots \times \mathrm{SL}(U_t)))\\ &= \mathfrak{Y}/\!\!/(T_1 \times T_2 \times (\mathrm{SL}(U_1) \times \cdots \times \mathrm{SL}(U_t)))\\ &= \mathfrak{Y}/\!\!/(\mathrm{GL}(U_1) \times \cdots \times \mathrm{GL}(U_t))\end{aligned}$$

exists as a projective scheme (for the equality in the second line, see Example 2.1.6, i), in [6]). By its construction, Remark 4.5.4, and our results on S-equivalence in Section 4.4, $\mathrm{M}(\rho)^{\mathrm{ss}}_{\underline{P}}$ is the moduli space we have been looking for. □

5. Moduli spaces of G-bundles for reductive groups via faithless representations

In the definition of singular principal G-bundles, we use a representation ρ which is assumed to be faithful. This rules out one important case: the adjoint representation. In this section, we show how we can extend the techniques to consider representations which are not faithful (for the precise assumptions we make on ρ, see the beginning of Section 5.5). To do this, we will introduce *principal ρ-sheaves*, which can be considered as generalizations of singular principal G-bundles.

5.1 *Sheaves of sets*

We denote by fl, et, and Zar the flat, étale, and Zariski topology on k-schemes, respectively. For a k-scheme Y, we denote by Y_E the big E-site for E = fl, et, Zar, i.e., the category of schemes locally of finite type over Y endowed with the topology E. Any group scheme G over k defines a constant group scheme $G_Y = Y \times_k G$ on Y. This can be treated as a sheaf of groups $\underline{G}_Y$ on Y_{fl} by sending $T \in \mathrm{Ob}(Y_{\mathrm{fl}})$ to $\underline{G}_Y(T) = \mathrm{Hom}_Y(T, G_Y)$. Sometimes, we denote this sheaf by $\underline{G}$ (or even G), if it is obvious which Y we mean. If G is an abelian group, then $\underline{G}$ denotes the sheaf associated to the constant group scheme G_k. Let $f\colon Y \longrightarrow S$ be a morphism of schemes. It induces a continuous morphism $f^E\colon Y_{\mathrm{fl}} \longrightarrow S_E$ between the big flat site on Y and the big E-site on S. For convenience, we usually write f instead of f^{Zar}. Let F be a sheaf of abelian groups on Y_{fl}. Then, $R^i f^E_\star F$ denotes the right derived functor of $f^E_\star$. For any group scheme G over k and any continuous morphism $f^E\colon Y_{\mathrm{fl}} \longrightarrow S_E$, we define $R^1 f^E_\star \underline{G}$ as the sheaf of pointed sets on S_E associated with the presheaf

$$(T \longrightarrow S) \longrightarrow H^1(T \times_S U, \underline{G})$$

where $H^1(\cdot, \underline{G})$ is the set of isomorphy classes of G-torsors pointed by the trivial G-torsor ([5], V, 2.1.1). If G is abelian, then, by [21], III, Proposition 1.13, Corollary 2.10, and Corollary 4.7, the above definition of $R^1 f_\star^E \underline{G}$ agrees with the previous definition of this sheaf as the derived functor. We also define $R^2 f_\star^E \underline{G}$ as the sheaf of pointed sets on S_E associated with the presheaf

$$(T \longrightarrow S) \longrightarrow H^2(T \times_S U, \underline{G})$$

where $H^2(\cdot, \underline{G})$ is the set of equivalence classes of gerbes with band $\underline{G}$ (see [5], IV, 3.1 or [21], IV, proof of Theorem 2.5). If G is an abelian group scheme, then $R^2 f_\star^E \underline{G}$ as defined above is canonically isomorphic to $R^2 f_\star^E \underline{G}$ computed as the derived functor of $f_\star^E$ (this is true already at the presheaf level; see [5], IV, 3.4, or [21], IV, proof of Theorem 2.5).

5.2 *Finite diagonalizable group schemes*

Let us recall that we have a well-defined functor Diag from the category of abelian groups to the category of k-group schemes (see [16], 2.5). A k-group scheme is called *diagonalizable*, if it is isomorphic to $\mathrm{Diag}(\Lambda)$ for some abelian group Λ. The functor Diag is an anti-equivalence between the category of abelian groups and the category of diagonalizable group schemes. Let us set $\mu_n = \mathrm{Diag}(\mathbb{Z}/n\mathbb{Z})$. It can also be defined as $\mathrm{Spec}\, k[T]/(T^n - 1)$ with the obvious group operations. Since any finite abelian group Λ is isomorphic to the product of groups $\mathbb{Z}_{q^k}$ for some prime numbers q, we see that $\mathrm{Diag}(\Lambda)$ is the product of group schemes of the form μ_{q^k}.

Assume that k has characteristic p. Let $\pi : Y \longrightarrow S$ be a smooth morphism where S is a scheme over k. Let F_S be the absolute Frobenius morphism of S and set $Y^{(1/S)} = Y \times_{F_S} S$. Then, F_Y is the composition of the relative Frobenius morphism $F = F_{Y/S} \colon Y \longrightarrow Y^{(1/S)}$ and the projection $W \colon Y^{(1/S)} \longrightarrow Y$. The differential d in the de Rham complex $\Omega^\bullet_{Y/S}$ is $\mathcal{O}_{Y^{(1/S)}}$-linear, so that

$$Z\Omega^1_{Y/S} = \ker(\Omega^1_{Y/S} \longrightarrow \Omega^2_{Y/S})$$

is an $\mathcal{O}_{Y^{(1/S)}}$-module. Then, there exists a map $C \colon Z\Omega^1_{Y/S} \longrightarrow \Omega^1_{Y^{(1/S)}/S}$ of $\mathcal{O}_{Y^{(1/S)}}$-modules, called the *Cartier operator* (see, e.g., [15]). By [15], 0, Corollaire 2.1.18, we have the following exact sequence of sheaves of abelian groups on Y_{et}:

$$0 \longrightarrow \mathcal{O}^\star_{Y^{(1/S)}} \xrightarrow{F_{Y/S}} \mathcal{O}^\star_Y \longrightarrow Z\Omega^1_{Y/S} \xrightarrow{W^\star - C} \Omega^1_{Y^{(1/S)}/S} \longrightarrow 0. \qquad (65)$$

Now, assume that X is a smooth variety defined over k, $Y = X \times_k S$, and π is the projection onto the second factor. Then, $Y^{(1/S)} = X^{(1/k)} \times_k S$ and $F_{Y/S} = F_{X/k} \times_k \mathrm{id}_S$. Since k is perfect, F_k is an isomorphism and we can identify $X^{(1/k)}$ with X. Then, $Y^{(1/S)}$ is identified with Y (note that, if U is an open subset of Y, then we cannot identify $U^{(1/S)}$ with U). Let $\iota \colon Y_{\mathrm{fl}} \longrightarrow Y_{\mathrm{et}}$ be the morphism induced by the identity morphism and let $\pi' \colon Y_{\mathrm{et}} \longrightarrow S_E$ for $E = \mathrm{Zar}$, et denote the continuous morphism. In our notation, we have $\pi^E = \pi' \circ \iota \colon Y_{\mathrm{fl}} \longrightarrow S_E$. It is easy to see that in our identifications $R^i\iota_\star\mu_p = 0$ for $i \neq 1$, and the sequence (65) gives rise to the following short exact sequence of abelian groups on Y_{et}:

$$0 \longrightarrow R^1\iota_\star\mu_p \longrightarrow Z\Omega^1_{Y/S} \stackrel{1-C}{\longrightarrow} \Omega^1_{Y/S} \longrightarrow 0. \tag{66}$$

By the Leray spectral sequence, we have $R^i\pi'_\star(R^1\iota_\star\mu_p) = R^{i+1}\pi^E_\star\mu_p$ for all integers i. Therefore, the above exact sequence yields the following long exact sequence of sheaves of abelian groups on S_E:

$$\cdots \longrightarrow R^{i+1}\pi^E_\star\mu_p \longrightarrow R^i\pi^E_\star Z\Omega^1_{Y/S} \longrightarrow R^i\pi^E_\star\Omega^1_{Y/S} \longrightarrow R^{i+2}\pi^E_\star\mu_p \longrightarrow \cdots. \tag{67}$$

Proposition 5.2.1. *Let X be a smooth projective variety over k and let D be a finite diagonalizable group scheme over k. Then, $H^1_{\mathrm{fl}}(X, D)$ is finite.*

Note that the above statement is non-trivial as there exist finite abelian group schemes for which $H^1_{\mathrm{fl}}(X, D)$ is infinite. For example, if α_p is the group scheme $\mathrm{Spec}(\mathbb{F}_p[T]/(T^p))$ and E is a supersingular elliptic curve (i.e., the Frobenius morphism acts trivially on $H^1(E, \mathcal{O}_E)$), then $H^1_{\mathrm{fl}}(E, \alpha_p)$ is equal to k. For surfaces, also $H^2_{\mathrm{fl}}(X, \mu_n)$ may contain a subgroup which is a vector space over k, so that the proposition works only for H^1.

Lemma 5.2.2. *$H^1_{\mathrm{fl}}(X, \mu_p)$ has the structure of a vector space over $\mathbb{F}_p$ of dimension $\leq h^0(X, \Omega_X)$.*

Note that, in positive characteristic, $h^0(X, \Omega_X) \neq h^1(X, \mathcal{O}_X)$, in general (see [31], Proposition 16).

Proof. Let $\Omega_{X,\mathrm{cl}} = Z\Omega^1_{X/k}$ denote the sheaf of closed 1-forms on X_{et}. The Cartier operator induces a p^{-1}-linear map $C\colon V := H^0(X, \Omega_{X,\mathrm{cl}}) \longrightarrow W := H^0(X, \Omega_X)$. Obviously, V is a k-vector subspace of W. Let us choose any projection $\pi : W \longrightarrow V$. Then $C' = \pi \circ C\colon V \longrightarrow V$ is a p^{-1}-linear operator. By [21], III, Proposition 4.14, we have

$$H^1_{\mathrm{fl}}(X, \mu_p) = \{\omega \in \Gamma(X, \Omega_X) | d\omega = 0,\ C(\omega) = \omega\}.$$

This means that $H^1_{\mathrm{fl}}(X, \mu_p)$ is contained in $V' = \{v \in V \mid C'(v) = v\}$. Now, we need a fact analogous to [21], III, Lemma 4.13, but for p^{-1}-linear operators. One can either check it directly or replace C' by its transpose endomorphism $D\colon V^\vee \longrightarrow V^\vee$ defined by

$$\langle v, Dv'\rangle = \langle C'v, v'\rangle^p,$$

for any $v \in V$ and $v' \in V^\vee$ (see [31], Section 9) and then apply [21], III, Lemma 4.13, to D. Therefore, V' is a vector space over $\mathbb{F}_p$ of dimension at most $\dim_k(V)$. One can easily see that $H^1(X_{\mathrm{fl}}, \mu_p)$ is an $\mathbb{F}_p$-vector subspace of V. □

Proof of Theorem 5.2.1. Obviously, the number of elements of $H^1_{\mathrm{fl}}(X, D)$ is bounded by the sum of the numbers of elements of $H^1_{\mathrm{fl}}(X, \mu_{q^k})$. If $q \neq p$, then there is no problem: $H^1_{\mathrm{fl}}(X, \mu_{q^k})$ corresponds to q^k-torsion points in the projective group scheme $\mathrm{Pic}^\tau X$. (In principle this is also true for $q = p$, but $\mathrm{Pic}^\tau X$ can be non-reduced.) If $q = p$, then the number of elements of $H^1_{\mathrm{fl}}(X, \mu_{p^k})$ is less or equal to k times the number of elements in $H^1_{\mathrm{fl}}(X, \mu_p)$ which is finite by Lemma 5.2.2. □

Let X be a smooth projective variety over k and let S be a k-scheme of finite type over k (in fact, we use only that S is a locally noetherian k-scheme). Let $j: U \subseteq Y$ be a Zariski open subset of $Y = S \times_k X$, such that $U \cap (X \times \{s\})$ is a big open subset of $X \times \{s\}$ for every geometric point $s \in S$. In this case, we say that $U \subset Y \longrightarrow S$ is *relatively big*. Let $q: Y \longrightarrow S$ and $p = q_{|U}: U \longrightarrow S$ be the canonical projections.

Proposition 5.2.3. *Let D be a finite diagonalizable group scheme defined over the field k. Then*

$$R^1 p_\star^E \underline{D} = \underline{H^1_{\mathrm{fl}}(X, D)}$$

for $E =$ fl, et, Zar.

Proof. For simplicity, we do not use E in the notation, but the proof works for each E (and not only for the Zariski topology as our convention would suggest). Obviously $R^1 p_\star \underline{(G_1 \times G_2)} = R^1 p_\star \underline{G_1} \times R^1 p_\star \underline{G_2}$ for any two group schemes G_1 and G_2 defined over k. Hence, it is sufficient to prove the proposition assuming that $D = \mu_n$ for some integer n. Let us consider a short exact sequence of sheaves of abelian groups on k_{fl}:

$$0 \longrightarrow \underline{\mu_n} \longrightarrow \underline{\mathbb{G}_m} \stackrel{n}{\longrightarrow} \underline{\mathbb{G}_m} \longrightarrow 0.$$

It induces the corresponding short exact sequences of sheaves on U_{fl} and Y_{fl} (but not on the étale sites; see [21], II.2.18). Let us recall that, for any abelian group scheme G defined over k, we have $j^\star(\underline{G}_Y) = \underline{G}_U$ (see [21], II, Remark 3.1 (d)). So, the map $\underline{G}_Y \longrightarrow j_\star j^\star(\underline{G}_Y)$ gives $\underline{G}_Y \longrightarrow j_\star(\underline{G}_U)$ which induces the corresponding maps $R^i q_\star(\underline{G}_Y) \longrightarrow R^i q_\star(j_\star \underline{G}_U)$ of higher direct images. Composing them with the maps $R^i q_\star(j_\star \underline{G}_U) \longrightarrow R^i p_\star(\underline{G}_U)$ coming from the Leray spectral sequence, we get the maps

$$\alpha^i_G : R^i q_\star(\underline{G}_Y) \longrightarrow R^i p_\star(\underline{G}_U).$$

By [21], III, Corollary 4.7, for any abelian group scheme G over k, the sheaf $R^1 q_\star(\underline{G}_Y)$ is equal to the sheaf associated with the presheaf

$$(T \longrightarrow S) \longrightarrow \check{H}^1_{\mathrm{fl}}(T \times_S Y, \underline{G}_{T \times_S Y}) = \check{H}^1_{\mathrm{fl}}(T \times_k X, \underline{G}_{T \times X})$$

where $\check{H}^1_{\mathrm{fl}}$ is the Čech cohomology. Since $H^1_{\mathrm{fl}}(X, G) = \check{H}^1_{\mathrm{fl}}(X, G)$ and $\check{H}^1_{\mathrm{fl}}$ can be computed from the Čech complex for an arbitrary covering (see [21], III, Corollary 2.10), we get

$$\underline{H^1_{\mathrm{fl}}(X, G)}_S \stackrel{\cong}{\longrightarrow} R^1 q_\star(\underline{G}_Y).$$

Hence, $R^i q_\star(\underline{\mu_n})_Y \cong \underline{H^i_{\mathrm{fl}}(X, \mu_n)}_S$, and we need only to show that $\alpha^1_{\mu_n}$ is an isomorphism. Applying the five lemma to the following commutative diagram

$$\begin{array}{ccccccccc}
q_\star(\underline{\mathbb{G}_m})_Y & \longrightarrow & q_\star(\underline{\mathbb{G}_m})_Y & \longrightarrow & R^1 q_\star(\underline{\mu_n})_Y & \longrightarrow & R^1 q_\star(\underline{\mathbb{G}_m})_Y & \longrightarrow & R^1 q_\star(\underline{\mathbb{G}_m})_Y \\
\downarrow{\scriptstyle \alpha^0_{\mathbb{G}m}} & & \downarrow{\scriptstyle \alpha^0_{\mathbb{G}m}} & & \downarrow{\scriptstyle \alpha^1_{\mu n}} & & \downarrow{\scriptstyle \alpha^1_{\mathbb{G}m}} & & \downarrow{\scriptstyle \alpha^1_{\mathbb{G}m}} \\
p_\star(\underline{\mathbb{G}_m})_U & \longrightarrow & p_\star(\underline{\mathbb{G}_m})_U & \longrightarrow & R^1 p_\star(\underline{\mu_n})_U & \longrightarrow & R^1 p_\star(\underline{\mathbb{G}_m})_U & \longrightarrow & R^1 p_\star(\underline{\mathbb{G}_m})_U,
\end{array}$$

we see that it is sufficient to show that $\alpha^0_{\mathbb{G}_m}$ and $\alpha^1_{\mathbb{G}_m}$ are isomorphisms. First, we show that $\alpha^0_{\mathbb{G}_m}$ is an isomorphism. If $T \longrightarrow S$ is a locally finite morphism, then, by Proposition 2.4.1,

$$\Gamma(T \times_S Y, \mathcal{O}_{T\times_S Y}) \longrightarrow \Gamma(T \times_S U, \mathcal{O}_{T\times_S U})$$

is an isomorphism, so that

$$\begin{aligned}\alpha^0_{\mathbb{G}_m}(T)\colon (q_\star(\underline{\mathbb{G}_m})_Y)(T) = \Gamma(T \times_S Y, \mathcal{O}_{T\times_S Y})^\star &\longrightarrow \Gamma(T \times_S U, \mathcal{O}_{T\times_S U})^\star \\ &= (p_\star(\underline{\mathbb{G}_m})_U)(T)\end{aligned}$$

is also an isomorphism. Similarly, using Proposition 2.4.1 and Hilbert's theorem 90 (see [21], III, Proposition 4.9), it is easy to see that, for any locally finite morphism $T \longrightarrow S$,

$$\begin{aligned}\check{H}^1_{\mathrm{fl}}(T \times_S Y, \underline{\mathbb{G}_m}) \cong H^1_{\mathrm{Zar}}(T \times_S Y, \mathcal{O}^\star_{T\times_S Y}) &\xrightarrow{\cong} H^1_{\mathrm{Zar}}(T \times_S U, \mathcal{O}^\star_{T\times_S U}) \\ &\cong \check{H}^1_{\mathrm{fl}}(T \times_S U, \underline{\mathbb{G}_m})\end{aligned}$$

where the map is determined by $j_V : V \times_S U \hookrightarrow V \times_S Y = V \times X$. Since this map induces $\alpha^1_{\mathbb{G}_m}$, we see that $\alpha^1_{\mathbb{G}_m}$ is an isomorphism already at the level of presheaves.

□

Note that, in the proof of the above proposition, we showed that $R^1 q^E_\star \underline{D} \cong R^1 p^E_\star \underline{D}$. This is an analogue of the Zariski-Nagata purity theorem (see [1], Exposé XVI, Théorème 3.3), but for non-reduced group schemes and in the flat topology instead of the étale.

Proposition 5.2.4. *For any i, the sheaves $R^i q^E_\star \underline{D}$ are representable by group schemes of finite type over S for $E =$ et and $E =$ Zar.*

Proof. For simplicity, we omit E in notation in the proof. First, assume that $D = \mu_p$. Let us recall that we have the exact sequence (65) of sheaves of abelian groups on S_{fl}

$$\cdots \longrightarrow R^{i+1} q_\star \mu_p \longrightarrow R^i q_\star Z\Omega^1_{Y/S} \longrightarrow R^i q_\star \Omega^1_{Y/S} \longrightarrow R^{i+2} q_\star \mu_p \longrightarrow \cdots . \quad (68)$$

The sheaves $R^i q_\star Z\Omega^1_{Y/S}$ and $R^i q_\star \Omega^1_{Y/S}$ are representable by S-group schemes of finite dimensional k-vector spaces (in fact, they are represented by $H^i(X, Z\Omega^1_{X,\mathrm{cl}})$ and $H^i(X, \Omega^1_X)$). So, $R^i q_\star \mu_p$ is also representable by an S-group scheme for every i. If n is not divisible by p, then μ_n is a smooth group scheme. Hence by [21], III, Theorem 3.9, the sheaf $R^i q_\star \underline{\mu_n}$ is associated with the presheaf

$$(T \longrightarrow S) \longrightarrow H^i_{\mathrm{et}}(T \times_S Y, \underline{\mu_n}).$$

By the proper base change theorem (see [21], VI, Corollary 2.3), we have an isomorphism

$$\underline{H^i(X, \mu_n)(T)} \xrightarrow{\cong} H^i_{\mathrm{et}}(T \times_k X, \underline{\mu_n}) = H^i_{\mathrm{et}}(T \times_S Y, \underline{\mu_n}).$$

This implies that $R^i q_\star \underline{\mu_n} = \underline{H^i_{\rm fl}(X, \mu_n)}$. Note that every finite diagonalizable group scheme D has a filtration whose sub quotients are of the form μ_q for some prime numbers q. Moreover, if $D'' = D/D'$ then

$$0 \longrightarrow \underline{D'} \longrightarrow \underline{D} \longrightarrow \underline{D''} \longrightarrow 0$$

is an exact sequence of sheaves of diagonalizable group schemes (see [16], I, 5.6 (4)). So, if the proposition holds for both D' and D'', then it also holds for D. Hence, the required assertion follows easily by induction on the length of D. □

Note that the proof of the above proposition fails for the flat topology on S-schemes. The above proposition is similar in spirit to the smooth specialization of étale cohomology groups (see [21], VI, Corollary 4.2).

Conjecture 5.2.5 (Relative purity of R^2). *The natural map $R^2 q_\star \underline{D} \longrightarrow R^2 p_\star \underline{D}$ is an isomorphism.*

We say that *Conjecture* 5.2.5 *holds for* X, if it holds for all k-schemes S and all relatively big subsets $U \subseteq S \times_k X \longrightarrow S$. If Conjecture 5.2.5 holds for X, then, by Proposition 5.2.4, the sheaf $R^2 p_\star \underline{D}$ is represented by a group scheme on S. The conjecture holds for all curves, because in this case $U = Y$ and $p = q$. If $k = \mathbb{C}$, then Conjecture 5.2.5 follows from [8], Lemma 3.2. Thus, by the Lefschetz principle, Conjecture 5.2.5 holds for varieties defined over any algebraically closed field of characteristic zero. In general, the conjecture is motivated by the purity theorems in the étale cohomology (unfortunately, such theorems are proved only for smooth pairs; see, e.g., [21], VI, Theorem 5.1 and Corollary 5.3). If D is a smooth diagonalizable group scheme and S is a point, then the conjecture follows from [21], VI, Lemma 9.1.

5.3 *Principal ρ-sheaves*

Let X be a smooth projective variety defined over an algebraically closed field k. For a torsion free sheaf $\mathscr{A}$, $U_{\mathscr{A}}$ denotes the maximal open subset where it is locally free. We fix a faithful representation $\rho \colon G \longrightarrow \mathrm{GL}(V)$ of the reductive group G with semisimple image G' in $\mathrm{SL}(V)$.

A *principal ρ-sheaf* on the smooth projective variety X is a triple $\mathfrak{P} = (\mathscr{P}, \mathscr{A}, \psi)$ consisting of a principal G-bundle $\mathscr{P}$ defined on the big open subset $U_{\mathscr{A}}$ of X, a torsion free sheaf $\mathscr{A}$ on X, and an isomorphism $\psi \colon \mathscr{P}(V^\vee) \longrightarrow \mathscr{A}_{|U_{\mathscr{A}}}$.

Remark 5.3.1. Note that if ρ is faithful (i.e., with trivial kernel group scheme), then a principal ρ-sheaf is the same thing as a singular principal G-bundle. Namely, $\sigma_{|U}$ corresponds to a reduction of the frame $\mathrm{GL}(V)$-bundle of $\mathscr{A}_{|U}$ to G and it uniquely extends to the corresponding section σ giving τ. Therefore, if ρ is faithful, then we can talk about (semi)stable principal ρ-sheaves.

A *family of principal ρ-sheaves* parameterized by a k-scheme S is a triple $\mathfrak{P}_S = (\mathscr{P}, \mathscr{A}, \psi)$ consisting of a principal G-bundle $\mathscr{P}$ defined on a big open subset U of $X \times S$ and an S-flat family $\mathscr{A}$ of torsion free sheaves, such that $U = U_{\mathscr{A}}$ and $\mathscr{A}$

extends $\mathscr{P}(V^\vee)$. Let $\mathscr{S}$ be a set of isomorphy classes of principal ρ-sheaves on X. We say that $\mathfrak{P}_S$ is a *family of principal ρ-sheaves in* $\mathscr{S}$, if, for every geometric point $s \in S$, the isomorphy class of $(\mathscr{P} \otimes k(s), \mathscr{A} \otimes k(s), \psi \otimes k(s))$ belongs to $\mathscr{S}$.

Proposition 5.3.2. *Let* $\rho\colon G \longrightarrow \mathrm{GL}(V)$ *be a faithful representation. Let* $\mathscr{P}$ *be a principal G-bundle defined on a big open subset* $i: U \hookrightarrow X$ *and set* $\mathscr{A} = i_\star \mathscr{P}(V^\vee)$. *Then,* $\mathscr{A}$ *is reflexive, and* $\mathscr{P}$ *can be uniquely extended to a principal G-bundle* $\mathscr{P}'$ *on* $U_{\mathscr{A}}$, *such that* $\mathscr{P}'(V^\vee) = \mathscr{A}_{|U_{\mathscr{A}}}$.

Proof. Let $\mathscr{E}$ be the frame $\mathrm{GL}(V)$-bundle on $U_{\mathscr{A}}$ associated with $\mathscr{A}^\vee_{|U_{\mathscr{A}}}$. Then, $\mathscr{P}$ gives a section $\sigma : U \longrightarrow \mathscr{E}/G$. Since $\mathscr{E}/G = \mathscr{E}(\mathrm{GL}(V)/G)$ is affine over $U_{\mathscr{A}}$, this section extends uniquely to a section defined on $U_{\mathscr{A}}$. This gives the required $\mathscr{P}'$. □

The above proposition shows that, in the definition of a principal ρ-sheaf, we do not need to assume that $U = U_{\mathscr{A}}$, if ρ is faithful.

Note that, by Proposition 4.5.1 and 4.5.2, we know that there exists a universal family for the set of isomorphy classes of semistable principal ρ-sheaves for $G \hookrightarrow \mathrm{GL}(V)$ (see Remark 5.3.1).

5.4 *Construction of universal spaces via central isogenies*

Let G and G' be two connected, reductive, linear algebraic groups defined over an algebraically closed field k. Let $\phi\colon G \longrightarrow G'$ be a *central isogeny*, i.e., a surjective homomorphism of algebraic groups with finite kernel group scheme $\ker(\phi)$ contained in the center group scheme $\mathscr{Z}(G)$ of G. For simplicity, we will assume that G' is semisimple. Since $\mathscr{Z}(G)$ is contained in a maximal torus of G ([16], Part 2, 1.6), $\ker(\phi)$ is a diagonalizable group scheme and, thus, we can write it as a product

$$\ker(\phi) = \mu_{q_1^{k_1}} \times \cdots \times \mu_{q_n^{k_n}}$$

for some prime numbers $q_1 < \cdots < q_n$ and positive integers $k_1, \ldots, k_n$.

Let $\mathfrak{P} = (\mathscr{P}, \mathscr{A}, \psi)$ be a family of principal G'-sheaves on X parameterized by S. Then, $\mathscr{P}$ is a principal G'-bundle on $U = U_{\mathscr{A}} \subseteq X \times S$. Let us define the functor of families of reductions of G'-bundles to G-bundles $\Gamma(\phi, \mathscr{P})\colon \underline{\mathrm{Sch}}_S \longrightarrow \underline{\mathrm{Set}}$ by sending $(t\colon T \longrightarrow S)$ to the set of isomorphy classes of reductions of the structure group of $\mathscr{P}_T = \bar{t}^\star \mathscr{P}$ to G via ϕ. On morphisms, $\Gamma(\phi, \mathscr{P})$ is defined by pull backs. Let $\widetilde{\Gamma}(\phi, \mathscr{P})$ denote the sheafification of $\Gamma(\phi, \mathscr{P})$ with respect to the Zariski (or étale) topology on S. In our case, $\widetilde{\Gamma}(\phi, \mathscr{P})$ is easy to describe, because $\Gamma(\phi, \mathscr{P})$ is a separated presheaf of sets. Hence, an element of $\widetilde{\Gamma}(\phi, \mathscr{P})(T)$ can be represented by $((s_i), (U_i))$ for some Zariski open covering (U_i) of T and some $s_i \in \Gamma(\phi, \mathscr{P})(U_i)$.

Proposition 5.4.1. *Assume that Conjecture 5.2.5 holds for X. Then, the functor* $\widetilde{\Gamma}(\phi, \mathscr{P})$ *is representable by a scheme which is flat and finite over a closed subscheme of S.*

Proof. We have a short exact sequence of sheaves of groups on U_{fl}

$$0 \longrightarrow \underline{D} \longrightarrow \underline{G} \longrightarrow \underline{G'} \longrightarrow 0.$$

In fact, $G \longrightarrow G'$ is a D-torsor, so that this sequence is exact already on k_{fl}. This can be seen from the construction of the central isogeny $G \longrightarrow G'$ from the corresponding map of tori $T \longrightarrow T'$ (see [16], II, 1.13-1.14) which is easily seen to be a D-torsor. Let $T \longrightarrow S$ be a locally finite morphism and let $U_T = T \times_S U$. By [5], IV, 4.2.7.3, Proposition 4.2.7.8 and Remarque 4.2.10, we have the following exact sequence of pointed sets:

$$H^1_{\text{fl}}(U_T, \underline{G}) \longrightarrow H^1_{\text{fl}}(U_T, \underline{G'}) \longrightarrow H^2_{\text{fl}}(U_T, \underline{D}).$$

It induces an exact sequence of sheaves of pointed sets on $\underline{\text{Sch}}_{S,\text{Zar}}$:

$$R^1 p_\star \underline{G} \longrightarrow R^1 p_\star \underline{G'} \longrightarrow R^2 p_\star \underline{D}. \tag{69}$$

The principal G'-bundle $\mathscr{P}$ on U gives a section σ of $R^1 p_\star \underline{G'}$ over S. By Proposition 5.2.4, $R^2 q_\star \underline{D}$ is representable by an S-group scheme K. Hence, by Conjecture 5.2.5, $R^2 p_\star \underline{D}$ is also representable by K. Let $\sigma' \in \text{Hom}(S, K) \cong R^2 p_\star \underline{D}(S)$ be the morphism corresponding to the image of σ. Then, the fiber of σ' over the unit element of K is a closed subscheme S' of S.

Lemma 5.4.2. *For any S-scheme T, the set $\widetilde{\Gamma}(\phi, \mathscr{P})(T)$ is non-empty, if and only if the structure morphism $t \colon T \longrightarrow S$ factorizes through $S' \hookrightarrow S$.*

Proof. If $\widetilde{\Gamma}(\phi, \mathscr{P})(T)$ is non-empty, then there exists a Zariski open covering (V_i) of T, such that the principal G'-bundles $\bar{t}^\star \mathscr{P}_{|U_{V_i}}$ have reductions of the structure group to G. This means that $\sigma_{|T} \in R^1 p_\star G'(T)$ is mapped to the unit element of $K(T) = \text{Hom}_S(T, K) \cong R^2 p_\star \underline{D}$. But $\sigma_{|T}$ goes to $\sigma' t$. Therefore, T factorizes through the fiber S' of σ' over the unit of K. To see the other implication, it is sufficient to prove that $\widetilde{\Gamma}(\phi, \mathscr{P})(S')$ is non-empty. But this is clear from the definition of S' and the sequence (69). □

To finish the proof of Proposition 5.4.1, note that, for any $t \colon T \longrightarrow S$, we have a short exact sequence:

$$H^1_{\text{fl}}(U_T, \underline{D}) \longrightarrow H^1_{\text{fl}}(U_T, \underline{G}) \longrightarrow H^1_{\text{fl}}(U_T, \underline{G'}).$$

Hence, the set of reductions of $\bar{t}^\star \mathscr{P}$ to a principal G-bundle is either empty or in bijection to $H^1_{\text{fl}}(U_T, \underline{D})$ (here, we use that D is central). Thus, by Lemma 5.4.2, $\widetilde{\Gamma}(\phi, \mathscr{P})(T) = R^1 p_\star \underline{D}(T)$, if t factorizes through $S' \hookrightarrow S$, and $\widetilde{\Gamma}(\phi, \mathscr{P})(T) = \varnothing$ otherwise. But, by Proposition 5.2.3, this means that $\widetilde{\Gamma}(\phi, \mathscr{P})$ is represented by an $H^1_{\text{fl}}(X, D)$-torsor over S'. Now, the required assertion follows from Proposition 5.2.1. □

Assume that we have a notion of semistability for principal G'-sheaves. Then, we extend it to principal ρ-sheaves by saying that a principal ρ-sheaf is *(semi)stable*, if the G'-sheaf obtained by extension of the structure group by $\Phi \colon G \longrightarrow G'$ is (semi)stable.

Proposition 5.4.3. *Assume that Conjecture* 5.2.5 *holds for X and that there exists a universal family for semistable principal G'-sheaves on X. Then, there exists a universal space for semistable principal ρ-sheaves.*

Proof. The theorem follows from [25], Lemma 4.10, and Proposition 5.4.1. □

5.5 *Construction of moduli spaces for reductive groups via faithless representations*

Let $\rho\colon G \longrightarrow \mathrm{SL}(V) \subset \mathrm{GL}(V)$ be a representation, such that the kernel group scheme of ρ is contained in the center group scheme $\mathscr{Z}(G)$. Set $G' = G/\ker\rho$ and let $\rho'\colon G' \longrightarrow \mathrm{SL}(V) \subset \mathrm{GL}(V)$ be the induced faithful representation. Again, we assume, for simplicity, that G' is semisimple. Let $\mathfrak{P} = (\mathscr{P}, \mathscr{A}, \psi)$ be a principal ρ-sheaf. Then, the *degree of* $\mathfrak{P}$ is the homomorphism $d_{\mathfrak{P}}\colon X^\star(G) \longrightarrow \mathbb{Z}$, such that, for any character χ, $d_{\mathfrak{P}}(\chi)$ is the degree of the unique line bundle extending the line bundle $\mathscr{P}(\chi)$ from $U_{\mathscr{A}}$ to X. By Remark 5.3.1, we can say when a principal G'-sheaf is semistable. We extend the definition of (semi)stability by saying that a principal ρ-sheaf $\mathfrak{P}$ is *(semi)stable*, if the corresponding principal G'-sheaf is (semi)stable. Equivalently, we can describe it in the following way: Let $\mathfrak{P} = (\mathscr{P}, \mathscr{A}, \psi)$ be a principal ρ-sheaf. Then, by a *reduction of* $\mathfrak{P}$ *to a parabolic subgroup* $P \subset G$, we mean a section $\beta\colon U' \longrightarrow \mathscr{P}(G/P)$ over a big open subset $U' \subseteq U_{\mathscr{A}}$. Then, $P = Q_G(\lambda)$ for some one parameter subgroup $\lambda\colon \mathbb{G}_m(k) \longrightarrow G$ (see (6) for the notation) and β is called a *reduction of* $\mathfrak{P}$ *to* λ. Let $\lambda'\colon \mathbb{G}_m(k) \longrightarrow G'$ be the one parameter subgroup of G' obtained by composing λ with the quotient map $G \longrightarrow G/\ker\rho = G'$ (it can be trivial, if the image of λ is contained in the kernel of ρ). Let us also set $P' = Q_{G'}(\lambda')$. Then, $G/P = G'/P'$, so that we get a reduction of the principal G'-bundle $\mathscr{P}'$ associated with $\mathscr{P}$ to a parabolic subgroup $P' \subset G'$. Since we assume G' to be semisimple, λ' yields a weighted filtration $(\mathscr{A}_\bullet(\beta), \alpha_\bullet(\beta))$. We say that $\mathfrak{P}$ is *(semi)stable*, if for every reduction of $\mathfrak{P}$ to a one parameter subgroup λ with non-trivial image in G' we have

$$M(\mathscr{A}_\bullet(\beta), \alpha_\bullet(\beta))(\succeq)0.$$

Remark 5.5.1. Since we assume G' to be semisimple, the map $\mathscr{D}(G) \longrightarrow G'$ has finite kernel, and the radical of G is contained in the Kernel of ρ. Thus, we may replace "one parameter subgroup with non-trivial image in G'" by "one parameter subgroup of $\mathscr{D}(G)$" in the above definition.

We now describe S-equivalence for principal ρ-sheaves. A one parameter subgroup $\lambda\colon \mathbb{G}_m(k) \longrightarrow G$ is called *admissible*, if there exists a reduction β of $\mathfrak{P}$ to λ, such that $M(\mathscr{A}_\bullet(\beta), \alpha_\bullet(\beta)) = 0$. Let L be the Levi subgroup of $P = Q_G(\lambda)$ and let $\mathscr{P}^P$ be the principal P-bundle on U' obtained from $\mathscr{P}$ by the reduction β. Let $\mathscr{P}^\lambda$ be the principal G-bundle on U' obtained from $\mathscr{P}^P$ by extension through $P \longrightarrow L \subset P \subset G$. Then, $(\mathscr{P}^\lambda, \oplus\mathscr{A}_{i+1}/\mathscr{A}_i)$ is a principal ρ-sheaf and we call it the *admissible deformation of* $\mathfrak{P}$ *associated with* λ. Any admissible deformation of a semistable principal ρ-sheaf is still semistable. Moreover, after a finite number of admissible deformations, we obtain a principal ρ-sheaf $\mathrm{gr}(\mathfrak{P})$, such that any of its admissible deformations is isomorphic to it. Such a principal ρ-sheaf is called *polystable*. Note that any stable principal ρ-sheaf is polystable. We say that two principal ρ-sheaves $\mathfrak{P}_1$ and $\mathfrak{P}_2$ are *S-equivalent*, if and only if $\mathrm{gr}(\mathfrak{P}_1)$ and $\mathrm{gr}(\mathfrak{P}_2)$ are isomorphic principal ρ-sheaves. Note that we cannot define S-equivalence of principal ρ-sheaves as S-equivalence of associated principal G'-sheaves, because there are several non-isomorphic stable principal ρ-sheaves that correspond to the same stable principal G'-sheaf, if $H^1_{\mathrm{fl}}(X, D)$ is non-trivial.

Let us fix a polynomial P and a homomorphism $d\colon X^\star(G) \longrightarrow \mathbb{Z}$. Let us recall that the moduli functor of principal ρ-sheaves

$$\underline{\mathrm{M}}(\rho)^{\mathrm{(s)s}}_{P,d}\colon \underline{\mathrm{Sch}_k} \longrightarrow \underline{\mathrm{Set}}$$

was defined by

$$S \longmapsto \left\{ \begin{array}{l} \text{Isomorphy classes of families of} \\ \text{(semi)stable principal } \rho\text{-sheaves of} \\ \text{degree } d \text{ with Hilbert polynomial } P \end{array} \right\}.$$

Theorem 5.5.2. *Assume that Conjecture 5.2.5 holds for X. Then, there exists a quasi-projective moduli scheme* $\mathrm{M}(\rho)^{\mathrm{ss}}_{P,d}$ *which is a coarse moduli scheme for* $\underline{\mathrm{M}}(\rho)^{\mathrm{ss}}_{P,d}$.

Proof. Let $\mathscr{S}$ be the family of isomorphy classes of semistable principal G'-sheaves. Since $G' \hookrightarrow \mathrm{GL}(V)$ is faithful, we can construct a universal family with group H parameterized by R for the set $\mathscr{S}$. Moreover, we know that the good quotient $R/\!\!/H$ exists, and it is the coarse moduli scheme for principal G'-sheaves [6]. There exists a torus $T = \mathbb{G}_m^q$ and a homomorphism $G \longrightarrow T$, such that the induced homomorphism $G \longrightarrow T \times G'$ is a central isogeny. Indeed, it is sufficient to take the derived group $\mathscr{D}(G)$ and then define T as the image of the connected component of the reduced part of the kernel of ρ under the projection $G \longrightarrow G/\mathscr{D}(G)$. The composition $d'\colon X^\star(T) \longrightarrow X^\star(G) \longrightarrow \mathbb{Z}$ determines the degrees $d_1, \ldots, d_q$ of line bundles corresponding to a principal T-bundle obtained by extension $G \longrightarrow T$ from a rational principal G-bundle of degree d. Using Poincaré line bundles on the Picard schemes J^{d_i} parameterizing line bundles of degree d_i on X, one can easily construct a universal family of T-bundles of degree d'. Set $R' = J^{d_1} \times \cdots \times J^{d_q} \times R$. Combining the above two universal families, one can easily construct a universal family for principal $(T \times G')$-sheaves with group H parameterized by R' (where T acts trivially on R', but acts as scalar multiplication on the product of Poincaré bundles parameterized by $J^{d_1} \times \cdots \times J^{d_q}$). Then, Proposition 5.4.3 implies existence of a universal space R'' for the set of semistable principal ρ-sheaves. By construction, R'' is an $H^1_{\mathrm{fl}}(X, D)$-torsor over a closed H-equivariant subscheme of R'. By Lemma 2.1.7 in [6], we know that there exists a good quotient $R''/\!\!/H$. This quotient uniformly corepresents our moduli functor $\underline{\mathrm{M}}(\rho)^{\mathrm{ss}}_{P,d}$, so that we can set $\mathrm{M}(\rho)^{\mathrm{ss}}_{P,d} = R''/\!\!/H$. If we take the quotient $\mathrm{M}(\rho)^{\mathrm{s}}_{P,d}$ of the preimage of R^{s}, then it uniformly corepresents the functor $\underline{\mathrm{M}}(\rho)^{\mathrm{s}}_{P,d}$, and the quotient is a geometric one, by Lemma 2.1.7 in [6]. Hence, the points of $\mathrm{M}(\rho)^{\mathrm{s}}_{P,d}$ correspond to isomorphy classes of stable ρ-sheaves. To finish proof of the theorem, we need to prove that there is a bijection between geometric points of $\mathrm{M}(\rho)^{\mathrm{ss}}_{P,d}$ and the S-equivalence classes of principal ρ-sheaves. First, let us show that two S-equivalent principal ρ-sheaves correspond to the same point in $\mathrm{M}(\rho)^{\mathrm{ss}}_{P,d}$. To prove this, it is sufficient to construct a family $\mathfrak{P}_T = (\mathscr{P}_T, \mathscr{A}_T, \psi_T)$ of principal ρ-sheaves parameterized by $T = \mathbb{A}^1$, such that $\mathfrak{P}_t \cong \mathfrak{P}$ for $t \neq 0$ and $\mathfrak{P}_0$ is an admissible deformation of $\mathfrak{P}$ associated with an admissible one parameter subgroup

$\lambda\colon \mathbb{G}_m(k) \longrightarrow G$. Set $P = Q_G(\lambda)$ and let $\mathscr{P}^P$ be the principal P-bundle corresponding to β. Let $g_{\alpha\beta}\colon U_{\alpha\beta} \longrightarrow P$ be a 1-cocycle describing $\mathscr{P}^P$. Then, we define $\mathscr{P}_T$ by the 1-cocycle $\lambda(t)\cdot g_{\alpha\beta}\cdot\lambda(t)^{-1}\colon U_{\alpha\beta}\times T \longrightarrow P \subset G$. Let $\lambda'\colon \mathbb{G}_m \longrightarrow \mathrm{GL}(V)$ be the composition of σ and λ. By the construction of the moduli space of decorated sheaves, we get a family $\mathscr{A}_T$ associated with λ' which extends $\mathscr{P}_T(V^\vee)$ and such that $\mathscr{A}_t \cong \mathscr{A}$ for $t \neq 0$ and $\mathscr{A}_0 \cong \mathrm{gr}(\mathscr{A}_\bullet(\beta))$. This finishes the sketch of the construction of the required family. Since $R'' \longrightarrow R''/\!\!/H$ is a good quotient, it separates closed H-invariant subschemes. Therefore, to finish the proof of the theorem, it is sufficient to show that the H-orbit of a based principal ρ-sheaf is closed in R'', if and only if it corresponds to a polystable principal ρ-sheaf. But we know that the H-orbit of a based principal ρ-sheaf is closed in R', if and only if it corresponds to a polystable principal G'-sheaf. But a principal ρ-sheaf is polystable, if and only if the corresponding principal G'-sheaf is polystable. So the required assertion follows from the fact that the map $R'' \longrightarrow R'$ is finite and H-equivariant. □

Corollary 5.5.3. *Assume that either X is a curve or k has characteristic* 0. *Then, there exists a quasi-projective coarse moduli scheme* $\mathrm{M}(\rho)^{(\mathrm{s})\mathrm{s}}_{P,d}$ *for the functor* $\underline{\mathrm{M}}(\rho)^{(\mathrm{s})\mathrm{s}}_{P,d}$.

Of course, $\mathrm{M}(\rho)^{(\mathrm{s})\mathrm{s}}_{P,d}$ is projective, if the ground field k has characteristic zero. If X is a curve, then, by the semistable reduction theorem of Heinloth [11], $\mathrm{M}(\rho)^{(\mathrm{s})\mathrm{s}}_{P,d}$ is projective.

If we apply the above corollary to the adjoint representation $G \longrightarrow \mathrm{GL}(\mathfrak{g})$, we get the Ramanathan-Gómez-Sols moduli spaces of principal ρ-sheaves (on curves) but in arbitrary characteristic.

6. Examples

As mentioned in the introduction, interesting definitions of (semi)stability and corresponding projective moduli are obtained for particular reductive groups G and representations ρ. These we sketch now, the details being mere exercises.

6.1 *Stable torsion free sheaves*

For $\mathrm{GL}(V)$, we may choose the representation $\rho\colon \mathrm{GL}(V) \longrightarrow \mathrm{SL}(V \oplus k)$, $m \longmapsto (m, \det(m)^{-1})$. Then, we are dealing with objects $(\mathscr{A}, \mathscr{L}, \tau)$ where $\mathscr{A}$ is a torsion free coherent sheaf of rank $\dim_k(V)$, and $\mathscr{L}$ is a line bundle. Note that it suffices to define τ over the maximal open subset $U_{\mathscr{A}}$ where $\mathscr{A}$ is locally free. Since $(\mathrm{Isom}(k^{\dim_k(V)}, V)\times \mathrm{Isom}(k,k))/\mathrm{GL}(V) \cong \mathbb{G}_m(k)$ by means of the map $\mathrm{Isom}(k^{\dim_k(V)}, V)\times\mathrm{Isom}(k,k) \longrightarrow k$, $(m,l)\longmapsto \det(m)\cdot l$, $\tau_{|U}$ corresponds to a section $\mathscr{O}_U \longrightarrow \det(\mathscr{A}_{|U})\otimes\mathscr{L}_{|U}$. Thus, the set of isomorphy classes of singular principal G-bundles $(\mathscr{A}, \mathscr{L}, \tau)$ maps bijectively onto the set of isomorphy classes of triples $(\mathscr{A}, \mathscr{L}, \varepsilon\colon \mathscr{L}^\vee \longrightarrow \det(\mathscr{A}))$, ε being an isomorphism and $(\mathscr{A}, \mathscr{L}, \varepsilon)$ being isomorphic to $(\mathscr{A}', \mathscr{L}', \varepsilon')$, if and only if there are isomorphisms $\psi\colon \mathscr{A} \longrightarrow \mathscr{A}'$ and

$\chi\colon \mathscr{L} \longrightarrow \mathscr{L}'$, such that $\varepsilon' = \det(\psi) \circ \varepsilon \circ \chi^{-1}$. Obviously, $(\mathscr{A}, \mathscr{L}, \varepsilon) \longmapsto \mathscr{A}$ yields a bijection between the set of isomorphy classes of triples and the set of isomorphy classes of torsion free sheaves on X. The concept of semistability for $(\mathscr{A}, \mathscr{L}, \tau)$ is that

$$\sum_{i=1}^{s} \alpha_i(\mathrm{rk}(\mathscr{A}_i)P(\mathscr{A}) - \mathrm{rk}(\mathscr{A})P(\mathscr{A}_i))(\succeq)0$$

for every weighted filtration $(\mathscr{A}_\bullet, \alpha_\bullet)$ of $\mathscr{A}$. This is obviously equivalent to the Gieseker-Maruyama (semi)stability of $\mathscr{A}$. Thus, we have a bijection between the set of isomorphy classes of (semi)stable singular principal $\mathrm{GL}(V)$-bundles and the set of isomorphy classes of Gieseker (semi)stable torsion free sheaves. If one considers appropriate sheafifications of the respective moduli functors, one can see that the above assignment induces indeed an isomorphism between $\mathrm{M}(\rho)^{(\mathrm{s})\mathrm{s}}_{\underline{P}}$ and the moduli space of Gieseker (semi)stable sheaves with Hilbert polynomial $\underline{P}$.

6.2 *Stable orthogonal and symplectic sheaves*

Let $G = \mathrm{Sp}_r(k)$ and $\rho\colon\ \mathrm{Sp}_r(k) \hookrightarrow \mathrm{SL}(V)$ be the standard faithful representation in an r-dimensional vector space V with non-degenerate skew-symmetric form $\varphi\colon V \otimes V \to k$. It is easy to check that a one parameter subgroup λ of $\mathrm{SL}(V)$, factoring through $\mathrm{Sp}_r(k)$, amounts to a decomposition

$$V = \bigoplus_{i \in \mathbb{Z}} V^i$$

with $\mathbb{G}_m(k)$ acting on the factors $V^\bullet$ with weights $\lambda_\bullet$ (order them increasingly), and each V^i being orthogonal to all V^j but $V^{-i} \cong (V^i)^\vee$, with $\lambda_{-i} = -\lambda_i$. The associated parabolic subgroup $Q_G(\lambda) \subseteq \mathrm{SL}(V)$ is the stabilizer of the symplectic weighted flag $(V_\bullet, \lambda_\bullet)$ in V defined by

$$V_i = \bigoplus_{j \le i} V^j.$$

Here, $(V_i)^\perp = V_{-i-1}$ and $\lambda_{-i} = -\lambda_i$. Giving a principal ρ-sheaf $\mathfrak{P} = (\mathscr{P}, \mathscr{A}, \psi)$ on X is equivalent to giving a symplectic sheaf $(\mathscr{A}, \varphi_{\mathscr{A}})$ of rank r, i.e., a torsion free sheaf $\mathscr{A}$ on X of this rank together with a skew-symmetric homomorphism $\varphi_{\mathscr{A}}\colon \mathscr{A} \otimes \mathscr{A} \to \mathscr{O}_X$, such that $\mathscr{A} \to \mathscr{A}^\vee$ is a monomorphism. (Namely, this is the homomorphism $\bigwedge^2 \mathscr{A}_{|U} \to \mathscr{O}_U$ amounting to the principal bundle $\mathscr{P}$ on the big open set U where $\mathscr{A}$ is locally free, then naturally extended to $\bigwedge^2 \mathscr{A} \to \iota_\star \bigwedge^2 \mathscr{A}_{|U} \to \iota_\star \mathscr{O}_U = \mathscr{O}_X$.) A reduction of $\mathscr{P}$ to the "weighted" parabolic subgroup $Q_G(\lambda) \subseteq G$ amounts, by extension to X, to a saturated filtration $\mathscr{A}_\bullet \subseteq \mathscr{A}$ which is an orthogonal weighted filtration with the weights $\lambda_\bullet$ (defining $\mathscr{A}_i^\perp$ as $\ker(\mathscr{A} \cong \mathscr{A}^\vee \to \mathscr{A}_i^\vee)$). Thus, our notion of (semi)stability becomes

$$\sum_{i=1}^{s} (\lambda_{i+1} - \lambda_i)(\mathrm{rk}\, \mathscr{A}_i \cdot P(\mathscr{A}) - \mathrm{rk}\, \mathscr{A} \cdot P(\mathscr{A}_i))(\succeq)0,$$

for all orthogonal weighted filtrations $(\mathscr{A}_\bullet, \lambda_\bullet)$ of $\mathscr{A}$. Since this expression is linear in the sequence $\lambda_\bullet$, this condition can be further expressed in terms of just orthogonal filtrations of two terms $\{0\} \subsetneq \mathscr{F} \subseteq \mathscr{F}^\perp \subsetneq \mathscr{A}$ with weights $\mathrm{rk}(\mathscr{F}) - r$ and $\mathrm{rk}(\mathscr{F})$, thus becoming the very simple condition announced in the introduction.

The same definition of (semi)stability and corresponding projective moduli are obtained for orthogonal sheaves $(\mathscr{A}, \varphi)$ of rank r, i.e., pairs made by a torsion free sheaf $\mathscr{A}$ of this rank and a symmetric form $\varphi\colon \mathscr{A} \otimes \mathscr{A} \to \mathscr{O}_X$ which amount to principal ρ-sheaves for the standard inclusion $\rho\colon m \longmapsto \mathrm{Diag}(m, \det(m)^{-1})$ of $\mathrm{O}_r(k)$ in $\mathrm{SL}_{r+1}(k)$ (this change of target slightly varies the details of the former proof).

The semistable singular principal G-bundles for the natural inclusion ρ of $\mathrm{SO}_r(k)$ in $\mathrm{SL}_r(k)$ turn out to be just semistable special orthogonal sheaves $(\mathscr{A}, \varphi_{\mathscr{A}}, \psi_{\mathscr{A}})$, i.e., semistable orthogonal sheaves $(\mathscr{A}, \varphi_{\mathscr{A}})$ together with an isomorphism $\psi\colon \det \mathscr{A} \cong \mathscr{O}_X$, such that $\psi_{\mathscr{A}}^2 = \det \varphi_{\mathscr{A}}$ (the proof is just as in the symplectic case, but then one has to check that the extra datum $\psi_{\mathscr{A}}$ does not alter the definition of semistability).

6.3 *Stable algebra sheaves*

Let $G \subseteq \mathrm{GL}(V)$ be the group $\mathrm{Aut}(V, \varphi)$ of automorphisms of a non-associative algebra structure $\varphi\colon V \otimes V \to V$ on a finite dimensional space V. Assume for simplicity that G lies in fact in $\mathrm{SL}(V)$ (for instance, if G is connected and the Killing form $(v, w) \longmapsto \mathrm{Tr}(\varphi(v, \_) \circ \varphi(w, \_))$ of V is non-degenerate), so that we can use the faithful embedding $\rho\colon G \hookrightarrow \mathrm{SL}(V)$ (otherwise, we arrive at the same definition of (semi)stability and corresponding projective moduli, but with the slight complication in the details caused by taking ρ as the standard embedding of G in $\mathrm{SL}(V \oplus k)$).

A one parameter subgroup of $\mathrm{SL}(V)$ amounts to a decomposition

$$V = \bigoplus_{i=1}^{s+1} V^i$$

into subspaces V^i acted on by $\mathbb{G}_m(k)$ with increasing integer weights λ_i which are balanced, i.e., $\sum \lambda_i \dim_k(V^i) = 0$. It factors through $G = \mathrm{Aut}(V, \varphi)$, if and only if $\varphi(V^i, V^j)$ is contained in V^k, if there is a k, such that $\lambda_i + \lambda_j = \lambda_k$, and otherwise is null. The corresponding parabolic subgroup $Q_G(\lambda) \subseteq G$ is the stabilizer of the filtration of V by subspaces $V_i = \bigoplus_{j \le i} V^j$ which we weight by the balanced sequence $\lambda_\bullet$, i.e., $\sum \lambda_i(\dim_k(V_i) - \dim_k(V_{i-1})) = 0$, so that it becomes an algebra weighted filtration, i.e., $\varphi(V_i, V_j) \subseteq V_k$ whenever $\lambda_i + \lambda_j \le \lambda_k$.

Just as in the former example, a singular principal G-bundle on X amounts to a φ-algebra sheaf $(\mathscr{A}, \varphi_{\mathscr{A}})$ of rank $\dim_k(V)$, as defined in the introduction. A reduction to the "weighted" parabolic subgroup $Q_G(\lambda) \subseteq G$ amounts to a balanced algebra filtration of $\mathscr{A}$, i.e., $\mathscr{A}_\bullet \subseteq \mathscr{A}$ with balanced weights $\lambda_\bullet$, such that $\varphi_{\mathscr{A}}(\mathscr{A}_i, \mathscr{A}_j) \subseteq \mathscr{A}_k^{\vee\vee}$ whenever $\lambda_i + \lambda_j \le \lambda_k$. The condition of (semi)stability of a φ-algebra sheaf $(\mathscr{A}, \varphi_{\mathscr{A}})$ then becomes

$$\sum_{i=1}^{s} (\lambda_{i+1} - \lambda_i)(\mathrm{rk}\, \mathscr{A}_i \cdot P(\mathscr{A}) - \mathrm{rk}\, \mathscr{A} \cdot P(\mathscr{A}_i))(\succeq)0,$$

for any balanced weighted algebra filtration $(\mathscr{A}_\bullet, \lambda_\bullet)$ of $\mathscr{A}$. This can be expressed more simply and equivalently in the usual terms of just algebra filtrations in the way mentioned in the introduction.

It can be applied, in particular, to the standard realizations of the exceptional groups as automorphism groups of some non-associative algebras. For instance, G_2 is the automorphism group of the octonions over k. In fact, any group G of adjoint type (i.e., with trivial center) over a field of characteristic zero can be identified with the component of identity $\mathrm{Aut}^0(\mathfrak{g}, \varphi)$ of the group of automorphisms of its Lie algebra $\varphi = [\_, \_]\colon \mathfrak{g} \otimes \mathfrak{g} \to \mathfrak{g}$. Since G has finite index in $\mathrm{Aut}(\mathfrak{g}, \varphi)$, principal ρ-sheaves, for the adjoint representation ρ, are just semistable φ-algebra sheaves together with a reduction to G over the big open set where they are locally free, so there is a projective moduli space for them.

In fact, this is also true for any reductive group, if the characteristic of k is 0. In fact, as a consequence of Section 5, the space parameterizing reductions from G/Z to G is a scheme over the space parameterizing principal G/Z-sheaves, but the notion of (semi)stability does not change, i.e., remains just (semi)stability of the adjoint algebra sheaf. *Thus, we recover the moduli of principal ρ-sheaves for the adjoint representation ρ which were called principal G-sheaves in* [8].

Acknowledgments

We thank Professor Mehta for suggesting this research while pointing out, in a conversation with the last author, that, since our previous work in characteristic 0 deals with the semistability of a tensor rather than with the semistability of a vector bundle, it was more suited for attacking the case of positive characteristic. Our thanks also go to N. Fakhruddin, Y. Holla, J. Milne, A. Premet, and S. Subramanian for discussions related to this work.

Tomás Gómez and Ignacio Sols are supported by grant number MTM2004-07090-C03-02 from the Spanish "Ministerio de Educación y Ciencia", and they thank the Polish Academy of Science (Warsaw) for hospitality. T.G. also thanks the Tata Institute of Fundamental Research (Mumbai) and the University of Duisburg-Essen for hospitality during his visits.

Adrian Langer was partially supported by a Polish KBN grant (contract number 1P03A03027). A.L. also thanks the DFG Schwerpunkt "Globale Methoden in der komplexen Geometrie" for supporting his visit to the University of Duisburg-Essen.

Alexander Schmitt acknowledges support by the DFG via a Heisenberg fellowship and via the "Schwerpunkt" program "Globale Methoden in der Komplexen Geometrie—Global Methods in Complex Geometry". A.S. was also supported by the DAAD via the "Acciones Integradas Hispano-Alemanas" program, contract number D/04/42257.

References

[1] M. Artin, A. Grothendieck and J. L. Verdier, Théorie des topos et cohomologie étale des schémas, Séminaire de Géométrie Algébrique du Bois-Marie 1963–1964 (SGA 4), Dirigé

par M. Artin, A. Grothendieck, et J.L. Verdier, avec la collaboration de N. Bourbaki, P. Deligne et B. Saint-Donat, Lecture Notes in Mathematics, 269, 270, 305, Springer-Verlag, Berlin-New York, 1972–73.

[2] V. Balaji and A. J. Parameswaran, Semistable principal bundles. II, Positive characteristics, *Transform. Groups*, **8** (2003) 3–36.

[3] A. Borel, Linear algebraic groups, Second edition, Graduate Texts in Mathematics 126, Springer-Verlag, New York (1991) xii+288 pp.

[4] M. Demazure, Sous-groupes paraboliques des groupes réductifs in Schémas en groupes III: Structure des schémas en groupes réductifs, Séminaire de Géométrie Algébrique du Bois Marie 1962/64 (SGA 3), Dirigé par M. Demazure et A. Grothendieck, Lecture Notes in Mathematics, 153, Springer-Verlag, Berlin-New York, 1970, viii+529 pp.

[5] J. Giraud, Cohomologie non abélienne, Die Grundlehren der mathematischen Wissenschaften, Band 179, Springer-Verlag, Berlin-New York (1971) ix+467 pp.

[6] T. Gómez, A. Langer, A. H. W. Schmitt and I. Sols, Moduli spaces for principal bundles in arbitrary characteristic, *Advances in Mathematics*, **219** (2008) 1177–1245.

[7] T. L. Gómez and I. Sols, Stable tensors and moduli space of orthogonal sheaves, math. AG/0103150 v4, 36 pp.

[8] T.L. Gómez and I. Sols, Moduli space of principal sheaves over projective varieties, *Ann. of Math.*, **161** (2005) 1033–88.

[9] A. Grothendieck, Techniques de construction et théorèmes d'existence en géométrie algébrique. IV. Les schémas de Hilbert, Séminaire Bourbaki, Vol. 6, Exp. No. 221, 249–76, Soc. Math. France, Paris (1995).

[10] A. Grothendieck and J. A. Dieudonné, Éléments de géométrie algébrique. I, Die Grundlehren der mathematischen Wissenschaften, Band 166, Springer-Verlag, Berlin-Heidelberg-New York (1971) ix+466 pp.

[11] J. Heinloth, Semistable reduction for G-bundles on curves, *J. Algebraic Geom.*, **17** (2008) 167–183; Addendum: *J. Algebraic Geom.*, **19** (2010) 193–197.

[12] W. H. Hesselink, Uniform instability in reductive groups, *J. reine angew. Math.*, **303/304** (1978) 74–96.

[13] D. Huybrechts and M. Lehn, Framed modules and their moduli, *Internat. J. Math.*, **6** (1995) 297–324.

[14] D. Huybrechts and M. Lehn, The geometry of moduli spaces of sheaves, Aspects of Mathematics E31, Friedr. Vieweg & Sohn, Braunschweig (1997) xiv+269 pp.

[15] L. Illusie, Complexe de de Rham-Witt et cohomologie cristalline, *Ann. Sci. École Norm. Sup. (4)*, **12** (1979) 501–661.

[16] J. C. Jantzen, Representations of algebraic groups, Second edition, Mathematical Surveys and Monographs 107, American Mathematical Society, Providence, RI (2003) xiv+576 pp.

[17] H. Kraft and C. Procesi, Classical Invariant Theory—a Primer, Lecture Notes, Version 2000, 128 pages. Available at http://www.math.unibas.ch.

[18] A. Langer, Semistable sheaves in positive characteristic, *Ann. of Math. (2)*, **159** (2004) 251–76: Addendum, *Ann. of Math (2)*, **160** (2004) 1211–3.

[19] A. Langer, Moduli spaces of sheaves in mixed characteristic, *Duke Math. J.*, **124** (2004) 571–86.

[20] M. Maruyama, On boundedness of families of torsion free sheaves, *J. Math. Kyoto Univ.*, **21** (1981) 673–701.

[21] J. S. Milne, Étale cohomology, Princeton Mathematical Series 33, Princeton University Press, Princeton, New Jersey (1980).

[22] D. Mumford et al., Geometric Invariant Theory, Third edition, Ergebnisse der Mathematik und ihrer Grenzgebiete (2) 34, Springer-Verlag, Berlin (1994) xiv+292 pp.

[23] S. Ramanan and A. Ramanathan, Some remarks on the instability flag, *Tohoku Math. J. (2)*, **36** (1984) 269–291.

[24] A. Ramanathan, Stable principal bundles on a compact Riemann surface, *Math. Ann.*, **213** (1975) 129–52.

[25] A. Ramanathan, Moduli for principal bundles over algebraic curves I–II, *Proc. Indian Acad. Sci. Math. Sci.*, **106** (1996) 301–28, 421–49.

[26] A. H. W. Schmitt, Singular principal bundles over higher-dimensional manifolds and their moduli spaces, *Int. Math. Res. Not.*, **23** (2002) 1183–209.

[27] A. H. W. Schmitt, A universal construction for moduli spaces of decorated vector bundles over curves, *Transformation Groups*, **9** (2004) 167–209.

[28] A. H. W. Schmitt, A closer look at semistability for singular principal bundles, *Int. Math. Res. Not.*, **62** (2004) 3327–66.

[29] A. H. W. Schmitt, Global boundedness for decorated sheaves, *Internat. Math. Res. Not.*, **68** (2004) 3637–71.

[30] A. H. W. Schmitt, Moduli for decorated tuples of sheaves and representation spaces for quivers, *Proc. Indian Acad. Sci. Math. Sci.*, **115** (2005) 15–49.

[31] J. P. Serre, Sur la topologie des variétés algébriques en caractéristique p, 1958, Symposium internacional de topología algebraica, International symposium on algebraic topology, pp. 24–53, Universidad Nacional Autónoma de México and UNESCO, Mexico City.

[32] C. S. Seshadri, Geometric reductivity over arbitrary base, *Advances in Math.*, **26** (1977) 225–74.

[33] C. T. Simpson, Moduli of representations of the fundamental group of a smooth projective variety I, *Inst. Hautes Études Sci. Publ. Math.*, **79** (1994) 47–129.

[34] T. A. Springer, Linear algebraic groups, Second edition, Progress in Mathematics, 9, Birkhäuser Boston, Inc., Boston, MA (1998) xiv+334 pp.

Teichmüller Theory and Moduli Problem No. 10, 2010, pp. 373–385.

Some Remarks About Curves on K3 Surfaces

Mihai Halic

Mathematisches Institut der Universität Basel, Rheinsprung 21, 4051 Basel, Switzerland
e-mail: Mihai.Halic@unibas.ch

Abstract. The article studies the geometry of curves on $K3$ surfaces. Its aim is to give an alternative proof to the fact that there are rational curves on every projective $K3$ surface.

2000 Subject Classification: 14J28, 14H10

Introduction

In this note I wish to present my attempt at understanding the geometry of the curves on projective $K3$ surfaces. Actually I shall restrict my attention to the Picard general $K3$ surfaces, that is those whose Picard group has rank one. The reason is that these form a dense subset in the Moduli space of projective $K3$ surfaces.

A classical result, due to Bogomolov and Mumford, states that on every projective $K3$ surface there are rational curves, and 1-dimensional families of elliptic curves. Its proof has been sketched in [10], and has been treated in full detail by Xi Chen in [1]. Actually, in *loc. cit.* the author proves more, namely:

Theorem. *For any integers $n \geqslant 3$ and $d > 0$, the linear system $|\mathcal{O}_S(d)|$ on a general $K3$ surface in $S \hookrightarrow \mathbb{P}^n$ contains a $(p(d)-\delta)$-dimensional family of irreducible nodal curves, whose arithmetic genus equals $p(d) := d^2(n-1)+1$, and which posses precisely δ nodes.*

The common feature of these approaches is that one first studies degenerations of the general $K3$ surfaces to special ones, namely to the so-called trigonal surfaces. On these latter one explicitely produces singular curves, usually a connected curve with two smooth rational components. The second step consists in deforming the trigonal surface, together with (smoothings of) the curve, back to the starting $K3$ surface.

While this method is very elegant and short, it does not give much insight into the geometry of curves on a general $K3$ surface. A geometric approach has been considered by M. Green and Ph. Griffiths in [6, Part B], where they studied the geometry of the dual variety $S^* \hookrightarrow \mathbb{P}^n$ of S. Their guiding idea is that curves with nodes are obtained as hyperplane sections $H \cap S$, such that the corresponding hyperplane H has simple tangency points with S. Due to technical difficulties, they have been able to carry over the study only for $K3$'s in $\mathbb{P}^n$, with $n = 3, 4$, and for 2-sheeted coverings of the plane. Another study concerning the deformations of nodal curves on $K3$ surfaces is done in [11], but no existence result is proved there.

The idea in this note is similar to that of M. Green and Ph. Griffiths but the methods are different: we investigate the stability of the restriction of the tangent bundle of the $K3$ surface S to the curves on it. The consequence of the stability is that curves on $K3$ surfaces are rigid, that is they can not be deformed *within their conformal class*. Therefore, a deformation of a nodal curve inside S induces a non-trivial deformation of the conformal structure of the normalization of the curve. Combining this with S. Diaz's upper bound in [3] on the dimension of complete subvarieties of the Deligne-Mumford space M_g, we get an inductive process which produces curves on S whose geometric genus is strictly decreasing. Our main result is stated in theorem 2.3: under some technical hypothesis (the assumption ($\star\star$) in section 2), we reprove the theorem above up a a certain number of nodes, by investigating the geometry of the stable map spaces in the sense of Kontsevich-Manin.

1. A stability result

In this section we investigate the properties of the restriction of the tangent bundle of a $K3$ surface to curves. Since the tangent bundle of a $K3$ surface is stable, the results of Flenner and Mehta/Ramanathan (see [4,8] or [7, chapter 7]) imply that its restriction to the general hypersurface section is stable.

What we wish to find are *effective bounds* for the degree, which guarantee the stability of the restriction, and also investigate the analogous issue for nodal curves. Observe that for a general $K3$ surface $S \xrightarrow{|\mathcal{O}_S(1)|} \mathbb{P}^n$, Flenner's bound implies that the restriction of T_S to the *general member* of $|\mathcal{O}_S(d)|$ is still stable for $d \geqslant 3(n-1)$. As we will see, this bound is quite weak. Our main tool is Bogomolov's effective restriction theorem (see [7, section 7.3]).

We make the following convention: for a connected nodal curve $\hat{C} \hookrightarrow S$, and a torsion free sheaf of rank one $\hat{Q} \to \hat{C}$, we define *the degree* $\deg_{\hat{C}} \hat{Q} := \chi(\hat{Q}) + p_a(\hat{C}) - 1$, where $p_a(\hat{C}) = h^1(\mathcal{O}_{\hat{C}})$ is the arithmetic genus of $\hat{C}$. We shall repeatedly use the following computational lemma.

Lemma 1.1.

(i) *Let $\hat{C}$ be a reduced, irreducible, nodal curve, and let $\nu : C \to \hat{C}$ be its normalization. Consider a rank 2 vector bundle $E \to \hat{C}$, and a quotient line bundle $q : \nu^*E \to Q$ of ν^*E. We denote by $\{x_{11}, x_{21}\}, \ldots, \{x_{1\delta}, x_{2\delta}\} \subset C$ the pairs of points which are identified by ν, and let $0 \leqslant \delta' \leqslant \delta$ be the number of pairs $\{x_{1j}, x_{2j}\}$ for which $q_{x_{1j}}$ and $q_{x_{2j}}$ are not collinear. There is a torsion-free quotient $\hat{q} : E \to \hat{Q}$ of E, of degree $\deg_{\hat{C}} \hat{Q} = \deg_C Q + \delta'$, such that $Q = \nu^*\hat{Q}/(\nu^*\hat{Q})_{\mathrm{tor}}$.*

(ii) *With the previous notations, assume that $\hat{C}$ is a curve on a $K3$-surface S, and that $E := T_S \otimes_{\mathcal{O}_S} \mathcal{O}_{\hat{C}}$. Denote by G the kernel of the composition $T_S \twoheadrightarrow E \twoheadrightarrow \hat{Q}$. Then G is a locally free, rank 2 sheaf on S, with*

$$c_1(G) = -[\hat{C}] \quad \textit{and} \quad c_2(G) = 24 + \delta' + \deg_C Q.$$

Proof.

(i) The idea is to construct a rank one, torsion free sheaf on $\hat{C}$ whose degree is as large as possible. For $j \in \{1, \dots, \delta\}$ we get the pair of linear applications

$$q_{1j} : T_{S,\hat{x}_j} = (\nu^* T_S)_{x_{1j}} \longrightarrow Q_{x_{1j}} \cong \mathbb{C}, \quad q_{2j} : T_{S,\hat{x}_j} = (\nu^* T_S)_{x_{2j}} \longrightarrow Q_{x_{2j}} \cong \mathbb{C}.$$

For each index j we distinguish between two cases: either q_{1j}, q_{2j} are collinear or they are not.

Step 1. If q_{1j}, q_{2j} are collinear, that is $\mathrm{Ker}(q_{1j}) = \mathrm{Ker}(q_{2j})$, we find an isomorphism τ_j such that the diagram below commutes:

$$\begin{array}{ccc} T_{S,\hat{x}_j} & \xrightarrow{q_{1j}} & Q_{x_{1j}} \\ \| & & \downarrow \tau_j \\ T_{S,\hat{x}_j} & \xrightarrow{q_{2j}} & Q_{x_{2j}} \end{array}$$

Identifying $Q_{x_{1j}}$ and $Q_{x_{2j}}$ using τ_j we get an invertible sheaf on the nodal curve $C/\{x_{1j} = x_{2j}\}$ whose degree equals $\deg_C(Q)$.
We repeat this procedure until we eliminate all the pairs of points such that q_{1j}, q_{2j} are collinear. The result consists of a nodal curve C', a naturally induced morphism $\nu' : C' \to \hat{C}$, and an invertible sheaf $Q' \to C'$ which is a destabilizing quotient for $(\nu')^* T_S \to C'$. We denote

$$\{x'_{11}, x'_{21}\}, \dots, \{x'_{1\delta'}, x'_{2\delta'}\} \subset C', \text{ where } 0 \leqslant \delta' \leqslant \delta,$$

the remaining points which are identified under the partial normalization $\nu' : C' \to \hat{C}$, and $q'_{11}, q'_{21}, \dots, q'_{1\delta'}, q'_{2\delta'}$ the corresponding projections.

Step 2. If q_{1j}, q_{2j} are not collinear, $\mathrm{Ker}(q_{1j}) \oplus \mathrm{Ker}(q_{2j}) = E_{\hat{x}_j}$, and is not possible perform the previous construction. Since $(\nu')^* E \twoheadrightarrow Q'$ is surjective, $\nu_* q : E \otimes \nu_* \mathcal{O}_C \twoheadrightarrow \hat{Q} := \nu_* Q'$ is the same. Observe that $\hat{Q} \to \hat{C}$ is a torsion free sheaf of rank one and degree $\deg_{\hat{C}} \hat{Q} = \deg_C Q + \delta'$.

We claim that the composition $\hat{q} : E \to \hat{Q}$ is still surjective. Indeed, consider a double point $\hat{x} \in \{\hat{x}_1, \dots, \hat{x}_{\delta'}\}$ of $\hat{C}$; since the homomorphisms $q_1 : E_{\hat{x}} \to Q_{x_1}$ and $q_2 : E_{\hat{x}} \to Q_{x_2}$ are not collinear, we get the following local description:

$$k[\![u,v]\!]^{\oplus 2} \bigotimes_{k[\![u,v]\!]} \frac{k[\![u,v]\!]}{\langle uv \rangle} \xrightarrow{\iota} k[\![u,v]\!]^{\oplus 2} \bigotimes_{k[\![u,v]\!]} \left(\frac{k[\![u,v]\!]}{\langle v \rangle} \oplus \frac{k[\![u,v]\!]}{\langle u \rangle} \right)$$

$$\xrightarrow{\nu_* q} k[\![u]\!] \oplus k[\![v]\!], \quad \text{with}$$

$$\iota(f,g) = (f(u,0) + f(0,v), g(u,0) + g(0,v)) \quad \text{and}$$

$$\times \hat{q}(h_1(u) + k_1(v), h_2(u) + k_2(v)) = h_1(u) + k_2(v).$$

Our claim becomes now clear.

(ii) The fact that G is locally free is a consequence of the fact that $\hat{C}$ is a complete intersection on S, and the formulae for the degree are obtained by straightforward computation (see for instance [7, Proposition 5.2.2]). □

We are going to use this lemma to prove the following result:

Proposition 1.2. *Let S be a Picard general $K3$ surface, and denote by A the positive generator of* $\mathrm{Pic}(S)$. *Consider a reduced, irreducible, nodal curve $\hat{C} \in |dA|$, for some $d \geqslant 1$, and let $\nu : C \to \hat{C}$ be its normalization. Let δ denote the number of double points of $\hat{C}$. If either*

- $d = 1, 2$ *and* $\delta \leqslant \frac{d^2A^2}{4} - 25$ *or*
- $d \geqslant 3$ *and* $\delta \leqslant (d-1)A^2 - 23$

holds, then $\nu^*T_S \to C$ is a stable, rank two bundle.

Proof. Let us denote by A the positive generator of $\mathrm{Pic}(S)$, and assume that $\hat{C} \in |dA|$. We denote as before $\{x_{11}, x_{21}\}, \dots, \{x_{1\delta}, x_{2\delta}\} \subset C$, with $\delta \leqslant p_a(\hat{C}) = (d^2A^2+1)/2$, the pairs of points which are identified by ν.

Let us assume that $\nu^*T_S \to C$ is not stable. Then it has a destabilizing quotient $\nu^*T_S \twoheadrightarrow Q$ of rank one; in particular $\deg_C Q \leqslant 0$. We consider the torsion free sheaf constructed in the previous lemma, and the kernel $G := \mathrm{Ker}(\hat{q} : T_S \to \hat{Q})$.

The assumption on δ implies that the discriminant

$$\Delta(G) := 4c_2(G) - c_1^2(G) = 4\left(24 + \delta' + \deg_C Q - \tfrac{d^2A^2}{4}\right)$$
$$\leqslant 4\left(24 + \delta' - \tfrac{d^2A^2}{4}\right) < 0. \tag{1.1}$$

Using [7, theorem 7.3.4] we deduce the existence of a subsheaf $G' \subset G$ of rank one with torsion free quotient, having the following properties:

$$\xi_{G',G} := c_1(G') - \frac{c_1(G)}{2} > 0, \quad \text{and} \quad \xi^2_{G',G} \geqslant \frac{-\Delta(G)}{4}. \tag{1.2}$$

The tangent bundle of a $K3$ surface is stable as it admits a Kähler-Einstein metric. This implies that $c_1(G') \cdot A < 0$; since S is a Picard general $K3$ surface, we find that $c_1(G') = -kA$ with $k \geqslant 1$. Further, the inequality $\xi_{G',G} > 0$ is equivalent to $d-2k \geqslant 1$. In particular $d \geqslant 3$. For $d = 1$ or $d = 2$, we have already our contradiction which comes from the assumption that $\nu^*T_S \to C$ is not stable.

In higher degrees, we must go further a little bit, and use the second inequality in (1.2):

$$\frac{(d-2)^2A^2}{4} \geqslant \frac{(d-2k)^2A^2}{4} \geqslant \frac{d^2A^2 - 4(\delta+24)}{4}.$$

This implies that $\delta + 24 \geqslant (d-1)A^2$, which disagrees again with our hypothesis. □

Remark 1.3. The result is unsatisfactory because it gives no insight into what happens for a larger number of nodes. For $d = 1, 2$, the bound $\frac{d^2A^2}{4} - 25$ corresponds roughly to the half of the arithmetic genus of $\hat{C}$. For $d \geqslant 3$ the upper bound is very weak and does not seem to have any geometrical meaning.

The weakness in the previous proof is that in (1.1) we use only that $\deg_C Q \leqslant 0$, and that we have no control on δ'. If we restrict our attention to $T_S|_{\hat{C}}$, we obtain more precise informations.

Proposition 1.4. *Let S be a Picard general $K3$ surface with* $\mathrm{Pic}(S) = \mathbb{Z}A$, *and consider a reduced, irreducible, nodal curve $\hat{C} \in |dA|$, for some $d \geqslant 1$. Let $n := \frac{A^2}{2} + 1$. If either*

- *$d = 1$ and $n \geqslant 50$ or*
- *$d \geqslant 2$ and $(d-1)(n-1) \geqslant 13$,*

is fulfilled, then the following statements hold:

(i) *$T_S|_{\hat{C}} \to \hat{C}$ has no non-trivial section;*
(ii) *$T_S|_{\hat{C}} \to \hat{C}$ has no torsion free quotient of negative degree;*
(iii) *$T_S|_{\hat{C}} \to \hat{C}$ is a simple, rank two bundle.*

The inequalities in the lemma leave uncovered only finitely many cases, enumerated below:

n	2	3	4	5	6, 7	8,...,13	14,...,49
d	1,...,13	1,...,7	1,...,5	1,...,4	1, 2, 3	1, 2	1

Secondly, we observe that the exact sequences

$$0 \longrightarrow T_S \otimes A^{-d} \longrightarrow T_S \longrightarrow T_S \otimes \mathcal{O}_{\hat{C}} \longrightarrow 0, \quad \text{and}$$

$$0 \longrightarrow \mathrm{End}(T_S) \otimes A^{-d} \longrightarrow \mathrm{End}(T_S) \longrightarrow \mathrm{End}(T_S) \otimes \mathcal{O}_{\hat{C}} \longrightarrow 0$$

imply that the statements are trivially satisfied for $d \gg 0$. The point of the proposition is that of making this result effective.

Proof.

(i) We use the same notations as before. Let us assume to the contrary that there is a non-trivial section $\hat{\sigma} : \mathcal{O}_{\hat{C}} \to T_S \otimes \mathcal{O}_{\hat{C}}$. The induced section $\sigma : \mathcal{O}_C \to \nu^* T_S$ defines the line bundle $L = \mathcal{O}_C(\{\sigma = 0\}) \subset \nu^* T_S$, and the quotient $Q = \mathcal{O}_C(-\{\sigma = 0\})$ of negative degree.
This data determines the rank one, torsion free sheaf $\hat{Q} \to \hat{C}$ constructed in lemma 1.1. We use that σ is induced by a section in $T_S|_{\hat{C}}$ as follows: we divide the double points $\hat{x}_1, \ldots, \hat{x}_\delta$ of $\hat{C}$ into those where the section $\hat{\sigma}$ does not vanish and those where $\hat{\sigma}$ does vanish. Let again δ' denote the number of this latter set. For each such double point, the degree of Q goes down by 2. It follows that

$$\deg_{\hat{C}} \hat{Q} = \deg_C(Q) + \delta' \leqslant -2\delta' + \delta' = -\delta' \leqslant 0,$$

and therefore $\Delta(G) \leqslant 96 - d^2 A^2 < 0$. As before, we deduce the existence of a rank one subsheaf $G' \subset G$ such that (1.2) holds. For $d = 1, 2$ this is impossible, while for $d \geqslant 3$ we get

$$\frac{(d-2)^2 A^2}{4} \geqslant \frac{d^2 A^2 - 96}{4} \quad \Longleftrightarrow \quad (d-1)A^2 \leqslant 24.$$

Again, this contradicts our assumption.

(ii) In this case the term $\delta' = 0$ in formula (1.1), so that $\Delta(G) < 0$. We conclude as before.

(iii) Let $\hat{\varphi} : T_S|_{\hat{C}} \to T_S|_{\hat{C}}$ be an automorphism; we wish to prove that is a multiple of the identity. Since $\det T_S \cong \mathcal{O}_S$, after multiplying $\hat{\varphi}$ with a constant, we may assume that $\det(\hat{\varphi}) = 1$. Let λ be an eigenvalue of $\hat{\varphi}$ at a smooth point $y \in \hat{C}$.

Consider the induced automorphism $\varphi : \nu^* T_S \to \nu^* T_S$, and define

$$E(\lambda) := \operatorname{Ker}(\varphi - \lambda \mathrm{id}) = \{w \in \nu^* T_S | \varphi(w) = \lambda w\} \subset \nu^* T_S.$$

It is a torsion free sheaf over C, and therefore locally free. We claim that it is the pull-back of a locally free sheaf on $\hat{C}$.

If $\operatorname{rk} E(\lambda) = 2$ we are done. So assume that $\operatorname{rk} E(\lambda) = 1$. Consider a double point $\hat{x} \in \hat{C}$, and $x_1, x_2 \in C$ with $\nu(x_1) = \nu(x_2) = \hat{x}$. We wish to prove that the diagram below commutes:

$$\begin{array}{ccccccc} E(\lambda)_{x_1} & \subset & (\nu^* T_S)_{x_1} & \xrightarrow{\varphi_{\hat{x}}} & (\nu^* T_S)_{x_1} & \supset & E(\lambda)_{x_1} \\ & & \tau_{\hat{x}} \| & & \| \tau_{\hat{x}} & & \\ E(\lambda)_{x_2} & \subset & (\nu^* T_S)_{x_2} & \xrightarrow{\varphi_{\hat{x}}} & (\nu^* T_S)_{x_2} & \supset & E(\lambda)_{x_1} \end{array}$$

Since $\varphi = \nu^* \hat{\varphi}$, certainly holds $\tau_{\hat{x}} \varphi_{\hat{x}} = \varphi_{\hat{x}} \tau_{\hat{x}}$. For $w \in E(\lambda)_{x_1}$ we have

$$\varphi_{\hat{x}}(\tau_{\hat{x}}(w)) = \tau_{\hat{x}}(\varphi_{\hat{x}}(w)) = \tau_{\hat{x}}(\lambda w) = \lambda \tau_{\hat{x}}(w).$$

So $E(\lambda)$ is preserved by the identifications $\tau_{\hat{x}}$, and descends to a locally free sheaf $\hat{E}$ on $\hat{C}$ of the same degree. Further, let $Q(\lambda) := \operatorname{Im}(\varphi - \lambda \mathrm{id}) \subset \nu^* T_S$. The $\tau_{\hat{x}}$'s preserve $Q(\lambda)$ too, so it descends to $\hat{Q} \to \hat{C}$ of the same degree. All together, we get the exact sequence

$$0 \longrightarrow \hat{E} \longrightarrow T_S|_{\hat{C}} \longrightarrow \hat{Q} \longrightarrow 0 \text{ over } \hat{C}, \text{ with } \deg_{\hat{C}} \hat{E} + \deg_{\hat{C}} \hat{Q} = 0.$$

Actually $\det(T_S|_{\hat{C}}) = \mathcal{O}_{\hat{C}}$ implies that $\hat{Q} = \hat{E}^\vee$. As $\hat{Q} \subset T_S|_{\hat{C}}$ is a sub-vector bundle, we deduce also the exact sequence $0 \to \hat{Q} \to T_S|_{\hat{C}} \to \hat{Q}^\vee = \hat{E} \to 0$ over $\hat{C}$. This means that both $\hat{E}$ and $\hat{Q}$ are quotients of $T_S|_{\hat{C}}$. Since the sum of their degrees is zero, at least one of $\hat{E}$ or $\hat{Q}$ has negative degree. We have already seen that this is impossible. □

The result that is missing, which I am unable to address is:

Question 1.5. Let S be a Picard general $K3$ surface, and consider a reduced, irreducible, nodal curve $\hat{C} \in |dA|$, for some $d \geqslant 1$. Let $\nu : C \to \hat{C}$ denote the normalization morphism, and assume that $C \ncong \mathbb{P}^1$. Is it true that $\nu^* T_S \to C$ has no non-trivial section?

Remark 1.6. Proposition 1.2 ties in with the work of Demailly, Peternell and Schneider in [2]. They call (see section 6 in *loc. cit.*) a vector bundle $\mathcal{V} \to X$ over a projective manifold *almost nef* if there is a countable family $A_i \hookrightarrow X$ of proper subvarieties of X such that $\mathcal{V}|_C \to C$ is nef for all curves $C \not\subset \cup_i A_i$. They have also introduced the somewhat stronger notion of *pseudo-effective* bundles; any pseudo-effective bundle is almost nef, but the converse is unclear.

For Calabi-Yau manifolds X, the authors proved in [2, corollary 6.12] that neither T_X nor Ω^1_X is pseudo-effective. However, they leave open the question whether T_X or Ω^1_X is almost nef or not (even for $K3$ surfaces); is speculated there that this should not be the case.

For a vector bundle $E \to C$ with $\det E \cong \mathcal{O}_C$ over a smooth curve, Miyaoka's criterion [9, theorem 3.1] says that E is nef if and only if E is semi-stable. Our proposition 1.2 can be interpreted as follows: assuming that the tangent bundle $T_S \to S$ of a $K3$ surface is not almost nef, the curves $\hat{C} \subset S$ on which it fails to be nef are quite singular. This might be viewed as a hint that T_S still enjoys some very week positivity property.

2. Rigidity of curves on $K3$ surfaces and applications

The purpose of this section is to prove that nodal curves on $K3$ surfaces are rigid *within their conformal structure*, and to deduce the existence of low genus curves on projective $K3$'s. This represented actually the main motivation for this study.

The heuristic idea behind it is exactly the same as in the work of M. Green and Ph. Griffiths: for a polarized $K3$ surface (S, A) with $\mathrm{Pic}(S) = \mathbb{Z}A$ holds

$$\dim |dA| = \frac{d^2A^2}{2} + 1 =: p(d) = \text{ genus of the general curve } C \in |dA|.$$

One expects that the (irreducible) nodal curves with δ double points form a δ-codimensional quasi-projective subvariety of $|dA|$, since each double point imposes (theoretically) one restriction. The difficulty consists in proving that things happen like this, and not else.

From now on, S will stay for a Picard general $K3$ surface, and A will denote the ample generator of $\mathrm{Pic}(S)$. We fix integers $d \geqslant 1$, $\delta \geqslant 0$, and denote by

$$|dA|_\delta := \{C \in |dA| \mid C \text{ is reduced, irreducible, and its only singularities are } \delta \text{ nodes}\}.$$

In [11, Theorem 2.6] is proved that, if non-empty, $|dA|_\delta$ is a smooth, quasi-projective variety of dimension $p(d) - \delta$.

I make the following *two assumptions*:

$$(\star) \qquad \text{it holds: either } d = 1, 2 \text{ and } \delta \leqslant \tfrac{d^2A^2}{4} - 25 =: \delta_{\max}$$
$$\text{or } d \geqslant 3 \text{ and } \delta \leqslant (d-1)A^2 - 23 =: \delta_{\max},$$

and

$(\star\star)$ *every* irreducible component of $\overline{|dA|_\delta} \setminus |dA|_\delta$ contains *reduced* curves.

The second condition is trivially satisfied for $d = 1$, but I am unable to remove it in general. Under these assumptions, I can prove:

Proposition 2.1. *The linear system $|dA|$ contains a $(p(d)-\delta)$-dimensional family corresponding to reduced and irreducible curves having precisely δ nodes.*

For proving the statement we are going to use two classical results: the first one is due to O. Zariski [12] (in the case of the projective plane), and gives an upper bound on the dimension of families of nodal curves on surfaces. The statement below is [1, Lemma 3.1].

Lemma. *Let $U_{d,g} \subset |dA|$ be the subset consisting reduced and irreducible curves of geometric genus g. Then $\dim U_{d,g} \leqslant g$. If $g > 0$ and W is an irreducible component of $U_{d,g}$ of dimension g, then the general member of W is a nodal curve.*

The second result is due to S. Diaz, and gives an upper bound on the dimension of complete subvarieties of the (non-compactified) Deligne-Mumford space. He proves in [3] the following:

Theorem. *The dimension of any complete subvariety of M_g is at most $g-2$.*

I am going to use the language of stable maps (at a very elementary level); for the necessary definitions, notations and properties the reader may consult [5]. The advantage of this approach is that there are two natural morphisms

$$\begin{array}{ccc} \overline{M}_g(S, dA) & \xrightarrow{\Pi_g} & |dA| \\ \downarrow{\scriptstyle \pi_g} & & \\ \overline{M}_g & & \end{array} \qquad \begin{array}{ccc} [u : C \to S] & \longmapsto & \text{section of } A^{\otimes d} \text{ defining } u_*C \\ \downarrow & & \\ [C_{\text{stab}}] & & \end{array}$$

which permit to combine the properties of the linear system $|dA|$ with the geometry of $\overline{M}_g$; this latter allows to control the genus of the normalization of the hyperplane sections. Observe there is no natural (*overall* defined) morphism $|dA| \to \overline{M}_{p(d)}$. We denote

$$\mathrm{D}_{g-\delta,2\delta} := \mathrm{Image}(\overline{M}_{g-\delta,\{x_{1j},x_{2j}\}_{j=1,\ldots,\delta}} \to \overline{M}_g),$$

$$\Delta_{g',g''} := \mathrm{Image}(\overline{M}_{g',1} \times \overline{M}_{g'',1} \to \overline{M}_{g'+g''})$$

and $\partial\overline{M}_g := \mathrm{D}_{g-1,2} \cup \bigcup_{g'+g''=g} \Delta_{g',g''}$ the 'boundary divisor'.

Further, we let $\overline{M}^*_g(S, dA) \hookrightarrow \overline{M}_g(S, dA)$ be the (possibly empty) union of those irreducible components whose generic points are non-multiply covered (also called simple) stable maps, with domain of definition a smooth, irreducible curve of genus g.

For an irreducible component $W \hookrightarrow \overline{M}^*_g(S, dA)$, we denote $W^{**} \subset W$ the (possibly empty) Zariski open subset corresponding to morphisms whose image in S is an irreducible nodal curve with $p(d)-g$ nodes.

Lemma 2.2. *Let $r, d, d_1, \ldots, d_r \geqslant 1$, with $d_1 + \ldots + d_r = d$.*

Denote $\overline{M}^*_{g_1,\ldots,g_r}(S, (d_1, \ldots, d_r)A) \hookrightarrow \overline{M}^*_g(S, dA)$ the union of those components whose generic points are simple stable maps, with domain of definition consisting of r curves of genera $g_1, \ldots, g_r$, and degree d_j on the jth component. Let $W \hookrightarrow \overline{M}^*_{g_1,\ldots,g_r}(S, (d_1, \ldots, d_r)A)$ be an irreducible component.

Then dim $W \leqslant g_1 + \ldots + g_r$, *with equality if the domain of definition of the generic point in* W *consists of* r *smooth curves, and its image is a nodal curve in* S.

Proof. One applies inductively, r times, the previous lemma. □

Proposition 2.1 is a consequence of the following more precise statement, which describes the geometry of curves on general $K3$ surfaces.

Theorem 2.3. *Assume that* $(\star)$ *and* $(\star\star)$ *holds. Then:*

(i) *The space* $\overline{M}^*_{p(d)-\delta}(S, dA)$ *is non-empty. It contains a component* W *of dimension* $p(d) - \delta$, *whose generic point represents a morphisms* $\nu : C \to S$ *with the property that* $\nu(C) \hookrightarrow S$ *is an irreducible nodal curve.*
(ii) *The projection* $\pi_{p(d)-\delta} : W \to \overline{M}_{p(d)-\delta}$ *is generically finite.*

Proof. We prove by induction with respect to δ. For $\delta = 0$, the general point of $\overline{M}^*_{p(d)}(S, dA)$ corresponds to the general hyperplane section of $A^{\otimes d}$. In particular, $\overline{M}^*_{p(d)}(S, dA)$ is irreducible, of dimension $p(d)$. The infinitesimal deformations of the general morphism $\nu : C \to S$ relative to $\overline{M}_g$ are contained in $H^0(C, T_S|_C)$. Since, by proposition 1.2, $T_S|_C \to C$ is a stable bundle of degree zero, it has no non-trivial sections. It follows that the projection $\pi_{p(d)}$ is generically finite.

Let us assume that both statements hold for $\delta < \delta_{\max}$. We are going to prove that they hold for $\delta + 1$. Write $g := p(d) - \delta$ for shorthand.

The first part of the induction hypothesis says that there is a component $W \hookrightarrow \overline{M}^*_g(S, dA)$ of dimension g; its generic point corresponds to a stable map whose image is a nodal curve on S. The second hypothesis is that the projection π_g is quasi-finite, and therefore $\dim \pi_g(W) = g$. Diaz's theorem implies that the intersection $\pi_g(W) \cap \partial \overline{M}_g \neq \emptyset$. It follows that $W \setminus W^{**} \neq \emptyset$, and its components are divisors on W.

At this point, the assumption $(\star\star)$ guarantees the existence of a component $V_{\mathrm{spl}} \hookrightarrow W \setminus W^{**}$ whose generic point represents a simple (that is non-multiply covered) stable map. The previous discussion implies that

$$\pi_g(V_{\mathrm{spl}}) = \begin{cases} \pi_g(W) \cap \mathrm{D}_{g-1,2} & \text{or} \\ \pi_g(W) \cap \Delta_{g',g''} & \text{with } g' + g'' = g. \end{cases}$$

Case 1. If $\pi_g(V_{\mathrm{spl}}) = \pi_g(W) \cap \mathrm{D}_{g-1,2}$, then $V_{\mathrm{spl}} \hookrightarrow \mathrm{Image}(\overline{M}_{g-1,2}(S, dA) \to \overline{M}^*_g(S, dA))$; let $\tilde{V}_{\mathrm{spl}} \hookrightarrow \overline{M}_{g-1,2}(S, dA)$ be the component which projects onto V_{spl}, and let $[(C, u)] \in \tilde{V}_{\mathrm{spl}}$ be the generic point: we wish to prove that C is irreducible and that $\hat{C} := u(C) \hookrightarrow S$ is a nodal curve. From the recursive construction of W, we deduce that we may view

$$\tilde{V}_{\mathrm{spl}} \hookrightarrow \overline{M}_{p(d)-(\delta+1),\{x_{1j},x_{2j}\}_{j=1,\ldots,\delta+1}}(S, dA),$$

that is $[(C, u)]$ is obtained by identifying $\delta + 1$ pairs of points. Let $\{C_j\}_{j=1,\ldots,r}$ be the irreducible components of C, and let $\nu : \sqcup_{j=1}^r C_j \to C$ be the partial normalization (glueing) map. We consider the exact sequences

$$0 \to \mathcal{O}_C \to \nu_* \mathcal{O}_{\sqcup_{j=1}^r C_j} \to \mathcal{S}_C \to 0 \quad \text{and} \quad 0 \to \mathcal{O}_{\hat{C}} \to (u \circ \nu)_* \mathcal{O}_{\sqcup_{j=1}^r C_j} \to \mathcal{S} \to 0.$$

$\mathcal{S}_C$ is a skyscraper sheaf, supported at the double points where the irreducible components of C are glued together; at each such point the length of $\mathcal{S}_C$ equals one.

Since u is a simple map, $\mathcal{S}$ is a skyscraper sheaf too. It is supported at the various intersections of the $u(C_j)$'s; among these points appear the images of the $\delta + 1$ pairs of marked points on C, and also the images of the points where the C_j's are glued together. It follows that $\text{length}(\mathcal{S}) \geqslant (\delta + 1) + \text{length}(\mathcal{S}_C)$.

Taking the cohomology of these sequences, we deduce:

$$g - 1 = h^1(\mathcal{O}_C) = \sum_{j=1}^{r} h^1(\mathcal{O}_{C_j}) + \text{length}(\mathcal{S}_C) - (r-1)$$

$$= \sum_{j=1}^{r} h^1(\mathcal{O}_{C_j}) + b_1(\Gamma_C), \quad \text{and}$$

$$\sum_{j=1}^{r} h^1(\mathcal{O}_{C_j}) = h^1(\mathcal{O}_{\hat{C}}) - \text{length}(\mathcal{S}) + (r-1) = p(d) - \text{length}(\mathcal{S}) + (r-1).$$

Here Γ_C denotes the graph of C, and $b_1(\Gamma_C)$ is its first Betti number. We obtain that $\text{length}(\mathcal{S}) = (\delta + 1) + \text{length}(\mathcal{S}_C)$. The previous remark implies that the equality holds precisely when the curves $u(C_j)$'s meet only at the (corresponding images) of the glueing points between the C_j's, and at the images of the marked points.

We claim now that $b_1(\Gamma_C) = 0$, that is there are no loops in Γ_C. Otherwise there was a component of $\overline{M}^*_{h^1(\mathcal{O}_{C_1}),\dots,h^1(\mathcal{O}_{C_r})}(S, (d_j A)_j)$ of dimension $g - 1 > \sum_{j=1}^{r} h^1(\mathcal{O}_{C_j})$, in contradiction to lemma 2.2. This argument also proves that none of $u|_{C_j}$ is constant, that is $d_j \geqslant 1$ for all j. If u is constant on some component, then we may omit it from the sum $\sum_{j=1}^{r} h^1(\mathcal{O}_{C_j})$.

So, Γ_C is a tree. The exact sequence $0 \to \mathcal{O}_{\hat{C}} \to u_*\mathcal{O}_C \to \mathcal{S}' \to 0$ implies that

$$p(d) - (\delta + 1) = h^1(\mathcal{O}_C) = h^1(\mathcal{O}_{\hat{C}}) - \text{length}(\mathcal{S}') \quad \Longrightarrow \quad \text{length}(\mathcal{S}') = \delta + 1.$$

We have seen that the intersection pattern of the $u(C_j)$'s in S coincides with the intersection pattern of the C_j's (which is encoded in Γ_C), and the intersections given by the identifications of the $\delta+1$ pairs of double points. Since Γ_C is a tree, the r components $C_1, \dots, C_r$ are glued together at $r-1$ points; at these points there is no contribution to $\text{length}(\mathcal{S}')$. Assuming that $r \geqslant 2$, we deduce that

$$\delta + 1 = \text{length}(\mathcal{S}') \geqslant \#\{\text{intersection points between the } u(C_j)\text{'s}\} - (r-1)$$

$$= \sum_{1 \leqslant i < j \leqslant r} d_i d_j A^2 - (r-1) = \frac{d^2 - \left(\sum_{j=1}^{r} d_j^2\right)}{2} A^2 - (r-1)$$

$$\geqslant \frac{d^2 - [(d-r)^2 + (r-1)]}{2} A^2 - (r-1) = \frac{r(2d - r - 1) + 1}{2} A^2 - (r-1)$$

$$\geqslant r\left[\frac{d-1}{2} A^2 - 1\right] + \frac{1}{2} A^2 + 1 \geqslant \frac{2d-1}{2} A^2 - 1.$$

This contradicts the assumption ($\star$) on δ. It follows that $r = 1$, that is C is irreducible, of genus $h^1(\mathcal{O}_C) = g - 1$. Since the map $[(C, u)]$ is the generic point of a $(g-1)$-dimensional family, C must be smooth and $u(C) \hookrightarrow S$ a nodal curve. The natural projection $\pi_{g-1} : \tilde{V}_{\text{spl}} \to \overline{M}_{g-1}$ is generically finite by proposition 1.2.

Case 2. If $\pi_g(V_{\text{spl}}) = \pi_g(W) \cap \Delta_{g',g''}$, then

$$V_{\text{spl}} \hookrightarrow \text{Image}(\overline{M}_{g',1}(S, d'A) \times \overline{M}_{g'',1}(S, d''A) \to \overline{M}^*_g(S, dA))$$

for some d', d'', with $d' + d'' = d$. Let again $\tilde{V}_{\text{spl}} \hookrightarrow \overline{M}_{g',1}(S, d'A) \times \overline{M}_{g'',1}(S, d''A)$ be the component which projects onto V_{spl}, and $[(C, u)]$ be the generic point of $\tilde{V}_{\text{spl}}$. From the recursive construction of W, we deduce that the stabilized curve $[C_{\text{stab}}]$ lies in the fibred product

$$\begin{array}{ccc}
\overline{M}_{g',\{\bullet\}} \times \overline{M}_{g'',\{\bullet\}} & \longleftarrow & \bigcup\limits_{\delta'+\delta''=\delta} \overline{M}_{g'-\delta',\{\bullet\}\cup\{x'_{1j},x'_{2j}\}_{j=1,\dots,\delta'}} \times \overline{M}_{g''-\delta'',\{\bullet\}\cup\{x''_{1j},x''_{2j}\}_{j=1,\dots,\delta''}} \\
\downarrow & & \downarrow \\
\overline{M}_g & \longleftarrow & \overline{M}_{g-2\delta,\{x_{1j},x_{2j}\}_{j=1,\dots,\delta}}
\end{array}$$

It follows that $C_{\text{stab}} = C'_{\text{stab}} \cup C''_{\text{stab}}$, and these components carry δ' and δ'' pairs of marked points respectively. The curve C is obtained by attaching (bubble) trees of $\mathbb{P}^1$'s at smooth points of C'_{stab} and C''_{stab}, or at their glueing point. If it happens that such a tree is inserted at a marked point, this one 'jumps' on the tree.

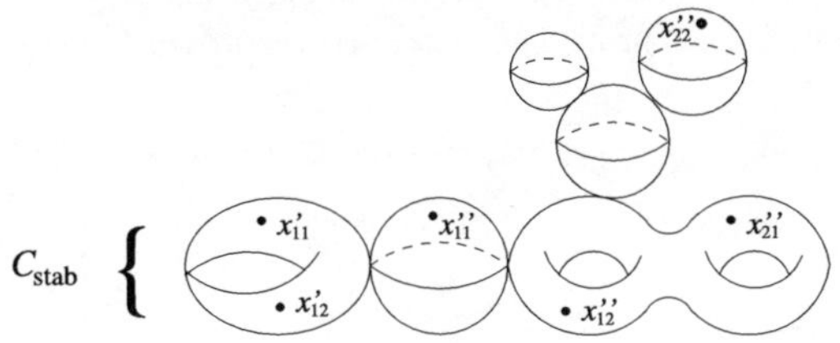

In any case, $C = C' \cup C''$ and the restrictions $[(C', u' := u|_{C'})]$ and $[(C'', u'' := u|_{C''})]$ are (simple) stable maps to S; let d' and d'' be their degrees respectively. The skyscraper sheaf $\mathcal{S}'$ defined by the exact sequence

$$0 \to \mathcal{O}_{u'm(C')} \to u'_*\mathcal{O}_{C'} \to \mathcal{S}' \to 0$$

is supported at least at the images of the δ' pairs of marked points. Observe that whenever different pairs of marked points are mapped to the same point in S, the length of $\mathcal{S}'$ increases. It follows that $g' = h^1(\mathcal{O}_{C'}) = p(d') - \text{length}(\mathcal{S}') \leqslant p(d') - \delta'$. The same holds for g''. All together we obtain:

$$p(d) - \delta = g = g' + g'' \leqslant (p(d') - \delta') + (p(d'') - \delta'') = p(d') + p(d'') - \delta \Rightarrow d'd''A^2 \leqslant 1.$$

This is possible only if either d' or d'' vanishes. Assume that $d'' = 0$, that is $d' = d$.

We have found a component of $\overline{M}_{g'}(S, dA)$, with $g' \leqslant (p(d) - \delta) - 1$, whose dimension equals $p(d) - (\delta + 1)$ and whose generic point represents a simple map. Lemma 2.2 implies that necessarily $g' = p(d) - (\delta + 1)$.

We are now in the previous case, and we conclude that the domain of definition of $[(C', u|_{C'})]$ is smooth and irreducible, and that $u(C') \hookrightarrow S$ is a nodal curve. The same proposition 1.2 implies that the projection $\pi_{g-1} : \tilde{V}_{\mathrm{spl}} \to \overline{M}_{g-1}$ is generically finite. □

Remark 2.4.

(i) For $d = 1$, the issue concerning the irreducibility of the domain of definition of the stable maps appearing in the proof is trivially satisfied. It would be very helpful to have a criterion to decide when is the limit of points in $|dA|_\delta$ a reduced and irreducible curve, or whether at least this situation appears generically.

(ii) A very simple example of a *theoretically possible* scenario, which we exclude assuming $(\star\star)$, is the following: suppose that for some genus g and (even) degree d, a component $W \hookrightarrow \overline{M}^{*}_{g}(S, dA)$ enjoys the properties in theorem 2.3; it is *a priori* possible that every boundary component of $W \setminus W^{**}$ (which is a divisor in W) consists 2-sheeted coverings.

The Riemann-Hurwitz formula implies that for a 2-sheeted covering $C \to D$ of smooth curves, ramified at ϱ points, holds: $g = g(C) = 2g(D) - 1 + \frac{\varrho}{2}$. On the other hand, the choice of the ramification points on D gives a ϱ-dimensional component in $\overline{M}_g(D, 2[D])$. So, its dimension grows faster than g.

Back to our situation, if $C \stackrel{u}{\to} S$ is a multiply-covered curve, it factorizes through $C \to D \stackrel{\bar{u}}{\to} S$. The phenomenon described above prevents to get informations about the genus of D (the domain of the underlying simple stable map) only in terms of the dimension of W, to which the multiply-covered map u belongs.

Acknowledgements

I wish to thank the organizers of the Workshop on Teichmüller Theory and Moduli Problems for their invitation to participate at the workshop, and for their hospitality. I thank Prof. Chris Peters for his comments on earlier drafts of the paper. This research has been partially supported by the Schweizerischer Nationalfonds.

References

[1] Xi Chen, Rational curves on $K3$ surfaces, *J. Algebr. Geom.*, **8** (1999) 245–278.

[2] J.-P. Demailly, Th. Peternell and M. Schneider, Pseudo-effective line bundles on compact Kähler manifolds, *Internat. J. Math.*, **12** (2001) 689–741.

[3] S. Diaz, A bound on the dimension of complete subvarieties of M_g, *Duke Math. J.*, **51** (1984) 405–408.

[4] H. Flenner, Restrictions of semistable bundles on projective varieties, *Comment. Math. Helv.*, **59** (1984) 635–650.

[5] W. Fulton and R. Pandharipande, Notes on stable maps and quantum cohomology, in J. Kollár et al. (Eds.), Algebraic Geometry Santa Cruz 1995, AMS (1997) Part 2, 45–96.

[6] M. Green and Ph. Griffiths, Two applications of algebraic geometry to entire holomorphic mappings, in The Chern Symposium, Springer Verlag New York Heidelberg Berlin (1979), 41–74.

[7] D. Huybrechts and M. Lehn, The geometry of moduli spaces of sheaves, Aspects of mathematics, Vieweg & Sohn (1997).

[8] V. B. Mehta and A. Ramanathan, Semistable sheaves on projective varieties and the restriction to curves, *Math. Ann.*, **258** (1982) 213–224.

[9] Y. Miyaoka, The Chern classes and Kodaira dimension of a minimal variety, *Adv. Stud. Pure Math.*, **10** (1985) 449–476.

[10] S. Mori and S. Mukai, The uniruledness of the moduli spaces of curves of genus 11, in Lecture Notes in Math., 1016, Springer (1983), 334–353.

[11] A. Tannenbaum, Families of curves with nodes on $K3$ surfaces, *Math. Ann.*, **260** (1982) 239–253.

[12] O. Zariski, Dimension theoretic characterisation of maximal irreducible algebraic systems of plane nodal curves of a given order n and with a number d of nodes, *Amer. J. Math.*, **104** (1982) 209–226.

Teichmüller Theory and Moduli Problem No. 10, 2010, pp. 387–394.

The Moduli Stack of Vector Bundles on a Curve

Norbert Hoffmann

Mathematisches Institut der Freien Universitaet, Arnimallee 3, D-14195 Berlin, Germany
e-mail: norbert.hoffmann@fu-berlin.de

Abstract. This expository text tries to explain briefly and not too technically the notions of stack and algebraic stack, emphasizing as an example the moduli stack of vector bundles on an algebraic curve.

Introduction

The aim of this text is to explain what an algebraic stack is, in particular what the moduli stack of vector bundles on an algebraic curve is. (At the Allahabad conference, a shorter form of this introduction to stacks was followed by a brief report on the paper [6] about the birational type of moduli stacks for vector bundles with some extra structure.)

Algebraic stacks were introduced by Deligne-Mumford [3] and by M. Artin [1]. In order to motivate them, we recall the notion of fine moduli scheme and the usual problem with its existence in the presence of automorphisms in section 1. Then we define stacks as some sort of sheaves of groupoids in section 2, and finally we discuss the algebraicity notions of Deligne-Mumford and of M. Artin in section 3.

Warning: In the present text, several technicalities are suppressed or oversimplified (in order not to obscure the basic ideas). For example, open coverings and gluing are used freely without discussing the topology, where the correct thing would be to specify an appropriate Grothendieck topology and replace gluing by descent. Moreover, we use a simplified notion of prestack in order to avoid the technically better, but less suggestive notion of fibered category; cf. [5, Exposé VI]. Also finiteness conditions are not systematically taken into account; for example, we omit the condition that an algebraic stack has to be quasi-separated.

Full technical details about algebraic stacks can be found in the textbook [7] of Laumon and Moret-Bailly, or in a book in preparation by Behrend, Conrad, Edidin, Fulton, Fantechi, Göttsche and Kresch [2]. Expository texts have also been written by Vistoli [10], Sorger [9] and Gomez [4].

1. Fine moduli schemes and why they don't exist

Let $k = \bar{k}$ be an algebraically closed field, and let C be a connected smooth projective algebraic curve of genus $g \geq 2$ over k. We consider (algebraic) vector bundles E of fixed rank r on C. Their basic discrete invariant is the degree $\deg(E) \in \mathbb{Z}$, defined as the degree of the line bundle $\det(E) := \Lambda^r E$.

(Recall that the degree of a line bundle L on C is by definition the number of zeros minus the number of poles of any nonzero rational section s of L, both counted with multiplicities; this difference does not depend on the choice of s.)

"Classifying" vector bundles E on C of given rank r and degree d means understanding the set of isomorphism classes

$$\mathrm{Bun}_{r,d} := \{E \text{ vector bdl. on } C \text{ of rank } r, \text{ deg. } d\}/\cong .$$

This set is roughly the set of points of some algebraic variety, the *moduli space* of such vector bundles E. More precisely, Seshadri [8] constructed in 1967 a connected smooth quasiprojective variety

$$\mathfrak{Bun}_{r,d}^{\mathrm{stab}}$$

of dimension $r^2(g-1)+1$ over k whose set of k-valued points is ... not quite $\mathrm{Bun}_{r,d}$, but at least the subset $\mathrm{Bun}_{r,d}^{\mathrm{stab}} \subseteq \mathrm{Bun}_{r,d}$ consisting of the isomorphism classes of stable vector bundles E.

(Recall that a nonzero vector bundle E on C is called *stable* if

$$\deg(E')/\mathrm{rank}(E') < \deg(E)/\mathrm{rank}(E)$$

holds for every proper subbundle $0 \neq E' \subsetneq E$.)

How to give the set $\mathrm{Bun}_{r,d}$ (or $\mathrm{Bun}_{r,d}^{\mathrm{stab}}$) the structure of an algebraic variety over k? One way to do this is to decide, for every variety S over k, which maps of the underlying point sets $S(k) \to \mathrm{Bun}_{r,d}$ are actually morphisms of k-varieties. The natural answer is this: such a map should be a morphism of k-varieties if and only if it comes from a vector bundle $\mathcal{E}$ on $C \times_k S$, in the sense that it sends every point $s \in S(k)$ to the isomorphism class of the restriction $\mathcal{E}_s := \mathcal{E}|_{C\times\{s\}}$. This line of thought motivates the notion of fine moduli scheme, which we recall now.

Definition 1.1. *Let S be a k-scheme. We denote by*

$$\mathrm{Bun}_{r,d}(S) := \{\mathcal{E} \text{ vector bdl. on } C \times_k S \text{ of rank } r, \text{ deg. } d\}/\cong$$

the set of isomorphism classes of rank r vector bundles $\mathcal{E}$ on $C \times_k S$ with constant degree d; here the degree $\deg(\mathcal{E})$ *is by definition the locally constant function on S which assigns to each point $s \in S$ the integer* $\deg(\mathcal{E}_s)$.

Every morphism of k-schemes $f : T \to S$ induces a map

$$f^* : \mathrm{Bun}_{r,d}(S) \longrightarrow \mathrm{Bun}_{r,d}(T), \quad [\mathcal{E}] \mapsto [f^*\mathcal{E}]$$

by pullback; thus we get a contravariant functor

$$\mathrm{Bun}_{r,d}(\_) : \mathsf{Schemes}/k \longrightarrow \mathsf{Sets}$$

from the category of schemes over k to the category of sets.

Definition 1.2. *A scheme M over k is a* fine moduli scheme *for vector bundles E of rank r and degree d on C if M represents the functor* $\mathrm{Bun}_{r,d}(\_)$.

More explicitly, M is such a fine moduli scheme of vector bundles if

$$\{\varphi : S \to M \text{ } k\text{-morphism}\} = \{\mathcal{E} \text{ vector bdl. on } C \times_k S \text{ of rank } r, \text{ deg. } d\}/\cong \quad (1)$$

where the equality sign means the existence of a bijection which is functorial in S.

By Yoneda's lemma, the fine moduli scheme M is unique up to unique isomorphisms if it exists. Unfortunately, it doesn't exist, i.e. the functor $\mathrm{Bun}_{r,d}(\_)$ is not representable. To see why, we compare how both sides of (1) behave under gluing. So assume given an open covering $S = \bigcup_i U_i$ of the k-scheme S.

- For any k-scheme M, a k-morphism $\varphi : S \to M$ is given by
  - a k-morphism $\varphi_i : U_i \to M$ for each i

  satisfying the condition
  - $\varphi_i = \varphi_j$ on the intersection $U_i \cap U_j$ for all i, j.
- A vector bundle $\mathcal{E}$ on $C \times_k S$ is given by
  - a vector bundle $\mathcal{E}_i$ on $C \times_k U_i$ for each i, and
  - an isomorphism $\alpha_{ij} : \mathcal{E}_i \xrightarrow{\sim} \mathcal{E}_j$ over $C \times_k (U_i \cap U_j)$ for each i, j,

  satisfying the condition
  - $\alpha_{il} = \alpha_{jl} \circ \alpha_{ij}$ over $C \times_k (U_i \cap U_j \cap U_l)$ for all i, j, l.

We see that both sides of (1) behave fundamentally different under gluing. To make this more precise, we recall some terminology.

Definition 1.3.

i) *A* presheaf *is a contravariant functor F* : Schemes$/k \to$ Sets.
ii) *A presheaf F is a* sheaf *if the following condition holds for every open covering of k-schemes $S = \bigcup_i U_i$:*

   Given a tuple $(\varphi_i)_i$ of elements $\varphi_i \in F(U_i)$ such that φ_i and φ_j have the same image in $F(U_i \cap U_j)$ for all i, j, there is a unique element $\varphi \in F(S)$ whose image in $F(U_i)$ is φ_i for all i.

Now the above observations about gluing can be summarized as follows. The presheaf represented by any scheme M over k is a sheaf, but it is easy to check that the presheaf $\mathrm{Bun}_{r,d}(\_)$ is not a sheaf. Hence the latter is not representable, i. e. there is no fine moduli scheme.

This is a typical problem. Functors of isomorphism classes like $\mathrm{Bun}_{r,d}(\_)$ are usually not representable, the standard exception being isomorphism classes of objects without automorphisms. (In that case, the above isomorphisms α_{ij} are unique if they exist, and the cocycle relation $\alpha_{il} = \alpha_{jl} \circ \alpha_{ij}$ is automatically satisfied, so the presheaf of isomorphism classes is actually a sheaf, and in fact in many cases representable.)

The classical solution for this problem is the following. Instead of representing the functor $\mathrm{Bun}_{r,d}(\_)$, we approximate it by a representable functor, as closely as possible in a specific sense. The representing scheme is then called a *coarse moduli scheme*. For example, Seshadri's variety $\mathfrak{Bun}_{r,d}^{\text{stab}}$ is a coarse moduli scheme of stable vector bundles E on C with rank r and degree d.

A more radical solution for this problem is given by the notion of stack. It is motivated by the above observation that objects (e. g. vector bundles) don't glue like maps to a set. Even if we are just interested in isomorphism classes, we need to keep track of actual isomorphisms (in particular of automorphisms) in order to understand gluing. This leads to the idea that a moduli space should not be an underlying *set* (of isomorphism classes) endowed with some geometric structure, it should be an underlying *groupoid* (of objects and their isomorphisms) endowed with some geometric structure. That's roughly what a stack is.

2. Stacks

Definition 2.1. *A* groupoid *is a category in which every morphism is invertible.*

Example 2.2. The vector bundles E on C of rank r and degree d, together with the isomorphisms of vector bundles as morphisms, form a groupoid.

Example 2.3. If a group G acts on a set X, then one has a quotient groupoid X/G: its objects are the elements $x \in X$, its morphisms from x to x' are the elements $g \in G$ satisfying $g \cdot x = x'$, and its composition law is the multiplication in G.

Note that the isomorphism class of the object x in X/G is precisely the orbit of the element x in X, whereas the automorphism group of the object x in X/G is precisely the stabilizer of the element x in X.

When replacing sets by groupoids, presheaves (in the sense of definition 1.3) get replaced by prestacks, and sheaves get replaced by stacks; more precisely:

Definition 2.4. *A* prestack $\mathcal{M}$ *over* k *consists of*

- *a groupoid $\mathcal{M}(S)$ for each k-scheme S,*
- *a functor $f^* : \mathcal{M}(S) \to \mathcal{M}(T)$ for each k-morphism $f : T \to S$, and*
- *an isomorphism of functors $c_{f,g} : g^* f^* \xrightarrow{\sim} (f \circ g)^*$ for each composable pair of k-morphism $g : U \to T$ and $f : T \to S$,*

such that the isomorphisms $c_{f,g}$ satisfy the following two compatibility conditions:

- *$c_{f,\mathrm{id}_T} = \mathrm{id}_{f^*}$ and $c_{\mathrm{id}_S,f} = \mathrm{id}_{f^*}$ for all k-morphisms $f : T \to S$.*
- *The diagram of isomorphisms of functors*

$$\begin{array}{ccc} h^* g^* f^* & \xrightarrow{c_{g,h} f^*} & (g \circ h)^* f^* \\ {\scriptstyle h^* c_{f,g}} \downarrow & & \downarrow {\scriptstyle c_{f, g\circ h}} \\ h^*(f \circ g)^* & \xrightarrow[c_{f\circ g,h}]{} & (f \circ g \circ h)^* \end{array}$$

commutes for all triples of k-morphisms $h : V \to U$, $g : U \to T$ and $f : T \to S$.

Example 2.5. The prestack $\mathcal{B}un_{r,d}$ of vector bundles E on C with rank r and degree d consists of the following data.

- Objects of the groupoid $\mathcal{B}un_{r,d}(S)$ are vector bundles $\mathcal{E}$ on $C \times_k S$ with rank r and (constant) degree d.
- Morphisms of the groupoid $\mathcal{B}un_{r,d}(S)$ are isomorphisms $\alpha : \mathcal{E} \xrightarrow{\sim} \mathcal{E}'$ of vector bundles on $C \times_k S$.
- The functor $f^* : \mathcal{B}un_{r,d}(T) \to \mathcal{B}un_{r,d}(S)$ is the usual pullback of vector bundles along the k-morphism $f : T \to S$.
- $c_{f,g}(\mathcal{E})$ is the usual canonical isomorphism between $g^* f^* \mathcal{E}$ and $(f \circ g)^* \mathcal{E}$ whenever $g : U \to T$ and $f : T \to S$ is a composable pair of k-morphisms.

If $\mathcal{M}$ is a prestack and $f : T \hookrightarrow S$ is an embedding of k-schemes, then we will write $\mathcal{E}|_T$ and $\alpha|_T$ for the images in $\mathcal{M}(T)$ of objects $\mathcal{E}$ and morphisms α in $\mathcal{M}(S)$. If $g : U \hookrightarrow T$ is another embedding, then the isomorphism $c_{f,g}$ allows us to identify $(\mathcal{E}|_T)|_U$ with $\mathcal{E}|_U$ and $(\alpha|_T)|_U$ with $\alpha|_U$.

Just like a sheaf is a presheaf whose sections can be glued, a stack is a prestack whose objects and morphisms can be glued; more precisely:

Definition 2.6. *A prestack $\mathcal{M}$ over k is a* stack *if the following two conditions hold for every open covering of k-schemes $S = \bigcup_i U_i$:*

- *Given two objects $\mathcal{E}, \mathcal{E}'$ of $\mathcal{M}(S)$ and for each i a morphism $\alpha_i : \mathcal{E}|_{U_i} \to \mathcal{E}'|_{U_i}$ in $\mathcal{M}(U_i)$ such that*

$$\alpha_i|_{U_i \cap U_j} = \alpha_j|_{U_i \cap U_j} \text{ in } \mathcal{M}(U_i \cap U_j) \text{ for all } i, j,$$

there is a unique morphism $\alpha : \mathcal{E} \to \mathcal{E}'$ in $\mathcal{M}(S)$ such that $\alpha_i = \alpha|_{U_i}$ for all i.
- *Given for each i an object $\mathcal{E}_i$ in $\mathcal{M}(U_i)$ and for each i, j an isomorphism $\alpha_{ij} : \mathcal{E}_i|_{U_i \cap U_j} \to \mathcal{E}_j|_{U_i \cap U_j}$ in $\mathcal{M}(U_i \cap U_j)$ such that*

$$\alpha_{il} = \alpha_{jl} \circ \alpha_{ij} \text{ in } \mathcal{M}(U_i \cap U_j \cap U_l) \text{ for all } i, j, l,$$

there is an object $\mathcal{E}$ in $\mathcal{M}(S)$ and for each i an isomorphism $\beta_i : \mathcal{E}|_{U_i} \to \mathcal{E}_i$ in $\mathcal{M}(U_i)$ such that $\beta_j = \alpha_{ij} \circ \beta_i$ in $\mathcal{M}(U_i \cap U_j)$ for all i, j.

Example 2.7. The gluing behavior of vector bundles described in section 1 means that the prestack $\mathcal{B}un_{r,d}$ of example 2.5 is actually a stack.

Both prestacks and stacks over k form 2-categories: A 1-morphism $\Phi : \mathcal{M} \to \mathcal{M}'$ is given by a functor $\Phi(S) : \mathcal{M}(S) \to \mathcal{M}'(S)$ for each k-scheme S, together with an isomorphism of functors $f^* \circ \Phi(S) \cong \Phi(T) \circ f^*$ for each k-morphism $f : T \to S$ satisfying appropriate compatibility conditions; a 2-morphism $\tau : \Phi_1 \Rightarrow \Phi_2$ is given by a natural transformation $\tau(S) : \Phi_1(S) \Rightarrow \Phi_2(S)$ for each k-scheme S satisfying appropriate compatibility conditions.

Recall that for every presheaf, one has an associated sheaf (obtained by the sheafification process). For every prestack, one similarly has an associated stack, obtained by an analogous process called stackification.

3. Algebraic stacks

A sheaf in the sense of definition 1.3 is not yet a very geometric object; it only yields a geometric object (namely a fine moduli scheme) if it is representable. Similarly, a stack is not yet a very geometric object; it only is so if it is "algebraic", i. e. if it satisfies some further condition which we explain next. There are two variants of this condition, one due to Deligne-Mumford [3] and one due to M. Artin [1].

Definition 3.1. *A* groupoid scheme *over k is a groupoid object in the category* $\mathsf{Schemes}/k$ *of schemes over k.*

More explicitly, a groupoid scheme over k consists of two k-schemes X, R and five k-morphisms

$$e : X \to R, \quad s : R \to X, \quad t : R \to X, \quad \circ : R \times_{s,X,t} R \to R \quad \text{and} \quad i : R \to R,$$

such that for each k-scheme S, the sets $X(S)$ and $R(S)$ of k-morphisms $S \to X$ and $S \to R$ together with the induced maps

$$e_* : X(S) \longrightarrow R(S), \qquad s_* : R(S) \longrightarrow X(S), \qquad t_* : R(S) \longrightarrow X(S),$$
$$\circ_* : R(S) \times_{s_*, X(S), t_*} R(S) \longrightarrow R(S) \quad \text{and} \quad i_* : R(S) \longrightarrow R(S)$$

form a (small) groupoid, in which $X(S)$ is the set of objects, $R(S)$ is the set of morphisms, e_* sends each object to its identity automorphism, s_* sends each morphism to its source, t_* to its target, $\circ_*$ sends each composable pair of morphisms to their composition, and i_* sends each morphism to its inverse.

Note that a groupoid scheme $(X, R, e, s, t, \circ, i)$ over k induces a prestack $\mathcal{M}$ over k, by sending each k-scheme S to the groupoid $\mathcal{M}(S)$ with objects $X(S)$ and morphisms $R(S)$ that has just been described.

Example 3.2. Let a group scheme G over k act on a scheme X over k. In analogy to example 2.3, one has a quotient groupoid scheme $(X, R := G \times_k X, e, s, t, \circ, i)$ over k where $s : G \times_k X \to X$ is the second projection, $t : G \times_k X \to X$ is the group action, and $\circ : G \times_k G \times_k X \to G \times_k X$ is given by the group multiplication.

Definition 3.3. *Let a group scheme G over k act on a scheme X over k. The* stack quotient *X/G is the stackification of the prestack $\mathcal{M}$ induced by the groupoid scheme $(X, G \times_k X, \dots)$ described in example 3.2.*

Remark 3.4. The stack quotient X/G contains information not only about the orbit sets, but also about the stabilizer groups, cf. example 2.3.

Example 3.5. Let $X = \mathrm{Spec}(k)$ be just one point, and let the group scheme G over k act trivially on X. The resulting stack quotient $\mathrm{B}G := X/G$ is called the *classifying stack* associated to G. It is easy to check that $\mathrm{B}G(S)$ is precisely the groupoid of locally trivial principal G-bundles over S for every k-scheme S.

Definition 3.6. *A stack $\mathcal{M}$ over k is* algebraic *in the sense of Artin (resp. in the sense of Deligne-Mumford) if $\mathcal{M}$ is (1-isomorphic to) the stackification of the prestack induced by a groupoid scheme $(X, R, e, s, t, \circ, i)$ over k in which the morphisms $s, t : R \to X$ are both smooth (resp. étale).*

A stack $\mathcal{M}$ that is algebraic in the sense of Artin (resp. of Deligne-Mumford) is usually called an Artin stack (resp. Deligne-Mumford stack).

Example 3.7. Let a group scheme G over k act on a scheme X over k.

i) Suppose that G is smooth over k. Then X/G is an Artin stack. In particular, the classifying stack $\mathrm{B}G$ is an Artin stack in this case.
ii) Suppose that G is étale over k. Then X/G is a Deligne-Mumford stack. In particular, the classifying stack $\mathrm{B}G$ is a Deligne-Mumford stack in this case.

Remark 3.8. One can guess how to apply geometric notions to a quotient stack X/G: the natural thing to do is to apply the same notions to the scheme X in a G-equivariant way. This actually works, no matter how bad the action of G on X is (maybe huge stabilizers, non-closed orbits, etc.). Thus the geometry of the quotient stack X/G is really just the G-equivariant geometry of X, even if this would not at all be the case for anykind of orbit space X/G because the action is bad.

Remark 3.9. The above way of applying geometric notions to quotient stacks X/G can often be generalized to stacks given by a groupoid scheme $(X, R, \dots)$ in a rather formal way. In some sense, this is one reason why *algebraic* stacks are quite geometric objects, even if general stacks aren't.

Proposition 3.10. *The stack $\mathcal{B}un_{r,d}$ is algebraic in the sense of Artin.*

Sketch of the proof. One can define what an open covering of a stack is. A stack is algebraic if it admits an open covering by algebraic stacks. Choose a very ample line bundle $\mathcal{O}(1)$ on C, and let

$$P(n) := d + r(1 - g + n \deg \mathcal{O}(1))$$

be the common Hilbert polynomial of all vector bundles E on C with rank r and degree d. An appropriate open subscheme in one of Grothendieck's Quot-schemes

$$X_n \subseteq \mathrm{Quot}_P(\mathcal{O}(-n)^{P(n)})$$

is a fine moduli scheme of pairs (E, B) consisting of a vector bundles E on C with

- $\mathrm{rank}(E) = r$ and $\deg(E) = d$,
- $E(n) := E \otimes \mathcal{O}(n)$ is generated by its global section,
- the Zariski sheaf cohomology $\mathrm{H}^1(C, E(n))$ vanishes,

and a basis B of $\mathrm{H}^0(E(n))$. $G_n := \mathrm{GL}_{P(n)}$ acts on X_n by changing this basis B. Because the above conditions on E are open conditions, X_n/G_n is an open substack of $\mathcal{B}un_{r,d}$. Since every vector bundle E on C with rank r and degree d satisfies them for $n \gg 0$, we obtain an open covering

$$\mathcal{B}un_{r,d} = \bigcup_n X_n/G_n$$

which shows that $\mathcal{B}un_{r,d}$ is an Artin stack because the X_n/G_n are. □

Remark 3.11. $\mathcal{B}un_{r,d}$ is not a Deligne-Mumford stack, because vector bundles have automorphism groups of positive dimension.

Example 3.12. The moduli stack $\mathcal{M}_g$ of smooth projective curves over k with fixed genus $g \geq 2$ is given by the following groupoid for each k-scheme S:

- The objects of $\mathcal{M}_g(S)$ are the smooth projective morphisms $\pi : \mathcal{C} \to S$ all of whose geometric fibers are connected curves of genus g.
- The morphisms in $\mathcal{M}_g(S)$ from $\pi : \mathcal{C} \to S$ to $\pi' : \mathcal{C}' \to S$ are the isomorphisms $\mathcal{C} \xrightarrow{\sim} \mathcal{C}'$ of schemes over S.

$\mathcal{M}_g$ is known to be a Deligne-Mumford stack [3] (and hence also an Artin stack).

Remark 3.13. An Artin stack $\mathcal{M}$ given by a groupoid scheme $(X, R, e, s, t, \circ, i)$ is called *smooth* if X is smooth. The *dimension* of $\mathcal{M}$ is by definition the dimension of X minus the relative dimension of R over X.

For example, the stack quotient X/G of a k-scheme X modulo a smooth group scheme G over k is smooth if and only if X is, and $\dim(X/G) = \dim(X) - \dim(G)$. In particular, $\mathrm{B}G$ is smooth of dimension $-\dim(G)$.

Remark 3.14. $\mathcal{B}un_{r,d}$ is known to be smooth of dimension $r^2(g-1)$, one less than the dimension of the coarse moduli scheme $\mathfrak{Bun}_{r,d}^{\mathrm{stab}}$. This difference comes from the one-dimensional (scalar) automorphism groups of stable vector bundles which $\mathfrak{Bun}_{r,d}^{\mathrm{stab}}$ does not see, but which the stack $\mathcal{B}un_{r,d}$ does take into account.

Remark 3.15. The stack $\mathcal{M}_g$ is known to be smooth of dimension $3g-3$.

References

[1] M. Artin, Versal deformations and algebraic stacks, *Invent. Math.*, **27** (1974) 165–189.

[2] K. Behrend, B. Conrad, D. Edidin, W. Fulton, B. Fantechi, L. Göttsche and A. Kresch, Algebraic stacks, In preparation. Drafts of chapters available at www.math.uzh.ch/index.php?pr_vo_det&key1=1287&key2=580

[3] P. Deligne and D. Mumford, The irreducibility of the space of curves of given genus, *Inst. Hautes Études Sci. Publ. Math.*, **(36)** (1969) 75–109.

[4] T. Gomez, Algebraic stacks, *Proc. Indian Acad. Sci., Math. Sci.*, **111(1)** (2001) 1–31.

[5] A. Grothendieck, SGA 1: Revêtements étales et groupe fondamental, vol. 224 of Lecture Notes in Mathematics, Springer-Verlag, Berlin (1971).

[6] N. Hoffmann, Rationality and Poincare families for vector bundles with extra structure on a curve, *Int. Math. Res. Not. 2007*, No. 3, Article ID rnm010, 30 p. (2007).

[7] G. Laumon and L. Moret-Bailly, Champs algébriques, vol. 39 of Ergebnisse der Mathematik und ihrer Grenzgebiete, Springer-Verlag, Berlin (2000).

[8] C. S. Seshadri, Space of unitary vector bundles on a compact Riemann surface, *Ann. of Math. (2)*, **85** (1967) 303–336.

[9] C. Sorger, Lectures on moduli of principal G-bundles over algebraic curves, In School on Algebraic Geometry (Trieste, 1999), vol. 1 of *ICTP Lect. Notes*, Trieste (2000).

[10] A. Vistoli, Intersection theory on algebraic stacks and on their moduli spaces, *Invent. Math.*, **97(3)** (1989) 613–670.

Teichmüller Theory and Moduli Problem No. 10, 2010, pp. 395–412.

The Klein-Maskit Combination Theorems

John Hubbard

Department of Mathematics, Cornell University, Ithaca, NY, USA
e-mail: jhh8@cornell.edu

The object of this paper is to give a presentation of the Klein-Maskit combination theorems. These theorems are fundamental to Thurston's proof of hyperbolization for 3-manifolds.

Both theorems assert that appropriate subsets of $\mathrm{PSL}_2\,\mathbb{C}$ generate discrete subgroups, they further give precise descriptions of the subgroups generated. Maskit's proof [6,7] is strictly 2-dimensional, and is quite difficult and involved, in part because it is quite difficult to see whether a subgroup of $\mathrm{PSL}_2\,\mathbb{C}$ is discrete by considering its action on the Riemann sphere. At about the same time, R. Kulkarni [4,5] proved some closely related results using 3-dimensional methods.

In the late 1970's, Thurston proposed a much simpler 3-dimensional proof; a subgroup of $\mathrm{PSL}_2\,\mathbb{C}$ is discrete if and only if its orbits in $\mathbb{H}^3$ are discrete, so discreteness is much easier to observe in the 3-dimensional setting. The proof given here is a variant of Thurston's proof. Full details will appear in [3]. A similar construction can be found in [1].

1. Amalgamated sums and HNN extensions

The statements of the combination theorems involve amalgamated sums and HNN extensions. These are fairly standard topics in combinatorial group theory, and they are also encountered in topology when studying Van Kampen's theorem. Here we present a review of these topics, in part because although amalgamated sums are quite familiar and standard, HNN extensions are less so, and it is easier to understand the relevant techniques if one has just seen them at work in a familiar context.

Covering spaces and fundamental groups

Our proofs are all topological, and we will need the fundamental theorem relating covering spaces to the fundamental group. Our presentation is essentially that of [2].

There is a difficulty right at the beginning of this subject: choosing the right class of topological spaces. As soon as a space is a bit complicated, like the Hawaiian earring, it doesn't have a universal covering space, and the relation between covering spaces and the fundamental group becomes murky. We will not be applying the theorems to such spaces, of course, but they are relevant to the context of the present paper: limit sets of Kleinian groups are usually just as pathological as the Hawaiian earring and often worse. We will work in the category of UC spaces, which by definition are the spaces which have universal covering spaces. All the spaces in this paper will be assumed to be Hausdorff (or proved to be Hausdorff if necessary).

A universal covering space of a pointed space (X, x) is a pointed space $(\tilde{X}, \tilde{x})$, together with a covering map $p : (\tilde{X}, \tilde{x}) \to (X, x)$, such that for any pointed covering map $p_1 : (Y, y) \to (X, x)$, there exists a unique covering map $p_2 :(\tilde{X}, \tilde{x}) \to (Y, y)$ with $p = p_2 \circ p_1$. It is then clear that if a universal covering space of (X, x) exists, it is unique up to unique isomorphism, so one can speak of *the universal covering space* $\widetilde{(X, x)}$. In elementary topology one proves that the category of UC spaces includes the path-connected, semi-locally simply connected spaces, and that the fundamental group can then be defined as homotopy classes of based loops. This will be called the *standard case*. Example 6.1 shows that there are UC-spaces that are not semi-locally simply connected.

Remark 1. We have deliberately not assumed that X is connected. As a consequence of the definition it is easy to see that $\widetilde{(X, x)}$ *is* connected, and that its image in X is the component of X containing x. Note that the empty set is trivially a cover of any space. △

Remark 2. Why work in this generality? One answer is that it applies to more spaces than the standard case. This is not a very convincing argument, as I don't know of any cases where the more general UC spaces play a significant role, though of course, this may change. A more serious argument is that it makes Van Kampen's theorem and the analog for HNN-extensions much more transparent. But the best argument I know is that this approach emphasizes the similarities of Galois theory and covering space theory. The choice of a universal covering space for a space X corresponds to the choice of a separable algebraic closure $\bar{k}$ for a field k; the absolute Galois group $\mathrm{Aut}_k(\bar{k})$ corresponds to the fundamental group. In Galois theory, it is a profinite group, and the category of étale algebras over k (i.e., finite direct sums of finite separable field extensions) is equivalent to the category of finite sets on which $\mathrm{Aut}_k(\bar{k})$ operates continuously. For a further development of these ideas, see [2]; they are further developed in the large but difficult literature on the étale fundamental group. △

A space X is a UC space if for every $x \in X$ the universal covering space $\widetilde{(X, x)}$ exists; the morphisms of the category of UC spaces are the continuous maps.

We can then define $\pi_1(X, x)) = \mathrm{Aut}_X \widetilde{(X, x)}$ to be the group of covering transformations of $\tilde{X}$. Of course, $\pi_1(X, x)$ maps to $p^{-1}_{(X,x)}(x)$ by $\alpha \mapsto \alpha(\tilde{x})$; we invite the reader to check that this map is bijective. Then π_1 is a functor on the category of pointed UC spaces: if $f :(X, x) \to (Y, y)$ is a continuous map, there is by the universal property a unique map $\tilde{f} : \widetilde{(X, x)} \to \widetilde{(Y, y)}$ such that $\tilde{f}(\tilde{x}) = \tilde{y}$. The map $\tilde{f}$ restricts to a map

$$\pi_1(f) : p^{-1}_{(X,x)}(x) = \pi_1(X, x) \to p^{-1}_{(Y,y)}(y) = \pi_1(Y, y);$$

we leave it to the reader that this map is a group homomorphism. Although $\pi_1(f)$ is the categorically correct name of the mapping, it is usually written $f_*: \pi_1(X, x) \to \pi_1(Y, y)$, and we will respect this usage.

Remark 3. Why the base-points, if we aren't using loops as our basic construction? Because without base points, the universal covering space is not a *functor*. We needed the base point $\tilde{x}$ to specify p_2 uniquely. Example 6.2 illustrates how dangerous it can be to speak loosely of "*the* universal covering space of X" without mentioning a base point. △

If $\tilde{y} \in \widetilde{(X, x)}$ is a point with $p_{(X,x)}(\tilde{y}) = y$, then $\pi_1(X, x)$ acts on $p_{(X,x)}^{-1}(y)$, and the map

$$\pi_1(X, x) \to p_{(X,x)}^{-1}(y) \quad \text{given by} \quad \alpha \mapsto \alpha(\tilde{y})$$

induces an isomorphism $[\tilde{y}]: \pi_1(X, x) \to \pi_1(X, y)$. This construction corresponds, in the standard case, to choosing a path $\tilde{\delta}$ in $\widetilde{(X, x)}$ connecting $\tilde{x}$ to $\tilde{y}$, and projecting this path to a path δ in X connecting x to y. Then $\gamma \mapsto \delta^{-1} \cdot \gamma \cdot \delta$ defines an isomorphism $\pi_1(X, x) \to \pi_1(Y, y)$; here $\cdot$ denotes concatenation. The isomorphism depends only on $\tilde{y}$ since any two paths in the univeral covering space connecting x to y are homotopic.

Let (X, x) be a path-connected topological space with a base point. If $p: Y \to X$ is a covering map, an action of $\pi_1(X, x)$ on $p^{-1}(x)$ can be defined as follows: for any $y \in p^{-1}(x)$ let $p_y: (\tilde{X}, \tilde{x}) \to (Y, y)$ be the map given by the universal property, and set $g * y := p_y(g * \tilde{x})$. In the standarde case, this is the action corresponding to lifting a loop in (X, x) to a path in (Y, y).

This construction defines the *fiber functor*

$$\mathrm{Fib}: Cov(X) \to \pi_1(X, x) - \textsc{Sets}$$

from the category of covering maps of X and deck transformations to the category $\pi_1(X, x) - \textsc{Sets}$, i.e., sets on which $\pi_1(X, x)$ acts. Theorem 4 is the central result of covering space theory.

Theorem 4. *Let* (X, x) *be a pointed* UC-*space. Then the functor*

$$\mathrm{Fib}: Cov(X) \to \pi_1(X, x) - \textsc{Sets}$$

is an isomorphism of categories.

Sketch of proof. We need to construct an inverse functor. Let Z be a set on which $\pi_1(X, x)$; decompose $Z = \sqcup Z_i$ into the orbits of $\pi_1(X, x)$, so that the Z_i are $\pi_1(X, x)$-invariant subsets on which $\pi_1(X, x)$ acts transitively. Set

$$Y_i = \widetilde{(X, x)}/G_i,$$

where $G_i \subset \pi_1(X, x) = \mathrm{Aut}_X\, \widetilde{(X, x)}$ is the subgroup acting trivially on Z_i. Then the obvious projection $\sqcup_i Y_i \mapsto X$ is a covering map whose fiber as a $\pi_1(X, x)$-set is Z. For details consult [2].

To use this theorem effectively, we need the following property of G-sets. If $\varphi: G_1 \to G_2$ is a group homomorphism, there is an obvious functor $\varphi^*: G_2 - \textsc{Sets} \to G_1 - \textsc{Sets}$.

Lemma 5. *Let G_1, G_2 be groups, and $\varphi : G_1 \to G_2$ be a homomorphism. Then the functor*

$$\varphi^* : G_2 - \text{SETS} \to G_1 - \text{SETS}$$

induced by φ is an isomorphism of categories if and only if φ is an isomorphism of groups.

The easy proof can also be found in [2].

Amalgamated sums and van Kampen's theorem

Definition 6. *Let G_1, G_2, H be groups, and $\alpha_1 : H \to G_1$, $\alpha_2 : H \to G_2$ be homomorphisms. Then the amalgamated sum $G_1 \sqcup_H G_2$ is a group G, together with homomorphisms $\beta_1 : G_1 \to G$, $\beta_2 : G_2 \to G$ such that $\beta_1 \circ \alpha_1 = \beta_2 \circ \alpha_2$, satisfying the universal property.*

If K is a group and $\gamma_1 : G_1 \to K$, $\gamma_2 : G_2 \to K$ are group homomorphisms such that $\gamma_1 \circ \alpha_1 = \gamma_2 \circ \alpha_2$, then there exists a unique homomorphism $\delta : G \to K$ such that $\delta \circ \beta_1 = \gamma_1$, $\delta \circ \beta_2 = \gamma_2$, as summarized in the following diagram.

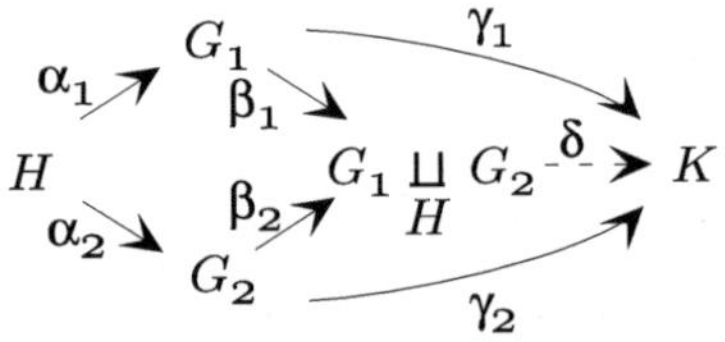

Remark. Amalgamated sums are often called *amalgamated products*, or *amalgamated free products*, or *free products with amalgamation*; indeed, in the topological literature this sort of terminology is nearly universal. This is unfortunate: the universal property above makes sense in any category, and in the categorical context it is always called an amalgamated sum. It seems wiser to stick to a single name for a single construction. △

It is easy to write generators and relations for amalgated sums, and in particular to show that amalgamated sums exist in the category of groups.

Amalgamated sums are usually first encountered in topology when studying Van Kampen's theorem.

Theorem 7. *Let (X, x) be a pointed* UC*-space, and U_1, U_2 be connected open subsets of X containing x, with $X = U_1 \cup U_2$ and $U := U_1 \cap U_2$ connected. Denote by i_k the inclusions $U_k \hookrightarrow X$, and j_k the inclusions $U \hookrightarrow U_k$. Then the homomorphisms $(i_k)_* : \pi_1(U_k, x) \to \pi_1(X)$ satisfy $(i_1)_* \circ (j_1)_* = (i_2)_* \circ (j_2)_*$, and the homomorphism*

$$\varphi : \left(\pi_1(U_1, x) \bigsqcup_{\pi_1(U,x)} \pi_1(U_2, x) \right) \to \pi_1(X, x)$$

coming from the universal property is an isomorphism of groups.

Proof. By Lemma 5, it is enough to show that

$$\varphi^* : \pi_1(X,x) - \text{SETS} \to \left(\pi_1(U_1,x) \coprod_{\pi_1(U,x)} \pi_1(U_2,x)\right) - \text{SETS}.$$

is an isomorphism of categories. A $(\pi_1(U_1,x) \sqcup_{\pi_1(U,x)} \pi_1(U_2,x))$-space with underlying set F is precisely an action of $\pi_1(U_1,x)$ and of $\pi_1(U_2,x)$ on the set F, such that the two induced actions of $\pi_1(U,x)$ coincide.

Under the inverse of the functor Fib, this data corresponds to a covering map $\pi_1 : Z_1 \to U_1$, a covering map $\pi_2 : Z_2 \to U_2$, both with fiber F, and an isomorphism of $\psi : Z_1|_U \to Z_2|_U$ inducing the identity on F. We can use ψ to glue Z_1 and Z_2 above U and construct a space Z and a covering map $\pi : Z \to X$. The functor Fib then gives us an action of $\pi_1(X,x)$ on F. We leave to the reader to prove that we have constructed an inverse of φ^*. □

HNN-extensions and an analog of Van Kampen's theorem

There is a second construction in group theory analogous to amalgamated sums: *HNN-extensions*, which is a way of embedding a group G in a bigger group G', such that two isomorphic subgroups of G are conjugate in G'. Note that even if the subgroups were already conjugate in G (or even identical), we still have $G' \neq G$.

Definition 8. *Let G, H be groups, and let $\alpha_1, \alpha_2 : H \to G$ be two group homomorphisms. The HNN extension $(G*_{(\alpha_1,\alpha_2)}, a)$ is a group $G*_{\alpha_1,\alpha_2}$ together with an element $a \in G*_{(\alpha_1,\alpha_2)}$ called the stable element and a group homomorphism $\varphi : G \to G*_{(\alpha_1,\alpha_2)}$, which satisfy*

$$\varphi \circ \alpha_2 = a^{-1}(\varphi \circ \alpha_1)a,$$

and which is universal for that property. This means that if K is a group, $b \in K$ is an element and $f : G \to K$ is a group homomorphism such that $f \circ \alpha_2 = b^{-1}(f \circ \alpha_1)b$, then there exists a unique $\tilde{f} : G*_{(\alpha_1,\alpha_2)} \to K$ with $g(a) = b$, such that $f = \tilde{f} \circ \varphi$. The following diagram illustrates the universal property:

$$\begin{array}{ccccc} H & \overset{\alpha_1}{\underset{\alpha_2}{\rightrightarrows}} & G & \overset{\varphi}{\longrightarrow} & G*_{(\alpha_1,\alpha_2)} \\ & & & \searrow f & \downarrow \tilde{f} \\ & & & & (K,b) \end{array}$$

It is not hard to show with generators and relations that the HNN-extension exists; it is of course unique up to unique isomorphism.

There is also an analog of Van Kampen's theorem, proved with a very similar proof.

Let X be a UC-space, such that there exists a connected UC-space Y, two disjoint connected UC-subsets Y_1, Y_2 and a homeomorphism $h : Y_1 \to Y_2$ such that X is homeomorphic to $Y/\sim$, where $\sim$ is the equivalence relation generated by $u \sim v$ if $h(u) = v$. Denote by $f : Y \to X$ the projection inducing this homeomorphism. Choose base points $y \in Y$, $y_1 \in Y_1$, $y_2 \in Y_2$ with $h(y_1) = y_2$ and points

$$\tilde{y}_1, \tilde{y}_2 \in \widetilde{(Y,y)} \quad \text{with} \quad p_{(Y,y)}(\tilde{y}_1) = y_1,\ p_{(Y,y)}(\tilde{y}_2) = y_2.$$

These points induce isomorphisms $[\tilde{y}_i] : \pi_1(Y,y) \to \pi_1(Y,y_i)$.

We then obtain a diagram

$$\begin{array}{ccccc} \pi_1(Y_1, y_1) & \overset{u_1}{\underset{u_2}{\rightrightarrows}} & \pi_1(Y, y) & \overset{\varphi}{\longrightarrow} & (\pi_1(Y, y)*_{(u_1,u_2)}, a) \\ & & & \searrow f_* & \downarrow \tilde{f}_* \\ & & & & (\pi_1(X, x), u) \end{array}$$

We have denoted by u_1 the composition

$$\pi_1(Y_1, y_1) \to \pi_1(Y, y_1) \overset{[\tilde{y}_1]^{-1}}{\to} \to \pi_1(Y, y)$$

and by u_2 the composition

$$\pi_1(Y_1, y_1) \overset{h_*}{\to} \to \pi_1(Y_2, y_2) \to \pi_1(Y, y_2) \overset{[\tilde{y}_2]^{-1}}{\to} \to \pi_1(Y, y).$$

The element $u \in \pi_1(X, x)$ is defined as follows. Denote by $z \in X$ the common image $f(y_1) = f(y_2)$. The pull-back covering map $f^*\widetilde{(X, x)} \to Y$ is a covering map of Y with a base point, so by the universal property of universal covering spaces, there is a natural map

$$g : \widetilde{(Y, y)} \to f^*\widetilde{(X, x)}.$$

Then $z_1 = g(y_1)$ is a point of the fiber of $f^*\widetilde{(X, x)}$ above y_1 and $z_2 = g(y_2)$ is a point of the fiber of $f^*\widetilde{(X, x)}$ above y_2. But these fibers are the same set: the fiber above z for $\widetilde{(X, x)} \to X$. Thus there is a unique automorphism $u \in \mathrm{Aut}_X \widetilde{(X, x)}$ such that $u(z_1) = z_2$.

The homomorphism $\tilde{f}_*$ is provided by the universal property of HNN extensions, which requires knowing that $u^{-1}u_2u = u_1$. This is seen as follows: the maps $f|_{Y_1} : Y_1 \to X$ and $f \circ h : Y_1 \to X$ coincide. In particular, they induce the same map $\pi_1(Y, y_1) \to \pi_1(X, z)$. But the two compositions $f_* \circ u_1$ and $f_* \circ u_2$ have codomain $\pi_1(X, x)$, not $\pi_1(X, z)$, and they each depend on the choice of an element of $p^{-1}_{(X,x)}(z)$, in fact z_1 and z_2 respectively. The formula $u^{-1}u_2u = u_1$ follows easily.

Theorem 9. *The homomorphism $\tilde{f}_*$ is an isomorphism.*

As in van Kampen's theorem, it is enough to prove that

$$(\tilde{f}_*)^* : \pi_1(X, x) - \text{SETS} \to (\pi_1(Y, y)*_{(u_1,u_2)}) - \text{SETS}$$

is an isomorphism of categories. We will construct an inverse, in the equivalent category of covering spaces. Let F be a set on which $\pi_1(Y, y)*_{(u_1,u_2)}$ operates. This means that $\pi_1(Y, y)$ operates on F, and that the actions of $\pi_1(Y_1, y_1)$ on F by u_1 and u_2 are conjugated by the stable element a.

In terms of covering spaces, this translates to a covering space $p : Z \to Y$ with fiber $p^{-1}(y) = F$, and an isomorphism $\tilde{a}$ of the two covering spaces

$$p : p^{-1}(Y_1) \to Y_1 \quad \text{and} \quad h \circ p : p^{-1}(Y_2) \to Y_1.$$

Equivalently, we can interpret $\tilde{a}$ as a homeomorphism $\tilde{h} : p^{-1}(Y_1) \to p^{-1}(Y_2)$ such that $h \circ p = p \circ \tilde{h}$.

Thus we can glue Z to itself by $\tilde{h}$; the resulting space $\bar{Z}$ is a covering space of X, corresponding to an action of $\pi_1(X, x)$ on F. We leave it to the reader to check that we have constructed an inverse of $(\tilde{f}_*)^*$. QED

2. Harmonic separators

Some terminology

First let us set up some terminology concerning Kleinian groups; a good place to learn this material is [9], and I hope that [3] will also soon be available.

Hyperbolic space will be denoted $\mathbb{H}^3$; often we will work with the ball model $\mathbf{D}^3$ of $\mathbb{H}^3$. The Euclidian boundary of $\mathbf{D}^3$ is the unit sphere, which can be identified with the Riemann sphere by stereographic projection. Since stereographic projection is conformal, this gives gives the "sphere at infinity $\partial\mathbb{H}^3$ of $\mathbb{H}^3$" a natural conformal structure. We will write $\bar{\mathbb{H}}^3 = \mathbb{H}^3 \sqcup \partial\mathbb{H}^3$, which is homeomorphic to a closed ball. The group $\operatorname{Aut}\mathbb{H}^3$ of orientation-preserving isometries of $\mathbb{H}^3$ is well-known to be isomorphic to $\mathrm{PSL}_2\,\mathbb{C}$; the isomorphism can be chosen so that if an isometry $\alpha \in \operatorname{Aut}\mathbb{H}^3$ corresponds to $\begin{bmatrix} a & b \\ c & d \end{bmatrix}$, then $\alpha : \mathbb{H}^3 \to \mathbb{H}^3$ extends continuously to the conformal automorphism

$$z \mapsto \frac{az+b}{cz+d}$$

of the Riemann sphere.

A *Kleinian group* is a discrete subgroup $G \subset \operatorname{Aut}\mathbb{H}^3$. The accumulation set in $\bar{\mathbb{H}}^3$ of the orbit Gx of any point $x \in \mathbb{H}^3$ is independent of x, and is called the *limit set* of G, denoted by Λ_G. A Kleinian group is *quasifuchsian* if it is quasiconformally conjugate to a Fuchsian group of the first kind, i.e., one for which the limit set is $\mathbb{R} \cup \{\infty\}$. Thus the limit set of a quasifuchsian group is always a quasicircle, and conversely any Fuchsian group whose limit set is a quasicircle is either quasifuchsian or has a subgroup of index 2 that is quasifuchsian.

A substitute for the convex hull

Thurston has taught us that often working with the 3-dimensional convex hull $\hat{\Lambda}_G \subset \bar{\mathbb{H}}^3$ often makes proofs very transparent that are quite obscure when working in the 2-dimensional sphere at infinity.

For our purposes, as we will explain, the convex hull of the limit set will not be convenient, and we will choose instead the *harmonic separator*. This is defined as follows: a quasi-Fuchsian group H has a limit set Λ_H that is a quasicircle, breaking up $\mathbb{P}^1$ into two components U_0 and U_1. Construct the harmonic function h_H on $\bar{\mathbb{H}}$ that extends the function f_H that is 0 on U_0 and 1 on U_1. Then the harmonic separator S_H is the set of equation $h_H = 1/2$.

This harmonic function is easy to construct, and easy to understand, especially in the ball model $\mathbf{D}^3$ of $\mathbb{H}$. Let μ be normalized round measure on $\mathbb{P}^1 = \partial\mathbf{D}^3$, i.e., $\mu = \frac{1}{4\pi} dS$. One way to define h_H is to set

$$h_H(x) = \int_{\mathbb{P}^1} A_x^* f_H(y)\mu(dy)$$

where A_x is an element of $\operatorname{Aut}\mathbf{D}^3$ that sends 0 to x. Any other B_x differs from A_x by precomposition with a rotation, and hence yields the same value since μ is invariant by rotations. In formulas this description becomes the Poisson integral formula. It is clear that h_H is natural in the sense that

$$h_{A^{-1}HA} = A^* h_H$$

for all $A \in \operatorname{Aut}\mathbf{D}^3$.

Usually S_H is a smooth real-analytic surface, but it is possible to construct examples where it has singularities. They need not concern us here; we construct an example in the last section. More relevant to us is the following proposition.

Proposition 11. *Let G be a Kleinian group, and $H \subset G$ be a quasi-Fuchsian subgroup which is the stabilizer of some component U_0 of $\mathbb{P}^1 - \Lambda_G$. Suppose that $\Lambda_G \neq \Lambda_H$. Then there exists a closed H-invariant neighborhood $\widetilde{S}_H$ of S_H such that if $g \in G$ satisfies $g(x) = y$ with $x, y \in \widetilde{S}_H$, then $g \in H$.*

Proof. Suppose $x, y \in S_H$ and $g(x) = y$. Then

$$\frac{1}{2} = h_H(y) = h_H(g(x)) = h_{gHg^{-1}}(x)$$

so $x \in S_{gHg^{-1}}$, and hence $S_H \cap S_{gHg^{-1}} \neq \emptyset$; let us see that this is impossible if $g \notin H$. Recall that f_H is the function which is 0 on U_0, and 1 on the other component U_1 of $\mathbb{P}^1 - \Lambda_H$. The hypotheses $\Lambda_G \neq \Lambda_H$ and U_0 a component of $\mathbb{P}^1 - \Lambda_G$ together imply that U_1 is not a component of $\mathbb{P}^1 - \Lambda_G$, and thus that $g^{-1}(U_0) \subset U_1$, and $U_1 - g^{-1}(U_0)$ has positive measure. The functions f_H and $1 - f_H \circ g$ satisfy $f_H \geq 1 - f_H \circ g$, and the inequality is strict on a set of positive measure. By naturality we have $h_H \circ g = h_{gHg^{-1}}$, and $h_H \circ g$ is the harmonic extension of $f_H \circ g$. Thus if $g \notin H$ the function $h_H - h_{gHg^{-1}}$ never vanishes.

Next, note that $S_H \subset \widehat{\Lambda}_H$ (recall that the hat indicates the convex hull). Otherwise there is a plane in $\mathbb{H}^3$ with Λ_H on one side, and points of S_H on the other; if such a point is moved to the origin then U_0 is entirely moved inside a hemisphere, and so $h_H > 1/2$ at such a point.

It follows that a compact neighborhood of a point $x \in S_H$ intersects $S_{H'}$ for only finitely many subgroups $H' \subset G$ conjugate to H. We may suppose x is at the origin; for any $\epsilon > 0$ there are only finitely many images of U_0 by elements $g \in G$ that have areas $> \epsilon$, hence only finitely many with diameter $> \epsilon$, and for any $R > 0$, there are only finitely many such images whose convex hulls can intersect the ball of radius R around the origin.

Thus there is a well-defined function

$$\rho : S_H \to \mathbb{R}, \quad \rho(x) := \inf_{y \in S_{H'}, H' \text{ conjugate to } H \text{ in } G, H' \neq H} d(x, y),$$

and $\rho(x) > 0$ everywhere on its domain, though its infimum may be 0. The union

$$\widetilde{S}_H := \bigcup_{x \in S_H} \bar{B}_{\rho(x)/2}(x)$$

is a neighborhood of the required sort. □

Remark. The analog of Proposition 11 is not true for the convex hull $\widehat{\Lambda}_H$: it is sometimes possible to find elements $g \in G - H$ and points $x, y \in \widehat{\Lambda}_H$ such that $g(x) = y$; as a consequence it is not always true that $\widehat{\Lambda}_H / H$ is an embedded subset of $\mathbb{H}^3/G$. We show an example in subsection 6.3. △

3. The combination theorem for amalgamated sums

Both of the Klein-Maskit combination theorems assert that if Kleinian groups have appropriate conjugate quasi-Fuchsian subgroups, then they generate a discrete subgroup; the first concerns the case where the quasi-Fuchsian subgroups are subgroups of different groups, and the second where they are subgroups of the same group.

Theorem 12 (Klein-Maskit Combination theorem for amalgamated sums). *Let G_1, G_2 be two Kleinian groups, such that $H := G_1 \cap G_2$ is quasi-Fuchsian. Denote by U_1, U_2 the components of $\mathbb{P}^1 - \Lambda_H$, and suppose that $\Lambda_{G_1} \cap U_2 = \Lambda_{G_2} \cap U_1 = \emptyset$. Then G_1 and G_2 generate a discrete subgroup $G \subset \mathrm{PSL}_2\,\mathbb{C}$, and the diagram*

$$\begin{array}{ccccc} & & H & & \\ & \swarrow & & \searrow & \\ G_1 & & & & G_2 \\ & \searrow & & \swarrow & \\ & & G & & \end{array}$$

is an amalgamated sum.

Proof. Here we will prove the result only when G_1 and G_2 are torsion-free. In Section 5 we will see how to recover the general case.

Define $X_i := \mathbb{H}^3/G_i$. The spaces X_1 and X_2 are complete hyperbolic manifolds, since neither G_1 or G_2 have torsion elements. By Proposition 11, both contain an embedded copy of $\widetilde{S}_H/H$ (this requires the hypothesis $\Lambda_{G_1} \cap U_2 = \Lambda_{G_2} \cap U_1 = \emptyset$) whose complement in each X_i has two components. Denote by V_i the component of $X_i - \widetilde{S}_H/H$ that has U_i/H in its closure, and set $Y_i = X_i - V_i$. Note that Y_1 and Y_2 both contain a copy of $\widetilde{S}_H/H$.

Finally define $X = (Y_1 \sqcup Y_2)/\sim$, where $\sim$ identifies a point of $\widetilde{S}_H/H \subset Y_1$ to the same point of $\widetilde{S}_H/H \subset Y_2$; denote by $p : (Y_1 \sqcup Y_2) \to X$ the quotient map.

We need to show that X is a complete hyperbolic manifold locally isometric to $\mathbb{H}^3$. Any point $x \in X$ has an inverse image x' in the interior of Y_1 or Y_2, say Y_1. Thus there

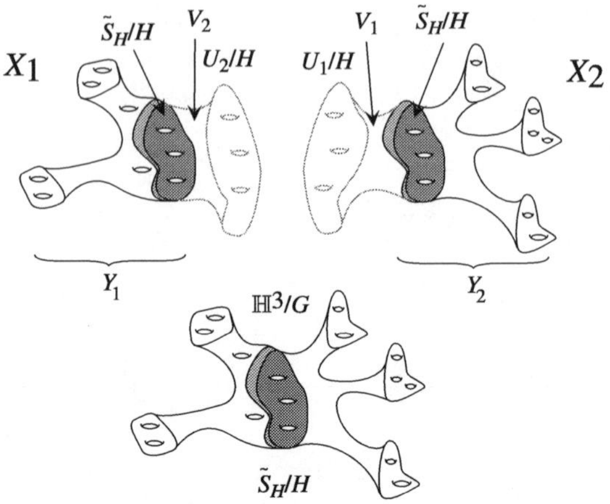

Figure 1. On the top, we see the quotients $X_1 = \mathbb{H}^3/G_1$ and $X_2 = \mathbb{H}^3/G_2$. Shaded in the middle of each is $\widetilde{S}_H/H$; it is the same in both figures. Thus we can glue together the "complicated part" Y_1 of X_1 with the complicated part of Y_2 of X_2 to make X, which is complicated on both sides.

exists $r > 0$ such that the ball of radius r around x' is isometric to a ball in $\mathbb{H}^3$ and contained in Y_1; on this ball p is injective, and we give the image the metric of the ball. This defines an infinitesimal metric unambiguously: If there are two such points x' and x'', one is in the interior of the copy of $\widetilde{S}_H/H$ in Y_1 and the other in the copy in Y_2, and the images of the balls of radius r around x' and x'' coincide, together with their metrics.

There is a fussy point to be checked: is the "metric" associated to the infinitesimal metric actually a metric: might there be distinct points that can be joined by arbitrarily short curves. □

Example 13. Consider $(\mathbb{R} \times \{A, B\})/\sim$, where $\sim$ identifies (t, A) with (t, B) for $t \in (0, 1)$. Then everything we said above goes through: every point of the quotient has a neighborhood isometric to an interval, and the quotient is a "non-Hausdorff manifold" with a "Riemann metric." But the points $(0, A)$ and $(0, B)$ are distance 0 apart: for $0 < \epsilon < 1$, the path that joins $(0, A)$ to $(\epsilon, A) \sim (\epsilon, B)$ to $(0, B)$ has length 2ϵ. Of course the problem is that we didn't identify points in a closed set: for the quotient of a space Z by an equivalence relation $\sim$ to be Hausdorff, the graph of the equivalence relation needs to be a closed subset of $Z \times Z$. Even this is not enough to guarantee that the quotient is Hausdorff, though it is if one adds that the equivalence classes are finite. It is easier to argue directly as below. △

At the moment our X has the same status as the quotient of Example 13. Still, looking at the example shows what to worry about; we leave to the reader to show that distinct points of X have disjoint neighborhoods isometric to balls in $\mathbb{H}^n$.

Moreover, X is a complete metric space. Indeed, if (x_n) is a Cauchy sequence in X, and y_n is a point of $Y_1 \sqcup Y_2$ in the equivalence class of x_n, then we can choose a subsequence (y_{n_i}) belonging entirely to Y_1 or Y_2. But these are are complete metric spaces, and the sequence (y_{n_i}) is still Cauchy. We actually need to think about this a bit; the Y_i are not convex, and the path metric in Y_i does not coincide with the induced metric from X_i. For the induced metric the sequence (y_{n_i}) is Cauchy, so it converges to a point in the equivalence class of the limit of the x_n.

Thus the universal covering space $\widetilde{X}$ is a complete simply connected manifold, and as such isometric to $\mathbb{H}^3$; clearly the isometry $\varphi : \mathbb{H}^3 \to \widetilde{X}$ can be chosen so that φ is the identity on $\widetilde{S}_H$. With this choice of φ, the fundamental group of $Y_1 \subset X$ is identifed to G_1, and the fundamental group of $Y_2 \subset X$ is identifed to G_2.

The theorem is now proved: G is a group of deck transformations, hence discrete, and the statement about amalgamated sums follows from van Kampen's theorem.

4. The combination theorem for HNN extensions

In the second Klein-Maskit combination theorem, instead of gluing two different hyperbolic manifolds to each other along a common subset $\tilde{S}_H/H$ for some quasi-Fuchsian group H, we glue a manifold to itself along two isometric $\tilde{S}_{H_1}/H_1$ and $\tilde{S}_{H_2}/H_2$ for two different quasi-Fuchsian goups H_1, H_2.

Theorem 15 (Klein-Maskit Combination theorem for HNN-extensions). *Let G be a Kleinian groups, and U_1, U_2 be two distinct components of $\mathbb{P}^1 - \Lambda_G$ with quasi-Fuchsian stabilizers H_1, H_2 in G. Suppose further that $\varphi \in \operatorname{Aut}\mathbb{P}^1$ satisfies $\varphi^{-1}H_2\varphi = H_1$, and that $\varphi(U_1) = \mathbb{P}^1 - \bar{U}_2$. Then G and φ generate a Kleinian group, and the natural map $G *_{(id,\varphi)} \to \langle G, \varphi\rangle$ is an isomorphism.*

Note that the hypothesis $\varphi(U_1) = \mathbb{P}^1 - \bar{U}_2$ implies that $\varphi \notin G$.

Proof. Again we will deal here only with the case where G is torsion-free, we show how to recover the full statement in Section 5. Define $Z := \mathbb{H}^3/G$. The quotient Z contains disjoint embedded subsets $\widetilde{S}_{H_1}/H_1$ and $\widetilde{S}_{H_2}/H_2$. The complement of $\widetilde{S}_{H_1}/H_1$ has components V_1, V_1', where V_1 is the component whose closure contains U_1/H_1. Similarly, let V_2, V_2', be the components of $X - \tilde{S}_{H_2}$, with U_2/H_2 in the closure of V_2. The map φ induces a homeomorphism

$$\bar{\varphi} : \tilde{S}_{H_2}/H_2 \to \tilde{S}_{H_1}/H_1;$$

this map is "inside out", i.e., it sends $V_2 \cap \tilde{S}_{H_2}/H_2$ to $V_1' \cap \tilde{S}_{H_1}/H_1$ and $V_2' \cap \tilde{S}_{H_2}/H_2$ to $V_1 \cap \tilde{S}_{H_1}/H_1$.

Let $Y := Z - (V_1 \cup V_2)$, and $X = Y/\sim$, where $\sim$ identifies a point $x \in S_{H_2}/G$ with $\bar{\varphi}(x) \in S_{H_1}$. The proof that X is a complete hyperbolic manifold is essentially identical as in the case of Theorem and is left to the reader.

Let us denote by $T \subset X$ the subset image of $\tilde{S}_{H_1}/H_1$ (which is also the image of $\tilde{S}_{H_2}/H_2$, since these have been glued to each other by $\bar{\varphi}$). The universal covering space $\tilde{X}$ contains many copies of $\tilde{T}$, and we can choose our isomorphism $\psi : \tilde{X} \to \mathbb{H}^3$ so that it maps one such copy T_1 to $\tilde{S}_{H_1}$, and so that if H_1' is the stabilizer of T_1 in the group of deck transformations, then $H_1 = \psi H_1' \psi^{-1}$.

Then ψ maps the group of deck transformations to the group generated by G and φ, and Theorem shows that this group is naturally isomorphic to the HNN-extension of G with stable element φ. □

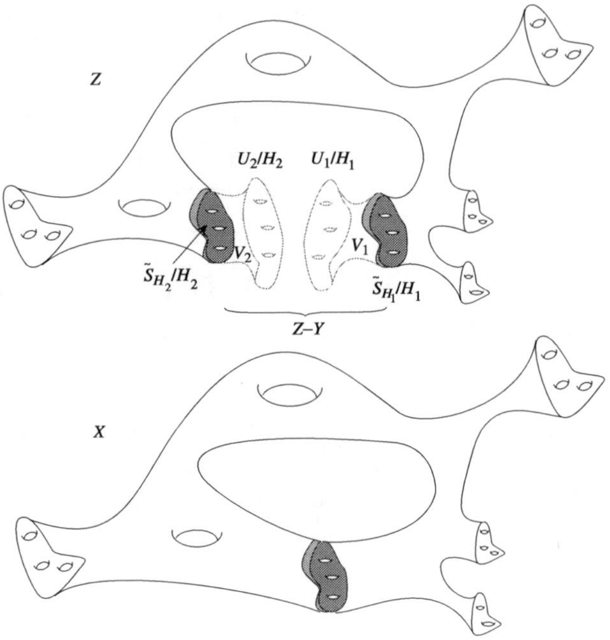

5. Orbifolds and torsion

In both combination theorems, we constructed a complete hyperbolic manifold X and identified its universal covering space with $\mathbb{H}^3$; if any of the groups involved have torsion elements, the space X is not a hyperbolic manifold, and this approach fails, at least if we take it literally. Instead, we should think of X as an *orbifold* and take $\widetilde{X}$ to be its *orbifold universal covering space*; in that case $\widetilde{X}$ is isometric to $\mathbb{H}^3$, and the proof continues exactly the same way.

The problem is that not all orbifolds have universal covering spaces that are manifolds; but our space X, and more generally any orbifold that carries a complete hyperbolic (or spherical, or Euclidian) structure, does have such a covering space, as we will prove using the geodesic flow.

First, we will define orbifolds and orbifold covering maps, and show that their geodesic flow is well defined. Although most of the generalities work for spaces modelled on any symmetric space, we will restrict our attention to hyperbolic orbifolds. Although our main interest is in 3-dimensional orbifolds, we will treat the n-dimensional case, as it appears to be non more difficult.

Hyperbolic orbifolds and their covering maps

An n-dimensional hyperbolic manifold is a metric space locally modelled on $\mathbb{H}^n$, with changes of coordinates isometric. Similarly, an n-dimensional orientable hyperbolic orbifold is a metric space locally modelled on the quotient of $\mathbb{H}^n$ by a finite group of orientation-preserving isometries of $\mathbb{H}^n$. We will require our metric spaces to be geodesics metric spaces, meaning that the distance between pairs of points is the infimum of the lengths of geodesics joining the points.

More precisely, an n-dimensional hyperbolic orbifold is a geodesic metric space X, such that every point $x \in X$ has a neighborhood U such that there exists a finite group $\Gamma \subset \operatorname{Aut} \mathbb{H}^n$, a Γ-invariant open subset $V \subset \mathbb{H}^n$ and a map $\varphi_U : V \to U$ inducing an isometry $\bar{\varphi} : V/\Gamma \to U$, where V/Γ carries the quotient metric from $\mathbb{H}^n$.

Note that we required $\Gamma \subset \operatorname{Aut}(\mathbb{H}^n)$, which by our definition consists only of orientation-preserving isometries so all elements of Γ as above are orientation-preserving; we could have allowed reflections too but that complicates the description.

Remark. There are many possible definitions of an orbifold, even more than for manifolds. In most cases, for instance differentiable or analytic orbifolds, we cannot simply choose X a metric space, as there is no natural metric on the model V/Γ. One needs then to define charts and changes of coordinates; the crucial issue is always what kinds of changes of coordinates are allowed. The difficulties involved are not immense, but they are substantial and unrelated to the issues in this paper, so we will ignore them here and stick to our definition. △

Hyperbolic orbifolds are easiest to understand when $n = 2$. The finite groups of orientation-preserving isometries of $\mathbb{H}^2$ are cyclic groups of rotations. Thus a point x of a 2-dimensional orbifold X has a neighborhood U isometrically modelled on D_r/C_m where $D_r = \{z \in \mathbf{D} | |z| < r\}$ for some $r < 1$, and C_m is the group generated by $z \mapsto e^{2\pi i/m} z$. If $x \in X$ corresponds to $0 \in \mathbf{D}$, then all points of $U - \{x\}$ are smooth. Thus the set of non-smooth points is discrete, and each can be assigned an integer, $m \geq 2$ in the case above. Note that we might think of the smooth points as marked with the integer 1. A 2-dimensional orbifold has *singular locus* which is a discrete set with each point marked with an integer ≥ 2; the complement is a hyperbolic manifold, and points of the singular locus are *cone points*: such a point x marked m has a neighborhood isometric to a neighborhood of 0 in $\mathbf{D}/C_m$.

When $n = 3$ the structure of a hyperbolic orbifold is a bit more elaborate, though still quite easily describable. Again we know what the elements of finite order in $\operatorname{Aut} \mathbb{H}^3$ are: rotations of some order m around some axis. If for a finite group of isometries the stabilizer of some point is neither trivial nor cyclic, then the placing the point at $0 \in \mathbf{D}^3$, the finite group must be conjugate to a finite subgroup of SO_3, hence dihedral, tetrahedral, octohedral, or icosahedral. It is then easy to see that the singular locus of a hyperbolic orbifold is an embedded graph. The points where the orbifold is modelled on $\mathbb{H}^3/C_m$ with $m \geq 2$ are edges, that we will mark with the integer m. These edges can meet at vertices, which are necessarily trivalent, and where the edges have quite specific structure. There are 4 possibilities:

- (dihedral case) three edges meet, marked 2, 2, m;
- (tetrahedral case) three edges meet, marked 2, 3, 3;
- (octohedral case) three edges meet, marked 2, 3, 4;
- (icosahedral case) three edges meet, marked 2, 3, 5.

A point of an edge point of the singular locus has a neighborhood isometric to $\mathbb{H}^3/C_m$, and the vertices have neighborhoods isometric to a neighborhood of the fixed point in $\mathbb{H}^3/\Gamma$ for Γ one of the non-cyclic finite subgroups of SO_3.

Covering maps of orbifolds

A map $f : X \to Y$ of hyperbolic orbifolds is a covering map if every $y \in Y$ has a neighborhood U such that if $V \subset \mathbb{H}^n$ is a subset invariant under a finite group $\Gamma \subset \operatorname{Aut}\mathbb{H}^n$ and there is an isometry $\varphi : V/\Gamma \to Y$, then for every component W of $f^{-1}(U)$, there is a subgroup $\Gamma' \subset \Gamma$ and an isometry $\varphi_W : V/\Gamma' \to W$ such that $\varphi = f \circ \varphi_W$.

A geodesic on a hyperbolic orbifold X is a parametrized curve $\gamma : I \to X$ with I some interval, which is locally near every point x the image of a geodesic arc in $\mathbb{H}^n$.

Proposition 16. *If $(x, \mathbf{v})$ is a unit tangent vector at a smooth point of a complete hyperbolic orbifold X, then there exists a unique geodesic $\gamma_{x,\mathbf{v}} : \mathbb{R} \to X$ with $\gamma_{x,\mathbf{v}}(0) = x$ and $\gamma'_{x,\mathbf{v}}(0) = \mathbf{v}$.*

The point $\gamma_{x,\mathbf{v}}(t)$ depends continuously on x, $\mathbf{v}$ and t.

Proof. Since x is a smooth point of X, the curve $\gamma_{x,\mathbf{v}}(t)$ exists and is unique for $|t| < \epsilon$ for some $\epsilon > 0$. Let t_0 be the supremum of such ϵ. Note that if $t_0 < \infty$, then $z := \lim_{t \nearrow t_0} \gamma_{x,\mathbf{v}}(t)$ certainly exists, since $\gamma_{x,\mathbf{v}}$ is Lipschitz with ratio 1 (this is where X complete is used). Choose a neighborhood U of z, a finite subgroup $\Gamma \subset \operatorname{Aut}\mathbb{H}^n$, a Γ-invariant neighborhood of 0 in $\mathbf{D}^n$ and a map $\varphi : V \to U$ inducing an isometry $V/\Gamma \to U$. There exists $\delta > 0$ such that $\gamma_{x,\mathbf{v}}[t_0 - \delta, t_0) \in U$.

Let $\tilde{\delta} : [t_0 - \delta, t_0) \to V$ be a lift of $\delta : [t_0 - \delta, t_0) \to U$. Then δ can be extended to $\bar{\delta} : [t_0 - \delta, t_0 + \delta']$ for some $\delta' > 0$, and $\varphi(\bar{\delta})$ extends δ. The extension is unique, since any two $\tilde{\delta}$ differ by an element of Γ. This contradicts $t_0 < \infty$. We leave the continuity with respect to initial conditions to the reader. □

Theorem 17. *Let X be a complete hyperbolic orbifold, and $x \in X$ be a smooth point. Choose $U \subset \mathbb{H}^n$ an open set, and $f : U \to X$ a hyperbolic local coordinate. Then the map $\tilde{f} : \mathbb{H}^n \to X$ that maps unit-speed geodesics in $\mathbb{H}$ through $f^{-1}(x)$ to unit-speed geodesics through x and that extends f is a universal covering map of orbifolds. In particular, every universal covering space of X is smooth, isometric to $\mathbb{H}^n$.*

Proof. Suppose that T is the supremum of the $t \geq 0$ such that there exists a unique geodesic arc $\gamma : [0, t] \to X$ with $\gamma(0) = x, \gamma'(0) = \mathbf{v}$. If $T = \infty$ we are done, so suppose $T < \infty$. Then since γ is Lipschitz with ratio 1 and X is complete, the point $y = \lim_{t \to T} \gamma(t)$ exists. Choose a chart $\varphi : D_r \to U$ at y, and a point $z \in D_r$ such that $\varphi(z) = \gamma(t)$ for some t with $T - t < r$.

Then by our definition of a geodesic, there exists $\epsilon > 0$ and a parametrized geodesic arc $\delta : (t - \epsilon, t + \epsilon) \to D_r$ such that $\varphi \circ \delta = \gamma$ on $(t - \epsilon, t + \epsilon)$. But δ can be continued to $\tilde{\delta} : (t - \epsilon, t + r) \to D_r$ (in fact, δ can be continued to all of $\mathbb{R}$, of course, if you allow values in $\mathbf{D}^n$), and then $\varphi \circ \tilde{\delta} : (t - \epsilon, t + r) \to X$ continues γ to $t + r > T$, which is a contradiction if we can show that the continuation is unique. The only ambiguity in the construction is the choice of y; any other choice y_1 differs from y by an element

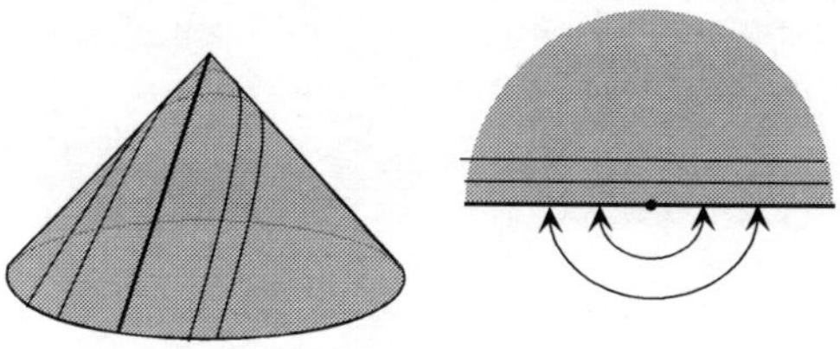

Figure 2. On a cone the straight lines to the vertex are geodesics, and if the cone opening is π they just bounce back. But a nearby geodesic just loops around the vertex and very nearly bounces back. There is no "opposite side."

$g \in \Gamma$, where Γ is the finite group such that D_r/Γ is isometric to U. But then the new δ_1 also satisfies $\delta_1 = g(\delta)$, and its continuation $\tilde{\delta}_1$ also satisfies $\tilde{\delta}_1 = g(\tilde{\delta})$. Thus the continuation of γ to $[0, t+r]$ is unique. □

One may well wonder what the geodesic flow looks like, and the result is paradoxical. In dimension 2, geodesics bounce off points marked 2; how do you get to the points "opposite?" This requires thinking of what geodesics on a cone look like.

Theorem 18. *The universal covering space of a complete hyperbolic n-orbifold X is always isomorphic to* $\mathbb{H}^n$. *Thus X is isometric to a quotient* $\mathbb{H}^n/G$ *for some discrete subgroup* $G \subset \mathrm{Aut}(\mathbb{H}^n)$.

Proof. Choose a smooth point $x \in X$, and a chart $\varphi : D_r \to U$ with $\varphi(0) = x$; since x is a smooth point we may take φ to be an isometry to its image.

Every point of $\mathbf{D}^n$ can be written $\delta_{\mathbf{v}}(t)$, where $\mathbf{v} \in T_0\mathbf{D}$ is a unit vector, δ is a geodesic in $\mathbf{D}^n$ with $\delta(0) = 0$, $\delta'(0) = \mathbf{v}$ and $t \geq 0$; this writing is unique if $t > 0$. Consider the map $F : \mathbf{D}^n \to X$ given by

$$F(\delta_{\mathbf{v}}(t)) = \gamma_{\bar{\mathbf{v}}}(t),$$

where $\gamma_{\bar{\mathbf{v}}}$ is the geodesic on X with $\gamma(0) = x$ and $\gamma'(0) = [D\varphi(0)]\mathbf{v}$.

We claim F is a covering map. So for every point $y \in X$, we need to find a neighborhood U that is "evenly covered." Choose an open neighborhood U of y and a chart $\psi : V \to U$ representing U as V/Γ for some finite subgroup $\Gamma \in \mathrm{Aut}^+(\mathbf{D}^n)$. Let $w \in \mathbf{D}^n$ satisfy $F(w) = y$, and let W be the component of $F^{-1}(U)$ containing W. What we need to show is that $F : W \to U$ lifts to a homeomorphism $\tilde{F} : W \to V$. We have $w = \delta_{\mathbf{v}}(t)$ for some $\mathbf{v}$ and t; by definition of the geodesic flow there exists a neighborhood I of t and a lift $\eta : I to V$ of $\delta|_I$. The two arcs $\delta_{\mathbf{v}}|_I$ and η are isometric parametrizations of arcs in open subsets hyperbolic space, and as such $\delta_{\mathbf{v}} \circ \eta^{-1}$ extends to an isometry $\mathbb{H}^n \to \mathbf{D}^n$.

There are many such isometries, differing by rotations around $\delta_{\mathbf{v}}$, but only finitely many are locally lifts of F, in fact, only one because we chose x smooth, so $\gamma_{\bar{\mathbf{v}}}$ has only a discrete set of points in the singular locus. The restriction of this extension to W is the lift we required to see that F is a covering map. It is clearly universal since $\mathbf{D}^n$ is simply-connected. □

Finally, we can complete the proofs of the Klein-Maskit theorems. The spaces X constructed in the proofs of Theorems 4 and 15 are clearly complete hyperbolic orbifolds, so their universal covering spaces are isometric to $\mathbb{H}^3$ by Theorem , Thus they are quotients of $\mathbb{H}^3$ by discrete subgroups of isometries.

6. Constructing some counterexamples

6.1 *A* UC-*space that is not semi-locally simply connected*

In the x, y-plane, let Y be the union of the x-axis, and of all the half-hyperbolas

$$y = \frac{1}{x-n}, \quad n \in \mathbb{Z}, \ y > 0.$$

Let $Z' \subset \mathbb{R}^3$ be obtained by rotating Y around the x-axis. To make Z path connected, add a path joining

$$\begin{pmatrix} n \\ 0 \\ 0 \end{pmatrix} \to \begin{pmatrix} n \\ 1 \\ 0 \end{pmatrix}$$

Clearly Z is invariant by translation τ by $\mathbf{e}_1$; let $X = Z/\mathbb{Z}\tau$.

Then the space X is UC but not semi-locally simply connected; more precisely, the projection map $p : Z \to X$ is a universal covering map. This takes a bit of checking, which we leave to the reader. Note that Z is itself not simply connected, and has further non-trivial covering spaces, but with these sorts of spaces, a composition of covering maps is not necessarily a covering map. Note also that Z has open subsets that are not UC-spaces, for instance the quotient of the subset of Z' consisting of points whose x-coordinate is not an integer.

6.2 *The universal covering space is not a functor*

Consider the Hopf fibration $h : S^3 \to S^2$, with fibers homeomorphic to circles. If there were a functorial universal covering space, there would be a "universal covering space fiber by fiber" $\tilde{h} : X \to S^2$ with some appropriate X and with fibers homeomorphic to $\mathbb{R}$. Since the fibers are contractible, such a fibration would be trivial, hence have a section $\sigma : S^2 \to X$, which would project to a section $\bar{\sigma} : S^2 \to S^3$ of h. But h is a principal bundle, if it had a section, it would be trivial.

6.3 *A Kleinian group with two quasi-Fuchsian subgroups whose convex hulls intersect*

One consequence of the Poincaré polyhedron theorem [7,3] is that for any set $\mathcal{D}$ of disjoint open (round) discs $D_i \subset \mathbb{P}^1$, $i \in I$, the subgroup of Aut $\mathbb{H}^3$ generated by reflections in the boundaries of the discs is a discrete subgroup $\Gamma_{\mathcal{D}}$ of Aut $\mathbb{H}^3$. The boundary

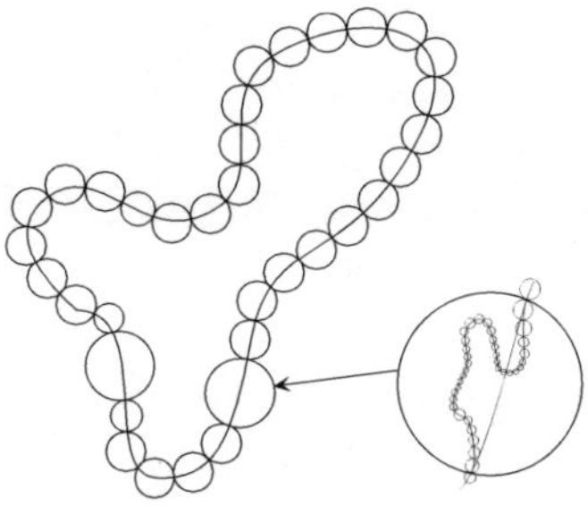

Figure 3. If you put a chain of beads along a simple closed curve, the group generated by reflections in the circles generates a discrete subgroup of $\operatorname{Aut}\mathbb{H}^3$, and a subgroup of index 2 of this group is quasi-Fuchsian, with limit set close to the original curve in the Hausdorff metric.

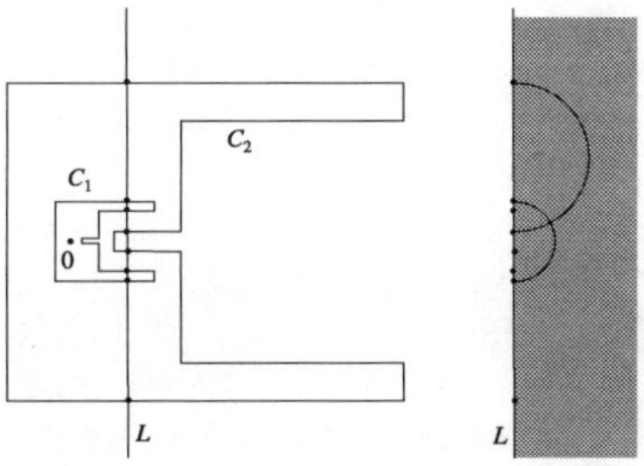

Figure 4. LEFT: The curve C_2 is the image of C_1 by $A: z \mapsto 4z$. If you daisy chain little circles along C_1, sufficiently small that if you apply $z \mapsto 4z$ to all of them the new set is disjoint from the previous one, you find a Kleinian group with two quasi-Fuchsian subgroups that are conjugate in $\operatorname{Aut}\mathbb{H}^3$ but not within the group itself. RIGHT: The vertical halfplane through L, viewed as the plane of the paper. Geodesics joining points of C_1 and C_2 intersect.

of each disc is the boundary of a plane, whose complement has two components H_i^+ and H_i^-; with D_i in the closure of H_i^+. The intersection

$$V = \bigcap_{i \in I} H_i^-$$

is a fundamental domain for $\Gamma_{\mathcal{D}}$.

One consequence is the following proposition.

Proposition 19. *Any simple closed curve $C \subset \mathbb{P}^1$ can be approximated in the Hausdorff topology by limit sets of finitely-generated quasi-Fuchsian groups.*

Proof. Put a chain $\mathcal{D}$ of disjoint, externally tangent discs along C. Figure 3 suggests why the limit set is homeomorphic to a circle. By the Ahlfors finiteness theorem, if G is a finitely generated Kleinian group whose limit set Λ_G is homeomorphic to a circle, then the subgroup stabilizing the components (either G or of index 2 in G) is quasi-Fuchsian. □

Next, construct two simple closed curves C_1, C_2 such that C_2 is the image of C_1 under $A : z \mapsto 4z$. as indicated in Figure 4

Put a daisychain of small circles along C_1, and move them by A to form a daisy-chain along C_2. Suppose that the circles are sufficiently small that the second batch of discs is disjoint from the first. By the Poincaré polyhedron theorem, the reflections in all these circles is a discrete group G, and it evidently has two quasi-Fuchsian subgroups G_1, G_2 such that $G_2 = AG_1A^{-1}$. One can check that the quasi-Fuchsian groups stabilize components U_1, U_2 of $\mathbb{P}^1 - \Lambda_G$, and A maps U_1 to $\mathbb{P}^1 - \overline{U_2}$. We are thus in a position to apply Theorem 15.

But the convex hulls in $\mathbb{H}^3$ of the Λ_{G_1} and Λ_{G_2} intersect. The line L intersects each of C_1 and C_2 in 4 points. These points do not lie in disjoint arcs, so in the plane of $\mathbb{H}^3$ which, in the upper half space model of $\mathbb{H}^3$, is the vertical plane through L, geodesics joining points of one curve intersect geodesics joining points of the other curve.

References

[1] J. W. Anderson and R. D. Canary, The visual core of a hyperbolic 3-manifold, *Mathematische Annalen*, **321** (2001) 989–1000.

[2] A. Douady and R. Douady, Algébre et théories Galoisiennes, Cassini, Paris (2005).

[3] J. H. Hubbard, Teichmüller theory and Applications to Topology, Geometry and Dynamics, Vol. 2 Matrix Editions, Ithaca, NY to appear.

[4] R. Kulkarni and U. Pinkall, Uniformization of Geometric Structures with Applicationsto Conformal Geometry, Differential Geometry, Peniscola (1985) 109–209, Lecture Notes in Mathematics 1209, Springer Verlag (1985).

[5] R. Kulkarni, On the Principle of Uniformization, *Journal of Differential Geometry*, **13** (1978) 109–138.

[6] B. Maskit, Kleinian Groups, Springer Verlag (1988).

[7] B. Maskit, On boundaries of Teichmüller spaces and on Kleinian groups, II, *Ann. Math.*, **91** (1970) 607–639.

[8] J. P. Otal, Thurston's hyperbolization theorem of Haken manifolds, Surveys in differential geometry, Vol. 3, Int. Press, Boston, MA (1998) 77–194.

[9] W. P. Thurston, edited by S. Levy, Three-Dimensional Geometry and Topology, Princeton University Press (1997).

[10] W. P. Thurston, The geometry and topology of 3-manifolds http://www.msri.org/publications/books/gt3m (1979).

Teichmüller Theory and Moduli Problem No. 10, 2010, pp. 413–435.

Function Models for Teichmüller Spaces and Dual Geometric Gibbs Type Measure Theory for Circle Dynamics

Yunping Jiang

Department of Mathematics, Queens College of the City University of New York,
Flushing, NY 11367-1597
and
Department of Mathematics, Graduate School of the City University of New York,
365 Fifth Avenue, New York, NY 10016
e-mail: yunping.jiang@qc.cuny.edu

Abstract. Geometric models and Teichmüller structures have been introduced for the space of smooth expanding circle endomorphisms and for the space of uniformly symmetric circle endomorphisms. The latter one is the completion of the previous one under the Techmüller metric. Moreover, the spaces of geometric models as well as the Teichmüller spaces can be described as the space of Hölder continuous scaling functions and the space of continuous scaling functions on the dual symbolic space. The characterizations of these scaling functions have been also investigated. The Gibbs measure theory and the dual Gibbs measure theory for smooth expanding circle dynamics have been viewed from the geometric point of view. However, for uniformly symmetric circle dynamics, an appropriate Gibbs measure theory is unavailable, but a dual Gibbs type measure theory has been developed for the uniformly symmetric case. This development extends the dual Gibbs measure theory for the smooth case from the geometric point of view. In this survey article, we give a review of these developments which combines ideas and techniques from dynamical systems, quasiconformal mapping theory, and Teichmüller theory. There is a measure-theoretical version which is called g-measure theory and which corresponds to the dual geometric Gibbs type measure theory. We briefly review it too.

Keywords. Scaling function, C^{1+} expanding circle endomorphism, uniformly symmetric circle endomorphism, Teichmüller space, symbolic dynamical system, dual symbolic dynamical system.

2000 Subject Classification: 58F23, 30C62

1. Introduction

Circle maps are basic elements in dynamical systems. The dynamics of a smooth expanding circle map presents many profound phenomena in mathematics and physics such as the structural stability theory, ergodic theory, probability theory, and, more recently, chaos theory. Teichmüller theory studies complex manifold structures of almost complex structures on Riemann surfaces. We have brought in some concepts and techniques in Teichmüller theory into the study of geometric structures of spaces

The research is partially supported by PSC-CUNY awards.

This article is prepared for the proceedings of the International Workshop on Teichmüller theory and moduli problems held at the Harish-Chandra Research Institute (HRI), Allahabad, India, January 5 to 14, 2006.

of circle expanding maps. In this survey article we review some development in this direction.

2. Circle endomorphisms

The theme here is an orientation-preserving covering map f from the unit circle $T = \{z \in \mathbb{C} \mid |z| = 1\}$ onto itself. Let d be the topological degree of f. We assume that $d \geq 2$. The universal cover of T is the real line $\mathbb{R}$ with a covering map

$$\pi(x) = e^{2\pi xi} : \mathbb{R} \to T$$

Then f can be lifted to an orientation-preserving homeomorphism F of $\mathbb{R}$ with the property that $F(x+1) = F(x) + d$. Since a covering map of degree ≥ 2 has a fixed point, we assume that $z = 1$ is a fixed point of f. Then by assuming $F(0) = 0$, we set up a one-to-one correspondence between degree d circle covering maps f with $f(1) = 1$ and real line homeomorphisms F with $F(x+1) = F(x) + d$. Thus we call in this paper f or the corresponding F a circle endomorphism. We use f^n (or F^n) to mean the composition of f (or F) by itself $n > 0$ times.

A circle endomorphism f is C^k for $k \geq 1$ if its kth-derivative $F^{(k)}$ is continuous and $C^{k+\alpha}$ for some $0 < \alpha \leq 1$ if, furthermore, $F^{(k)}$ is α-Hölder continuous, that is,

$$\sup_{x \neq y \in \mathbb{R}} \frac{|F^{(k)}(x) - F^{(k)}(y)|}{|x-y|^\alpha} = \sup_{x \neq y \in [0,1]} \frac{|F^{(k)}(x) - F^{(k)}(y)|}{|x-y|^\alpha} < \infty.$$

A C^1 circle endomorphism f is called expanding if there are constants $C > 0$ and $\lambda > 1$ such that

$$(F^n)'(x) \geq C\lambda^n, \quad n = 1, 2, \ldots.$$

3. Topological models

The topological classification of smooth expanding circle endomorphisms was first considered by Shub [24] in 1960's. He proved that two C^2 expanding circle endomorphisms f and g are topologically conjugate if and only if they have the same degree. Here f and g are topologically conjugate if there is a homeomorphism h of T such that

$$f \circ h = h \circ g.$$

By also considering the lift G of g and the lift H of h, we have an equivalent definition that f and g are topologically conjugate if there is a homeomorphism H of $\mathbb{R}$ with $H(x+1) = H(x) + 1$ such that

$$F \circ H = H \circ G \pmod 1.$$

Shub's proof is an application of the contracting fixed point theorem in functional analysis. Consider the space $\mathcal{C}$ of all continuous functions ϕ on $\mathbb{R}$ with $\phi(x+1) = \phi(x) + 1$ with the maximum norm

$$\|\phi\| = \sup_{x \in [0,1]} |\phi(x)|.$$

Then $\mathcal{C}$ is a Banach space. Define an operator $\mathcal{L} = \mathcal{L}_{F,G}$ as

$$\mathcal{L}\phi(x) = F^{-1} \circ \phi \circ G : \mathcal{C} \to \mathcal{C}.$$

(Note that $F^{-1}(x+d) = F^{-1}(x) + 1$.) Without loss of generality, we assume that $C = 1$. Then one can check that

$$\|\mathcal{L}\phi - \mathcal{L}\psi\| \leq \frac{1}{\lambda}\|\phi - \psi\|.$$

So $\mathcal{L}$ is a contracting functional from the Banach space $\mathcal{C}$ into itself. And thus it has a unique fixed point H, that is,

$$F^{-1} \circ H \circ G = H$$

Similarly, $\mathcal{L}_{G,F}$ has a unique fixed point $\tilde{H}$, that is,

$$G^{-1} \circ \tilde{H} \circ F = \tilde{H}.$$

This implies that

$$H \circ \tilde{H} = id.$$

Therefore, H is a homeomorphism of $\mathbb{R}$ such that

$$F \circ H = H \circ G \pmod{1}.$$

There is a more general theorem if we bring in the consideration of Markov partitions. Consider a partition of [0, 1] by

$$I_i = I_{i,f} = F^{-1}([i, i+1]), \quad 0 \leq i \leq d-1.$$

It is a Markov partition in the following sense: If we consider a corresponding partition of T, which we still denote as $\{I_i\}_{i=0}^{d-1}$, then

(1) the union of these intervals is T;
(2) all the intervals in the partition have pairwise disjoint interiors;
(3) the restriction of f to the interior of every interval in the partition is injective.

We use

$$\eta_0 = \eta_{0,f} = \{I_i\}_{i=0}^{d-1}$$

to denote this initial Markov partition. We then have a sequence of Markov partitions

$$\eta_n = \eta_{n,f} = f^{-n}(\eta_0), \quad n = 0, 1, 2, \ldots$$

on the unit circle T as well as the unit interval [0, 1]. We can label each interval in η_n as follows. Define

$$g_i(x) = F^{-1}(x+i) : [0,1] \to I_i, \quad i = 0, 1, \ldots, d-1.$$

Each g_i is a homeomorphism. Given a word $w_n = i_0 \cdots i_k \cdots i_{n-1}$ of $\{0, \ldots, d-1\}$ of length $n \geq 1$, define

$$g_{w_n} = g_{i_0} \circ g_{i_1} \circ \cdots \circ g_{i_{n-1}}.$$

Let

$$I_{w_n} = I_{w_n, f} = g_{w_n}([0, 1]).$$

Then

$$\eta_n = \{I_{w_n} \mid w_n = i_0 \dots i_k \dots i_{n-1},\ i_k \in \{0, \dots, d-1\}\}.$$

One can check that for a word $w = i_0 \dots i_{n-1} i_n \dots$ of infinite length, and with $w_n = i_0 \dots i_{n-1}$, then

$$\dots \subset I_{w_n} \subset I_{w_{n-1}} \subset \dots I_{w_1} \subset [0, 1].$$

Since each I_{w_n} is compact,

$$I_w = \cap_{n=1}^{\infty} I_{w_n} \neq \emptyset.$$

Consider the space

$$\Sigma^+ = \Sigma_d^+ = \prod_{n=0}^{\infty} \{0, 1, \dots, d-1\}$$

$$= \{w = i_0 i_1 \dots i_k \dots i_{n-1} \dots \mid i_k \in \{0, 1, \dots, d-1\},\ k = 0, 1, \dots\}$$

with the product topology. Then it is a compact topological space.

If each $I_w = \{x_w\}$ contains only one point, then we define the projection $\pi_+ = \pi_{+,f}$ from Σ^+ onto T as

$$\pi_+(w) = x_w.$$

The projection π_+ is 1-1 except for a countable set B consisting of all labellings w of endpoints in the partitions $\eta_n = \{I_{w_n}\}, n = 0, 1, \dots$.

Let

$$\sigma^+(w) = i_1 \dots i_{n-1} i_n \dots$$

be the left shift map. Then (Σ^+, σ^+) is called a symbolic dynamical system. From our construction, one can check that

$$\pi_+ \circ \sigma^+(w) = f \circ \pi_+(w), \quad w \in \Sigma^+.$$

Let

$$\varepsilon_n = \varepsilon_{n,f} = \max_{w_n} |I_w|$$

where w_n runs over all words of length n of $\{0, 1, \dots, d-1\}$. Then we have a more general Shub type theorem.

Theorem 1. *Let f and g be two circle endomorphisms such that both $\varepsilon_{n,f}$ and $\varepsilon_{n,g}$ tend to zero as $n \to \infty$. Then f and g are topologically conjugate if and only if their topological degrees are the same.*

Proof. Since both sets $I_{w,f} = \{x_w\}$ and $I_{w,g} = \{y_w\}$ contain only a single point for each w, we define

$$h(x_w) = y_w.$$

One can check that h is a homeomorphism with the inverse $h^{-1}(y_w) = x_w$. □

Therefore, for a fixed degree $d > 1$, there is only one topological model (Σ^+, σ^+) for the dynamics of all circle endomorphisms of degree d with $\varepsilon_n \to 0$.

4. Geometric models, part I

The next theme is the study of geometric models. A result analogous to Mostow's rigidity theorem for closed hyperbolic 3-manifolds was proved by Shub and Sullivan [25]. The result can be stated as follows: Suppose f and g are two topologically conjugate real analytic expanding circle endomorphisms. If the conjugacy h is absolutely continuous, it must be also real analytic. Later, this result was proved for a more general case: Suppose f and g are two topologically conjugate $C^{k+\alpha}$ expanding circle endomorphisms for $1 \leq k \leq \omega$ and $0 < \alpha \leq 1$. If the conjugacy h is absolutely continuous, it must be also $C^{k+\alpha}$. Smooth invariants of a circle endomorphism have also been investigated. A quantity is called a smooth invariant if it is the same for f and g as long as f and g are smoothly conjugate (this means that the conjugacy is C^k for $k \geq 1$). A point p of f is called a periodic point of period $n \geq 1$ if $f^i(p) \neq p$ for $0 \leq i \leq n-1$ but $f^n(p) = p$. The eigenvalue at a periodic point p of period n is defined as $e_p = (f^n)'(p)$. The eigenvalue e_p is a smooth invariant. The set of all eigenvalues of a $C^{1+\alpha}$ expanding circle endomorphism is actually a set of complete smooth invariants, where $0 < \alpha \leq 1$. This means that two $C^{1+\alpha}$ expanding circle endomorphisms f and g of degree $d > 1$ are smoothly conjugate if and only if their eigenvalues at the corresponding periodic points are the same. Therefore, one can use the set of all eigenvalues to classify geometric models of smooth expanding circle endomorphisms of the same degree. (Research in this direction has been extended to a larger class which even allows one to include maps with critical points. The reader who is interested in the smooth classification of one-dimensional dynamical systems in this direction may refer to [10–14] for more details.)

However, the structure of the set of all eigenvalues is not clear. In what follows, we define a function which is called a scaling function and will contain full information about the set of all eigenvalues in this context. (The name of the scaling function in this context was first used by Feigenbaum [8] in describing the universal geometric structure of attractors of infinitely period doubling folding maps. It was then used by Sullivan [26] for Cantor sets on the line to describe differential structures for fractal sets. The present form of the definition was formulated in [10] for any Markov map and then used to study the smooth classification of one-dimensional maps which have certain Markov properties. The reader may also refer to [11,12] for more details.)

As we have already seen, given a circle endomorphism of degree $d > 1$, there is an interval system

$$\{\eta_n\}_{n=1}^{\infty} = \{\{I_{w_n}\}_{w_n}\}_{n=1}^{\infty}$$

where w_n runs over all words of length n of $\{0, 1, \ldots, d-1\}$. When we constructed the topological model $(\Sigma^+, \sigma^+\}$ from this interval system, we read each w_n from the left to the right, i.e., $w_n = i_0 i_1 \ldots i_{n-1}$. From the topological point of view, this means that we consider the set of all left cylinders

$$[w_n] = [w_n]_l = [i_0 i_1 \ldots i_{n-1}]_l = \{w' = i'_0 i'_1 \ldots i'_{n-1} i'_n \ldots \mid i'_0 = i_0, \ldots, i'_{n-1} = i_{n-1}\}$$

as a basis for the topology.

Now let us consider another topology which has a basis consisting of all right cylinders. We read w_n from the right to the left, $\kappa_n = w_n = j_{n-1} j_{n-2} \ldots j_0$ and define

$$\Sigma^- = \Sigma_d^- = \{\kappa = \ldots j_{n-1} \ldots j_k \ldots j_1 j_0 \mid j_k \in \{0, 1, \ldots, d-1\},\ k = 0, 1, \ldots\}.$$

It is a topological space with a basis for the topology consisting of all right cylinders

$$[\kappa_n] = [\kappa_n]_r = [j_{n-1} \ldots j_0]_r = \{\kappa' = \ldots j'_n j'_{n-1} \cdots j'_0 \mid j'_{n-1} = j_{n-1}, \ldots, j'_0 = j_0\}.$$

Consider the right shift map

$$\sigma^- : \ldots j_{n-1} \ldots j_1 j_0 \mapsto \cdots j_{n-1} \ldots j_1.$$

Then we call (Σ^-, σ^-) the dual symbolic dynamical system for f.

Another way to view the symbolic dynamical system and the dual symbolic dynamical system is to consider the inverse limit of $f : T \to T$. This inverse limit can be viewed as a solenoid with the symbolic representation

$$\Sigma = \Sigma^- \times \Sigma^+.$$

Then Σ^- represents the transversal direction and Σ^+ represents the leaf direction.

On the transversal direction Σ^-, we define a function called the scaling function for f as follows. For any $\kappa = \ldots j_{n-1} \ldots j_1 j_0 \in \Sigma^-$, let $\kappa_n = j_{n-1} \ldots j_1 j_0$, $\sigma^-(\kappa_n) = j_{n-1} \ldots j_1$. Then

$$I_{\kappa_n} \subset I_{\sigma^-(\kappa_n)}.$$

Define

$$S(\kappa_n) = S_f(\kappa_n) = \frac{|I_{\kappa_n}|}{|I_{\sigma^-(\kappa_n)}|}.$$

Definition 1. *If for every* $\kappa \in \Sigma^-$,

$$S(\kappa) = S_f(\kappa) = \lim_{n \to \infty} S(\kappa_n)$$

exists, then we have a function

$$S = S_f : \Sigma^- \to \mathbb{R}^+.$$

We call this function the scaling function of f.

The space Σ^- is a metric space with the metric

$$d(w, w') = \sum_{k=0}^{\infty} \frac{|i_k - i'_k|}{d^k}.$$

A function S on Σ^- is called Hölder continuous if there are constants $C > 0$ and $0 < \beta \leq 1$ such that

$$|S(\kappa) - S(\kappa')| \leq C(d(\kappa, \kappa'))^\beta, \quad \kappa, \kappa' \in \Sigma^-.$$

Let $\mathcal{C}^{1+}$ denote the space of all $C^{1+\alpha}$ expanding circle endomorphisms for some $0 < \alpha \leq 1$. We have:

Theorem 2. *The scaling function S for $f \in \mathcal{C}^{1+}$ exists and is a Hölder continuous function. Furthermore, S is a completely smooth invariant. This means that $f, g \in \mathcal{C}^{1+}$ are C^1 conjugate if and only if they have the same scaling functions, i.e., $S_f = S_g$.*

This result is actually proved for a larger class of one-dimensional maps which may have critical points. The reader who is interested in this direction can refer to [11–14].

Thus geometric models of $\mathcal{C}^{1+}$ can be represented by degrees $d > 1$ and scaling functions S as follows. We say $f \sim_s g$ if f and g are C^1 conjugate. It is an equivalence relation in $\mathcal{C}^{1+}$. Then we have that

$$\mathcal{C}^{1+}/\sim_s = \{d, S\}.$$

For a fixed $d > 1$, let $\mathcal{C}_d^{1+}$ be the space of $f \in \mathcal{C}^{1+}$ with the degree d. Then

$$\mathcal{C}_d^{1+}/\sim_s = \{S\}.$$

There are two natural problems now. One is to study the geometric structure on $\mathcal{C}_d^{1+}/\sim_s$. The other is to characterize a scaling function. We will discuss these two problems.

5. Teichmüller structures, part I

A discussion of the first problem follows a similar idea to that of Teichmüller theory for Riemann surfaces with the help of the following theorem (refer to [10,11,15]).

Theorem 3. *Suppose f and g are two maps in $\mathcal{C}_d^{1+}$. Suppose h is the topological conjugacy between f and g. Then h is a quasisymmetric homeomorphism.*

A homeomorphism h of T is called quasisymmetric (see [1]) if there is a constant $K \geq 1$ such that

$$K^{-1} \leq \frac{|H(x+t) - H(x)|}{|H(x) - H(x-t)|} \leq K$$

for all $x \in \mathbb{R}$ and all $t > 0$, where H is a lift of f to the real line.

Take $q_d(z) = z^d$ as a basepoint in $\mathcal{C}_d^{1+}$. For any $f \in \mathcal{C}_d^{1+}$, let h_f be the conjugacy from f to q_d, i.e.,

$$f \circ h_f = h_f \circ q_d.$$

Thus, we can think of $\mathcal{C}_d^{1+}$ as pairs (f, h_f). Two pairs satisfy $(f, h_f) \sim_t (g, h_g)$ if $h_f \circ h_g^{-1}$ is a C^1-diffeomorphism. Then $\sim_t$ is an equivalence relation. The Teichmüller space

$$\mathcal{T}\mathcal{C}_d^{1+} = \{[(f, h_f)] \mid f \in \mathcal{C}_d^{1+}, \text{ with the basepoint } [(q_d, id)]\}$$

is the space of all $\sim_t$-equivalence classes $[(f, h_f)] = [(f, h_f)]_t$ with the basepoint $[(q_d, id)]$. This space has a Teichmüller metric $d_T(\cdot, \cdot)$ as we describe below.

We first consider the universal Teichmüller space. Let $\mathcal{QS}$ be the set of all quasisymmetric homeomorphisms of the unit circle T factored by the space of all Möbius transformations of the circle. (Then $\mathcal{QS}$ may be identified with the set of all quasi-symmetric homeomorphisms of the unit circle fixing three points). For any $h \in \mathcal{QS}$, let $\mathcal{E}_h$ be the set of all quasiconformal extensions of h into the unit disk. Let $K_{\tilde{h}}$ be the quasiconformal dilatation of $\tilde{h} \in \mathcal{E}_h$. Using quasiconformal dilatation, one defines a distance in $\mathcal{QS}$ by

$$d_T(h_1, h_2) = \frac{1}{2} \inf\{\log K_{\tilde{h}_1 \tilde{h}_2^{-1}} \mid \tilde{h}_1 \in \mathcal{E}_{h_1}, \tilde{h}_2 \in \mathcal{E}_{h_2}\}.$$

Here $(\mathcal{QS}, d)$ is called the universal Teichmüller space. It is a complete metric space and a complex manifold with complex structure compatible with the Hilbert transform (see, for example, [1]).

A quasisymmetric homeomorphism h is called symmetric if there is a bounded positive function $\epsilon(t)$ such that $\epsilon(t) \to 0^+$ as $t \to 0^+$ and

$$1 - \epsilon(t) \leq \frac{|H(x+t) - H(x)|}{|H(x) - H(x-t)|} \leq 1 + \epsilon(t)$$

for any x in $\mathbb{R}$, where $H(x+1) = H(x) + 1$ is a lift of h. A C^1-diffeomorphism of the unit circle is symmetric. However, a symmetric homeomorphism of the unit circle could be very singular. Let $\mathcal{S}$ be the subset of $\mathcal{QS}$ consisting of all symmetric homeomorphisms of the unit circle. The space $\mathcal{S}$ is a closed subgroup of $\mathcal{QS}$. The topology coming from the metric d_T on $\mathcal{QS}$ induces a topology on the factor space $\mathcal{QS}$ mod $\mathcal{S}$. Given two cosets $\mathcal{S}f$ and $\mathcal{S}g$ in this factor space, define a metric by

$$\bar{d}_T(\mathcal{S}f, \mathcal{S}g) = \inf_{A, B \in \mathcal{S}} d(Af, Bg).$$

The factor space $\mathcal{QS}$ mod $\mathcal{S}$ with this metric is a complete metric space and a complex manifold. The topology on $(\mathcal{QS} \bmod \mathcal{S}, \bar{d}_T)$ is the finest topology which makes the projection $\pi : \mathcal{QS} \to \mathcal{QS}$ mod $\mathcal{S}$ continuous, and π is also holomorphic. An equivalent topology can be defined as follows. For any $h \in \mathcal{QS}$, let $\tilde{h}$ be a quasiconformal extension of h to a small neighborhood U of T in the complex plane. Let

$$\mu_{\tilde{h}}(z) = \frac{\tilde{h}_{\bar{z}}(z)}{\tilde{h}_z(z)}, \quad k_{\tilde{h}} = \|\mu_{\tilde{h}}\|_\infty \quad \text{and} \quad B_{\tilde{h}} = \frac{1 + k_{\tilde{h}}}{1 - k_{\tilde{h}}}.$$

Then the boundary dilatation is defined as

$$B_h = \inf B_{\tilde{h}}$$

where the infimum is taken over all quasiconformal extensions of h near the unit circle. It is known that h is symmetric if and only if $B_h = 1$. Define

$$\tilde{d}(h_1, h_2) = \frac{1}{2} \log B_{h_2^{-1} h_1}.$$

The two metrics $\bar{d}$ and $\tilde{d}$ on $\mathcal{QS}$ mod $\mathcal{S}$ are equal. The reader may refer to [9] for this. The Teichmüller metric on $\mathcal{TC}_d^{1+}$ is defined similarly. Let τ and τ' be two points in $\mathcal{TC}_d^{1+}$. Then

$$d_T(\tau, \tau') = \frac{1}{2} \log B_{h_f^{-1} \circ h_g}$$

where $(f, h_f) \in \tau$ and $(g, \tau_g) \in \tau'$.

Since the space of geometric models can be represented by the space of scaling functions, *the Teichmüller space can also be represented by the space of Hölder continuous scaling functions* S_f *for* $f \in \mathcal{C}_d^{1+}$ *with the basepoint* $S(\kappa) = 1/d$.

6. Characterizations, part I

The characterization of a scaling function has been done for $d = 2$, which is the most interesting case for us. First by the definition of a scaling function, one can easily check that

$$S(\kappa 0) + S(\kappa 1) = 1, \quad \kappa \in \Sigma^- = \Sigma_2^-.$$

We call this the summation condition. In addition to this condition, a scaling function also enjoys another non-trivial condition which we call the compatibility condition,

$$\prod_{n=0}^{\infty} \frac{S(\kappa 1 \overbrace{0 \cdots 0}^{n})}{S(\kappa 0 \underbrace{1 \cdots 1}_{n})} = \text{const.}, \quad \kappa \in \Sigma^-.$$

Actually, this infinite product converges to the constant exponentially. (Its general term must tend to 1 as n goes to ∞. This implies that $S(\cdots 000) = S(\cdots 111)$). We showed that the converse is also true as follows.

Theorem 4. *Let S be a positive Hölder continuous function on* Σ^-. *Then S is the scaling function of a map in* $\mathcal{C}_2^{1+}$ *if and only if S satisfies the summation and compatibility conditions.*

The original proof of this theorem is given in [4] and uses the Gibbs measure theory and some constructions in [22]. A proof without using the Gibbs measure theory can be founded in [16]. In the proof, we find the connection between the scaling function and the solenoid function and the linear model for a circle endomorphism and use some constructions in [5] (refer to Theorem 7). We would like to note that the solenoid function and the linear model are also interesting geometric invariants for a circle endomorphism. The solenoid function for a circle endomorphism is defined in [27] and used to describe an affine structure along leave directions Σ^+ of the solenoid $\Sigma^- \times \Sigma^+$. It has been studied in [21]. The linear model for a circle endomorphism is defined in [6]. A linear model can be thought of as a nonlinear coordinate on the unit circle. In [6], a question about what kind of nonlinear coordinate can be realized by a smooth expanding circle endomorphism arose. This question was studied in [3] by employing some results in quasiconformal theory (refer to [1]). A much simpler understanding was given in [16] by employing the naive distortion property (see [11]).

Therefore, *the Tecihmüller space* $\mathcal{TC}_2^{1+}$ *is represented by the space of positive Hölder continuous functions on* Σ^- *satisfying the summation and compatibility conditions with the basepoint* $S(\kappa) = 1/2$.

It is clear that the summation condition is true for any $d > 2$. This means that the scaling function S of $f \in \mathcal{C}_d^{1+}$ satisfies

$$S(\kappa 0) + S(\kappa 1) + \cdots + S(\kappa(d-1)) = 1, \quad \kappa \in \Sigma^- = \Sigma_d^-.$$

The compatibility condition for $d > 2$ should be similar. However, the proof of the characterization of the scaling function for a map in $\mathcal{C}_d^{1+}$ for $d > 2$ should be slightly more complicated than the case $d = 2$, but it is a promising problem.

7. Teichmüller structures, part II

The Teichmüller space $(\mathcal{TC}_d^{1+}, d_T(\cdot,\cdot))$ is not complete. Its completion is an interesting subject to be studied. A circle endomorphism f of degree d is called uniformly symmetric if all its inverse branches for f^n, $n = 1, 2, \ldots$, are symmetric uniformly. More precisely, there is a bounded positive function $\epsilon(t)$ with $\epsilon(t) \to 0^+$ as $t \to 0^+$ such that

$$1 - \epsilon(t) \le \frac{|F^{-n}(x+t) - F^{-n}(x)|}{|F^{-n}(x) - F^{-n}(x-t)|} \le 1 + \epsilon(t), \quad x \in \mathbb{R},\ t > 0,\ n = 1, 2, \ldots.$$

By the naive distortion lemma (see, for example, [11]) we have

Proposition 1. *Any map $f \in \mathcal{C}^{1+}$ is uniformly symmetric.*

Let $\mathcal{US}$ be the space of all uniformly symmetric circle endomorphisms of degree $d \ge 2$. The above proposition says that $\mathcal{C}^{1+} \subset \mathcal{US}$. However, a map in $\mathcal{US}$ can be quite different. For example, it may not be differentiable and may not be absolutely continuous. However, we have shown that from the dual point of view, it has a lot of similarity to what we have studied for a map in $\mathcal{C}^{1+}$. (However, a C^1 expanding circle endomorphism is very different from what we have studied for a map in $\mathcal{C}^{1+}$ (see [22]).)

For a fixed $d \ge 2$, let $\mathcal{US}_d$ be the space of all uniformly symmetric circle endomorphisms of degree d. For $f \in \mathcal{US}_d$, it is certainly uniformly M-quasisymmetric for a fixed constant $M > 1$, that is,

$$M^{-1} \le \frac{|F^{-n}(x+t) - F^{-n}(x)|}{|F^{-n}(x) - F^{-n}(x-t)|} \le M, \quad x \in \mathbb{R},\ t > 0,\ n = 1, 2, \ldots.$$

We have:

Proposition 2. *If $f \in \mathcal{US}_d$, then there is a constant $C > 0$ such that*

$$S(\kappa_n) = \frac{|I_{\kappa_n}|}{|I_{\sigma^-(\kappa_n)}|} \ge C$$

for all finite words $\kappa_n = j_{n-1} \cdots j_1 j_0$ of $\{0, 1, \ldots, d-1\}$.

The above proposition means that the sequence of nested partitions

$$\{\eta_n\}_{n=1}^{\infty} = \{\{I_{w_n}\}_{w_n}\}_{n=1}^{\infty}$$

has bounded geometry. Using the above proposition and the summation condition, we have constants $D > 0$ and $0 < \tau < 1$ such that

$$\varepsilon_n \leq D\tau^n, \quad n = 1, 2, \ldots.$$

Therefore, just like in the proofs of Theorem 1 and Theorem 3, we have:

Theorem 5. *Any two maps $f, g \in \mathcal{US}_d$ are topologically conjugate and, furthermore, the conjugacy h is quasisymmetric.*

With this proposition, we can define the Teichmüller space for $\mathcal{US}_d$ as we did for $\mathcal{C}_d^{1+}$. Take $q_d(z) = z^d$ as a basepoint in $\mathcal{US}_d$. For any $f \in \mathcal{US}_d$, let h_f be the conjugacy from f to q_d, i.e.,

$$f \circ h_f = h_f \circ q_d.$$

Thus we can think of $\mathcal{US}_d$ as pairs (f, h_f). Two pairs satisfy $(f, h_f) \sim_t (g, h_g)$ if $h_f \circ h_g^{-1}$ is symmetric. Then $\sim_t$ is an equivalence relation. The Teichmüller space

$$\mathcal{TUS}_d = \{[(f, h_f)] \mid f \in \mathcal{US}_d, \text{ with the basepoint } [(q_d, id)]\}$$

is the space of all $\sim_t$-equivalence classes $[(f, h_f)] = [(f, h_f)]_t$ equipped with a Teichmüller metric

$$d_T(\tau, \tau') = \frac{1}{2} \log B_{h_f^{-1} \circ h_g}$$

where $(f, h_f) \in \tau$ and $(g, h_g) \in \tau'$.

If $f, g \in \mathcal{C}_d^{1+}$ and if the conjugacy h between f and g is symmetric, then h must be C^1. The reason is that if the conjugacy between f and g is symmetric, then their scaling functions S_f and S_g must be the same (refer to Theorem 7). Therefore they are C^1-conjugate and the conjugacy h must be C^1 (see Theorem 2). (A related easy but interesting fact is that the ratio of eigenvalues $e_f(p)$ and $e_g(h(p))$ of f and g at corresponding periodic point p and $h(p)$ determines the local quasisymmetric constant of h at p. If h is symmetric, its local quasisymmetric constant at p is 1, so the ratio $e_f(p)/e_g(h(p))$ is 1.) This implies that the Teichmüller space $\mathcal{TC}_d^{1+}$ is indeed a subspace of the Teichmüller space $\mathcal{TUS}_d$. Furthermore, we have (refer to [4,5]):

Theorem 6. *The space $(\mathcal{TUS}_d, d_T(\cdot, \cdot))$ is a complete complex Banach manifold and is the completion of the space $(\mathcal{TC}_d^{1+}, d_T(\cdot, \cdot))$.*

The local model of the complex Banach manifold can be thought of as the set of Beltrami coefficients on the upper-half plane $\mathbb{H}$ (complex L^∞ functions $\mu(z)$ on the upper-half plane $\mathbb{H}$ with $\|\mu(z)\|_\infty < 1$) such that $\mu(dz) = \mu(z)$ and $|\mu(z+n) - \mu(z)| \to 0$ uniformly for n as $\Im(z) \to 0$ (refer to [3–5]).

8. Geometric models, part II

The geometric models of maps in $\mathcal{US}_d$ can also be represented by their scaling functions. Two maps $f, g \in \mathcal{US}_d$ are called symmetrically conjugate if the conjugacy between them is symmetric. This is an equivalence relation which we denote as $f \sim_{sy} g$. The space $\mathcal{US}_d/\sim_{sy}$ of geometric models for maps in $\mathcal{US}_d$ is the space of all equivalence classes. We have (refer to [5]):

Theorem 7. *Suppose* $f \in \mathcal{US}_d$. *Then its scaling function*

$$S = S_f : \Sigma^- \to \mathbb{R}^+.$$

exists and is a continuous function. Furthermore, it is a complete symmetric invariant for $\mathcal{US}_d$*; this means* f *and* g *are symmetrically conjugate if and only if their scaling functions are the same, i.e.,* $S_f = S_g$.

Thus $\mathcal{TUS}_d$ can be represented by scaling functions S_f, i.e.,

$$\mathcal{US}_d / \sim_{sy} = \{S_f \mid f \in \mathcal{US}_d\}$$

and

$$\mathcal{TUS}_d = (\{S_f \mid f \in \mathcal{US}_d, \text{ with the basepoint } S = \tfrac{1}{d}\}, d_T(\cdot, \cdot)).$$

9. Characterizations, part II

The characterization of the scaling functions for $\mathcal{US}_2$ has been given as

Theorem 8. *Let* S *be a positive continuous function on* $\Sigma^- = \Sigma_2^-$. *Then* S *is the scaling function of a map in* $\mathcal{US}_2$ *if and only if* S *satisfies the summation and compatibility conditions.*

The proof of this theorem can be founded in [5]. In this case, the infinite product

$$\prod_{n=0}^{\infty} \frac{S(\kappa 1 \overbrace{0 \cdots 0}^{n})}{S(\kappa 0 \underbrace{1 \cdots 1}_{n})} = const., \qquad \kappa \in \Sigma^-,$$

in the compatibility condition converges uniformly to a constant.

Therefore, *the Tecihmüller space* $\mathcal{TUS}_2$ *is represented by the space of positive continuous functions on* Σ^- *satisfying the summation and compatibility conditions.*

Just as in the end of §6, the characterization of a scaling function of a map in $\mathcal{US}_d$ should be slightly more complicated than the case $d = 2$, but it is a promising problem.

10. Invariant measures and dual invariant measures

Consider the symbolic dynamical system (Σ^+, σ^+) and a positive Hölder continuous function $\psi = \psi(w)$. The standard Gibbs theory (refer to [2] or [7]) implies that there is a number $P = P(\log \psi)$ called the pressure and a σ^+-invariant probability measure $\mu_+ = \mu_{+,\psi}$ such that

$$C^{-1} \leq \frac{\mu_+([i_0 \cdots i_{n-1}])}{\exp(-Pn + \sum_{i=0}^{n-1} \log \psi((\sigma^+)^i(w)))} \leq C$$

for any left cylinder $[i_0 \cdots i_{n-1}]$ and any $w = i_0 \dots i_{n-1} \cdots \in [i_0 \dots i_{n-1}]$, where C is a fixed constant. Here μ_+ is a σ^+-invariant measure means that

$$\mu_+((\sigma^+)^{-1}(A)) = \mu_+(A)$$

for all Borel sets of Σ^+. A σ^+-invariant probability measure satisfying the above inequalities is called the Gibbs measure with respect to the given potential function $\log \psi$.

Two positive Hölder continuous functions ψ_1 and ψ_2 are said to be cohomologous equivalent if there is a continuous function $u = u(w)$ on Σ^+ such that

$$\log \psi_1(w) - \log \psi_2(w) = u(\sigma^+(w)) - u(w).$$

If two functions are cohomologous to each other, they have the same Gibbs measure. Therefore, the Gibbs measure can be thought of as a representation of a cohomologous class.

The Gibbs measure is also an equilibrium state. Consider the measure-theoretical entropy $h_{\mu_+}(\sigma^+)$. Since the Borel σ-algebra of Σ^+ is generated by all left cylinders, then $h_{\mu_+}(\sigma^+)$ can be calculated as

$$\begin{aligned} h_{\mu_+}(\sigma^+) &= \lim_{n\to\infty} \frac{1}{n} \sum_{w_n} (-\mu_+([w_n]) \log \mu_+([w_n])) \\ &= \lim_{n\to\infty} \sum_{w_n} \left(-\mu_+([w_n]) \log \left(\frac{\mu_+([w_n])}{\mu_+(\sigma^+([w_n]))} \right) \right), \end{aligned}$$

where w_n runs over all words $w_n = i_0 \dots i_{n-1}$ of $\{0, 1, \dots, d-1\}$ of length n. Then μ_+ is an equilibrium state in the sense that

$$P(\log \psi) = h_{\mu_+}(\sigma^+) + \int_{\Sigma^+} \log \psi(w) d\mu(w) = \sup \left\{ h_\nu(\sigma^+) + \int_{\Sigma^+} \log \psi(w) d\nu(w) \right\}$$

where ν runs over all σ^+-invariant probability measures. The measure μ_+ is unique in this case.

There is a natural way to transfer a σ^+-invariant probability measure μ_+ (not necessarily a Gibbs measure) to a σ^--invariant probability measure μ_- as follows. Given any right cylinder $[j_{n-1} \dots j_0]_r$ in Σ^-, let $i_0 \dots i_{n-1} = j_{n-1} \dots j_0$ define a left cylinder

$$[i_0 \dots i_{n-1}]_l = \{w' = i'_0 \dots i'_{n-1} i'_n \dots \mid i'_0 = i_0, \dots, i'_{n-1} = i_{n-1}\}.$$

Then define

$$\mu_-([j_{n-1} \dots j_0]_r) = \mu_+([i_0 \dots i_{n-1}]_l).$$

Then

$$\mu_-([j_{n-1}\dots j_0]_r) = \mu_+([i_0\dots i_{n-1}]_l) = \mu_+((\sigma^+)^{-1}([i_0\dots i_{n-1}]_l))$$
$$= \mu_+\left(\bigcup_{i=0}^{d-1}[ii_0\dots i_{n-1}]_l\right) = \sum_{i=0}^{d-1}\mu_+([ii_0\dots i_{n-1}]_l)$$
$$= \sum_{j=0}^{d-1}\mu_-([jj_{n-1}\dots j_0]).$$

This implies that μ_- satisfies the finite additive law for all cylinders, i.e., if $A_1, \dots, A_k$ are finitely many pairwise disjoint right cylinders in Σ^-, then

$$\mu_-\left(\bigcup_{l=1}^{k} A_k\right) = \sum_{l=1}^{k}\mu_-(A_l).$$

Also μ_- satisfies the continuity law in the sense that if $\{A_n\}_{n=1}^{\infty}$ is a decreasing sequence of cylinders and tends to the empty set (this means $A_{n+1} \subset A_n$ and $\cap_{n=1}^{\infty} A_n = \emptyset$), then $\mu_-(A_n)$ tends to zero as n goes to ∞. The reason is that since a cylinder of Σ^- is a compact set, a decreasing sequence of cylinders tending to the empty set must be eventually all empty. The Borel σ-algebra in Σ^- is generated by all right cylinders. So μ_- extends to measure on Σ^-. We have the following proposition.

Proposition 3. *μ_- is a σ^--invariant probability measure.*

Proof. We have seen that μ^- is a measure on Σ^-. Since $\mu_-(\Sigma^-) = 1$, it is a probability measure. For any right cylinder $[j_{n-1}\dots j_0]_r$,

$$\mu_-((\sigma^-)^{-1}([j_{n-1}\dots j_0]_r) = \mu_-\left(\bigcup_{j=0}^{d-1}[j_{n-1}\dots j_0 j]_r\right)$$
$$= \sum_{j=0}^{d-1}\mu_-([j_{n-1}\dots j_0 j]_r) = \sum_{i=0}^{d-1}\mu_+([i_0\dots i_{n-1} i]_l)$$
$$= \mu_+\left(\bigcup_{i=0}^{d-1}[i_0\dots i_{n-1} i]_l\right)$$
$$= \mu_+([i_0\dots i_{n-1}]_l) = \mu_-([j_{n-1}\dots j_0]_r).$$

So μ_- is σ^--invariant. □

We call μ_- a dual invariant measure. A natural question now is as follows. *Is a dual invariant measure a Gibbs measure with respect to some continuous or Hölder continuous function on Σ^-?*

A more interesting geometric question is the following. Consider a metric induced from the dual invariant measure μ_- (in the case that μ_- is supported on the whole Σ^- and is non-atomic), that is,

$$d(\kappa, \kappa') = \mu_-([j_{n-1} \cdots j_0])$$

where $[j_{n-1} \ldots j_0]$ is the smallest right cylinder containing both $\kappa = \ldots j_n j_{n-1} \ldots j_0$ and $\kappa' = \ldots j'_n j_{n-1} \ldots j_0$, $j_n \neq j'_n$. *Is σ^- differentiable with a continuous or Hölder continuous derivative under this metric?* More precisely, *does the limit*

$$\frac{d\sigma^-}{dx}(\kappa) = \lim_{n\to\infty} \frac{\mu_-(\sigma^-([j_{n-1} \cdots j_1 j_0]))}{\mu_-([j_{n-1} \cdots j_1 j_0])} = \lim_{n\to\infty} \frac{\mu_-([j_{n-1} \cdots j_1])}{\mu_-([j_{n-1} \cdots j_1 j_0])}$$

exist for every $\kappa = \ldots j_{n-1} \ldots j_1 j_0 \in \Sigma^-$? If it exists, is the limiting function continuous or Hölder continuous on Σ^-?

Actually, there is a measure-theoretical version related to these questions. I will first give a brief review of this theory.

11. *g*-measures

Let X be Σ^- (or Σ^+) and let f be σ^- (or σ^+). Let $\mathcal{B}$ be the Borel σ-algebra of X. Let $\mathcal{M}(X)$ be the space of all finite Borel measures on X. Let $\mathcal{M}(X, f)$ be the space of all f-invariant probability measures in $\mathcal{M}(X)$. Let $\mathcal{C}(X)$ be the space of all continuous real functions on X. Then $\mathcal{M}(X)$ is the dual space of $\mathcal{C}(X)$. Denote

$$\langle \phi, \mu \rangle = \int_X \phi(x) d\mu, \quad \phi \in \mathcal{C}(X) \text{ and } \mu \in \mathcal{M}(X).$$

A real non-negative continuous function ψ on X is called a g-function (the historic reason to call such a function a g-function is because of Keane's paper [19]) if

$$\sum_{f(y)=x} \psi(y) = 1.$$

For a function ψ, define the transfer operator $\mathcal{L}_\psi$ from $\mathcal{C}(X)$ into itself as

$$\mathcal{L}_\psi \phi(x) = \sum_{f(y)=x} \phi(y)\psi(y), \quad \phi \in \mathcal{C}(X).$$

One can check that $\mathcal{L}_\psi \phi = \mathcal{L}_1(\psi\phi)$ and if ψ is a g-function, then $\mathcal{L}_\psi 1 = 1$. Let $\mathcal{L}^*_\psi$ be the dual operator of $\mathcal{L}_\psi$, that is, $\mathcal{L}^*_\psi$ is the operator from $\mathcal{M}(X)$ into itself satisfying

$$\langle \phi, \mathcal{L}^*_\psi \mu \rangle = \langle \mathcal{L}_\psi \phi, \mu \rangle, \quad \forall \phi \in \mathcal{C}(X) \text{ and } \forall \mu \in \mathcal{M}(X).$$

Suppose ψ is a g-function. Then a probability measure $\mu \in \mathcal{M}(X)$ is called a g-measure if it is a fixed point of $\mathcal{L}_\psi$, that is,

$$\mathcal{L}^*_\psi \mu = \mu.$$

A g-measure is a f-invariant measure because

$$\mu(f^{-1}(B)) = \langle 1_{f^{-1}(B)}, \mu \rangle = \langle 1_B \circ f, \mathcal{L}^*_\psi \mu \rangle$$
$$= \langle \mathcal{L}_\psi 1_B \circ f, \mu \rangle = \langle 1_B, \mu \rangle = \mu(B), \ \forall B \in \mathcal{B}.$$

For any $\mu \in \mathcal{M}(X)$, let $\tilde{\mu} = \mathcal{L}^*_1 \mu$.

Proposition 4.

$$\tilde{\mu}(B) = \sum_{j=0}^{d-1} \mu(f(B \cap [j]))$$

where B is any Borel subset in $\mathcal{B}$ and $[j]$ is the right cylinder of j. Moreover, if $\mu \in \mathcal{M}(X, f)$, μ is absolutely continuous with respect to $\tilde{\mu}$.

Proof. For any Borel subset $B \in \mathcal{B}$,

$$\tilde{\mu}(B) = \langle 1_B, \mathcal{L}_1^* \mu \rangle = \langle \mathcal{L}_1 1_B, \mu \rangle.$$

But

$$\mathcal{L}_1 1_B(x) = \sum_{j=0}^{d-1} 1_B(xj) = \sum_{j=0}^{d-1} 1_{f(B\cap[j])}(x).$$

So we have that

$$\tilde{\mu}(B) = \sum_{j=0}^{d-1} \mu(f(B \cap [j])).$$

If μ is f-invariant, then we have

$$\tilde{\mu}(B) = \sum_{j=0}^{d-1} \mu(f(B \cap [j])) = \sum_{j=0}^{d-1} \mu(f^{-1}(f(B \cap [j]))) \geq \sum_{j=0}^{d-1} \mu(B \cap [j]) = \mu(B).$$

Therefore, $\mu(B) = 0$ whenever $\tilde{\mu}(B) = 0$. So μ is absolutely continuous with respect to $\tilde{\mu}$. □

Suppose $\mu \in \mathcal{M}(X, f)$. Then μ is absolutely continuous with respect to $\tilde{\mu}$. So the Radon-Nikodým derivative

$$D_\mu(x) = \frac{d\mu}{d\tilde{\mu}}(x), \quad \tilde{\mu} - a.e.\ x$$

of μ with respect to $\tilde{\mu}$ exists $\tilde{\mu}$-a.e. and is a $\tilde{\mu}$-measurable function. We would like to note that $\tilde{\mu}$ may not be absolutely continuous with respect to μ.

The following theorem was proved by Ledrappier in [20] and was used by Walters in [28] in the study of a generalized version of Ruelle's theorem.

Theorem 9. *Suppose ψ is a g-function and $\mu \in \mathcal{M}(X)$ is a probability measure. The followings are equivalent:*

i) *μ is a g-measure, i.e., $\mathcal{L}_\psi^* \mu = \mu$.*
ii) *$\mu \in \mathcal{M}(X, f)$ and $D_\mu(x) = \psi(x)$ for $\tilde{\mu}$-a.e. x.*
iii) *$\mu \in \mathcal{M}(X, f)$ and*

$$E[\phi | f^{-1}(\mathcal{B})](x) = \mathcal{L}_\psi \phi(fx) = \sum_{fy=fx} \psi(y)\phi(y), \text{ for } \mu\text{-a.e. } x$$

where $E[\phi | f^{-1}(\mathcal{B})]$ is the conditional expectation of ϕ with respect to $f^{-1}(\mathcal{B})$.

iv) $\mu \in \mathcal{M}(X, f)$ *and is an equilibrium state in the meaning that*

$$0 = h_\mu(f) + \int_X \log \psi \, d\mu = \sup \left\{ h_\nu(f) + \int_X \log \psi \, d\nu \mid \nu \in \mathcal{M}(X, f) \right\}.$$

(Note that the pressure $P(\log \psi) = 0$ for a g-function ψ.)

For any σ^+-invariant probability measure μ_+, let μ_- be the dual σ^--invariant probability measure which we have constructed in the previous section. Then we have a $\tilde{\mu}_+$-measurable function

$$D_{\mu_+}(w) = \lim_{n\to\infty} \frac{\mu_+([i_0 i_1 \ldots i_{n-1}])}{\mu_+([i_1 \ldots i_{n-1}])}, \quad \text{for } \tilde{\mu}_+ \text{-a.e. } w = i_0 i_1 \ldots i_{n-1} \ldots$$

and a $\tilde{\mu}_-$-measurable function

$$D_{\mu_-}(\kappa) = \lim_{n\to\infty} \frac{\mu_-([j_{n-1} \ldots j_1 j_0])}{\mu_-([j_{n-1} \ldots j_1])}, \quad \text{for } \tilde{\mu}_- \text{-a.e. } \kappa = \ldots j_{n-1} \ldots j_0.$$

Now the question related to those at the end of the previous section is as follows. Can we extend D_{μ_-} as well as D_{μ_+} to a continuous g-function or a Hölder continuous g-function?

The Borel σ-algebra of Σ^+ (or of Σ^-) is generated by all left cylinders (or all right cylinders). The measure-theoretical entropy $h_{\mu_+}(\sigma^+)$ can be calculated as

$$\begin{aligned} h_{\mu_+}(\sigma^+) &= \lim_{n\to\infty} \frac{1}{n} \sum_{w_n} (-\mu_+([w_n]) \log \mu_+([w_n])) \\ &= \lim_{n\to\infty} \sum_{w_n} \left(-\mu_-([w_n]) \log \left(\frac{\mu_-([w_n])}{\mu_-(\sigma^+([w_n]))} \right) \right), \end{aligned}$$

where w_n runs over all words $w_n = i_0 \ldots i_{n-1}$ of $\{0, 1, \ldots, d-1\}$ of length n. The measure-theoretical entropy $h_{\mu_-}(\sigma^-)$ can be calculated as

$$\begin{aligned} h_{\mu_-}(\sigma^-) &= \lim_{n\to\infty} \frac{1}{n} \sum_{\kappa_n} (-\mu_-([\kappa_n]) \log \mu_-([\kappa_n])) \\ &= \lim_{n\to\infty} \sum_{\kappa_n} \left(-\mu_-([\kappa_n]) \log \left(\frac{\mu_-([\kappa_n])}{\mu_-(\sigma^-([\kappa_n]))} \right) \right), \end{aligned}$$

where κ_n runs over all words $\kappa_n = j_{n-1} \ldots j_0$ of $\{0, 1, \ldots, d-1\}$ of length n. We would like to know when is μ_+ (or μ_-) an equilibrium state? We have studied these questions for $\mathcal{C}^{1+}$ and for $\mathcal{US}$.

12. Geometric Gibbs measures and dual geometric Gibbs measures

Consider $f \in \mathcal{C}^{1+}$. Then $1/f'(x)$ can be lifted to a positive Hölder continuous function $\psi(w) = \psi_f(w) = 1/f'(\pi_+(w))$ on the symbolic space Σ^+. By thinking of

$\log \psi$ as a potential function for the dynamical system (Σ^+, σ^+), there is a unique σ^+-invariant measure $\mu_+ = \mu_{+,\psi}$ as we have mentioned in the previous section such that

$$C^{-1} \leq \frac{\mu_+([i_0 \ldots i_{n-1}])}{\prod_{i=0}^{n-1} \psi((\sigma^+)^i(w))} \leq C$$

for any left cylinder $[i_0 \ldots i_{n-1}]$ and any $w = i_0 \ldots i_{n-1} \ldots \in [i_0 \ldots i_{n-1}]$, where C is a fixed constant. (Note that $P = P(\log \psi) = 0$ in this case.)

The geometric model $[f]_s$ in $\mathcal{C}^{1+}$ can also be represented by the Gibbs measure μ_+ with respect to $\psi(w) = 1/f'(\pi_+(w))$. The reason is that any $g \in [f]_s$ is smoothly conjugate to f, so there is a C^1 diffeomorphism h of T such that $f(h(x)) = h(g(x))$. Then $f'(h(x))h'(x) = h'(g(x))g'(x)$. Therefore,

$$\log \psi_f(w) - \log \psi_g(w) = \log h'(w) - \log h'(\sigma^+(w)).$$

So ψ_g and ψ_f are cohomologous to each other. We call this μ_+ a geometric Gibbs measure because it enjoys the following geometric property too: The push-forward measure $\mu = (\pi_+)_*\mu_+$ is a smooth f-invariant measure. This means there is a continuous function ρ on T such that

$$\mu(A) = \int_A \rho(x)dx, \quad \text{for all Borel subsets } A \text{ on } T.$$

There is another way to find the density ρ. First it is a standard method to find an invariant measure for a dynamical system f. Let ν_0 be the Lebesgue measure. Consider the push-forward measure $\nu_n = (f^n)_*\nu_0$ by the nth iterates of f. Sum up these measures to get

$$\mu_n = \frac{1}{n}\sum_{k=0}^{n-1} \nu_n$$

Any limit μ of a subsequence of $\{\mu_n\}$ will be an f-invariant measure. Since we start with an $f \in \mathcal{C}^{1+}$, we can prove that the sequence $\{\mu_n\}$ is actually convergent in C^1 topology. This means that each $\nu_n = (f^n)_*\nu_0$ has a Hölder continuous density

$$\rho_n(x) = \sum_{f^n(y)=x} \frac{1}{(f^n)'(y)}.$$

Following the theory of transfer operators (refer to [17,7]), $\rho_n(x)$ converges uniformly to a continuous function $\rho(x)$. The density of μ_n is just

$$\frac{1}{n}\sum_{k=0}^{n-1} \rho_n.$$

So it also converges to ρ uniformly. Thus $\mu(A) = \int_A \rho(x)dx$ is the limit of μ_n and is a smooth f-invariant measure.

Let $y = h(x) = \mu([0, x])$. Then $y = h(x)$ is a C^1-diffeomorphism of T. Let

$$g(y) = h \circ f \circ h^{-1}(y), \quad x = h^{-1}(y)$$

(Note that g here means a circle endomorphism not a g-function!) Then g preserves the Lebesgue measure dy (which means that $g_*(dy) = dy$, or equivalently, the Lebesgue measure is g-invariant). Since the Lebesgue measure is an ergodic g-invariant measure, g is unique in the geometric model $[f]_s$.

By considering $\psi(w) = 1/g'(\pi(w))$, it is a g-function on Σ^+ and μ_+ is a g-measure. Thus μ_+ is an equilibrium state. It follows that μ is also an equilibrium state, that is,

$$0 = P(-\log f'(x)) = h_\mu(f) - \int_T \log f'(x) d\mu = h_\mu(f) - \int_T \log f'(x)\rho(x)dx$$

$$= \sup\left\{h_\nu(f) - \int_T \log f'(x)d\nu \mid \nu \text{ is an } f\text{-invariant propbability measure}\right\}$$

$$= h_{Leb}(g) - \int_T \log g'(y)dy,$$

where $h_{Leb}(g)$ denotes the measure-theoretical entropy with respect to the Lebesgue measure. The equilibrium state μ is unique in this case.

Now by considering the dual invariant measure μ_- for this geometric Gibbs measure μ_+, we have:

Theorem 10. *Suppose $f \in C^{1+}$. Consider Σ^- with the metric $d(\cdot,\cdot)$ induced from μ_-. Then the right shift σ^- is a C^{1+} differentiable with respect to d. The derivative is one over the scaling function S_f, i.e.,*

$$\frac{d\sigma^-}{dx}(\kappa) = \frac{1}{S(\kappa)}, \quad \kappa \in \Sigma^-.$$

The proof follows the proof of Theorem 2 and the definition of μ_-. Note that by the definition of the derivative for $\kappa = \ldots j_{n-1}\ldots j_1 j_0 \in \Sigma^-$,

$$\frac{d\sigma^-}{dx}(\kappa) = \lim_{n\to\infty} \frac{\mu_-(\sigma^-([j_{n-1}\ldots j_1 j_0]))}{\mu([j_{n-1}\ldots j_1 j_0])} = \lim_{n\to\infty} \frac{\mu_-([j_{n-1}\ldots j_1])}{\mu_-([j_{n-1}\ldots j_1 j_0])}.$$

The theorem says that it equals to $1/S(\kappa)$ pointwise. Moreover, this convergence is exponentially fast. Then, following the fact that Σ^- is a compact space, we have automatically the Gibbs inequality that

$$C^{-1} \le \frac{\mu_-([j_{n-1}\cdots j_0])}{\prod_{l=0}^{n-1} S((\sigma^-)^l(\kappa))} \le C$$

for any right cylinder $[j_{n-1}\ldots j_0]$ and any κ in this cylinder, where $C > 0$ is a fixed constant. Thus μ_- is a Gibbs measure with respect to the potential function $\log S_f$. We call μ_- a dual geometric Gibbs measure.

Corollary 1. *The dual geometric Gibbs measure μ_- is a g-measure with respect to the g-function S_f whose pressure $P(\log S_f) = 0$. Moreover, $D_{\mu_-} = S_f$ for $\tilde{\mu}$-a.e. κ and μ_- is a unique equilibrium state in the sense that*

$$0 = P(\log S) = h_{\mu_-}(\sigma^-) + \int_{\Sigma^-} \log S(\kappa) d\mu_-(\kappa)$$

$$= \sup \left\{ h_\nu(\sigma^-) + \int_{\Sigma^-} \log S(\kappa) d\nu(\kappa) \mid \nu \textit{ is a } \sigma^-\textit{-invariant measure} \right\}.$$

So following Theorem 4 and Theorem 10, we have:

Theorem 11. *Suppose $S(\kappa)$ is a Hölder continuous function on Σ^- satisfying the summation and compatibility conditions. Then there is a unique measure μ_- and the metric $d(\cdot,\cdot)$ induced from μ_- on Σ^- such that $1/S$ is the derivative of the right shift σ^- with respect to this metric. Moreover, μ_- is an equilibrium state for the dynamical system $\sigma^- : (\Sigma^-, d(\cdot,\cdot)) \to (\Sigma^-, d(\cdot,\cdot))$ from a metric space into itself with the potential* $\log S$.

13. Dual geometric Gibbs type measures

A map $f \in \mathcal{US}$ may not be differentiable everywhere (it may not be even be absolutely continuous). There is no suitable Gibbs theory to be used in the study of geometric properties of a σ^+-invariant measure. We have turned to the dual symbolic dynamical system (Σ^-, σ^-) and produced a similar dual geometric Gibbs type measure theory.

An f-invariant measure μ can be found as we did in the previous section. Let ν_0 be the Lebesgue measure. Consider the push-forward measure $\nu_n = (f^n)_*\nu_0$ and sum them up to get

$$\mu_n = \frac{1}{n} \sum_{i=0}^{n-1} \nu_n.$$

Take a weak limit μ of a subsequence of $\{\mu_n\}$. Then μ is an f-invariant measure.

Each $h_n(x) = \mu_n([0, x])$ defines a homeomorphism on T. Since f is uniformly symmetric, the sequence $\{h_n\}$ is also uniformly symmetric. The space of all quasisymmetric homeomorphisms with a fixed quasisymmetric constant is a normal family (refer to [1]). So there is a subsequence of $\{h_n\}$ that converges uniformly to a function which is a symmetric homeomorphism $h(x)$ in this case. Furthermore, we have

$$h(x) = \mu([0, x]).$$

Moreover, by considering

$$g = h \circ f \circ h^{-1},$$

we see that g is a uniformly symmetric circle endomorphism in the geometric model $[f]_{sy}$ preserving the Lebesgue measure.

We can lift μ to Σ^+ to get a σ^+-invariant measure μ_+ as follows. For any finite word $w_n = i_0 \cdots i_{n-1}$, consider the cylinder $[w_n]$. Define

$$\mu_+([w_n]) = \mu(I_{w_n}),$$

where I_{w_n} is the interval in the interval system labeled by w_n. One can check that it satisfies the finite additive law and the continuity law. So it can be extended to a σ^+-invariant probability measure μ_+ on Σ^+ such that $(\pi_+)_*\mu_+ = \mu$. For μ_+, we can construct its dual invariant measure μ_- on Σ^- as we did in the previous two sections. Then we have the following geometric Gibbs type property as we had before in the smooth case:

Theorem 12. *Suppose $f \in \mathcal{US}$. Consider Σ^- with the metric $d(\cdot,\cdot)$ induced from μ_-. Then the right shift σ^- is C^1 differentiable. The derivative is one over the scaling function S_f, i.e.,*

$$\frac{d\sigma^-}{dx}(\kappa) = \frac{1}{S_f(\kappa)}, \qquad \kappa \in \Sigma^-.$$

The proof of the theorem follows the proof of Theorem 7 and the definition of μ_-.

Definition 2. *Suppose $\psi(\kappa)$ is a positive continuous function on Σ^-. A σ^--invariant measure ν is called a geometric Gibbs type measure with the potential $-\log\psi(\kappa)$ if*

$$\lim_{n\to\infty} \frac{\nu([j_{n-1}\dots j_1 j_0])}{\nu([j_{n-1}\dots j_1])} = \psi(\kappa), \quad \forall\, \kappa = \dots j_{n-1}\dots j_0 \in \Sigma^-.$$

Corollary 2. *The measure μ_- in Theorem 12 is a geometric Gibbs type measure with the potential $\log S_f$. Furthermore, μ_- is a g-measure with respect to the g-function S_f and $D_{\mu_-}(\kappa) = S_f(\kappa)$ for $\tilde{\mu}_-$-a.e. κ. Moreover, μ_- is an equilibrium state in the sense that*

$$0 = P(-\log S) = h_{\mu_-}(\sigma^-) + \int_{\Sigma^-} \log S(\kappa)d\mu_-(\kappa)$$
$$= \sup\left\{h_\nu(\sigma^-) + \int_{\Sigma^-} \log S(\kappa)d\nu(\kappa)\right\}$$

where ν runs over all σ^--invariant probability measures.

So following Theorem 8 and Theorem 12, we have:

Theorem 13. *Suppose $S(\kappa)$ is a continuous function on Σ^- satisfying the summation and compatibility conditions. Then there is a geometric Gibbs type measure μ_- with the potential $\log S$. Moreover, μ_- is an equilibrium state for the dynamical system σ^- with the potential $\log S$.*

The Teichmüller space $\mathcal{TUS}$ is a complex Banach manifold and consists of certain positive functions on the dual symbolic space Σ^-. It is an interesting problem now to study the change of μ_- when the potential $\log S$ is changed in the manifold. The reader who is interested in this direction may refer to [18,23] for some results about differentiating the absolutely continuous invariant measure of a map with respect to this map.

Acknowledgement

During my research in this direction and related topics, I have had many interesting conversations with Professors Guizhen Cui, Aihua Fan, Fred Gardiner, Jihua Ma, Sudeb Mitra, Huyi Hu, Anthony Quas, David Ruelle, and Sheldon Newhouse. Professor Dennis Sullivan introduced me to this interesting topic. I would like to express my thanks to them all.

References

[1] L. V. Ahlfors, Lectures on Quasiconformal Mappings, Van Nostrand Mathematical studies 10, D. Van Nostrand Co. Inc., Toronto-New York-London (1966).

[2] R. Bowen, Equilibrium States and the Ergodic Theory of Anosov Diffeomorphisms, Springer-Verlag, Berlin (1975).

[3] G. Cui, Circle expanding maps and symmetric structures, *Ergod. Th. & Dynamical Sys.*, **18** (1998) 831–842.

[4] G. Cui, Y. Jiang and A. Quas, Scaling functions, g-measures, and Teichmüller spaces of circle endomorphisms, *Discrete and Continuous Dynamical Sys.*, **3** (1999) 534–552.

[5] G. Cui, F. Gardiner and Y. Jiang, Scaling functions for circle endomorphisms, *Contemporary Mathematics, AMS Series*, **355** (2004) 147–163.

[6] A. Douady and J. H. Hubbard, On the dynamics of polynomial-like mappings, *Ann. Sci. Éc. Norm. Sup.*, **18** (1985) 287–343.

[7] A. Fan and Y. Jiang, On Ruelle-Perron-Frobenius operators. I. Ruelle's theorem and II. Convergence speed, *Commun. Math. Phys.*, **223** (2001) no. 1, 125–141 and 143–159.

[8] M. Feigenbaum, The universal metric properties of non-linear transformations, *J. Stat. Phys.*, **21** (1979), 669-706.

[9] F. Gardiner and D. Sullivan, Symmetric and quasisymmetric structures on a closed curve, *Amer. J. of Math.*, **114** (1992) no. 4, 683–736.

[10] Y. Jiang, Generalized Ulam-von Neuman Transformations, Ph.D Thesis (1990) CUNY Graduate Center.

[11] Y. Jiang, Renormalization and Geometry in One-Dimensional and Complex Dynamics, World Scientific, Singapore-New Jersey-London-Hong Kong 10 (1996).

[12] Y. Jiang, Smooth classification of geometrically finite one-dimensional maps, *Trans. of the Amer. Math. Soc.*, **348(6)** (1996) 2391–2412.

[13] Y. Jiang, On rigidity of one-dimensional maps, *Contemporary Mathematics, AMS series*, **211** (1997) 319–341.

[14] Y. Jiang, Differentiable rigidity and smooth conjugacy, *Annales Academiæ Scientiarum FennicæMathematica*, **30** (2005) 361–383.

[15] Y. Jiang, Geometry of geometrically finite one-dimensional mappings, *Commun. in Math. Phys.*, **156** (1993) 639–647.

[16] Y. Jiang, Metric invariants in dynamical systems. *Journal of Dynamics and Differentiable Equations*, **17** (2005) no. 1, 51–71.

[17] Y. Jiang, A Proof of existence and simplicity of a maximal eigenvalue for Ruelle-Perron-Frobenius operators, *Letters in Mathematical Physics*, **48** (1999) 211–219.

[18] Y. Jiang and D. Ruelle, Analyticity of the susceptibility function for unimodal Markovian maps of the interval, *Nonlinearity*, **18** (2005) 2447–2453.

[19] M. Keane, Strongly mixing g-measures, *Invent. Math.*, **16** (1972) 309–324.

[20] F. Ledrappier, Principe variationel et systèmes symboliqué, *Comm. Math. Phys.*, **33** (1973) 119–128.

[21] A. Pinto and D. Sullivan, Dynamical systems applied to asymptotic geometry, preprint.

[22] A. Quas, Non-ergodicity for C^1 expanding maps and g-measures, *Erg. Th. & Dynam. Sys.*, **16** (1996) no. 3, 531–543.
[23] D. Ruelle, Differentiating the absolutely continuous invariant measure of an interval map f with respect to f, *Commun. Math. Phys.*, **258** (2005) 445–453.
[24] M. Shub, Endomorphisms of compact differentiable manifolds, *Amer. J. Math.*, **91** (1969) 129–155.
[25] M. Shub and D. Sullivan, Expanding endomorphisms of the circle revisited, *Ergod. Th & Dynam. Sys.*, **5** (1985) 285–289.
[26] D. Sullivan, Differentiable structure on fractal like sets determined by intrinsic scaling functions on dual Cantor sets, *The Proceedings of Symposia in Pure Mathematics*, **48** (1988).
[27] D. Sullivan, Linking the universalities of Milnor-Thurston, Feigenbaum and Ahlfors-Bers, *Topological Methods in Modern Mathematics*, (1993) 543–563.
[28] P. Walters, Ruelle's operator theorem and g-measures, *Trans. Amer. Math. Soc.*, **214** (1975) 375–387.

Teichmüller Theory and Moduli Problem No. 10, 2010, pp. 437–449.

Stable Maps into the Classifying Space of the General Linear Group

Ivan Kausz

NWF I - Mathematik, Universität Regensburg, 93040 Regensburg, Germany
e-mail: ivan.kausz@mathematik.uni-regensburg.de

Abstract. We define a moduli stack which parametrizes morphisms from semistable curves to BGL_r, the classifying stack of GL_r. Furthermore, for each combinatorial type we define a stack parametrizing morphisms of that type together with a boundary morphism. All this is in close analogy to Kontsevich's moduli stack of stable maps into a projective algebraic variety.

1. Introduction

In this note we give a definition of stable maps into the classifying stack BGL_r of the general linear group. To support our belief that the definition is the correct one, we show that there are natural boundary morphisms between the moduli groupoids parameterizing stable maps to BGL_r. We warn the reader beforehand that we do not as yet have applications of our constructions. Our purpose in this note is mainly to explain the combinatorics of stable maps into BGL_r.

Before giving an idea of what stable maps to BGL_r are, we first recall the case of stable maps into a complex projective variety V. Let $g \in \mathbb{N}_0$, and let S be a finite set. A stable S-pointed map of genus g consists of a prestable curve C together with nonsingular points $(x_s)_{s\in S}$ on it and a morphism $f : C \to V$ of algebraic varieties such that the collection of these data has at most finitely many automorphisms.

Stable maps into V are a fundamental notion for defining Gromov-Witten invariants for V. Namely, there is a Deligne-Mumford stack $\bar{M}_{g,S}(V)$ parameterizing all S-pointed stable maps of genus g, and Gromov-Witten invariants of V are defined by means of intersection theory on $\bar{M}_{g,S}(V)$.

The moduli stack $\bar{M}_{g,S}(V)$ breaks up into connected components $\bar{M}_{g,S}(V,\beta)$ where β runs through some monoid. Stable maps which belong to that component are called of class β. Given two stable maps, say $S_i \cup \{*_i\}$-pointed, of genus g_i and class β_i for $i = 1, 2$ such that the points $*_i$ are mapped to the same point in V, one obtains a new stable map which is $S_1 \cup S_2$-pointed, of genus $g_1 + g_2$ and class $\beta_1 + \beta_2$, by clutching together the two curves at the points $*_i$. This yields a morphism

$$\bar{M}_{g_1,S_1\cup\{*_1\}}(V,\beta_1) \times_V \bar{M}_{g_2,S_2\cup\{*_2\}}(V,\beta_2) \to \bar{M}_{g_1+g_2,S_1\cup S_2}(V,\beta_1+\beta_2)$$

which is an instance of a boundary morphism. A general point in its image belongs to a stable map whose underlying curve has two smooth irreducible components joined in one double point such that the restriction of the map to the i-th component is a S_i-pointed stable map of genus g_i and class β_i. The combinatorial type of such a stable map is a decorated graph containing two vertices labeled by (g_1, β_1) and (g_2, β_2) and half-edges (or "tails") labeled by elements of S_1 and S_2 respectively:

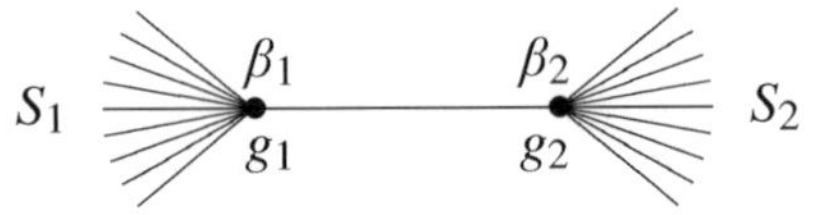

More generally, to each S-pointed stable map of genus g and class β one can associate some decorated graph τ, its combinatorial type. For each combinatorial type there is a stack $\bar{M}(V, \tau)$, defined as fiber product of some stacks $\bar{M}_{g_i,S_i}(V, \beta_i)$, and a boundary morphism

$$\bar{M}(V, \tau) \to \bar{M}_{g,S}(V, \beta)$$

generalizing the clutching described above. The collection of boundary morphisms is responsible for the rich algebraic structure of quantum cohomology of V (cf. [M]).

By the very definition of the stack BGL_r a morphism from a curve C to BGL_r is nothing else but a vector bundle of rank r over C. Thus it is clear that the generic stable map to BGL_r should be a pair consisting of a smooth curve together with a vector bundle of rank r on it. Now it is well known that given a family of stable curves which is generically smooth and has a singular special fiber, a vector bundle defined on the complement of that fiber does not necessarily extend to a vector bundle over the whole family, not even if one allows for an étale base change of the family. This is just like in the situation of morphisms into a variety V, where it may happen that a morphism to V, defined on the complement of the special fiber, does not extend to a morphism from the whole family to V, even after base change.

However there is always a family of *semistable* curves which outside the special fiber coincides with the given one, such that the morphism extends to that new family, at least after an étale base change. The same solution applies to vector bundles, i.e. to morphisms into BGL_r. Namely there exists a semistable family such that the given vector bundle extends after étale base change. Furthermore the new family and the extension can be chosen in such a way that the restriction of the vector bundle on those chains of projective lines in the semistable special fiber which are contracted under stabilization satisfies a certain condition which we call "admissibility" (cf. Definition 2.2). Accordingly, we define S-pointed stable maps of genus g and degree d to BGL_r as semistable S-pointed curves of genus g together with vector bundles of rank r and degree d satisfying that condition.

It is straightforward to define families of stable S-pointed maps of genus g and degree d to BGL_r, and pull back and isomorphisms of these. Thus we obtain a groupoid over $\mathbb{C}$ which we denote by $\bar{M}_{g,S}(\mathrm{BGL}_r, d)$.

To give an idea of what the boundary morphisms are, let us first consider the analogue of the example above. Thus we are given two curves, together with vector bundles of rank r on them, and on each of the curves one special point. Now if we join the curves by identifying the two special points, we need an additional datum to obtain a vector bundle on the new curve. This datum consists in an isomorphism between the respective fibers of two given bundles at the special points. However isomorphisms between two vector spaces may degenerate in families. In other words, after chosing a bases of the two vector spaces, an isomorphism between them is given by an element in GL_r which is noncompact and therefore in general a family of isomorphisms cannot be extended to special points. Our solution of this problem is a certain compactification KGL_r of GL_r which we have constructed in an earlier paper [K1].

To be more precise, let us give a definition of the groupoid $\bar{M}(\mathrm{BGL}_r, \boldsymbol{\tau})$, where $\boldsymbol{\tau}$ is the decorated graph

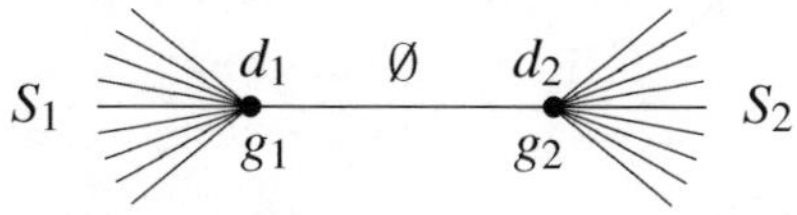

which corresponds to the example considered above. First we observe that one can interpret the diagonal morphism $\Delta : \mathrm{BGL}_r \to \mathrm{BGL}_r \times \mathrm{BGL}_r$ as follows. There is a commutative diagram

$$\begin{array}{ccc} \mathrm{BGL}_r & \xrightarrow{\cong} & \mathrm{Isom}(\mathrm{pr}_1^*\mathbb{E}, \mathrm{pr}_2^*\mathbb{E}) \\ & {\scriptstyle\Delta}\searrow \quad \swarrow & \\ & \mathrm{BGL}_r \times \mathrm{BGL}_r & \end{array}$$

where the horizontal arrow is an isomorphism, $\mathbb{E}$ is the universal vector bundle over BGL_r, and $\mathrm{pr}_i : \mathrm{BGL}_r \times \mathrm{BGL}_r \to \mathrm{BGL}_r$ is the i-th projection. We define

$$\overline{\mathrm{BGL}_r} := \mathrm{KGL}(\mathrm{pr}_1^*\mathbb{E}, \mathrm{pr}_2^*\mathbb{E})$$

as the fiberwise compactification of the locally trivial GL_r-bundle $\mathrm{Isom}(\mathrm{pr}_1^*\mathbb{E}, \mathrm{pr}_2^*\mathbb{E}) \to \mathrm{BGL}_r \times \mathrm{BGL}_r$ and let

$$\bar{\Delta} : \overline{\mathrm{BGL}_r} \to \mathrm{BGL}_r \times \mathrm{BGL}_r$$

be the natural projection. With this notation the groupoid $\bar{M}(\mathrm{BGL}_r, \boldsymbol{\tau})$ is defined by the Cartesian diagram

$$\begin{array}{ccc} \bar{M}(\mathrm{BGL}_r, \boldsymbol{\tau}) & \longrightarrow & \bar{M}_{g_1, S_1\cup\{*_1\}}(\mathrm{BGL}_r, d_1) \times \bar{M}_{g_2, S_2\cup\{*_2\}}(\mathrm{BGL}_r, d_2) \\ \downarrow & & \downarrow {\scriptstyle \mathrm{ev}_1 \times \mathrm{ev}_2} \\ \overline{\mathrm{BGL}_r} & \xrightarrow{\bar{\Delta}} & \mathrm{BGL}_r \times \mathrm{BGL}_r \end{array}$$

The existence of a natural boundary morphism

$$\bar{M}(\mathrm{BGL}_r, \boldsymbol{\tau}) \to \bar{M}_{g,S}(\mathrm{BGL}_r, d).$$

where $g = g_1 + g_2$, $S = S_1 \cup S_2$ and $d = d_1 + d_2$ then follows from results in [K2] and [K3].

Under this boundary morphism, a point of $\bar{M}(\mathrm{BGL}_r, \boldsymbol{\tau})$ which lies over the complement of BGL_r in $\overline{\mathrm{BGL}_r}$ is mapped to a stable map to BGL_r whose underlying curve is not stable but only semistable, and the stabilizing map contracts a chain of projective lines. The restriction of the underlying bundle of the stable map to this chain of projective lines yield a tuple of degrees and we in §5 below we give a careful description of which strata of $\overline{\mathrm{BGL}_r}$ (or of KGL_r) correspond to which tuple of degrees.

It should be mentioned that A. Schmitt [Sch] and J. Li [Li] have constructed varieties parameterizing a certain subset of all stable maps to BGL_r. However they do not define boundary morphisms and we doubt whether that would be possible at all in their context.

We would also like to mention work of D. Abramovich, A. Corti and A. Vistoli [AV,ACV] where stable maps to BG for finite groups G are defined. In [K4] we have established some connection between their construction an ours, but the exact relationship remains mysterious.

2. Stable maps

First we recall some definitions from [K2]

Definition 2.1. *A* chain of projective lines *is a connected prestable curve over a field, whose components are projective lines and whose associated graph is linear. An irreducible component of R is called* extremal, *if it intersects only one other component.*

Definition 2.2. *A vector bundle E of rank r on a chain R of projective lines is called* admissible, *if the following two properties are satisfied:*

(1) *For each irreducible component R_i of R there exists a number $d_i \geq 1$ such that the restriction of E to $R_i \cong \mathbb{P}^1$ is isomorphic to the bundle $d_i\mathcal{O} \oplus (r - d_i)\mathcal{O}(1)$.*
(2) *If a global section of E vanishes in two smooth points lying in different extremal components of R, then it vanishes everywhere.*

It is not difficult to see that if there exists an admissible vector bundle E of rank r on a chain R of projective lines, then the length of that chain is bounded by r (cf. section 3 in [K2]).

Definition 2.3. *Let C' be a prestable curve over a field k. A* Gieseker vector bundle of rank r on C' *is a pair $(C \to C', E)$, where C is a prestable curve over k, $C \to C'$ is a morphism and E is a rank r vector bundle on C which satisfy the following properties:*

(1) *The morphism $C \to C'$ is an isomorphism over the complement of the singular points of C'.*
(2) *The fiber of $C \to C'$ over a singular point of C' is either a point or a chain of projective lines.*
(3) *The restriction of E to each of the nontrivial fibers of $C \to C'$ is admissible.*

Now we come to the main definition of this note:

Definition 2.4. *Let S be a finite set. A* stable S-pointed map to BGL_r *is a tupel $(C, x_i \mid i \in S, E)$, where $(C, x_i \mid i \in S)$ is a prestable S-pointed curve over a field and E is a vector bundle on C such that if $(C, (x_i)) \to (C^{st}, (y_i))$ denotes the stabilization morphism (cf. [M] V.1.6) then $(C \to C^{st}, E)$ is a Gieseker vector bundle of rank r on C^{st}.*

3. Combinatorial types

Before defining the combinatorial type of a stable S-pointed map to BGL_r, let us first recall some definitions from [M].

Definition 3.1. *A* (finite) graph τ *consists of the data* $(F_\tau, V_\tau, \partial_\tau, j_\tau)$, *where* F_τ *is a (finite) set (of flags),* V_τ *is a finite set (of vertices),* $\partial_\tau : F_\tau \to V_\tau$ *is the boundary map, and* $j_\tau : F_\tau \to F_\tau$ *is an involution.*

An isomorphism $\tau \xrightarrow{\sim} \sigma$ between two graphs consists of two bijections $F_\tau \xrightarrow{\sim} F_\sigma$, $V_\tau \xrightarrow{\sim} V_\sigma$, compatible with ∂ and j.

The fixed points of j_τ form the set S_τ of tails, the pairs (f, f') with $f \in F_\tau$ and $f' = j_\tau(f) \neq f$ form the set E_τ^{or} of oriented edges of τ. If $e = (f, f') \in E^{\mathrm{or}}$ is an oriented edge, then $j_\tau(e) := (j_\tau(f), j_\tau(f'))$ is the edge with the opposite orientation. The two-element orbits of j_τ (viewed as an inversion of F_τ or of E_τ^{or}) form the set E of (un-oriented) edges of τ. For $v \in V_\tau$ let $F_\tau(v) := \partial^{-1}(v)$ denote the set of flags starting from v, and let $|v| := |F_\tau(v)|$ denote the valence of v.

The topological realization of a graph τ is denoted by $||\tau||$.

Definition 3.2. *A* modular graph *is a graph* τ *together with a map* $g : V_\tau \to \mathbb{N}_0$, $v \mapsto g_\tau$. *A modular graph* τ *is called* stable, *if* $|v| \geq 3$ *for all* v *with* $g_v = 0$ *and* $|v| \geq 1$ *for all* v *with* $g_v = 1$.

There is a standard way to associate a modular graph (τ, g) to a prestable S-pointed curve $(C, x_i \mid i \in X)$ (cf. [M], III. 2). The vertices v of τ correspond to the irreducible components C_v of C, and g_v is the genus of C_v. The un-oriented edges e of τ correspond to the singular points x_e of C. A prestable S-pointed curve is stable if and only if (τ, g) is stable.

Definition 3.3. *A* chain-type *is a tupel* $\mathbf{d} = (d_1, \ldots, d_q)$ *of integers* $d_i \geq 1$, *where* $q \geq 0$. *The number* q *is called the* length *of the chain-type* d *and the integer* $|\mathbf{d}| := \sum_{i=1}^{q} d_i$ *is called its* degree. *The (empty) chain-type* $\mathbf{d} = ()$ *of length zero will also be denoted by the symbol* $\emptyset$.

For a chain-type $\mathbf{d} = (d_1, \ldots, d_q)$ we denote by $j(\mathbf{d})$ the reversed chain-type $(d_q, \ldots, d_1)$. The reversed empty chain-type is again empty.

Definition 3.4. *A* chain-graph *(for rank r) is a tupel* $(\tau, g, d, \mathbf{d})$ *where* (τ, g) *is a stable modular graph and* d, $\mathbf{d}$ *are maps*

$$d : V_\tau \to \mathbb{Z}$$
$$\mathbf{d} : E_\tau^{\mathrm{or}} \to \{\text{chain-types of degree} \leq r\}$$

where $\mathbf{d}$ *has the property that* $\mathbf{d}(j_\tau(e)) = j(\mathbf{d}(e))$ *for every oriented edge* e *of* τ.

Definition 3.5. *The* combinatorial type *of an S-labeled stable map* $(C, x_i \mid i \in S, E)$ *to* BGL_r *is the chain-graph* $(\tau, g, d, \mathbf{d})$, *where* (τ, g) *is the stable modular graph associated to the stabilization* C^{st} *of* C *(in particular,* S_τ *is identified with* S*) and the maps* d, $\mathbf{d}$ *are characterized by the following properties.*

(1) *Let $v \in V_\tau$ and let C_v be the associated component of the stabilized curve C^{st}. There exists a unique component of C which is mapped birationally onto C_v. By a slight abuse of notation we denote that component again by the symbol C_v. We require that the restriction of E to C_v is of degree $d(v)$.*
(2) *Let $e = (f, f')$ be an oriented edge of τ and let $x_e \in C^{st}$ be the corresponding singular point. If the fiber of $C \to C^{st}$ over x_e consists of just one point, then $\mathbf{d}(e) = \emptyset$. Else the fiber R is a chain of length $q \geq 1$ of projective lines. To the flags f and f' there are associated components C_f $C_{f'}$ of C^{st} together with points $x_f \in C_f$ and $x_{f'} \in C_{f'}$. By abuse of notation we denote by (C_f, x_f) and $(C_{f'}, x_{f'})$ also the pointed components of C which are mapped birationally to the corresponding pointed components of C^{st}. Let $R_1, \ldots, R_q$ be the irreducible components of the chain R numbered in such a way that R_1 meets the point x_f and R_q meets the point $x_{f'}$ and such that $R_i \cap R_j$ is empty for $|i - j| > 1$. Then $\mathbf{d}(e) = (\deg(E|_{R_1}), \ldots, \deg(E|_{R_q}))$.*

It is clear that for any chain-graph $(\tau, g, d, \mathbf{d})$ for rank r there exists an S_τ-labeled stable map to BGL_r whose combinatorial type is $(\tau, g, d, \mathbf{d})$.

4. The groupoid of stable maps to $\mathbf{BGL}_r$

Let S be a finite set and let $r \geq 1$. An *S-labeled stable map over a scheme T to BGL_r* is a tupel $(C/T, x_i \mid i \in S, E)$, where $(C/T, x_i \mid i \in S)$ is a prestable S-pointed curve and E is a vector bundle of rank r on C such that for each geometric point t of T the fiber $(C_t, x_i(t) \mid i \in S, E_t)$ is a stable map to BGL_r.

A morphism

$$(C/T, x_i \mid i \in S, E) \to (C'/T', x_i' \mid i \in S, E')$$

between two S-labeled stable maps to BGL_r consists of the following data:

(1) A morphism $T \to T'$,
(2) An isomorphism $f : C \xrightarrow{\sim} C' \times_{T'} T$ which for each $i \in S$ maps the section x_i to the pull back of the section x_i',
(3) An isomorphism $E \xrightarrow{\sim} f^*(\mathcal{O}_T \otimes_{\mathcal{O}_{T'}} E')$.

For integers $g \geq 0$ and d let $\bar{M}_{g,S}(\mathrm{BGL}_r, d)$ be the groupoid (over the category of schemes) whose objects are S-labeled stable maps $(C/T, x_i \mid i \in S, E)$ to BGL_r where C/T is a T-curve of genus g and the restriction of the vector bundle E to each fiber of $C \to T$ is of degree d.

5. Generalized isomorphisms and Gieseker vector bundle data

This section and the following are preparatory to section 7 below, where we will define an analogue of the groupoid $\bar{M}(V, \tau)$ and the boundary morphism $\bar{M}(V, \tau) \to \bar{M}_{g,S}(V, \beta)$ (cf. [M] V. 4.7) for the case when the variety V is replaced by the stack BGL_r.

Let F_1 and F_2 be two r-dimensional k-vector spaces. In [K1] we have defined a smooth compactification $\mathrm{KGL}(F_1, F_2)$ of the variety $\mathrm{Isom}(F_1, F_2) \cong \mathrm{GL}_r$ of linear isomorphisms from F_1 to F_2 which has properties analogous to the so called "wonderful compactification" of adjoint linear groups introduced by De Concini and Procesi. We call the k-valued points of $\mathrm{KGL}(F_1, F_2)$ *generalized isomorphisms* from F_1 to F_2. The complement of $\mathrm{Isom}(F_1, F_2)$ inside $\mathrm{KGL}(F_1, F_2)$ is a normal crossing divisor with smooth irreducible components

$$Y_i = Y_i(F_1, F_2) \quad \text{and} \quad Z_i = Z_i(F_1, F_2) \quad (i \in [0, r-1]).$$

For subsets I, J of the set $[0, r-1]$ we write $\bar{\mathbf{O}}_{I,J}(F_1, F_2)$ for the intersection of all Z_i and Y_j, where i and j run through the set I and J respectively. The variety $\bar{\mathbf{O}}_{I,J}(F_1, F_2)$ is non-empty if and only if

$$\min(I) + \min(J) \geq r$$

(here we use the convention that the minimum of the empty subset of $[0, r-1]$ is r). A pair (I, J) of subsets of $[0, r-1]$ which satisfies this property is called a *GI-type* (for the rank r). A k-rational point of $\bar{\mathbf{O}}_{I,J}(F_1, F_2)$ is called a *generalized isomorphism of type* (I, J). There is an isomorphism

$$\mathrm{inv} : \begin{cases} \mathrm{KGL}(F_1, F_2) & \to \mathrm{KGL}(F_2, F_1) \\ \Phi & \mapsto \Phi^{-1} \end{cases}$$

with $\mathrm{inv}^2 = \mathrm{id}$, whose restriction to $\mathrm{Isom}(F_1, F_2)$ is the inversion. We have $\mathrm{inv}(Y_i(F_1, F_2)) = Z_i(F_2, F_1)$ for all $i \in [0, r-1]$. Thus if Φ is a generalized isomorphism of type (I, J) then Φ^{-1} is a generalized isomorphism of type (J, I).

Definition 5.1. *Let (D, x_1, x_2) be a two-pointed prestable curve over a field k. A* Gieseker vector bundle datum *on (D, x_1, x_2) is a pair (F, Φ) where F is a vector bundle on D with fibers F_1 and F_2 in x_1 and x_2 respectively, and Φ is a generalized isomorphism from F_1 to F_2.*

Let (D, x_1, x_2) be a two-pointed prestable curve and let C' be the prestable curve which arises by identifying the two points x_1 and x_2. Thus we have a morphism $D \to C'$ which is the partial normalization of the curve C' at a singular point x.

In our paper [K2] we have shown that a Gieseker vector bundle datum (F, Φ) on D induces a Gieseker vector bundle $(C \to C', E)$ on C'. Moreover the construction of $(C \to C', E)$ yields a distinguished singular point $y \in C$ which is mapped to the point x, and one can recover the data (D, x_1, x_2, F, Φ) from the data $(C \to C', E, y)$.

Let $(C \to C', E)$ be the Gieseker vector bundle associated to the Gieseker vector bundle datum (F, Φ). Let R be the fiber over x of the morphism $C \to C'$. Then R is either a single point or a chain of projective lines. In the second case we may consider D as a subscheme of C and may number the components $R_1, \ldots, R_m$ of R such that R_1 meets D in x_1 and R_m meets D in x_2 and $R_i \cap R_j = \emptyset$ if $|i-j| > 1$. Let $\mathbf{d} = \emptyset$ if R is reduced to a point and $\mathbf{d} = (\deg(E|_{R_1}), \ldots, \deg(E|_{R_m}))$ else.

Now what can be said about the chain-type **d**, if the generalized isomorphism Φ is chosen in $\bar{\mathbf{O}}_{I,J}(F_1, F_2)$ for some GI-type (I, J)? To answer this question we define a mapping

$$\{\text{GI-types}\} \to \{\text{chain-types}\}$$

as follows: Let (I, J) be a GI-type, with $I = \{i_1, \dots, i_p\}$, $J = \{j_1, \dots, j_q\}$ where $p, q \geq 0$ and

$$0 \leq i_1 \leq \cdots \leq i_p < i_{p+1} := r$$
$$0 \leq j_1 \leq \cdots \leq j_q < j_{q+1} := r$$

The GI-type (I, J) is mapped to the chain-type $(d_1, \dots, d_{p+q})$, where

$$d_m = \begin{cases} i_{m+1} - i_m & \text{if } m \in [1, p] \\ j_{p+q-m+2} - j_{p+q-m+1} & \text{if } m \in [p+1, p+q] \end{cases}$$

The mapping can be visualized by the following picture

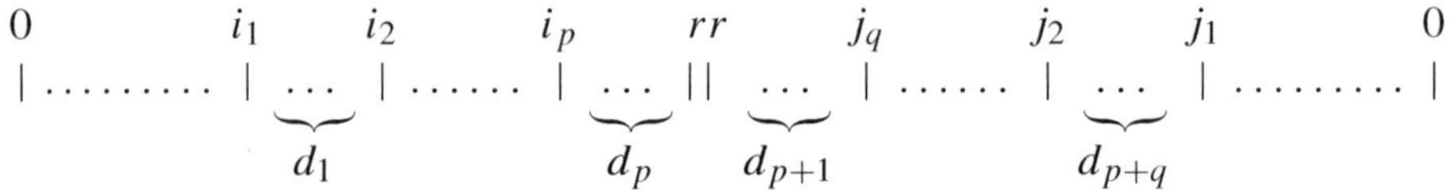

The answer to the above question can now be formulated as follows: If Φ is generic in $\bar{\mathbf{O}}_{I,J}(F_1, F_2)$ (more precisely, if Φ is contained in $\bar{\mathbf{O}}_{I,J}(F_1, F_2)$ but not in $\bar{\mathbf{O}}_{I',J'}(F_1, F_2)$ for I' strictly larger than I or J' strictly larger than J) then the chain-type **d** is the image of the GI-type (I, J) by the mapping

$$\{\text{GI-types}\} \to \{\text{chain-types}\}$$

defined above.

6. The groupoid of Gieseker vector bundle data

Let T be a scheme and let F_1 and F_2 be two vector bundles of rank r on T. In [K1] we have defined a smooth T-scheme $\mathrm{KGL}(F_1, F_2)$, whose fiber over at a point $t \in T$ is a compactification of the variety $\mathrm{Isom}(F_{1,t}, F_{2,t})$, which is isomorphic to $\mathrm{KGL}(F_{1,t}, F_{2,t})$. A section of $\mathrm{KGL}(F_1, F_2) \to T$ is called a *generalized isomorphism from F_1 to F_2*. For each GI-type (I, J) there exists a natural closed subscheme $\bar{\mathbf{O}}_{I,J}(F_1, F_2)$ of $\mathrm{KGL}(F_1, F_2)$ whose fiber over $t \in T$ is the variety $\bar{\mathbf{O}}_{I,J}(F_{1,t}, F_{2,t})$. A section of $\bar{\mathbf{O}}_{I,J}(F_1, F_2) \to T$ is called a *generalized isomorphism of type (I, J) from F_1 to F_2*. Also the isomorphism $\mathrm{inv} : \mathrm{KGL}(F_1, F_2) \xrightarrow{\sim} \mathrm{KGL}(F_2, F_1)$, $\Phi \mapsto \Phi^{-1}$ generalizes to the relative situation and has the properties $\mathrm{inv}^2 = \mathrm{id}$ and $\mathrm{inv}(\bar{\mathbf{O}}_{I,J}(F_1, F_2)) = \mathrm{inv}(\bar{\mathbf{O}}_{J,I}(F_2, F_1)$. All these objects commute (in an obvious sense) with base change $T' \to T$.

Definition 6.1. *Let* $(D/T, x_1, x_2)$ *be a two-pointed prestable curve over a scheme* T *and let* (I, J) *be a GI-type for a rank* $r \in \mathbb{N}$. *A* Gieseker vector bundle datum of rank r and type (I, J) on $(D/T, x_1, x_2)$ *is a pair* (F, Φ), *where* F *is a vector bundle of rank* r *on* D *and* Φ *is a section of* $\bar{\mathbf{O}}_{I,J}(x_1^*F, x_2^*F) \to T$.

An isomorphism

$$(F, \Phi) \xrightarrow{\sim} (F', \Phi')$$

between two Gieseker vector bundle data of type (I, J) on $(D/T, x_1, x_2)$ is an isomorphism $F \xrightarrow{\sim} F'$ such that the induced isomorphism $\bar{\mathbf{O}}_{I,J}(x_1^*F, x_2^*F) \xrightarrow{\sim} \bar{\mathbf{O}}_{I,J}(x_1^*F', x_2^*F')$ carries Φ to Φ'.

We denote by $\mathrm{GVBD}_{I,J}$ the groupoid whose objects are Gieseker vector bundle data of rank r and type (I, J) on $(D/T, x_1, x_2)$ and whose arrows are isomorphisms between Gieseker vector bundle data.

Let (I, J) and (I', J') be two GI-types which are mapped to the same chain-type. Then by similar arguments as used in the proof of [K3] 4.2 it can be shown that there is a canonical isomorphism of groupoids

$$\beta_{I',J'}^{I,J} : \mathrm{GVBD}_{I,J} \xrightarrow{\sim} \mathrm{GVBD}_{I',J'}.$$

Moreover, we have $\beta_{I,J}^{I,J} = \mathrm{id}$ and if (I'', J'') is a third GI-type with the same associated chain-type as (I, J) and (I', J'), then $\beta_{I'',J''}^{I,J} = \beta_{I'',J''}^{I',J'} \circ \beta_{I',J'}^{I,J}$.

Let $\mathbf{d}$ be a chain-type. The family of isomorphisms $(\beta_{I',J'}^{I,J})_{(I,J),(I',J') \mapsto \mathbf{d}}$ enables us to define a new groupoid $\mathrm{GVBD}_{\mathbf{d}}$ as follows. The objects of $\mathrm{GVBD}_{\mathbf{d}}$ are the union of all objects of the groupoids $\mathrm{GVBD}_{I,J}$ where (I, J) runs through all GI-types which are mapped to the chain-type $\mathbf{d}$. A morphism from an object (F, Φ) of $\mathrm{GVBD}_{I,J}$ to an object (F', Φ') of $\mathrm{GVBD}_{I',J'}$ is a family $(f_{A,B})_{(A,B) \mapsto \mathbf{d}}$ where

$$f_{A,B} : \beta_{A,B}^{I,J}(F, \Phi) \to \beta_{A,B}^{I',J'}(F', \Phi')$$

is a morphism in $\mathrm{GVBD}_{A,B}$ such that $\beta_{A',B'}^{A,B}(f_{A,B}) = f_{A',B'}$ for all GI-types (A, B), (A', B') mapped to $\mathbf{d}$.

Remark 6.2. Let (I, J) be a GI-type and let $\mathbf{d}$ be its associated chain-type. Then the inclusion $\mathrm{GVBD}_{I,J} \hookrightarrow \mathrm{GVBD}_{\mathbf{d}}$ is an equivalence of categories. The advantage of $\mathrm{GVBD}_{\mathbf{d}}$ over $\mathrm{GVBD}_{I,J}$ is that its definition depends only on the chain-type $\mathbf{d}$ and does not involve the choice of of a GI-type over $\mathbf{d}$.

Definition 6.3. *An object* (F, Φ) *of the groupoid* $\mathrm{GVBD}_{\mathbf{d}}$ *is called a* Gieseker vector bundle datum of type $\mathbf{d}$ *on the two-pointed curve* $(D/T, x_1, x_2)$.

Let C' be the prestable T-curve constructed from D by identifying the two sections x_1 and x_2 and let x be the section of C' given by the composition $T \xrightarrow{x_i} D \to C'$. Let GVB be the following groupoid: The objects of GVB are Gieseker vector bundles $(C \to C', E)$ of rank r on C' such that $C \to C'$ is an isomorphism over the complement of $x(T)$. An arrow $(C_1 \to C', E_1) \to (C_2 \to C', E_2)$ between two such objects consists of a C'-isomorphism $f : C_1 \xrightarrow{\sim} C_2$ and an isomorphism $E_1 \xrightarrow{\sim} f^*E_2$.

In section 5 we have already mentioned that a Gieseker vector bundle datum on a two-pointed prestable curve over a field induces a Gieseker vector bundle on the prestable curve obtained by identifying the two marked points. This construction works also in families over an arbitrary base scheme T (cf. [K2]) and gives for each GI-type (I, J) a natural morphism of groupoids

$$\nu_{I,J} : \mathrm{GVBD}_{I,J} \to \mathrm{GVB}.$$

The morphisms $\nu_{I,J}$ are compatible with the isomorphisms $\beta^{I,J}_{I',J'}$ in the sense that all triangles

$$\begin{array}{ccc} \mathrm{GVBD}_{I,J} & \xrightarrow{\beta^{I,J}_{I',J'}} & \mathrm{GVBD}_{I',J'} \\ & {\scriptstyle \nu_{I,J}} \searrow \quad \swarrow {\scriptstyle \nu_{I',J'}} & \\ & \mathrm{GVB} & \end{array}$$

commute.

It follows that for each chain-type $\mathbf{d}$ the family $(\nu_{I,J})_{(I,J)\mapsto \mathbf{d}}$ induces a morphism of groupoids

$$\nu_{\mathbf{d}} : \mathrm{GVBD}_{\mathbf{d}} \to \mathrm{GVB}.$$

7. Boundary morphisms

In this section we will define for each chain-graph (cf. section 3)

$$\boldsymbol{\tau} = (\tau,\ g_v \mid v \in V_\tau,\ d_v \mid v \in V_\tau,\ \mathbf{d}_e \mid e \in E^{\mathrm{or}}_\tau)$$

a groupoid (over the category of schemes) $\bar{M}(\mathrm{BGL}_r, \boldsymbol{\tau})$ together with a boundary morphism

$$\nu_{\boldsymbol{\tau}} : \bar{M}(\mathrm{BGL}_r, \boldsymbol{\tau}) \to \bar{M}_{g,S}(\mathrm{BGL}_r, d),$$

where $g = \sum g_v + \dim H_1(||\tau||)$ is the genus of the combinatorial type, $d = \sum d_v + \sum |\mathbf{d}_e|$ is its total degree and $S = S_\tau$ is the set of its tails.

We start with an auxiliary definition.

Definition 7.1. *A* GI-graph *(for rank r) is a tupel*

$$\boldsymbol{\gamma} = (\tau,\ g_v \mid v \in V_\tau,\ \delta_v \mid v \in V_\tau,\ I_f \mid f \in F_\tau \backslash S_\tau)$$

where $(\tau, g_v \,|v \in V_\tau)$ is a stable modular graph, where the δ_v are integers and where for each oriented edge (f, f') the pair $(I_f, I_{f'})$ is a GI-type for the rank r.

Before defining the groupoid $\bar{M}(\mathrm{BGL}_r, \boldsymbol{\tau})$ we will first define the groupoid $\bar{M}(\mathrm{BGL}_r, \boldsymbol{\gamma})$ for a GI-graph $\boldsymbol{\gamma}$. We will see below that to $\boldsymbol{\gamma}$ there is associated a chain-graph $\boldsymbol{\tau}$, and the relationship between the groupoids $\bar{M}(\mathrm{BGL}_r, \boldsymbol{\gamma})$ and $\bar{M}(\mathrm{BGL}_r, \boldsymbol{\tau})$ is analogous to the relationship between $\mathrm{GVBD}_{I,J}$ and $\mathrm{GVBD}_{\mathbf{d}}$ described in section 6.

Let

$$\boldsymbol{\gamma} = (\tau,\ g_v \mid v \in V_\tau,\ \delta_v \mid v \in V_\tau,\ I_f \mid f \in F_\tau \backslash S_\tau)$$

be a GI-graph and let T be a scheme. An object of $\bar{M}(\mathrm{BGL}_r, \boldsymbol{\gamma})$ over T is a family

$$((C_v/T,\ (x_i)_{i\in F_\tau(v)},\ \mathcal{F}_v) \mid v \in V_\tau;\ \Phi_e \mid e \in E_\tau^{\mathrm{or}})$$

where

(1) $(C_v/T,\ (x_i)_{i\in F_\tau(v)},\ \mathcal{F}_v)$ is an $F_\tau(v)$-labeled stable map over T to BGL_r such that C_v/T is a curve of genus g over T.
(2) For every oriented edge $e = (f, f')$ of τ with $f \in F_\tau(v)$ and $f' \in F_\tau(v')$, Φ_e is a generalized isomorphism of type $(I_f, I_{f'})$ from $x_f^*\mathcal{F}_v$ to $x_{f'}^*\mathcal{F}_{v'}$.
(3) For every oriented edge e the equality $\Phi_{j(e)} = \Phi_e^{-1}$ holds.
(4) For each vertex $v \in V_\tau$ we have $\deg \mathcal{F}_v = \delta_v$.

An arrow

$$\begin{aligned}&((C_v/T,\ (x_i)_{i\in F_\tau(v)},\ \mathcal{F}_v) \mid v \in V_\tau;\ \Phi_e \mid e \in E_\tau^{\mathrm{or}})\\ &\quad\to ((C'_v/T,\ (x'_i)_{i\in F_\tau(v)},\ \mathcal{F}'_v) \mid v \in V_\tau;\ \Phi'_e \mid e \in E_\tau^{\mathrm{or}})\end{aligned}$$

between two objects of $\bar{M}(\mathrm{BGL}_r, \boldsymbol{\gamma})$ over T is a family of isomorphisms

$$(C_v/T,\ (x_i)_{i\in F_\tau(v)},\ \mathcal{F}_v) \xrightarrow{\sim} (C'_v/T,\ (x'_i)_{i\in F_\tau(v)},\ \mathcal{F}'_v)$$

of stable maps over T to BGL_r (cf. section 4), where v runs through the set V_τ of vertices of τ, such that for each oriented edge $e = (f_1, f_2)$ with $f_1 \in F_\tau(v_1)$, $f_2 \in F_\tau(v_2)$ the induced isomorphism

$$\mathrm{KGL}(x_{f_1}^*\mathcal{F}_{v_1}, x_{f_2}^*\mathcal{F}_{v_2}) \xrightarrow{\sim} \mathrm{KGL}((x'_{f_1})^*\mathcal{F}'_{v_1}, (x'_{f_2})^*\mathcal{F}'_{v_2})$$

maps the generalized isomorphism Φ_e to Φ'_e.

If $T' \to T$ is a morphism of schemes then an object of $\bar{M}(\mathrm{BGL}_r, \boldsymbol{\gamma})$ over T pulls back to one over T' and an arrow between objects over T induces one between their respective pull back over T'. This defines $\bar{M}(\mathrm{BGL}_r, \boldsymbol{\gamma})$ as a groupoid over the category of schemes.

To a GI-graph $\boldsymbol{\gamma}$ we associate the chain-graph

$$\boldsymbol{\tau} = (\tau,\ g_v \mid v \in V_\tau,\ d_v \mid v \in V_\tau,\ \mathbf{d}_e \mid e \in E_\tau^{\mathrm{or}})$$

where $d_v = \delta_v - \sum_{f\in F_\tau(v)}(r - \min(I_f))$, and for each oriented edge $e = (f, f')$ the chain-type $\mathbf{d}_e$ is the image of the GI-type $(I_f, I_{f'})$. This yields a mapping

$$\{\text{GI-graphs}\} \longrightarrow \{\text{chain-graphs}\}$$

which obviously is surjective and finite to one.

Let $\boldsymbol{\tau}$ be a chain-graph and let $\boldsymbol{\gamma}$ and $\boldsymbol{\gamma}'$ be two GI-graphs mapped to $\boldsymbol{\tau}$. Using arguments similar as in the proof of [K3] 4.2, one shows that there is a canonical isomorphism

$$\beta_{\boldsymbol{\gamma}'}^{\boldsymbol{\gamma}} : \bar{M}(\mathrm{BGL}_r, \boldsymbol{\gamma}) \to \bar{M}(\mathrm{BGL}_r, \boldsymbol{\gamma}').$$

We have $\beta_{\boldsymbol{\gamma}}^{\boldsymbol{\gamma}} = \mathrm{id}$ and $\beta_{\boldsymbol{\gamma}''}^{\boldsymbol{\gamma}'} \circ \beta_{\boldsymbol{\gamma}'}^{\boldsymbol{\gamma}} = \beta_{\boldsymbol{\gamma}''}^{\boldsymbol{\gamma}}$ for any further GI-graph $\boldsymbol{\gamma}''$ mapped to $\boldsymbol{\tau}$.

We use the family $(\beta^{\gamma}_{\gamma'})_{\gamma,\gamma'\mapsto\tau}$ to define the groupoid $\bar{M}(\mathrm{BGL}_r,\tau)$ over the category of schemes. The construction is completely analogous to the construction of $\mathrm{GVBD}_{\mathbf{d}}$ in section 6: The objects of $\bar{M}(\mathrm{BGL}_r,\tau)$ are the union of objects of $\bar{M}(\mathrm{BGL}_r,\gamma)$ where γ runs through all GI-graphs mapped to τ. Let $\mathcal{E}$ and $\mathcal{E}'$ be two objects in $\bar{M}(\mathrm{BGL}_r,\tau)$ and assume that $\mathcal{E}$ belongs to $\bar{M}(\mathrm{BGL}_r,\gamma)$ and $\mathcal{E}'$ to $\bar{M}(\mathrm{BGL}_r,\gamma')$. An arrow of $\bar{M}(\mathrm{BGL}_r,\tau)$ from $\mathcal{E}$ to $\mathcal{E}'$ is then a family of arrows

$$f_\pi : \beta^{\gamma}_{\pi}(\mathcal{E}) \to \beta^{\gamma'}_{\pi}(\mathcal{E}')$$

in $\bar{M}(\mathrm{BGL}_r,\pi)$ where π runs through all GI-graphs over τ such that for all pairs π, π' of GI-graphs over τ we have $\beta^{\pi}_{\pi'}(f_\pi) = f_{\pi'}$.

Remark 7.2. Of course for each GI-graph γ mapped to a chain-graph τ the inclusion $\bar{M}(\mathrm{BGL}_r,\gamma) \hookrightarrow \bar{M}(\mathrm{BGL}_r,\tau)$ is an equivalence of categories and the reason why we defined $\bar{M}(\mathrm{BGL}_r,\tau)$ in the way we did is that we want to make apparent that it does not depend on any choice of a GI-graph γ over τ.

The boundary morphism

$$\nu_\tau : \bar{M}(\mathrm{BGL}_r,\tau) \to \bar{M}_{g,S}(\mathrm{BGL}_r,d)$$

is now defined as follows: With the help of the constructions in [K2] we get a canonical morphism $\nu_\gamma : \bar{M}(\mathrm{BGL}_r,\gamma) \to \bar{M}_{g,S}(\mathrm{BGL}_r,d)$ for every GI-graph γ over τ such that for any two GI-graphs γ and γ' the diagram

$$\begin{array}{ccc} \bar{M}(\mathrm{BGL}_r,\gamma) & \xrightarrow{\beta^{\gamma}_{\gamma'}} & \bar{M}(\mathrm{BGL}_r,\gamma') \\ & {\scriptstyle\nu_\gamma}\searrow \quad \swarrow{\scriptstyle\nu_{\gamma'}} & \\ & \bar{M}_{g,S}(\mathrm{BGL}_r,d) & \end{array}$$

Therefore the family $(\nu_\gamma)_{\gamma\mapsto\tau}$ induces a morphism $\nu_\tau : \bar{M}(\mathrm{BGL}_r,\tau) \to \bar{M}_{g,S}(\mathrm{BGL}_r,d)$.

References

[AV] D. Abramovich and A. Vistoli, Compactifying the space of stable maps, *J. Amer. Math. Soc.*, **15** (2002) no. 1, 27–75.

[ACV] D. Abramovich, A. Corti and A. Vistoli, Twisted bundles and admissible covers. Special issue in honor of Steven L. Kleiman, *Comm. Algebra*, **31** (2003) no. 8, 3547–3618.

[G] Gieseker, A degeneration of the moduli space of stable bundles, *J. Differential Geom.*, **19** (1984) no. 1, 173–206.

[K1] I. Kausz, A Modular Compactification of the General Linear Group, *Documenta Math.*, **5** (2000) 553–594.

[K2] I. Kausz, A Gieseker Type Degeneration of Moduli Stacks of Vector Bundles on Curves, math.AG/0201197.

[K3] I. Kausz, A canonical decomposition of generalized theta functions on the moduli stack of Gieseker vector bundles. math.AG/0305034.

[K4] I. Kausz, Twisted vector bundles on pointed nodal curves, *J. Indian Acad. Sc.*

[Li] J. Li

[M] Y. I. Manin, Frobenius Manifolds, Quantum Cohomology, and Moduli Spaces, AMS Colloquium Publications (1999).

[NS] D. S. Nagaraj and C. S. Seshadri, Degenerations of the moduli spaces of vector bundles on curves II, *Proc. Indian Acad. Sci. Math. Sci.*, **109** (1999) no. 2, 165–201.

[Sch] A. Schmitt, The Hilbert Compactification of the Universal Moduli Space of Semistable Vector Bundles over Smooth Curves, *J. Diff. Geom.*

[Se1] C. S. Seshadri, Degenerations of the moduli spaces of vector bundles on curves, School on Algebraic Geometry (Trieste, 1999), 205–265, ICTP Lect. Notes, 1, Abdus Salam Int. Cent. Theoret. Phys., Trieste, 2000.

Teichmüller Theory and Moduli Problem No. 10, 2010, pp. 451–463.

Limit Points of Iterated Function Systems

Linda Keen and Nikola Lakic

Department of Mathematics, Lehman College and Graduate Center, CUNY Bronx, NY 10468
e-mail: linda.keen@lehman.cuny.edu; lakinik@aol.com

Abstract. This is a written version of a lecture given at the *International Workshop on Teichmüller theory and moduli problems* held at the Harish Chandra Research Institute, Allahabad, U.P., India, in January 2006. It is based on work that originally appeared in [3], [5], and [6].

2000 Subject Classification: 32G15, 30C60, 30C70, 30C75

1. Introduction

Let Δ denote the unit disk and X denote a subdomain of Δ. Let $f_1, f_2, f_3, \ldots$ be a randomly chosen sequence of holomorphic maps $f_i : \Delta \to X \subset \Delta$ and consider the compositions

$$F_n = f_1 \circ f_2 \circ \cdots f_{n-1} \circ f_n.$$

Definition 1. *The sequence* $\{F_n\}$ *is called the* iterated function system IFS *coming from the sequence* $f_1, f_2, f_3, \ldots$.

The sequence F_n forms a normal family and the limit functions F of the family are either open maps or constant maps. The constant limit points may be located either inside X or on ∂X.

Note that the compositions are taken in the *reverse* of the usual order; that is, backwards. There is a theory for forward iterated function systems that is somewhat simpler (see [4]).

Iterated function systems can also be defined for families of holomorphic functions from any plane domain Ω onto a subdomain $X \subseteq \Omega$. We will show what kinds of limit functions occur and whether a given sequence admits more than one limit function depends only on the relative geometry of X in Ω.

Clearly if Ω is the whole complex plane and X has more than two boundary points, the only functions from Ω to X are constants and the iterated function systems are not interesting. We therefore will assume, in this paper, that all the domains we consider have at least two boundary points. Since such a domain admits a hyperbolic metric, we call it *a hyperbolic domain*. The results here originally appeared in [3], [5] and [6]. The reader is also referred to [7] for background and other results.

Paritally supported by a PSC-CUNY Grant. Thanks to the HRI, Allahabad, U.P. India. Partially supported by NSF grant DMS 0200733.

2. History

The simplest situation occurs when all of the functions in the family are the same. Then we have the theorem of Denjoy.

Theorem (Denjoy). *Given an IFS such that $f_n = f$ for all n and f is not a Möbius transformation. Then*

$$\lim_{n\to\infty} F_n = \lim_{n\to\infty} f^n \equiv k$$

$k \in \bar{X}$ is unique.

The first generalization of the Denjoy theorem was given independently by Gill and Lorentzen.

Theorem (Gill (88)–Lorentzen (90)). *If X is relatively compact in Δ then*

$$\lim_{n\to\infty} F_n \equiv k$$

$k \in X$, $(k \notin \partial X)$ is unique.

This completely settles the question of what happens when the subdomain X is relatively compact in Δ. The next natural case to consider is case when the subdomain $X \subset \Delta$ is not relatively compact in Δ. To this end, we recall the following concept for subdomains of the complex plane.

Definition 2. *A domain $E \subset \mathbb{C}$ is a* Bloch domain *if*

$$r_a = \sup_r(|z-a| < r \subset X) \textit{ is bounded for all } a \in X.$$

An example of a non-compact Bloch domain is an infinite strip whose boundary is a pair of parallel lines.

In [2], Beardon, Carne, Minda and Ng generalize this concept as follows. Let ρ_Δ denote the hyperbolic metric on Δ.

Definition 3. *A domain $X \subset \Delta$ is a* hyperbolic Bloch subdomain *of Δ, if*

$$r_a = \sup_r(\rho_\Delta(z,a) < r \subset X) \textit{ is bounded for all } a \in X.$$

Since, in this paper we are only concerned with hyperbolic domains, we will abuse notation and use the term Bloch subdomain to mean a hyperbolic Bloch subdomain.

Clearly if X is relatively compact in Δ, it is a Bloch subdomain. There are, however, lots of non-relatively compact subdomains that are also Bloch subdomains. Let us consider some examples:

- The interior of a polygon of finite hyperbolic area is a Bloch domain. Such a polygon may have cusps on the boundary of Δ where the adjacent sides are tangent. Fundamental domains of finitely generated Fuchsian groups that represent Riemann surfaces with punctures are thus Bloch subdomains.

- The hyperbolic analog of the Euclidean infinite strip is a domain with the property that those sides that touch the boundary of Δ meet there at a finite angle. Note that such a polygon has infinite area but it is a Bloch subdomain.
- A more interesting example is the following. Let G be a finitely generated Fuchsian group G whose quotient $S = \Delta/G$ is a compact Riemann surface. Let D be a fundamental polygon for G and let $V = \{v_1, \ldots, v_n\}$ be the set of vertices of D. For any $g \in G$ let $g(V) = \{g(v_1), \ldots, g(v_n)\}$. Then

$$X = \Delta \backslash \cup_{g \in G} g(V)$$

is an example of a non-simply connected Bloch subdomain.

We have the following theorem about Bloch domains.

Theorem ([2],[3]). *All limit functions of any IFS from Δ to X are constant if and only if X is a Bloch subdomain.*

The "If" statement was proved by Beardon, Carne, Minda and Ng in [2] and the "Only if" statement was originally proved in [3]. We prove it here as theorem 2 in the next section.

3. Uniqueness of limits

If X is relatively compact in Δ the Gill–Lorentzen theorem says that every IFS has a unique limit. It is natural then to ask whether the same is for true limit functions of iterated function systems coming from non-relatively compact subdomains X. Answers to this question are the content of the following theorems:

Theorem 1 ([6]). *If X is non-relatively compact in Δ and n is any integer, then given any n points $c_1, c_2, \ldots, c_n \in X$, there is an IFS with at least n limit functions G_i, $i = 1, \ldots, n$ such that $G_i(0) = c_i$. If X is Bloch $G_i \equiv c_i$ and these are the only limit functions.*

When X is non-Bloch we have the following results:

Theorem 2 ([3]). *If X is non-Bloch then there always exist non-constant limit functions – the Bloch condition is necessary for all limits to be constant.*

Theorem 3 ([5]). *If X is non-Bloch in Δ and $f : \Delta \to X$ is any holomorphic map. Then f is the limit of an IFS.*

4. Proof of theorem 1

With no loss of generality we may assume that $0 \in X$. The idea of the proof is to construct functions f_k such that the set $S = \{c_0 = 0, c_1 = F_1(0) = f_1(0), c_2 = F_2(0) = f_1 \circ f_2(0), \ldots c_{n-1} = F_{n-1}(0) = f_1 \circ f_2 \circ \cdots f_{n-1}(0)\}$ consists of distinct points and such that the *cycle relation*

$$f_i \circ f_{i+1} \circ \cdots \circ f_{i+n-1}(0) = 0 \tag{1}$$

holds for all integers i. From this relation we see that $F_n(0) = 0$ and if $m = qn + r$, $0 \leq r < n$, $F_m(0) = c_r$. The point here is that because the iteration is backwards, we have to find successive pre-images of the points c_i.

Once we have done this, for any subsequence $F_{n_k} = f_1 \circ f_2 \circ \cdots \circ f_{n_k}$, we see that $F_{n_k}(0) \in S$ for all k. It follows that any limit function must map 0 to a point in S. Choosing subsequences appropriately, we can find n distinct limit functions G_i such that $G_i(0) = c_i$, $i = 0 \ldots n-1$.

If X is Bloch, all limit functions must be constant so these are all the limit functions.

We divide the proof into three cases, $n = 1, n = 2$ and $n = 4$. It will be clear from the $n = 4$ case how the proof for any $n > 4$ should go.

$n = 1$: This is the most trivial case. Take $f_i : \Delta \to X$ as a covering map with fixed point c, $f_i(c) = c$. Then clearly $F_n(c) = c$ for all n and any limit function satisfies $F(c) = c$.

$n = 2$: We can use covering maps again in this case. Assume $0 \in X$ and $c_0 = 0$. Choose a covering map

$$f_1 : \Delta \to X, \quad \text{such that} \quad f_1(0) = c_1.$$

The map f_1 defined up to a rotation. Choose it so that for some x_1 with $f_1(x_1) = 0$ we have $x_1 \in X$. Now choose f_2 so that $f_2(0) = x_1$ and for some x_2

$$f_2(x_2) = 0, \quad \text{and} \quad x_2 \in X.$$

Continue to find f_n and x_n such that

$$f_n(0) = x_{n-1}, x_n \in X, f_n(x_n) = 0.$$

We can always do this by the non-compactness of X. Then for all n we have

$$f_1 \circ f_2 \circ \cdots \circ f_{2n}(0) = 0 \quad \text{and}$$

$$f_1 \circ f_2 \circ \cdots \circ f_{2n+1}(0) = c_1.$$

For $n \geq 2$ we need more than covering maps. In order to find the functions f_n and their successive pre-images we need a preparatory lemma.

Lemma 1. *Given distinct points $a_1, \ldots a_n \in \Delta \backslash \{0\}$, there $\exists f : \Delta \to \Delta$ and $x_i \in X$, such that*

$$f(x_i) = \frac{a_i}{x_i}.$$

The function f is rational of degree $n-1$.

Proof. Assume $n \geq 2$ and set

$$A(a, z) = \frac{z - a}{1 - \bar{a}z}, \quad A(a, A(-a, z)) = z.$$

Let $g_1 : \Delta \to \Delta$ be a holomorphic function to be determined later. Choose $x_1 \in X$, $|x_1| > |a_1|$. Set

$$f(z) = \frac{A(x_1, z)g_1(A(x_1, z)) + \frac{a_1}{x_1}}{1 + \frac{\bar{a}_1}{\bar{x}_1}A(x_1, z)g_1(A(x_1, z))}$$

Then $f(x_1) = a_1/x_1$.

Rewrite this as

$$A(x_1, z)g_1(A(x_1, z)) = A\left(\frac{a_1}{x_1}, f(z)\right).$$

Want to choose x_2, $|x_2| > |a_2|$ such that

$$A(x_1, x_2)g_1(A(x_1, x_2)) = A\left(\frac{a_1}{x_1}, \frac{a_2}{x_2}\right)$$

and so on. For $1 \leq j \leq k \leq n$ set

$$a_{jk} = A(x_j, x_k).$$

Next, for $k = 2, \ldots n$ set

$$b_{1k} = A\left(\frac{a_1}{x_1}, \frac{a_k}{x_k}\right) \tag{2}$$

For $j = 2, \ldots n - 1$ and $k = j, j + 1, \ldots n$ set

$$b_{jk} = A\left(\frac{b_{(j-1)j}}{a_{(j-1)j}}, \frac{b_{(j-1)k}}{a_{(j-1)k}}\right). \tag{3}$$

In order that our construction work we need to choose the x_i so that the following inequalities hold:

$$\left|\frac{a_i}{x_i}\right| < 1, \quad i = 1, \ldots n. \tag{4}$$

In step 1 we chose x_1 so this holds for $i = 1$.

For all j, k such that $j < k$ we also need to have

$$\left|\frac{b_{jk}}{a_{jk}}\right| < 1 \tag{5}$$

We use the non-compactness of X to prove we can choose x_i satisfying these conditions. Note first that for fixed j, and all $k > j$, $|x_k| \to 1$ implies $|a_{jk}| \to 1$.

Next, as $|x_j| \to 1$,

$$\limsup |b_{1j}| \leq \left|A\left(\frac{a_1}{x_1}, a_j e^{\theta_j}\right)\right| = B_{1j} < 1$$

where θ_j is chosen so that $\arg a_j e^{\theta_j} = \arg \frac{a_1}{x_1} + \pi$ and B_{1j} is maximal.

Since X is not relatively compact we get conditions on x_i, $i = 2, \ldots, n$ so that all the inequalities (4) and all the inequalities (5) hold with $j = 1$.

Now fix x_2 so that (4) and (5) with $j = 1$ hold, assuming the remaining $|x_i|$ are close enough to 1.

We now find bounds

$$\limsup_{|x_j| \to 1} |b_{2j}| \leq \left|A\left(\frac{b_{12}}{a_{12}}, B_{1j} e^{\theta_j}\right)\right| = B_{2j} < 1$$

where again θ_j is chosen to maximize B_{2j}.

We repeat this process, choosing $x_3, \ldots x_{n-1}, x_n$, in turn so that all the inequalities above hold.

Next, define the functions $g_k(z) : \Delta \to \Delta, k = 2, \ldots n$ recursively by

$$A(x_k, z) g_k(A(x_k, z)) = A\left(\frac{b_{(k-1)k}}{a_{(k-1)k}}, g_{(k-1)}(A(x_{(k-1)}, z))\right) \tag{6}$$

Now take the function $g_n(z) \equiv 0$. Then work back through equations (6) to obtain the functions g_1 and f. Note that f is rational of degree $n-1$.

We check that $f(x_i) = \frac{a_i}{x_i}$ for $i = 1, \ldots, n$, so that we have the required points x_i and the function f.

We show how the theorem is proved for $n = 4$.

Assume $0 \in X$ and $c_0 = 0, c_1, c_2, c_3 \in X$. We want the cycle relation

$$f_i \circ f_{i+1} \circ f_{i+2} \circ f_{i+3}(0) = 0$$

to hold for all i.

Let π_1 be a covering map such that $\pi_1(0) = c_1$. Choose $y_1, b_2', b_3' \in \Delta$ such that

$$\pi_1(y_1) = 0, \quad \pi_1(b_2') = c_2, \quad \pi_1(b_3') = c_3.$$

That is,

$$\begin{array}{ccccc} & y_1 & 0 & b_2' & b_3' \\ \pi_1 & \downarrow & \downarrow & \downarrow & \downarrow \\ & 0 & c_1 & c_2 & c_3 \end{array}$$

Use the lemma to find g and $x_1, b_2, b_3 \in X$ so that

$$g(x_1) = \frac{y_1}{x_1}, \quad g(b_2) = \frac{b_2'}{b_2}, \quad g(b_3) = \frac{b_3'}{b_3}.$$

Let $g_1(x) = xg(x)$. Then

$$\begin{array}{ccccc} & x_1 & 0 & b_2 & b_3 \\ g_1 & \downarrow & \downarrow & \downarrow & \downarrow \\ & y_1 & 0 & b_2' & b_3' \\ \pi_1 & \downarrow & \downarrow & \downarrow & \downarrow \\ & 0 & c_1 & c_2 & c_3 \end{array}$$

Set $f_1 = \pi_1 g_1$. Choose a covering map π_2 such that $\pi_2(0) = b_2$ and points y_{21}, y_2, b'_{32} so that

$$
\begin{array}{ccccc}
 & y_{21} & y_2 & 0 & b'_{32} \\
\pi_2 & \downarrow & \downarrow & \downarrow & \downarrow \\
 & x_1 & 0 & b_2 & b_3 \\
g_1 & \downarrow & \downarrow & \downarrow & \downarrow \\
 & y_1 & 0 & b'_2 & b'_3 \\
\pi_1 & \downarrow & \downarrow & \downarrow & \downarrow \\
 & 0 & c_1 & c_2 & c_3
\end{array}
$$

Now use the lemma to find g_2 and points x_{21}, x_2, b_{32} so that

$$
\begin{array}{ccccc}
 & x_{21} & x_2 & 0 & b_{32} \\
g_2 & \downarrow & \downarrow & \downarrow & \downarrow \\
 & y_{21} & y_2 & 0 & b'_{32} \\
\pi_2 & \downarrow & \downarrow & \downarrow & \downarrow \\
 & x_1 & 0 & b_2 & b_3 \\
g_1 & \downarrow & & \downarrow & \downarrow \\
 & y_1 & 0 & b'_2 & b'_3 \\
\pi_1 & \downarrow & \downarrow & \downarrow & \downarrow \\
 & 0 & c_1 & c_2 & c_3
\end{array}
$$

Set $f_2 = \pi_2 g_2$. Now find a covering map π_3 and points y_{321}, y_{32}, y_3 so that

$$
\begin{array}{ccccc}
 & y_{321} & y_{32} & y_3 & 0 \\
\pi_3 & \downarrow & \downarrow & \downarrow & \downarrow \\
 & x_{21} & x_2 & 0 & b_{32} \\
g_2 & \downarrow & \downarrow & \downarrow & \downarrow \\
 & y_{21} & y_2 & 0 & b'_{32} \\
\pi_2 & \downarrow & \downarrow & \downarrow & \downarrow \\
 & x_1 & 0 & b_2 & b_3 \\
g_1 & \downarrow & \downarrow & \downarrow & \downarrow \\
 & y_1 & 0 & b'_2 & b'_3 \\
\pi_1 & \downarrow & \downarrow & \downarrow & \downarrow \\
 & 0 & c_1 & c_2 & c_3
\end{array}
$$

Then use the lemma again to find g_3 and points x_{321}, x_{32}, x_3 so that

$$\begin{array}{ccccc}
 & x_{321} & x_{32} & x_3 & 0 \\
g_3 & \downarrow & \downarrow & \downarrow & \downarrow \\
 & y_{321} & y_{32} & y_3 & 0 \\
\pi_3 & \downarrow & \downarrow & \downarrow & \downarrow \\
 & x_{21} & x_2 & 0 & b_{32} \\
g_2 & \downarrow & \downarrow & \downarrow & \downarrow \\
 & y_{21} & y_2 & 0 & b'_{32} \\
\pi_2 & \downarrow & \downarrow & \downarrow & \downarrow \\
 & x_1 & 0 & b_2 & b_3 \\
g_1 & \downarrow & \downarrow & \downarrow & \downarrow \\
 & y_1 & 0 & b'_2 & b'_3 \\
\pi_1 & \downarrow & \downarrow & \downarrow & \downarrow \\
 & 0 & c_1 & c_2 & c_3
\end{array}$$

Set $f_3 = \pi_3 g_3$. Finally, find a covering map π_4 and use the lemma to find g_4 so that

$$\begin{array}{ccccc}
 & 0 & x_{432} & x_{43} & x_4 \\
g_4 & \downarrow & \downarrow & \downarrow & \downarrow \\
 & y_{4321} & y_{432} & y_{43} & y_4 \\
\pi_4 & \downarrow & \downarrow & \downarrow & \downarrow \\
 & x_{321} & x_{32} & x_3 & 0 \\
g_3 & \downarrow & \downarrow & \downarrow & \downarrow \\
 & y_{321} & y_{32} & y_3 & 0 \\
\pi_3 & \downarrow & \downarrow & \downarrow & \downarrow \\
 & x_{21} & x_2 & 0 & b_{32} \\
g_2 & \downarrow & \downarrow & \downarrow & \downarrow \\
 & y_{21} & y_2 & 0 & b'_{32} \\
\pi_2 & \downarrow & \downarrow & \downarrow & \downarrow \\
 & x_1 & 0 & b_2 & b_3 \\
g_1 & \downarrow & \downarrow & \downarrow & \downarrow \\
 & y_1 & 0 & b'_2 & b'_3 \\
\pi_1 & \downarrow & \downarrow & \downarrow & \downarrow \\
 & 0 & c_1 & c_2 & c_3
\end{array}$$

Set $f_4 = \pi_4 g_4$. Note that $f_4(0) = 0$. Note $f_1 \circ f_2 \circ f_3 \circ f_4(0) = 0$. This is the first cycle. We continue in this way to add new points to the cycles: to find f_5, f_6, f_7, we replace c_1, c_2 and c_3 with x_{432}, x_{43} and x_4, respectively; for the rest of the f_n, we repeat the process, changing the indices in the obvious way.

Next we find universal covering maps π_n and apply lemma 1 to the preimages in Δ under π_n of points in the incomplete cycles to get new points in X and maps h_n and g_n. The map $f_n = \pi_n g_n$ completes the next cycle and we have added points to the incomplete cycles. In the limit, for all $i \geq 1$ we have the cycle relation

$$f_i \circ f_{i+1} \circ f_{i+2} \circ f_{i+3}(0) = 0$$

for all i.

Let

$$F_k = f_1 \circ \cdots f_k.$$

Then, $F_k(0) = c_r$ where $r = k$ mod 4. The accumulation points are limits of subsequences $\{F_{n_k}\}$. For any such limit F, $F(0) = c_r$ for some $r = 0, \ldots, 3$. Because the c_r are distinct, we have at least 4 distinct accumulation points.

If X is Bloch, all the limit functions of this IFS are constant. Since $F(0) = c_k$ for some k for every limit function there are exactly 4 possible constant functions.

5. Proofs of theorems 2 and 3

Theorem 2 follows from theorem 3, but we give a separate proof that makes it easier to understand how the Bloch condition controls the dynamics.

Theorem 2 ([4]). *X is Bloch in Δ only if all limit functions of any IFS are constant.*

The idea of the proof is: Given X, non-Bloch in Δ, construct an IFS with a limit function that is not constant using covering maps and Blaschke products of degree 2. We will need two lemmas. The proofs are relatively straightforward and we omit them.

Lemma 2 (Geometric Lemma). *Given $a \in X$ and*

$$\rho_X(a, z) < 1 < C = \text{ Bloch rad at } a,$$

then there is a function $\epsilon(C)$ such that

$$\rho_X(a, z) \leq (1 + \epsilon(C))\rho(a, z)$$

and $\epsilon(C) \to 0$ as $C \to \infty$.

This lemma says that in a relatively compact subset of X, ρ and ρ_X are equivalent:

$$\rho(z, w) < \rho_X(z, w) < K\rho(z, w)$$

and "deep inside X", K is very close to 1.

Lemma 3 (Blaschke product lemma). *Let $c \neq 0$ be any point in Δ such that $\rho(0,c) < 1$. If*

$$A_a(z) = \frac{z(z-a)}{1-\bar{a}z},$$

then $A^{-1}(c) = \{z_1, z_2\}$ (that is, $A_a(z_1) = A_a(z_2) = c$) and

$$\rho(0, z_1) = \rho(a, z_2) \to \rho(0, c) \text{ as } |a| \to 1.$$

This lemma says that Blaschke products contract, but the contraction constant is close to 1 near the boundary.

Proof of Theorem 1. Pick any two distinct points $a_0, w_0 \in X$ such that

$$\rho_X(a_0, w_0) < 1/2.$$

Now, using the lemmas we want to recursively find f_n, and compositions $F_n \to F$ satisfying

$$F(0) = a_0 \quad \text{and} \quad F(\tilde{w}) = w_0$$

for some $\tilde{w}$, $\rho(0, \tilde{w}) < 1$. To do this we will need a sequence $\epsilon_n \to 0$ such that $\Pi_1^\infty (1+\epsilon_n)^2 \leq 2$.

To get the first function, let $\pi_1 : \Delta \to X$, $\pi_1(0) = a_0$ be a universal covering map. Then $\exists c_0 \in \Delta$, such that

$$\pi_1(c_0) = w_0 \quad \text{and} \quad \rho(0, c_0) = \rho_X(a_0, w_0)$$

For any choice of $a_1 \in X$, find $w_1, \tilde{w}_1 \in \Delta$ with

$$A_{a_1}(\tilde{w_1}) = A_{a_1}(w_1) = c_0 \quad \text{and} \quad \rho(0, \tilde{w_1}) = \rho(a_1, w_1)$$

Define $f_1 = \pi_1 \circ A_{a_1}$ so that

$$f_1(0) = f_1(a_1) = a_0 \quad \text{and} \quad f_1(w_1) = f_1(\tilde{w}_1) = w_0.$$

We need to make sure that w_1 belongs to X. By the Blaschke product lemma, if $|a_1|$ is close enough to 1

$$\begin{aligned} \rho(0, \tilde{w_1}) &= \rho(a_1, w_1) \\ &< (1+\epsilon_1)\rho(0, c_0) \\ &= (1+\epsilon_1)\rho_X(a_0, w_0) \end{aligned}$$

Moreover, we may assume $R(X, \Delta, a_1) \gg 1$ so that w_1 is inside a disk in X "deeply enough" that by the geometric lemma

$$\begin{aligned} \rho_X(a_1, w_1) &< (1+\epsilon_1)\rho(a_1, w_1) \\ &< (1+\epsilon_1)^2 \rho_X(a_0, w_0) \\ &< 1 \end{aligned}$$

We find the rest of the functions inductively. By our choice of ϵ_n, there exist points $a_n, w_n \in X$ and $\tilde{w}_n \in \Delta$ such that

$$f_n(0) = f_n(a_n) = a_{n-1} \quad \text{and} \quad f_n(w_n) = f_n(\tilde{w}_n) = w_{n-1}, \tag{7}$$

with

$$\rho(a_n, w_n) = \rho(0, \tilde{w}_n) < (1 + \epsilon_n)\rho_X(a_{n-1}, w_{n-1})$$

and

$$\begin{aligned}\rho_X(a_n, w_n) &< (1 + \epsilon_n)\rho(a_n, w_n)\\ &< (1 + \epsilon_n)^2 \rho_X(a_{n-1}, w_{n-1}).\end{aligned}$$

Therefore

$$\rho(0, \tilde{w}_n) < \Pi_1^n (1 + \epsilon_i)\rho(0, c_0) < 1 \quad \text{and}$$

$$\rho_X(a_n, w_n) < \Pi_1^n (1 + \epsilon_i)^2 \rho(0, c_0) < 1.$$

Now if $F_n = f_1 \circ f_2 \circ f_3 \circ \cdots f_n$, formulas (7) yield

$$F_n(0) = a_0 \quad \text{and} \quad F_n(\tilde{w}_n) = w_0.$$

By Montel's theorem, F_n is a normal family so that any subsequence F_{n_j} converges locally uniformly to a holomorphic limit function F. Therefore,

$$F(0) = a_0 \quad \text{and} \quad F(\tilde{w}) = w_0,$$

where $\tilde{w}$ is an accumulation point of the sequence $\tilde{w}_{n_j}$. Since, we have $\rho(0, \tilde{w}_n) < 1$ for all n, the point $\tilde{w}$ belongs to Δ. This implies that F is a nonconstant function concluding the proof of theorem 2.

Theorem 3 ([5]). *If X is non-Bloch in Δ and $f : \Delta \to X$ is any holomorphic map. Then f is the limit of an IFS.*

Sketch of proof of Theorem 3

The reader can find the full details in [6]. The main tool is the use of infinite Blaschke products:

Let $a_1, a_2, a_3, \ldots$ be a sequence of points in Δ and let k be a positive integer. The finite Blaschke products

$$A_n(z) = z^k \frac{z - a_1}{1 - \overline{a_1}z} \frac{z - a_2}{1 - \overline{a_2}z} \cdots \frac{z - a_n}{1 - \overline{a_n}z},$$

form a normal family. Thus, some subsequence $A_{n(l)}$ of A_n converges locally uniformly to an infinite Blaschke product

$$A(z) = z^k \frac{z - a_1}{1 - \overline{a_1}z} \frac{z - a_2}{1 - \overline{a_2}z} \cdots$$

If $|a_n| \to a > 0$, then $A^{(k)}(0) \neq 0$ for any accumulation point A and $A : \Delta \to \Delta$ is open.
A rough idea of the proof is:

Construct infinite Blaschke products $B_i : X \to \Delta$. Interleave the B_i with covering maps $\pi_i : \Delta \to X$. Set $f_i = \pi_{i-1} B_i$ for $i \geq 2$ and show that

$$B_1 f_2 f_3 \dots f_n \to e^{i\theta} z.$$

Next set $f_1 = f(e^{-i\theta} B_1(z))$.

Because X is non-Bloch we can find $c_n \in X$, $|c_n| \to 1$ and $D(c_n, n) \subset X$. We need the following lemmas.

Lemma 4. *Given $C > 1$, $\exists a_n = c_{i_n}$ such that a limit A of the Blaschke product for the a_n satisfies*

$$|A(z)| \geq e^{-k/2^{n-2}C} \tanh(\rho(a_n, z))$$

$\forall z \in D(a_n, (n+4)/2)$.

Lemma 5. *For all $(p, q) \in \Delta$ with $\rho(p, q) < 1$, and any $\sigma > 0$, $\exists C > 1$ such that if n is sufficiently large, $\exists p_n, q_n \in D(a_n, (n+4)/2)$ with*

$$A(p_n) = p, \quad A(q_n) = q, \quad \rho(p_n, q_n) \leq (1 + \sigma)\rho(p, q).$$

For the construction pick $z_0 \in X$ with $z_0 \in D(0, 1)$ and pick ϵ_n with $(1 + \epsilon_n)^{2n+1} \to 0$. Find a_n such that $D(a_n, n) \subset X$.

To obtain B_1, take subsequence of a_n and form the infinite Blaschke product. Using lemmas 4 and 5 find

$$z_1 \in D(0, 1) \text{ with } B_1(z_1) = z_0$$

$$a_{n_1}, q_1 \in D(a_{n_1}, (n_1 + 4)/2)$$

with

$$B_1(a_{n_1}) = 0, \; B_1(q_1) = z_0$$

and

$$\rho(0, z_1) \leq (1 + \epsilon_1)\rho(0, z_0)$$

$$\rho_X(a_{n_1}, q_1) \leq (1 + \epsilon_2)\rho(0, z_0).$$

Observe that by lemma 4, B_1 maps X onto Δ.

Define $\pi_1 : \Delta \to X$, $\pi_1(0) = a_{n_1}$ and find w_1 with $\pi_1(w_1) = q_1$ and

$$\rho(0, w_1) = \rho_X(a_{n_1}, q_1) \leq (1 + \epsilon_2)\rho(0, z_0)$$

To obtain B_2, take a sub-subsequence and form its infinite Blaschke product. Using lemmas 4 and 5 find $z_2 \in D(0, 1)$ with $B_2(z_2) = w_1$. Now pull back by B_1 and π_1 to get $a_{n_2}, q_2 \in D(a_{n_2}, (n_2 + 4)/2)$ with

$$B_1 \pi_1 B_2(a_{n_2}) = 0, \; B_1 \pi_1 B_2(q_2) = z_0$$

$$\text{and} \quad \rho(0, z_2) \leq (1 + \epsilon_2)^3 \rho(0, z_0).$$

Note that you get an extra factor of $(1+\epsilon_2)$ for each pullback.

Set $f_2 = \pi_1 B_2$.

Continuing in this way, we obtain sequences of maps π_{n-1} and B_n with

$$f_n = \pi_{n-1} \circ B_n$$

and a sequence of points z_n with $\rho(0, z_n) \to \rho(0, z_0)$ such that for

$$G_n = B_1 \circ f_2 \circ f_3 \cdots \circ f_n$$

we have

$$G_n(z_n) = z_0, \; G_n(0) = 0$$

and

$$\rho(0, z_n) \leq (1+\epsilon_n)^{2n+1} \rho(0, z_0).$$

Each G_n is holomorphic, maps the unit disk to itself and satisfies $G_n(0) = 0$ and $G_n(z_n) = z_0$. Therefore any accumulation point G of G_n is a holomorphic self map of the unit disk that maps an accumulation point of z_n to z_0. Thus, $G(z) = e^{i\theta}z$. Now to complete the proof of theorem 3, set $f_1 = f(e^{-i\theta}z)$. Then the IFS f_n with $F_n = f_1 \circ \cdots f_n$ has $f(z) = f_1 \circ G$ as a limit function.

References

[1] L. Ahlfors, Lectures on Quasiconformal Mappings, Van Nostrand Studies, 10 (1966).

[2] A. F. Beardon, T. K. Carne, D. Minda and T. W. Ng, Random iteration of analytic maps, *Ergodic Th. and Dyn. Systems*, **24** (2004) no. 3, 659–675.

[3] L. Keen and N. Lakic, Random holomorphic iterations and degenerate subdomains of the unit disk, *Proceedings of Amer. Math. Soc.*, **134** (2006) no. 2, 371–378.

[4] L. Keen and N. Lakic, Forward Iterated Function Systems, Complex Dynamics and Related Topics, Lectures at the Morningside Center of Mathematics, New Studies in Advanced Mathematics, IP Vol. 5 (2003).

[5] L. Keen and N. Lakic, Accumulation points of iterated function systems, in Complex Dynamics: 25 years after the appearance of the Mandelbrot set, Contemp. Mathc 396, (ed.) R. Devaney and L. Keen, A.M.S. (2006).

[6] L. Keen and N. Lakic, Accumulation constants of iterated function systems with Bloch target domains, *Annales Acad. Sci. Finn.*, Vol. 32, Fasc. (2007) no. 1, 73–82.

[7] L. Keen and N. Lakic Hyperbolic Geometry from a local viewpoint Cambridge Univ. Press (2007).

Teichmüller Theory and Moduli Problem No. 10, 2010, pp. 465–479.

SaddleDrop: A Tool for Studying Dynamics in $\mathbb{C}^2$

Sarah C. Koch*

Department of Mathematics, Cornell University, Ithaca, NY, USA
e-mail: kochs@math.cornell.edu

Abstract. In this expository note, we discuss the mathematics behind the computer program, SaddleDrop. Based upon ideas of a group at Cornell University, this program draws parameter space pictures for the complex Hénon map. The difficulty of this task is explained as we contrast the well-developed theory of one variable complex dynamics with the two variable Hénon case.

Introduction

Some of the most beautiful results of complex dynamics arise from studying families of maps which depend on parameters. One such example is the quadratic family $f_c(x) = x^2 + c$, which is parameterized by the complex plane; for each $c \in \mathbb{C}$, there is a map $f_c : \mathbb{C} \to \mathbb{C}$. We can color each point c in the parameter space according to different dynamical properties of the map f_c. For this particular example, coloring the parameter space generates the Mandelbrot set, one of the most important objects in complex dynamics. There is a simple algorithm to draw this set, but it requires a crucial theorem from one variable dynamics.

The Hénon map $H_{a,c}(x, y) = (x^2 + c - ay, x)$ is a polynomial diffeomorphism of $\mathbb{C}^2$, where $(a, c) \in \mathbb{C}^2$, $a \neq 0$, are complex parameters. We can color the parameter space for this family just as before for the quadratic polynomials f_c: we color each pair $(a, c) \in \mathbb{C}^2$ based upon certain dynamical properties of the map $H_{a,c}$. However, it is much more difficult to know what to draw in this case. We not only face the issue of drawing in $\mathbb{C}^2$, but the theory from one variable dynamics that was so important in the Mandelbrot set example, begins to break.

As an undergraduate at Cornell University, K. Papadantonakis worked with J. H. Hubbard, J. Smillie, and E. Bedford to write a computer program called SaddleDrop which was one of the first to draw parameter space pictures for the Hénon map. In this note, we explain the mathematics behind the algorithm used by the program.

We first introduce some preliminaries from the field of complex dynamics, and a key result from the one variable case. We then discuss the Mandelbrot set in greater detail and explain the simple algorithm used to draw it. The remainder of the paper is devoted to two variable complex dynamics: we discuss the analogs of the theory of one variable dynamics when they exist, the challenges that arise when they do not, and the program SaddleDrop which conquers some of these difficulties.

*Author is supported by a National Physical Science Consortium Fellowship.

The author would like to heartily thank J. H. Hubbard for his contagious enthusiasm about Hénon maps, J. Smillie for invaluable SaddleDrop discussions, and the organizers of the International Workshop on Teichmüller Theory and Moduli Problems at the Harish-Chandra Research Institute in Allahabad, India. Special thanks to Sudeb Mitra for the invitation to participate in the workshop, and for the opportunity to visit India which was a truly remarkable experience.

One variable preliminaries

We are interested in iterating maps and asking questions about what happens to certain points in the domain as we continue to iterate. For this section we suppose that $f : \mathbb{C} \to \mathbb{C}$ is a polynomial. Many of the following definitions and remarks hold for arbitrary spaces and maps, but we will not discuss those here.

To any point $p \in \mathbb{C}$, we can associate the sequence of all forward images of p: $f^n(p)$, where f^n is the composition of f with itself n times.

Definition 1. *Let $p \in \mathbb{C}$. The sequence of all forward images of p is called the orbit of p under f, and we denote it as $O^+(p)$. That is, $O^+(p) := \{f^n(p) : n \geqslant 0\}$. If f is invertible, we can define $O^-(p)$ to be the sequence of inverse images of the point p.*

It is quite reasonable to wonder how the sequences $O^+(p)$ behave for a given p. For example, there may be some special values of p where $O^+(p)$ is periodic. These points are naturally called periodic points of the map f. To every periodic point, we associate the corresponding periodic cycle.

Definition 2. *A point $p \in \mathbb{C}$ is periodic if $f^n(p) = p$ for some $n \geqslant 1$. The smallest such n is called the period of p.*

Definition 3. *If $p \in \mathbb{C}$ is periodic of period n, then the set $\{p, \ldots, f^{n-1}(p)\}$ is called a periodic cycle of length n, or a periodic n-cycle.*

There are different types of periodic n-cycles. Their dynamical properties are classified by the value of the derivative of f^n, at a point p_i in the cycle. Note that the chain rule guarantees that the value of $(Df^n)(p_i) = (Df^n)(p_j)$, where p_i and p_j both belong to the periodic cycle. Hence this classification of periodic cycles is well defined.

Definition 4. *If p is a periodic point of period n and $|(Df^n)(p)| < 1$, then $\{p, \ldots, f^{n-1}(p)\}$ is called an attracting cycle.*

An attracting cycle attracts an open set of points called a basin. Not all cycles are attracting of course; some are repelling.

Definition 5. *If p is a periodic point of period n and $|(Df^n)(p)| > 1$, then $\{p, \ldots, f^{n-1}(p)\}$ is called a repelling cycle.*

We omit the case where p is a periodic point of period n and $|(Df^n)(p)| = 1$. This is significantly different from those listed above and will not be discussed here.

As p varies, we can partition $\mathbb{C}$ into two subsets based upon whether the sequence $O^+(p)$ is bounded.

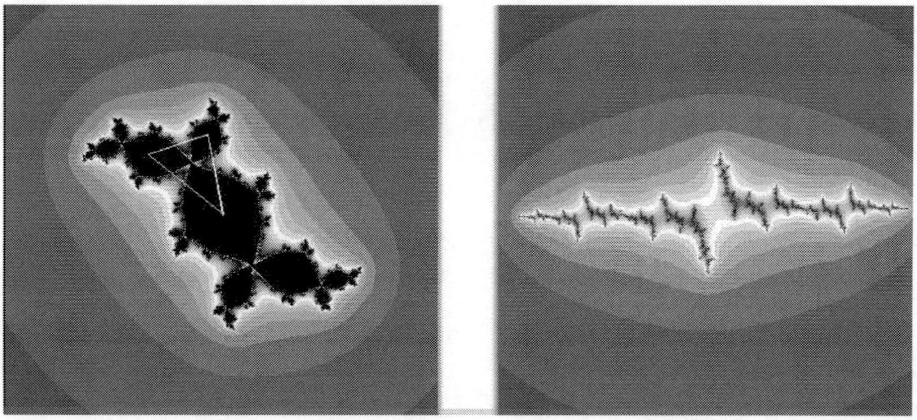

Figure 1. Above are the sets $K^+(f)$ for two different quadratic polynomials. One is connected and the other is a Cantor set. The one on the left contains an attracting 3-cycle given by the vertices of the triangle. The set on the left is called the "Douady rabbit".

Dynamical space

Definition 6. *Let $K^+(f) := \{p \in \mathbb{C} : O^+(p)$ is bounded$\}$. If f is a polynomial, the set $K^+(f)$ is called the filled Julia set.*

The filled Julia set and its complement are dynamically interesting; these sets are both forward invariant; that is, $f(K^+(f)) \subset K^+(f)$, and $f(\mathbb{C}\backslash K(f)) \subset \mathbb{C}\backslash K(f)$, so we can consider f restricted to each as a separate dynamical system.

Example 1. Let $f(z) = z^2$, and suppose $p \in \mathbb{C}$. We can compute a formula for $|f^n(p)|$, namely

$$f^n(p) = p^{2^n}, \text{ and if } p = re^{i\theta}, \text{ then } |f^n(p)| = r^{2^n}.$$

As $n \to \infty$ there are three possible outcomes:

- If $p \in \mathbb{D}$, then $r < 1$ and $|f^n(p)| \to 0$
- If $p \in \partial\mathbb{D}$, then $r = 1$ and $|f^n(p)| = 1$
- If $p \in \mathbb{C}\backslash\overline{\mathbb{D}}$, then $r > 1$ and $|f^n(p)| \to \infty$

We see that if $p \in K^+(f) \iff p \in \overline{\mathbb{D}}$, so $K^+(f) = \overline{\mathbb{D}}$. Note for this example that $p = 0$ is an attracting fixed point of f, and $p = 1$ is a repelling fixed point. The basin of attraction for the fixed point at $p = 0$ is $\mathbb{D}$.

One may be tempted to think that the formula for $f^n(p)$ always lends itself so nicely to determining $K^+(f)$; this example is deceptive. The set can be amazingly complicated and almost always has fractal boundary. We can easily draw pictures of K^+ on the computer. For each point $p \in \mathbb{C}$ we ask if $O^+(p)$ is bounded. If yes, we color the p black, and if no, we color p grey[1]. This generates what is known as a dynamical space picture for f.

---

[1] In this paper, we use grey for our pictures, but the computer program used to generate the pictures uses other colors.

It is quite natural to wonder what $K^+(f)$ looks like, and we can ask questions of a topological nature: Is it connected? If not, how many connected components does it have? We could just draw the computer picture to get a visual answer, but there is a very important theorem (due to Fatou) which is the key to rigorously answering the connectedness question question.

Theorem 1. *Let f be a polynomial of degree $d \geqslant 2$. If the filled Julia set $K^+(f)$ contains all of the finite critical points of f then both $K^+(f)$ and $\partial K^+(f)$ are connected.*

This theorem is incredibly useful. In the case where f is a quadratic polynomial, there is only one critical point to consider, so determining whether $K^+(f)$ is connected in that case is precisely asking what happens to the critical point under forward iteration. This lays the foundation for drawing parameter space pictures for polynomials; we will focus on the quadratics.

The Mandelbrot set, *M*

We would like to draw a parameter space picture for the family of quadratic polynomials. It may appear that a quadratic polynomial, $f(x) = a_1x^2 + a_2x + a_3$, depends on three complex parameters, however, this polynomial is topologically conjugate to a unique polynomial of the form $g(x) = x^2 + c$ through an affine change of variables. Dynamically, it suffices to consider conjugacy classes of maps since orbits are preserved under conjugation. Hence the parameter space for the entire family of quadratic polynomials is $\mathbb{C}$.

Let $f_c(x) = x^2 + c$. As discussed in the introduction, it is desirable to color parameter space according to the dynamical properties of the map f_c. In particular, we would like to color the parameter plane according to properties of $K^+(f_c)$. Color

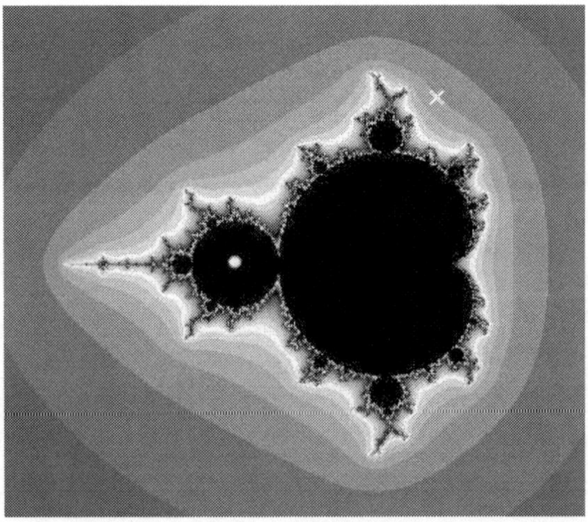

Figure 2. The Mandelbrot Set M with two parameters marked.

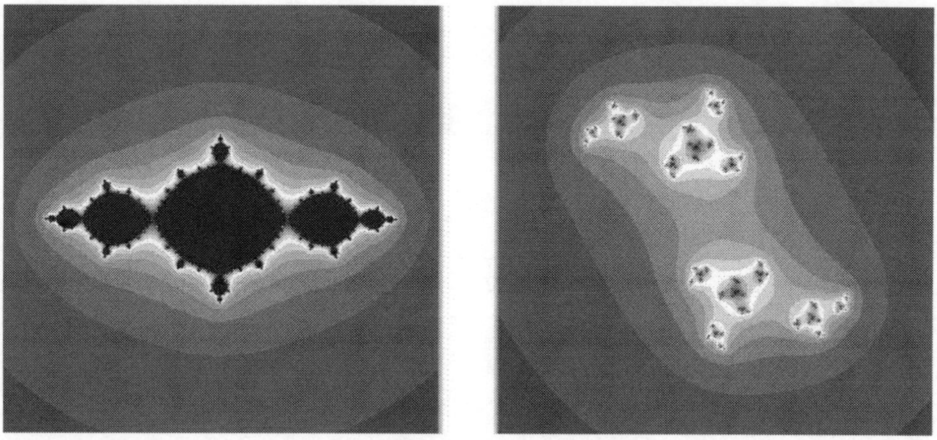

Figure 3. The set on the left is $K^+(f_c)$ for the polynomial $f_c(x) = x^2 - 1$. There is a white dot marking this parameter value in figure 2. The set on the right is $K^+(f_c)$ for the polynomial $f_c(x) = x^2 + 0.23 + 0.91i$. There is a white 'x' marking this parameter value in $\mathbb{C}\backslash M$ above.

$c \in \mathbb{C}$ black if the corresponding $K^+(f_c)$ is connected and color $c \in \mathbb{C}$ grey if the corresponding $K^+(f_c)$ is disconnected. The theorem above provides an easy algorithm for this task: we must follow the critical point of f_c, which is $z_0 = 0$. For each $c \in \mathbb{C}$ we have a polynomial f_c, and we must determine if $O^+(z_0) \subset K^+(f_c)$. This simple procedure generates the Mandelbrot set $M := \{c \in \mathbb{C} : K^+(f_c) \text{ is connected}\}$.

Notice that there are grey curves in $\mathbb{C}\backslash M$ drawn above in figure 2. If $c \in \mathbb{C}\backslash M$, then $O^+(z_0) \to \infty$; we can color c according to how quickly z_0 escapes to infinity. Color $c \in \mathbb{C}\backslash M$ dark grey if z_0 escapes to infinity quickly, and color $c \in \mathbb{C}\backslash M$ light grey if z_0 escapes relatively slowly. We therefore have a "rate of escape" function defined on $\mathbb{C}\backslash M$ in parameter space. The grey curves in $\mathbb{C}\backslash M$ are level curves of this function.

In the quadratic case, there is a nice dichotomy: either $K^+(f_c)$ is connected or it is a Cantor set. If the filled Julia set is a Cantor set, then the dynamics of f_c on $K^+(f_c)$ is conjugate to the horseshoe dynamics of the one-sided shift map σ on the space of infinite sequences of two symbols. We mention this as an interesting fact; the theory behind it is omitted, and we will not require this result for any future discussion. More information can be found in [BuS] or [BH].

The green function

We refer the reader to figure 3 for this portion of the discussion. Consider the complement of $K^+(f_c)$; it is shaded in grey in both pictures. This set contains all points p that escape to infinity under iteration of f_c. Similar to the case of $\mathbb{C}\backslash M$ above, it is natural to ask how quickly these points escape: the darker the grey, the faster the point escapes to infinity. We therefore have a function, $G : \mathbb{C}\backslash K^+(f) \to (0, \infty)$, which measures the rate of escape of these points. This function is called a Green function, or a potential function for $K^+(f_c)$. Notice that there are curves faintly drawn in the grey regions of both figures; these are level curves of G. If we set $G := 0$ on $K^+(f_c)$, the Green function is defined on all of $\mathbb{C}$. For the polynomial $f_c(x) = x^2 + c$, the following gives a formula for G.

Definition 7. *The map* $G : \mathbb{C} \to [0, \infty)$ *associated to the filled Julia set* $K^+(f_c)$ *is defined as*

$$G(x) = \lim_{n\to\infty} \frac{1}{2^n} \log^+ |f_c^n(x)|, \text{ where } \log^+ |z| = \max\{0, \log |z|\}$$

and is the Green function of $K^+(f_c)$*. This function is subharmonic everywhere and harmonic on* $\mathbb{C}\backslash\partial K^+(f_c)$*. The level curves of this function discussed above are called equipotentials.*

Critical points of *G*

From the formula for G, one can derive the equation $G(f_c(x)) = 2G(x)$, which implies that f_c maps equipotentials to equipotentials. The function G has critical points, and this equation reveals exactly what they are. The critical points we consider are contained in $\mathbb{C}\backslash K^+(f_c)$. We do not consider critical points of G for $z \in K^+(f_c)$ since $G = 0$ there.

$$G(f_c(z)) = 2G(z) \implies DG(f_c(z))f_c'(z) = 2DG(z)$$

Suppose $z \in \mathbb{C}\backslash K^+(f_c)$ is a critical point of G. We then have $DG(z) = 0$. Therefore if z is a critical point of G, then either of the following holds:

- $f_c'(z) = 0$, which implies $z = z_0$
- $DG(f_c(z)) = 0$, which implies $f_c(z)$ is also a critical point of G

Note that the first possibility implies $z_0 = 0$ is a critical point of G; this is indeed true if $z_0 \notin K^+(f_c)$.

Proposition 1. *If* $z \in \mathbb{C}\backslash K^+(f_c)$ *is a critical point of* G*, then there exists* $m \geqslant 0$ *such that* $f_c^m(z) = z_0$.

Proof. As mentioned above, if z is a critical point of G, then either $z = z_0$, or $f_c(z)$ is also a critical point of G. If $z = z_0$, then the proposition is true with $m = 0$. Suppose then that $z \neq z_0$. We therefore must have that $f_c(z)$ is also a critical point of G. Since $f_c(z)$ is a critical point of G, then either $f_c(z) = z_0$, and the proposition holds for $m = 1$, or $f_c(z) \neq z_0$, and $f_c(f_c(z))$ is a critical point by the second possibility above. We can continue in this way, and we see that if there is no m such that $f_c^m(z) = z_0$, then we must have that $f_c^m(z)$ is a critical point of G for every m. Since $z \in \mathbb{C}\backslash K^+(f_c)$, $O^+(z)$ is unbounded and $f_c^m(z) \to \infty$ as $m \to \infty$. We therefore have a sequence of critical points of G which are arbitrarily close to infinity. However, analysis of the formula for G reveals that $G(w) \sim \log |w|$ as $w \to \infty$. The function $\log |w|$ does not have critical points arbitrarily close to infinity, and we have arrived at a contradiction. □

The equation $G(f_c^n(z)) = 2^n G(z)$, and proposition 1 imply that the critical points of G are precisely the critical point, $z_0 = 0$ of f_c as well as all inverse images of this

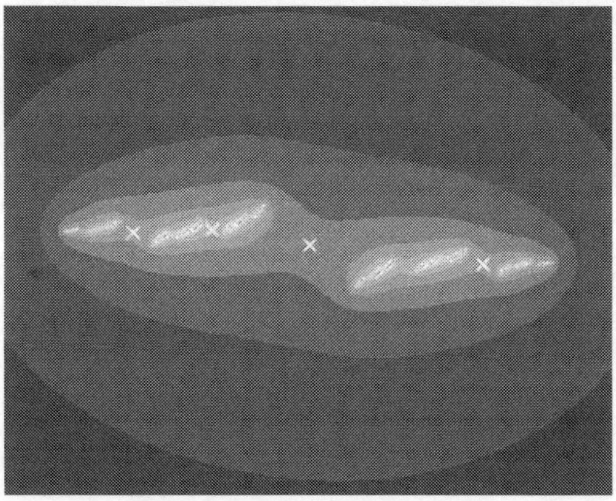

Figure 4. An example of a disconnected $K^+(f_c)$. The white marks are placed at a few critical points of G.

point under the polynomial f_c as long as these points are not in $K^+(f_c)$. If one of these points were contained in $K^+(f_c)$, then all of them would be, and G would have no critical points. We therefore observe that for the case $f_c(x) = x^2 + c$, $K^+(f_c)$ is connected if and only if G has no critical points, by theorem 1. This is a very important fact.

Two variable preliminaries

We now advance to complex dynamics in two variables. The map we consider is the Hénon map, given by the formula

$$H_{a,c} : \mathbb{C}^2 \to \mathbb{C}^2, \quad \text{where} \quad H_{a,c}\begin{pmatrix} x \\ y \end{pmatrix} = \begin{pmatrix} x^2 + c - ay \\ x \end{pmatrix}.$$

The Jacobian of $H_{a,c}$ equals a, and H^{-1} is given by

$$H_{a,c}^{-1}\begin{pmatrix} x \\ y \end{pmatrix} = \begin{pmatrix} y \\ \frac{y^2+c-x}{a} \end{pmatrix}.$$

Note that this map is a polynomial diffeomorphism of $\mathbb{C}^2$ as long as $a \neq 0$. We can define two variable analogs for the Hénon map of the one variable objects previously discussed.

Definition 8. *Let* $\mathbf{p} \in \mathbb{C}^2$. *The sequence of all forward images of* $\mathbf{p}$ *is called the orbit of* $\mathbf{p}$ *under* $H_{a,c}$ *, and we denote it as* $O^+(\mathbf{p})$. *The Hénon map is invertible, so we can define* $O^-(\mathbf{p})$ *to be the sequence of all inverse images of* $\mathbf{p}$.

Definition 9. *A point* $\mathbf{p}$ *is periodic if* $H_{a,c}^n(\mathbf{p}) = \mathbf{p}$ *for some* $n \geqslant 1$. *The smallest such* n *is called the period of* $\mathbf{p}$.

Definition 10. *If* $\mathbf{p}$ *is periodic of period* n, *then the set* $\{\mathbf{p}, \ldots\ H_{a,c}^{n-1}(\mathbf{p})\}$ *is called a periodic cycle of length* n.

Definition 11. *Let* **p** *be a periodic point of period n, and let* λ_1 *and* λ_2 *be the eigenvalues of* $DH^n_{a,c}(\mathbf{p})$. *If* $|\lambda_1| < 1$ *and* $|\lambda_2| < 1$, *then* $\{\mathbf{p}, \ldots\ H^{n-1}_{a,c}(\mathbf{p})\}$ *is an attracting cycle.*

Definition 12. *Let* **p** *be a periodic point of period n, and let* λ_1 *and* λ_2 *be the eigenvalues of* $DH^n_{a,c}(\mathbf{p})$. *If* $|\lambda_1| < 1$ *and* $|\lambda_2| > 1$, *then* $\{\mathbf{p}, \ldots\ H^{n-1}_{a,c}(\mathbf{p})\}$ *is a saddle cycle.*

We can also define repelling cycles as those periodic cycles where the eigenvalues λ_1 and λ_2 are greater than 1 in modulus. We shall be most concerned with saddle cycles. We can again partition our dynamical space into different sets depending upon the orbits of points.

Definition 13. *Let* $K^+(H_{a,c}) := \{\mathbf{p} \in \mathbb{C} : O^+(\mathbf{p}) \text{ is bounded}\}$.

Definition 14. *Let* $K^-(H_{a,c}) := \{\mathbf{p} \in \mathbb{C} : O^-(\mathbf{p}) \text{ is bounded}\}$.

Definition 15. *Let* $K(H_{a,c}) := K^+(H_{a,c}) \cap K^-(H_{a,c})$.

As in the one variable case the sets above and their complements are invariant sets and are therefore dynamically interesting, and we can ask topological questions: for a given pair of parameters (a, c), which of these sets above is connected? The answers to these questions are a bit complicated. Previously, we saw pictures of $K^+(f_c)$ for some quadratic polynomial examples (see figures 3 and 4). In the one variable case, the sets $K^+(f_c)$ and its complement were contained in $\mathbb{C}$; we could draw this copy of $\mathbb{C}$ on a computer screen, and observe visually whether these sets were connected.

The algorithm: dynamical space

The dynamical space for $H_{a,c}$ is $\mathbb{C}^2$. We face two challenges. The first is how do we draw computer pictures in $\mathbb{C}^2$? One possible way is to slice $\mathbb{C}^2$ with a complex line, say l, and just as in the one variable case, we could examine which points in this

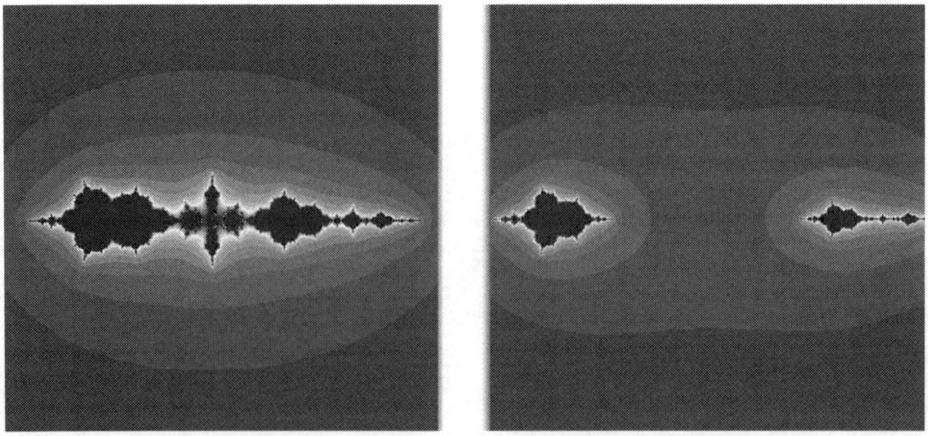

Figure 5. Both pictures depict the set $K^+(H_{a,c})$ for the parameter values $a = 0.3$, $c = -1.17$. On the left is a slice of $\mathbb{C}^2$ with the line $y = 1$ and on the right is a slice of $\mathbb{C}^2$ with the line $y = 4$.

line escape to infinity under forward iteration of the Hénon map. This line is a one-dimensional object and thus lends itself nicely to computer pictures. We could then color the points in our computer picture based on whether or not they escape. This provides us with a picture of the intersection of $K^+(H_{a,c})$ with the complex line.

The pictures above are a little disconcerting. In one picture, $K^+(H_{a,c})$ appears to be connected, but in the other picture it does not. It is possible for a set to be disconnected in a complex line l but to be connected in $\mathbb{C}^2$. And it is equally possible for a set to be connected in a complex line l but to be disconnected in $\mathbb{C}^2$. These pictures reveal nothing about the connectedness of $K^+(H_{a,c})$. This is our second challenge: is there a one-dimensional object in $\mathbb{C}^2$ we can draw, which provides topological information about any of the sets above? We will focus our attention on the set $K(H_{a,c})$.

Definition 16. *If* $\mathbf{p}$ *is a saddle fixed point of* $H_{a,c}$*, then the unstable manifold of* $\mathbf{p}$ *is the set* $W^u(\mathbf{p}) := \left\{\binom{x}{y} \in \mathbb{C}^2 : H_{a,c}^{-n}\binom{x}{y} \to \mathbf{p} \text{ as } n \to \infty\right\}$.

This set is an immersed one-dimensional manifold in $\mathbb{C}^2$. Moreover, note that it is an invariant set for $H_{a,c}$ and for $H_{a,c}^{-1}$, and is therefore a dynamical object of interest. Such a set exists for all $H_{a,c}$ that have a saddle fixed point.

Proposition 2. *For every pair of parameters* (a, c)*, with* $|a| \neq 1$*,* $H_{a,c}$ *has a saddle fixed point, except for those on the curve* $(a+1)^2 = 4c$.

This result in [HP] guarantees that there is a saddle fixed point for practically any parameter pair (a, c), and for each saddle fixed point $\mathbf{p}$, there is an unstable manifold $W^u(\mathbf{p})$. This is a one-dimensional object, but it is all wound up in $\mathbb{C}^2$; drawing it with a computer would be rather difficult if it were not for the following result due to J. H. Hubbard.

Proposition 3. *Let* $\mathbf{p}$ *be a saddle fixed point of* $H_{a,c}$*. Let* $(\lambda, \mathbf{v})$ *be the eigenvalue-eigenvector pair of* $DH_{a,c}(\mathbf{p})$*, where* $|\lambda| > 1$ *is the expanding eigenvalue. The map* $\gamma : \mathbb{C} \to W^u(\mathbf{p})$*, given by*

$$\gamma(z) = \lim_{n\to\infty} H_{a,c}^n\left(\mathbf{p} + \frac{z}{\lambda^n}\mathbf{v}\right)$$

is an analytic injective immersion, where $H_{a,c}(\gamma(z)) = \gamma(\lambda z)$.

This has been a tremendously useful fact. For any saddle fixed point $\mathbf{p}$, the map γ gives a parameterization of $W^u(\mathbf{p})$ by $\mathbb{C}$. We can draw this copy of $\mathbb{C}$ on a computer. For each point $z \in \mathbb{C}$, we calculate $\gamma(z) \in W^u(\mathbf{p})$. We then examine $H_{a,c}^n(\gamma(z))$ as $n \to \pm\infty$ (since we are attempting to draw $K(H_{a,c})$, we must consider both forward and backward orbits). We color the original point z grey if either $O^+(\gamma(z))$ or $O^-(\gamma(z))$ is unbounded, and color z black if $O^+(\gamma(z))$ and $O^-(\gamma(z))$ are both bounded, or equivalently if $\gamma(z) \in K(H_{a,c})$. Note however, that $W^u(\mathbf{p}) \subset K^-(H_{a,c})$, so $O^-(\gamma(z))$ is necessarily bounded; in fact, $O^-(\gamma(z)) \to \mathbf{p}$ since $\gamma(z) \in W^u(\mathbf{p})$. We therefore need only investigate whether $O^+(\gamma(z))$ is bounded.

We have found another one-dimensional object to draw in $\mathbb{C}^2$. Observe that in the figure above $K(H_{a,c}) \cap W^u(\mathbf{p})$ is disconnected. But is $K(H_{a,c})$ disconnected inside $\mathbb{C}^2$? How does the connectedness of $K(H_{a,c}) \cap W^u(\mathbf{p})$ relate to the connectedness of

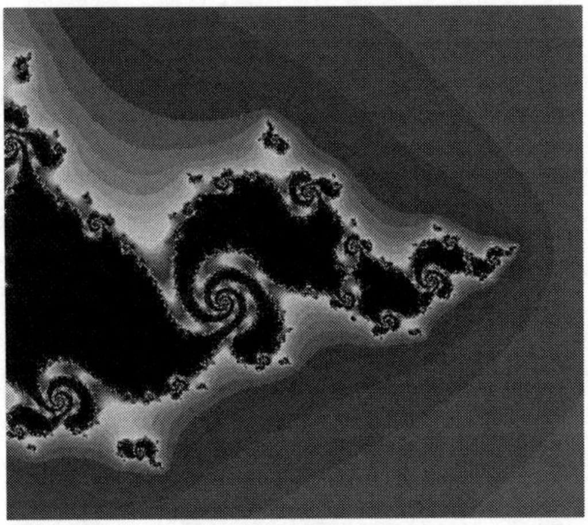

Figure 6. This is a copy of $\mathbb{C}$ colored as described above. In black we see $\gamma^{-1}(K(H_{a,c}))$, for the parameter values $a = 0.3$, $c = -0.375 + 0.6125i$.

$K(H_{a,c})$ inside $\mathbb{C}^2$? Moreover, it is certainly possible for a Hénon map, $H_{a,c}$ to have two saddle fixed points. We may wonder how the connectedness of $K(H_{a,c}) \cap W^u(\mathbf{p}_1)$ compares with the connectedness of $K(H_{a,c}) \cap W^u(\mathbf{p}_2)$, where $\mathbf{p}_1$ and $\mathbf{p}_2$ are the two points in question. The following two theorems by Bedford and Smillie provide very promising answers to these questions in [BS6].

Theorem 2. *The set $K(H_{a,c})$ is connected if and only if $K(H_{a,c}) \cap W^u(\mathbf{p})$ is connected for some saddle point* $\mathbf{p}$.

Theorem 3. $K(H_{a,c}) \cap W^u(\mathbf{p}_1)$ *is connected for the saddle point* $\mathbf{p}_1$ *if and only if* $K(H_{a,c}) \cap W^u(\mathbf{p}_2)$ *is connected for the other saddle point* $\mathbf{p}_2$.

We have succeeded: we found a one-dimensional object to draw that provides topological information about $K(H_{a,c})$. The first part of the algorithm is complete.

The algorithm: parameter space

Drawing a parameter space picture for the Hénon map is quite difficult. It is unclear where to even begin. The parameter space is $\mathbb{C}^2$, so again we face challenges of drawing in this space on a computer. In this case, slicing the parameter space $\mathbb{C}^2$ with a complex line will suffice. We choose to fix $a = a_0$, (where $0 < |a_0| < 1$) and examine a c parameter plane, which is just a copy of $\mathbb{C}$. We color each $c \in \mathbb{C}$ according to the properties of $K(H_{a_0,c})$. Alternatively, we could have fixed $c = c_0$ and examined a parameter planes as well, however, the c parameter planes provide analogs of the Mandelbrot set. Recall the formula for the Hénon map

$$H_{a,c}\begin{pmatrix} x \\ y \end{pmatrix} = \begin{pmatrix} x^2 + c - ay \\ x \end{pmatrix}.$$

Notice that if we fix $a = a_0$, and vary c, the parameter we are changing is the translation; this translation parameter space is the analog of the parameter space $\mathbb{C}$ for the polynomials $f_c(x) = x^2 + c$. Indeed, if $a = 0$, we obtain the Mandlebrot set M in the c parameter plane.

Ideally, we would like to color points in $\mathbb{C}^2$ according to the properties of $K(H_{a_0,c})$. We could color a point $(a_0, c) \in \mathbb{C}^2$ black if $K(H_{a_0,c})$ is connected, and we could color it grey if $K(H_{a_0,c})$ is disconnected. But is there a systematic way to determine if $K(H_{a_0,c})$ is connected?

This is where we encounter one of the fundamental differences between complex dynamics in one variable and complex dynamics in several variables. Theorem 1 provided a very simple algorithm for drawing the Mandelbrot set: the connectedness of $K^+(f_c)$ was precisely related to the orbit of the critical point z_0 of the polynomial f_c. Unfortunately, we cannot even hope that such a statement will be as helpful to us for $H_{a,c}$. One immediate observation is that $H_{a,c}$ is a diffeomorphism and therefore has no critical points. Regardless of this, there is no direct analog of theorem 1 in higher dimensions. We must find another criterion with which to draw parameter space.

Fix a parameter pair, (a, c). Consider $W^u(\mathbf{p})$; let $\mathbf{z} \in W^u(\mathbf{p})$. Either $O^+(\mathbf{z})$ is bounded or it is not, so either $\mathbf{z} \in K(H_{a,c})$ or it is not. If it is not, then $\|H^n_{a,c}(\mathbf{z})\| \to \infty$ as $n \to \infty$, and we can therefore measure how quickly $\mathbf{z}$ escapes, providing an analog of the Green function.

Definition 17. *The map $G^+ : \mathbb{C}^2 \to [0, \infty)$ associated to $K^+(H_{a,c})$ defined as*

$$G^+\begin{pmatrix} x \\ y \end{pmatrix} = \lim_{n\to\infty} \frac{1}{2^n} \log^+ \left\| H^n_{a,c}\begin{pmatrix} x \\ y \end{pmatrix} \right\|$$

is the Green function of $K^+(H_{a,c})$. This function is plurisubharmonic everywhere and pluriharmonic (harmonic on any complex line) on $\mathbb{C}^2 \backslash \partial K^+(H_{a,c})$.

Note that this function is identically 0 on $K^+(H_{a,c})$. We can examine $G^+ \circ \gamma$ where γ is the parameterization of $W^u(\mathbf{p})$ by $\mathbb{C}$. Observe that $K^+(H_{a,c}) \cap W^u(\mathbf{p}) \subset K(H_{a,c})$; in fact, $K^+(H_{a,c}) \cap W^u(\mathbf{p}) = K(H_{a,c}) \cap W^u(\mathbf{p})$, so G^+ restricted to

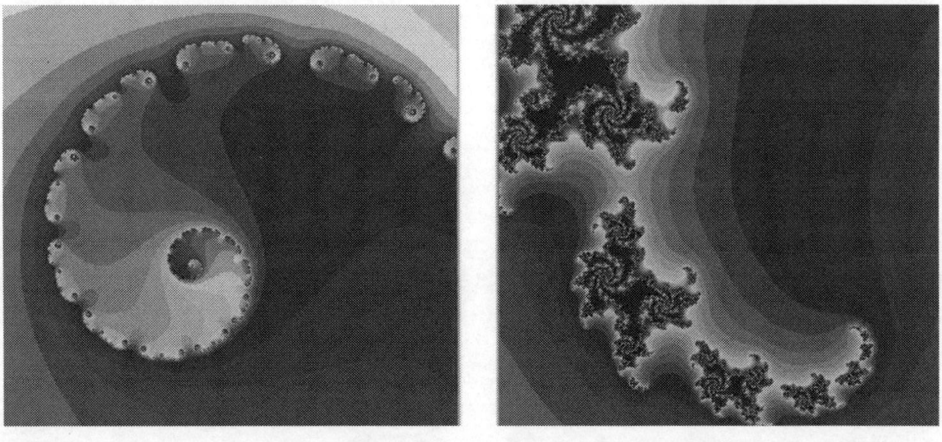

Figure 7. These are unstable manifolds for two different pairs of parameter values. For the map on the left, it appears that $K(H_{a,c}) \cap W^u(\mathbf{p})$ is a Cantor set, and for the map on the right, $K(H_{a,c}) \cap W^u(\mathbf{p})$ is disconnected but not a Cantor set.

$W^u(\mathbf{p})$ is defined on $W^u(\mathbf{p})\backslash K^+(H_{a,c}) = W^u(\mathbf{p})\backslash K(H_{a,c})$. In our computer pictures, we color $\gamma(z) \notin K(H_{a,c})$ according to how fast it escapes under forward iteration. In the pictures below, we see faint grey curves drawn in the complement of $K(H_{a,c})$. These are level curves of the function G^+. Remember that these pictures are really $\gamma^{-1}(W^u(\mathbf{p}))$.

Although there are no critical points of our map $H_{a,c}$, there are sometimes critical points of $G^+ \circ \gamma$; that is, for some parameter values there are critical points of G^+ restricted to $W^u(\mathbf{p})$. Recall that for the case $f_c(x) = x^2 + c$, $K^+(f_c)$ is connected if and only if there are no critical points of the Green function G in the dynamical space $\mathbb{C}$. This suggests that the critical points of G^+ in $W^u(\mathbf{p})$ may be related to the connectedness of $K(H_{a,c})$. This is indeed the case; Bedford and Smillie assert in [BS6] that $K(H_{a,c})$ is connected if and only if for μ almost every $\mathbf{p}$, $G^+|_{W^u(\mathbf{p})\backslash K^+(H_{a,c})}$ contains no critical points. The measure μ is a special harmonic measure that will not be discussed here.

This result is somewhat encouraging; it provides a preliminary criterion for drawing a parameter space picture. Fix a parameter pair (a_0, c_0) and find a saddle fixed point, $\mathbf{p}$. Draw the dynamical space $W^u(\mathbf{p})$. In $W^u(\mathbf{p})$, find a critical point of G^+ if there are any. Color (a_0, c_0) in parameter space according to the 'behavior' of this critical point. Note that these critical points of G^+ are contained in $W^u(\mathbf{p})\backslash K(H_{a_0,c_0}) = W^u(\mathbf{p})\backslash K^+(H_{a_0,c_0})$, which means that they have unbounded forward orbits, but bounded backward orbits. We can therefore measure how quickly these critical points escape forward to infinity. More specifically, color (a_0, c_0) according to how fast the critical point escapes. Move through parameter space, coloring each parameter pair (a, c) according to how fast a critical point of G^+ in $W^u(\mathbf{p})\backslash K(H_{a,c})$ escapes.

There are some issues that arise with this algorithm, namely, if there are any critical points of G^+ in $W^u(\mathbf{p})$, then there are infinitely many. How can we be sure that this algorithm is well defined? If α_1 and α_2 are two critical points in $W^u(\mathbf{p})$, how do we color (a_0, c_0) according to behavior of the critical points if the critical points display two different behaviors? The way we color the parameter pair fundamentally depends on which critical point we choose; the algorithm is not well defined. In fact, essentially different parameter space pictures can arise if a different critical point is chosen. This is actually part of the richness of the subject: we do not fully understand this phenomenon yet, but it is very tantalizing and continues to create more and more questions about $H_{a,c}$.

Another issue that surfaces with this algorithm surrounds the idea of 'moving around in parameter space', coloring each pair (a, c) according to how fast a critical point of G^+ escapes. Let $\mathbf{p}_1$, $\mathbf{p}_2$ be saddle fixed points for the parameter pairs (a_1, c_1) and (a_2, c_2) respectively. As we move from parameters (a_1, c_1) to (a_2, c_2) does the critical point we select for (a_1, c_1) in $W^u(\mathbf{p}_1)$ influence the choice of the one we select for (a_2, c_2) in $W^u(\mathbf{p}_2)$? Is there any continuity of the critical point selection as the parameter pair is varied? Is there any systematic way of choosing a corresponding critical point as we move in parameter space? The answer is yes: one systematic approach is to use Newton's method. This is the algorithm in the program SaddleDrop.

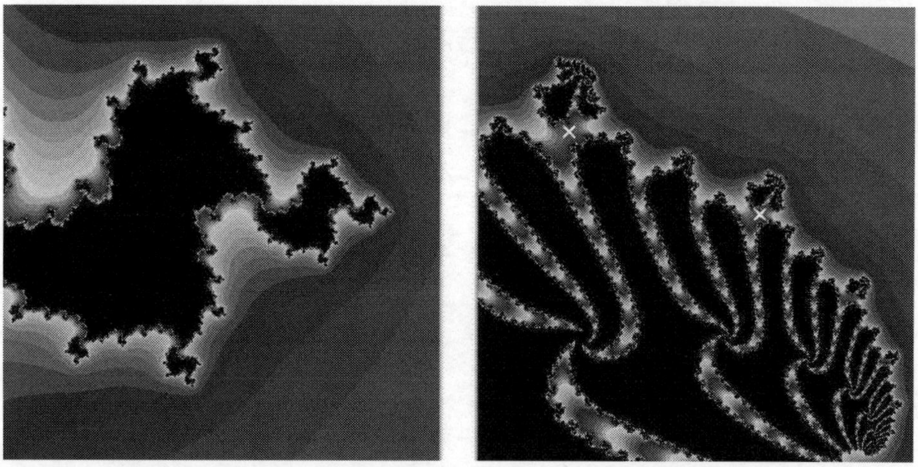

Figure 8. On the left is a picture of $W^u(\mathbf{p})$ for a parameter pair (a, c) such that $W^u(\mathbf{p})\backslash K(H_{a,c})$ contains no critical points of G^+. On the right is a picture of $W^u(\mathbf{p})$ for a parameter pair (a, c) such that $W^u(\mathbf{p})\backslash K(H_{a,c})$ contains infinitely many critical points of G^+. Two of them are marked with a white 'x'.

SaddleDrop

The program SaddleDrop was written by Karl Papadantonakis in 2000 at Cornell University. He collaborated with J. H. Hubbard, J. Smillie, and E. Bedford to draw a parameter space for the Hénon map using the algorithm outlined above. To summarize the points above and to be a bit more precise, we outline the algorithm:

GOAL : Draw parameter space pictures for the Hénon map.

- Fix a parameter. Suppose we fix $a = a_0$.
- Choose a single value of the other parameter; suppose we pick $c = c_0$.
- For the parameter pair (a_0, c_0), find a saddle fixed point $\mathbf{p}$ of H_{a_0,c_0}.
- Draw $W^u(\mathbf{p})$ as discussed in the dynamical space algorithm.
- Find a critical point, α, of G^+ in $W^u(\mathbf{p})\backslash K(H_{a_0,c_0})$. If there are no critical points, change the initial parameter pair recalculating $\mathbf{p}$ and $W^u(\mathbf{p})$ until a critical point of G^+ is found in $W^u(\mathbf{p})\backslash K(H_{a_0,c_0})$.
- For the parameter pair (a_0, c_0), measure how fast α escapes to ∞. Color (a_0, c_0) accordingly.
- Move to new parameter pair (a_0, c'), keeping a_0 but changing c_0 to c'. Let $\mathbf{p}'$ be the saddle fixed point for this new parameter pair.
- Use Newton's method to find a corresponding critical point α' in $W^u(\mathbf{p}')$.
- Color (a_0, c') according to how fast α' escapes.
- Do this for all parameter pairs (a_0, c), where $c \in \mathbb{C}$.

The end result will be a c parameter space picture for $H_{a_0,c}$.

Notice that the above picture somewhat resembles a squished Mandelbrot set. This is not so surprising as the Hénon map can be viewed as a two variable perturbation of the polynomial $f_c(x) = x^2 + c$, for small enough values of $|a|$. The picture is colored according to the behavior of one particular critical point. For the grey c values

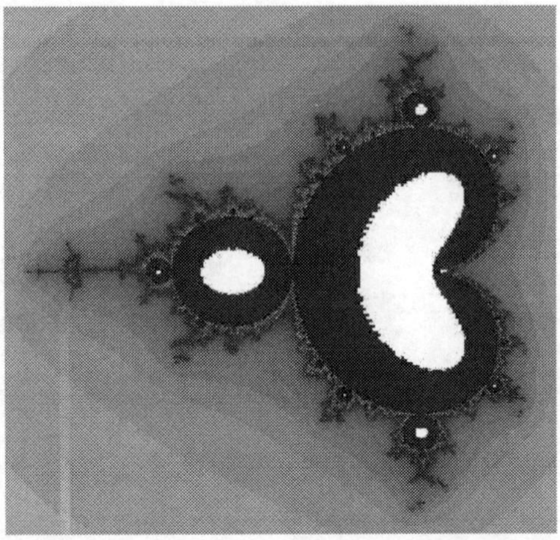

Figure 9. Above is the c parameter space picture for $a = -0.2$ obtained with SaddleDrop.

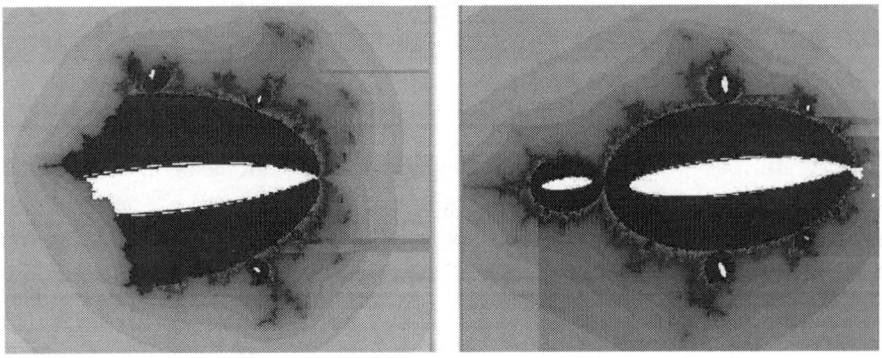

Figure 10. Above are two c parameter space pictures for $a_0 = 0.37 + 0.04i$. Each was obtained by following around a critical point of G^+ in $W^u(\mathbf{p})$.

surrounding the black set in the middle, the critical point found by Newton's method escapes to infinity.

The black and white c parameters are rather interesting. For the black c values in the middle, Newton's method converged to a critical point that escaped very slowly; slower than some set tolerance. For the parameter values that are colored white, Newton's method did not converge to a critical point of G^+, so the program could not color these parameter values according to the behavior of a critical point since it could not find one. Perhaps two critical points of G^+ coalesced for these c parameter values, thus confusing Newton's method there. Note also there are lines of discontinuity visible in part of the grey region; these are byproducts of Newton's method which is used as we move through parameter space.

The pictures above appear to be very different. The picture on the left was obtained by following around a critical point in $W^u(\mathbf{p})$, while the picture on the right was obtained by following around a different critical point. The parameter pictures vary greatly with the critical point originally chosen. How the parameter space picture depends on the critical point is quite interesting. We do not understand this yet but

Figure 11. A zoom-in of a c parameter space; note the finger-like appearance of the black set.

have some conjectures. This is one of the many questions that have arisen as a result of this program. We have also found some strange phenomena in parameter space; an example of this is fingering, see figure 11.

SaddleDrop has been an extremely important contribution to the study of Hénon maps. There is still much to understand about the program; it has given us some of the first glimpses of parameter space for $H_{a,c}$ and continues to raise many fruitful questions about complex dynamics in more than one variable.[2]

References

[BH] X. Buff and J. H. Hubbard, Dynamics in one Complex Variable, to appear.

[BS5] E. Bedford and J. Smillie, Polynomial diffeomorphisms of $\mathbb{C}^2$ V: Critical points and Lyapunov exponents, *J. Geom. Anal.*, **8** (1998) 349–383.

[BS6] E. Bedford and J. Smillie, Polynomial diffeomorphisms of $\mathbb{C}^2$ VI: Connectivity of J, *Ann. of Math.*, **148** (1998), 695–735.

[BuS] G. T. Buzzard and J. Smillie, Complex dynamics in several variables, in Flavors of Geometry (S. Levy, ed.) CUP (1997), 117–150.

[DH] A. Douady and J. H. Hubbard Etude dynamique des polynômes complexes I & II, *Publ. Math. d'Orsay*, (1984–85).

[H1] J. H. Hubbard, Parametrizing Unstable and Very Unstable Manifolds, *Moscow Math. J.*, **5** (2005).

[H2] J. H. Hubbard, The Hénon mapping in the complex domain, in Chaotic Dynamics and Fractals, ed. Barnsley and Demko Academic Press, Orlando (1986), 101–111.

[HO1] J. H. Hubbard and R. W. Oberste-Vorth, Hénon mappings in the complex domain I: The global topology of dynamical space, *Pub. Math. IHES*, **79** (1994) 5–46.

[HO2] J. H. Hubbard and R. W. Oberste-Vorth, Hénon mappings in the complex domain II: Projective and inductive limits of polynomials, in Real and Complex Dynamical Systems, ed. Branner and Hjorth, Kluwer Academic Publishers (1995), 89–132.

[HP] J. H. Hubbard and K. Papadantonakis, Exploring the parameter space of complex Hénon mappings, to appear.

[M] J. Milnor, Dynamics in one complex variable. Introductory lectures, Wiesbaden: Vieweg (1999).

[2] SaddleDrop is available for download at http://www.math.cornell.edu/~dynamics.

Teichmüller Theory and Moduli Problem No. 10, 2010, pp. 481–488.

The Action of Elliptic Modular Transformations on Asymptotic Teichmüller Spaces

Katsuhiko Matsuzaki

Department of Mathematics, Okayama University, Okayama 700-8530, Japan
e-mail: matsuzak@math.okayama-u.ac.jp

Abstract. A modular transformation of the Teichmüller space is called elliptic if it has a fixed point on it. We consider its action on an invariant fiber over the asymptotic Teichmüller space and prove that its orbit is always discrete but it does not necessarily have a fixed point on the fiber.

Keywords. Analytically infinite Riemann surface, quasiconformal mapping class group, Teichmüller modular group.

2000 Subject Classification: Primary 30F60, 37F30; Secondary 32G15

1. Introduction

In this note, we investigate the action of elliptic modular transformations of an infinite dimensional Teichmüller space on a fiber over its asymptotic Teichmüller space. In this section, we explain the terminology in brief and state our theorems. For details on Teichmüller spaces, modular transformations and asymptotic Teichmüller spaces, one can consult a monograph [4] and articles [1], [2] and [3].

Let R be an analytically infinite Riemann surface. The Teichmüller space $T(R)$ is the set of all Teichmüller equivalence classes $[f]$ of quasiconformal homeomorphisms f of R, which is endowed with a complex structure and a metric structure. The asymptotic Teichmüller space $AT(R)$ is a certain quotient space of the Teichmüller space $T(R)$ with the quotient map $\alpha : T(R) \to AT(R)$. This is also endowed with the complex structure and the metric structure compatible with those of $T(R)$ under α. Asymptotic equivalence of quasiconformal homeomorphisms of R is defined similarly to Teichmüller equivalence by using asymptotically conformal homeomorphisms. The projection α maps each Teichmüller equivalence class $p = [f] \in T(R)$ to the asymptotic equivalence class $\alpha(p) = [[f]] \in AT(R)$. The fiber over any $\alpha(p) \in AT(R)$ is denoted by T_p, which is a closed submanifold of $T(R)$.

The quasiconformal mapping class is a homotopy equivalence class $[g]$ (relative to the ideal boundary of R) of quasiconformal automorphisms $g : R \to R$ and the quasiconformal mapping class group $\mathrm{MCG}(R)$ is the group of all these mapping classes. Every element $[g] \in \mathrm{MCG}(R)$ induces a biholomorphic and isometric automorphism of $T(R)$ by $[f] \mapsto [f \circ g^{-1}]$. In this way, $\mathrm{MCG}(R)$ acts on $T(R)$ and a representation $\iota : \mathrm{MCG}(R) \to \mathrm{Aut}(T(R))$ is given. The image of ι is called the Teichmüller modular group and is denoted by $\mathrm{Mod}(R)$. Each element of $\mathrm{Mod}(R)$ is called a *modular transformation*. Similarly, $[g] \in \mathrm{MCG}(R)$ induces a biholomorphic and isometric automorphism of $AT(R)$ by $[[f]] \mapsto [[f \circ g^{-1}]]$. Thus we have a representation $\iota_A : \mathrm{MCG}(R) \to \mathrm{Aut}(AT(R))$. The image of ι_A is denoted by $\mathrm{Mod_A}(R)$.

The action of $\mathrm{MCG}(R)$ preserves the fibers of the projection $\alpha : T(R) \to AT(R)$. Hence a surjective homomorphism $A : \mathrm{Mod}(R) \to \mathrm{Mod_A}(R)$ is well-defined by $\iota_A \circ \iota^{-1}$, which sends $\gamma \in \mathrm{Mod}(R)$ to $\gamma_A := A(\gamma) \in \mathrm{Mod_A}(R)$. In other words, we have $\gamma_A \circ \alpha(p) = \alpha \circ \gamma(p)$ for every $p \in T(R)$ and for every $\gamma \in \mathrm{Mod}(R)$. We say that a modular transformation $\gamma \in \mathrm{Mod}(R)$ is *asymptotically trivial* if γ belongs to the kernel of A, that is, $\gamma_A = \mathrm{id}$. For instance, a modular transformation corresponding to a mapping class caused by the Dehn twist along a simple closed curve in R is asymptotically trivial.

In our companion paper [7], we have given a certain classification of the modular transformations. We say that $\gamma \in \mathrm{Mod}(R) - \{\mathrm{id}\}$ is *elliptic* if γ has a fixed point in $T(R)$ and *asymptotically elliptic* if γ_A has a fixed point in $AT(R)$. A mapping class $[g] \in \mathrm{MCG}(R)$ has a conformal representative g on a Riemann surface R_p determined by $p \in T(R)$ if and only if the corresponding modular transformation $\gamma = \iota([g])$ fixes p. Elliptic modular transformations can be of infinite order in the case where R is of analytically infinite type. We denote the fixed point locus of γ in $T(R)$ by $\mathrm{Fix}(\gamma)$ and the fixed point locus of γ_A in $AT(R)$ by $\mathrm{Fix}(\gamma_A)$.

In this note, we are concerned with discreteness of the orbit of an elliptic modular transformation. Our first result is the following.

Theorem 1. *Suppose that an elliptic modular transformation $\gamma \in \mathrm{Mod}(R)$ keeps a fiber T_p invariant, namely, $\alpha(p) \in \mathrm{Fix}(\gamma_A)$. Then the orbit $\langle\gamma\rangle(p)$ is a discrete set in T_p.*

Note that, in the above theorem, the invariant fiber T_p can have no intersection with the fixed point locus $\mathrm{Fix}(\gamma)$, as is seen in Theorem 5 below.

It is proved in [7] that, for an asymptotically elliptic but not elliptic modular transformation γ, the orbit $\langle\gamma\rangle(p)$ is a discrete set in $T(R)$ for every $p \in T(R)$. Combining Theorem 1 with this result, we have the following.

Corollary 2. *If $\alpha(p) \in \mathrm{Fix}(\gamma_A)$ for $p \in T(R)$ and $\gamma \in \mathrm{Mod}(R)$, then the orbit $\langle\gamma\rangle(p)$ is a discrete set in T_p.*

It is also proved in [7] that, for an elliptic modular transformation γ of infinite order, there exists a point $p \in T(R)$ such that the orbit $\langle\gamma\rangle(p)$ is not a discrete set in $T(R)$. Then we obtain another corollary to Theorem 1.

Corollary 3. *An elliptic modular transformation $\gamma \in \mathrm{Mod}(R)$ of infinite order is not asymptotically trivial.*

Note that Petrovic [8] has constructed an asymptotically elliptic modular transformation that is neither elliptic nor asymptotically trivial.

For an elliptic modular transformation $\gamma \in \mathrm{Mod}(R)$, Theorem 1 asserts that, if $\alpha(p) \in \mathrm{Fix}(\gamma_A)$, then $\langle\gamma\rangle(p)$ is a discrete set. Since $\alpha(\mathrm{Fix}(\gamma)) \subset \mathrm{Fix}(\gamma_A)$, the assumption $\alpha(p) \in \alpha(\mathrm{Fix}(\gamma))$ also gives the same conclusion. We extend this claim in the following way. The distances on $T(R)$ and $AT(R)$ are denoted by d and d_A respectively.

Theorem 4. *There exists a continuous strictly-increasing function $h : [0, \infty) \to [0, \infty)$ with $h(0) = 0$ that holds the following property: if a point $p \in T(R)$ and an elliptic modular transformation $\gamma \in \mathrm{Mod}(R)$ satisfy*

$$h(d_A(\alpha(p), \alpha(\mathrm{Fix}(\gamma)))) \leq d(p, \mathrm{Fix}(\gamma)),$$

then $\langle\gamma\rangle(p)$ is a discrete set in $T(R)$.

Finally, we compare the projection $\alpha(\mathrm{Fix}(\gamma))$ on $AT(R)$ with the fixed point locus $\mathrm{Fix}(\gamma_A)$ for an elliptic modular transformation $\gamma \in \mathrm{Mod}(R)$. Note that $\mathrm{Fix}(\gamma)$ is a closed submanifold of $T(R)$ and $\mathrm{Fix}(\gamma_A)$ and $\alpha(\mathrm{Fix}(\gamma))$ are closed submanifolds of $AT(R)$. This is easily seen if we realize them in the Bers embedding; γ and γ_A are regarded as linear transformations of certain Banach spaces and α is a surjective linear projection between them. See Section 2.

Theorem 5. *Let $\gamma \in \mathrm{Mod}(R)$ be an elliptic modular transformation of infinite order such that $\mathrm{Fix}(\gamma)$ is not a singleton. Then $\alpha(\mathrm{Fix}(\gamma))$ is a proper submanifold of $\mathrm{Fix}(\gamma_A)$ in $AT(R)$.*

On the contrary, if γ is of finite order, then $\alpha(\mathrm{Fix}(\gamma)) = \mathrm{Fix}(\gamma_A)$. This is because, for every $\alpha(p) \in \mathrm{Fix}(\gamma_A)$, the average of the finite orbit $\langle\gamma\rangle(p) \subset T(R)$ gives a fixed point $q \in \mathrm{Fix}(\gamma)$ satisfying $\alpha(q) = \alpha(p)$.

2. The Bers embedding

The proof of Theorem 1 is carried out by using the Bers embedding of the Teichmüller space. Let R^* be the complex conjugate of the Riemann surface R and denote its hyperbolic metric by $\rho(z)|dz|$. Let $B(R^*)$ be the complex Banach space of all holomorphic quadratic differentials φ on R^* with the norm $\|\varphi\|_\infty = \sup_{z \in R^*} \rho(z)^{-2}|\varphi(z)|$. By the Bers embedding $\beta : T(R) \to B(R^*)$, the Teichmüller space $T(R)$ is identified with a bounded contractible domain in $B(R^*)$ with the base point $o \in T(R)$ corresponding to the origin $0 \in B(R^*)$. For every $p \in T(R)$, we set $\varphi_p := \beta(p) \in B(R^*)$.

Each fiber T_p over the asymptotic Teichmüller space $AT(R)$ is identified with the intersection of $\beta(T(R))$ and an affine subspace $\varphi_p + B_0(R^*)$ of $B(R^*)$. Here $B_0(R^*)$ is the closed subspace of $B(R^*)$ consisting of all those φ vanishing at infinity, for which $\rho(z)^{-2}|\varphi(z)|$ tends to zero as z goes to the point at infinity of R^*. Let $\hat{B}(R^*)$ be the quotient Banach space $B(R^*)/B_0(R^*)$ and let $a : B(R^*) \to \hat{B}(R^*)$ the quotient map for the correspondence $\varphi \to \hat{\varphi}$. Then we have the asymptotic Bers embedding $\hat{\beta} : AT(R) \to \hat{B}(R^*)$ satisfying $a \circ \beta = \hat{\beta} \circ \alpha$. See [1] and [3].

Let g be a conformal automorphism of R. By reflection, it extends to a conformal automorphism of R^*. Then g acts on $B(R^*)$ linearly and isometrically by $(g_*\varphi)(z)dz^2 = \varphi(g^{-1}(z))dg^{-1}(z)^2$. For the modular transformation $\gamma = \iota([g])$ and for every $p \in T(R)$, we see that $\varphi_{\gamma(p)} = g_*\varphi_p$. Since $B_0(R^*)$ is invariant under g_*, it induces the linear isometry $\hat{g}_*$ of the quotient Banach space $\hat{B}(R^*)$ such that $a(g_*\varphi) = \hat{g}_*\hat{\varphi}$ for every $\varphi \in B(R^*)$.

Proof of Theorem 1. We may assume that the base point $o \in T(R)$ is a fixed point of γ. Then the mapping class corresponding to γ has a conformal representative g on R and on its complex conjugate R^*.

The fiber T_p is represented by $\beta(T(R)) \cap (\varphi_p + B_0(R^*))$ in the Bers embedding. Since γ preserves T_p, we have $g_*\varphi_p = \varphi_p + \psi$ for some $\psi \in B_0(R^*)$. Then, using this formula inductively, we see that

$$g_*^n \varphi_p = \varphi_p + \sum_{i=0}^{n-1} g_*^i \psi$$

for every integer $n \geq 1$.

We have only to consider the case where γ is of infinite order. Suppose that the orbit $\langle \gamma \rangle(p)$ is not a discrete set. Then there exists an increasing sequence of positive integers $\{n_k\}$ such that $\gamma^{n_k}(p) \to p$ as $k \to \infty$, or equivalently $g_*^{n_k}\varphi_p \to \varphi_p$. This implies that $\sum_{i=0}^{n_k-1} g_*^i \psi \to 0$ and, by operating g_* once more, $\sum_{i=1}^{n_k} g_*^i \psi \to 0$ as well. Hence, by subtracting the first one from the second, we have $g_*^{n_k}\psi \to \psi$ as $k \to \infty$.

Take an arbitrary point $z \in R^*$ and consider

$$\rho^{-2}(z)|(g_*^{n_k}\psi)(z)| = \rho^{-2}(g^{-n_k}(z))|\psi(g^{-n_k}(z))|,$$

which converge to $\rho^{-2}(z)|\psi(z)|$. Since ψ vanishes at infinity and since the group of conformal automorphisms $\langle g \rangle$ acts on R discontinuously, we see that the above quantities converge to zero as $k \to \infty$. Hence we have $\psi = 0$ and thus $g_*\varphi_p = \varphi_p$. However, this implies that $\langle \gamma \rangle(p) = \{p\}$, contradicting the assumption. □

Theorem 1 and the following result proved in [7] yield Corollaries 2 and 3 concerning asymptotically elliptic modular transformations.

Theorem 6. *An asymptotically elliptic modular transformation $\gamma \in \mathrm{Mod}(R)$ satisfies one of the following alternative conditions:*

(1) *γ is elliptic of finite order;*
(2) *γ is elliptic of infinite order and the orbit $\langle \gamma \rangle(p)$ is not a discrete set for some $p \in T(R)$;*
(3) *γ is not elliptic and the orbit $\langle \gamma \rangle(p)$ is a discrete set for every $p \in T(R)$.*

Proof of Corollary 2. By Theorem 6, we have only to consider the case where γ is elliptic. Then by Theorem 1, the orbit $\langle \gamma \rangle(p)$ is a discrete set. □

Proof of Corollary 3. By Theorem 6, there exists some $p \in T(R)$ such that the orbit $\langle \gamma \rangle(p)$ accumulates. If γ is asymptotically trivial, then γ keeps the fiber T_p invariant and Theorem 1 implies that $\langle \gamma \rangle(p)$ is a discrete set. This contradiction proves that γ is not asymptotically trivial. □

3. The cross ratio and the quasisymmetric distance

For the proof of Theorem 4, we introduce another distance on the Teichmüller space, which is defined by the variation of the cross ratio under quasisymmetric automorphisms.

Let $f_1 : R \to R_1$ and $f_2 : R \to R_2$ be quasiconformal homeomorphisms of R and consider their lifts $\tilde{f}_1$ and $\tilde{f}_2$, which are quasiconformal automorphisms of the unit disk Δ extendable to the unit circle $\partial\Delta$ as quasisymmetric automorphisms. Let $X = (x_1, x_2, x_3, x_4)$ be any ordered quadruple of distinct points on $\partial\Delta$ whose cross ratio $c(X)$ belongs to the interval $(1, \infty)$. Then the *quasisymmetric distance* $d_{qs}([f_1], [f_2])$ between the Teichmüller classes $[f_1]$ and $[f_2]$ in $T(R)$ is defined by

$$\sup_X d_{0,1}(c(\tilde{f}_1(X)), c(\tilde{f}_2(X))),$$

where $d_{0,1}$ denotes the hyperbolic distance on the three-punctured sphere $\mathbb{C} - \{0, 1\}$. It is known that

$$\lambda(d([f_1], [f_2])) \leq d_{qs}([f_1], [f_2]) \leq d([f_1], [f_2])$$

for a certain continuous strictly-increasing function $\lambda : [0, \infty) \to [0, \infty)$ with $\lambda(0) = 0$. See [5] and [6].

Proof of Theorem 4. By using the inverse function λ^{-1} of λ, we define a continuous strictly-increasing function $h(t) := 2t + \lambda^{-1}(2t)$. We will prove that this function h satisfies the required properties. Set $\delta := d_A(\alpha(p), \alpha(\mathrm{Fix}(\gamma)))$. If $\delta = 0$, then $\alpha(p) \in \alpha(\mathrm{Fix}(\gamma))$ and Theorem 1 yields the assertion. Hence we may assume that $\delta > 0$.

Since $\alpha(\mathrm{Fix}(\gamma))$ is a closed set in $AT(R)$, there exists some point in $\mathrm{Fix}(\gamma)$ that attains the distance δ and, without loss of generality, we may assume that it is the base point $o \in T(R)$. Namely, $d_A(\alpha(p), \alpha(o)) = \delta$ and $o \in \mathrm{Fix}(\gamma)$. Then, since $d(p, T_o) = \delta$ and since T_o is closed in $T(R)$, there exists some $p_0 \in T_o$ such that $d(p, p_0) = \delta$.

The assumption yields

$$d(p, o) \geq d(p, \mathrm{Fix}(\gamma)) \geq h(d_A(\alpha(p), \alpha(\mathrm{Fix}(\gamma)))) = 2\delta + \lambda^{-1}(2\delta).$$

By $d(p_0, o) \geq d(p, o) - \delta$, we have $d(p_0, o) > \lambda^{-1}(2\delta)$ and hence

$$d_{qs}(p_0, o) \geq \lambda(d(p_0, o)) > 2\delta.$$

Here we set $\varepsilon = \{d_{qs}(p_0, o) - 2\delta\}/3 > 0$.

Let f be an asymptotically conformal homeomorphism of R such that $[f] = p_0$ and let $\tilde{f}$ be its lift to Δ. Since $d_{qs}(p_0, o) = 2\delta + 3\varepsilon$, there exists an ordered quadruple X of distinct points on $\partial\Delta$ such that

$$d_{0,1}(c(\tilde{f}(X)), c(X)) \geq 2\delta + 2\varepsilon. \tag{1}$$

We have only to consider the case where γ is of infinite order. Let g be a conformal automorphism of R such that $[g] = \gamma$ and let $\tilde{g}$ be its lift to Δ. Then $\gamma^n(p_0) = [f \circ g^{-n}]$ and $\tilde{f} \circ \tilde{g}^{-n}$ is a lift of $f \circ g^{-n}$. Since f is asymptotically conformal, we see that the complex dilatation $\mu(\tilde{f} \circ \tilde{g}^{-n})$ converge to zero almost everywhere as $n \to \pm\infty$. Choose a conformal automorphism j_n of Δ for each n such that $j_n \circ \tilde{f} \circ \tilde{g}^{-n}$

fixes three particular points on $\partial\Delta$. Then $j_n \circ \tilde{f} \circ \tilde{g}^{-n}$ converge uniformly to id and hence $c(f \circ \tilde{g}^{-n}(X)) = c(j_n \circ \tilde{f} \circ \tilde{g}^{-n}(X))$ converge to $c(X)$. This implies that

$$d_{0,1}(c(\tilde{f} \circ \tilde{g}^{-n}(X)), c(X)) \leq \varepsilon \tag{2}$$

for all n whose absolute values are greater than some n_0.

By inequalities (1) and (2), we have

$$d_{0,1}(c(\tilde{f} \circ \tilde{g}^{-n}(X)), c(\tilde{f}(X))) \geq 2\delta + \varepsilon$$

and thus

$$d(\gamma^n(p_0), p_0) \geq d_{qs}(\gamma^n(p_0), p_0) \geq 2\delta + \varepsilon$$

for all n with $|n| > n_0$. By

$$d(\gamma^n(p), p) \geq d(\gamma^n(p_0), p_0) - d(p, p_0) - d(\gamma^n(p), \gamma^n(p_0)) = d(\gamma^n(p_0), p_0) - 2\delta,$$

we finally have $d(\gamma^n(p), p) \geq \varepsilon > 0$ for all n with $|n| > n_0$. This means that the orbit $\langle\gamma\rangle(p)$ does not accumulate to p. From this, we see that $\langle\gamma\rangle(p)$ is a discrete set. □

4. Infinitesimally asymptotically trivial Beltrami differentials

Let R be a Riemann surface that admits a conformal automorphism g of infinite order. Let Belt(R) be the complex Banach space of all Beltrami differentials μ on R with the supremum norm. We consider the following three Banach subspaces of Belt(R). Let Belt(R, g) be the subspace consisting of all Beltrami differentials invariant under g, let N be the subspace consisting of all infinitesimally trivial Beltrami differentials, and let $\hat{N}$ be the subspace consisting of all infinitesimally asymptotically trivial Beltrami differentials. Here "infinitesimally" means that μ is regarded as a tangent vector at $o \in T(R)$.

To define these spaces precisely, let $Q(R)$ be the complex Banach space of all integrable holomorphic quadratic differentials φ on R with the norm $\|\varphi\|_1 = \int_R |\varphi|$. Then the tangent space of $T(R)$ at o is identified with the dual space $Q(R)^*$. Each element $\mu \in$ Belt(R) induces a bounded linear functional $v_\mu \in Q(R)^*$ by $v_\mu(\varphi) = \int \mu\varphi$. We say that $\mu \in$ Belt(R) is *infinitesimally trivial* if $v_\mu = 0$, that is, $\int \mu\varphi = 0$ for every $\varphi \in Q(R)$. A *degenerating sequence* is a sequence $\{\varphi_n\} \subset Q(R)$ such that $\|\varphi_n\|_1 = 1$ and φ_n converge locally uniformly to zero. We say that $\mu \in$ Belt(R) is *infinitesimally asymptotically trivial* if $\lim_{n\to\infty} |v_\mu(\varphi_n)| = 0$ for every degenerating sequence $\{\varphi_n\} \subset Q(R)$. It is clear that N is contained in $\hat{N}$. Actually, for the Banach subspace Belt$_0(R) \subset$ Belt(R) consisting of all Beltrami differentials vanishing at infinity, we know that $\hat{N} = N +$ Belt$_0(R)$. Here we say that $\mu \in$ Belt(R) vanishes at infinity if, for every $\varepsilon > 0$, there exists a compact subset $V \subset R$ such that ess. $\sup_{z\in R-V} |\mu(z)| < \varepsilon$.

Let $M(R)$ be the open unit ball of the complex Banach space Belt(R). The Teichmüller projection $\Phi : M(R) \to T(R)$ is defined by $\Phi(\mu) = [f^\mu]$ where $[f^\mu] \in T(R)$ is the Teichmüller class of the quasiconformal homeomorphism f^μ of R having the complex dilatation μ. The Teichmüller projection Φ is a holomorphic split submersion. The kernel of the derivative $d\Phi_0$ at the origin is coincident with the Banach subspace $N \subset \mathrm{Belt}(R)$. The asymptotic Teichmüller projection $\hat{\Phi} : M(R) \to AT(R)$ is defined by $\hat{\Phi} = \alpha \circ \Phi$. The kernel of the derivative $d\hat{\Phi}_0$ at the origin is coincident with the Banach subspace $\hat{N} \subset \mathrm{Belt}(R)$. This is given in [1]. See also [4] for the facts mentioned above.

Proof of Theorem 5. We may assume that the base point $o \in T(R)$ belongs to $\mathrm{Fix}(\gamma)$. Take the conformal representative g of γ and consider the Riemann surface $R_g = R/\langle g \rangle$. Then the assumption $\mathrm{Fix}(\gamma) = T(R_g) \neq \{o\}$ is equivalent to saying that R_g is not a three-punctured sphere.

Take a Beltrami differential $\check{\mu} \in \mathrm{Belt}(R_g)$ that is not infinitesimally trivial as follows. Since $T(R_g) \neq \{o\}$, there exists a non-trivial element $\psi \in Q(R_g)$ such that $\sup_{z \in R_g} \rho_g^{-2}(z)|\psi(z)| < \infty$, where ρ_g is the hyperbolic density on R_g. For a closed disk W in R_g, we set $\check{\mu}(z) = \rho_g^{-2}(z)\overline{\psi(z)}$ on W and $\check{\mu}(z) = 0$ elsewhere.

Let $\mu \in \mathrm{Belt}(R)$ be the lift of $\check{\mu}$ to R, which belongs to $\mathrm{Belt}(R, g) - N$. Consider the inverse image of W under the covering projection $R \to R_g$ and denote it by $\bigsqcup_{i \in \mathbb{Z}} W_i$ with $W_i = g^i(W_0)$. For any $n \in \mathbb{Z}$, we set $\mu_n = \mu$ on $\bigsqcup_{i \geq n} W_i$ and $\mu_n = 0$ elsewhere.

We will show that $\mu_0 \notin \mathrm{Belt}(R, g) + \hat{N}$. Suppose to the contrary that we can write $\mu_0 = \nu + \lambda$ where $\nu \in \mathrm{Belt}(R, g)$ and $\lambda \in \hat{N}$. Take any $\varphi \in Q(R)$ with $\|\varphi\| = 1$ and set $\varphi_n = g_*^n \varphi$ for every $n \in \mathbb{N}$. Then $\{\varphi_n\}$ is a degenerating sequence. A simple calculation shows that

$$\int_R \mu_{-n}\varphi = \int_R g_*^n(\mu_{-n}\varphi) = \int_R \mu_0\varphi_n = \int_R \nu\varphi_n + \int_R \lambda\varphi_n.$$

Since $\mu_{-n} \to \mu$ as $n \to \infty$, the left-hand side converges to $\int_R \mu\varphi$. By $\nu \in \mathrm{Belt}(R, g)$, we have $\int_R \nu\varphi_n = \int_R \nu\varphi$, and by $\lambda \in \hat{N}$, we have $\int_R \lambda\varphi_n \to 0$ as $n \to \infty$; the right-hand side converges to $\int_R \nu\varphi$. Hence $\int_R \mu\varphi = \int_R \nu\varphi$, which implies that $\mu - \nu \in N \subset \hat{N}$. From this and $\mu_0 = \nu + \lambda$, we have $\mu - \mu_0 = \lambda'$ for another $\lambda' \in \hat{N}$.

Here we set $\varphi_{-n} = g_*^{-n}\varphi$ for every $n \in \mathbb{N}$ and take another degenerating sequence $\{\varphi_{-n}\}$. Similar to the above, it holds

$$\int_R (\mu - \mu_n)\varphi = \int_R (\mu - \mu_0)\varphi_{-n} = \int_R \lambda'\varphi_{-n}$$

and the left-hand side converges to $\int_R \mu\varphi$ and the right-hand side converges to 0 as $n \to \infty$. Hence $\int_R \mu\varphi = 0$ for every $\varphi \in Q(R)$, which implies that $\mu \in N$. However, this contradicts the fact that μ is chosen so that $\mu \notin N$.

Consider an arc $\Phi(t\mu_0)$ in $T(R)$ for sufficiently small $t > 0$ and its projection $\hat{\Phi}(t\mu_0)$ on $AT(R)$. Since $g_*\mu_0 - \mu_0 = \mu_1 - \mu_0$ vanishes at infinity, we see that $\hat{\Phi}(t\mu_0)$ is contained in $\mathrm{Fix}(\gamma_A) \subset AT(R)$. On the other hand, since $\mu_0 \notin \mathrm{Belt}(R, g) + \hat{N}$, the tangent vector $d\hat{\Phi}_0(\mu_0)$ of the arc $\hat{\Phi}(t\mu_0)$ at $\alpha(o) \in AT(R)$ does not lie along the submanifold $\alpha(\mathrm{Fix}(\gamma))$ in $AT(R)$. Hence, for some $p = \Phi(t\mu_0)$ with sufficiently small $t > 0$, $\alpha(p) = \hat{\Phi}(t\mu_0)$ does not belong to $\alpha(\mathrm{Fix}(\gamma))$. Since $\alpha(p) \in \mathrm{Fix}(\gamma_A)$, we obtain the assertion that $\alpha(\mathrm{Fix}(\gamma))$ is a proper subset of $\mathrm{Fix}(\gamma_A)$. □

References

[1] C. Earle, F. Gardiner and N. Lakic, Asymptotic Teichmüller space, Part I: the complex structure, In the tradition of Ahlfors and Bers, *Contemp. Math.*, **256**, AMS (2000) 17–38.

[2] C. Earle, F. Gardiner and N. Lakic, Asymptotic Teichmüller space, Part II: the metric structure, In the tradition of Ahlfors and Bers, III, *Contemp. Math.*, **355**, AMS (2004) 187–219.

[3] C. Earle, V. Markovic and D. Saric, Barycentric extension and the Bers embedding for asymptotic Teichmüller space, Complex manifolds and hyperbolic geometry, *Contemp. Math.*, **311** AMS (2002) 87–105.

[4] F. Gardiner and N. Lakic, Quasiconformal Teichmüller theory, SRUV 76, American Mathematical Society (2000).

[5] I. Kra, On Teichmüller's theorem on the quasi-invariance of cross ratios, *Israel. J. Math.*, **30** (1978) 152–158.

[6] O. Lehto, Univalent functions and Teichmüller spaces, GTM 109, Springer (1986).

[7] K. Matsuzaki, A classification of the modular transformations of infinite dimensional Teichmüller spaces, In the tradition of Ahlfors and Bers, IV, *Contemp. Math.*, **432**, AMS (2007) 167–178.

[8] I. Petrovic, A Teichmüller model for period doubling, In the tradition of Ahlfors and Bers, III, *Contemp. Math.*, **355**, AMS (2004) 333–351.

Teichmüller Theory and Moduli Problem No. 10, 2010, pp. 489–511.

Cannon-Thurston Maps and Bounded Geometry

Mahan Mj

RKM Vivekananda University, Belur Math, India
e-mail: mahan@rkmvu.ac.in; mahan.mj@gmail.com

Abstract. This is an expository paper. We prove the Cannon-Thurston property for bounded geometry surface groups with or without punctures. We prove three theorems, due to Cannon-Thurston, Minsky and Bowditch. The proofs are culled out of earlier work of the author.

2000 Subject Classification: 20F32, 57M50

1. Introduction

Let M be a closed hyperbolic 3-manifold fibering over the circle with fiber F. Let $\widetilde{F}$ and $\widetilde{M}$ denote the universal covers of F and M respectively. Then $\widetilde{F}$ and $\widetilde{M}$ are quasi-isometric to $\mathbb{H}^2$ and $\mathbb{H}^3$ respectively. Now let $\mathbb{D}^2 = \mathbb{H}^2 \cup \mathbb{S}^1_\infty$ and $\mathbb{D}^3 = \mathbb{H}^3 \cup \mathbb{S}^2_\infty$ denote the standard compactifications. In [CT85] Cannon and Thurston show that the usual inclusion of $\widetilde{F}$ into $\widetilde{M}$ extends to a continuous map from $\mathbb{D}^2$ to $\mathbb{D}^3$. This was extended to Kleinian surface groups of bounded geometry without parabolics by Minsky [Min94]. Bowditch [Bow02a] [Bow02b] proved the Cannon-Thurston property for bounded geometry surface groups with parabolics.

In [Mit98a], [Mit98b], [Mj05b], [Mj05c], [Mj05a], we have given a different approach to the Cannon-Thurston problem. Though the theorems of Cannon-Thurston, Minsky and Bowditch can be deduced from ours, it might be instructive to write down a complete proof of these results. In some ways, the proof here is easier and more minimalistic. Another reason for writing this paper is that Cannon and Thurston's original result [CT85] is unpublished. It seems only appropriate that the theorem that motivated all the above results be available.

Much of what follows is true in the setting of hyperbolic metric spaces in the sense of Gromov. We shall often state results in this generality.

2. Preliminaries

2.1 *Hyperbolic metric spaces*

We start off with some preliminaries about hyperbolic metric spaces in the sense of Gromov [Gro85]. For details, see [CDA90], [GdlH90]. Let (X, d) be a hyperbolic metric space. The **Gromov boundary** of X, denoted by ∂X, is the collection of equivalence classes of geodesic rays $r : [0, \infty) \to X$ with $r(0) = x_0$ for some fixed

$x_0 \in X$, where rays r_1 and r_2 are equivalent if $sup\{d(r_1(t), r_2(t))\} < \infty$. Let $\widehat{X} = X \cup \partial X$ denote the natural compactification of X topologized the usual way (cf. [GdlH90] pg. 124).

The **Gromov inner product** of elements a and b relative to c is defined by

$$(a, b)_c = 1/2[d(a, c) + d(b, c) - d(a, b)].$$

Definitions. *A subset Z of X is said to be k-**quasiconvex** if any geodesic joining $a, b \in Z$ lies in a k-neighborhood of Z. A subset Z is **quasiconvex** if it is k-quasiconvex for some k. A map f from one metric space (Y, d_Y) into another metric space (Z, d_Z) is said to be a (K, ϵ)-**quasi-isometric embedding** if*

$$\frac{1}{K}(d_Y(y_1, y_2)) - \epsilon \leq d_Z(f(y_1), f(y_2)) \leq K d_Y(y_1, y_2) + \epsilon$$

*If f is a quasi-isometric embedding, and every point of Z lies at a uniformly bounded distance from some $f(y)$ then f is said to be a **quasi-isometry**. A (K, ϵ)-quasi-isometric embedding that is a quasi-isometry will be called a (K, ϵ)-quasi-isometry.*

*A (K, ϵ)-**quasigeodesic** is a (K, ϵ)-quasi-isometric embedding of a closed interval in $\mathbb{R}$. A $(K, 0)$-quasigeodesic will also be called a K-quasigeodesic.*

Let (X, d_X) be a hyperbolic metric space and Y be a subspace that is hyperbolic with the inherited path metric d_Y. By adjoining the Gromov boundaries ∂X and ∂Y to X and Y, one obtains their compactifications $\widehat{X}$ and $\widehat{Y}$ respectively.

Let $i : Y \to X$ denote inclusion.

Definition. *Let X and Y be hyperbolic metric spaces and $i : Y \to X$ be an embedding. A **Cannon-Thurston map** $\hat{i}$ from $\widehat{Y}$ to $\widehat{X}$ is a continuous extension of i.*

The following lemma says that a Cannon-Thurston map exists if for all $M > 0$ and $y \in Y$, there exists $N > 0$ such that if λ lies outside an N ball around y in Y then any geodesic in X joining the end-points of λ lies outside the M ball around $i(y)$ in X. For convenience of use later on, we state this somewhat differently. The proof is similar to Lemma 2.1 of [Mit98a].

Lemma 2.1. *A Cannon-Thurston map from $\widehat{Y}$ to $\widehat{X}$ exists if the following condition is satisfied:*

Given $y_0 \in Y$, there exists a non-negative function $M(N)$, such that $M(N) \to \infty$ as $N \to \infty$ and for all geodesic segments λ lying outside an N-ball around $y_0 \in Y$ any geodesic segment in X joining the end-points of $i(\lambda)$ lies outside the $M(N)$-ball around $i(y_0) \in X$.

Proof. Suppose $i : Y \to X$ does not extend continuously. Since i is proper, there exist sequences $x_m, y_m \in Y$ and $p \in \partial Y$, such that $x_m \to p$ and $y_m \to p$ in $\widehat{Y}$, but $i(x_m) \to u$ and $i(y_m) \to v$ in $\widehat{X}$, where $u, v \in \partial X$ and $u \neq v$.

Since $x_m \to p$ and $y_m \to p$, any geodesic in Y joining x_m and y_m lies outside an N_m-ball $y_0 \in Y$, where $N_m \to \infty$ as $m \to \infty$. Any bi-infinite geodesic in X joining $u, v \in \partial X$ has to pass through some M-ball around $i(y_0)$ in X as $u \neq v$. There exist constants c and L such that for all $m > L$ any geodesic joining $i(x_m)$ and $i(y_m)$ in

X passes through an $(M+c)$-neighborhood of $i(y_0)$. Since $(M+c)$ is a constant not depending on the index m this proves the lemma. □

The above result can be interpreted as saying that a Cannon-Thurston map exists if the space of geodesic segments in Y embeds properly in the space of geodesic segments in X.

We shall be needing the notion of a tree of hyperbolic metric spaces [BF92].

Definition. *A tree T of hyperbolic metric spaces satisfying the* **q(uasi) i(sometrically) embedded condition** *is a metric space (X, d) admitting a map $P : X \to T$ onto a simplicial tree T, such that there exist δ, ϵ and $K > 0$ satisfying the following:*

1. *For all vertices $v \in T$, $X_v = P^{-1}(v) \subset X$ with the induced path metric d_v is a δ-hyperbolic metric space. Further, the inclusions $i_v : X_v \to X$ are uniformly proper, i.e. for all $M > 0$, there exists $N > 0$ such that for all $v \in T$ and $x, y \in X_v$, $d(i_v(x), i_v(y)) \leq M$ implies $d_v(x, y) \leq N$.*
2. *Let e be an edge of T with initial and final vertices v_1 and v_2 respectively. Let X_e be the pre-image under P of the mid-point of e. Then X_e with the induced path metric is δ-hyperbolic.*
3. *There exist maps $f_e : X_e \times [0, 1] \to X$, such that $f_e|_{X_e \times (0,1)}$ is an isometry onto the pre-image of the interior of e equipped with the path metric.*
4. *$f_e|_{X_e \times \{0\}}$ and $f_e|_{X_e \times \{1\}}$ are (K, ϵ)-quasi-isometric embeddings into X_{v_1} and X_{v_2} respectively. $f_e|_{X_e \times \{0\}}$ and $f_e|_{X_e \times \{1\}}$ will occasionally be referred to as f_{v_1} and f_{v_2} respectively.*

d_v and d_e will denote path metrics on X_v and X_e respectively. i_v, i_e will denote inclusion of X_v, X_e respectively into X.

A few general lemmas about hyperbolic metric spaces will be useful. We reproduce the proofs here from [Mit98b].

Nearest point projections

The following Lemma says nearest point projections in a δ-hyperbolic metric space do not increase distances much. It is a standard fact that any geodesic metric space is quasi-isometric to a graph [BH99]. In what follows, we shall often implicitly identify spaces with their graph approximations. Further, graphs are declared to have edge length one. Also, without loss of generality, we may look at functions restricted to the vertex sets of these graphs.

Lemma 2.2. *Let (Y, d) be a δ-hyperbolic metric space and let $\mu \subset Y$ be a geodesic segment. Let $\pi : Y \to \mu$ map $y \in Y$ to a point on μ nearest to y. Then $d(\pi(x), \pi(y)) \leq C_3 d(x, y)$ for all $x, y \in Y$ where C_3 depends only on δ.*

Proof. Let $[a, b]$ denote a geodesic edge-path joining vertices a, b. Recall that the Gromov inner product $(a, b)_c = 1/2[d(a, c) + d(b, c) - d(a, b)]$. It suffices by repeated use of the triangle inequality to prove the Lemma when $d(x, y) \leq 1$. Let u, v, w be points on $[x, \pi(x)]$, $[\pi(x), \pi(y)]$ and $[\pi(y), x]$ respectively such that

$d(u, \pi(x)) = d(v, \pi(x))$, $d(v, \pi(y)) = d(w, \pi(y))$ and $d(w, x) = d(u, x)$. Then $(x, \pi(y))_{\pi(x)} = d(u, \pi(x))$. Also, since Y is δ-hyperbolic, the diameter of the inscribed triangle with vertices u, v, w is less than or equal to 2δ (See [ABC+91]).

$$\begin{aligned} d(u, x) + d(u, v) \geq d(x, \pi(x)) &= d(u, x) + d(u, \pi(x)) \\ \Rightarrow d(u, \pi(x)) &\leq d(u, v) \leq 2\delta \\ \Rightarrow (x, \pi(y))_{\pi(x)} &\leq 2\delta \end{aligned}$$

Similarly, $(y, \pi(x))_{\pi(y)} \leq 2\delta$.

$$\begin{aligned} \text{i.e.}\quad & d(x, \pi(x)) + d(\pi(x), \pi(y)) - d(x, \pi(y)) \leq 4\delta \\ \text{and}\quad & d(y, \pi(y)) + d(\pi(x), \pi(y)) - d(y, \pi(x)) \leq 4\delta \end{aligned}$$

Therefore,

$$\begin{aligned} 2d(\pi(x), \pi(y)) &\leq 8\delta + d(x, \pi(y)) - d(y, \pi(y)) + d(y, \pi(x)) - d(x, \pi(x)) \\ &\leq 8\delta + d(x, y) + d(x, y) \\ &\leq 8\delta + 2 \end{aligned}$$

Hence $d(\pi(x), \pi(y)) \leq 4\delta + 1$. Choosing $C_3 = 4\delta + 1$, we are through. □

Lemma 2.3. *Let (Y, d) be a δ-hyperbolic metric space. Let μ be a geodesic segment in Y with end-points a, b and let x be any vertex in Y. Let y be a vertex on μ such that $d(x, y) \leq d(x, z)$ for any $z \in \mu$. Then a geodesic path from x to y followed by a geodesic path from y to z is a k-quasigeodesic for some k dependent only on δ.*

Proof. As in Lemma 2.2, let u, v, w be points on edges $[x, y]$, $[y, z]$ and $[z, x]$ respectively such that $d(u, y) = d(v, y)$, $d(v, z) = d(w, z)$ and $d(w, x) = d(u, x)$. Then $d(u, y) = (z, x)_y \leq 2\delta$ and the inscribed triangle with vertices u, v, w has diameter less than or equal to 2δ (See [ABC+91]). $[x, y] \cup [y, z]$ is a union of 2 geodesic paths lying in a 4δ neighborhood of a geodesic $[x, z]$. Hence a geodesic path from x to y followed by a geodesic path from y to z is a k-quasigeodesic for some k dependent only on δ. □

Lemma 2.4. *Suppose (Y, d) is a δ-hyperbolic metric space. If μ is a (k_0, ϵ_0)-quasigeodesic in Y and p, q, r are 3 points in order on μ then $(p, r)_q \leq k_1$ for some k_1 dependent on k_0, ϵ_0 and δ only.*

Proof. $[a, b]$ will denote a geodesic path joining a, b. Since p, q, r are 3 points in order on μ, $[p, q]$ followed by $[q, r]$ is a (k', ϵ')-quasigeodesic in the δ-hyperbolic metric space Y where (k', ϵ') depend only on δ, k_0, ϵ_0. Hence there exists a k_1 dependent on k_0, ϵ_0 and δ alone such that $d(q, [p, r]) \leq k_1$. Let s be a point on $[p, r]$ such that $d(q, s) = d(q, [p, r]) \leq k_1$. Then

$$\begin{aligned} (p, r)_q &= 1/2(d(p, q) + d(r, q) - d(p, r)) \\ &= 1/2(d(p, q) + d(r, q) - d(p, s) - d(r, s)) \\ &\leq d(q, s) \leq k_1. \end{aligned}$$

□

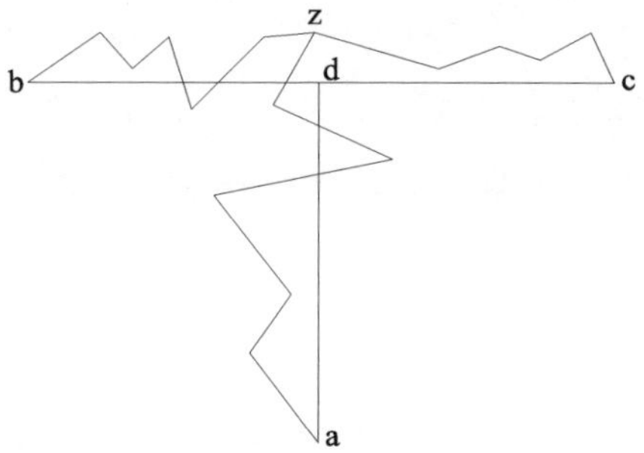

Figure 1. Quasi-isometries and nearest point projections almost commute.

2.2 *Stability of tripods, or NPP's and QI's almost commute*

A crucial property of hyperbolic metric spaces is *stability of quasigeodesics*, i.e. any quasigeodesic (in particular an image of a geodesic under a quasi-isometry) lies in a bounded neighborhood of a geodesic. This property can be easily extended to quasi-convex sets [GdlH90] [CDA90]. Here we are interested in a particular kind of quasi-convex set, the **tripod**. In general, a tripod is a union of three geodesic segments, all of which share a common end-point. It is easy to see (for instance by thinness of triangles) that a tripod is quasiconvex. We shall be interested in a special kind of a tripod. Let $[a, b]$ be a geodesic in a hyperbolic metric space X. Let $x \in X$ be some point. Let p be a nearest point projection of x onto $[a, b]$. We shall look at tripods of the form $[a, b] \cup [x, p]$. We shall show that such tripods are stable under quasi-isometries.

However, we shall interpret this differently to say that nearest point projections and quasi-isometries in hyperbolic metric spaces 'almost commute'. The following Lemma says precisely this: nearest point projections and quasi-isometries in hyperbolic metric spaces 'almost commute'.

Lemma 2.5. *Suppose (Y, d) is δ-hyperbolic. Let μ_1 be some geodesic segment in Y joining a, b and let p be any vertex of Y. Also let q be a vertex on μ_1 such that $d(p, q) \leq d(p, x)$ for $x \in \mu_1$. Let ϕ be a (K, ϵ)-quasi-isometry from Y to itself. Let μ_2 be a geodesic segment in Y joining $\phi(a)$ to $\phi(b)$ for some $g \in S$. Let r be a point on μ_2 such that $d(\phi(p), r) \leq d(\phi(p), x)$ for $x \in \mu_2$. Then $d(r, \phi(q)) \leq C_4$ for some constant C_4 depending only on K, ϵ and δ.*

Proof. Since $\phi(\mu_1)$ is a (K, ϵ)-quasigeodesic joining $\phi(a)$ to $\phi(b)$, it lies in a K'-neighborhood of μ_2 where K' depends only on K, ϵ, δ. Let u be a point in $\phi(\mu_1)$ lying at a distance at most K' from r. Without loss of generality suppose that u lies on $\phi([q, b])$, where $[q, b]$ denotes the geodesic subsegment of μ_1 joining q, b. [See Figure 1 below.]

Let $[p, q]$ denote a geodesic joining p, q. From Lemma 2.3 $[p, q] \cup [q, b]$ is a k-quasigeodesic, where k depends on δ alone. Therefore $\phi([p, q]) \cup \phi([q, b])$ is a (K_0, ϵ_0)-quasigeodesic, where K_0, ϵ_0 depend on K, k, ϵ. Hence, by Lemma 2.4 $(\phi(p), u)_{\phi(q)} \leq K_1$, where K_1 depends on K, k, ϵ and δ alone. Therefore,

$$\begin{aligned}(\phi(p), r)_{\phi(q)} &= 1/2[d(\phi(p), \phi(q)) + d(r, \phi(q)) - d(r, \phi(p))]\\ &\leq 1/2[d(\phi(p), \phi(q)) + d(u, \phi(q)) + d(r, u)\\ &\quad - d(u, \phi(p)) + d(r, u)]\\ &= (\phi(p), u)_{\phi(q)} + d(r, u)\\ &\leq K_1 + K'\end{aligned}$$

There exists $s \in \mu_2$ such that $d(s, \phi(q)) \leq K'$.

$$\begin{aligned}(\phi(p), r)_s &= 1/2[d(\phi(p), s) + d(r, s) - d(r, \phi(p))]\\ &\leq 1/2[d(\phi(p), \phi(q)) + d(r, \phi(q)) - d(r, \phi(p))] + K'\\ &= (\phi(p), r)_{\phi(q)} + K'\\ &\leq K_1 + K' + K'\\ &= K_1 + 2K'\end{aligned}$$

Also, as in the proof of Lemma 2.2 $(\phi(p), s)_r \leq 2\delta$

$$\begin{aligned}d(r, s) &= (\phi(p), s)_r + (\phi(p), r)_s\\ &\leq K_1 + 2K' + 2\delta\\ d(r, \phi(q)) &\leq K_1 + 2K' + 2\delta + d(s, \phi(q))\\ &\leq K_1 + 2K' + 2\delta + K'\end{aligned}$$

Let $C_4 = K_1 + 3K' + 2\delta$. Then $d(r, \phi(q)) \leq C_4$ and C_4 is independent of a, b, p. □

3. Rays of spaces and hyperbolic ladders

3.1 *Trees of hyperbolic metric spaces*

In this section, we shall consider trees T of hyperbolic metric spaces satisfying the qi-embedded condition. (X, d_X) will denote the space that is a tree of hyperbolic metric spaces. Our trees will be rather special.

- **T will either be a ray $[0, \infty)$ or a bi-infinite geodesic $(-\infty, \infty)$.**

The vertices of T will be the integer points $j \in \{0\} \cup \mathbb{N}$ or $\mathbb{Z}$ according as T is $[0, \infty)$ or $(-\infty, \infty)$. The edges will be of the form $[j, j+1]$. All the vertex and edge spaces will be abstractly isometric and identified with a fixed hyperbolic metric space Y. The vertex space over j will be denoted as Y_j, and the edge space over $[j, j+1]$ will be denoted as Y_{ej}. There are two qi-embeddings of Y_{ej}, one into Y_j (denoted i_{j0}), and the other into Y_{i+1} (denoted i_{j1}). We shall demand:

1. i_{j0} is the **identity** map for all j, i.e the identification of Y_j and Y_{ej} with Y is the same
2. There exist K, ϵ such that for all j, i_{j1} is a (K, ϵ) quasi-isometry between Y_{ej} and Y_{j+1}.

Thus, the tree T is assumed rooted at 0, and maps are described on this basis.

This induces a map ϕ_j from Y_j to Y_{j+1} which is a uniform (K, ϵ) quasi-isometry for all j. Let Φ_j denote the induced map on geodesics. So $\Phi_j([a, b])$ is a geodesic in Y_{j+1} joining $\phi_j(a), \phi_{j+1}(b)$. ϕ_j^{-1} and Φ_j^{-1} will denote the quasi-isometric inverse of ϕ_j and the induced map on geodesics respectively. We shall assume that the quasi-isometric inverse ϕ_j^{-1} is also a (K, ϵ) quasi-isometry. *(Note that ϕ_j^{-1} is just a notation for the quasi-isometric inverse of ϕ and is not necessarily a set-theoretic inverse.)*

- The space Y in question will also be quite special. **Y will either be the universal cover of a closed hyperbolic surface (hence the hyperbolic plane), or the universal cover of a finite volume hyperbolic surface minus cusps.**

The first case, where Y is $\mathbb{H}^2$ will be necessary when we prove the results of Cannon-Thurston[CT85] and Minsky [Min94]. The second case, where Y is $\mathbb{H}^2$ minus an equivariant collection of horodisks, will be useful while proving a result of Bowditch [Bow02a].

3.2 *Hyperbolic ladders*

Given a geodesic $\lambda = \lambda_0 \subset Y_0$, we shall now construct a hyperbolic ladder $B_\lambda \subset X$ containing λ_0. We shall then prove that B_λ is uniformly quasiconvex (independent of λ). Inductively define:

$$\lambda_{j+1} = \Phi_j(\lambda_j) \quad \text{for} \quad j \geq 0$$

$$\lambda_{j-1} = \Phi_j^{-1}(\lambda_j) \quad \text{for} \quad j \leq 0$$

$$B_\lambda = \bigcup_j \lambda_j$$

[See Figure 2 below.]

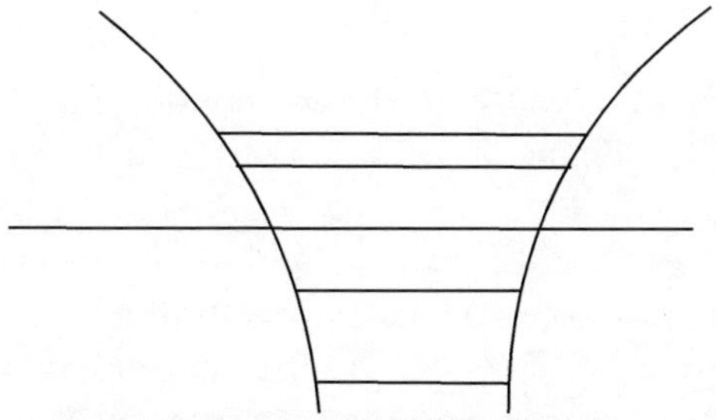

Figure 2. The hyperbolic ladder.

To prove quasiconvexity, we construct a retraction Π_λ from $\bigcup_j Y_j$ onto B_λ that fixes B_λ and does not stretch distances much. So Π_λ is a 'quasi-Lipschitz' map.

On Y_j define $\pi_j(y)$ to be a nearest-point projection of y onto λ_j. Thus, $d_{Y_j}(y, \pi_j(y)) \leq d_{Y_j}(y, z)$ for all $z \in \lambda_j$.

Next, define

$$\Pi_\lambda(y) = \pi_j(y) \quad \text{for} \quad y \in Y_j.$$

We shall choose a collection of **admissible paths** in X, such that any path in X can be uniformly approximated by these. An admissible path in X is a union of paths of the following forms:

1. Paths lying in Y_j for some j.
2. Paths of length one joining $y \in Y_j$ to $\phi_j(y) \in Y_{j+1}$ for $j \geq 0$.
3. Paths of length one joining $y \in Y_j$ to $\phi_j^{-1}(y) \in Y_{j-1}$ for $j \leq 0$.

Note. Any path in X can be uniformly approximated by an admissible path.

Definition. *An admissible geodesic is an admissible path that minimizes distances amongst admissible path.*

We are now in a position to prove the main technical theorem of this paper.

Theorem 3.1. *There exists* $C_0 \geq 0$ *such that* $d_X(\Pi_\lambda(x), \Pi_\lambda(y)) \leq C_0 d_X(x, y)$ *for* $x, y \in \bigcup_i Y_i$. *Further, given* $\delta \geq 0$, *there exists* $C \geq 0$ *such that if* X *is* δ*-hyperbolic, then* B_λ *is* C*-quasiconvex.*

Proof. It suffices to prove the the theorem when $d_X(x, y) \leq 1$.

Case (a). $x, y \in Y_j$ for some j:

From Lemma 2.2, there exists C_3 such that $d_{Y_j}(\pi_j(x), \pi_j(y)) \leq C_3$. Since embeddings of Y_j in X are 1-Lipschitz (edges go to edges), $d_X(\Pi_\lambda(x), \Pi_\lambda(y)) \leq C_3$.

Case (b). $x \in Y_j$ and $y = \phi_j(x) \in Y_{j+1}$ for some $j \geq 0$:

By Lemma 2.5, there exists $C_4 \geq 0$ such that

$$d_Y(\phi_j(\pi_j(x)), \pi_{j+1}(\phi_j(x))) \leq C_4$$

Unravelling definitions, and using the uniformity of the embedding of Y_j in X,

$$d(\Pi_\lambda(x), \Pi_\lambda(y)) \leq C_4 + 1.$$

Case (c). $x \in Y_j$ and $y = \phi_j^{-1}(x) \in Y_{j-1}$ for some $j \leq 0$:

The argument is just as in Case(b) above.

Finally, to prove quasiconvexity of B_λ, we consider any two points $a, b \in B_\lambda$. Let μ be a geodesic in X joining a, b. Then by the above argument, $\Pi_\lambda(\mu \cap \bigcup_j Y_j)$ is a 'dotted (K, ϵ)-quasigeodesic' for some K, ϵ, i.e. a quasigeodesic which is not necessarily connected. Note that by our definition of a quasigeodesic (viz. a quasi-isometric image of an interval in $\mathbb{R}$) a dotted quasigeodesic is a quasigeodesic. Since X is δ-hyperbolic, there exists $C \geq 0$ such that the geodesic μ lies in a C neighborhood of $\Pi_\lambda(\mu \cap \bigcup_j Y_j)$. Further, since the latter lies on B_λ, we conclude that B_λ is C-quasiconvex. ∎

Note. Suppose instead of being geodesics λ_i are all (K_Y, ϵ_Y)-quasigeodesics in each Y_i. Then observe that the Theorem 3.1 still goes through, and the union B_λ (in this case of **quasigeodesics**) of λ_i is still a retract of $\bigcup_i Y_i$.

The following lemma gives us a method of finding an admissible path from λ_j to λ_0, whose length is of the order of j.

Lemma 3.2. *There exists $A > 0$, such that if $a \in \lambda_j$ for some j then there exists $b \in i(\lambda) = \lambda_0$ such that $d_X(a, b) \leq Aj$.*

Proof. It suffices to prove that for $j \geq 1$, there exists $A > 0$ independent of j such that if $p \in \lambda_j$, there exists $q \in (\lambda_{j-1})$ with $d(p, q) \leq A$. A symmetric argument works for $j \leq -1$.

By construction, $\lambda_j = \Phi_j(\lambda_{j-1})$. Since ϕ_j is a (K, ϵ) quasi-isometry for all j, there exists C_1 such that $\phi_j^{-1}(p)$ lies in a C_1 neighborhood of λ_j. Hence, there exists $q \in \lambda_j$ such that $d(q, p) \leq 1 + C_1$. Choosing $A = C_1 + 1$ we are through. □

Remark. Note that Lemma 3.2 also goes through when each λ_i is a (K_Y, ϵ_Y)-quasigeodesic in Y_i for all i.

The main theorem of this section follows:

Theorem 3.3. *Let (X, d) be a tree (T) of hyperbolic metric spaces satisfying the qi-embedded condition. Suppose in addition that T is either $\mathbb{R}$ or $[0, \infty)$ with the usual vertex and edge sets Y_j, for j integers. If X is hyperbolic, then $i : Y_0 \to X$ extends continuously to $\hat{i} : \widehat{Y_0} \to \widehat{X}$.*

Proof. To prove the existence of a Cannon-Thurston map, it suffices to show (from Lemma 2.1) that for all $M \geq 0$ and $y_0 \in Y_0$ there exists $N \geq 0$ such that if a geodesic segment λ lies outside the N-ball around $y_0 \in Y_0$, B_λ lies outside the M-ball around $y_0 \in X$.

To prove this, we show that if λ lies outside the N-ball around $y_0 \in Y_0$, B_λ lies outside a certain $M(N)$-ball around $y_0 \in Y_0 \subset X$, where $M(N)$ is a proper function from $\mathbb{N}$ into itself.

Since Y_0 is properly embedded in X there exists $f(N)$ such that $i(\lambda)$ lies outside the $f(N)$-ball around y_0 in X and $f(N) \to \infty$ as $N \to \infty$.

Let p be any point on B_λ. Then $p \in Y_j$ for some j. There exists $y \in \lambda = \lambda_0$ such that $d_X(y, p) \leq Aj$ by Lemma 3.2. Therefore,

$$\begin{aligned} d_X(y_0, p) &\geq d_X(y_0, y) - Aj \\ &\geq f(N) - Aj \end{aligned}$$

By our choice of metric on X,

$$d_X(y_0, p) \geq j$$

Hence

$$\begin{aligned} d(x_0, p) &\geq \max(f(N) - Aj, j) \\ &\geq \frac{f(N)}{A+1} \end{aligned}$$

From Theorem 3.1 there exists C independent of λ such that B_λ is a C-quasiconvex set containing λ. Therefore any geodesic joining the end-points of λ lies in a C-neighborhood of B_λ.

Hence any geodesic joining end-points of λ lies outside a ball of radius $M(N)$ where

$$M(N) = \frac{f(N)}{A+1} - C.$$

Since $f(N) \to \infty$ as $N \to \infty$ so does $M(N)$. □

4. Closed surface groups of bounded geometry: Theorems of Cannon-Thurston and Minsky

4.1 *Three manifolds fibering over the circle*

Now, let M be a closed hyperbolic 3-manifold, fibering over the circle with fiber F. Then the universal cover $\tilde{M}$ may be regarded as a tree T of spaces, where $T = \mathbb{R}$, and where each vertex and edge set is $\tilde{F}$. Thus X is $\tilde{M}$, which is quasi-isometric to $\mathbb{H}^3$ and Y is $\tilde{F}$ which is quasi-isometric to $\mathbb{H}^2$. Further, ϕ_i can all be identified with $\tilde{\phi}$, where ϕ is the pseudo-anosov monodromy of the fibration. Then as a direct consequence of Theorem 3.3, we obtain the following theorem of Cannon and Thurston, which motivated much of the present work.

Theorem 4.1 (Cannon-Thurston [CT85]). *M be a closed hyperbolic 3-manifold, fibering over the circle with fiber F. Let $i : \mathbb{H}^2 \to \mathbb{H}^3$ denote the inclusion of $\tilde{F}$ into $\tilde{M}$ (where $\tilde{F}$ and $\tilde{M}$ are identified with $\mathbb{H}^2$ and $\mathbb{H}^3$ respectively. Then i extends continuously to a map $\hat{i} : \mathbb{D}^2 \to \mathbb{D}^3$, where $\mathbb{D}^2$ and $\mathbb{D}^3$ denote the compactifications of $\mathbb{H}^2$ and $\mathbb{H}^3$ respectively.*

4.2 *Closed surface Kleinian groups*

In this section we apply Theorem 3.3 to geometrically tame closed groups of bounded geometry.

The *convex core* of a hyperbolic 3-manifold N (without cusps) is the smallest convex submanifold $C(N) \subset N$ for which inclusion is a homotopy equivalence. If an ϵ neighborhood of $C(N)$ has finite volume, N is said to be *geometrically finite*. Suppose $N = \mathbb{H}^3/\rho(\pi_1(S))$ for a closed surface S. We say that an end of N is *geometrically finite* if it has a neighborhood missing $C(N)$. N is *simply degenerate* if it has only one end E; if a neighborhood of E is homeomorphic to $S \times \mathbb{R}$; and if, in addition, there is a sequence of pleated surfaces homotopic in this neighborhood to the inclusion of S, and exiting every compact set (This last condition is automatic by Bonahon [Bon86]. N is called *doubly degenerate* if it has two ends, both of which are simply degenerate. For a more detailed discussion of pleated surfaces and geometrically tame ends, see [Thu80] or [Min92].

Let $inj_N(x)$ denote the injectivity radius at $x \in N$. N is said to have **bounded geometry** if there exists $\epsilon_0 > 0$ such that $inj_N(x) > \epsilon_0$ for all $x \in N$. In order to apply Theorem 3.3 we need some preliminary Lemmas.

Let E be a simply degenerate end of N. Then E is homeomorphic to $S \times [0, \infty)$ for some closed surface S of genus greater than one.

Lemma 4.2 ([Thu80]). *There exists $D_1 > 0$ such that for all $x \in N$, there exists a pleated surface $g : (S, \sigma) \to N$ with $g(S) \cap B_{D_1}(x) \neq \emptyset$.*

The following Lemma follows easily from the fact that $inj_N(x) > \epsilon_0$:

Lemma 4.3 ([Bon86], [Thu80]). *There exists $D_2 > 0$ such that if $g : (S, \sigma) \to N$ is a pleated surface, then $dia(g(S)) < D_2$.*

The following Lemma due to Minsky [Min92] follows from compactness of pleated surfaces.

Lemma 4.4 ([Min92]). *Fix S and $\epsilon > 0$. Given $a > 0$ there exists $b > 0$ such that if $g : (S, \sigma) \to N$ and $h : (S, \rho) \to N$ are homotopic pleated surfaces which are isomorphisms on π_1 and $inj_N(x) > \epsilon$ for all $x \in N$, then*

$$d_N(g(S), h(S)) \leq a \Rightarrow d_{Teich}(\sigma, \rho) \leq b,$$

where d_{Teich} denotes Teichmüller distance.

Definition. *The* **universal curve** *over $X \subset Teich(S)$ is a bundle whose fiber over $x \in X$ is x itself. [McM01]*

Lemma 4.5. *There exist K, ϵ and a homeomorphism h from E to the universal curve S_γ over a Lipschitz path γ in Teichmüller space, such that $\tilde{h}$ from $\tilde{E}$ to the universal cover of S_γ is a (K, ϵ)-quasi-isometry.*

Proof. We can assume that $S \times \{0\}$ is mapped to a pleated surface $S_0 \subset N$ under the homeomorphism from $S \times [0, \infty)$ to E. We shall construct inductively a sequence of 'equispaced' pleated surfaces $S_i \subset E$ exiting the end. Assume that $S_0, \ldots, S_n$ have been constructed such that:

1. If E_i be the non-compact component of $E \backslash S_i$, then $S_{i+1} \subset E_i$.
2. Hausdorff distance between S_i and S_{i+1} is bounded above by $3(D_1 + D_2)$.
3. $d_N(S_i, S_{i+1}) \geq D_1 + D_2$.
4. From Lemma 4.4 and condition (2) above there exists D_3 depending on D_1, D_2 and S such that $d_{Teich}(S_i, S_{i+1}) \leq D_3$

Next choose $x \in E_n$, such that $d_N(x, S_n) = 2(D_1 + D_2)$. Then by Lemma 4.2, there exists a pleated surface $g : (S, \tau) \to N$ such that $d_N(x, g(S)) \leq D_1$. Let $S_{n+1} = g(S)$. Then by the triangle inequality and Lemma 4.3, if $p \in S_n$ and $q \in S_{n+1}$,

$$D_1 + D_2 \leq d_N(p, q) \leq 3(D_1 + D_2).$$

This allows us to continue inductively. The Lemma follows. □

Note that in the above Lemma, pleated surfaces are not assumed to be embedded. This is because immersed pleated surfaces with a uniform lower bound on injectivity radius are uniformly quasi-isometric to the corresponding Riemann surfaces.

Since there are exactly one or two ends E_i, we have thus shown:

Lemma 4.6. *The hyperbolic metric space $\widetilde{C(N)}$ is quasi-isometric to a tree (T) of hyperbolic metric spaces satisfying the qi-embedded condition, where T is either $[0, \infty)$ (simply degenerate) or $\mathbb{R}$ (doubly degenerate).*

Applying Theorem 3.3, we obtain the following theorem of Minsky:

Theorem 4.7 (Minsky [Min94]). *Let $\Gamma = \rho(\pi_1(S))$ be a closed surface Kleinian group, such that $\mathbb{H}^3/\Gamma = M$ has injectivity radius uniformly bounded below by some $\epsilon > 0$. Then there exists a continuous map from the Gromov boundary of Γ (regarded as an abstract group) to the limit set of Γ in $\mathbb{S}^2_\infty$.*

Since a continuous image of a compact locally connected set is locally connected [HY61], and since the limit set of Γ is a continuous image of the circle, by Theorem 3.3, we have:

Corollary 4.8. *Suppose Γ is a closed surface Kleinian group, such that $N = \mathbb{H}^3/\Gamma$ has bounded geometry, i.e. $inj_N(x) > \epsilon_0$ for all $x \in N$. Then the limit set of Γ is locally connected.*

5. Punctured surface groups of bounded geometry: A theorem of Bowditch

5.1 *Outline of proof*

In [Bow02a] [Bow02b], Bowditch proved the existence of Cannon-Thurston maps for punctured surface groups of bounded geometry using some of the ideas from [Mit98b]. We give below a different proof of the result, which is in some ways simpler. First, a sketch.

Let N^h be a bounded geometry 3-manifold corresponding to a representation of the fundamental group of a punctured surface S^h. Excise the cusps (if any) of N^h leaving us a manifold that has one or two ends. Let N denote N^h minus cusps. Then $\widetilde{N}$ is quasi-isometric to the universal curve over a Lipschitz path in Teichmüller space from which cusps have been removed. This path is semi-infinite or bi-infinite according as N is one-ended or two-ended. Fix a reference finite volume hyperbolic surface S^h. Let S denote S^h minus cusps. Then $\widetilde{S}$ is quasi-isometric to the Cayley graph of $\pi_1(S)$ which is (Gromov) hyperbolic. We fix a base surface in N and identify it with S. Now look at $\widetilde{S} \subset \widetilde{N}$. Let $\lambda = [a, b]$ be a geodesic segment in $\widetilde{S}$. We 'flow' λ out the end(s) of $\widetilde{N}$ to generate the **hyperbolic ladder-like set** B_λ as in the proof of Theorem 3.1.

Theorem 3.1 ensures that there is a retraction from $\widetilde{N}$ to B_λ which does not increase distances much. From this it follows that B_λ is quasi-isometrically embedded in $\widetilde{N}$. Recall that for the construction of B_λ, we only require the hyperbolicity of $\tilde{S}$ and not that of $\tilde{N}$.

Now if λ lies outside a large ball about a fixed reference point p in $\widetilde{S}$, then so does B_λ in $\widetilde{N}$. Since B_λ is q.i. embedded in $\widetilde{N}$, there exists an ambient $\widetilde{N}$-quasigeodesic μ lying in a bounded neighborhood of B_λ joining the end-points of λ. If S^h had no cusps, as in the case for closed surfaces, we could immediately conclude that for any geodesic segment λ in $\widetilde{S^h}$ lying outside large balls around p, there is a quasigeodesic in $\widetilde{N^h}$ joining its endpoints and lying outside a large ball around p in $\widetilde{N^h}$. However, since S^h has cusps, $\widetilde{S^h}$ and $\widetilde{S}$ are different. So a little more work is necessary. Suppose as before that λ_0 is a geodesic in $\widetilde{S^h}$ lying outside a large ball around p. For ease of exposition we assume that the end-points of λ_0 lie outside cusps. Let $\lambda \subset \widetilde{S}$ be the geodesic in (the path-metric on) $\widetilde{S}$ joining the same pair of points. Then off horodisks, λ_0 and λ track each other. Construct B_λ as before, and let μ be an ambient quasigeodesic in $\widetilde{N}$ lying in a bounded neighborhood of B_λ joining the end-points of λ. Then, modulo horoballs in $\widetilde{N^h}$, μ lies outside a large ball around p. Let μ_0 be the hyperbolic geodesic joining the end points of μ. Off horoballs, μ and μ_0 track each other. Hence, off horoballs, μ_0 lies outside large balls about p. The points at which μ_0 enters and leaves a particular horoball therefore lie outside large balls about p. But then the hyperbolic segment joining them must do the same. This shows that μ_0 must itself lie outside large balls around p. As before we conclude that there exists a continuous extension of the inclusion of $\widetilde{S^h}$ into $\widetilde{N^h}$ to the boundary. The remaining part of this section fleshes out this argument.

5.2 *Relative hyperbolicity*

We shall be requiring certain properties of hyperbolic spaces minus horoballs. These were studied by Farb [Far98] under the garb of 'electric geometry'. We combine Farb's results with a version that is a (slight variant of) theorem due to McMullen (Theorem 8.1 of [McM01]).

Definition. *A path $\gamma : I \to Y$ to a path metric space Y is an ambient K-quasigeodesic if we have*

$$L(\beta) \leq KL(A) + K$$

for any subsegment $\beta = \gamma|[a, b]$ and any path $A : [a, b] \to Y$ with the same end-points.

The following definitions are adapted from [Far98]

Definition. *Let M be a convex hyperbolic manifold. Let Y be the universal cover of M minus cusps and $X = \widetilde{M}$. γ is said to be a K-quasigeodesic in X* **without backtracking** *if*

- *γ is a K-quasigeodesic in X*
- *γ does not return to any horoball* **H** *after leaving it.*

Definition. *γ is said to be an ambient K-quasigeodesic in Y* **without backtracking** *if*

- *γ is an ambient K-quasigeodesic in Y.*
- *γ is obtained from a K-quasigeodesic without backtracking in X by replacing each maximal subsegment with end-points on a horosphere by a quasigeodesic lying on the surface of the horosphere.*

Note that in the above definition, we allow the behavior to be quite arbitrary on horospheres (since Euclidean quasigeodesics may be quite wild); however, we do not allow wild behavior off horoballs.

Remark. Our definition of *ambient quasigeodesic without backtracking* does not allow a path to follow a horosphere for a long distance without entering it. This is a point where the definition differs from the usual definition of an ambient quasi-geodesic.

$B_R(Z)$ will denote the R-neighborhood of the set Z.

Let $\mathcal{H}$ be a locally finite collection of horoballs in a convex subset X of $\mathbb{H}^n$ (where the intersection of a horoball, which meets ∂X in a point, with X is called a horoball in X). The following theorem is due to McMullen [McM01].

Theorem 5.1 ([McM01]). *Let $\gamma : I \to X \setminus \bigcup \mathcal{H} = \mathcal{Y}$ be an ambient K-quasigeodesic for a convex subset X of $\mathbb{H}^n$ and let $\mathcal{H}$ denote a collection of horoballs. Let η be the hyperbolic geodesic with the same endpoints as γ. Let $\mathcal{H}(\eta)$ be the union of all the horoballs in $\mathcal{H}$ meeting η. Then $\eta \cup \mathcal{H}(\eta)$ is (uniformly) quasiconvex and $\gamma(I) \subset B_R(\eta \cup \mathcal{H}(\eta))$, where R depends only on K.*

Theorem 5.1 is similar in flavor to certain theorems about relative hyperbolicity *a la* Gromov [Gro85], Farb [Far98] and Bowditch [Bow97]. We give below a related theorem that is derived from Farb's 'Bounded Horosphere Penetration' property.

Let $\gamma_1 = \overline{pq}$ be a hyperbolic K-quasigeodesic without backtracking starting from a horoball $\mathbf{H_1}$ and ending within (or on) a *different* horoball $\mathbf{H_2}$. Let $\gamma = [a, b]$ be the hyperbolic geodesic minimizing distance between $\mathbf{H_1}$ and $\mathbf{H_2}$. Following [Far98] we put the zero metric on the horoballs that γ penetrates. The resultant pseudo-metric is called the electric metric. Let $\widehat{\gamma}$ and $\widehat{\gamma_1}$ denote the paths represented by γ and γ_1 respectively in this pseudometric. It is shown in [Far98] that γ, $\widehat{\gamma}$ and $\widehat{\gamma_1}$ have similar intersection patterns with horoballs, i.e. there exists C_0 such that

- •1 If only one of γ and $\widehat{\gamma_1}$ penetrates a horoball $\mathbf{H}$, then it can do so for a distance $\leq C_0$.
- •2 If both $\widehat{\gamma_1}$ and γ enter (or leave) a horoball $\mathbf{H}$ then their entry (or exit) points are at a distance of at most C_0 from each other. [Here by 'entry' (resp. 'exit') point of a quasigeodesic we mean a point at which the path switches from being in the complement of or 'outside' (resp. in the interior of or 'inside') a closed horoball to being zinside (resp. outside) it].

The point to observe here is that quasigeodesics without backtracking in our definition gives rise to quasigeodesics without backtracking in Farb's sense. Since this is true for arbitrary γ_1 we give below a slight strengthening of this fact. Further, by our construction of ambient quasigeodesics without backtracking, we might just as well consider ambient quasigeodesics without backtracking in place of quasigeodesics.

Theorem 5.2 ([Far98]). *Given $C > 0$, there exists C_0 such that if*

- •1 *either two quasigeodesics without backtracking γ_1, γ_2 in X, OR*
- •2 *two ambient quasigeodesics without backtracking γ_1, γ_2 in Y, OR*
- •3 *γ_1 – an ambient quasigeodesic without backtracking in Y and γ_2 – a quasigeodesic without backtracking in X,*

start and end

- •1 *either on (or within) the same horoball OR*
- •2 *a distance C from each other then they have similar intersection patterns with horoballs (except possibly the first and last ones), i.e. there exists C_0 such that*
- •1 *If only γ_1 penetrates or travels along the boundary of a horoball* **H**, *then it can do so for a distance $\leq C_0$.*
- •2 *If both γ_1 and γ_2 enter (or leave) a horoball* **H** *then their entry (or exit) points are at a distance of at most C_0 from each other.*

5.3 *Horo-ambient quasigeodesics*

A special kind of quasigeodesic without back-tracking will be necessary. We start with a hyperbolic geodesic λ^h in $\widetilde{S}^h$. Fix a neighborhood of the cusps lifting to an equivariant family of horoballs in the universal cover $\mathbb{H}^2 = \widetilde{S}^h$. Since λ^h is a hyperbolic geodesic in $\widetilde{S}^h$ there are unique entry and exit points for each horoball that λ^h meets and hence unique Euclidean geodesics joining them on the corresponding horosphere. Replacing the segments of λ^h lying inside Z-horoballs by the corresponding Euclidean geodesics, we obtain an ambient quasigeodesic λ in $\widetilde{M_0}$ as a consequence of Theorem 5.1 (See Corollary 5.3 and figure below):

Ambient quasigeodesics obtained by this kind of a construction will be termed **horo-ambient quasigeodesics** to distinguish them from *electro-ambient quasigeodesics* defined earlier.

The following Corollary of Theorem 5.1 justifies the terminology.

Corollary 5.3. *There exists K, ϵ, such that any horo-ambient quasigeodesic in Y is indeed a K, ϵ ambient quasigeodesic in Y.*

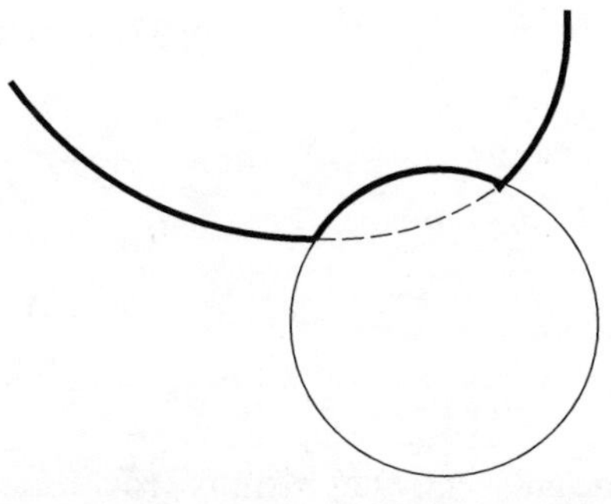

Figure 3. Horo-ambient quasigeodesics.

Proof. Let $[a,b]_h$ be a hyperbolic geodesic in X joining a,b, where a,b lie outside horoballs. Let $[a,b]_{ha}$ be the horo-ambient quasigeodesic in Y joining a,b. Let $[a,b]_a$ be the ambient geodesic in Y joining a,b. Then by Theorem 5.1,there exists $R \geq 0$, such that $[a,b]_a$ lies in an R-neighborhood of $[a,b]_h \cup \mathcal{H}(\eta)$. Project $[a,b]_a$ onto $[a,b]_h \cup \mathcal{H}(\eta)$ using the nearest point projection in X. Removing the backtracking induced, we get some K,ϵ depending on R such that the image is an ambient quasigeodesic without backtracking in our sense. Clearly, such an ambient quasigeodesic coincides with $[a,b]_{ha}$ off horoballs $\mathcal{H}(\eta)$. Since the interpolating segments in $[a,b]_{ha}$ on any horoball in $\mathcal{H}(\eta)$ are Euclidean *geodesics*, the result follows. □

5.4 *Trees of spaces*

We want to first show that the universal cover of N^h minus Z-cusps is quasi-isometric to a tree of hyperbolic metric spaces.

Let E^h be a simply degenerate end of N^h. Then E^h is homeomorphic to $S^h \times [0,\infty)$ for some surface S^h of negative Euler characteristic. Cutting off a neighborhood of the cusps of S^h we get a surface with boundary denoted as S. Let E denote E^h minus a neighborhood of the Z-cusps. We assume that each Z-cusp has the standard form coming from a quotient of a horoball in $\mathbb{H}^3$ by Z. Also, we shall take our pleated surfaces to be such that the pair $(S, cusps)$ is mapped to the pair $(E, cusps)$ for each pleated S^h. We shall now show that each $\widetilde{E}$ is quasi-isometric to a ray of hyperbolic metric spaces satisfying the q-i embedded condition. Each edge and vertex space will be a copy of $\widetilde{S}$ and the edge to vertex space inclusions shall be quasi-isometries. Note that each $\widetilde{S}$ can be thought of as a copy of $\mathbb{H}^2$ minus an equivariant family of horodisks. The following Lemmas are generalizations to punctured surfaces of Lemmas 4.2, 4.3, 4.4.

Lemma 5.4 ([Thu80]). *There exists $D_1 > 0$ such that for all $x \in E$, there exists a pleated surface $g : (S^h,\sigma) \to E^h$ with $g(S) \cap B_{D_1}(x) \neq \emptyset$. Also g maps $(S, cusps)$ to $(E, cusps)$.*

Lemma 5.5 ([Bon86], [Thu80]). *There exists $D_2 > 0$ such that if $g : (S^h,\sigma) \to N^h$ is a pleated surface, then the diameter of the image of S is bounded, i.e. $dia(g(S)) < D_2$.*

The following Lemma due to Thurston (Theorems 9.2 and 9.6.1 of [Thu80]) and Minsky [Min92] follows from compactness of pleated surfaces.

Lemma 5.6 ([Min92]). *Fix S^h and $\epsilon > 0$. Given $a > 0$ there exists $b > 0$ such that if $g : (S^h,\sigma) \to E^h$ and $h : (S^h,\rho) \to E^h$ are homotopic pleated surfaces which are isomorphisms on π_1 and E^h is of bounded geometry, then*

$$d_E(g(S), h(S)) \leq a \Rightarrow d_{Teich}(\sigma,\rho) \leq b,$$

where d_{Teich} denotes Teichmüller distance.

In [Min92] a specialization of this statement is proven for closed surfaces. However, the main ingredient, a Theorem due to Thurston is stated and proven in [Thu80]

(Theorems 9.2 and 9.6.1 – *'algebraic limit is geometric limit'*) for finite area surfaces. The arguments given by Minsky to prove the above Lemma from Thurston's Theorems (Lemma 4.5, Corollary 4.6 and Lemma 4.7 of [Min92]) go through with very little change for surfaces of finite area.

Construction of equispaced pleated surfaces exiting the end

We next construct a sequence of **equispaced** pleated surfaces $S^h(i) \subset E^h$ exiting the end as before. Assume that $S^h(0), \ldots, S^h(n)$ have been constructed such that:

1. $(S(i), cusps)$ is mapped to $(E, cusps)$
2. If $E(i)$ be the component of $E \backslash S(i)$ for which $E(i)$ is non-compact, then $S(i+1) \subset E(i)$.
3. Hausdorff distance between $S(i)$ and $S(i+1)$ is bounded above by $3(D_1 + D_2)$.
4. $d_E(S(i), S(i+1)) \geq D_1 + D_2$.
5. From Lemma 5.6 and condition (3) above there exists D_3 depending on D_1, D_2 and S such that $d_{Teich}(S(i), S(i+1)) \leq D_3$.

Next choose $x \in E(n)$, such that $d_E(x, S_n) = 2(D_1 + D_2)$. Then by Lemma 5.4, there exists a pleated surface $g : (S^h, \tau) \rightarrow E^h$ such that $d_E(x, g(S)) \leq D_1$. Let $S^h(n+1) = g(S^h)$. Then by the triangle inequality and Lemma 5.5, if $p \in S(n)$ and $q \in S(n+1)$,

$$D_1 + D_2 \leq d_E(p, q) \leq 3(D_1 + D_2).$$

This allows us to continue inductively. $S(i)$ corresponds to a point x_i of *Teich*(S). Joining the x_i's in order, one gets a Lipschitz path in *Teich*(S).

Definition. *A sequence of pleated surfaces satisfying conditions (1–5) above will be called an* **equispaced sequence of pleated surfaces**. *The corresponding sequence of* $S(i) \subset E$ *will be called an* **equispaced sequence of truncated pleated surfaces**.

Each $S(i)$ being compact (with or without boundary), $\widetilde{S(i)}$ is a hyperbolic metric space. We can think of the universal cover $\widetilde{E}$ of E as being quasi-isometric to a ray T of hyperbolic metric spaces by setting $T = [0, \infty)$, with vertex set $\mathcal{V} = \{n : n \in \mathbb{N} \cup \{0\}\}$, edge set $\mathcal{E} = \{[n-1, n] : n \in \mathbb{N}\}$, $X_n = \widetilde{S(n)} = X_{[n-1,n]}$. Further, by Lemma 5.6 this tree of hyperbolic metric spaces satisfies the quasi-isometrically embedded condition. We have thus shown

Lemma 5.7. *If E^h be a simply degenerate end of a hyperbolic 3 manifold N^h with bounded geometry, then there is a sequence of equispaced pleated surfaces exiting E^h and hence a sequence of truncated pleated surfaces exiting $\widetilde{E}$. Further, $\widetilde{E}$ is quasi-isometric to a ray of hyperbolic metric spaces satisfying the q.i. embedded condition.*

5.5 *Construction of B_λ: modifications for punctured surfaces*

Now, let N^h be a bounded geometry 3-manifold corresponding to a representation of the fundamental group of a punctured surface S^h. Excise the cusps (if any) of N^h

leaving us a manifold that has one or two ends. Let N denote N^h minus cusps. Fix a reference finite volume hyperbolic surface S^h. Let S denote S^h minus cusps. Then $\widetilde{S}$ is quasi-isometric to the Cayley graph of $\pi_1(S)$ which is (Gromov) hyperbolic, in fact quasi-isometric to a tree. We fix a base surface in N and identify it with S. Now look at $\widetilde{S} \subset \widetilde{N}$.

Then, by Lemma 5.7 $\widetilde{N}$ is quasi-isometric to a tree T of hyperbolic metric spaces. Each of the vertex and edge spaces is a copy of $\widetilde{S}$. Also, the map ϕ_i induced from $\widetilde{S} \times \{i\}$ to $\widetilde{S} \times \{i+1\}$ is a (K, ϵ)-quasi-isometry for all i as are their quasi-isometric inverses. Further, T is a semi-infinite or bi-infinite interval in $\mathbb{R}$ according as N is one-ended or two-ended. So far, this is exactly like the case for closed surfaces. But here, we can assume in addition that each ϕ_i is the restriction of a map ϕ_i^h from $\widetilde{S^h} \times \{i\}$ to $\widetilde{S^h} \times \{i+1\}$ which preserves horodisks. Let Φ_i^h denote the induced map on hyperbolic geodesics and let Φ_i denote the induced map on horo-ambient geodesics.

Let λ^h be a hyperbolic geodesic segment in $\widetilde{S^h}$. Let λ be the horo-ambient quasi-geodesic in $\widetilde{S}$ joining the end-points of λ^h.

Starting with a horo-ambient quasigeodesic $\lambda \subset \widetilde{S}$, we can now proceed as in the proof of Theorem 3.1 to construct the **hyperbolic ladder-like set** B_λ.

There is only one difference: **Each λ_i in this situation is a horo-ambient quasigeodesic, and not necessarily a hyperbolic geodesic.** Thus, we set $\lambda = \lambda_0$ to be some horo-ambient quasigeodesic in $\widetilde{S} = \widetilde{S} \times \{0\}$. Next, (for $i \geq 0$), inductively, set λ_{i+1} to be the (unique) horo-ambient quasigeodesic in $\widetilde{S} \times \{i+1\}$ joining the end-points of $\phi_i(\lambda_i)$. That is to say, $\lambda_{i+1} = \Phi_i(\lambda_i)$. Similarly, for $i \leq 0$.

Note that Corollary 5.3 ensures that there exist K_0, ϵ_0 such that each λ_i is a (K_0, ϵ_0)-quasigeodesic in $\widetilde{S} \times \{i\}$.

Then Theorem 3.1 (or more precisely, the Note following it) ensures that there is a retraction from $\widetilde{N}$ to B_λ which does not increase distances much. From this it follows that B_λ is quasi-isometrically embedded in $\widetilde{N}$. Recall that for the construction of B_λ, we only require the hyperbolicity of $\tilde{S}$ and not that of $\tilde{N}$.

As before, (by projecting a geodesic in $\widetilde{N}$ onto B_λ) we obtain an ambient quasi-geodesic contained in B_λ joining the end-points a, b of λ.

Let

- β^h = geodesic in $\widetilde{N^h}$ joining a, b
- β^0_{amb} = horo-ambient quasigeodesic in $\widetilde{N}$ obtained from β^h by replacement of hyperbolic by 'Euclidean' geodesic segments for horoballs in $\widetilde{N^h}$
- $\beta_{amb} = \Pi_\lambda(\beta^0_{amb})$

5.6 *Quasigeodesic rays*

Let λ_i^c denote the union of the segments of λ_i which lie along horocycles and let $\lambda_i^b = \lambda_i - \lambda_i^c$. Let

$$B_\lambda^c = \cup_i \lambda_i^c$$

$$B_\lambda^b = \cup_i \lambda_i^b$$

We want to show that for all $x \in B^b_\lambda$ there exists a C-quasigeodesic $r_x : \{0\} \cup \mathbb{N} \to B^b_\lambda$ such that $x \in r_x(\{0\} \cup \mathbb{N})$ and $r_x(i) \in \lambda^b_i$. Suppose $x \in \lambda^b_k \subset B^b_\lambda$. We define r_x by starting with $r_x(k) = x$ and construct $r_x(k-i)$ and $r_x(k+i)$ inductively (of course $(k-i)$ stops at 0 for T a semi-infinite ray but $(k+i)$ goes on to infinity). For the sake of concreteness, we prove the existence of such a $r_x(k+1)$. The same argument applies to $(k-1)$ and inductively for the rest.

Lemma 5.8. *There exists $C > 0$ such that if $r_x(k) = x \in \lambda^b_k$ then there exists $x' \in \lambda^b_{k+1}$ such that $d(x, x') \leq C$. We denote $r_x(k+1) = x'$.*

Proof. Let $[a, b]$ be the maximal connected component of λ^b_k on which x lies. Then there exist two horospheres $\mathbf{H_1}$ and $\mathbf{H_2}$ such that $a \in \mathbf{H_1}$ (or is the initial point of λ_k) and $b \in \mathbf{H_2}$ (or is the terminal point of λ_k). Since ϕ_k preserves horocycles, $\phi_k(a)$ lies on a horocycle (or is the initial point of λ_{k+1}) as does $\phi_k(b)$ (or is the terminal point of λ_{k+1}). Further, the image of $[a, b]$ under ϕ_k is a quasigeodesic in $\widetilde{S} \times \{k+1\}$ **which we now denote as** $\phi_k([a, b])$. Recall that $\Phi_k([a, b])$ is the horo-ambient geodesic in $\widetilde{S} \times \{k+1\}$ joining $\phi_k(a)$ and $\phi_k(b)$.

Therefore by (Gromov) hyperbolicity of $\widetilde{S}$, $\Phi_k([a, b])$ lies in a bounded neighborhood of $\phi_k([a, b])$ (which in turn lies at a bounded distance from $\Phi^h_k([a, b])$) and hence by Theorem 5.2 there exists an upper bound on how much $\Phi^h_k([a, b])$ can penetrate horoballs, i.e. there exists $C_1 > 0$ such that for all $z \in \Phi^h_k([a, b])$, there exists $z' \in \Phi^h_k([a, b])$ lying outside horoballs with $d(z, z') \leq C_1$. Further, since ϕ^h_k is a quasi-isometry there exists $C_2 > 0$ such that $d(\phi_k(x), \Phi^h_k([a, b])) \leq C_2$. Hence there exists $x' \in \Phi^h_k([a, b])$ such that

- $d(\phi_k(x), x') \leq C_1 + C_2$
- x' lies outside horoballs.

Again, $\Phi_k([a, b])$ lies at a uniformly bounded distance $\leq C_3$ from λ_{k+1} and so, if $c, d \in \lambda_{k+1}$ such that $d(a, c) \leq C_3$ and $d(b, d) \leq C_3$ then the segment $[c, d]$ can penetrate only a bounded distance into any horoball. Hence there exists $C_4 > 0$ and $x'' \in [c, d] \subset \lambda_{k+1}$ such that

- $d(x', x'') \leq C_4$
- x'' lies outside horoballs.

Hence, $d(\phi_k(x), x'') \leq C_1 + C_2 + C_4$. Since $d(x, \phi_{k+1}(x)) = 1$, we have, by choosing $r_{k+1}(x) = x''$,

$$d(r_k(x), r_{k+1}(x)) \leq 1 + C_1 + C_2 + C_4.$$

Choosing $C = 1 + C_1 + C_2 + C_4$, we are through. □

Using Lemma 5.8 repeatedly (inductively replacing x with $r_x(k+i)$ we obtain the values of $r_x(i)$ for $i \geq k$. By an exactly similar symmetric argument, we get $r_x(k-1)$ and proceed down to $r_x(0)$. Now for any $i, z \in \widetilde{S} \times \{i\}$ and $y \in \widetilde{S} \times \{i+1\}, d(z, y) \geq 1$. Hence, for any $z \in \widetilde{S} \times \{i\}$ and $y \in \widetilde{S} \times \{j\}, d(z, y) \geq |i - j|$. This gives

Corollary 5.9. *There exist $K, \epsilon > 0$ such that for all $x \in \lambda_k^b \subset B_\lambda^b$ there exists a (K, ϵ) quasigeodesic ray r_x such that $r_x(k) = x$ and $r_x(i) \in \lambda_i^b$ for all i.*

To fix and recall notation:

- λ^h = hyperbolic geodesic in $\widetilde{S}^h$ joining a, b
- λ = horo-ambient quasigeodesic in $\widetilde{S}$ constructed from $\lambda^h \subset \widetilde{S}^h$
- β^h = geodesic in N^h joining a, b
- β^0_{amb} = horo-ambient quasigeodesic in $\widetilde{N}$ obtained from β^h by replacement of hyperbolic by 'Euclidean' geodesic segments for horoballs in $\widetilde{N^h}$
- $\beta_{amb} = \Pi_\lambda(\beta^0_{amb}) \cap B_\lambda$

5.7 *Proof of theorem for surfaces with punctures*

By construction, the hyperbolic geodesic β^h and the ambient quasigeodesic β^0_{amb} agree exactly off horoballs. β_{amb} is constructed from β^0_{amb} by projecting it onto B_λ and so by Theorem 3.1, it is an ambient quasigeodesic. But it might 'backtrack'. Hence, we need to modify it such that it satisfies the no backtracking condition. First, observe by Theorem 5.1 that all three β^h, β^0_{amb}, β_{amb} track each other off some K-neighborhood of horoballs.

The advantage of working with β_{amb} is that it lies on B_λ. However, it might backtrack.

Lemma 5.10. *There exists $C > 0$ such that for all $x \in \lambda_i^b \subset B_\lambda^b \subset B_\lambda$ if λ^h lies outside $B_n(p)$ for a fixed reference point $p \in \widetilde{S}^h$ and also assuming that in fact the reference point lies in $\widetilde{S}$, then x lies outside an $\frac{n-C}{C+1}$ ball about p in $\widetilde{N}$.*

Proof. Since λ^b is a part of λ^h, therefore $r_x(0)$ lies outside $B_n(p)$. By Corollary 5.9, there exists $C > 0$ such that for all $i, j \in \{0, 1, 2, \dots\}$,

$$|i - j| \leq d(r_x(i), r_x(j)) \leq C|i - j|$$

Also, $d(x, p) \geq i$ since $x \in \mu_i^b$. (Here distances are all measured in $\widetilde{N}$ regarded as a tree of spaces with "successive" vertex spaces separated by distance one.) Hence,

$$d(x, p) \geq \min\{i, n - C(i + 1)\}$$
$$\geq \frac{n - C}{C + 1}$$

This proves the result. □

If $x \in B_\lambda$, then $x \in B_\lambda^b$ or $x \in B_\lambda^c$ for some μ. Hence $x \in B_\lambda$ implies that either x lies on some horosphere bounding some $\mathbf{H} \in \mathcal{H}$ or, from Lemma 5.10 above, $d(x, p) \geq \frac{n-C}{C+1}$. Since β_{amb} lies on B_λ, we conclude that β_{amb} is an ambient quasigeodesic in $\widetilde{N}$ such that every point x on β_{amb} either lies on some horosphere bounding some $\mathbf{H} \in \mathcal{H}$ or, from Lemma 5.10 above, $d(x, p) \geq \frac{n-C}{C+1}$.

McMullen [McM01] shows (cf Theorem 5.1) that in $\widetilde{N^h}$, any such ambient quasigeodesic β_{amb} lies in a bounded neighborhood of $\beta^h \cup \mathcal{H}(\beta^h)$. We do not as yet know that β_{amb} does not backtrack, but we can convert it into one without much effort. Let Π denote nearest point projection of $\widetilde{N^h}$ onto $\beta^h \cup \mathcal{H}\,(\beta^h)$. Then $\Pi(\beta_{amb}) = \beta_1$ is again an ambient quasigeodesic in $\widetilde{N}$. Further, β_1 tracks β_{amb} throughout its length, since Π moves points through a uniformly bounded distance. Now β_1 might backtrack, but it can do so in a trivial way, i.e. if β_1 re-enters a horoball after leaving it, it must do so at exactly the point where it leaves it. Removing these 'trivial backtracks', we obtain an **ambient quasigeodesic without backtracking** β which tracks β_{amb} throughout its length.

Note. On the one hand β is an ambient quasigeodesic without backtracking. Hence, it reflects the intersection pattern of β^h with horoballs. On the other hand, it **tracks** β_{amb} whose properties we already know from Corollary 5.9 above.

Since, of β and β^h, one is an ambient quasigeodesic without backtracking, and the other a hyperbolic geodesic joining the same pair of points, we conclude from Theorem 5.2 that they have similar intersection patterns with horoballs, i.e. there exists C_0 such that

- If only of β and β^h penetrates or travels along the boundary of a horoball $\mathbf{H}$, then it can do so for a distance $\leq C_0$.
- If both β and β^h enter (or leave) a horoball $\mathbf{H}$ then their entry (or exit) points are at a distance of at most C_0 from each other.

Again, since β tracks β_{amb}, we conclude that there exists $C > 0$ such that β lies in a C-neighborhood of β_{amb} and hence from Lemma 5.10

- Every point x on β either lies on some horosphere bounding some $\mathbf{H} \in \mathcal{H}$ or, $d(x, p) \geq \frac{n-C}{C+1} - C$.

The above three conditions on β and β^h allow us to deduce the following condition for β^h.

Proposition 5.11. *Let x, λ^h satisfy the hypotheses of Lemma 5.10. Every point x on β^h either lies inside some horoball $\mathbf{H} \in \mathcal{H}$ or, $d(x, p) \geq \frac{n-C}{C+1} - C = m(n)$*

We split β^h into two parts. β^c consists of those points of β^h which lie within horoballs. We set β^b to be the closure of $\beta^h - \beta^c$.

We have denoted $\frac{n-C}{C+1} - C$ by $m(n)$, so that $m(n) \to \infty$ as $n \to \infty$. The above Proposition asserts that the geodesic β^h lies outside large balls about p modulo horoballs. By Lemma 2.1 this is almost enough to guarantee the existence of a Cannon-Thurston map. The rest of the necessary work is given below.

Theorem 5.12. *Suppose S^h is a hyperbolic surface of finite volume. Suppose that N^h is a hyperbolic manifold corresponding to a representation of $\pi_1(S^h)$ without accidental parabolics. Let $i : S^h \to N^h$ be a proper homotopy equivalence.*

Then $\tilde{i} : \widetilde{S^h} \to \widetilde{N^h}$ *extends continuously to the boundary* $\hat{i} : \widehat{S^h} \to \widehat{N^h}$*. If* Λ *denotes the limit set of* $\widetilde{M}$*, then* Λ *is locally connected.*

Proof. Let λ^h be a geodesic segment in $\widetilde{S^h}$ lying outside $B_n(p)$ for some fixed reference point p. Fix neighborhoods of the cusps and lift them to the universal cover. Let $\mathcal{H}$ denote the set of horoballs. Assume without loss of generality that p lies outside horoballs. Let β^h be the hyperbolic geodesic in $\widetilde{N^h}$ joining the endpoints of λ^h. Further, let $\beta^h = \beta^b \cup \beta^c$ as above. Then by Proposition 5.11, β^b lies outside an $m(n)$ ball about p, with $m(n) \to \infty$ as $n \to \infty$.

Next, let $\mathbf{H}_1$ be any one of the equivariant collection of horoballs that β^h meets. By shrinking the cusps in the quotient manifold slightly if necessary, we may assume without loss of generality that none of the equivariant collection of horoballs contain p. The entry and exit points u and v of β^h into and out of $\mathbf{H}_1$ lie outside an $m(n)$ ball about p. Let z be the point on the boundary sphere that $\mathbf{H}_1$ is based at. Then for any sequence $x_i \in \mathbf{H}_1$ with $d(p, x_i) \to \infty$, it follows that $x_i \to z$. Let $\{x_i\}$ and $\{y_i\}$ denote two such sequences on any horoball $\mathbf{H}_1$ not containing p. Then the visual diameter of the set $\{x_i, y_i\}$ must go to zero. Hence, if $[x_i, y_i]$ denotes the geodesic joining x_i, y_i then $d(p, [x_i, y_i]) \to \infty$. Since, u, v lie outside an $m(n)$ ball, there exists some function ψ (independent of $\mathbf{H}_1$), such that the geodesic $[u, v]$ lies outside a $\psi(m(n))$ ball around p, where $\psi(k) \to \infty$ as $k \to \infty$.

Since the choice of this function does not depend on $\mathbf{H}_1$, which is chosen at random, we conclude that there exists such a function for all of β^c. We have thus established:

- β^b lies outside an $m(n)$ ball about p.
- β^c lies outside a $\psi(m(n))$ ball about p.
- $m(n)$ and $\psi(m(n))$ tend to infinity as $n \to \infty$

Define $f(n) = min(m(n), \psi(m(n)))$. Then β^h lies outside an $f(n)$ ball about p and $f(n) \to \infty$ as $n \to \infty$.

By Lemma 2.1 $i : \widetilde{S^h} \to \widetilde{N^h}$ extends continuously to the boundary $\hat{i} : \widehat{S^h} \to \widehat{N^h}$. This proves the first statement of the theorem.

Now, the limit set of $\widetilde{S^h}$ is the circle at infinity, which is locally connected. Further, the continuous image of a compact locally connected set is locally connected [HY61]. Hence, if Λ denotes the limit set of $\widetilde{N^h}$, then Λ is locally connected. This proves the theorem. □

References

[ABC+91] J. Alonso, T. Brady, D. Cooper, V. Ferlini, M. Lustig, M. Mihalik, M. Shapiro and H. Short, Notes on word hyperbolic groups, Group Theory from a Geometrical Viewpoint (E. Ghys, A. Haefliger, A. Verjovsky eds.) (1991) 3–63.

[BF92] M. Bestvina and M. Feighn, A Combination theorem for Negatively Curved Groups, *J. Diff. Geom.*, **35** (1992) 85–101.

[BH99] M. Bridson and A. Haefliger, Metric spaces of non-positive curvature, Springer (1999).

[Bon86] F. Bonahon, Bouts de varietes hyperboliques de dimension 3, *Ann. Math.*, **124** (1986) 71–158.

[Bow97] B. H. Bowditch, Relatively hyperbolic groups, preprint, Southampton (1997).

[Bow02a] B. H. Bowditch, The Cannon-Thurston map for punctured surface groups, preprint, Southampton (2002).

[Bow02b] B. H. Bowditch, Stacks of hyperbolic spaces and ends of 3 manifolds, preprint, Southampton (2002).

[CDA90] M. Coornaert, T. Delzant and A. Papadopoulos, Geometrie et theorie des groupes, Lecture Notes in Math., Vol. 1441, Springer Verlag (1990).

[CT85] J. Cannon and W. P. Thurston, Group Invariant Peano Curves, preprint, Princeton (1985).

[Far98] B. Farb, Relatively hyperbolic groups, *Geom. Funct. Anal.*, **8** (1998) 810–840.

[GdlH90] E. Ghys and P. de la Harpe (eds.), Sur les groupes hyperboliques d'apres Mikhael Gromov, Progress in Math., Vol. 83, Birkhauser, Boston Ma. (1990).

[Gro85] M. Gromov, Hyperbolic Groups, in Essays in Group Theory, ed. Gersten, MSRI Publ., Vol. 8, Springer Verlag (1985) 75–263.

[HY61] J. G. Hocking and G. S. Young, Topology, Addison Wesley (1961).

[McM01] C. McMullen, Local connectivity, Kleinian groups and geodesics on the blow-up of the torus, *Invent. math.*, **97** (2001) 95–127.

[Min92] Y. N. Minsky, Teichmüller Geodesics and Ends of 3-Manifolds, *Topology* (1992) 1–25.

[Min94] Y. N. Minsky, On Rigidity, Limit Sets, and End Invariants of Hyperbolic 3-Manifolds, *J. A.M.S.*, **7** (1994) 539–588.

[Mit98a] Mahan Mitra, Cannon-Thurston Maps for Hyperbolic Group Extensions, *Topology*, **37** (1998) 527–538.

[Mit98b] Mahan Mitra, Cannon-Thurston Maps for Trees of Hyperbolic Metric Spaces, *Jour. Diff. Geom.*, **48** (1998) 135–164.

[Mj05a] Mahan Mj, Cannon-Thurston Maps for Closed Surface Groups I and II, preprint arXiv:math/0512539, arXiv:math/0607509.

[Mj05b] Mahan Mj, Cannon-Thurston Maps for Pared Manifolds of Bounded Geometry, *Geometry & Topology*, **13** (2009) 189–245.

[Mj05c] Mahan Mj, Cannon-Thurston Maps, i-bounded Geometry and a Theorem of McMullen, preprint, arXiv:math.GT/0511041 (2005).

[Thu80] W. P. Thurston, The Geometry and Topology of 3-Manifolds, Princeton University Notes (1980).

Teichmüller Theory and Moduli Problem No. 10, 2010, pp. 513–517.

Computing a Generating Set of Arithmetic Kleinian Groups

Gregory Muller

Department of Mathematics, Cornell University, Ithaca, NY, USA
e-mail: gpm23@cornell.edu; morilac@gmail.com

Abstract. The goal of this paper is to demonstrate the use of techniques from hyperbolic geometry to compute generating sets of certain subgroups of $SL^+(2,\mathbb{C})$; specifically, $SO^+(Q,\mathbb{Z})$ for Q some integral quadratic form of signature (3, 1) that does not represent 0.

Introduction

Hyperbolic space $\mathbb{H}^3$ can be thought of as the subset of $\mathbb{R}^4$ that satisfies the equation $Q_1(\mathbf{x}) = x_1^2 + x_2^2 + x_3^2 - x_4^2 = -1$ and the condition $x_4 > 0$, together with the Riemannian metric induced by Q_1. The group of isometries of $\mathbb{H}^3$ is the space of 4 by 4 real matrices A such that $A^t Q_1 A = Q_1$ and $e_4 \cdot Ae_4 > 0$, where e_4 is the unit vector in the x_4 direction.[1] It is denoted $SO^+(Q_1,\mathbb{R})$. This group contains the discrete subgroup $SO^+(Q_1,\mathbb{Z})$ of integer valued matrices.

Now consider a different quadratic form on $\mathbb{R}^4$; we will be particularly interested in

$$Q_7(x_1,x_2,x_3,x_4) = x_1^2 + x_2^2 + x_3^2 - 7x_4^2.$$

Over $\mathbb{R}$, Q_7 and Q_1 are isomorphic; that is, there exists an isometry $M:\mathbb{R}^4 \to \mathbb{R}^4$ such that $M^t Q_7 M = Q_1$. Therefore, $SO^+(Q_1,\mathbb{R})$ is isomorphic to $SO^+(Q_7,\mathbb{R})$. However, there is no reason for there to be an isomorphism between $SO^+(Q_7,\mathbb{Z})$ and $SO^+(Q_1,\mathbb{Z})$. What is true is that $SO^+(Q_7,\mathbb{Z})$ does act on $\mathbb{H}^3$ as a subgroup of $SO^+(Q_7,\mathbb{R}) = SO^+(Q_1,\mathbb{R})$.

All of the above is true for any integral quadratic form of signature (3, 1). Why, then, focus on the form Q_7? The following theorem sheds some light on this.

Theorem 1 ([4]). *$Q(\mathbf{x}) = 0$ has a non-zero integral solution if and only if $\mathbb{H}^3/SO^+(Q,\mathbb{Z})$ is not compact.*

Q_7 does not have such a solution. To see this, assume it does, and furthermore, assume that the coordinates of $\mathbf{x}$ have no common factor (no loss of generality because a common factor can just be pulled out). Then consider the above equation mod 8.

$$0 \equiv x_1^2 + x_2^2 + x_3^2 - 7x_4^2 \equiv x_1^2 + x_2^2 + x_3^2 + x_4^2 \bmod 8$$

[1]The expression $A^t Q_1 A = Q_1$ is using the fact that any quadratic form Q can be written as a symmetric matrix $\tilde{Q}$ such that $Q(\mathbf{x}) = \mathbf{x}^t \tilde{Q}\mathbf{x}$. Throughout this paper, the same notation will be used for a quadratic form and its associated matrix.

However, mod 8, there are only three possible squares: 0, 1 and 4. In the above equation, none of the summands can be 1, because then the rest would have to also be 1 to satisfy the equation mod 4, which is not a solution mod 8. Therefore, all the summands are either 0 or 4, meaning x_1, x_2, x_3 and x_4 are all even, contradicting the assumption of having no common factor.

Thus, the quotient orbifold $\mathbb{H}^3/SO^+(Q_7, \mathbb{Z})$ is compact, and the goal now is to use this to understand the algebraic structure of $SO^+(Q_7, \mathbb{Z})$.

It is important to note that the techniques used here to compute a generating set for this group can be used to find similar generating sets for $SO^+(Q_n, \mathbb{Z})$, with $Q_n = x_1^2 + x_2^2 + x_3^2 - nx_4^2$ and $Q_n(\hat{x}) = 0$ having no non-zero solutions[2]. The basic algorithm will, in fact, work for any integral quadratic form of signature (3, 1) which yields a compact quotient orbifold, although the details will require some work.

Dirichlet fundamental domains

To further understand $\mathbb{H}^3/SO^+(Q_7, \mathbb{Z})$, the next step will be to create a fundamental domain for it. The fundamental domain of choice will be a Dirichlet fundamental domain.

Constructing a Dirichlet fundamental domain D for the action of a group G begins with choosing a point x in hyperbolic space. Then, determine two open sets, U and V.

Let U be a fundamental domain for the stabilizer of x. In the disc model, if the point x is the central point, then the stabilizer consists entirely of rotations, and so the fundamental domain U can be taken to be a cone.

Let V be the set of points in hyperbolic space that are closer to x than any other point in Gx, the orbit of x. This is often called the 'Voronoi cell' of x. In practice, Voronoi cells can be constructed in the following way. Let $A \in G$, and let H_A denote the half-space consisting of points that are closer to x than to Ax. Then, the Voronoi cell is the intersection of H_A over all $A \in G$.

Then, the Dirichlet fundamental domain D is $U \cap V$.

When the quotient orbifold is compact, a Dirichlet fundamental domain can be computed algorithmically.

Step 1. Choose a fundamental domain U for the stabilizer of x.

Step 2. Produce G_L, the set of elements of G that move the point x less than some distance L.

Step 3. Let V_L be the intersection of H_A over all $A \in G_L$. If V_L is compact, then there is some distance M such that no point in V_L is more than M away from x.

Step 4. If $M < L/2$, then V_L is the same as V, the Voronoi cell of all of G, because $V_L \subset H_A$ for all A that move x more that $2M$. If $M \geq L/2$ or V_L is not compact, increase L and do it again.

This will work eventually, because it must work when L is greater than the diameter of the orbifold.

[2]The proof that Q_7 was such a form also proves that Q_n will also work, for any $n = 8m - 1$, for $m > 0$.

How does this look for $\mathbb{H}^3/SO^+(Q_7,\mathbb{Z})$? The point x will be $(0,0,0,7^{-1/2})$, and its stabilizer consists of $SO^+(3,\mathbb{Z})$ acting on the first 3 coordinates. It is a group of order 24, and consists of compositions of permutations and reflections along axes which are orientation preserving. Fortunately, their action on the disc model with x in the center is exactly the same as their action on the first 3 coordinates of the hyperboloid model, and so creating a fundamental domain is easy. We will take the fundamental domain U to be the set of points in the disc which have positive Euclidean inner-product with the following three vectors:

$$v_1 = (1,-1,0,0); \quad v_1 = (0,1,-1,0); \quad v_2 = (0,0,1,0)$$

Generating elements of $SO^+(Q_7,\mathbb{Z})$. The hard part about producing the rest of the fundamental domain is producing all elements of $SO^+(Q_7,\mathbb{Z})$ which move x less than some fixed distance. To do this, several facts about elements of $SO^+(Q_7,\mathbb{Z})$ will be helpful. Let A be an element of $SO^+(Q_7,\mathbb{Z})$, and let a_{ij} denote the ijth coefficient of the matrix A.

Fact 1. The first three columns of the matrix A have pseudolength[3] 1, the last column has pseudolength -7, and they are all pairwise orthogonal. Furthermore, any integral matrix that satisfies these properties is automatically in $SO(Q_7,\mathbb{Z})$.

Fact 2. Let x be an element of $\mathbb{H}^3$. The distance in $\mathbb{H}^3$ from x to Ax is $\cosh(a_{44})$, where a_{44} is the bottom right entry in the matrix.

Fact 3. The entries a_{14}, a_{24} and a_{34}, the right column except for the bottom right entry, are all divisible by 7.

Fact 4. The entry a_{44} is either 1 more or 1 less than a multiple of 7.

Fact 5. The absolute values of a_{41}, a_{42} and a_{43}, the bottom entries save the bottom right entry, are all bounded above by $a_{44}/\sqrt{7}$.

Proof.

1) This is just a different way of saying $A^t Q_7 A = Q_7$.
3) A satisfies the equation $A^t Q_7 A = Q_7$. $(Q_7)^{-1} A^t Q_7 A = 1$, and so $A^{-1} = (Q_7)^{-1} A^t Q_7$. Given the simple nature of Q_7 written as a symmetric matrix, this has a simple form:

$$A^{-1} = \begin{pmatrix} a_{11} & a_{21} & a_{31} & -7a_{41} \\ a_{12} & a_{22} & a_{32} & -7a_{42} \\ a_{13} & a_{23} & a_{33} & -7a_{43} \\ -a_{14}/7 & -a_{24}/7 & -a_{34}/7 & a_{44} \end{pmatrix}$$

Since this matrix also consists of integers, the specified entries are divisible by 7.
4) The right column has pseudolength -7, which means $a_{14}^2+a_{24}^2+a_{34}^2-7a_{44}^2 = -7$. Using their divisiblity by 7, let $a_{14} = 7\alpha$, $a_{24} = 7\beta$ and $a_{34} = 7\gamma$. Then $49(\alpha^2 + \beta^2 + \gamma^2) - 7a_{44}^2 = -7$, and so $7(\alpha^2+\beta^2+\gamma^2) = a_{44}^2 - 1 = (a_{44}-1)(a_{44}+1)$. 7 must divide $a_{44} - 1$ or $a_{44} + 1$, proving the statement.

[3] The term 'pseudolength' of a vector $\mathbf{x}$ in $\mathbb{R}^4$ just means $Q_7(\mathbf{x})$. The term emphasizes the idea of Q_7 giving $\mathbb{R}^4$ a pseudometric.

5) Since A is the inverse of some other matrix in $SO^+(q,\mathbb{Z})$, its entries satisfy the equation $7(a_{41}^2 + a_{42}^2 + a_{43}^2) = a_{44}^2 - 1$. Therefore, $\max(|a_{41}|, |a_{42}|, |a_{43}|) \leq a_{44}/\sqrt{7}$. □

The algorithm for generating such elements. With these facts established, elements of $SO^+(Q_7,\mathbb{Z})$ which move x less than some fixed distance $\cosh(L)$ can be generated.

Step 1. Generate a list of pseudolength 1 vectors whose fourth component is bounded above by $L/\sqrt{7}$. This list is finite since the sum of the squares of the first three coordinates is bounded above by $L^2 + 1$.

Step 2. For every $0 \leq n := (7i \pm 1) \leq L$ (for i an integer) find all the solutions of $(7a)^2+(7b)^2+(7c)^2-7n^2 = -7$, and correspondingly, all the possible right columns of elements of $SO^+(Q_7,\mathbb{Z})$. Again, there are only a finite number of possibilities.

Step 3. For each possible right column found in Step 2, compare it to every vector on the list from Step 1, to find the vectors which are orthogonal to the chosen right column. Of the vectors which are orthogonal, compare them to each other and find which pairs are orthogonal to each other. Then, any triple of vectors which are orthogonal to the chosen right column and each other constitutes an element of $SO^+(Q_7,\mathbb{Z})$.

Once the elements have been generated, the half planes they generate must be drawn, and compactness and maximum distance must be checked.

The generators

So what does this algorithm yield? First, in the interest of coming up with a generating set for $SO^+(Q_7,\mathbb{Z})$, we need a generating set of the stabilizer of the distinguished point. As was mentioned before, this group is $SO^+(3,\mathbb{Z})$, which is known to be isomorphic to S_4.

To see this, notice that this group is the symmetry group of the cube, and that a symmetry of the cube is determined by its permutation of the diagonals of the cube. There are four diagonals in a cube, and every permutation of the diagonals is possible. Generators for this group as a subgroup of $SO^+(Q_7,\mathbb{Z})$ are:

$$(12) = \begin{pmatrix} 0 & 1 & 0 & 0 \\ 1 & 0 & 0 & 0 \\ 0 & 0 & -1 & 0 \\ 0 & 0 & 0 & 1 \end{pmatrix}, \quad (1234) = \begin{pmatrix} 0 & 0 & 1 & 0 \\ 0 & 1 & 0 & 0 \\ -1 & 0 & 0 & 0 \\ 0 & 0 & 0 & 1 \end{pmatrix}$$

The above algorithm yields the following as a sufficient set of generators. There is some ambiguity in which matrices to choose, up to multiplication by an element of the stabilizer. These particularly matrices have been chosen so that they move the Dirichlet fundamental domain to an adjacent region.

$$A = \begin{pmatrix} -2 & -2 & 0 & 7 \\ -5 & -2 & 0 & 14 \\ 0 & 0 & -1 & 0 \\ -2 & -1 & 0 & 6 \end{pmatrix} \quad A^{-1} = \begin{pmatrix} -2 & -5 & 0 & 14 \\ -2 & -2 & 0 & 7 \\ 0 & 0 & -1 & 0 \\ -1 & -2 & 0 & 6 \end{pmatrix}$$

$$B = \begin{pmatrix} 0 & -1 & 0 & 0 \\ -8 & 0 & 0 & 21 \\ 0 & 0 & -1 & 0 \\ -3 & 0 & 0 & 8 \end{pmatrix} \quad C = \begin{pmatrix} -4 & -3 & -2 & 14 \\ -3 & -4 & -2 & 14 \\ -2 & -2 & 0 & 7 \\ -2 & -2 & -1 & 8 \end{pmatrix} \quad B^{-1} = \begin{pmatrix} 0 & -8 & 0 & 21 \\ -1 & 0 & 0 & 0 \\ 0 & 0 & -1 & 0 \\ 0 & -3 & 0 & 8 \end{pmatrix}$$

$$D = \begin{pmatrix} -4 & -3 & -2 & 14 \\ -9 & -4 & -4 & 28 \\ -4 & -2 & -3 & 14 \\ -4 & -2 & -2 & 13 \end{pmatrix} \quad D^{-1} = \begin{pmatrix} -4 & -9 & -4 & 28 \\ -3 & -4 & -2 & 14 \\ -2 & -4 & -3 & 14 \\ -2 & -4 & -2 & 13 \end{pmatrix}$$

$$E = \begin{pmatrix} -7 & -8 & 0 & 28 \\ -8 & -7 & 0 & 28 \\ 0 & 0 & -1 & 0 \\ -4 & -4 & 0 & 15 \end{pmatrix}$$

$$F = \begin{pmatrix} -10 & -12 & -10 & 49 \\ -2 & -4 & -3 & 14 \\ -3 & -4 & -2 & 14 \\ -4 & -5 & -4 & 20 \end{pmatrix} \quad F^{-1} = \begin{pmatrix} -10 & -2 & -3 & 28 \\ -12 & -4 & -4 & 35 \\ -10 & -3 & -2 & 28 \\ -7 & -2 & -2 & 20 \end{pmatrix}$$

Notice that a left-right reflection of this list sends matrices to their inverses.

Which of these matrices are necessary to generate $SO^+(Q_7, \mathbb{Z})$? The following relations demonstrate that only A, C and the stabilizer group are necessary.

$$B = (1234)^2(12)A^{-2}(12)$$

$$D = (1234)^3(12)C(1234)^3$$

$$E = A^{-1}(1234)^2(12)A^{-1}(12)$$

$$F = C(12)(1234)^2A^{-1}(12)$$

Therefore, the four matrices (12), (1234), A and C are enough to generate the entire group $SO^+(Q_7, \mathbb{Z})$.

References

[1] Hubbard and H. John, Teichmüller Theory, Vol. 1 (Ithaca, NY: Matrix Editions, 2006).
[2] Maclachlan, Colin and Reid, Alan The Arithmetic of Hyperbolic 3-Manifolds (New York, Springer, 2002).
[3] T. Y. Lam, Introduction to Quadratic Forms over Fields (New York, Springer, 2004).
[4] Borel and Armand Introduction aux Groupes Arithetiques (Paris, Hermann, 1969).

Teichmüller Theory and Moduli Problem No. 10, 2010, pp. 519–528.

On the Spectrum of Asymptotic Slopes

A. J. Parameswaran and S. Subramanian

School of Mathematics, Tata Institute of Fundamental Research, Homi Bhabha Road, Mumbai 400 005, India
e-mail: param@math.tifr.res.in; subramnn@math.tifr.res.in

Abstract. The slopes of maximal subbundles of rank s divided by the degree of the map under various pull backs form a bounded collection of numbers called the s-spectrum of the bundle. We study the supremum of the s-spectrum and determine it in terms of the Harder Narasimhan filtration of the bundle.

1. Introduction

Line subbundles of maximal degree in a rank 2 bundle on a curve have been studied in [LN] and many other subsequent papers. In [BP] it was shown that a vector bundle is strongly semistable if and only if the slope of the maximal subbundle of a given rank in a finite pull back is bounded by the slope of the pull back bundle.

In this note we study the behaviour of maximal subbundles of vector bundles on curves after finite pull backs. These slopes of maximal subbundles of rank s divided by the degree of the map under various pull backs form a bounded collection of numbers called the s-spectrum of the bundle. We study the supremum of the s-spectrum and determine it in terms of the Harder Narasimhan filtration of the bundle. We also give a criterion for a spectrum value to be isolated.

2. Preliminaries

Let k be an algebraically closed field of arbitrary characteristic. Let C be an irreducible smooth projective curve over k of genus g. For a vector bundle V over C we will denote by $\mathbb{G}r(s, V)$ the Grassmann bundle over C defined by the space of all s dimensional quotients of V. The tautological line bundle over $\mathbb{G}r(s, V)$ will be denoted by $\mathcal{O}_{\mathbb{G}r(s,V)}(1)$. This is the determinant of the universal quotient bundle on $\mathbb{G}r(s, V)$. Tensor powers of this tautological line bundle are denoted by $\mathcal{O}_{\mathbb{G}r(s,V)}(n)$. The slope of V is defined as

$$\mu(V) := \frac{\deg(V)}{\mathrm{rank}(V)}$$

V is called *semistable* if for all subbundles $W \subset V$ of positive rank $\mu(W) \leq \mu(V)$. V is called *strongly semistable* if the iterated Frobenius pull back, $F_C^{n*}(V)$, is semistable for all $n > 0$, where $F_C : C \to C$ denotes the Frobenius morphism. If V is a strongly semistable vector bundle of rank r, then for any representation of GL_r into GL_n, the induced rank n vector bundle is strongly semistable (cf. [RR]).

Given a vector bundle V, there is a unique filtration (the Harder Narasimhan filtration) $V_\bullet := \{0 = V_0 \subset \cdots \subset V_l = V\}$ such that each V_i/V_{i-1} is semistable and the slopes of successive quotients are strictly decreasing, i.e., $\mu_i := \mu(V_i/V_{i-1}) > \mu_{i+1} := \mu(V_{i+1}/V_i)$. The subbundles V_i are defined inductively as the inverse image of the maximal subbundle of maximal slope in V/V_{i-1}. We call the bundles V_i/V_{i-1}, the *Harder Narasimhan factors* of the bundle V. The bundle V_1 is called the *maximal subbundle* of V and denoted by $V_{\max}$. The slope $\mu(V_1) = \mu(V_{\max})$ is called the maximal slope of V and denoted by $\mu_{\max}(V)$. We recall the following result whose proof we will omit.

Lemma 2.1. *If V and W are semistable vector bundles over a smooth curve C, with $\mu(V) > \mu(W)$, then $Hom_C(V, W) = H^0(C, V^* \otimes W) = 0$.*

Now we have the following result which gives enough complete intersection curves in the projective bundle.

Lemma 2.2. *Let V be a strongly semistable vector bundle of rank r over C and L a line bundle of degree $> 2g - ns\mu(V)$ with $n > 0$. Then $\mathcal{O}_{\mathbb{G}r(s,V)}(n) \otimes \pi^*L$ separates points on $\mathbb{G}r(s, V)$.*

Proof. Since $\pi : \mathbb{G}r(s, V) \to C$ is a smooth fibration with Grassmann varieties as fibres and higher cohomologies of ample bundles vanish on Grassmannians, it follows that

$$H^1(\mathbb{G}r(s, V), \mathcal{O}_{\mathbb{G}r(s,V)}(n) \otimes \pi^*L) \cong H^1(C, \pi_*(\mathcal{O}_{\mathbb{G}r(s,V)}(n) \otimes \pi^*L))$$
$$\cong H^1(C, V_{s,n} \otimes L)$$

where $V_{s,n}$ is the vector bundle associated to V by the Weyl module with highest weight $n\omega_s$ with ω_s as the fundamental weight corresponding to the s-th exterior power representation. Notice that the slope of $\wedge^s V$ is equal to $s\mu(V)$.

Hence for any two points $x, y \in C$, Serre duality implies

$$H^1(C, V_{s,n} \otimes L(-x-y)) = H^0(C, V_{s,n}^* \otimes L^{-1}(x+y) \otimes K_C)^*$$

To show this cohomology group vanishes, it suffices to show the semistable bundle $V_{s,n}^* \otimes L^{-1}(x+y) \otimes K_C$ has negative slope. Now we have:

$$\mu(V_{s,n}^* \otimes L^{-1}(x+y) \otimes K_C) = \mu(V_{s,n}^*) - \deg L + 2g = -ns\mu(V) - \deg L + 2g$$

which is negative if and only if $\deg L > 2g - ns\mu(V)$. Hence for two distinct points $x, y \in C$,

$$H^0(C, V_{s,n} \otimes L) \to H^0(C, V_{s,n}|_{x+y})$$

is surjective whenever $\deg L > 2g - ns\mu(V)$. This implies $\mathcal{O}_{\mathbb{G}r(s,V)}(n) \otimes \pi^*L$ separate points and surjects onto sections of $\mathcal{O}_{\mathbb{G}r(V_x)}(n)$ on the Grassmannian $\mathbb{G}r(s, V_x)$. □

Consider the universal exact sequence on $\mathbb{G}r(s, V)$:

$$0 \to \mathcal{S} \to \pi^*V \to \mathcal{Q} \to 0 \tag{2.1}$$

Then $\det \mathcal{Q} \cong \mathcal{O}_{\mathbb{G}r(s,V)}(1)$ is the tautological line bundle. For any line bundle on $\mathbb{G}r(s, V)$ let [-] denote the corresponding cycle class. The Chow group of 0 cycles on $\mathbb{G}r(s, V)$ is canonically isomorphic to the Chow group of 0 cycles of the curve C (cf. [F], Prop. 14.6.5,). Then we have the following:

Lemma 2.3. *Given any line bundle L on C and any fibre F of $\pi : \mathbb{G}r(s, V) \to C$ we have*

$$[\mathcal{O}_{\mathbb{G}r(s,V)}(1)]^{s(r-s)+1} = (s(r-s)+1)s\mu(V)([\mathcal{O}_{\mathbb{G}r(s,V)}(1)]^{s(r-s)} \cdot F) \quad (2.2)$$

$$[\mathcal{O}_{\mathbb{G}r(s,V)}(1)]^{s(r-s)} \cdot [\pi^*L] = \mathrm{degL}([\mathcal{O}_{\mathbb{G}r(s,V)}(1)]^{s(r-s)} \cdot F) \quad \text{and}$$

$$\times [\pi^*L] \cdot [\pi^*L] = 0 \quad (2.3)$$

Proof. If we pull back the bundle under a finite map, both sides of the formulae get multiplied by the degree of the map. Hence we may assume, after a finite pull back if necessary, that there exists a line bundle L such that $L^{\otimes r} \cong \det V$., i.e., det V has an rth root. Then $\oplus_1^r L \cong \mathcal{O}_C^{\oplus r} \otimes L$. Let $\mathcal{L}^{-1}$ be a very ample line bundle on C such that $V \otimes \mathcal{L}^{-1}$ and $L \otimes \mathcal{L}^{-1}$ are globally generated and have vanishing first cohomology.

Let $Quot_P(\mathcal{O}_C^{\oplus N} \otimes \mathcal{L})$ denote the quot scheme over C of quotient of fixed Hilbert polynomial P (the degree and rank determine the polynomial) of the trivial vector bundle of rank $N \gg 0$ twisted by the line bundle $\mathcal{L}$. Let $U \subset Quot_P(\mathcal{O}_C^{\oplus N} \otimes \mathcal{L})$ be the open set where the universal quotient sheaf is locally free and $H^0(C, V \otimes \mathcal{L}^{-1}) \cong H^0(C, \mathcal{O}_C^{\oplus N})$. Then by [Ne] (Remark 5.5, page 140) U is smooth and irreducible and hence connected. Then $\oplus_1^r L \cong \mathcal{O}_C^{\oplus r} \otimes L$ and V belong to the same quot scheme, in fact U.

Let $\mathcal{V}$ be the universal bundle over $C \times U$ and $\mathbb{G}r(s, \mathcal{V})$ be the corresponding Grassmannian bundle over $C \times U$. For each vector bundle W representing a closed point $[W] \in U$, the restriction $\mathcal{V}|_{C \times [W]}$ is canonically isomorphic to W. Hence the restriction of the Grassmannian bundle $\mathbb{G}r(s, \mathcal{V})$ is canonically $\mathbb{G}r(s, W)$. Since the isomorphism of $CH_0(\mathbb{G}r(s, V))$ with $CH_0(C)$ is canonical, it suffices to check the formula for any closed point (vector bundle) of the open subset U of the quot scheme $Quots_P(\mathcal{O}_C^{\oplus N} \otimes \mathcal{L})$.

Hence we obtain,

$$\begin{aligned}
[\mathcal{O}_{\mathbb{G}r(s,V)}(1)]^{s(r-s)+1} &\cong [\mathcal{O}_{\mathbb{G}r(s,\oplus L)}(1)]^{s(r-s)+1} \cong [\mathcal{O}_{\mathbb{G}r(s,\oplus \mathcal{O}_C \otimes L)}(1)]^{s(r-s)+1} \\
&\cong \{[\mathcal{O}_{\mathbb{G}r(s,\oplus \mathcal{O}_C)}(1)] + [L]^{\otimes s}\}^{s(r-s)+1} \\
&= (s(r-s)+1)s\deg L(\mathcal{O}_{\mathbb{G}r(s,\oplus \mathcal{O}_C)}(1).F) \\
&= (s(r-s)+1)s\mu(V)(\mathcal{O}_{\mathbb{G}r(s,\oplus L)}(1).F) \\
&= (s(r-s)+1)s\mu(V)(\mathcal{O}_{\mathbb{G}r(s,V)}(1).F) \qquad \square
\end{aligned}$$

Choose constants $0 < \epsilon_n \leq 1$ and line bundles L_n on C such that $\deg L_n = 2g - ns\mu(V) + \epsilon_n$. Then $\mathcal{O}_{\mathbb{G}r(s,V)}(n) \otimes \pi^*L_n$ separates points by Lemma 2.2 and hence defines enough smooth complete intersection curves in $\mathbb{G}r(s, V)$ by Bertini's theorem. Now we have the following result.

Lemma 2.4. *Let D be an irreducible complete intersection curve in $\mathbb{G}r(s, V)$ defined by sections of $\mathcal{O}_{\mathbb{G}r(s,V)}(n) \otimes \pi^* L_n$. Then*

$$\mu(\mathcal{S}|_D) = n^{s(r-s)}([\mathcal{O}_{\mathbb{G}r(s,V)}(1)]^{s(r-s)} \cdot F)\left(\mu(V) - \frac{s(2g+\epsilon_n)}{n}\right)$$

Proof. Since D is a complete intersection, the degree of $\mathcal{O}_{\mathbb{G}r(s,V)}(1)$ on D can be calculated as the cup product (denoted by $\cdot$ in the Chow ring) of the cycle classes of the corresponding divisors (line bundles) with the class of $\mathcal{O}_{\mathbb{G}r(s,V)}(1)$. Note that $[\mathcal{O}_{\mathbb{G}r(s,V)}(n)] = n[\mathcal{O}_{\mathbb{G}r(s,V)}(1)]$ as the tensor product of line bundles gives the sum of the corresponding classes. Hence we can interpret deg $\mathcal{O}_{\mathbb{G}r(s,V)}(1)|_D$ as

$$\begin{aligned}
\deg \mathcal{O}_{\mathbb{G}r(s,V)}(1)|_D &= [D] \cdot [\mathcal{O}_{\mathbb{G}r(s,V)}(1)] \\
&= ([\mathcal{O}_{\mathbb{G}r(s,V)}(n)] + [\pi^* L_n])^{s(r-s)} \cdot [\mathcal{O}_{\mathbb{G}r(s,V)}(1)] \\
&= \{[\mathcal{O}_{\mathbb{G}r(s,V)}(n)]^{s(r-s)} \\
&\quad + s(r-s)[\mathcal{O}_{\mathbb{G}r(s,V)}(n)]^{s(r-s)-1} \cdot [\pi^* L_n]\} \cdot [\mathcal{O}_{\mathbb{G}r(s,V)}(1)] \\
&= n^{s(r-s)}[\mathcal{O}_{\mathbb{G}r(s,V)}(1)]^{s(r-s)+1} \\
&\quad + s(r-s)n^{s(r-s)-1}[\mathcal{O}_{\mathbb{G}r(s,V)}(1)]^{s(r-s)} \cdot [\pi^* L_n]
\end{aligned}$$

Now the degree of the universal subbundle on D,

$$\begin{aligned}
\deg \mathcal{S}|_D &= \deg \pi^* V|_D \quad - \quad \deg \mathcal{Q}|_D \\
&= n^{s(r-s)}([\mathcal{O}_{\mathbb{G}r(s,V)}(1)]^{s(r-s)} \cdot F) \deg V \\
&\quad - n^{s(r-s)}(s(r-s)+1)s\mu(V)([\mathcal{O}_{\mathbb{G}r(s,V)}(1)]^{s(r-s)} \cdot F) \\
&\quad - n^{s(r-s)-1}s(r-s)\deg L_n \, ([\mathcal{O}_{\mathbb{G}r(s,V)}(1)]^{s(r-s)} \cdot F)
\end{aligned}$$

Now by substituting for deg L_n in the above and simplifying this expression we get

$$\mu(\mathcal{S}|_D) = n^{s(r-s)}([\mathcal{O}_{\mathbb{G}r(s,V)}(1)]^{s(r-s)} \cdot F)\left(\mu(V) - \frac{s(2g+\epsilon_n)}{n}\right) \qquad \square$$

Note that the degree of $\pi_D : D \to C$ is equal to cardinality of a general fibre of π_D over x which equals $[\mathcal{O}_{\mathbb{G}r(s,V_x)}(n)]^{s(r-s)} \cdot F = n^{s(r-s)}([\mathcal{O}_{\mathbb{G}r(s,V)}(1)]^{s(r-s)} \cdot F)$.

3. Genuinely ramified maps

Let us begin with the following definition.

Definition 3.1. *Let $f : D \to C$ be a finite morphism of integral curves. Then f is said to be* genuinely ramified *if f is separable and does not factor through an étale cover of C.*

Lemma 3.2. *A separable morphism $f : D \to C$ is genuinely ramified if and only if $(f_* \mathcal{O}_D)_{\max} \cong \mathcal{O}_C$.*

Proof. By projection formula we have

$$H^0(D, f^*W) = H^0(C, W \otimes f_* \mathcal{O}_D)$$

Hence for a semistable bundle S on C of positive slope

$$Hom(S, f_* \mathcal{O}_D) = Hom(f^*S, \mathcal{O}_D) = 0$$

because f^*S remains semistable of positive slope as f is separable. This shows that $\mu_{\max} f_* \mathcal{O}_D = 0$.

If rank $(f_* \mathcal{O}_D)_{\max} > 1$, then it forms a sheaf of subalgebras of $f_* \mathcal{O}_D$ on C. Hence by taking the spectrum of $f_* \mathcal{O}_{D \max}$ we obtain an étale cover of C factoring $f : D \to C$. Hence f is not genuinely ramified. □

In [N], Madhav Nori has constructed the fundamental group scheme $\pi(X)$ of any complete variety X. This is constructed as a Tannaka category whose objects are essentially finite vector bundles on X. Further he has also constructed a principal $\pi(X)$ bundle $\tilde{X} \to X$. Let $\pi_1^{\mathrm{alg}}(X)$ denote the étale fundamental group of a complete scheme X. Then one can show that $\pi_1^{\mathrm{alg}}(X)$ is a quotient of $\pi(X)$ whose objects in the Tannaka category are the vector bundles that are trivial on a finite étale cover of X. This quotient morphism induces a $\pi_1^{\mathrm{alg}}(X)$-bundle $\mathcal{X} \to X$ over X.

Lemma 3.3. *A separable morphism of curves $D \to C$ is genuinely ramified if and only if the induced map $\pi_1^{\mathrm{alg}}(D) \to \pi_1^{\mathrm{alg}}(C)$ is an epimorphism.*

Proof. Clearly an epimorphism on the fundamental group implies genuine ramification. This follows from the fact that for any finite étale morphism $f : D \to C$ of degree d the index of the image of the étale fundamental group is equal to d.

Any finite morphism $f : D \to C$ induces the map on étale fundamental groups whose image $f_*(\pi_1^{\mathrm{alg}}(D)) = \Gamma$ is of finite index. Define $\tilde{D} := \mathcal{C}/\Gamma$. Then the induced map $\tilde{D} \to C$ is étale such that $f : D \to C$ factors through this. This proves the converse. □

Lemma 3.4. *If D is a general complete intersection curve in $\mathbb{G}r(s, V)$ defined by an ample line bundle, then the induced projection $D \to C$ is genuinely ramified*

Proof. Notice that the algebraic fundamental group of $\mathbb{G}r(s, V)$ is naturally isomorphic to the algebraic fundamental group of C as Grassmannians are algebraically simply connected. If D is a complete intersection curve in $\mathbb{G}r(s, V)$, then the algebraic fundamental group of D surject onto the algebraic fundamental group of $\mathbb{G}r(s, V)$ by the algebraic analogue of Lefshetz Theorem. Hence $\pi_1^{\mathrm{alg}}(D) \to \pi_1^{\mathrm{alg}}(C)$ is surjective. Now the result follows from Lemma 3.3. □

Lemma 3.5. *Let $f : D \to C$ be genuinely ramified morphism of smooth projective curves. Then*

(a) *If V and W are two semistable bundles on C of same slope, then*

$$Hom_C(V, W) \cong Hom_D(f^*V, f^*W)$$

(b) *If V is a stable bundle on C, then f^*V is stable on D.*
(c) *If V is a semistable bundle on C and $F \subset f^*V$ is a subbundle of same slope as f^*V, then F is isomorphic to the pull back of a subbundle of V.*

Proof.

(a) Given two semistable bundles V, W of same slope on C, we have

$$\begin{aligned} Hom_D(f^*V, f^*W) &\cong Hom_C(V, f_*f^*W) \\ &\cong Hom_C(V, W \otimes f_*\mathcal{O}_D) \cong Hom_C(V, W) \end{aligned}$$

The last equality follows from the fact that $Hom_C(F, f_*\mathcal{O}_D/\mathcal{O}_C) = 0$ for any semistable bundle F of slope ≥ 0 as genuine ramification of f implies that $f_*\mathcal{O}_D/\mathcal{O}_C$ has negative maximal slope (see Lemma 3.2).
(b) Since the socle (maximal subbundle that is a direct sum of stable bundles (cf. [MR])) is unique, it follows that the socle of f^*V descends to the socle of V when f is separable. Since V is stable, this descended bundle has to be V itself. This shows that the pull back of a stable bundle is polystable under any finite separable map. Now the stability of f^*V for genuinely ramified maps follows from (a) since projections to direct summands are endomorphisms which do not come from below.
(c) Let V be a semistable bundle over C. Let $F \subset f^*V$ be a subbundle of same slope. Then the socle S_F of F is contained in the socle S_{f^*V} of f^*V and hence a direct summand of S_{f^*V}. But by uniqueness of the socle, S_{f^*V} is $f^*(S_V)$. Since stable bundles pull back to stable bundles, S_F coincides with some factors of $f^*(S_V)$ and hence is a pull back. Now the assertion follows by induction on the rank applied to the bundle $F/S_F \subset f^*V/S_F$ □

Corollary 3.6. *Let $f : D \to C$ be a finite separable morphism of smooth projective irreducible curves. Then for any semistable vector bundle W on D we have*

(a) $\mu_{\max}(f_*W) \leq \frac{\mu(W)}{\deg f}$
(b) *If* $\mu_{\max}(f_*W) = \frac{\mu(W)}{\deg f}$, *then* $\mathrm{rank}(f_*W)_{\max} \leq \mathrm{rank}\, W \cdot \mathrm{rank}(f_*\mathcal{O}_D)_{\max}$

Proof. First assertion follows from the fact that $Hom_C(F, f_*W) \cong Hom_D(f^*F, W)$. Hence semistable bundles of slope $> \frac{\mu(W)}{\deg f}$ have no morphism to f_*W.

If $\mathrm{rank}(f_*\mathcal{O}_D)_{\max} > 1$, then the morphism $f : D \to C$ factors through an étale morphism $\pi : \tilde{C} \to C$ such that $\tilde{f} : D \to \tilde{C}$ satisfies $\tilde{f}_*\mathcal{O}_D = \mathcal{O}_{\tilde{C}}$ (see Lemma 3.2).

If W' is any semistable bundle on $\tilde{C}$, then $\chi(W') = \chi(\pi_*W')$ as the cohomologies do not change by taking direct images under finite maps. By Riemann-Roch theorem we get

$$\chi(W') = (\mathrm{rank}\,(W'))(1 - g_{\tilde{C}} + \mu(W')) = (\mathrm{rank}\,(W'))(\deg\,(\pi)(1 - g_C) + \mu(W'))$$

and

$$\chi(\pi_*(W')) = \text{rank } (\pi_*(W'))(1 - g_C + \mu(\pi_*(W')))$$
$$= (\deg (\pi)\text{rank } (W'))(1 - g_C + \mu(\pi_*(W')))$$

since π is étale, $1 - g_{\tilde{C}} = (\deg \pi)(1 - g_C)$. Hence by comparing the terms above we conclude that $\mu(W') = (\deg (\pi))(\mu(\pi_*(W')))$. That $\pi_* W'$ is semistable of slope $\frac{\mu(W')}{\deg \pi}$ follows from (a).

Hence it suffices to prove (b) for the case when f is genuinely ramified.

On the contrary, assume there is a semistable bundle W with $\mu_{\max}(f_* W) = \frac{\mu(W)}{\deg f}$ and $\text{rank}(f_* W)_{\max} > \text{rank} W$, then consider the natural map, $f^*(f_* W)_{\max} \to Q \subset W$ with image Q. By taking direct image we obtain $(f_* Q)_{\max} = (f_* W)_{\max}$. Since $f^*(f_* W)_{\max}$ and W are semistable of same slope, Q is a semistable vector bundle of same slope $\frac{\mu(W)}{\deg f}$. By Lemma 3.5(c), $Q \cong f^* Q'$ is itself a pull back. Then $(f_* Q)_{\max} = (f_* f^* Q')_{\max} = Q'$, hence has the same rank as Q, which is at most rank W, a contradiction. □

4. Asymptotic slopes and strong semistability

Let C be a smooth curve defined over an algebraically closed field k of arbitrary characteristic. Let V be a vector bundle of rank r over C. For each $1 \leq s < r$, we denote the slope of maximal subbundle by $e_s(V)$.

$$e_s(V) := Max \left\{ \frac{\deg(W)}{s} | W \subset V \text{ is a subbundle of rank } s \right\}$$

Define the asymptotic s-spectrum $\mathcal{AS}_s(V)$ and the asymptotic s-slope $\nu_s(V)$ as follows:

$$\mathcal{AS}_s(V) := \left\{ \frac{e_s(f^*(V))}{\deg f} \right\}$$

$$\nu_s(V) := Limsup \frac{e_s(f^*(V))}{\deg f} = Limsup\, \mathcal{AS}_s(V)$$

where the supremum is taken over all finite morphisms $f : D \to C$. Now we have the following criterion for strong semistability in terms of the asymptotic slopes.

Theorem 4.1. *A vector bundle V is strongly semistable if and only if $\nu_s(V) = \mu(V)$ for some s. Then the asymptotic s-slopes $\nu_s(V)$ are equal to the usual slope $\mu(V)$ for all s.*

Proof. Let s be a given integer such that $1 \leq s < r$. In [BP], it is proved that a vector bundle is strongly semistable if and only if for every morphism $f : D \to C$, and for every subbundle $W \subset f^*V$ of rank s, the slope of W is at most the slope of f^*V. Hence if V is not strongly semistable then there exists a finite morphism

$f : D \to C$ and a subbundle $W \subset f^*V$ of rank s such that $\mu(W) > \mu(f^*(V))$. Hence $\nu_s(V) > \mu(V)$.

Assume V is strongly semistable. Then for any given finite map $f : D \to C$, f^*V is semistable and hence for every subbundle $W \subset f^*V$ of rank s, the slope of W is at most the slope of f^*V. Hence $\nu_s(V) \leq \mu(V)$.

Given $\epsilon > 0$, choose $n \gg 0$ such that $s(2g + \epsilon_n) < n\epsilon$. Then for the line bundle L_n, $\mathcal{O}_{\mathbb{G}r(s,V)}(n) \otimes \pi^*L$ separate points by Lemma 2.2. The kernel of the universal sequence on a general complete intersection D on $\mathbb{G}r(s, V)$ defined by $\mathcal{O}_{\mathbb{G}r(s,V)}(n) \otimes \pi^*L$ (which exists as it separates points) has slope $n^{s(r-s)}([\mathcal{O}_{\mathbb{G}r(s,V)}(1)]^{s(r-s)} \cdot F)\big(\mu(V) - \frac{s(2g+\epsilon_n)}{n}\big)$, by Lemma 2.4. The degree of $\pi_D : D \to C$ is $n^{s(r-s)}$ $([\mathcal{O}_{\mathbb{G}r(s,V)}(1)]^{s(r-s)} \cdot F)$. Hence dividing by the degree of $D \to C$, we obtain a number whose difference with $\mu(V)$ is less than ϵ. Hence $\nu_{r-s}(V) = \mu(V)$ for semistable bundles. □

Remark 4.2. Let V be a strongly semistable vector bundle over a smooth curve C. Then there is a sequence of genuinely ramified maps $f_i : D_i \to C$ such that

$$Lim \frac{e_s(f_i^*(V))}{\deg f_i} = mu(V)$$

follows from Lemma 3.4.

Let $V_\bullet := \{0 = V_0 \subset \cdots \subset V_l = V\}$ be the Harder Narasimhan filtration of V. Let d_i, r_i denote the degree and rank of V_i for $i = 0, 1, 2, \ldots l$. Assume that the Harder Narasimhan factors V_i/V_{i-1} are strongly semistable if the characteristic is positive. By [L], any vector bundle on a curve in positive characteristic has such a Harder Narasimhan Filtration after a finite number of Frobenius pull backs. Now we can determine the asymptotic slopes of V for each s.

Theorem 4.3. $s\nu_s = d_i + (s - r_i)\mu_{i+1}$ *where* $r_i < s \leq r_{i+1}$

Proof. First we note that if L is any line bundle, then $e_s(V \otimes L) = e_s(V) + \deg L$ and the spectrum $\mathcal{AS}_s(V \otimes L) = \mathcal{AS}_s(V) + \deg L$. Hence $\nu_s(V \otimes L) = \nu_s(V) + \deg L$. Since the Harder-Narasimhan filtration of $V \otimes L$ is given by $V_i \otimes L$, $d_i(V \otimes L) = d_i(V) + r_i \cdot \deg L$. Now assuming the formula for $V \otimes L$, we get:

$$s(\nu_s(V \otimes L)) = d_i(V \otimes L) + (s - r_i)\mu_{i+1}(V \otimes L)$$

$$s(\nu_s(V)) + s(\deg L) = d_i(V) + r_i(\deg L) + (s - r_i)(\mu_{i+1}(V)) + (s - r_i)(\deg L)$$

By simplifying, we obtain the formula for V.

Hence by taking deg L to be sufficiently large we may assume that all μ_i's are positive.

Let $f : D \to C$ be any finite map and $W \subset f^*(V)$ be a subbundle of rank s. Let $W_j \subset f^*(V_j/V_{j-1})$ be the saturation of the image of $W \cap f^*V_j$ in $f^*(V_j/V_{j-1})$. Let s_j be the rank of W_j and δ_j be equal to $\frac{\deg W_j}{\deg f}$. Then we have $s_j \leq r_j - r_{j-1}$, $r_i < \sum s_j = s \leq r_{i+1}$ and $\frac{\text{degree } W}{\deg f} \leq \sum \delta_j$. Since V_j/V_{j-1} is strongly semistable, we also have $\delta_j \leq s_j\mu_j \leq (r_j - r_{j-1})\mu_j$ for all $j \geq 1$. Now by comparing the first i terms and the rest of the terms we get the inequality:

$$\frac{\deg W}{\deg f} \leq \sum_{j=1}^{l} s_j\mu_j = \sum_{j=1}^{i} s_j\mu_j + \sum_{j=i+1}^{l} s_j\mu_j$$

$$\leq \sum_{j=1}^{i}(r_j - r_{j-1})\mu_j + \sum_{j=i+1}^{l} s_j\mu_{i+1} \leq d_i + (s - r_i)\mu_{i+1}$$

To show the equality we produce subbundles of rank s in coverings with degree divided by the degree of the covering arbitrarily close to $d_i+(s-r_i)\mu_{i+1}$ where $r_i < s \leq r_{i+1}$. By Theorem 4.1, we can find a covering $f : D \to C$ and a subbundle $W_i \subset f^*(V_i/V_{i-1})$ of rank $s - r_i$ such that $\mu(V_i/V_{i-1}) - \mu(W_i) < \epsilon$. Let $W := \pi^{-1}(W_i)$ be the inverse image of W_i in $V_i \subset V$ by the projection $\pi : V_i \to V_i/V_{i-1}$. Then it shows that $d_i + (s - r_i)\mu_{i+1} - \mu(W)/\deg \mathrm{f} < \epsilon$. Hence the theorem. □

Theorem 4.4. *Let $f : Y \to X$ be a morphism and $W \subset f^*V$ be a subbundle of rank s with $r_i < s \leq r_{i+1}$, such that*

$$\frac{\deg (W)}{\deg f} > d_i + (s - r_i)\mu_{i+1} - (\mu(V_i) - \mu_{i+1})$$

Then $f^*V_i \subset W \subset f^*V_{i+1}$.

Proof. From the proof of Theorem 4.3, $s\nu_s = d_i + (s - r_i)\mu_{i+1}$, the inequalities become equality if and only if $s_j = r_j - r_{j-1}$ for all j, and hence $W \cap V_i = V_i$, proving $V_i \subset W$. □

5. The geometry of the spectrum

Notice that the s-spectrum $\mathcal{AS}_s(V) \subset [e_s(V), \bar{\mu}_{\max}(V)] \subset \mathbb{R}$ is a subset of the bounded interval. Hence it has maximum, minimum, and cluster points. We have described the supremum (asymptotic slopes) of the spectrum for each s. This leads to the following natural question.

Question. Are asymptotic s-slopes the only limit points of the spectrum? Or is it likely to be dense in the interval $[e_s(V), \mu_{\max}(V)]$?

Now we give a criterion for the asymptotic slopes to be an isolated value for strongly semistable vector bundles.

Lemma 5.1. *Let V be a strongly semistable vector bundle. Then $\mu(V)$ is an isolated point of the asymptotic s-spectrum $\mathcal{AS}_s(V)$ if and only if $\mathcal{AS}_s(V) = \{ \mu(V) \}$, i.e., there exists a subbundle $W \subset V$ of rank s such that $\mu(W) = \mu(V)$.*

Proof. Assume $W \subset V$ is a subbundle of slope $\mu(V)$. Then for any map $f : D \to C$, $f^*(W) \subset f^*(V)$ is a maximal subbundle and hence $\frac{\mu(f^*W)}{\deg f} = \mu(W) = \mu(V)$. Hence the spectrum is a singleton, proving $\mathcal{AS}_s(V) = \{ \mu(V) \}$.

Let V be a strongly semistable vector bundle such that $\mu(V)$ is isolated in the spectrum $\mathcal{AS}_s(V)$.

From Remark 4.2 and the hypothesis that $\mu(V)$ is isolated, it follows that there is a genuinely ramified map such that the pull back of V has a subbundle of same slope. By Lemma 3.5 (c), this subbundle descends to a subbundle of same slope as V. □

References

[BP] I. Biswas and A. J. Parameswaran, A criterion for strongly semistable principal bundles over a curve in positive characteristic, *Bull. Sci. Math.*, **128(9)**, 761–773.

[F] W. Fulton, Intersection Theory, Springer Verlag, New York (1984).

[L] A. Langer, Semistable sheaves in positive characteristic, *Ann. of Math.*, **159** (2004) 251–276.

[LN] H. Lange and M. S. Narasimhan, Maximal subbundles of rank two vector bundles on curves, *Math. Ann.*, **266** (1983) 55–72.

[MR] V. B. Mehta and A. Ramanathan, Restriction of stable sheaves and representations of the fundamental group, *Inv. Math.*, **77** (1984) 163–172.

[M] J. P. Murre, Lectures on an introduction to Grothendieck's theory of the fundamental group, T.I.F.R. Lecture notes (1967).

[Ne] Newstead, Lectures on Introduction to moduli problems and orbit spaces, TIFR Lecture notes, Narosa Publishing House, New Delhi (1978).

[N] M. V. Nori, The fundamental group scheme, *Proc. Indian Acad. Sci. (Math. Sci.)*, **91** (1982) 73–122.

[RR] S. Ramanan and A. Ramanathan, Some remarks on the instability flag, *Tohoku Math. Jour.*, **36** (1984) 269–291.

Teichmüller Theory and Moduli Problem No. 10, 2010, pp. 529–548.

Rigidity, Past and Present

Chris Peters

Institut Fourier, Université de Grenoble I, St.-Martin d'Hères, France
e-mail: chris.peters@ujf-grenoble.fr

Abstract. The Hodge theoretic aspects of the finiteness theorem of Arakelov are reviewed. Earlier results by the author as well as more recent results by Zucker, M.-H. Saito and others are explained to the non-specialist. New and shorter proofs of some of these recent results are provided.

Introduction

This note is an elaborated version of the talk with the same title I gave in the January 2006 workshop at the HRI. The talk was aimed at (complex) algebraic geometers with little background in Hodge theory. In the talk, I have tried to give as much details from Hodge theory so as to allow the audience to follow the main arguments. Much of these details are presented here as well. I try to avoid to become too technical by substituting references to the existing literature.

The point of departure is a beautiful result of Arakelov [Arak] which is explained below in § 1.2. A brief sketch of the original proof is given in § 1.3. Since its appearance, this theorem motivated many people to look for analogs in the number theoretic setting. Analogs which eventually led Faltings to his proof [Fa83b] of the Mordell conjecture. Indeed, a version of Arakelov's theorem for Abelian varieties appears by the same author [Fa83a] in the same journal: it is the article which is just preceding the famous article in which the Mordell conjecture has been proven.

In [Pe90] I proved a generalization of Faltings' result for variations of Hodge structures. At the request of the organizers of the conference I explained this proof and added a few recent developments. Among the latter I mention the classification of non-rigid variations of K3-surfaces [S-Zu] due to M.-H. Saito and S. Zucker, as well as for Abelian varieties [Sa], due to M.-H. Saito. The reader will find a few new proofs of some recent results as well. For example, one of the main results of [L-T-Y-Z]: any family of Calabi-Yau's with maximally unipotent local monodromy at some point of the boundary must be rigid (Cor 3.5). The same approach gives an easy proof of a result from [VZ]: a variation with maximal Higgs field is rigid (Prop. 3.7).

Contrary to when I wrote [Pe90], nowadays some good introductory works to Griffiths' theory have appeared such as [V] and [C-P-M]. The latter treats the Lie-theoretic background which I need. For that reason I often refer to it for details omitted in the presentation below.

At the conference H. Shiga pointed out to me that he together with Y. Imayoshi in [I-S] proved Arakelov's rigidity result by analytic means.

1. Variations on Arakelov's theorem

1.1 *Families*

The statement of Arakelov's theorem uses the concept of a family of complex algebraic varieties:

Definition 1.1.

1) A family of algebraic varieties *is a proper and flat morphism* $f : X \to Y$ *between complex manifolds whose fibres are (irreducible) algebraic varieties. If* $\dim Y = n$ *one speaks of an* n-dimensional family. *Note that neither* X *nor* Y *are assumed to be algebraic. Indeed,* Y *could be a disc or a polydisc.*
2) *The locus over which* f *is* not *of maximal rank is called* discriminant locus *and is denoted* $\Delta(f)$*; on the complement the fibres are smooth. If the discriminant locus is empty one says that* f *is a* smooth family.
3) *A morphism from the family* $f' : X' \to Y'$ *to the family* $f : X \to Y$ *is given by a commutative square*

$$\begin{array}{ccc} X' & \xrightarrow{\tilde{f}} & X \\ \downarrow f' & & f \downarrow \\ Y' & \xrightarrow{f} & X \end{array}$$

If $f = \mathrm{id}$ *and* $\tilde{f}$ *is an* isomorphism, *one says that* f *and* f' *are isomorphic; if* $f = \mathrm{id}$ *but* $\tilde{f}$ *is only a bimeromorphic map, we say that* f *and* f' *are* bimeromorphic.

Obvious examples are given by products: $X_0 \times Y \to Y$, where X_0 is some smooth algebraic variety. Any family isomorphic to such a family is called *trivial.*

Whenever we are given a holomorphic map between manifolds $g : Y' \to Y$, there is a standard procedure to obtain a new family over Y': one takes the fibre product $X' := X \times_Y Y' = \{(x, y') \mid f(x) = g(y')\}$ which comes with a natural morphism $X' \to Y'$. This family is called the *pull back of* f *by* g. A family is called *isotrivial* if it becomes bimeromorphically trivial after pulling back through some *finite* covering map $Y' \to Y$.

1.2 *Arakelov's theorem*

Consider a class $\mathcal{C}$ of smooth projective varieties and fix a smooth projective variety $\bar{S}$, the *base manifold* and a closed subvariety $\Sigma \subset \bar{S}$, the *degeneracy locus*. The *Shafarevich problem* for the triple $(\mathcal{C}, \bar{S}, \Sigma)$ consists in determining the isomorphism classes of families over a given base $\bar{S}$ of which the fibres over $\bar{S} - \Sigma$ belong to $\mathcal{C}$. In other words, one asks how many families f there are over $\bar{S}$ with smooth members in $\mathcal{C}$ and with $\Delta(f) \subset \Sigma$, possibly with strict inclusion.

Arakelov's theorem answers this question for 1-dimensional families of curves of a fixed genus $g \geq 2$.

Theorem ([Arak]). *Let $\bar{S}$ be a complete curve and a finite set Σ of points on $\bar{S}$. There are at most finitely many non-isotrivial families of curves of given genus $g \geq 2$ over $\bar{S}$ that are smooth over $\bar{S} - \Sigma$.*

In order to sketch Arakelov's proof let us return to the Shafarevich problem for a class $\mathcal{C}$ of varieties which has a good moduli space M. For simplicity of exposition this shall mean that M is some quasi-projective variety, possibly singular, such that the points of M are in one to one correspondence with isomorphism classes of varieties in $\mathcal{C}$. In addition, given a family $f : X \to S$, there is an induced algebraic *moduli map* $\mu_f : S \to M$ which sends $s \in S$ to the isomorphism class of the fibre $X_s := f^{-1}s$ in M. Usually, some more functorial properties are needed which are formalised in the concept of a *coarse moduli space*. For details, see [Nst]. In many cases these exist.

Examples 1.2.

1. Fix an integer $g \geq 0$. Smooth projective curves of genus g (= compact Riemann surfaces of genus g) have a quasi-projective moduli space M_g, as is well known.
2. Fix an integer $g \geq 1$. Abelian varieties (complex tori embeddable in projective space) with a fixed (principal) polarization have a quasi-projective moduli space A_g as is equally well known.
3. A K3-surface is a simply connected compact complex surface with trivial canonical bundle. More generally, a Calabi-Yau manifold is a simply connected compact complex manifold with trivial canonical bundle. Projective Calabi-Yau's with fixed cohomology ring, fixed Chern classes and a fixed polarization admit a coarse moduli space. This is less well known. See for example [L-T-Y-Z, Thm. 8].

It would be even more ideal if there exists a universal family $\mathcal{C}_M \to M$ of varieties in $\mathcal{C}$ so that f is isomorphic to its pull back under the moduli map. In this situation one speaks of a *fine moduli space*. This rarely occurs; for instance, μ_f then would be the constant map if and only if the family f is trivial, but it is easy to construct 1-dimensional non-trivial families of curves all of which are isomorphic (see e.g. [B-H-P-V, Ch. V.§5]). This is related to the existence of automorphisms of the fibres which in fact make the family a fibre-bundle with non-trivial monodromy. The fact that a fine moduli space of curves does not exist also is related to the fact that iso-trivial families enter the statement of Arakelov's theorem: it is another way of saying that the moduli map is constant. In terms of moduli maps, the theorem states that there are only finitely many non-constant maps from a quasi-projective curve to M_g.

Generalizing to any algebraic variety M, one is led to consider

$$\mathrm{Mor}(S, M) = \{\mu : S \to M \mid \mu \text{ algebraic.}\}$$

This set has a natural structure of a topological space with possibly infinitely many components, and one expects that each of these has an algebraic structure. In the curve case it is not hard to see that $\mathrm{Mor}(S, M_g)$ indeed has such a structure.

The first step in the proof of Arakelov's theorem is to show that there are in fact only *finitely* many components. This is a *boundedness* statement. The second step consists

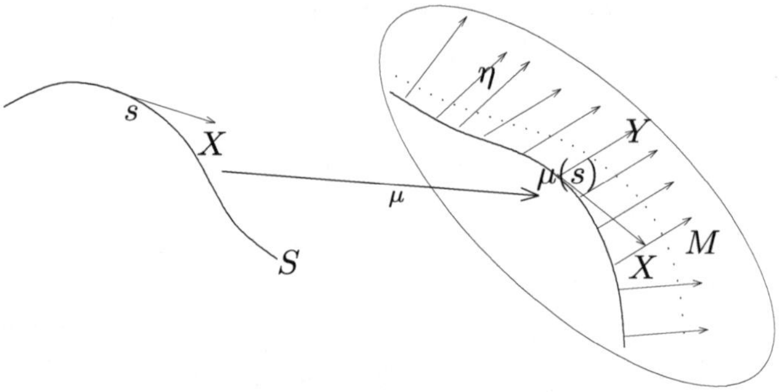

Figure 1. Infinitesimal deformations.

of establishing *rigidity*, i.e. to show that the individual components corresponding to non-isotrivial families are points. In this survey I'll concentrate on this second aspect.

1.3 *Deforming maps and rigidity*

Details of what follows can be found in [Pe90, § 2]. Let me start with the basic

Definition 1.3. *Let $\mu : S \to M$ be a morphism between complex varieties. A* deformation *of μ (keeping base and target fixed) parametrized by a germ of a complex space (T, t) is a holomorphic map $m : S \times T \to M$ such that $m|S \times \{t\} = \mu$.*

Taking for T the point with ring of functions the dual numbers $\mathbb{C}[\epsilon]/\epsilon^2$ we get the *infinitesimal deformations* which are classified by the vector space $H^0(S, \mu^* T_M)$.

Intuitively, infinitesimally small deformations of μ are given by a vector field in M only defined at points of the image $\mu(M)$ pointing in a potential deformation direction. The set of deformations of μ can be shown to be a complex analytic space whose Zariski tangent space at μ is the space of infinitesimal deformations:

$$T_\mu \mathrm{Def}(\mu) = H^0(S, \mu^* T_M). \tag{1}$$

However, in the situations we are interested in, S is quasi-projective (we speak of an *algebraic family*) and in this case not all deformations are of interest: only the analytic subspace

$$\mathrm{Def}(\mu)^{\mathrm{alg}} = \mathrm{Def}(\mu) \cap \mathrm{Mor}(S, M) \subset \mathrm{Def}(\mu) \tag{2}$$

of *algebraic* deformations count. To detail this further, in this situation S can be compactified by some complex projective manifold $\bar{S}$ such that $\Sigma = \bar{S} - S$ is a divisor with normal crossings and the algebraic infinitesimal deformations of f correspond to global sections of some suitable extension of $\mu^* T_M$ with certain controlled growth conditions "near infinity", i.e. near the points of Σ. Usually the bundle $\mu^* T_M$ will extend to a bundle on all of $\bar{M}$ and the sections of $\mu^* T_M$ having the desired growth conditions correspond precisely to sections of this extension.

Let me come back to the case of the situation in Arakelov's theorem where the moduli map $\mu = \mu_f : S \to M_g$ comes from the restriction to S of a family $\bar{f} : \bar{X} \to \bar{S}$ of curves of genus g over the compact curve $\bar{S}$. There is a canonical identification $[\mu^* T_{M_g}]_t = H^1(X_t, T_{X_t})$ where, I recall $X_t = f^{-1}t$. This globalizes to $\mu^* T_{M_g} = R^1 f_* T_{X/S}$ and one aims to calculate its global algebraic sections over S. In this case $T_{X/S}$ is a line bundle on X which extends naturally to a line bundle

$$L := T_{\bar{X}/\bar{S}}$$

and the looked for space of infinitesimal deformations is $H^0(\bar{S}, R^1 \bar{f}_* L)$. The Leray spectral sequence for $\bar{f}$ also involves $H^1(\bar{S}, \bar{f}_* L)$, but since $g \geq 2$ the fiber of $f_* L$ at $t \in S$ which is $H^0(X_t, T_{X_t})$ vanishes. The upshot is that

$$T_\mu \mathrm{Def}(\mu)^{\mathrm{alg}} = H^0(\bar{S}, R^1 \bar{f}_* L) = H^1(\bar{X}, L).$$

Now L^{-1} is the relative canonical bundle on $\bar{X}$ (with respect to $\bar{f}$) and to show rigidity, by Kodaira vanishing, it suffices to show that L^{-1} is ample. Arakelov does this by a clever geometric argument involving the locus of Weierstraß points on the fibers.

1.4 *Towards Hodge theory*

Starting with the family of curves $f : X \to S$, one may look at the associated family of Jacobians $J_t = \mathrm{Jac}(X_t)$. Recall (example 1.2.2) that g-dimensional Abelian varieties have a moduli space A_g. This leads to a commutative diagram in which the moduli map μ_f and the *period map* p_f figure:

$$\begin{array}{ccc} S & \xrightarrow{\ \mu_f\ } & M_g \\ & {\scriptstyle p_f}\searrow \quad \swarrow{\scriptstyle \iota} & \\ & A_g. & \end{array}$$

Torelli's theorem implies that ι is an immersion and so rigidity for the moduli map is implied by rigidity for the period map.

To explain the transition to Hodge theory, recall that there is a Hodge decomposition for 1-cohomology $H^1(X_t) \otimes_{\mathbb{Z}} \mathbb{C} = H^{1,0}(X_t) \oplus H^{0,1}(X_t)$ and the position of the g-dimensional subspace $H^{1,0}(X_t)$ in $H^1(X_t) \otimes_{\mathbb{Z}} \mathbb{C} \simeq \mathbb{C}^{2g}$ is described by a point $\widetilde{p_f}(t)$ in the Siegel upper half space $\mathfrak{h}_g$. The assignment $t \mapsto \widetilde{p_f}(t)$ is ambiguous because of the monodromy action of the fundamental group $\pi_1(S)$ through the properly discontinuous action of $\mathrm{Sp}(2g)$ on $\mathfrak{h}_g$ and the quotient $A_g = \mathfrak{h}_g/\mathrm{Sp}(2g)$ is the moduli space considered in 1.2.2 and $p_f(t)$ is the equivalence class of $\widetilde{p_f}(t)$.

There is a description of the tangent space to $\mathrm{Def}(p_f)$ entirely in terms of Hodge theory as follows. The 1-dimensional cohomology of a genus g curve is modelled on a symplectic lattice $H := (\mathbb{Z}^{2g}, Q)$ carrying a weight one Hodge structure. More generally, recall:

Definition 1.4. *Let H be a free $\mathbb{Z}$-module of finite rank. A* weight k Hodge structure *on H is a direct sum decomposition on its complexification*

$$H_{\mathbb{C}} := H \otimes_{\mathbb{Z}} \mathbb{C} = \bigoplus_{p+q=k} H^{p,q}$$

such that $H^{q,p} = \overline{H^{p,q}}$ where complex conjugation is with respect to the natural underlying real structure.

In the curve case there is an extra ingredient coming from the cup-product pairing $Q = \langle -, - \rangle$. It is an example of a polarization:

Definition 1.5. *A non-degenerate integral bilinear form Q on H, symmetric for k even and skew-symmetric for k odd* polarizes *the Hodge structure on H if the two Riemann bilinear relations hold:*

$$Q(H^{p,q}, H^{r,s}) = 0 \quad \text{if } (r,s) \neq (q,p) \tag{3}$$

$$Q(Cu, \bar{u}) > 0 \quad \text{if } u \neq 0 \quad \text{and where} \quad C|H^{p,q} = \mathrm{i}^{p-q}. \tag{4}$$

The pair (H, Q) is called a weight k *polarized Hodge structure*. The example of H^1 for a curve is by no means special:

Example 1.6. Let X be a projective manifold. Cup product with the hyperplane class in $H^2(X)$ defines the Lefschetz operator L on integral cohomology and it is well known that there is a splitting (over $\mathbb{Q}$) of the integral cohomology $H^k(X)$ into basic building blocks of the form $L^m H^{k-2m}_{\text{prim}}(X)$, where the subscript denotes *primitive cohomology*. Primitive cohomology is defined over $\mathbb{Z}$ and comes with a polarization Q. The Hodge structure on primitive cohomology is polarized by Q.

Given a polarized Hodge structure (H, Q), its endomorphism group $\mathrm{End}(H, Q)$ consists of integral endomorphisms which are skew-symmetric with respect to Q. Its complexification has a canonical decomposition:

$$\mathrm{End}(H_{\mathbb{C}}, Q_{\mathbb{C}}) = \bigoplus_j \mathrm{End}^{-j,j} H_{\mathbb{C}}$$

where $\mathrm{End}^{-j,j} H_{\mathbb{C}}$ is the complex vector space of Q-skew endomorphisms X of H_C for which $XH^{p,q} = H^{p-j,q+j}$, $j \in \mathbb{Z}$. It is not hard to see that this is a weight 0 Hodge structure.

After this digression, coming back to the curve case, we note that the vector space $\mathrm{End}(H^1_{\mathbb{C}}(X_t), Q)$ decomposes in three pieces: a real piece $\mathrm{End}^{0,0} H^1_{\mathbb{C}}(X_t)$, which corresponds essentially to the endomorphisms of the Jacobian of X_t, $\mathrm{End}^{-1,1} H^1_{\mathbb{C}}(X_t)$ and $\mathrm{End}^{1,-1} H^1_{\mathbb{C}}(X_t)$ which is the complex conjugate of $\mathrm{End}^{-1,1} H^1_{\mathbb{C}}(X_t)$.

The next step is to let t vary over the base S. The $\mathbb{Z}$-modules $H^1_{\mathbb{C}}(X_t)$ give a local system $\underline{H}_S$ on S and the subspaces $H^{1,0}(X_t)$ form a holomorphic subbundle of the associated vector bundle $\mathcal{H} = \underline{H}_S \otimes_{\mathbb{Z}} \mathcal{O}_S$. The holomorphicity is equivalent to the period map $S \to A_g$ being holomorphic.The bundle $\mathcal{E}nd(\mathcal{H}, Q)$ is the bundle of Q-skew endomorphisms of $\mathcal{H}$ and it splits further into types; it is not hard to show that

$$p_f^* T_{A_g} = \mathcal{E}nd^{-1,1}(\mathcal{H}, Q).$$

As I noted before, one really only is interested in those deformations that give families of curves over $\bar{S}$ and so one needs some growth conditions as I now explain briefly.

For simplicity let me suppose that $\dim S = 1$ so that $S = \bar{S} - \{p_1, \dots, p_N\}$. Each of the punctures p_j is called a "point at infinity". So locally at such a point the smooth part of the family lives over a punctured disc Δ^*. Now, quite generally, if we have a smooth family of varieties $\{X_t\}$ over the punctured disc, going once around the puncture in the positive direction induces on $H^k(X_t)$, $t \in \Delta^*$ the *local monodromy operator* T. The crucial result is the

Theorem 1.7 (Monodromy Theorem). *The operator T acts quasi-unipotently on $H^k(X_t)$: for some integers ℓ, M, $(T^\ell - I)^M = 0$. Moreover $M \le k+1$.*

Choose ℓ such that T^ℓ is unipotent. The smallest number M for which $(T^\ell - I)^M = 0$ is called the *order of unipotency* of T, which by the previous theorem at most equals $k+1$. The quasi-unipotency of T makes it possible to extend the vector bundle $\mathcal{H}^k$ whose fiber at $t \in \Delta^*$ is $H^k(X_t)$ to a fiber bundle on all of Δ, the so-called *quasi-canonical extension* $\mathcal{H}^k_{\text{can}}$ of $\mathcal{H}$. For details I refer to [Del70, p. 94].

In the case at hand there is a vector bundle $\mathcal{H}_{\text{can}}$ on $\bar{S}$ and one can show that the Hodge spaces $H^{1,0}(X_t)$, which make up a holomorphic subbundle, also extend to a subbundle $\mathcal{H}^{1,0}_{\text{can}}$ of $\mathcal{H}_{\text{can}}$. Likewise $\mathcal{E}nd(\mathcal{H}, Q)$ and $\mathcal{E}nd(\mathcal{H}, Q)^{-1,1}$ extend to $\mathcal{E}nd(\mathcal{H}, Q)_{\text{can}}$ and $\mathcal{E}nd(\mathcal{H}, Q)^{-1,1}_{\text{can}}$ respectively. The crucial observation now is:

Lemma 1.8. *Any holomorphic section of the* vector bundle $\mathcal{E}nd(\mathcal{H}, Q)_{\text{can}}$ *over $\bar{S}$ restricts over S to a global section of the* local system $\underline{\text{End}}(\underline{H}_S, Q)$ *of S. Such a global section is precisely a Q-skew endomorphism of any of its fibers $H^1(X_t)$ commuting with the monodromy.*

This is by no means trivial and depends on the existence of a special metric on the bundle $\mathcal{E}nd(\mathcal{H}, Q)$ as will be explained below (Proposition 2.9) which it inherits from the period domain. The metric in this case is any invariant metric on the Siegel upper half space (see Cor. 2.10).

From the preceding discussion a criterion for rigidity emerges:

Criterion 1.9. One has

$$\begin{aligned} T_{p_f}\text{Def}^{\text{alg}}(p_f) &= T_{p_f}\text{Mor}(S, A_g) = H^0(\bar{S}, \mathcal{E}nd^{-1,1}(\mathcal{H}, Q)_{\text{can}}). \\ &= Q\text{-skew endomorphisms of } H^1_{\mathbb{C}}(X_t) \text{ of type} \\ &\quad\ (-1,1) \text{ commuting with the monodromy group.} \end{aligned} \tag{5}$$

In particular p_f is rigid if the bundle $\mathcal{E}nd(\mathcal{H}, Q)_{\text{can}}$ only admits global endomorphisms over $\bar{S}$ of type $(0,0)$. It follows that p_f is rigid if and only if the global (skew) endomorphisms of the bundle of 1-cohomology of X_t consists of the endomorphisms of any smooth member of the family of Jacobians J_t which commute with monodromy.

1.5 *Rigidity results for Abelian varieties*

The criterion 1.9 is essentially due to Faltings [Fa83a]. Indeed, there is nothing peculiar to Jacobians here and the criterion applies to any family of Abelian varieties.

In [Fa83a] an example is given of a family of smooth 8-dimensional Abelian varieties which is not isotrivial but still non-rigid. In fact, the example has no non-isotrivial factors.

M.-H. Saito in [Sa] has taken up a systematic study of rigidity for Abelian varieties *parametrized by a curve* and he shows that families of g-dimensional Abelian varieties without non-isotrivial factors are rigid if $g < 7$ so that Faltings' example is in the lowest possible dimension. In [Sa] Faltings' example is shown to belong to a wide class of Kuga fiber spaces. In the following description non-compact Hermitian symmetric spaces are needed. One type already came up: type III spaces are just the bounded incarnations of the Siegel upper half spaces $\mathfrak{h}_g$ and have dimension $\frac{1}{2}g(g+1)$. Type II symmetric spaces depend on a parameter $h \geq 2, h \in \mathbb{Z}$ and will be denoted II_h. They have dimension $\frac{1}{2}h(h-1)$ (see [Sa, §6]). Type I symmetric spaces depend on two parameters $p, q \in \mathbb{N}$ and will be denoted I_{pq}; they have dimension pq.

Examples 1.10.

1. Fix natural numbers t, t', n and m, and consider the arithmetic quotients
$$\mathfrak{h}_n^{t-t'}/\Gamma, \quad II_m^{t'}/\Gamma.$$
Over the product there is a Kuga fiber space which is a particular family of Abelian varieties of relative dimension $g = 2tmn$. Fixing a point q in the first factor gives a Kuga fiber space over $III_m^{t'}/\Gamma$ which has no isotrivial factor if $m \geq 2$ and $t' \geq 1$. Moreover, all such spaces obtained by varying q are isomorphic. So, if $n \geq 1$ and $t - t' \geq 1$ they are not rigid. Similarly, fixing a point in the second factor gives a Kuga fiber space over $\mathfrak{h}_n^{t-t'}/\Gamma$ which has no isotrivial factor if $n \geq 1$ and $t-t' \geq 1$, and it is non-rigid if $m \geq 2$. In conclusion, this gives examples for all even $g \geq 8$.
2. A similar construction is possible for type I spaces. Again integers t, t', n, m are given with $t' \leq t$. In addition one has pairs of integers (p_j, q_j) $j = 1, \ldots, t$ for which $p_j + q_j = n$ for $j \leq t'$ and $p_j + q_j = m$ for $j > t'$. As base one takes the product of the two spaces
$$(I_{p_1 q_1} \times \cdots \times I_{p'_t q'_t})/\Gamma, \quad (I_{p'_{t'+1} q'_{t'+1}} \times \cdots \times I_{p_t q_t})/\Gamma.$$
There is a Kuga fiber space over the product of relative dimension $g = tmn$ which, over each the two factors give families without isotrivial factors as soon as the base dimension is > 0, and deformable as soon as the other factor has positive dimension. This gives examples of all dimensions $g \geq 8$ which factor as $g = tnm$, $t, n, m \geq 2$.

As to rigidity results, Saito shows that his general classification implies:

Theorem.

- *A 1-parameter family of Abelian varieties of prime dimension whose generic fiber is simple and has no isotrivial factors must be rigid.*
- *Assume S is a non-compact curve, the local monodromy around at least one point at infinity has infinite order and the generic fiber is a simple Abelian variety. Then the family is rigid.*

Referring to the Monodromy theorem 1.7, in the weight one case the order of unipotency is at most 2 and so either T is finite or has maximal order of unipotency 2, which puts the previous result into perspective.

Remark 1.11. The results in [Sa] have been explained geometrically by Ben Moonen [Mo] in terms of Shimura theory.

1.6 *Rigidity results for K3-surfaces*

It is easy to construct examples starting from products of modular families of elliptic curves, say $\{E_\tau \times E_\sigma \mid (\sigma, \tau) \in C \times D\}$, C and D modular curves. Fixing $\tau = \tau_0$ gives a family of abelian surfaces over C with deformation parameter $\tau \in D$. This does not contradict the previous non-rigidity results for Abelian varieties, since the family has a trivial factor. However, the associated family of Kummer surfaces is a genuine 1-parameter family of K3's with a deformation parameter. The results of [S-Zu] generalize this construction. The upshot is that the base of any non-rigid families of K3's must be of the form $\mathfrak{h}/\Gamma$, where Γ is a discrete subgroup of $\mathrm{SL}(2, \mathbb{R})$ acting freely on $\mathfrak{h}$ and there is one deformation parameter as in the preceding example. This is deduced from the very special structure of the $\mathbb{Q}$-algebra

$$\mathrm{End} = \left\{ \begin{array}{l} \text{endomorphisms of } H^2(X_t; \mathbb{Q}) \\ \text{commuting with the monodromy group.} \end{array} \right\} \tag{6}$$

Note that one takes *all* endomorphisms instead of just the Q-skew endomorphisms.

To explain the main result concerning this algebra, I briefly have to say something about the structure of the cohomology groups $H^2(X_t, \mathbb{Q})$ of the fibres. The cup product form on this $\mathbb{Q}$-vector space induces an orthogonal direct sum decomposition

$$H^2(X_t, \mathbb{Q}) = S_t \oplus T_t, \quad S_t = \mathrm{NS}(X_t) \otimes \mathbb{Q},\ T_t = S_t^\perp,$$

where $\mathrm{NS}(X_t)$ is the Néron-Severi group of the surface X_t. It can be shown that there is a local system $\underline{S}$ over S, the *algebraic part* of the local system

$$H_S := \{H^2(X_t, \mathbb{Q})\}_{t \in S}$$

such that $\underline{S}_t = \mathrm{NS}(X_t)$ for t in the complement of a countable set of points in S. The rank of the system $\underline{S}$ is therefore called the *generic Picard number* ρ of the family of K3's. The orthogonal complement of $\underline{S}$ is called the *transcendental part* $\underline{T}$ of the local system H_S.

Theorem 1.12. *The algebra* (6) *is a quaternion-algebra over a totally real number field* Z. *If* $[Z : \mathbb{Q}] = g$, *the local system* $\underline{T}$ *over* S *is a rank-1* End-*module and hence in particular has dimension* $4g$.

Since $\dim H^2(X_t, \mathbb{Q}) = 22$, one has in particular

Corollary 1.13. *If* $22 - \rho$ *is not divisible by* 4, *the family is rigid.*

The monodromy group Γ of the family acts always as a finite group on the algebraic part (the algebraic part has signature $(1, \rho - 1)$, but a polarization is always fixed). However, it might act non-finitely on the transcendental part $\underline{T}$. The induced group is a subgroup of the algebra End and one has [S-Zu, The. 5.6.2]:

Theorem 1.14. *Suppose that the family of K3's is non-isotrivial and non-rigid. Then the monodromy group* Γ *either acts as finite group on* $\underline{T}$ *or as an arithmetic subgroup of* $\mathrm{SL}(2, \mathbb{R})$ *commensurable with* $\mathrm{SL}(2, \mathbb{Z})$. *In particular, if there is an infinite order local monodromy operator, one has* $\mathrm{End} = M_2(\mathbb{Q})$ *and* $\rho = 18$.

Recalling the Monodromy Theorem 1.7 the maximal order of unipotency on 2-cohomology is 3. So a local monodromy operator T can be finite, or has order of unipotency 2 or 3. In the situation of Theorem 1.14 the order of unipotency must be at most 2 and so one has:

Corollary 1.15. *If there is a local monodromy operator at infinity* T *having maximal order of unipotency* 3, *the family must be rigid.*

2. A Hodge theoretic approach

2.1 *Period maps*

Fix a free $\mathbb{Z}$-module H of finite rank, an integer k, the *weight*, and positive integers $h^{p,q}$, $p+q = k$, the *Hodge numbers* adding up to $\mathrm{rank}(H)$ and assembled in a *Hodge vector* $\vec{h} = (\dots, h^{p,k-p}, h^{p+1,k-1-p}, \dots)$.

The set of all possible Hodge structures on H whose Hodge numbers make up the fixed Hodge vector $\vec{h}$ forms a nice algebraic variety, a partial flag variety. In order to make this identification it is essential to pass from the Hodge *decomposition* $H = \bigoplus H^{r,s}$ to the corresponding Hodge *filtration* $\cdots F^p \supset F^{p-1} \supset \cdots$, where

$$F^p = \bigoplus_{r \geq p} H^{r,k-r}. \tag{7}$$

The condition $H^{p,q} = \overline{H^{q,p}}$ translates into $H \otimes \mathbb{C} = F^p \oplus \overline{F^{k-p+1}}$. On can also go in the reverse direction since $H^{p,q} = F^p \cap \overline{H^q}$. This shows that instead of Hodge decompositions, one may equivalently speak of Hodge filtrations.

The Hodge structures coming from geometry are polarized in the sense of Definition 1.5. So assume that a non-degenerate $\mathbb{Z}$-valued form Q on H is given which is symmetric for k even, and skew-symmetric otherwise and which satisfies the two Riemann bilinear relations. These imply a condition on the signature of Q: it should be equal to $\sum(-1)^p h^{p,k-p}$.

The group G of isometries (with respect to Q) of the real vector space $H_{\mathbb{R}} := H \otimes_{\mathbb{Z}} \mathbb{R}$ acts on the *period domain* $D = D(H, Q, \vec{h})$ consisting of possible Hodge structures of Hodge vector $\vec{h}$ polarized by Q. The pair $(\vec{h}, Q)$ will be called the *polarization type*. The form Q induces a hermitian metric on $H_{\mathbb{C}} =: H \otimes_{\mathbb{Z}} \mathbb{C}$, the *Hodge metric* defined by

$$h(u, v) := Q(Cu, \bar{v}), \quad u, v \in H_{\mathbb{C}}. \tag{8}$$

It turns out that G acts transitively on D. At a reference point $F_0 \in D$ corresponding to a Hodge decomposition $\oplus H^{p,q}$ the isotropy group is

$$V = \{g \in G \mid gH^{p,q} \subset H^{p,q}\}.$$

Hence the period domain is the homogeneous space $D = G/V$ and so it is in particular a manifold, but one cannot expect this to be an algebraic variety because of non-algebraic nature of the second Riemann relation (4). However, it embeds naturally in a projective manifold, the *compact dual* $\check{D}$ of D which by definition is the set $\check{D} = \check{D}(H, Q, \vec{h})$ of Hodge decompositions on H with given Hodge vector $\vec{h}$ which only satisfy the first Riemann equation (3). The "complex" group

$$G_{\mathbb{C}} = \mathrm{Aut}(H \otimes_{\mathbb{Z}} \mathbb{C}, Q \otimes 1)$$

turns out to act transitively on $\check{D}$ with isotropy group (at a reference point $F_0 \in \check{D}$ corresponding to a Hodge decomposition $\oplus H^{p,q}$) the subgroup

$$B = \{g \in G_{\mathbb{C}} \mid gH^{p,q} \subset H^{p,q}\}.$$

So, again, the compact dual

$$\check{D} = G_{\mathbb{C}}/B,$$

a homogeneous variety, and in this case $\check{D}$ is even a smooth projective variety in which D sits as an open (but not Zariski open) subset.

Note that $\check{D}$ is the base of a principal bundle $B \to G_{\mathbb{C}} \to \check{D}$ and the holomorphic tangent bundle $T(\check{D})$ is the vector bundle which is associated to the adjoint representation of B on $\mathrm{Lie}\, G_{\mathbb{C}}/\mathrm{Lie}\, V$. The tangents coming from geometry turn out to be very special. To describe these, introduce

$$\mathfrak{g} : \mathrm{Lie}\, G_{\mathbb{C}}$$
$$\mathfrak{g}^{j,-j} := \{X \in \mathfrak{g} \mid XH^{p,k-p} \subset H^{p+j,k-p-j} \quad \text{for all } p \in \mathbb{Z}.\}$$

Clearly, $\mathfrak{g} = \mathrm{Lie}\, G \otimes \mathbb{C} = \oplus \mathfrak{g}^{j,-j}$ is a Hodge structure on $\mathrm{Lie}\, G$ which depends on the point $F_0 \in \check{D}$. The subbundle $T^{\mathrm{hor}} \subset T(\check{D})$ is given at F_0 by

$$T^{\mathrm{hor}}_{F_0} := \mathrm{Lie}\, B + \mathfrak{g}^{-1,1}/\mathrm{Lie}\, B.$$

This $G_{\mathbb{C}}$-equivariant subbundle of $T(\check{D})$ is called the *horizontal bundle*. We have a natural adjoint action of $\mathrm{Lie}\, B$ on $\mathfrak{g}^{-1,1}$ as well and hence an associated vector bundle on $\check{D}$ with fiber at F_0 given by $\mathfrak{g}^{-1,1}$ and we fix an identification of this bundle with T^{hor}:

$$T^{\mathrm{hor}}_{F_0} \simeq \mathfrak{g}^{-1,1}. \tag{9}$$

Let S be a complex manifold and suppose that a representation of $\pi_1(S, *)$ in $G_{\mathbb{Z}}$, the group of isometries of (H, Q) is given. So over some unramified cover $\tilde{S} \to S$ the representation becomes trivial. The minimal such cover is called the *monodromy cover*, S^{mon}. The image G^{mon} of $\pi_1(S, *)$ in $G_{\mathbb{Z}}$ is called the *monodromy group*.

By definition, a *variation of Hodge structures* on S of polarization type $(\vec{h}, Q)$ is a holomorphic map $p : S^{\mathrm{mon}} \to D$ which is *horizontal* in the sense that $p_*[T_s S^{\mathrm{mon}}] \subset T_F^{\mathrm{hor}} D$, $F = p(s)$. Such a map is called *period map*. Equivalently, it is a holomorphic map $p : S \to D/G^{\mathrm{mon}}$ lifting to a horizontal map $\tilde{S} \to D$ on a suitable unramified cover $\tilde{S} \to S$. There is also a direct definition:

Definition 2.1. *A* variation of Hodge structure *on S of weight k is a local system $\underline{H}_S$ of free $\mathbb{Z}$-modules of finite rank on S such that each fiber over $t \in S$ of the complexification admits a Hodge structure of weight k and such that*

— *the associated Hodge flag $F_t^\bullet$ depends holomorphically on t (this is the holomorphicity of the period map)*
— *the flat connection ∇ satisfies* Griffiths' horizontality condition*:*

$$\nabla_\xi F_t^q \subset F_t^{q-1}, \quad \xi \text{ a germ of a holomorphic tangent field at } t.$$

(this last condition is the horizontality of the period map).

The Hodge structure is polarized *by a flat bilinear integral form Q if Q induces a polarization on the Hodge structures on each fibers of $\underline{H}_S$.*

This reformulation leads to the basic examples provided by the primitive cohomology group of fixed rank of smooth projective varieties varying in a family:

Example 2.2. Let $p : X \to S$ be a smooth family of projective varieties. For simplicity, assume that every fiber X_t can be embedded in the same projective space so that it makes sense to speak of "the" cohomology class $h \in H^2(X_t)$ of a hyperplane section and hence there for each positive integer k there is an associated local system of primitive cohomology groups $H^k_{\mathrm{prim}}(X_t)$. For later reference, let me observe that since these form a local system, there is a canonically defined flat connection ∇, the *Gauss-Manin*-connection.

By Example 1.6 each of the modules $H^k_{\mathrm{prim}}(X_t)$ carries a polarized Hodge structure. Pulling back to the monodromy cover, this local system trivializes and a deep result of Griffiths states that Hodge filtration (7) varies holomorphically, i.e. the associated period map $p : S^{\mathrm{mon}} \to D$ is holomorphic and, horizontal. See [C-P-M] and the references therein.

To a polarized variation of Hodge structures on a local system H_S at any point $t \in S$ one can associate its *infinitesimal variation*, consisting of the defining polarized Hodge structure on (H, Q) together with the tangents at t to the period map, i.e.

$$\sigma : T = T_t S \to \mathrm{End}^{1,1}(H, Q) \otimes \mathbb{C}. \tag{10}$$

Pick $\eta \in T$ and write

$$\sigma_{p,q}(\eta) = \sigma | H^{p,q} \to H^{p-1,q+1}.$$

One can show (cf. [C-P-M, Theorem 5.3.4]) that in the geometric situation (example 2.2) this map just comes from cup-product with the image of η under the Kodaira-Spencer map $\rho : T \to H^1(X_t, X_t)$:

$$\sigma_{p,q} : H^q(X_t, \Omega^p_{X_t}) \xrightarrow{\cup\rho(\eta)} H^{q+1}(X_t, \Omega^{p-1}_{X_t}) \tag{11}$$

Infinitesimal variations have a rich multi-linear algebra structure. The maps $\sigma_{p,q}(\eta)$ can be composed into

$$\tau(\eta_1, \ldots, \eta_k) = \sigma_{0,k}(\eta_1) \circ \cdots \circ \sigma_{k,0}(\eta_k) : H^{k,0} \to H^{0,k}$$

By [C-P-M, § 5.5] this endomorphism is Q-symmetric and depends symmetrically on the η_j, thereby producing a homomorphism

$$\tau_t : \operatorname{Sym}^k T \to \operatorname{Sym}^2(H^{k,0})^\vee \subset \operatorname{End}^{-k,k}(H \otimes \mathbb{C}) \tag{12}$$

Remark 2.3. In the geometric situation this map, the *Griffiths-Yukawa coupling* can be defined in concrete terms using the Gauss-Manin connection (cf. Examples 2.2) as follows. Recall that the Gauss-Manin connection is the flat connection ∇ on the local system of the primitive cohomology groups $H^k_{\text{prim}}(X_t; \mathbb{C})$. Assume that S is a polydisc with local coordinates $(t_1, \ldots, t_n)$. Then $\nabla_{\partial/\partial t_j}$ acts as differentiation on cohomology classes. It will be abbreviated as $\frac{\partial}{\partial t_j}$. With this convention, at the origin, for any class $\omega \in H^{k,0}_{\text{prim}}(X_t; \mathbb{C})$ there is the following formula for the Yukawa-coupling in terms of the polarisation Q:

$$\tau\left(\frac{\partial}{\partial t_1}, \ldots, \frac{\partial}{\partial t_k}\right)\omega = Q\left(\omega, \frac{\partial}{\partial t_1} \circ \ldots \circ \frac{\partial}{\partial t_k}\omega\right) \tag{13}$$

2.2 *Higgs bundles*

An important property of variation of Hodge structures is their complete reducibility [Del71, 4.2.6]:

Theorem 2.4. *A polarized variation of Hodge structures over a quasi-projective manifold is direct sum of irreducible ones.*

It turns out that a further reduction is often needed. For this, one needs *complex systems of Hodge bundles*. To explain this notion, recall that a Hodge structure on a real vector space leads to a grading on the associated complex vector space and hence to a grading $H_S \otimes \mathcal{O}_S = \oplus_p \mathcal{H}^{p,k-p}_S$ on the bundle associated to a variation of Hodge structure. This grading is real in the sense that $\mathcal{H}^{p,k-p}_S = \overline{\mathcal{H}^{k-p,p}_S}$. One can identify $\mathcal{H}^{p,k-p}_S$ with the *holomorphic* bundle $\mathcal{F}^p/\mathcal{F}^{p+1}$. The horizontality just means that ∇ induces $\mathcal{O}_S$-linear bundle maps $\sigma_p : \mathcal{H}^{p,p-k}_S \to \mathcal{H}^{p-1,p+1-k}_S \otimes \Omega^1_S$. The pair

$$(\mathcal{H}_S, \sigma), \quad \text{with} \quad \mathcal{H}_S := \bigoplus_p \mathcal{H}^{p,k-p}_S, \quad \text{and} \quad \sigma := \bigoplus_p \sigma_p$$

is a *complex system of Hodge bundles*. It is an example of a Higgs bundle:

Definition. *A* Higgs bundle *over S is a couple $(\mathcal{H}, \sigma)$ consisting of a holomorphic vector bundle $\mathcal{H}$ over S and*

$$\sigma \in \mathrm{End}^{-1,1}(\mathcal{H}) \otimes \Omega^1_S \iff \sigma : TS \to \mathrm{End}^{-1,1}(\mathcal{H}), \quad \textit{the } \text{Higgs field}$$

having the he property that $\sigma \wedge \sigma = 0$ which is just the flatness of the connexion ∇.

Remarks 2.5.

1. Note that since there is no a priori real structure on a Higgs bundle so that the additional reality constraint on a complexified variation of Hodge structure is an extra ingredient.
2. In general an irreducible variation gives a complex system of Hodge bundles which splits further into Higgs bundles which themselves are complex systems of Hodge bundles.

A flat bilinear *hermitian* form $Q = \sum Q_p$ on a complex system of Hodge bundles $\mathcal{H} = \bigoplus \mathcal{H}^p$ is called a *polarization* if the splitting is Q-orthogonal and if $(-1)^p Q_p$ is positive definite on $\mathcal{H}^p$. If $\mathcal{H}$ comes from a polarized variation of Hodge structures, the polarization indeed gives a polarization for $\mathcal{H}$.

The complete reducibility result stated for polarized variations of Hodge structures remains true for polarized complex systems of Hodge bundles [Del87, § 1.11]:

Theorem 2.6. *A polarized system of complex systems of Hodge bundles over a quasi-projective manifold is an orthogonal direct sum of irreducible polarized complex systems of Hodge bundles.*

This implies that the complexification of an irreducible variation of Hodge structures may further split into irreducible Higgs bundles.

Example 2.7. Suppose that H_S is a local system of $\mathbb{Q}$-vector spaces underlying a variation of Hodge structures of weight k. Let H be the typical fiber of H_S. Let $A \in \mathrm{End}(H, Q) \otimes \mathbb{C}$ be a non-zero flat endomorphism of H. Then $\mathrm{Ker}\, A$ defines a non-trivial subsystem of H_S. As remarked above, complex systems of Hodge bundles are completely reducible and so $H_S \otimes \mathbb{C} = \mathrm{Ker}(A) \oplus M_S$ where M_S is a complex variation of Hodge structures.

Suppose that A is pure of type $(-1, 1)$. If $H_S \otimes \mathbb{C}$ has a non-trivial part of type $(0, k)$, the preceding splitting is a non-trivial splitting since $AH_S^{0,k} \subset H_S^{-1,k+1} = 0$. This splitting in general is not a splitting of variations of Hodge structures since $H_S^{k,0}$ does not have to belong to $\mathrm{Ker}(A)$.

2.3 *Curvature*

The goal is to describe the curvature of a suitable connection on the horizontal tangent bundle $T^{\mathrm{hor}} D$. It is a subbundle of the full tangent bundle and thus one can calculate the curvature of the hermitian metric defined by the classical *trace form* on $\mathfrak{g}$

$$B(X, Y) = \mathrm{Tr}\, X \circ Y, \qquad X, Y \in \mathfrak{g}.$$

This a symmetric form, invariant under the adjoint action of $G_{\mathbb{C}}$ on $\mathfrak{g}$ and it is real on Lie (G). The involution θ on $\mathfrak{g}$ which is just multiplication with $(-1)^j$ on $\mathfrak{g}^{-j,j}$ can be shown to be real on Lie G and the expression

$$h(X, Y) := -B(\theta X, \bar{Y}), \qquad X, Y \in \mathfrak{g} \tag{14}$$

being invariant under the adjoint action of V defines a sesquilinear form on $T_{F_0}D =$ Lie $G/$Lie V. It turns out (see [C-P-M, Prop. 12.2.5]) that this form is in fact a hermitian metric on $T_{F_0}D$, which by construction extends to a G-invariant hermitian metric on TD.

Recall that the *Chern connection* D_h is the unique metric connection on TD whose (0, 1)-part is just the operator $\bar{\partial}$. The curvature of this connection is an End(TD)-valued 2-form F_h and when evaluated on germs of holomorphic vector fields X and Y at $F_0 \in D$ yields $F_h(X, Y)$, an endomorphism of $T_{F_0}D$.

To describe it on horizontal directions, one resorts to the identification (9). To start, note that for $X \in \mathfrak{g}$, by definition $Q(Xu, v) + Q(u, Xv) = 0$ for all $u, v \in H_{\mathbb{C}}$, i.e. $\mathfrak{g}$ consists of the Q-skew endomorphisms. This does not mean that these are skew with respect to the Hodge metric h (8), since here the Weil-operator C intervenes which depends entirely on the Hodge structure. Indeed, for $X \in \mathfrak{g}$ we may define $X^* \in \mathfrak{g}$ as the h-adjoint of X, i.e. $h(Xu, v) = h(u, X^*v)$ for all $u, v \in H_C$. If $\bar{X}$ is conjugation with respect to the real structure coming from $\mathfrak{g} =$ Lie $G \otimes \mathbb{C}$, i.e. $\bar{X}(u) = \overline{X(\bar{u})}$, $u \in H_{\mathbb{C}}$ we have $X^* = \bar{X}$ if $X \in \mathfrak{g}^{-j,j}$ with j odd, while $X^* = -\bar{X}$ if j is even (see [C-P-M, Cor. 12.2.3], the transpose-sign should be deleted, it has no meaning). This clarifies the relation between the Hodge metric and the global complex structure on $\mathfrak{g}$.

Coming back to the curvature calculation, assume now that X and Y are germs of horizontal vector fields near F_0 so that $Y^* = \bar{Y} \in \mathfrak{g}^{1,-1}$ and $[X, Y^*] \in \mathfrak{g}^{0,0} =$ Lie V. The adjoint action of this element on $\mathfrak{g}^{-1,1}$ gives an endomorphism on $T^{\text{hor}}_{F_0}D = \mathfrak{g}^{-1,1}$ and by [C-P-M, § 13.3] we have

$$F_h(X, Y) = -\frac{1}{2}\,\text{ad}[X, Y^*]. \tag{15}$$

On horizontal directions X, Y the metric (14) becomes $h(X, Y) = B(X, Y^*) =$ Tr $X \circ Y^*$. Using (15) and the Jacobi-identity, as in [Pe90, Corr. 1.8] one finds

Lemma 2.8. *Let* $X, Y \in \mathfrak{g}^{-1,1}$ *then*

$$h(F_h(X, X^*)Y, Y) = 2\|[X, Y]\|_h - \|[X^*, Y]\|_h$$

so in particular, if $[X, Y] = 0$, *one has* $h(F_h(X, X^*)Y, Y) \leq 0$.

2.4 *Deforming period maps*

Let $p : S^{\text{mon}} \to D$ be a period map. To simplify notation, assume that $S^{\text{mon}} = S$ so that the source of the period map remains algebraic. In particular, S has a suitable smooth compactification $\bar{S}$. As $p^*T^{\text{hor}}D$ is contained as the $(-1, 1)$-part of an endomorphism bundle of Hodge structures, one can take its canonical extension

$$E := [f^*T^{\text{hor}}D]_{\text{can}},$$

a holomorphic object E on $\bar{S}$ (see the end of § 1.3). Actually E generally only is a coherent sheaf, but I neglect details like this and I assume for simplicity that E is a vector bundle. The metric (14) then is defined on $E|S$.

Notation. A metric h on a holomorphic vector bundle E can be used to contract a pair of E-valued differential forms Ω, Θ to an ordinary differential form denoted $\mathrm{Tr}(\Omega \wedge \Theta)$: on decomposables $\Omega = e \otimes \alpha$, $\Theta = f \otimes \beta$ we put $\mathrm{Tr}(\Omega \wedge \Theta)(e \otimes \alpha, f \otimes \beta) := h(e, f)\alpha \wedge \bar{\beta}$; in general extend this in a sesquilinear fashion.

As an example, if $\xi \in H^0(\bar{S}, E)$ is a global section, the curvature form with respect to h is an E-valued $(1, 1)$-form $F_h\xi$ on S so that $\mathrm{Tr}(F_h\xi \wedge \xi)$ is an ordinary $(1, 1)$-form on S. It can be evaluated on the pair (X, Y), where $X \in \mathfrak{g}^{1,1}$ comes from a tangent vector $X' \in T_sS$ and the tangent Y at the point $F_0 = p(s)$ comes from the deformation. One has:

$$\mathrm{Tr}(F_h\xi \wedge \xi)(X, Y) = h(F_h(X, X^*)Y, Y). \tag{16}$$

Of course period maps can a priori be deformed into maps $S \to D$ that are no longer horizontal, but these are not coming from geometry. For this reason only those deformations $P : S \times T \to D$ are considered which are horizontal themselves. This implies by [C-T, Prop. 5.2.] that any two vectors in $P_*(T_{(s,t)})$ when considered in the Lie-algebra $\mathfrak{g}^{1,1}$ (see (9)) must commute. This applies to X and Y so that from Lemma 2.8 and (16) one concludes that $\mathrm{Tr}(F_h s \wedge s)$ is negative semi-definite. We now invoke the following principle (see [C-P-M, 11.1.9]:

Proposition 2.9 (Principle of Plurisubharmonicity). *Let (E, h) be a Hermitian holomorphic vector bundle over a compact complex manifold and let s be a* holomorphic *section of E. Suppose that the* $(1, 1)$*-form* $\mathrm{Tr}(F_h s \wedge s)$ *is negative semi-definite. Then s is* flat *with respect to the Chern connection D_h, i.e. $D_h s = 0$.*

So from Lemma 2.8 one concludes that the section ξ is flat with respect to D_h and this proves the analog of (5) in the Hodge theoretic setting of a period map $\mu = \tilde{p} : \tilde{S} \to D$:

Corollary 2.10. *The tangent space at μ to the space of period map deformations is*

$$T_\mu \mathrm{Def}(\mu) = \left\{ \begin{array}{c} Q\text{-skew endomorphisms of } H \text{ of type } (-1, 1) \\ \text{commuting with the monodromy group.} \end{array} \right\}$$

The left hand side is also equal to the algebra of global flat endomorphisms of the underlying local system (H_S, Q) which are of type $(-1, 1)$. To the variation one has associated a Higgs bundle (see § 2.2). Its Higgs field at every point $t \in S$ can be shown to be just the infinitesimal variation of Hodge structure (10) at that point. So one can reformulate Cor. 2.10 in terms of Higgs bundles as follows:

Theorem 2.11. *Tangents to algebraic deformations of a given variation of Hodge structure on S correspond to flat global type $(-1, 1)$ endomorphisms of the underlying local system (H_S, Q) which are at the same time Higgs fields for the associated Higgs bundle.*

3. Back to geometry

Let me now come back to the setting of the Shafarevich problem. The class $\mathcal{C}$ of projective manifolds is now assumed to have a nice quasi-projective coarse moduli space M. The base manifold $\bar{S}$ is fixed as well as the degeneracy locus Σ which will be assumed to be a strictly normal crossing divisor. Take a family $\bar{X} \to \bar{S}$ with fibers $X_t, t \in S = \bar{S} - \Sigma$ in $\mathcal{C}$ so that, by assumption, there is a moduli map $\mu_f : S \to M$. On the other hand, one can apply Example 2.2 to $f : X \to S$, the restriction of $\bar{f}$ to $X = f^{-1}S$ so that there is a period map $p_f : S \to D/\Gamma$, where D is the period domain associated to the primitive cohomology groups of the fibers of some fixed rank, and where Γ is the monodromy group. These two maps fit into a commutative diagram

$$\begin{array}{ccc} S & \xrightarrow{\mu_f} & M \\ & {\scriptstyle p_f}\searrow \quad \swarrow{\scriptstyle \iota} & \\ & D/\Gamma. & \end{array}$$

Clearly, if one wishes to draw geometric conclusions from properties of the variation of Hodge structure, one likes to know that the variation determines the family, at least locally. If this is the case, one says that the *(local) Torelli theorem holds* for $\mathcal{C}$. It means that the map ι is an immersion.

Examples 3.1. The Torelli theorem holds for curves. The local Torelli theorem holds for hypersurfaces in projective $(n+1)$-space of degree $\geq n+1$. It also holds for any projective manifold with trivial canonical bundle such as an Abelian variety, a K3-surface, or any other Calabi-Yau manifold. In fact, if n is the dimension, it suffices to look at the variation of $H^{n,0}$ inside primitive cohomology. See [C-P-M] for an explanation and references to the original works.

Under the hypothesis that ι is an immersion, the above diagram gives an inclusion $\mu_f^* T_M \subset p_f^* T_D^{\text{hor}}(D/\Gamma))$. Recall that the right hand side is just the $(-1,1)$-part of the endomorphism bundle of the variation of Hodge structure and has a quasi-canonical extension (see directly below the statement of Thm. 1.7). By definition, the quasi-canonical extension of $\mu_f^* T_M$ is just the closure of this bundle inside the quasi-canonical extension.

As shown before, the tangent space to deformations of p_f as a variation of Hodge structure, is just

$$\text{End}^Q = \left\{ \begin{array}{l} \text{endomorphisms of } (H^k_{\text{prim}}(X_t; \mathbb{Q}), Q) \\ \text{commuting with the monodromy group,} \end{array} \right\}$$

i.e.,

$$H^0(\bar{S}, [\mu_f^* T_M]_{\text{can}}) = [\text{End}^Q \otimes \mathbb{C}]^{-1,1}.$$

To bring the geometric deformations into the picture, note that being a coarse moduli space implies that at any smooth point $m \in M$ corresponding to a variety X_m, there is a canonical identification

$$T_m M \simeq H^1(X_m, TX_m)$$

and hence $\mu_f^* T_M$ can be identified as the bundle whose fiber at t equals $H^1(X_s, TX_t)$. Any geometric deformation η of f thus defines at each point $t \in S$ an element $\theta_t \in H^1(X_t, TX_t)$, the associated *infinitesimal Kodaira-Spencer class* of f at t. By (11) cup product with it gives an infinitesimal variation of Hodge structure at the point t:

$$\sigma(\eta_t) \in \mathrm{End}^{-1,1} H^k_{\mathrm{prim}}(X_t; \mathbb{C}).$$

Such an infinitesimal variation can be called the infinitesimal variation in the *geometric direction defined by* η. Tangents to geometric deformations correspond to global flat endomorphisms inducing such infinitesimal variations at each point $t \in S$ and vice versa. Summarizing, one has:

Proposition 3.2. *If the (local) Torelli property holds for* $\mathcal{C}$, *tangents* η *to geometric deformations correspond to those flat endomorphisms* $[\mathrm{End}^{\mathcal{Q}} \otimes \mathbb{C}]^{-1,1}$ *which at each point define an infinitesimal variation of the Hodge structure in the geometric direction defined by* η*, and vice versa.*

Let me now consider the kernel of such flat endomorphisms $\sigma(\eta)$. In Example 2.7 we have seen that the complexified variation splits as $\mathrm{Ker}(\sigma) \oplus M_S$ and $\mathrm{Ker}(\sigma)$ contains the $(0,k)$-part of the variation. If $K = M_S^{k,0}$ and if $A_1, \ldots A_k \in \mathrm{End}\, H \otimes \mathbb{C}$ all have pure type $(-1,1)$, their composition $A_1 \circ \ldots \circ A_k$ maps K to the $(0,k)$-part in M_S which is zero by construction. Applying this to the maps which compose into the Yukawa coupling (13), we conclude that any non-trivial deformation forces the Yukawa-coupling to vanish on K. If for instance $h^{k,0} = 1$ the entire $(k,0)$-part of the variation is contained in K and so:

Corollary 3.3. *In the above situation, if* $h^{k,0} = 1$*, any non-trivial geometric deformation forces the Griffiths-Yukawa coupling to vanish.*

If I combine this result with

Proposition 3.4. *Let there be given a weight* k *polarized variation of Hodge structures over a non-compact algebraic curve* S *with* $h^{k,0} = h^{0,k} = 1$*, and suppose hat there is a point at infinity where the local monodromy operator has maximal order of unipotency* $(k+1)$ *(see Theorem 1.7). Then the Griffiths-Yukawa coupling is non-zero.*

I deduce the following criterion:

Corollary 3.5. *Let* $f : X \to S$ *be a non-isotrivial family of Calabi-Yau's over a non-compact curve* S *and suppose hat there is a point at infinity where the local monodromy operator has maximal order of unipotency* $(k+1)$ *(see Theorem 1.7). Then* f *is rigid.*

Sketch of the proof of Prop. 3.4. Suppose (Δ, t) is a local coordinate disc centered at a puncture where the local monodromy T has order of unipotency $(k+1)$. For simplicity assume that the monodromy is unipotent of order $(k+1)$ and introduce

$$N = \log(T - I) = \sum \frac{(-1)^{\ell+1}}{\ell} T^{\ell}$$

an operator with $N^k \neq 0$, $N^{k+1} = 0$. Schmid in [Sch] has introduced a mixed Hodge structure on the typical fiber H of the local system H_S underlying the variation of Hodge structure whose weight filtration $W_0 \subset \cdots \subset W_{2n}$ is determined by N and whose Hodge structure is given by the fiber at 0 of the quasi-canonical extensions of the Hodge filtration bundles. In the case at hand $[\mathcal{H}^{k,0}]_{\mathrm{can}}$ is a line bundle trivialized by a section whose value at 0 can thus be seen as an element $\omega \in H \otimes \mathbb{C}$. The map N is a type $(-1,-1)$-morphism of $H \otimes \mathbb{C}$. In particular $N^k : W_{2k}/W_{2k-1} \xrightarrow{\sim} W_0$ maps the pure (k,k)-type Hodge structure of the left hand side to the $(0,0)$-type Hodge structure on the right hand side. Since $N^k \neq 0$ both sides are 1-dimensional and necessarily $\omega \in W_{2k}$ represents a generator. Now one needs to use that N is essentially the residue at 0 of the logarithmic connection on the canonical extension $\mathcal{H}_{\mathrm{can}}$ and after a suitable renormalization one then has $N\omega = \sigma(t[\partial/\partial t])\omega$ and in particular, the Griffiths-Yukawa coupling $\sigma^k(t[\partial/\partial t])$ does not vanish. □

Remark 3.6. In [L-T-Y-Z, Theorem 38] the proof appears to have a gap. The preceding correction has been communicated to me by Kang Zuo. I also want to remark that this gives another proof of Cor. 1.15 in the special case of K3-surfaces.

I want to finish by showing how the above ideas give a simpler proof of [VZ, Lemma 4.3]. First a couple of definitions. Let S be a curve. At any point $t \in S$, the Higgs field gives an endomorphism σ of the variation which is well defined up to a multiplicative constant. A variation of Hodge structure of weight k is said to be *of Hodge-Lefschetz type a* if

- for some integer a with $0 \leq a \leq 2k$ one has $h^{k,0} = \cdots = h^{k-a+1,a+1} = 0$
- the maps $\sigma : H^{k-j,j} \to H^{k-j-1,j+1}$ $j = a, \ldots, k-a$ are isomorphisms at a generic point $t \in S$.

This implies that the length of the Hodge filtration is exactly $k - 2a + 1$ and all non-zero Hodge numbers are equal. Any variation of Hodge structures of weight k which is the direct sum

$$H_S \otimes \mathbb{C} = F_0 \oplus \cdots \oplus F_k, \qquad F_a \text{ Hodge-Lefschetz of type } a$$

is called (strictly) *maximal*.

Proposition 3.7 ([VZ, Lemma 4.3]). *A variation over a curve with strictly maximal Higgs field is rigid.*

Proof. If A is a flat type $(-1,1)$ endomorphism of such a system, $\ker A$ contains the $(0, k-a)$-part of F_a, hence, since $\ker A$ is a Higgs subfield, it must contain the entire Hodge-Lefschetz subsystem F_a, hence $H_S \otimes \mathbb{C} = \ker A$, i.e. $A = 0$ □

References

[Arak] A. Arakelov, Families of algebraic curves with fixed degeneracies, *Izv. Ak. Nauk. S.S.S.R, ser. Math.*, **35** (1971) 1277–1302.

[B-H-P-V] W. K. Barth, K. Hulek, C. Peters and A. Van de Ven, Compact complex surfaces, Second Enlarged Edition, Erg. Math., 3. Folge Band 4, 433+X pag., Springer Verlag (2003).

[C-P-M] J. Carlson, S. Mueller-Stach and C. Peters, Period Mappings and Period Domains, *Cambridge University Press, Cambridge Studies in advanced mathematics*, **85** (2003) 430 pp.

[C-T] J. Carlson and D. Toledo, Integral manifolds, harmonic mappings, and the abelian subspace problem, in Algebra—some current trends (Varna, 1986) 60–74, Lecture Notes in Math., **1352**, Springer, Berlin (1988).

[Del70] P. Deligne Équations différentielles à points singuliers réguliers, Springer Lecture Notes in Math., **163** (1970).

[Del71] P. Deligne, Théorie de Hodge II, *Publ. Math. I.H.E.S*, **40** (1971) 5–58.

[Del87] P. Deligne, Un théorème de finitude pour la monodromie, in *Discrete groups in geometry and analysis*, Progr. in Math. 67, Birkh. Verlag (1987) 1–19.

[Fa83a] G. Faltings, Arakelov's theorem for abelian varieties, *Invent. Math.*, **73** (1983) 337–348.

[Fa83b] G. Faltings, Endlichkeitssätze für abelsche Varietäten über Zahlkörpern, *Invent. Math.*, **73** (1983) 349–366.

[I-S] Y. Imayoshi and H. Shiga, A finiteness theorem for holomorphic families of Riemann surfaces. In Holomorphic Functions and Moduli II, 207–219, Springer-Verlag: MSRI publications volume **11** (1988).

[L-T-Y-Z] K. Liu, A. Todorov, S-T. Yau and K. Zuo, Shafarevich's conjecture for CY manifolds, I, preprint (2004).

[Mo] B. Moonen, Linearity properties of Shimura varieties, *I. J. Algebraic Geom.*, **7** (1998) 539–567.

[Nst] P. E. Newstead, Introduction to Moduli Problems and Orbit Spaces, Springer-Verlag, Berlin/New York (1978).

[Pe84] C. A. M. Peters, A criterion for flatnes of Hodge bundles over curves and geometric applications, *Math. Ann.*, **268** (1984) 1–19.

[Pe90] C. A. M. Peters, Rigidity for variations of Hodge structure and Arakelov-type finiteness theorems, *Comp. Math.*, **75** (1990) 113–126.

[Pe00] Peters, Chris, Arakelov-type inequalities for Hodge bundles, prépublication de l'Institut Fourier, 511 (2000).

[Sa] M.-H. Saito, Classification of non-rigid families of abelian varieties, *Tohoku Math. J.*, **45** (1993) 159–189.

[S-Zu] M.-H. Saito and S. Zucker, Classification of non-rigid families of K3-surfaces and a finiteness theorem of Arakelov type, *Math. Ann.*, **289** (1991) 1–31.

[Sch] W. Schmid, Variation of Hodge structure: the singularities of the period mapping, *Invent. Math.*, **22** (1973) 211–319.

[Si90] C. Simpson, Harmonic bundles on non-compact curves, *J. Am. Math. Soc.*, **3** (1990) 713–770.

[Si92] C. Simpson, Higgs bundles and local systems, *Publ. Math. I.H.E.S.*, **75** (1992) 5–95.

[V] C. Voisin, Théorie de Hodge et géométrie algébrique complexe, Cours spécialisés, 10, Soc. Math. Fr. (2002).

[VZ] E. Viehweg and K. Zuo, Families over curves with strictly maximal Higgs field, arXiv:math.AG/03/03103.

Teichmüller Theory and Moduli Problem No. 10, 2010, pp. 549–576.

On the Art of Calculating Accessory Parameters of Conformal Mappings of Circular Arc Polygons—General Considerations and Special Situations

R. Michael Porter*

Department of Mathematics, Centro de Investigación y de Estudios Avanzados del I.P.N., Apdo. Postal 1-798, Arteaga 5, 76001 Santiago de Querétaro, Qro, México
e-mail: mike@math.cinvestav.mx

Abstract. A brief history of the computation of conformal mappings for regions bounded by circular arcs is presented and general considerations regarding this art are discussed. The particular problem of mappings to horocycle domains (formed by removing internally tangent disks from a base disk) is detailed.

Keywords. Conformal mapping, accessory parameter, circle arc polygon, cusp region, horocycle domain, Schwarzian derivative, Kleinian group.

Subject Classification: Primary 30C30, Secondary 30C20

1. Introduction

The field of numerical conformal mapping has a long history and is rich in beauty. In this report we will view one tiny corner of this field, the calculation of accessory parameters of conformal mappings of circular polygon domains and in particular domains with cusp angles.

Numerical conformal mapping means the quest for approximations to the function whose existence is given by the Riemann mapping theorem [25,46,54] (or analogous statements for multiply-connected domains). Let $D \subseteq \mathbb{C}$ be bounded by a nonpathological Jordan curve ∂D. Let $\mathbb{D} = \{z \in \mathbb{C} : |z| < 1\}$ and let $f : \mathbb{D} \to D$ be a conformal (Riemann) mapping. Since Carathéodory [17] it is known that there is a homeomorphic extension of f to the closures of the domains. The boundary map $\partial\mathbb{D} \to \partial D$ has a Fourier series which in fact has the same coefficients as the power series of f centered at the origin. When we know this boundary map we can recover f in the interior of $\mathbb{D}$ from these Fourier coefficients or alternatively by the Cauchy integral formula. For this reason the two-dimensional problem of finding $f(z)$ for all $z \in \mathbb{D}$ is often reduced to the one-dimensional problem of finding $f(z)$ for all $z \in \partial\mathbb{D}$. In the case of polygonal domains, however, the reduction is carried even further: the problem is to find a *zero-dimensional* set of *accessory parameters*.

This article is intended to be accessible to non-specialists. After some generalities on conformal mapping, we will discuss the peculiarities of the accessory parameter

*Research partially supported by CONACyT grants 60160 and 80503.

problem for mappings to simply connected domains bounded by circular arcs. Numerical work for straight-edged polygons and the classical Schwarz-Christoffel formula was pioneered in the 1960s by L. N. Trefethen, after initial steps by Kufarev [41] relating to domains slit from infinity. The first numerical work on circular polygons may have been in 1987 by Bjørstad and Grosse [10], and it seems that little more was done until [27]. We will discuss in some detail some recent work by P. Brown [13] on circular quadrilaterals. Then the major part of this article will explain a procedure given in [61] for calculating mappings to horocycle domains.

2. Context of numerical conformal mapping

We begin with some general background to clarify what we will and will not be discussing in detail. An *accessory-parameter* conformal mapping method is a numerical scheme for approximating a finite list of parameters which appear in a particular formula and permit one to calculate a conformal mapping within a predetermined family of mappings. Generally accessory parameters are constants which appear in differential equations. The most frequent applications are to polygonal-type domains such as we will discuss here, but this is not strictly necessary.

In contrast, a *continuous method* is any scheme for calculating conformal mappings which is not an accessory-parameter method. Continuous methods fall into a few roughly defined classes. There are methods which approximate $f(e^{2\pi ik/N}) \in \partial D$ $(0 \leq k < N)$ via some property of holomorphic functions related to power series, Fourier series, integral representations, or other analytic properties. These methods are generally the most competitive in terms of speed of convergence towards the desired mapping, but pay the price that one must have a sufficiently close initial guess to start the process. A detailed survey is to be found in [68]; see also [19,25,42]. Another class, the osculation methods, approximate f^{-1} beginning with a mapping $h_0 : D \to D_0 \subseteq \mathbb{D}$ and composing succesively with $h_j : D_{j-1} \to D_j \subseteq \mathbb{D}$ in such a way that ∂D_j approximates $\partial\mathbb{D}$ for large j. For these methods, which do not require a close initial guess but which generally converge slowly, see for example [23,25,60]. Finally, there are methods which resist classification such as W. Thurston's celebrated circle-packing method [62,64] and the Marshall-Rohde zipper algorithm [47]. It is surprising that there can be so many essentially different ways to get a hold on f numerically.

2.1 *Straight polygons*

For an ordinary n-gon, composed of straight segments β_k joining pairs $w_k, w_{k+1} \in \mathbb{C}$ of successive vertices (we will assume cyclic indexing, that is, $w_{n+1} = w_0$), the basic tool is the Schwarz-Christoffel formula which states that the derivative of f is of the form

$$f'(z) = \frac{c}{\prod_{k=1}^{n}(z - z_k)^{1-\theta_k/\pi}}. \tag{1}$$

Here c is a constant and z_k are the *prevertices* of the conformal mapping f, that is, $z_k \in \partial\mathbb{D}$, $f(z_k) = w_k$. The arc $\alpha_k \subseteq \partial\mathbb{D}$ from w_k to w_{k+1} is mapped by f to

β_k. The fractional powers appearing in (1) are well defined in the simply connected region $\mathbb{D}$. When the polygon ∂D is given, the angles θ_k are given and the mapping problem reduces to finding only the prevertices (or their arguments), which are thus the accessory parameters for this mapping problem. When they are known, integration of (1) gives values of f in the interior as well as on the boundary of $\mathbb{D}$. The definitive work on Schwarz-Christoffel mappings is [20].

Crowding An obstacle which conformal mapping methods must deal with is the crowding phenomenon, reflected in the fact that $|f'(z)|$ may become very large or very small as z tends to certain parts of $\partial\mathbb{D}$. In consequence small changes in z on or near the boundary may correspond to large changes in w or vice versa. Since numerical work generally restricts the calculations to approximately equally spaced points along $\partial\mathbb{D}$ or along ∂D, when the crowding factor $\max(|f'|, |f'|^{-1})$ is very large one will need a large number of points on one boundary to have a detectable difference between corresponding points on the other. The consequence will be either a loss of precision or a high cost of calculation. For example, a conformal mapping from $\mathbb{D}$ to a rectangle D of base a and height 1 has a crowding factor which grows exponentially with a. Specifically, suppose the vertices of D are $f(\pm e^{\pm it})$. Then the relation of a and t is approximately as follows.

a	t
5.0	10^{-13}
10.0	10^{-26}
15.0	10^{-41}
20.0	10^{-54}

A breakthrough in dealing with the crowding phenomenon in the mapping of polygons was presented in [21] with the CDRT (Cross Ratios of the Delaunay Triangulation) algorithm. In this method, the polygonal domain D is first triangulated in a reasonably homogeneous way (Delaunay triangulation). The pairs of adjacent triangles in this triangulation form a collection of $n-3$ quadrilaterals with vertices on ∂D. The conformal modules (see [20, p. 19–20]) of these quadrilaterals (or equivalently, the cross ratios of their vertices) provide a description of D which in many ways is more natural than the positions of the vertices w_k. The preimages under f of the vertices of the quadrilaterals are quadruples of prevertices, and the cross ratios of these prevertices give a convenient charactarization of the positions of the prevertices themselves. The CRDT method solves a system of nonlinear equations to find these cross ratios. Since the cross ratio of four points is invariant under Möbius transformations, one may freely map any triangle near $\partial\mathbb{D}$ to one near the center (and thus far from the boundary), solve the nonlinear equations, and then map back near the boundary. In this way the crowding difficulty is resolved in virtually any interesting mapping problem of polygons. There are few other conformal mapping methods which relate nicely to changes by Möbius transformations (circle packing is one).

While in principle any piecewise smooth curve can be approximated by a polygon and can thus be approached with the Schwarz-Christoffel method, in practice the number of vertices may have to be very large and the formula (1) becomes unwieldy.

2.2 *Circular polygons*

From now on D will be a circular polygon, that is, a Jordan curve formed of circular arcs β_k. The notation will be as for straight polygons: the edges $\beta_k = f(\alpha_k)$ comprising $\partial D = f(\partial\mathbb{D})$ meet at vertices $w_k = f(z_k)$ with interior angles θ_k, and β_k joins w_k to w_{k+1}.

Circular polygons are ubiquitous in the study of Kleinian groups [48] and figure prominently in the work of Poincaré, Koebe, Ahlfors, Bers, Maskit and many others. A number of applications of circular polygons in the proofs of important theorems in complex function theory are also mentioned in the introduction to [13] and we will discuss some in Section 4. Nonetheless, there have been few advances in our general understanding of circular polygon conformal mapping beyond the general structure, analogous to the Schwarz-Christoffel formula (1) given in the text [54], a version of which we give in Proposition 1.

The basic tool for studying f is the representation formula for its Schwarzian derivative $\mathcal{S}_f = (f''/f')' - (1/2)(f''/f')^2$. The invariance of S_f with respect to Möbius transformations applied to f is mirrored in the fact the Möbius transformations send circular arcs to circular arcs. The structure theorem states that for circular polygons the Schwarzian derivative is a rational function of degree two.

Proposition 1. *The Schwarzian derivative S_f of a conformal mapping from $\mathbb{D}$ to a circular-polygon domain D is of the form*

$$\mathcal{S}_f(z) = z^{-2} \sum_{k=1}^{n} \left(\frac{a_k z_k z}{(z - z_k)^2} + i r_k \frac{z + z_k}{z - z_k} \right), \tag{2}$$

where the constants $a_k \in [-3/2, 1/2]$ are related to the interior angles θ_k of D by

$$a_k = \frac{1}{2}(1 - (\theta_k/\pi)^2), \tag{3}$$

and where the accessory parameters $z_k = e^{it_k}$, r_k satisfy the linear relations

$$\sum_{k=1}^{n} r_k = 0, \quad \sum_{k=1}^{n} z_k(a_k + 2ir_k) = 0, \tag{4}$$

and $r_k \in \mathbb{R}$.

This formula is quite similar to the version given in [54] for the case of a mapping from the upper half-plane to $\mathbb{D}$. The relations (4) come out slightly different. The general principle is described in detail in [54,20], first for the preschwarzian f''/f', which is directly related to straight polygons, and then for $\mathcal{S}_f$. The present version, for f defined in the unit disk, was given in [27, p. 19] and a direct proof (rather than by change of variable from the upper half plane) suggested by Ch. Pommerenke was included in [61, p. 16–17] (note however that (3) here corrects a typographical error).

Once we know the accessory parameters $\vec{z}$ and $\vec{r} = (r_1, \ldots, r_n)$ we can solve for the mapping function f when suitably normalized (more on this in Section 3). As a partial converse, we can only say that if the Schwarzian derivative of a holomorphic function f defined in $\mathbb{D}$ is of the form (2) with its accessory parameters satisfying (4), then f is a *local* homeomorphism, and that its image domain is a (not necessarily schlicht) region bounded by a (not necessarily simple) circular polygon with interior angles $\theta_1, \ldots, \theta_n$.

Let us examine the case of a circular quadrilateral ∂D. Since we are interested in knowing the conformal mapping for D up to Möbius transformations, we assume that w_1, w_2, and w_3 are 0, 1, and ∞ respectively. Thus w_4 is left to wander with two real degrees of freedom, and by "flexing" the edges (taking care not to let them cross) we have four additional parameters (Figure 1). On the other hand, we could count the parameters instead by means of (2). The difference is that the viewpoint shifts to the domain of f rather than the codomain. In the formula there are four each of $z_k = e^{it_k}$, a_k, and r_k. By normalization of the prevertices (i.e., a self mapping of $\mathbb{D}$), we should disregard t_1, t_2, and t_3, and since (7) consists essentially of three real constraints, we find once again that the space of normalized circular quadrilaterals has six real dimensions.

In the same way it is not hard to see that the space $\mathcal{P}_n$ of circular n-gons (not necesarily simple), modulo equivalence under mappings by Möbius transformations, has dimension $3n - 6$, while the submanifold of polygons with the n angles θ_k specified has dimension $2n - 6$.

It could be reasonably argued that, due to the inherent limitations of computational schemes, every imaginable conformal mapping method must really be hunting for the desired solution as a point in some finite-dimensional space. For example, the methods such as Theodorsen [25], Symm [25], Fornberg [25], Wegmann [25], Porter [59] and

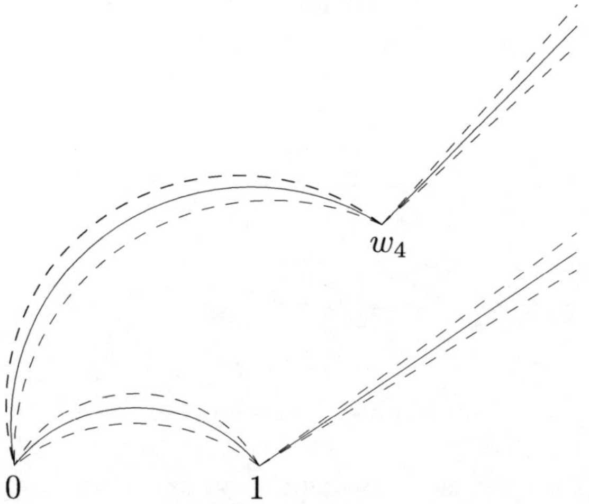

Figure 1. Degrees of freedom of circular quadrilaterals.

many others fix a value N and iterate some operation on a vector $(w_1, \ldots, w_N)$ which converges to a point in C^N which approximates $(f(e^{2\pi i/N}), \ldots, f(e^{2\pi Ni/N}))$ to the extent that the truncation error implied by the choice of N permits. Likewise the osculations methods [25] begin with a polygonal discretization of the boundary of D and approximate the preimages of this finite set of points. Nonetheless, this N must grow arbitrarily large as more precision is desired; the spirit of these methods is to approximate an intrinsically continuous phenomenon. Even for a domain as simple as an ellipse, we are hunting for a solution in an infinite-dimensional space (unless somehow we can formulate the map in terms of a class of particular functions, such as perhaps elliptic functions, and then look for our mapping in that class). This is in stark contrast to the accessory-parameter approaches to conformal mapping .

3. Numerical considerations for circular polygons

In this section we discuss the question of calculating the vertices w_k when all the parameters in (4) are provided. We will generally suppose that $0 \in D$ and that $f(0) = 0$. A full normalization

$$f(0) = 0, \quad f'(0) = 1 \tag{5}$$

will occasionally be required. The set of interest is naturally

$$\{(\vec{z}, \vec{r}, \vec{a}) \in (\partial\mathbb{D})^n \times \mathbb{R}^n \times [-3/2, 1/2]^n : \\ \text{(2) is the Schwarzian derivative of a schlicht function in } \mathbb{D}\}. \tag{6}$$

3.1 *Poles of* $\mathcal{S}_f$

The first matter to be dealt is that although the function $\mathcal{S}_f(z)$ is holomorphic at $z = 0$ (since f is conformal), this fact is not immediately obvious from (2). In fact, this formula would cause serious difficulties for computer work since every summand blows up at the origin. It is the extra conditions (4) that make the poles in the summands cancel. We can take advantage of this by the following simple trick. The summands of (2) take the values $-ir_k z^{-2}$ at $z = 0$. The first equality of (4) says that we can subtract these off without altering the sum. We find that

$$\mathcal{S}_f(z) = \sum_{k=1}^{n} z^{-2} \left(\frac{a_k z_k z}{(z - z_k)^2} + ir_k \frac{z + z_k}{z - z_k} + ir_k \right)$$

$$= \sum_{k=1}^{n} z^{-1} \left(\frac{a_k z_k}{(z - z_k)^2} + \frac{2ir_k}{z - z_k} \right)$$

and the order of the pole at 0 has diminished. In the last sum, the value of each summand at $z = 0$ is $(a_k - 2ir_k)z^{-1}/z_k$, so the second equality of (4) comes into play (together with the fact that $\overline{z_k} = 1/z_k$), and we find that

$$
\begin{aligned}
\mathcal{S}_f(z) &= \sum_{k=1}^{n} z^{-1}\left(\frac{a_k z_k}{(z-z_k)^2} + \frac{2ir_k}{z-z_k} - \frac{a_k - 2ir_k}{z_k}\right) \\
&= \sum_{k=1}^{n}\left(\frac{a_k}{(z-z_k)^2} + \frac{2ir_k - a_k}{z_k(z-z_k)}\right).
\end{aligned} \tag{7}
$$

The apparent singularity at $z = 0$ has been removed; the individual summands in (7) have no singularities in $\mathbb{D}$.

3.2 *Calculation of f from $\mathcal{S}_f$*

From (7) it is possible to calculate the Taylor series of f centered at the origin, and from this one could calculate $f(z)$ for z in $\mathbb{D}$. However, this series will converge very slowly near $\partial\mathbb{D}$ and is not of much use for calculating ∂D with any accuracy.

We describe a method for calculating $f(e^{it})$. We first assume that t is distinct from all the t_k; i.e., we are not calculating a vertex. Abbreviating (7) as $\mathcal{S}_f(z) = R(z)$, we restrict to the radius of $\mathbb{D}$ of argument t by introducing $h(r) = f(e^{it})$ $(0 \le r \le 1)$. Then by the Chain Rule for the Schwarzian derivative, we find

$$\mathcal{S}_h(r) = e^{2it} R(re^{it}).$$

One way to integrate this real Schwarzian differential equation is by using the classical fact [54] that the solution can be expressed

$$h = y_2/y_1 \tag{8}$$

as the quotient of two linearly independent solutions of the second order ordinary differential equation

$$y''(r) + \frac{1}{2}e^{2it}R(re^{it})y(r) = 0, \quad 0 \le r \le 1, \tag{9}$$

normalized, say, by $y_1(0) = 1 = y_2'(0)$, $y_2(0) = 0 = y_1'(0)$ to correspond with (5). There are many software packages available for solving such differential equations for complex functions of a real parameter (although generally one cannot provide a formal proof of the accuracy of the results obtained this way). Then $f(e^{it}) = h(1)$.

Some of the pros and cons of integrating two linear equations (9) as opposed to a single nonlinear equation (2) are mentioned in [27].

Of course this procedure would not work if $t = t_k$ because of the singularity of the Schwarzian derivative at the poles of R, yet the vertices are the points of greatest interest. To calculate the $f(e^{it_k})$ we proceed as follows. Within each interval (t_k, t_{k+1}) we choose three values of t, which may be approximately equally spaced and not too close to the endpoints, and obtain by radial integration three points on the circular arc $\beta_k \subseteq \mathbb{C}$ as just described. This is sufficient information for determining the center and radius of the circle containing β_k. Once we obtain such data for β_{k-1} and β_k we can find w_k as the first point of intersection when one runs along β_{k-1} in the appropriate direction. Thus f is effectively calculable on $\partial\mathbb{D}$.

3.3 *Alternative approaches to finding* $f(\partial\mathbb{D})$.

As is the case for straight polygons, the integration process also furnishes $f(z)$ for z interior to $\mathbb{D}$ without the need for applying the Cauchy integral formula to the boundary values.

In search for efficiency, one may look for alternatives which do not require three integrations from the origin to find a single boundary arc. However, some essential difficulties must be dealt with. If we start from $z = e^{it}$ with $t_{k-1} < t < t_k$ (abandoning the normalization (5)), we can solve the equation analogous to (9) with independent variable t instead of r. That is, the Schwarzian derivative with respect to t is

$$\frac{1}{2} - e^{2it}R(e^{it}).$$

Arbitrarily chosen initial conditions at the initial value of t will determine the arc β_{k-1}. The challenge is then to find the endpoint w_k, as one cannot integrate too close to the singularity at t_k. A technique for eliminating the singularity when the angle θ_k is nonzero is described in [27]. We will be interested in $\theta_k = 0$. The Möbius transformation $c_1/(w - w_k)$, for appropriate $c_1 \in \mathbb{C}$, takes the cusp region of D near w_k to a leftmost region in a horizontal band of width $\pi/2$, and we may then choose θ_0 so that the variable

$$u = e^{-i\theta_0 + \frac{c_1}{w - w_k}} \tag{10}$$

approaches 0 from negative values as t increases to t_k. By the Reflection Principle the correspondence $t \mapsto u$ is holomorphic near $t = t_k$ and we write

$$r_0 = \frac{du}{dt}(t_k) > 0.$$

Even though the singularity has been eliminated via the formal change of variable, it appears useless to try to solve a Schwarzian equation for u as a function of t since the unknown value of w_k appears in the formula for $S_{t\mapsto u}$. One way around this could be to solve for w_k and use the approximation

$$u \approx r_0(t_k - t)$$

from which we find that

$$w_k \approx w - \frac{c_1}{\log c_0 + \log(t_k - t)} \tag{11}$$

for t near t_k, where $c_0 = r_0 e^{i\theta_0}$. We can calculate three pairs (t, w) by integrating the corresponding Schwarzian differential equation, and then equate the resulting right hand sides of (11). This gives an estimate for $\log c_0$ and c_1.

It is not clear that one can always get close enough to t_k for these values to stabilize before the singularity makes integration unfeasable. Therefore a more reasonable strategy would be to integrate around a semicircle in the t-plane (dotted curve in Figure 2) to arrive at the following edge of the polygon. The would seem to be the closest one could come to "integrating around the boundary" of a circular polygon.

All the examples calculated in this paper are by the three-radius approach given in 3.2 above.

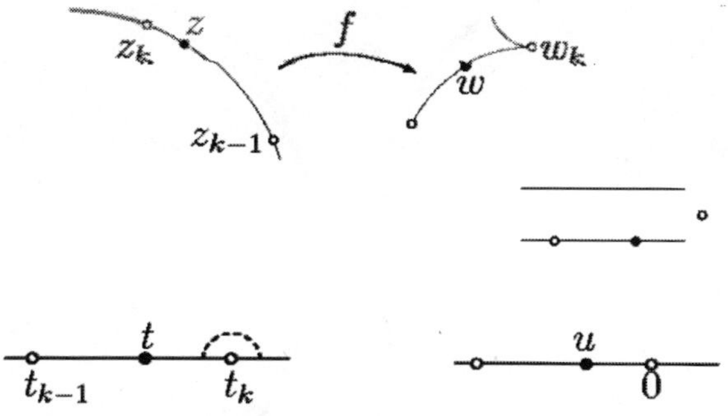

Figure 2. Scheme for integrating towards a vertex.

4. Polygons with symmetries

Much more so than is the case for the continuous methods, the accessory-parameter methods can benefit greatly from symmetries of D. A prime example is the regular n-gon, for which one can solve explicitly for the Schwarz-Christoffel parameters [9, p. 117]. In [46, vol. 3, p. 334–335] the parameters are obtained for the symmetric region obtained by removing from $\mathbb{C}$ the closed segments from the origin to the n-th roots of unity. More interesting examples are produced when there is a bit less symmetry. It is safe to say that for circular polygons, most of what has been done numerically to date is in the presence of some kind of symmetry. Perhaps the only exceptions are [10,27] and in a certain sense [61].

Right quadrilaterals Recent work of P. Brown [13] deals with a number of cases in which the symmetries of D leave precisely one accessory parameter which cannot be determined analytically. Such examples are particularly interesting because the conformal geometry defines real-valued functions of this parameter which lend themselves to detailed study. The domains considered by Brown are circular quadrilaterals and hexagons; we will discuss only the the first case here.

Let us look first at the circular quadrilateral ∂D with right angles at the vertices at $\pm 1 \pm i$, symmetric in the real and imaginary axes and, and also symmetric with respect to the rotation $z \mapsto iz$. Write $a = e^{i\pi/4}$. Because of the symmetry the conformal mapping $f : \mathbb{D} \to D$ normalized by (5) satisfies $f(iz) = if(z)$ and therefore $f(\pm a \pm ai) = \pm 1 \pm i$. Thus we have identified the four prevertices $z_1, \ldots, z_4$ given by $z_1 = a$, $z_k = iz_{k-1}$ (indices mod 4).

Let us apply (2) to this. The interior angles are θ_k with $\theta_1 = \cdots = \theta_4 = \pi/2$, translating into $a_1 = \cdots = a_4 = 3/8$ by (3). Then

$$(iz^2)S_f(iz) = -\frac{3}{8}\sum_{k=1}^{4}\left(\frac{z_k(iz)}{(iz - z_k)^2} + ir_k\frac{iz + z_k}{iz - z_k}\right)$$

$$= -\frac{3}{8}\sum_{k=1}^{4}\left(\frac{-z_{k-1}z}{-(z - z_{k-1})^2} + ir_k\frac{z + z_{k-1}}{z - z_{k-1}}\right).$$

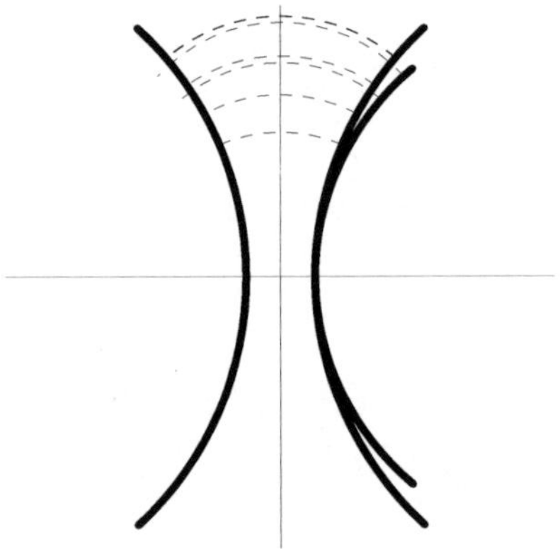

Figure 3. Degrees of freedom in symmetric right-angled quadrilaterals.

Since this is equal to $z^2\mathcal{S}_f(z)$ it is obvious that $r_k = -r_{k-1}$, so the four auxiliary parameters $r_1, \ldots, r_4$ are determined by the single real value r_1. This is a "universal constant" which can be approximated.

In the more general and more difficult situation studied by Brown, the vertices of D are as $\pm 1 \pm ib$ for some $b > 0$. Figure 3 illustrates the degrees of freedom available: b alone does not determine D. When one edge is flexed, the adjacent ones must also flex to maintain the right angle, which is possible since the total number of edges is even. Then the prevertices are $z_k = \pm e^{\pm it}$ where t has to be determined.

The symmetries of f are slightly weaker than before, $f(\overline{z}) = \overline{f(z)}$ and $f(-\overline{z}) = -\overline{f(z)}$, so $\mathcal{S}_f(z) = \mathcal{S}_f(\overline{z}) = \mathcal{S}_f(-\overline{z})$. We find using (7) that

$$\mathcal{S}_f(z) = \frac{3}{8}\sum_{k=1}^{4}\frac{1}{(z-z_k)^2} + \sum_{k=1}^{4}\frac{c_k}{z-z_k} \tag{12}$$

where $c_k = (2ir_k - 3/8)/z_k$, and then the symmetries reduce the c_k to a single complex value c.

Brown applies the fact that $\mathcal{S}_f$ is holomorphic at ∞ to find that $|c| = -3/(8\cos(t + \operatorname{Im} c))$ and with this relation expresses $\mathcal{S}_f(z)$ in terms of the two parameters t and $s = \arg c$. Let us write $f_{s,t}$ for the normalized solution. The remarkable fact is that the parameters s, t, representing arguments, are limited to a bounded interval.

It is after this preliminary setup that the real challenge of numerical conformal mapping begins. Even when we know t, it is not clear at the outset what values of s will make $f_{s,t}$ univalent. Thus Brown must resort to the classical theorems of Nehari [53,55] and others [43] relating univalence of f to the norm of S_f. The method adopted is to examine the curvature of the edges of $\partial f_{s,t}(\mathbb{D})$. For t fixed, let $O(s) \in \mathbb{R}$ be the center of the rightmost edge. It is shown that $O(s)$ is continuous and strictly increasing (it can cross ∞), which justifies the numerical approximations based on (12). Computer calculations can now seek the values of s, t which determine a mapping to the desired right-quadrilateral D.

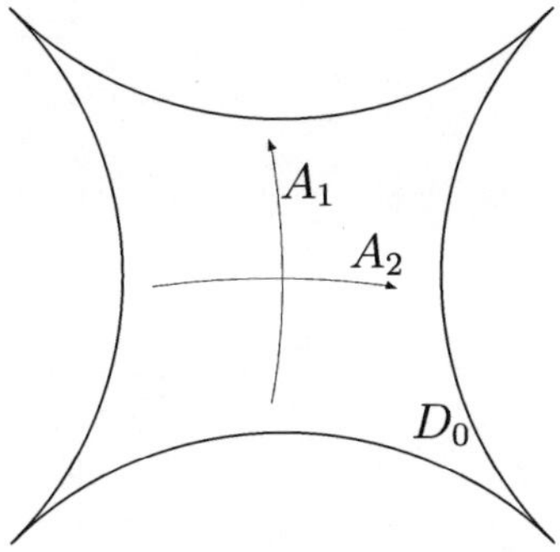

Figure 4. Symmetric cusp quadrilateral.

Cusp quadrilaterals Another example of the exploitation of symmetry is found in [56,57]. This appeared in the context of fundamental domains for certain two-generator quasifuchsian groups. We explain it here because of its connection with Teichmüller theory.

As in the example of Brown, we will look only at quadrilaterals, but this time with zero angles. Any circular polygon with zero angle at each vertex, i.e., $a_k = 1/2$ according to (3), will be called a *cusp polygon*.

The domain D_0 of Figure 4 is bounded by four complete geodesics in the standard Poincaré model of hyperbolic geometry. Hyperbolic isometries A_1 and A_2 identify pairs of points lying on the horizontal and on the vertical edges of D_0, respectively. The Fuchsian group G_0 which they generate is isomorphic to the fundamental group of the once punctured torus $S_0 = \mathbb{D}/G_0$. The Teichmüller space $\mathcal{T}(S_0)$ can be identified [43] with a domain in the space of bounded holomorphic quadratic differentials on S_0,

$$\mathcal{T}(S_0) \subseteq \mathcal{Q}(S_0) = \{\phi \in H(S_0) : \phi = (\phi \circ A_i)(A_i')^2,\ i = 1, 2\}.$$

Each $\phi \in \mathcal{Q}(S_0)$ determines (up to normalization) a holomorphic function $f = f_\phi \in H(\mathbb{D})$ by the relation $\mathcal{S}_f = \phi$, and the closure of $\mathcal{T}(S_0)$ in $\mathcal{Q}(S_0)$ is precisely the collection of those ϕ for which f_ϕ is univalent. In loose terms, as ϕ becomes larger, f_ϕ becomes more complicated, and by the results of Nehari mentioned earlier, bounds are not difficult to obtain for $\|\phi\|$ beyond which f_ϕ is definitely not 1-to-1. The question posed is to find some ϕ lying on the boundary $\partial\mathcal{T}(S_0)$.

The symmetry here is related to the fact that every element of $\mathcal{Q}(S_0)$ is a complex constant multiple of a single ϕ_0 which satisfies $\phi_0(\overline{z}) = \overline{\phi_0(z)}$. Thus if we limit the discussion to $\phi = \lambda\phi_0$ for real λ, the symmetry is passed along to f_ϕ and then to the domain $D_\lambda = f_\phi(D)$. When $\phi \in \mathcal{T}(S_0)$, this is a fundamental domain for a quasifuchsian group. As $\lambda > 0$ grows, the curvature of the top and bottom edges of D_λ shrinks, until for the boundary point $\lambda^*\phi_0$ they are parallel lines. Therefore the problem is to map D_0 to a domain of the type D_{λ^*} shown in Figure 5.

There are in fact infinitely many positive λ for which the top and bottom edges of D_λ are straight. Only the smallest one λ^* satisfies $\lambda\phi_0 \in \partial\mathcal{T}(S_0)$; for the others $f_{\lambda\phi_0}$ is not univalent. Therefore any numerical scheme for calculating λ^* based on the straightness of these edges must guarantee that it is not finding the Schwarzian

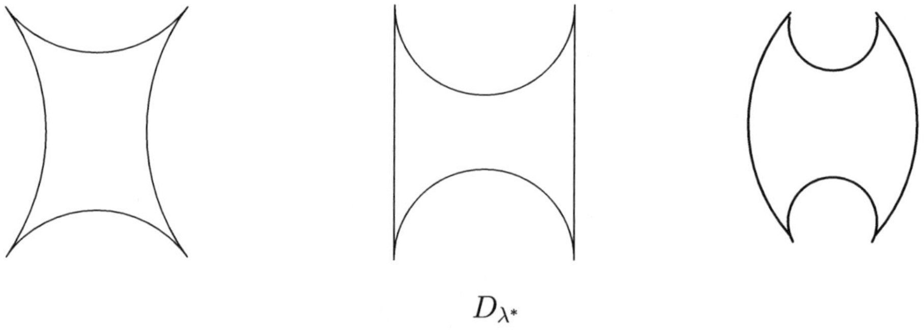

Figure 5. Symmetrically deformed cusp quadrilaterals.

derivative of some non-schlict function. To control this we can apply a change of independent variable $z = h(t)$ where h is a (symmetric!) conformal mapping from the unit square $E = \{t : 0 < \operatorname{Re} t < 1, 0 < \operatorname{Im} t < 1\}$ to D_0. The pullback ψ_0 of ϕ_0 via h is readily expressed in terms of the Weierstrass $\wp$-function having basic periods 1 and i, and we can focus our attention on the restriction to the bottom (real) edge of E. When we write the mapping $f \circ h(t)$ as a quotient $y_2(t)/y_1(t)$ as in (8), and normalize appropriately, we can interpret the values λ for the "straight-edged" mappings (Figure 5) as the eigenvalues of the Sturm-Liouville system

$$y_2'' + \lambda \psi_0 y_2 = 0, \quad y_2(0) = 0, \quad y_2'(0) = 1.$$

The classical theory of ordinary differential equations enables one to estimate the smallest few eigenvalues with considerable precision, and thus isolate the first one numerically. This effectively solves the mapping problem.

The basic quadratic differential ϕ_0 also satisfies the symmetry $\phi(i\,\bar{z}) = i\,\overline{\phi(z)}$, a symmetry along the diagonal of E. Calculations were made in the diagonal direction without benefit of the Sturm-Liouville theory, and a provisional value for the boundary point of $\mathcal{T}(S_0)$ in that direction was published in [57]. In the following decades there was some scattered further work concerning the boundary of $\mathcal{T}(S_0)$ [1,3,4,28–36,38,39,51,52,69]. Recently T. Sugawa [65] observed that the calculations simplify considerably when instead of the once-punctured torus one looks at the 4-times punctured sphere, whose uniformizing group is commensurable with G_0. Via "pleating coordinates" Sugawa was able to calculate all the rational points on $\partial\mathcal{T}(S_0)$ (they are known to be dense, see [7,8,63]), and in particular found that the estimate for the second point mentioned in [57] was incorrect.

5. Horocycle domains

In the rest of this article we will discuss how to calculate the accessory parameters for a certain family of circular polygons which need not be symmetric: D is a *horocycle domain* when ∂D is a cusp polygon and D is obtained by removing from a fixed "base" disk (not necessarily $\mathbb{D}$) a finite union of closed horodisks with disjoint interiors. Here we mean by *horodisk* a disk tangent internally to the base disk. All horocyclic domains

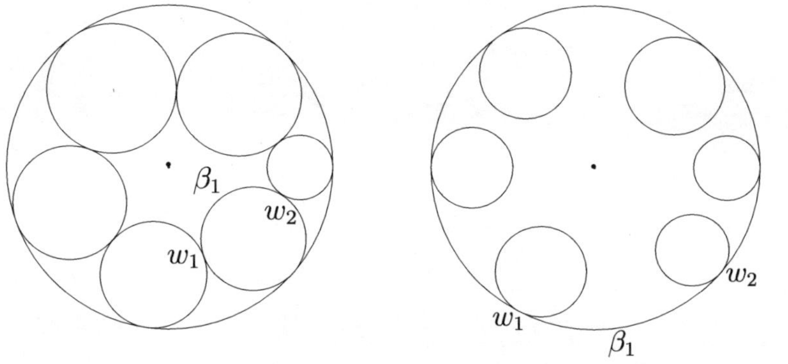

Figure 6. Singly cusped and doubly cusped horocycle domains.

are simply connected. Since all internal angles θ_k are zero, whenever we refer to (2) we will henceforth assume that $a_k = 1/2$ for all k. The material of this section was described in detail in [61].

The boundary ∂D is composed of arcs of horocycles, which we will call *inner edges*, and possibly also arcs of the base disk, which we will call *outer edges*. For evident reasons horocycle domains with no outer edges will be called *singly-cusped* while those with equal numbers of outer and inner edges are *doubly-cusped*. In the latter case the closed horodisks are disjoint. The region D_0 of the previous section is a particular case of a singly cusped horocycle domain.

Horocycle domains are convex in the hyperbolic metric of the base disk. Thus numerical data on horocycle domains can be of interest for studying the class of horocyclically convex domains. Work on conformal mappings for this class can be found in [5,13,37,49,50]. The natural question is to describe

$$\{(\vec{z}, \vec{r}) \in (\partial\mathbb{D})^n \times \mathbb{R}^n : \text{(2) with } a_k = 0 \text{ is the Schwarzian derivative of a schlicht function in } \mathbb{D}\}. \tag{13}$$

Also, given D, one would wish to find the corresponding values of $\vec{z}, \vec{r}$ for its Riemann mapping.

We will limit ourselves to the problem of finding $r_1, \ldots, r_n$ once we are given $z_1, \ldots, z_n$, so that the mapping f with $\mathcal{S}_f$ given by (2) will have a horocycle domain as its image. Thus we want to study the slice of (13) with fixed $\vec{z}$. Note that at the outset it is not even clear how to find values of $\vec{r} = (r_1, \ldots, r_n) \in \mathbb{R}^n$ such that the mapping is injective. At least we know that when (4) holds, the image of $\partial\mathbb{D}$ is a cusp-polygonal curve, not necessarily simple. Consider the following example. In Figure 7(a) we show what happens when z_n are equally spaced (roots of unity) and we take all $r_k = 0$. Note that this satisfies (4). As is often done in depicting conformal mappings, we draw the images of equally spaced concentric circles and radii in $\mathbb{D}$. Next for illustration we move z_k as in Figure 7(b). The mapping for $\vec{r} = 0$ is shown in (c) However, since we moved $\vec{z}$, the vector $\vec{r} = 0$ does not satisfy (4), so the boundary curves in (c) are not circular. Let us adjust $\vec{r}$ to a nearby value $\vec{r} = (0.0209308, 0.0602679, 0, -0.0602679, -0.0209308)$ as discussed in 5.4 below,

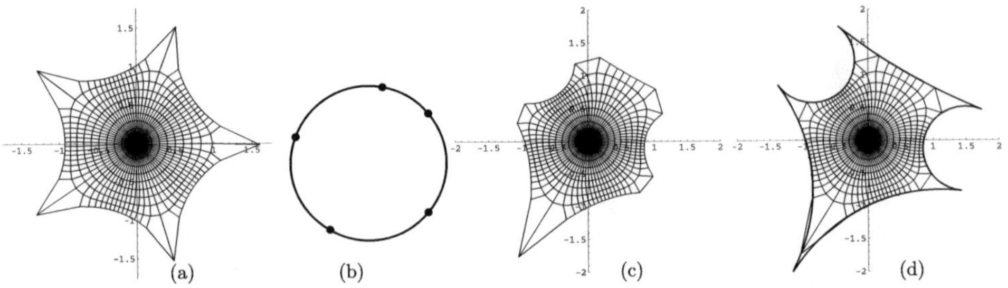

Figure 7. (a) Image of $\partial\mathbb{D}$ for $r_k = 0$ and equally spaced prevertices. (b) "budged" prevertices which produce image (c), (d).

so that (4) holds. The new image turns out to be nearly indistinguishable from (c); in particular there appears to be no problem with the mapping being 1-to-1. Since $f'(z)$ tends to infinity as z approaches a prevertex, the picture in (c) does not show the cusps clearly. When we draw in the boundary arcs all the way to the vertices as in (d) we see that the mapping is in $\mathcal{P}_n$ ($n = 5$) but is far from being a horocycle domain.

Before looking at strictly numerical questions a bit of qualitative description is necessary.

5.1 *Degeneration of horocycle domains and maps*

In order to understand the space of horocycle domains (modulo Möbius transformations) it is important to see what happens to the Schwarzian derivatives of the Riemann mappings when the number of edges is reduced due to a deformation which makes horocycles shrink to a point or touch each other.

Thus consider a family $D(\epsilon)$ in $\mathcal{P}_n$ of horocycle domains with n edges, sharing a common base circle and depending continuously on a parameter $\epsilon > 0$.

Assume that each $D(\epsilon)$ contains the origin. We do not require that 0 be the center of the base circle; in fact, the center need not lie in D. Recall that by the Carathéodory convergence theorem [46], when the domains $D(\epsilon)$ converge in the sense of Carathéodory to a horocycle domain $D(0)$ also containing the origin, we have convergence of their Riemann mappings $f_\epsilon\colon \mathbb{D} \to D(\epsilon)$ normalized by (5), that is, f_ϵ converges to $f_0 : \mathbb{D} \to D(0)$ uniformly on compact subsets of $\mathbb{D}$.

Following the notation we have been using, for $\epsilon > 0$ we have the edge $\alpha_k(\epsilon)$ of $\partial\mathbb{D}$ joining the prevertices $z_k(\epsilon)$, $z_{k+1}(\epsilon)$, and $\beta_k(\epsilon) = f_\epsilon(\alpha_k(\epsilon))$ the image edge of $D(\epsilon)$. When $\beta_k(\epsilon)$ is a full horocycle, there is a double cusp $w_k(\epsilon) = w_{k+1}(\epsilon)$. The edges $\beta_{k-1}(\epsilon)$ and $\beta_{k+1}(\epsilon)$ both have w_k as an endpoint, and both endpoints of $\beta_k(\epsilon)$ are w_k.

We will say that $\beta_k(\epsilon)$ *shrinks to a point* as $\epsilon \to 0$ when the diameter of $\beta_k(\epsilon)$ tends to 0 and its center tends to a point $w_k^* \in \partial D(0)$. Assuming that the adjacent edges $\beta_{k-1}(\epsilon)$ and $\beta_{k+1}(\epsilon)$ do not also degenerate, they will join to form a single edge $\beta_j(0)$ in the limit, as shown in Figure 8(a).

On the other hand, we will say that the edges $\beta_k(\epsilon)$ and $\beta_{k'}(\epsilon)$ (not necessarily adjacent) *touch in the limit* when they converge to arcs of the tangent circles containing

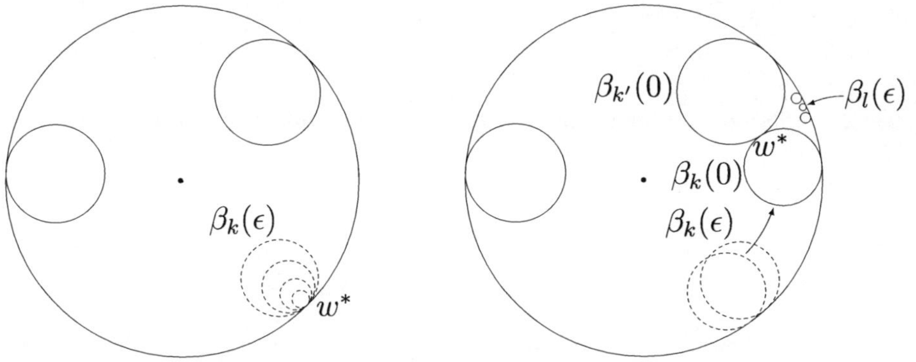

Figure 8. (a) Shrinking and (b) touching of horocycles.

consecutive inner edges $\beta_j(0)$ and $\beta_{j+1}(0)$ of $D(0)$. When this occurs, the limits of the intervening arcs $\beta_{k+1}(\epsilon), \ldots, \beta_{k'-1}(\epsilon)$, if any, are contained in a region of the base disk disjoint from $D(0)$ and effectively "disappear" in the limit, as shown in Figure 8(b).

The following two results, corresponding to degeneration by shrinking and degeneration by touching, are proved in detail in [61] as consequences of the Carathéodory convergence theorem.

Proposition 2 (Shrinking horocycles). *Let k be fixed, and suppose that for all* $\epsilon > 0$, *the inner edge* $\beta_k(\epsilon)$ *is a full horocycle which shrinks to a point* w^* *as* $\epsilon \to 0$, *where* w^* *is on the circumference of the base disk. Suppose further that* w^* *is an interior point of some edge* β_j *of the limiting horocycle domain* $D(0)$, *and the double vertex at which* $\beta_k(\epsilon)$ *ends is the only vertex of* $D(\epsilon)$ *in* β_j *for small* ϵ. *Then the preimage arcs* $\alpha_k(\epsilon)$ *converge to the point* $z^* = f_0^{-1}(w^*) \in \partial\mathbb{D}$ *as* $\epsilon \to 0$.

Proposition 3 (Touching horocycles). *Fix indices* $k \neq k'$. *Suppose that for* $\epsilon > 0$, *the edges* $\beta_k(\epsilon)$ *and* $\beta_{k'}(\epsilon)$ *of* $D(\epsilon)$ *are full horocycles, and that they converge to the tangent circles containing consecutive edges* β_j *and* β_{j+1} *of the limiting domain* $D(0)$, *which meet in a simple cusp* $w^* = w_{j+1} = f_0(z_{j+1})$ *of* $D(0)$. *Suppose further that for small* $\epsilon > 0$, *some edge* $\beta_l(\epsilon)$ *is exterior to* $D(0)$. *Then* $\alpha_l(\epsilon) \to z^* = z_{j+1}$ *as* $\epsilon \to 0$.

Proposition 3 states that the entire boundary preimage between α_k and $\alpha_{k'}$ degenerates to a single point z^*.

We can also describe quite precisely the behavior of specific terms in the Schwarzian $\mathcal{S}_{f_\epsilon}$, $\epsilon > 0$, when horocycles degenerate. We have observed that when a horocycle shrinks to a point as in Proposition 2, the adjacent outer edges combine to form a single new outer edge and so the number of vertices drops by two. Comparing this with (2) we see that the formulas for $\mathcal{S}_{f_\epsilon}$ must lose one pair of summands as they converge uniformly on compact subsets to $\mathcal{S}_{f_0}$ as $\epsilon \to 0$. Suppose the indexing is such that $k = 1$ in Proposition 2, that is, $w_1(\epsilon) = f_\epsilon(z_1(\epsilon))$, $w_2(\epsilon) = f_\epsilon(z_2(\epsilon))$ are the endpoints of the shrinking horocycle, that their prevertices $z_1(\epsilon)$, $z_2(\epsilon)$ converge to $z^* \in \partial\mathbb{D}$. Suppose further that there is no other degeneration: for $k \geq 3$ we have $z_k(\epsilon) \to z_k(0)$ which are distinct points on $\partial\mathbb{D}$. We wish to relate the $z_k(\epsilon)$ to the remaining auxiliary parameters $r_1(\epsilon), \ldots, r_n(\epsilon)$. When $\epsilon = 0$, condition (4) is $\sum_{k=3}^n r_k(0) = \sum_{k=3}^n z_k(0)(1/2 + 2ir_k(0)) = 0$.

The assumption that there is no qualitative degeneration of the domains near the w_k for $k \geq 3$ implies that $r_k = r_k(\epsilon) \to r_k(0)$ for these indices. This is simply because all of the formulas (2) for varying ϵ contain poles whose nature is completely specified. Now subtract the corresponding conditions for $\epsilon > 0$ to find that

$$r_1 + r_2 \to 0,$$

$$r_1 z_1 + r_2 z_2 \to \frac{iz^*}{2}.$$

as $\epsilon \to 0$. Thus we see $(r_1 + r_2)(z_1 + z_2) \to 0$ and therefore

$$r_2 z_1 + r_1 z_2 \to -\frac{iz^*}{2},$$

$$(r_1 - r_2)(z_1 - z_2) \to iz^*.$$

In terms of the arguments t_k, this says $(z_1 - z_2)/(t_1 - t_2) \to iz^*$ as $t_1, t_2 \to t^*$, so

$$(r_1 - r_2)(t_1 - t_2) \to 1.$$

From this we deduce the asymptotic behavior

$$r_1 \sim -\frac{1}{2(t_2 - t_1)}, \qquad (14)$$

$$r_2 \sim \frac{1}{2(t_2 - t_1)}.$$

In particular r_1, r_2 tend to infinity when a horocycle shrinks to a point.

Next consider two horocycles touching as in Proposition 3. Say that for $\epsilon > 0$ the horocycles join w_n to $w_1 = w_n$ and w_2 to $w_3 = w_2$. In the degeneration the vertices w_1, w_2 will be lost while a new one w^* will be created at the touching point while $z_1, z_2 \to z^*$ as $\epsilon \to 0$. Observe that $r_n(\epsilon) \to r_n(0)$ and $r_3(\epsilon) \to r_3(0)$ since there is no degeneration of domains near w_n, w_3. At $\epsilon = 0$ we have

$$r^* + \sum_{k=3}^{n} r_k(0) = z^*\left(\frac{1}{2} + 2ir^*\right) + \sum_{k=3}^{n} z_k(0)\left(\frac{1}{2} + 2ir_k(0)\right).$$

Upon subtracting from the analogous equation for $\epsilon > 0$, taking the limit, and cancelling, we have $r_1 + r_2 \to r^*$ and $r_1 z_1 + r_2 z_2 \to (r^* + i/4)z^*$, so $r_2 z_1 + r_1 z_2 \to (r^* - i/4)z^*$ and then

$$(r_1 - r_2)(z_1 - z_2) \to \frac{iz^*}{2}.$$

In terms of $t_1, t_2 \to t^*$ this can be expressed as

$$r_1 \sim \frac{r^*}{2} - \frac{1}{4(t_2 - t_1)}, \qquad (15)$$
$$r_2 \sim \frac{r^*}{2} + \frac{1}{4(t_2 - t_1)},$$

and as was previously mentioned,

$$r_1 + r_2 \sim r^*. \qquad (16)$$

The importance of (14) and (15) is that they provide approximate values for r^* belonging to qualitatively distinct horocycle domains and thus simplify certain mapping problems. For an application of this see [61].

5.2 *Accessory-parameter correspondence for horocycle domains*

Recall that the $2n$ accessory parameters $\vec{z} = (z_1, \ldots, z_n)$ and $\vec{r} = (r_1, \ldots, r_n)$ can vary essentially independently (that is, subject only to (4)) to define cusp polygons. We are interested in keeping these values from moving so far that univalence of f is lost, and further, we wish to find ways to assure that there is always a base disk. It is a remarkable fact that within the subspace of horocycle domains of fixed combinatorical type, $\vec{r}$ is determined by $\vec{z}$. In this section we prove this fact, first for the somewhat simpler case of doubly-cusped domains (Proposition 4) and then for arbitrary combinations of inner and outer edges (Proposition 5).

Proposition 4 ([61]). *Given an even number of distinct points $z_1, \ldots, z_n \in \partial\mathbb{D}$ there is a unique $\vec{r} \in \mathbb{R}^n$ such that (2) is the Schwarzian derivative of a conformal mapping from $\mathbb{D}$ to a doubly-cusped horocycle domain with prevertices precisely the points of $\vec{z}$.*

Proof. The existence proof is a continuity argument. Let $\mathrm{Cyc}_n \subseteq \mathbb{C}^n$ denote the collection of n-tuples of distinct points located along $\partial\mathbb{D}$ with increasing arguments. By relabeling we may assume that $\vec{z} = (z_1, \ldots, z_n) \in \mathrm{Cyc}_n$. Note that Cyc_n is a connected space.

Consider the subset $E \subseteq \mathrm{Cyc}_n$ of prevertex sets of doubly-cusped horocyclic conformal mappings. Since doubly-cusped domains exist, E is not empty by Proposition 1. If E were not all of Cyc_n, then it would have a boundary element $\vec{z} = \lim \vec{z}^{(i)} \in \mathrm{Cyc}_n$ where each $\vec{z}^{(i)}$ is the prevertex set of some doubly-cusped horocyclic map f_i but $\vec{z}$ is not. We normalize the mappings so that all $D_i = f_i(\mathbb{D})$ have the same base disk. Then the f_i form a normal family and we have Carathéodory convergence of D_i to a domain $D = f(\mathbb{D})$ (on a subsequence) and $f = \lim f_i$ is a conformal mapping, if we can show that f is not constant. However, this is easy to see: each f_i can be extended by the Reflection Principle across the arc $\beta_k^{(i)}$ from $z_k^{(i)}$ to $z_{k+1}^{(i)}$. Since $z_k^{(i)} \to z_k$, the f_i are in fact defined on a common domain containing $\mathbb{D}$ together with a neighborhood of an arc of $\partial\mathbb{D}$. The convergence of f_i is uniform on this compact arc, so f maps this arc into the base circle. Since $f(0) \neq 0$, f is not constant.

It is now clear that D is a doubly-cusped domain, being a limit of such domains. If it had fewer than n edges, then it is not hard to see that either some horocycles must have shrunk to points or some horocycles must have touched in the limiting process. It follows from Propositions 2 and 3 that certain sequences $z_k^{(i)}$ and $z_{k'}^{(i)}$ must have a common limit where $k \neq k'$. But this contradicts the meaning of convergence of $\vec{z}^{(i)}$ in Cyc_n. This contradiction implies that the cusp polygon ∂D has precisely n edges, and the prevertices under f are given by $\vec{z}$. This shows that $\vec{z} \in E$, contrary to assumption, and the contradiction proves the existence of $\vec{r}$ corresponding to every $\vec{z}$.

It remains to verify uniqueness. Consider two mappings f_1, f_2 with the same prevertices $\vec{z} \in \mathrm{Cyc}_n$ and accessory parameters $\vec{r}^{(1)}, \vec{r}^{(2)}$. Write $D_i = f_i(\mathbb{D})$. Fix i and consider the collection of horocycles of D_i together with the complement in $\widehat{\mathbb{C}}$ of the base disk of D_i. This is a collection of disks in $\widehat{\mathbb{C}}$ with disjoint interiors. Let G_i be the collection of compositions of inversions in the circumferences. It is well known [48] that G_i is an index-two extension of a Kleinian group (i.e., contains conformal and anticonformal transformations) with fundamental domain D_i. Via the action of G_i and the Reflection Principle we extend the map $f_2^{-1} \circ f_1 : D_1 \to D_2$ to a conformal mapping of the complements of the limit sets of G_1 and G_2. Then we extend the correspondence to the limit points and thus obtain a homeomorphism T of $\widehat{\mathbb{C}}$. Since the limit sets have measure zero, T is conformal almost everywhere, so it is a quasiconformal mapping of dilatation equal to 1. This implies [44] that T is conformal everywhere, and hence is a Möbius transformation. Therefore f_1 and $f_2 = T \circ f_1$ have the same Schwarzian derivative, from which it follows that $\vec{r}^{(1)} = \vec{r}^{(2)}$. This completes the proof of uniqueness. $\square$

Because of Proposition 4, we may define an *accessory-parameter function*

$$\varphi\colon \mathrm{Cyc}_n \to \mathbb{R}^n$$

as follows. Given $\vec{z} \in \mathrm{Cyc}_n$, write $\vec{r} = \varphi(\vec{z})$. With this value of $\vec{r}$, the rational function on the right hand side of (2) is the Schwarzian derivative of the Riemann mapping of some doubly-cusped horocycle domain with prevertices $\vec{z}$.

Signatures Now we apply the idea of Proposition 4 to more general horocycle domains. Let us associate the symbol $\sigma_k = 0$ with each outer edge β_k of D and $\sigma_k = 1$ with each inner edge β_k. This gives a word $\vec{\sigma} \in \{0, 1\}^n$, $n \geq 3$, generally to be regarded cyclically, which we call the *signature* $\vec{\sigma}$ of D (Figure 9).

Since every outer edge must adjoin two inner edges, $\vec{\sigma}$ cannot contain two consecutive 0's. This is the only restriction on σ. The two extreme cases are the singly-cusped domain with signature $\vec{\sigma} = 11\ldots1$, and the doubly-cusped domain (n even) with $\vec{\sigma} = 0101\ldots01$, as in Figure 4. Recall that we are assuming $a_k = 0$ in (2).

Proposition 5. *Let $\vec{\sigma}$ be a signature of length n, and let $\vec{z} \in \mathrm{Cyc}_n$. Then there exists a unique $\vec{r} \in \mathbb{R}^n$ such that (2) is the Schwarzian derivative of a Riemann mapping f from $\mathbb{D}$ to a horocycle domain with signature $\vec{\sigma}$.*

The necessary modification of the proof of Proposition 4 is given in [61].

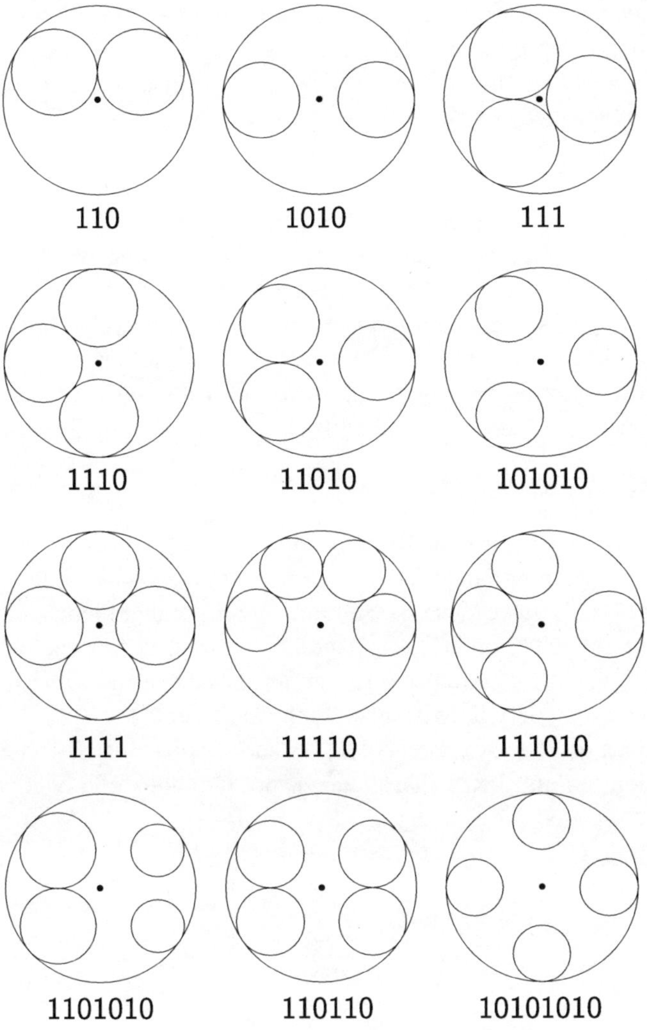

Figure 9. Signatures of horocycle domains.

By Proposition 5 there is a continuous map

$$\varphi_{\vec{\sigma}} : \mathrm{Cyc}_n \to \mathbb{R}^n \tag{17}$$

which associates to $\vec{z}$ the accessory parameters $\vec{r} = \varphi_{\vec{\sigma}}(\vec{z})$ for the Riemann mapping of an essentially unique horocycle domain with prevertices $\vec{z}$ and signature $\vec{\sigma}$. In the next section we will see how to calculate $\varphi_{\vec{\sigma}}(\vec{z})$.

5.3 *Equations relating horocycles to the base disk*

Let us fix a signature $\vec{\sigma}$ of length n, and take a fixed $\vec{z} \in (\partial\mathbb{D})^n$. Imagine that we have an approximate guess for $\varphi_{\vec{\sigma}}(\vec{z})$. If we use it as $\vec{r}$ in (2) with all $a_k = 1/2$ and if $\vec{r}$ satisfies (4), then $f(\partial\mathbb{D})$ will be a cusp polygon close to the boundary of a horocycle domain. We need to construct a measure of the discrepancy, which should reflect the amount in which the proposed outer edges (as specified by $\vec{\sigma}$) fail to form part of a single base circle. Once this discrepancy is made precise, we can look for a way to make it equal to zero.

Thus for now we suppose only that ∂D is an n-cusp polygon. How far is it from the set defined in (13)? Let $a_k \in \mathbb{C}$ (no relation to the notation in (2)) denote the center of the circle containing the edge β_k and let $b_k > 0$ be its radius.

If D is singly cusped, then there are no outer edges, and we will postpone for a moment the discussion of this situation. Suppose it is not, $\vec{\sigma} \neq 11\ldots1$, and choose one outer edge β_{k_0} to serve as a reference. If D were horocyclic, the remaining outer edges would be cocircular with β_{k_0}. Thus we can use the distances from the centers of the remaining outer edges to a_{k_0} as a measure of the mutual concentricity, which gives us a first set of equations in real numbers

$$\operatorname{Re} a_k - \operatorname{Re} a_{k_0} = 0, \quad \operatorname{Im} a_k - \operatorname{Im} a_{k_0} = 0 \quad (\textit{concentricity}) \tag{18}$$

that a horocyclic image must satisfy, as k ranges over the indices of the outer edges distinct from k_0. (A minor modification must be made for the doubly-cusped horocyclic structure $01\ldots01$ because the equations will not be independent in this case.)

With the concentricity conditions satisfied, to make the proposed outer edges lie on one same circle we still need to assure that all they have the same radius. Then the single circle on which they lie will serve as the base circle.

Consider a part of $\vec{\sigma}$ of the form $A11B11C$ and suppose that B contains at least one 0, say at position k. Suppose further that k_0 is not the index of any edge in B. We add the equation

$$b_k - b_{k_0} = 0 \quad (\textit{equal-radius}). \tag{19}$$

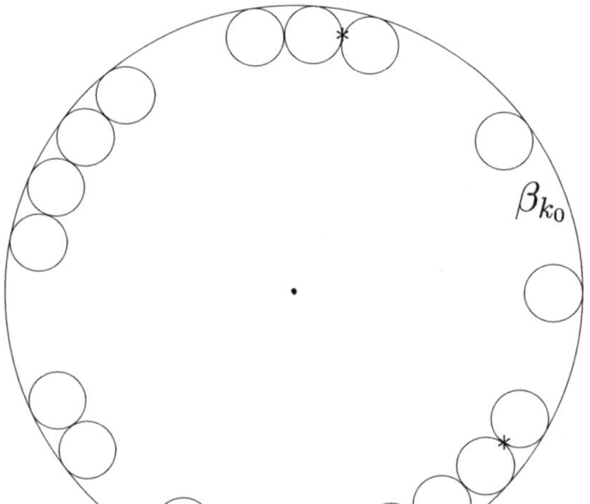

Figure 10. Combinatorial arrangement of horocycle domain. There are 18 edges, 6 concentricity equations, 3 equal-radius equations, and 6 tangency equations. The endpoints of the maximal chain $010\ldots10$ containing β_{k_0} are indicated by *.

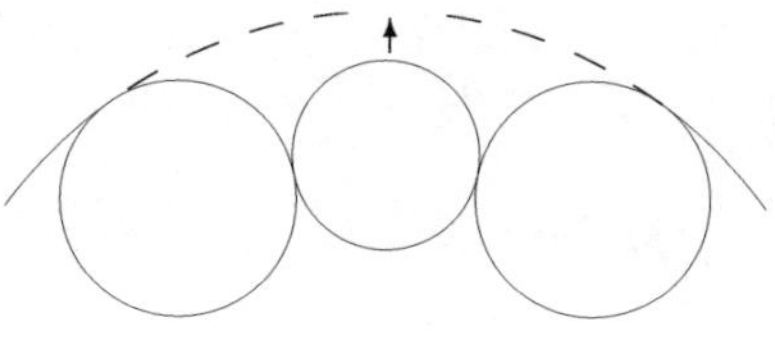

Figure 11. Tangency condition.

It turns out [61] that if B is of the form $01\ldots010$ (no two consecutive 1's), then the concentricity conditions imply that all of the outer edges included within B will automatically have the same radius. Therefore we include precisely one such equation for each maximal chain B of this type not containing β_{k_0}.

Finally, we must insure that the inner edges which are adjacent to two other inner edges are tangent to the base circle (see Figure 11). Each one is assigned an equation of the form

$$|a_k - a_{k_0}| + b_k - b_{k_0} = 0 \quad (\textit{tangency}). \tag{20}$$

(We associate no equations to the inner edges which are adjacent to an outer edge.)

Singly cusped domains must be dealt with by a different procedure, since we have no outer edges with which to describe the base circle K. If $n = 3$, then D is a cusp triangle, and all such are conformally equivalent to the domain

$$\{w : -1 < \operatorname{Re} w < 1, \ \operatorname{Im} w > (1 - (\operatorname{Re} w)^2)^{-1/2}\}.$$

Next let $\vec{\sigma} = 11\ldots1$ have length $n > 3$. In general there is a one-parameter family of circles tangent externally to β_1, β_2, and only one of them is tangent to β_3. This is clearly the base circle K. It is not difficult to write down formulas for the center a^* and radius b^* of K in terms of the centers and radii of β_1, β_2, β_3. For the remaining edges $\beta_4, \ldots, \beta_n$ we substitute these formulas into the tangency conditions

$$|a_k - a^*| + b_k - b^* = 0,$$

to obtain the desired relations among the centers and radii of the edges of ∂D.

In [61] it is verified that this description produces precisely $n - 3$ equations and whenever they are satisfied by a simple n-cusp polygon, then this polygon is the boundary of a horocycle domain whose signature is $\vec{\sigma}$.

5.4 *Computational procedure*

Now we know enough about horocycle domains to begin to calculate their parameters numerically. We first describe a mapping $\psi_{\vec{\sigma}}$ related to $\varphi_{\vec{\sigma}}$ which is more directly useful for numerical computation. Let us write the conditions described in the previous section as $A_1 = 0, \ldots, A_{n-3} = 0$. Here $A_1, \ldots, A_n$ are formulas (18),(19),(20) expressed in terms of centers and radii, and thus can be regarded also as functions on the space $\mathcal{P}_n$ of n-cusp polygons. Collectively they form a nonlinear differentiable mapping $A_{\vec{\sigma}} : \mathcal{P}_n \to \mathbb{R}^{n-3}$.

Let $\vec{z}$ be fixed. Recall that we noted in Section 3 that for any $\vec{r} \in \mathbb{R}^n$ we can solve the Schwarzian differential equation (2) outwardly along radii of $\mathbb{D}$, and then determine the center and radius of each β_k. From these the w_k are found. Thus we can say that the n-cusp polygon $B_n(\vec{r}) \in \mathcal{P}_n$ is computable as a function of the accessory parameter $\vec{r}$.

We also know that $\vec{r} = \phi_{\vec{\sigma}}(\vec{z})$ must lie in a certain codimension-3 affine linear subspace $E^{n-3} = E^{n-3}(\vec{z}) \subseteq \mathbb{R}^n$ defined by (4). Let $L_n : \mathbb{R}^{n-3} \to E^{n-3}$ be an affine linear isomorphism.

We compose these maps to define a differentiable mapping $\psi_{\vec{\sigma}} : \mathbb{R}^{n-3} \to \mathbb{R}^{n-3}$ as follows:

$$\psi_{\vec{\sigma}} : \mathbb{R}^{n-3} \xrightarrow{L_n} E^{n-3} \subseteq \mathbb{R}^n \xrightarrow{B_n} \mathcal{P}_n \xrightarrow{A_{\vec{\sigma}}} \mathbb{R}^{n-3} \tag{21}$$

The relationship between $\varphi_{\vec{\sigma}}$ and $\psi_{\vec{\sigma}}$ is quite natural. Suppose $\vec{u} \in \mathbb{R}^{n-3}$ satisfies $\psi_{\vec{\sigma}}(\vec{u}) = 0$. By (21) we have $A_{\vec{\sigma}}((B_n(L_n(\vec{u}))) = 0$; i.e., the edges of the polygon $B_n(L_n(\vec{u}))$ satisfy the $n-3$ circle conditions defined above in 5.4, so the the solution of (2) is horocyclic mapping with signature $\vec{\sigma}$ except that we cannot guarantee that the boundary curve is simple without further information. Now recall the uniqueness condition in Proposition 5. This implies that $L_n(\vec{u}) = \phi_{\vec{\sigma}}(\vec{z})$. The solution to the mapping problem is summarized as follows.

Let $\vec{\sigma}$ be an admissible signature for a horocycle domain with n vertices, and let $\vec{z} \in Cyc_n$. Suppose $\vec{u} \in \mathbb{R}^{n-3}$ is a solution of

$$\psi_{\vec{\sigma}}(\vec{u}) = 0 \tag{22}$$

where $\psi_{\vec{\sigma}}$ is defined by (21). Let $\vec{r} = L_n(\vec{u})$, and then let f be a holomorphic solution in $\mathbb{D}$ to the Schwarzian differential equation (2). Suppose additionally that the closed curve $f(\partial\mathbb{D})$ is simple. Then $f(\mathbb{D})$ is a horocycle domain with signature $\vec{\sigma}$ and prevertices $\vec{z}$.

It remains then to solve the equation (22). Finding $\vec{z}$ is the same to finding the real vector $\vec{t} = \arg \vec{z}$ and we will suppose

$$t_1 < t_2 < \cdots < t_n < t_1 + 2\pi.$$

The specific numerical method employed to solve (22) for $\vec{u}$ may be a matter of taste. We have found it feasible to use Broyden's method [15] (see also [16] for a pleasant account of the discovery of this simple and useful method). This algorithm, which has been termed a "secant method," commences with an initial guess $\vec{u}^{(0)}$ and an approximation B_0 for the Jacobian matrix of $\psi_{\vec{\sigma}}$ at $\vec{u}^{(0)}$. Iteratively the vectors

$$\vec{u}^{(k+1)} = \vec{u}^{(k)} - B_k^{-1}\psi_{\vec{\sigma}}(\vec{u}^{(k)}).$$

are defined and approach the solution. (In practice one calculates this without inverting B_k explicitly, solving the linear system $B_k\vec{s}^{(k)} = -\psi_{\vec{\sigma}}(\vec{u}^{(k)})$ for the difference

$$\vec{s}^{(k)} = \vec{u}^{(k+1)} - \vec{u}^{(k)}$$

by Gaussian reduction or any other of many well-known methods.) The matrix B_{k+1} to be used in the succeeding iteration is the following perturbation of B_k:

$$B_{k+1} = B_k + \frac{(\psi_{\vec{\sigma}}(\vec{u}^{(k+1)}) - \psi_{\vec{\sigma}}(\vec{u}^{(k)}) - B_k\vec{s}^{(k)}) \cdot \vec{s}^{(k)}}{\|\vec{s}^{(k)}\|_2^2}.$$

Broyden's idea is to avoid the cost of evaluating the Jacobian at each new point, using instead a matrix B_{k+1} which coincides with B_k on the subspace orthogonal to $\vec{s}^{(k)}$.

As with many numerical schemes of this type, the challenge is to find a close enough starting value for $\vec{r}$. If the starting value is not close enough, then the method may not converge, and when it does converge, the solution need not correspond to a univalent mapping.

There is still some leeway in the definition of $\psi_{\vec{\sigma}}$ since the linear isomorphism L_n in (21) is not uniquely specified. The simplest approach is to complete a given vector $\vec{u} \in R^{n-3}$ by defining

$$L_n(\vec{u}) = (u_1, \ldots, u_{n-3}, r_{n-2}, r_{n-1}, r_n)$$

with the last three entries easily found via (4). Numerical experiments suggest that this works, but the results are much better when one defines $L_n(\vec{u})$ as

$$T(u_1, \ldots, u_{n-3}, 0, 0, 0) + \vec{r}^{\,0}$$

where $\vec{r}^{\,0} \in E^{n-3}$ is the closest point of E^{n-3} to the origin in R^n, and T is an orthogonal mapping sending $R^{n-3} \times \{(0, 0, 0)\}$ to E^{n-3}.

5.5 *Applying the algorithm*

In a few special cases one knows a priori the value of $\vec{r}$. For example, when D has n-fold rotational symmetry about the origin, then so does $\vec{z}$, and it follows from (2) that $r_k = r_{k+1}$ and then from (4) that $r_k = 0$ for all k. Therefore D is close to a rotationally symmetric domain, we can use the value $(0, \ldots, 0)$ as an initial approximation for $\vec{r}$.

One may find a domain which is close to D in the sense of Carathéodory, but of a different signature, related by shrinking or touching of horocycles as in Section 5.1. Using (14) and (15), new horocycles may be created and nearby horocycles may be joined or separated as desired. In particular when the horocycles (the edges for which $\sigma_k = 1$) are all small, the corresponding differences $t_{k+1} - t_k$ are also small. An approximation to r_k is given by (14).

Say we are looking for a doubly-cusped domain as in Figure 12(a) with prevertices given as in (b). First we simplify the problem by supposing the horocycles to be very

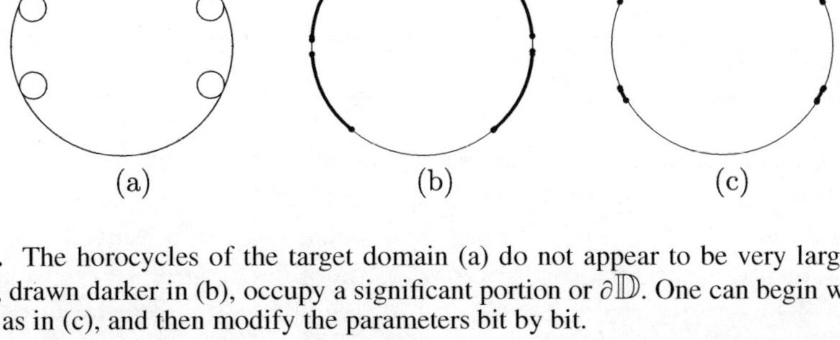

(a) (b) (c)

Figure 12. The horocycles of the target domain (a) do not appear to be very large, but their preimages, drawn darker in (b), occupy a significant portion or $\partial\mathbb{D}$. One can begin with smaller preimages as in (c), and then modify the parameters bit by bit.

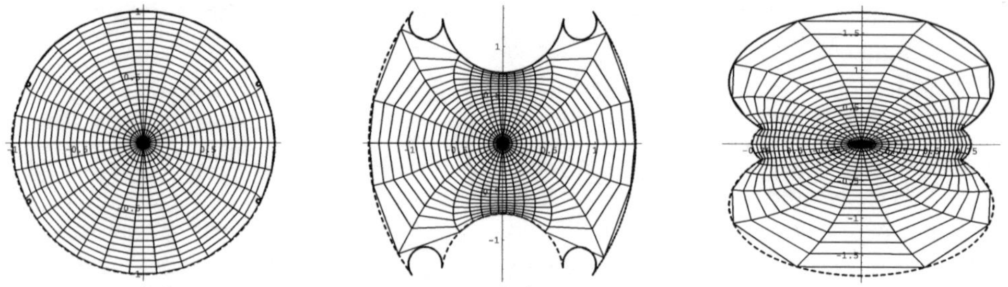

(a) (b) (c)

Figure 13. (a) The image with prevertices given in Figure 12(c) has extremely small horocycles. In (b),(c) the non-horocyclic images corresponding to perturbed accessory parameters r_k are shown.

small, which means that we move the prevertices e^{it_k} corresponding to their endpoints close together. In (c) we have taken the separation to be

$$t_{2k} - t_{2k-1} = dt = 0.1$$

and we investigate whether this is small enough to apply formula (14), which tells us that we should try

$$\vec{r} = (-5, 5, -5, 5, -5, 5, -5, 5).$$

Because of the symmetry we have $\vec{r}$ satisfies (4), so there is no need to project to E^{n-3}. The resulting map is shown in Figure 13(a), in which the four horocycles are barely visible. Using this value of $\vec{r}$ as the starting point in Broyden's method gives convergence to 7 decimals after three iterations, and the value 5 above is replaced by 4.995765. This tells us that the choice of dt above was sufficiently small. Figures (b) and (c) show the resulting image domains when one uses the entries of $\vec{r}$ decreased and increased by 50%, respectively, to give one a sense of the sensitivity of the Schwarzian differential equation to this parameter in horocycle mapping problems. Starting from (b) Broyden's method does not converge, while from (c) it converges to the accessory parameters of a nonunivalent mapping.

With the problem solved for small horocycles, one now increases the value of dt slightly to separate the prevertices of each horocycle a bit, the value of $\vec{r}$ previously obtained is used as the starting Broyden value, and the result is the new value of $\vec{r}$. This process is repeated until the spacing of the z_k in the original problem is reached, giving the domain we showed in Figure 12(a).

6. Closing remarks

We hope that this article has succeeded in giving a bit of the flavor of numerical conformal mapping and in particular the nature of the elusive accessory parameters. The question of calculating the mappings corresponding to a Schwarzian derivative and then adjusting to find a closer approximation of the desired accessory parameters is to some degree solved, although it will no doubt be refined in important ways in coming years. What is still very unclear is how to go about obtaining a reasonable initial guess for these parameters. The essence of the problem is that the constants in (2) do not seem to say very explicitly what the geometry is of the resulting circular polygon.

There may be some hope of attacking the problem via osculation methods. It has been noted [25] that these methods give surprisingly nice approximations in the first few iterations, although the convergence is finally excruciatingly slow. Thus it may be of interest to apply an osculation method to a given circular polygon. This would give a low-cost initial approximation of the location of the prevertices z_k of the generalized Schwarz-Christoffel mapping, and perhaps also some information about the parameters r_k.

Another method of attack, probably much more ambitious, would be to adapt the approach of CRDT to deal with circular polygons. On the one hand, circular polygons are much more naturally related to Möbius transformations than are straight polygons, so one would expect that cross ratios could naturally enter into their descriptions. On the other hand, the question of natural triangulations of circular polygons could be quite difficult, inasmuch as two points no longer determine an edge.

References

[1] R. Abe, On correspondences between once punctured tori and closed tori: Fricke groups and real lattices, *Tokyo J. Math.*, **23** (2000) 269–293.

[2] L. Ahlfors, Conformal invariants: topics in geometric function theory, McGraw-Hill Book Co., New York-Düsseldorf-Johannesburg (1973).

[3] H. Akiyoshi, H. Miyachi and M. Sakuma, A refinement of McShane's identity for quasi-fuchsian punctured torus groups, In the tradition of Ahlfors and Bers, III, *Contemp. Math.*, **355** (2004) 21–40.

[4] H. Akiyoshi, M. Sakuma, M. Wada, and Y. Yamashita, Jørgensen's picture of punctured torus groups and its refinement, Kleinian groups and hyperbolic 3-manifolds (Warwick, 2001) London Math. Soc. Lecture Note Ser. 299, Cambridge Univ. Press, Cambridge, (2003) 247–273.

[5] R. W. Barnard, L. Cole, K. Pearce and B. Williams, The sharp bound for the deformation of a disc under a hyperbolically convex map, *Proc. London Math. Soc.*, **93** (2006) 395–417.

[6] R. W. Barnard, K. Pearce, and G. B. Williams, Three extremal problems for hyperbolically convex functions, *Comput. Methods Funct. Theory*, **4** (2004) 97–109.

[7] L. Bers, The action of the modular group on the complex boundary, Riemann surfaces and related topics: Proceedings of the 1978 Stony Brook Conference, *Ann. of Math. Stud.*, **97**, Princeton Univ. Press, Princeton, N.J (1981) 33–52.

[8] L. Bers, On iterates of hyperbolic transformations of Teichmüller space, *Amer. J. Math.*, **105** (1983) 1–11.

[9] L. Bieberbach, Conformal Mapping, Chelsea Publishing Co., New York (1953).

[10] P. Bjørstad and E. Grosse, Conformal mapping of circular arc polygons, *SIAM J. Sci. Statist. Comput.*, **8** (1987) 19–32.

[11] D. Bonciani and F. Vlacci, Some remarks on Schwarz-Christoffel transformations from the unit disk to a regular polygon and their numerical computation, *Complex Var. Theory Appl.*, **49** (2004) 271–284.

[12] P. Bjørstad and E. Grosse, Conformal mapping of circular arc polygons, *SIAM J. Sci. Stat. Comput.*, **8** (1987) 19–32.

[13] P. R. Brown, Mapping onto circular arc polygons, *Complex Variables*, **50** (2005) 131–154.
[14] P. R. Brown, An analysis of a two parameter problem for conformal maps onto circular arc quadrilaterals, *Complex Var. Elliptic Equ.*, **53** (2008) 23–51.
[15] C. G. Broyden, A class of methods for solving nonlinear simultaneous equations, *Math. Comp.*, **19** (1965) 577–593.
[16] C. G. Broyden, On the discovery of the 'good Broyden' method, *Math. Program.*, **87** (2000) 209–213.
[17] C. Carathéodory, Theory of functions, Chelsea Publishing Company, New York (1954).
[18] T. K. DeLillo and J. A. Pfaltzgraff, Numerical conformal mapping methods for simply and doubly connected regions, Special issue on iterative methods (Copper Mountain, CO, 1996), *SIAM J. Sci. Comput.*, **19** (1998) 155–171.
[19] T. K. DeLillo, The accuracy of numerical conformal mapping methods: a survey of examples and results, *SIAM J. Numer. Anal.*, **31** (1994) 788–812.
[20] T. A. Driscoll and L. N. Trefethen, Schwarz-Christoffel mapping, Cambridge Monographs on Applied and Computational Mathematics, Cambridge University Press, Cambridge (2002).
[21] T. A. Driscoll and S. A. Vavasis, Numerical conformal mapping using cross-ratios and Delaunay triangulation, *SIAM J. Sci. Comput.*, **19** (1998) 1783–1803.
[22] D. Gaier, Konstructive methoden der konformen abbildung, Springer Tracts in Natural Philosophy 3 (1964).
[23] E. Grassmann, Numerical experiments with a method of successive approximation for conformal mapping, *Z. Angew. Math. Phys.*, **30** (1979) 873–884.
[24] H. Haddar and R. Kress, Conformal mappings and inverse boundary value problems, *Inverse Problems*, **21** (2005), 935–953.
[25] P. Henrici, Applied and Computational Complex Analysis, Vol. 3, Wiley, New York (1986).
[26] E. Hille, Analytic Function Theory, Vol. 2, Blaisdell Publishing Company, New York (1965).
[27] L. H. Howell, Numerical conformal mapping of circular arc polygons, *J. Comput. Appl. Math.*, **46** (1993) 7–28.
[28] T. Jørgensen, On pairs of once-punctured tori, Kleinian groups and hyperbolic 3-manifolds (Warwick, 2001) London Math. Soc. Lecture Note Ser. 299, Cambridge Univ. Press, Cambridge, (2003) 183–207.
[29] L. Keen, Accessory parameters and the uniformization of punctured tori, Modular functions in analysis and number theory, *Lecture Notes Math. Statist.* **5** Univ. Pittsburgh, Pittsburgh (1983) 132–149.
[30] L. Keen, Teichmüller spaces of punctured tori. I, II. *Complex Variables Theory Appl.*, **2** (1983) 199–211, 213–225.
[31] L. Keen, H. E. Rauch and A. T. Vasquez, Moduli of punctured tori and the accessory parameter of Lamé's equation, *Trans. Amer. Math. Soc.*, **255** (1979) 201–230.
[32] L. Keen and C. Series, Pleating coordinates for the Maskit embedding of the Teichmüller space of punctured tori, *Topology*, **32** (1993), 719–749.
[33] L. Keen and C. Series, How to bend pairs of punctured tori, Lipa's legacy, *Contemp. Math.*, **211** (1995) 359–387.
[34] L. Keen and C. Series, Pleating invariants for punctured torus groups, *Topology*, **43** (2004) 447–491.
[35] H. C. Kim, Matrix presentations of the Teichmüller space of a punctured torus, *J. Korea Soc. Math. Educ. Ser. B Pure Appl. Math*, **11** (2004) 73–88.
[36] H. C. Kim, Punctured torus representations using the gluing method, *Bull. Korean Math. Soc.*, **41** (2004) 709–721.
[37] S.-A. Kim and T. Sugawa, Characterizations of hyperbolically convex regions, *J. Math. Anal. Appl.*, **309** (2005) 37–51.

[38] Y. Komori, On the boundary of the Earle slice for punctured torus groups, Kleinian groups and hyperbolic 3-manifolds (Warwick, 2001), London Math. Soc. Lecture Note Ser. 299, Cambridge Univ. Press, Cambridge, (2003) 293–304.

[39] Y. Komori and T. Sugawa, Bers embedding of the Teichmüller space of a once-punctured torus, *Conform. Geom. Dyn.*, **8** (2004) 115–142.

[40] Y. Komori, T. Sugawa, M. Wada,and Y. Yamashita, Drawing Bers embeddings of the Teichmüller space of once-punctured tori, *Experiment. Math.*, **15** (2006) 51–60.

[41] P. P. Kufarev, On a method of numerical determination of the parameters of the Schwarz-Christoffel integral, (Russian) *Doklady Adad. Nauk SSSR*, **57** (1947) 535–357.

[42] P. K. Kythe, Computational conformal mapping, Birkhäuse, Boston (1998).

[43] O. Lehto, Univalent functions and Teichmüller spaces, Springer-Verlag, New York (1995).

[44] O. Lehto and K. I. Virtanen, Quasiconformal mappings in the plane, Second edition. Die Grundlehren der mathematischen Wissenschaften 126 Springer-Verlag, New York-Heidelberg (1973).

[45] W. Ma, D. Minda and D. Mejía, Hyperbolically 1-convex functions, *Ann. Polon. Math.*, **84** (2004) 185–202.

[46] A. I. Markushevich, Theory of functions of a complex variable Vols. I, II, III. Second English edition. Chelsea Publishing Co., New York (1977).

[47] D. E. Marshall and S. Rohde, Convergence of the zipper algorithm for conformal mapping, preprint available at `http://www.math.washington.edu/~marshall/preprints/preprints.html`.

[48] B. Maskit, Kleinian groups, Grundlehren der Mathematischen Wissenschaften 287, Springer-Verlag, Berlin (1988).

[49] D. Mejía and Ch. Pommerenke, Hyperbolically convex functions, Analysis and Applications—ISAAC 2001 (Berlin), *Int. Soc. Anal. Appl. Comput.*, 10, Kluwer Acad. Publ., Dordrecht (2003) 89–95.

[50] D. Mejía and Ch. Pommerenke, Horocyclically convex univalent functions, *Michigan Math. J.*, **53** (2005) 483–496.

[51] Y. N. Minsky, The classification of punctured-torus groups, *Ann. of Math.*, **149** (1999) 559–626.

[52] H. Miyachi, On the horocyclic coordinate for the Teichmüller space of once punctured tori, Travaux de la Conférence Internationale d'Analyse Complexe et du 8e Séminaire Roumano-Finlandais (Iassy, 1999), *Math. Rep. (Bucur.)*, **2** (2001) 527–532.

[53] Z. Nehari, The Schwarzian derivative and schlicht functions, *Bull. Amer. Math. Soc.*, **55** (1949) 552–553.

[54] Z. Nehari, Conformal mapping, McGraw-Hill, New York (1952).

[55] Z. Nehari, Some criteria of univalence, *Proc. Amer. Math. Soc.*, **5** (1949) 700–704.

[56] R. M. Porter, An accessory-parameter determination of a regular boundary point of Teichmüller space, doctoral thesis Northwestern University (1978).

[57] R. M. Porter, Computation of a boundary point of Teichmüller space, *Bol. Soc. Mat. Mexicana*, **24** (1979), 15–26.

[58] R. M. Porter, On computing some Bers boundary points for $T(1,1)$, Colloquium on dynamical systems (Guanajuato, 1983), Aportaciones Mat., Soc. Mat., Mexicana, México, (1985) 92–107.

[59] R. M. Porter, An interpolating polynomial method for numerical conformal mapping, *SIAM J. Sci. Comput.*, **23** (2001), 1027–1041.

[60] R. M. Porter, An accelerated osculation method and its application to numerical conformal mapping, *Complex Var. Theory Appl.*, **48** (2003) 569–582.

[61] R. M. Porter, Numerical calculation of conformal mapping to a disk minus finitely many horocycles, *Comput. Methods Funct. Theory*, **5** (2005) 471–488.

[62] B. Rodin and D. Sullivan, The convergence of circle packings to the Riemann mapping, *J. Differential Geom.*, **26** (1987) 349–360.

[63] H. Shiga, On Teichmüller spaces and modular transformations, *J. Math. Kyoto Univ.*, **25** (1985) 619–626.
[64] K. Stephenson, Introduction to circle packing. The theory of discrete analytic functions, Cambridge University Press, Cambridge (2005).
[65] T. Sugawa, Estimates of hyperbolic metric with applications to Teichmüller spaces, *Kyungpook Math. J.*, **42** (2002), 51–60.
[66] L. N. Trefethen, Numerical computation of the Schwarz-Christoffel transformation, *SIAM J. Sci. Statist. Comput.*, **1** (1980) 82–102.
[67] T. A. Driscoll and L. N. Trefethen, Schwarz-Christoffel mapping in the computer era, Proceedings of the International Congress of Mathematicians III, Doc. Math., extra, Vol. III (1988) 533–542.
[68] R. Wegmann, Methods for numerical conformal mapping, Handbook of complex analysis: geometric function theory, Vol. 2, Elsevier, Amsterdam (2005) 351–477.
[69] P. Will, Lagrangian decomposability of some two-generator subgroups of $PU(2, 1)$, *C. R. Math. Acad. Sci. Paris*, **340** (2005) 353–358.

Teichmüller Theory and Moduli Problem No. 10, 2010, pp. 577–582.

Denjoy-Wolff Theorem on Riemann Surfaces

Hiroshige Shiga

Department of Mathematics, Tokyo Institute of Technology
e-mail: shiga@math.titech.ac.jp

Abstract. The Denjoy-Wolff theorem is known as a theorem on iterations of holomorphic mappings of the unit disk to itself. We discuss the theorem for holomorphic mappings on Riemann surfaces. We shall also give a discussion from a view point of Fuchsian groups.

Keywords. Denjoy-Wolff theorem, Riemann surfaces.

2001 Subject Classification: 30F25, 37F

1. Introduction

Let Δ be the unit disk on the complex plane and f a holomorphic mapping from Δ to itself. The mapping f is called *escaping* if $\{f^n(K)\}_{n=1}^{\infty}$ diverges to $\partial\Delta$ for every compact subset K of Δ, where f^n means the n times iterations of f. The Denjoy-Wolff theorem says that if f is *escaping*, then $\{f^n\}_{n=1}^{\infty}$ converges to a point $\zeta \in \partial\Delta$ uniformly on every compact subset of Δ. The Denjoy-Wolff theorem plays an important role in complex dynamics (cf. [7], [9]). M. Heins([4]) gives a generalization of the theorem for holomorphic mappings on compact bordered Riemann surfaces.

In this article, we shall give an alternative proof for the Denjoy-Wolff-Heins theorem. We shall also give a generalization of the theorem and some discussion from a view point of Fuchsian groups. Finally, we will give examples related to these results. In particular, we will present an example of Riemann surfaces of infinite genus where the Denjoy-Wolff theorem hold in the Martin compactifications.

2. Heins' theorem and its proof

An open Riemann surface R is called *compact bordered* if it is of finite genus and the boundary ∂R of R consists of finitely many closed Jordan curves. Then, we shall show the following theorem which is a generalization of the Denjoy-Wolff theorem.

Theorem 2.1 (Heins[4]). *Let R be a compact bordered Riemann surface and f a holomorphic self-mapping of R. Suppose that f is escaping, that is, $\{f^n(K)\}_{n=1}^{\infty}$ diverges to ∂R for every compact subset K of R. Then, there exists $q \in \partial R$ such that $\{f^n\}_{n=1}^{\infty}$ converges to the point q uniformly on every compact subset of R.*

Proof. Take $p_0 \in R$ as a base point and denote by I_0 an arc in R from p_0 to $f(p_0)$. For $n \in \mathbb{N}$, set $I_n = f^n(I_0)$ and $I = \bigcup_{n=0}^{\infty} I_n$.

Since f is escaping, I is a curve from p_0 to ∂R. We put $E = \bar{I} \cap \partial R$. We show that E consists of only one point.

Suppose that E contains more than two points. Then E is a continuum with $m(E) > 0$, where $m(E)$ is the harmonic measure of E with respect to p_0. Here, we note the following lemma.

Lemma 2.1. *Let R be a compact bordered Riemann surface and f a holomorphic self-mapping of R. Then f has non-tangential limits almost everywhere on ∂R.*

Proof. Let $F : \Delta \to \Delta$ be a lift of f to the unit disk Δ which is the universal covering of R. Namely, $R = \Delta / \Gamma$ for some Fuchsian group Γ acting on Δ. Since R is a compact bordered Riemann surface, $|\Lambda_\Gamma| = 0$, where Λ_Γ denotes the limit set of Γ and $|\Lambda_\Gamma|$ is the Lebesgue measure on the unit circle $\partial\Delta$.

Indeed, let $\tilde{u}$ be the Dirichlet solution on Δ for the characteristic function of Λ_Γ. Since Λ_Γ is invariant under the action of Γ, $\tilde{u}$ is automorphic for Γ. Hence $\tilde{u}$ is projected to a harmonic function u on the quotient space R. From the construction, it is easy to see that u is continuous on $R \cup \partial R$ and $u|\partial R = 0$. Thus, we have $u \equiv 0$ and $\tilde{u} \equiv 0$ from the maximal principle. This implies $|\Lambda_\Gamma| = 0$ as we desired.

On the other hand, it follows from Fatou theorem (cf. [3] p. 94) that F has non-tangential limits almost everywhere on $\partial\Delta$. Let F^* denote the non-tangential boundary function on $\partial\Delta$. Since F is non-constant, we see that $F^{*-1}(\Lambda_\Gamma)$ is measure zero on $\partial\Delta$. Hence $F^*(e^{i\theta}) \in \Delta \cup (\partial\Delta \backslash \Lambda_\Gamma)$ for almost all $e^{i\theta}$. Noting that $\partial\Delta \backslash \Lambda_\Gamma$ corresponds to ∂R, we verify that f has non-tangential limits almost everywhere on ∂R. □

Now, we proceed to prove Theorem 2.1. Let E_0 be the set of points of E where f has non-tangential limits. From the above lemma, $m(E_0) > 0$. For every $\zeta \in E_0$, there exists a sequence $\{p_i\}_{i=1}^{\infty}$ such that it converges to ζ non-tangentially and $p_i \in I_{n(i)}$ for some $n(i) \in \mathbb{N}$.

Since $I_n = f^n(I)$ $(n \in \mathbb{N})$, we see that the hyperbolic diameter of I_n is bounded by that of I_0 from Schwartz's lemma. Therefore, both $f^{n(i)}(p_0)$ and $f^{n(i)+1}(p_0)$, the end points of $I_{n(i)}$, converge to ζ non-tangentially as $i \to \infty$. Hence we have

$$\zeta = \lim_{i\to\infty} f^{n(i)+1}(p_0) = \lim_{i\to\infty} f(f^{n(i)}(p_0)) = f^*(\zeta), \tag{2.1}$$

for every $\zeta \in E_0$, where f^* is the non-tangential boundary function of f on ∂R. (2.1) implies that non-tangential boundary function of f fixes every point on E_0. Let F denote a lift of f as in the proof of the above lemma. Then, for almost all $e^{i\theta}$ there exists $\gamma_\theta \in \Gamma$ such that

$$F^* \circ \gamma_\theta(e^{i\theta}) = e^{i\theta}.$$

The element γ_θ may depends on $e^{i\theta}$. However, Γ consists of countably many elements. Therefore, there exist a subset E' of $\partial\Delta$ with positive measure and $\gamma \in \Gamma$ such that

$$F^* \circ \gamma\,(e^{i\theta}) = e^{i\theta}$$

for any $e^{i\theta} \in E'$. Since $m(E') > 0$, it follows from Riesz theorem (cf.[3] Chap. II, Corollary 4.2) that $F \circ \gamma = id$ and hence f is the identity map. Thus, we have a contradiction because f is an escaping map. Therefore, E consists of only one point, say q, on ∂R.

Now, take a compact subset K of R. We may assume that K contains p_0. From the above argument, $\lim_{n\to\infty} f^n(p_0) = q$. On the other hand, the hyperbolic diameter of $f^n(K)(\ni f^n(p_0))$ is bounded by that of K from Schwartz's lemma. Hence $f^n(K)$ must converge to q as $n \to \infty$. □

Remark 2.1.

(1) From a classification theorem of holomorphic mappings of hyperbolic Riemann surfaces (cf. [7] Theorem 5.2), it is seen that f is escaping if for some $p \in R$, $\{f^n(p)\}_{n=1}^{\infty}$ contains a subsequence which diverges to the boundary.
(2) In [4], Heins uses the Martin compactification in the proof. On further development of the Denjoy-Wolff theorem from this point of view, see [5] and [8].
(3) The above proof is also an alternative proof of the Denjoy-Wolff theorem on the unit disk and it is different from the proof given in [7].
(4) A simple example $f(z) = z|z|^k$ $(0 < k)$ shows that the Denjoy-Wolff theorem does not hold for quasiconformal mappings.

The above proof also works for certain Riemann surfaces possibly of infinite type.

A Riemann surface S is called *parabolic* if it has no Green's function. The set of parabolic Riemann surfaces is denoted by O_G. We say that a subregion R of a Riemann surface S is an SO_{HB}-region if S belongs to O_G and the relative boundary ∂R consists of countably many analytic Jordan arcs which do not accumulate in S. Any compact bordered Riemann surface is an SO_{HB}-region because it is regarded as a subregion of a compact Riemann surface, which is in O_G.

An escaping holomorphic mapping $f : R \to R$ of an SO_{HB}-region R is called *escaping to border* if there exists a subsequence $\{f^{n_i}(p_0)\}_{i=1}^{\infty}$ of the orbit $\{f^n(p_0\}_{n=1}^{\infty}$ of some $p_0 \in R$ such that $f^{n_i}(p_0)$ converges to a point in ∂R as $i \to \infty$. It is easy to see that an escaping holomorphic mapping on a compact bordered Riemann surface is a map escaping to border.

Since it is not hard to see that the harmonic measure of ∂R is full, the same argument as in the proof of Theorem 2.1 yields the following;

Theorem 2.2. *Let R be an SO_{HB}-region and f a holomorphic self-mapping of R escaping to border. Then, there exists a point $q \in \partial R$ such that $\{f^n\}_{n=1}^{\infty}$ converges to q uniformly on every compact subset of R.*

Remark 2.2. The above definition of an SO_{HB}-region is slightly different from a standard one such as in [2]. We define it because of simplicity. Of course, the statement of the above theorem is also true for a "standard" SH_{HB}-region.

3. Iterations of geometric homomorphisms of Fuchsian groups

Let $\theta : \Gamma \to \Gamma$ be a homomorphism of a (torsion free) Fuchsian group Γ acting on the unit disk Δ. The homomorphism θ is called *geometric* if it is induced by a continuous self-mapping f of $R = \Delta/\Gamma$. More precisely, f defines θ as follows.

Take a point $p_0 \in R$ as a base point and consider the fundamental group $\pi_1(R, p_0)$ of R with a base point p_0. We may identify Γ with $\pi_1(R, p_0)$ by lifting closed curves passing p_0 to Δ. The continuous map f gives a homomorphism of $\pi_1(R, p_0)$ to $\pi_1(R, f(p_0))$. Thus, a lift F of f on Δ defines a homomorphism θ from Γ to itself with

$$F \circ \gamma = \theta(\gamma) \circ F \tag{3.1}$$

for any $\gamma \in \Gamma$. Inductively, we have

$$F^n \circ \gamma = \theta^n(\gamma) \circ F \tag{3.2}$$

for $n \in \mathbb{N}$.

Theorem 3.1. *Let Γ be a Fuchsian group on Δ representing an SO_{HB}-region R and $\theta : \Gamma \to \Gamma$ a geometric homomorphism. Suppose that θ is induced by a holomorphic mapping $f : R \to R$ escaping to border. Then there exists $N \in \mathbb{N}$ for each $\gamma \in \Gamma$ such that for any $n > N$, $\theta^n(\gamma) = id$ holds.*

Proof. We denote by π a natural projection from Δ to $R = \Delta/\Gamma$. Let $\tilde{c}$ be the hyperbolic geodesic arc connecting 0 and $\gamma(0)$. Then, $c := \pi(\tilde{c})$ is a closed curve passing through $\pi(0)$. From Theorem 2.2, there exist $e^{i\phi} \in \partial\Delta$ such that

$$e^{i\phi} = \lim_{n\to\infty} F^n(0) = \lim_{n\to\infty} F^n(\gamma(0)) = \lim_{n\to\infty} F^n(\tilde{c}).$$

By using Theorem 2.2 again, we see that the point $e^{i\phi}$ is projected to a point q_ϕ on ∂R and $c_n := \pi(F^n(\tilde{c}))$ is a closed curve which is freely homotopic to a closed curve for $\theta^n(\gamma)$ because of (3.2).

On the other hand, the hyperbolic length of c_n is bounded by that of c by Schwartz's lemma. Noting that c_n is passing through $f^n(\pi(0))$ and $f^n(\pi(0))$ is close to $q_\phi \in \partial R$, we see that c_n must be a trivial curve and $\theta^n(\gamma) = id$ for a sufficiently large n. □

4. Examples

In Theorem 3.1, the number N depends on $\gamma \in \Gamma$. Hence, it is a natural question whether we may take $N \in \mathbb{N}$ independent of $\gamma \in \Gamma$. Obviously, we may take N independent of $\gamma \in \Gamma$ if Γ is finitely generated. However, the following example shows us that we cannot take such N independent of $\gamma \in \Gamma$ in general.

Example 4.1. Let $f(z) = kz$ with $k > 1$ and $I_n = \{z = \sqrt{-1}y | k^{-n-1/2} \le y \le k^{-n}\}$. Since $\mathbb{C}^* = \mathbb{C}\setminus\{0\}$ is a parabolic Riemann surface, $R := \mathbb{H}\setminus \bigcup_{n=0}^{\infty} I_n = \Delta/\Gamma$ is an SO_{HB}-region and $f|R$ is a holomorphic mapping from R to itself escaping to border. However, $f^n(I_n) = I_0$ for each $n \in \mathbb{N}$. Thus, we may find $\gamma_n \in \Gamma$ for each $n \in \mathbb{N}$ such that $\theta^n(\gamma_n) \neq id$.

It is interesting to see what happens in Theorem 2.2 if f is a holomorphic mapping but not escaping to border. In this case, $\{f^n(p)\}_{n=1}^{\infty}$ diverges to the ideal boundary of R. Thus, we need to take a compactification of R to analyze the behavior of the divergence sequence. We consider the Martin compactification of R. As for the Martin compactification, see [2].

P. Poggi-Corradini [8] gives a planar domain Ω and a holomorphic self mapping ϕ such that the Denjoy-Wolff theorem does not hold in the Martin compactification of Ω. Here, we consider an SO_{HB}-region R of infinite genus whose relative boundary consists of finitely many analytic Jordan curves. We also assume that R has only one ideal boundary component. We call such an SO_{HB}-region a *Heins end.*

Definition 4.1. *Let R be a Heins end. The cardinality of the minimal points in the ideal boundary of R is called the harmonic dimension of R and we denote it by $hd(R)$.*

From Theorem 2.2, we obtain the following.

Corollary 4.1. *Let R be a Heins end with $hd(R) = 1$. Then, for any escaping holomorphic mapping $f : R \to R$ the Denjoy-Wolff theorem holds in the Martin compactification of R. Namely, on any compact subset K of R, $\{f^n\}_{n=1}^{\infty}$ uniformly converges to either a point in ∂R or the minimal point of the ideal boundary of R.*

Here, we give a Riemann surface R of infinite genus and a holomorphic mapping f such that R satisfies the condition in Corollary 4.1 and f^n converges to the minimal point as $n \to \infty$.

Example 4.2. Let $G = \Delta^* \setminus \bigcup_{n=1}^{\infty} I_n$, where $\Delta^* = \{0 < |z| < 1\}$ and $I_n = [a^{n+1/2}, a^n]$ for some a with $0 < a < 1$. Take two copies G_-, G_+ of G and make a Riemann surface R by identifying the same I_n of $\partial G_-, \partial G_+$ crosswisely for all n. Then, the Riemann surface R is an SO_{HB}-region of infinite genus. Thus, we have to show that $hd(R) = 1$.

To this end, first we note that $z = 0$ is a regular point of ∂G with respect to the Dirichlet problem. Indeed, from Wiener's criterion(cf. [10]) it suffices to show that

$$\sum_{n=1}^{\infty} \frac{n}{-\log cap(I_n)} = +\infty, \tag{4.1}$$

where $cap(E)$ is the logarithmic capacity of E. On the other hand, it is well known that $cap(I_n) = |I_n|/4 = a^n(1 - \sqrt{a})/4$. Hence, we see that (4.1) holds and $z = 0$ is a regular point. Now, we use a theorem in [6] on the harmonic dimension of a Heins end and we see that $hd(R) = 1$.

Here we consider a holomorphic function $f(z) = az$. Since $f(G) \subset G$ and $f(I_n) = I_{n+1}$, f is lifted to a holomorphic mapping F of R but not escaping to border. Hence from Corollary 4.1 we verify that $\{F^n\}$ converges to the minimal point of the ideal boundary.

References

[1] A. F. Beardon, Iteration of Rational Functions, Springer-Verlag (1991).
[2] C. Constantinescu and A. Cornea, Ideale Ränder Riemannscher Flächen, Springer (1963).

[3] J. B. Garnett, Bounded Analytic Functions, Academic Press (1981).
[4] M. Heins, A theorem of Wolff-Denjoy type, in "Complex Analysis", Birkhäuser, Basel (1988) 81–86.
[5] F. Lárusson, A Wolff-Denjoy theorem for infinitely connected Riemann surfaces, *Proc. Amer. Math. Soc.*, **124** (1996) 2745–2750.
[6] H. Masaoka and S. Segawa, Harmonic dimensions of covering Riemann surfaces, *Kodai Math. J.*, **17** (1994) 351–359.
[7] J. Milnor, Dynamics in One Complex Variable: Introductory Lectures, Vieweg (1999).
[8] P. Poggi-Corradini, On the failure of a generalized Denjoy-Wolff theorem, Conformal Geometry and Dynamics, **6** (2002) 12–32.
[9] N. Steinmetz, Rational iteration : complex analytic dynamical systems, de Gruyter (1993).
[10] M. Tsuji, Potential theory in modern function theory, Maruzen (1959).

Teichmüller Theory and Moduli Problem No. 10, 2010, pp. 583–596.

Cardioids and Teichmüller Spaces

Toshiyuki Sugawa

Department of Mathematics, Graduate School of Science, Hiroshima University,
Higashi-Hiroshima 739 8526, Japan
e-mail: sugawa@math.sci.hiroshima-u.ac.jp

Abstract. In this note, we give an expository account on the role played by cardioids in both Teichmüller spaces and in complex dynamics. Especially, we will observe that the shape of the Bers embedding of the Teichmüller space of a once-punctured rectangular torus converges to a cardioid in the sense of Carathéodory when the base surface is pinched along the meridian. A more general and complete result will be included in a forthcoming paper [27] of the author.

Keywords. Cardioid, Teichmüller space, holomorphic motion, once-punctured torus.

2000 Subject Classification: Primary 30F60; Secondary 30C62, 37F45

1. Introduction

A cardioid is a plane curve similar to $\{e^{i\theta} - e^{2i\theta}/2 : \theta \in \mathbb{R}\}$, see Figure 1. In this note, however, a *cardioid* means the (open) Jordan region bounded by a cardioid curve. We will call the *standard cardioid* the domain bounded by the above curve and denote

Figure 1. The cardioid.

The author was partially supported by the JSPS Grant-in-Aid for Scientific Research (B), 17340039.

it by C_0. Note that the function $f_0(z) = z - z^2/2$ maps the unit disk univalently onto C_0. Conventionally, we denote by $aC_0 + b$ the image of C_0 under the similarity map $z \mapsto az + b$.

It is an accidental coincidence that the cardioid appears both in the Kleinian group theory and in complex dynamics. However, according to Sullivan's dictionary, the Bers embedding of the Teichmüller space of a Riemann surface (or a Fuchsian group) corresponds to the Mandelbrot set (or its generalization). Therefore, this coincidence might not be a surprising fact. As is well known, the Mandelbrot set contains a cardioid (the so-called main cardioid) as a connected component of its interior. On the other hand, as Kalme [12] pointed out, the cardioid appears in the universal Teichmüller space in a natural way.

In this note, we point out similarities between the cardioid in the Mandelbrot set and cardioids in the Bers embedded Teichmüller spaces. Our emphasis will be put on the role played by holomorphic motions to give a unified aspect on these matters. Moreover, we will observe that the shape of a one-dimensional Teichmüller space tends to that of a cardioid when the base Riemann surface goes to the boundary of the moduli space, at least in a special case. It appears that this is a phenomenon which has not been observed in the literature.

2. Holomorphic motions

We briefly summarize the basic properties of holomorphic motions for convenience of the reader.

Let E be a subset of the Riemann sphere $\widehat{\mathbb{C}} = \mathbb{C} \cup \{\infty\}$ and D be a domain (or a complex manifold in general) with base point t_0. A holomorphic motion of E over (D, t_0) is a map $H : D \times E \to \widehat{\mathbb{C}}$ satisfying the following three conditions:

(i) $H_t = H(t, \cdot) : E \to \widehat{\mathbb{C}}$ is injective for each $t \in D$,
(ii) $H(\cdot, z) : D \to \widehat{\mathbb{C}}$ is holomorphic for each $z \in E$, and
(iii) $H(t_0, z) = z$ for $z \in E$.

Mañé-Sad-Sullivan [15] revealed the following remarkable facts about holomorphic motions over the unit disk (see also [3]): Let H be a holomorphic motion of E over $(\mathbb{D}, 0)$. Then

(a) H extends to a holomorphic motion of $\bar{E}$ uniquely in such a way that the extended $H : \mathbb{D} \times \bar{E} \to \widehat{\mathbb{C}}$ is (jointly) continuous, and
(b) H_t is $|t|$-quasiconformal on each component of $\operatorname{Int} \bar{E}$ for each $t \in \mathbb{D}$.

Here and hereafter, $\mathbb{D}$ denotes the unit disk $\{z \in \mathbb{C} : |z| < 1\}$. Note that we do not assume any continuity in the variable z for the holomorphic motion $H(t, z)$. For a constant $0 \le k < 1$, a sense-preserving homeomorphism $f : \Omega \to \Omega'$ between domains in $\widehat{\mathbb{C}}$ is called k-quasiconformal if f has locally integrable partial derivatives on $\Omega \setminus \{\infty, f^{-1}(\infty)\}$ (in the sense of distribution) such that $|f_{\bar{z}}| \le k |f_z|$ holds a.e.

Slodkowski [25] proved another remarkable fact: *every holomorphic motion H of E over the unit disk can be extended to that of $\widehat{\mathbb{C}}$. In particular, H_t is a restriction of a $|t|$-quasiconformal homeomorphism of $\widehat{\mathbb{C}}$ for each $t \in \mathbb{D}$.*

Readers interested in holomorphic motions may consult recent textbooks on Teichmüller theory such as Gardiner-Lakic [5] and Hubbard [10] for details.

3. Cardioid in mandelbrot set

Even polynomials of the simple form $P_c(z) = z^2 + c$ are not fully understood in the context of iterations, that is, in the theory of complex dynamics. We denote by P_c^n the n-th iterate of P_c, namely, $P_c^1 = P_c$ and $P_c^n = P_c^{n-1} \circ P_c$ for $n = 2, 3, \ldots$. Here c is a complex parameter. The boundary of the set $K_c = \{z \in \mathbb{C} : \{P_c^n(z)\}_{n=1,2,\ldots}$ is bounded$\}$ is called the Julia set and denoted by J_c. The Mandelbrot set M is defined to be the set of $c \in \mathbb{C}$ for which J_c is connected (see Figure 2). It is known that J_c is either connected or a Cantor set and that $c \in M$ iff $0 \in K_c$.

An *attracting fixed point* of P_c is a point $\alpha \in \mathbb{C}$ such that $P_c(\alpha) = \alpha$ and $|P_c'(\alpha)| < 1$. It is easily seen that a neighbourhood of an attracting fixed point of P_c is contained in K_c. In particular, $c \in M$. A fixed point α of P_c must satisfy the relation $P_c(\alpha) - \alpha = \alpha^2 - \alpha + c = 0$. If we set $\lambda = P_c'(\alpha) = 2\alpha$, then $c = \lambda/2 - (\lambda/2)^2$. We now conclude that P_c has an attracting fixed point iff $c \in D = \{\lambda/2 - (\lambda/2)^2 : \lambda \in \mathbb{D}\}$. Therefore, D is contained in the Mandelbrot set. Indeed, it is easy to see that D is a component of the interior of M and $D = (1/2)C_0$. The component D is called the main cardioid of M.

We now see that the Julia set J_c varies holomorphically (in a sense) on the parameter $c \in D$. A point z is called *periodic* for P_c if $P_c^n(z) = z$ for an integer n and it is called *repelling* further if $|(P_c^n)'(z)| > 1$. It is well known that the set of repelling periodic points of P_c is a dense subset of J_c. It is easy to see that $R_0 = \{e^{2\pi i\theta} : 2^n\theta \in \mathbb{Z}$ for some $n \in \mathbb{N}\}$ is the set of repelling fixed points of $P_0(z) = z^2$. Since repelling

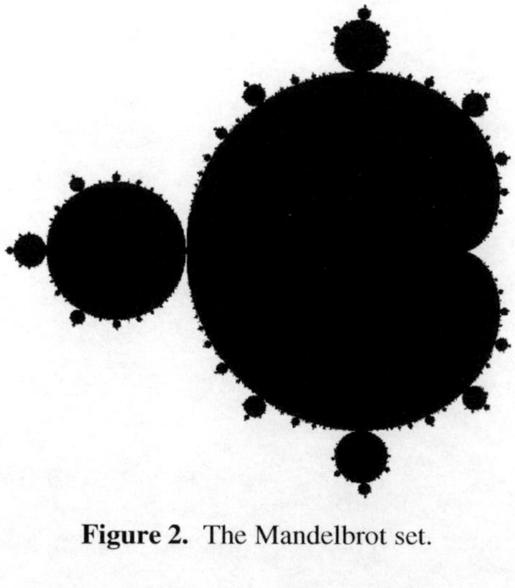

Figure 2. The Mandelbrot set.

periodic points do not collide for $c \in D$, for each $z_0 \in R_0$ one can take a repelling periodic point z_c of P_c, $c \in D$, in such a way that z_c is a holomorphic function of $c \in D$ assuming z_0 for $c = 0$. In this way, we obtain a function $H_c(z_0) = z_c$ on R_0 for each $c \in D$. The same reasoning as above yields that the function H_c is injective on R_0. Thus the map $H : (c, z) \mapsto H_c(z)$ gives a holomorphic motion of R_0 over $(D, 0)$. Since R_0 is dense in $\partial\mathbb{D}$, by the Mañé-Sad-Sullivan theorem, the holomorphic motion of R_0 extends to that of $\partial\mathbb{D}$ which will be still denoted by the same symbol $H(c, z) = H_c(z)$. Thus, one has the relation $J_c = H_c(\partial\mathbb{D})$ for $c \in D$. By the Slodkowski theorem, the map H_c can be extended to a quasiconformal map of $\widehat{\mathbb{C}}$ and thus $J_c = H_c(\partial\mathbb{D})$ is known to be a quasidisk for $c \in D$. For details and interesting figures, see Astala and Martin [2].

It is known that near the cusp point $1/4$ of the main cardiod D rather complicated phenomena of phase transition are observed, which are called the parabolic implosion. See [4] for details and references.

4. Bers embedding of Teichmüller space

We give minimal basics in the theory of Teichmüller spaces to state and to prove our result. For details, see [18] or [11].

In the following, we denote by $\mathbb{H}$ the upper half-plane $\{z \in \mathbb{C} : \operatorname{Im} z > 0\}$ and by $\mathbb{H}^*$ the lower half-plane $\{z \in \mathbb{C} : \operatorname{Im} z < 0\}$. Let Γ be a Fuchsian group acting on $\mathbb{H}$, namely, Γ is a discrete subgroup of $\mathrm{PSL}(2, \mathbb{R})$. We denote by $\mathrm{Belt}(\mathbb{H}, \Gamma)$ the space of Beltrami coefficient for Γ on $\mathbb{H}$, more precisely,

$$\mathrm{Belt}(\mathbb{H}, \Gamma) = \left\{ \mu \in L^\infty(\mathbb{H}) : \|\mu\|_\infty < 1,\ (\mu \circ \gamma)\frac{\overline{\gamma'}}{\gamma'} = \mu \text{ for all } \gamma \in \Gamma \right\}.$$

For $\mu \in \mathrm{Belt}(\mathbb{H}, \Gamma)$, we denote by f^μ the quasiconformal map f of $\widehat{\mathbb{C}}$ which is determined by the Beltrami equations $f_{\bar{z}} = \mu f_z$ in $\mathbb{H}$ and $f_{\bar{z}} = 0$ in $\mathbb{H}^*$ and normalization conditions $f(0) = 0$, $f(1) = 1$ and $f(\infty) = \infty$. Note that $f^\mu \Gamma (f^\mu)^{-1}$ is a Kleinian group acting on $f^\mu(\mathbb{H})$ and $f^\mu(\mathbb{H}^*)$ properly discontinuously.

Noting that f^μ is conformal on $\mathbb{H}^*$, we define a holomorphic function $\Phi(\mu)$ on $\mathbb{H}^*$ by $\Phi(\mu) = S_{f^\mu|\mathbb{H}^*}$, where S_f stands for the Schwarzian derivative of f :

$$S_f = \left(\frac{f''}{f'}\right)' - \frac{1}{2}\left(\frac{f'}{f}\right)^2 = \frac{f'''}{f'} - \frac{3}{2}\left(\frac{f'}{f}\right)^2.$$

It is convenient to measure the Schwarzian derivative by the norm

$$\|\varphi\|_{\mathbb{H}^*} = \sup_{z \in \mathbb{H}^*} (-2 \operatorname{Im} z)^2 |\varphi(z)|.$$

We denote by $B_2(\mathbb{H}^*)$ the Banach space consisting of analytic functions φ on $\mathbb{H}^*$ with $\|\varphi\|_{\mathbb{H}^*} < \infty$. Let

$$B_2(\mathbb{H}^*, \Gamma) = \{\varphi \in B_2(\mathbb{H}^*) : (\varphi \circ \gamma)(\gamma')^2 = \varphi \text{ for all } \gamma \in \Gamma\},$$

which is a closed subspace of $B_2(\mathbb{H}^*)$.

It is known that $\Phi(\mu) \in B_2(\mathbb{H}^*, \Gamma)$ for $\mu \in \mathrm{Belt}(\mathbb{H}, \Gamma)$ and that the image $\Phi(\mathrm{Belt}(\mathbb{H}, \Gamma))$ coincides with the Bers embedding of the Teichmüller space of Γ. We set $\mathrm{Teich}(\Gamma) = \Phi(\mathrm{Belt}(\mathbb{H}, \Gamma))$ and we identify it with the Teichmüller space of Γ (or equivalently, of the orbifold $\mathbb{H}/\Gamma$). It is known that $\mathrm{Teich}(\Gamma)$ is a bounded contractible domain in $B_2(\mathbb{H}^*, \Gamma)$. The map $\Phi_\Gamma = \Phi : \mathrm{Belt}(\mathbb{H}, \Gamma) \to \mathrm{Teich}(\Gamma)$ is called the Bers projection and known to be a holomorphic split submersion. In the case when Γ is the trivial group 1, the set $\mathrm{Teich}(1)$ is called the *universal Teichmüller space*.

We denote by $o(\Gamma)$ and $i(\Gamma)$ the outer and inner radii of the Teichmüller space $\mathrm{Teich}(\Gamma)$, in other words, $o(\Gamma)$ is the smallest number r with $\mathrm{Teich}(\Gamma) \subset \{\varphi \in \mathrm{B}_2(\mathbb{H}^*, \Gamma) : \|\varphi\|_{\mathbb{H}^*} < r\}$ and $i(\Gamma)$ is the largest number r with $\mathrm{Teich}(\Gamma) \supset \{\varphi \in \mathrm{B}_2(\mathbb{H}^*, \Gamma) : \|\varphi\|_{\mathbb{H}^*} < r\}$.

It is well known that $i(1) = 2$ and $o(1) = 6$ for trivial group 1, and hence, $2 \le i(\Gamma) \le o(\Gamma) \le 6$ for an arbitrary Fuchsian group Γ unless $\mathrm{Teich}(\Gamma)$ is a singleton. For a cofinite Fuchsian group Γ, it is also known that $i(\Gamma) > 2$ (cf. [20]) and $o(\Gamma) < 6$ (due to Sekigawa [24]). Furthermore, Nakanishi and Yamamoto [22] and Nakanishi and Velling [21] gave characterizing conditions for Γ to satisfy $o(\Gamma) = 6$ and $i(\Gamma) = 2$, respectively. It is remarkable that their conditions are coincident. In particular, $o(\Gamma) = 6$ iff $i(\Gamma) = 2$. One can see that the present investigation lie in the same line as above-mentioned studies.

5. Kalme's observation

Kalme [12] made an interesting observation on a special holomorphic family of quasiconformal maps of the Riemann sphere. Following [12], we present some facts related to our investigation.

For a complex number $\alpha \in \mathbb{C}$, consider the function

$$F_\alpha(z) = z^\alpha = e^{\alpha \log z}$$

on $\mathbb{H}^*$, where the branch of $\log z$ is taken so that $-\pi < \arg z = \mathrm{Im} \log z < 0$ for $z \in \mathbb{H}^*$. Since $F_\alpha'(z) = \alpha z^{\alpha-1} \neq 0$, the map F_α is locally univalent for $\alpha \neq 0$. A characterization of univalence of F_α is known.

Lemma 1 (Royster [23]). *Let $\alpha \in \mathbb{C}\backslash\{0\}$. Then, F_α is univalent in $\mathbb{H}^*$ if and only if either $|\alpha - 1| \le 1$ or $|\alpha + 1| \le 1$.*

An easy computation gives

$$S_{F_\alpha}(z) = \frac{1-\alpha^2}{2z^2} = 2(1-\alpha^2)\varphi_0^*(z),$$

where $\varphi_0^* \in B_2(\mathbb{H}^*)$ is given by $\varphi_0^*(z) = 1/(4z^2)$. Note that $\|\varphi_0^*\|_{\mathbb{H}^*} = 1$.

An interesting fact is that the holomorphic motion $f_t = F_{1-t}$ $(|t| < 1)$ of $\mathbb{H}^*$ can be extended to that of the Riemann sphere in an explicit way:

$$f_t(z) = \begin{cases} z\bar{z}^{-t} & z \in \overline{\mathbb{H}} \\ z^{1-t} & z \in \mathbb{H}^*. \end{cases}$$

Note that f_t is normalized and has the Beltrami coefficient

$$\mu_{f_t}(z) = -t\frac{z}{\bar{z}} = -t\frac{|\varphi_0(z)|}{\varphi_0(z)}$$

for $z \in \mathbb{H}$, where $\varphi_0(z) = \overline{\varphi_0^*(\bar{z})} = 1/(4z^2)$ for $z \in \mathbb{H}$. This is of the form of a Teichmüller differential on $\mathbb{H}$ and, by definition of Φ, one has the relation

$$\Phi(-t\mu_0) = 2(1-(1-t)^2)\varphi_0^* = g_0(t)\varphi_0^*, \tag{5.1}$$

where

$$\mu_0 = \frac{|\varphi_0|}{\varphi_0} \quad \text{and} \quad g_0(t) = 4\left(t - \frac{t^2}{2}\right).$$

We can now see that the intersection of Teich(1) with the linear span of φ_0^* is precisely the cardioid $4C_0$ (times φ_0^*).

Hille [9] observed that for $\varepsilon \in \mathbb{R}\backslash\{0\}$, $F_{i\varepsilon}$ is a universal covering projection onto an annulus, and hence, is never univalent whereas $S_{F_{i\varepsilon}}(z) = (1+\varepsilon^2)/(2z^2)$ and thus the norm $\|S_{F_{i\varepsilon}}\|_{\mathbb{H}^*} = 2(1+\varepsilon^2)$ tends to 2 as $\varepsilon \to 0$. We point out that the point $S_{F_{i\varepsilon}}$ approaches to the inward cusp of the above cardioid $4C_0\varphi_0^*$ from the outside.

6. One-dimensional Teichmüller spaces

From now on, we restrict ourselves to one-dimensional Teichmüller spaces. For simplicity, we assume Fuchsian groups to be torsion-free. Then $\dim \mathrm{Teich}(\Gamma) = 1$ iff the signature of Γ is (0,4) or (1,1). Since to each Fuchsian group Γ of signature (0,4), there corresponds a Fuchsian group Γ' of signature (1,1) such that $\mathrm{Teich}(\Gamma) = \mathrm{Teich}(\Gamma')$ (cf. [13]), without loss of generality we may further assume that Γ is of signature (1,1). Then the quotient Riemann surface $\mathbb{H}/\Gamma$ is represented by a once-punctured torus of the form $T = (\mathbb{C}\backslash\Omega)/\Omega$, where

$$\Omega = \langle\omega_1, \omega_2\rangle_{\mathbb{Z}} = \{m\omega_1 + n\omega_2 : m, n \in \mathbb{Z}\}$$

and $\omega_1, \omega_2 \in \mathbb{C}$ with $\mathrm{Im}\,(\omega_2/\omega_1) > 0$. We also write Teich($T$) for Teich($\Gamma$) when $T = \mathbb{H}/\Gamma$.

In recent years, much progress has been made in the study of one-dimensional Teichmüller spaces. For instance, Minsky [16] proved that the Bers embedding of a one-dimensional Teichmüller space is a Jordan domain. By using Minsky's method, Miyachi [17] showed that the one-dimensional Teichmüller space is "cusp-shaped" at every boundary point corresponding to a cusp. Goodman [6] even observed a spiraling shape of the boundary of the one-dimensional Teichmüller space. Computer graphics of the Bers embeddings are presented by [13] and [14]. Also, the inner and outer radii of the Teichmüller space of a once-punctured square torus are numerically computed in [26].

In [27], the following result is proved.

Theorem 2. *The Bers embedding* Teich(T) *of the Teichmüller space of a once-punctured torus* T *converges to a cardioid in the sense of Carathéodory (under suitable indentification of* $B_2(T) = B_2(\mathbb{D}, \Gamma)$ *with* $\mathbb{C}$*), when* T *goes to infinity in the moduli space*

There is some technicality with the suitable choice of the basis of the vector space $B_2(T)$ so that one can identify $B_2(T)$ with $\mathbb{C}$. Therefore, in this note, we restrict ourselves to the case when T is a rectangular torus for the sake of simplicity. See [27] for the general case.

Let Ω_λ be the lattice generated by $\lambda > 0$ and $\mathrm{i} = \sqrt{-1}$ over $\mathbb{Z}$, namely, $\Omega_\lambda = \{m\lambda + n\mathrm{i} : m, n \in \mathbb{Z}\}$. We set $T_\lambda = (\mathbb{C}\backslash\Omega_\lambda)/\Omega_\lambda$. We denote by $\pi_\lambda(\zeta) = [\zeta]$ the canonical projection $\mathbb{C}\backslash\Omega_\lambda \to T_\lambda$.

Take a holomorphic universal covering projection p_λ of the upper half-plane $\mathbb{H}$ onto $\mathbb{C}\backslash\Omega_\lambda$ so that $p_\lambda(\mathrm{i}) = a_\lambda$ and $p'_\lambda(\mathrm{i})/\mathrm{i} > 0$. Then $q_\lambda = \pi_\lambda \circ p_\lambda$ is a holomorphic universal covering projection of $\mathbb{H}$ onto T_λ. Let Γ_λ be the covering transformation group of $q_\lambda : \mathbb{H} \to T_\lambda$. Note that $\mathbb{H}/\Gamma_\lambda = T_\lambda$.

Let $\tilde{\varphi}_\lambda$ be the pullback of the quadratic differential $\mathrm{d}\zeta^2$ on $\mathbb{C}\backslash\Omega_\lambda$ under the map p_λ, namely, $\tilde{\varphi}_\lambda = (p'_\lambda)^2$. Further let

$$\varphi_\lambda = (\|\tilde{\varphi}_\lambda\|_{\mathbb{H}})^{-1}\tilde{\varphi}_\lambda.$$

Then, by definition, $\varphi_\lambda \in B_2(\mathbb{H}, \Gamma_\lambda)$ and $\|\varphi_\lambda\|_{\mathbb{H}} = 1$.

Set also

$$\varphi^*_\lambda(z) = \overline{\varphi_\lambda(\bar{z})}.$$

Then $\varphi^*_\lambda \in B_2(\mathbb{H}^*, \Gamma_\lambda)$ and $\|\varphi^*_\lambda\|_{\mathbb{H}^*} = 1$.

Since $\mathrm{d}\zeta^2$ on $\mathbb{C}\backslash\Omega_\lambda$ projects to a nontrivial holomorphic quadratic differential on T_λ via π_λ, the vector space $B_2(\mathbb{H}^*, \Gamma_\lambda)$ is spanned by φ^*_λ. Let

$$U_\lambda = \{w \in \mathbb{C} : w\varphi^*_\lambda \in \mathrm{Teich}(\Gamma_\lambda)\}.$$

Then, our result can be stated as in the following.

Theorem 3. *The domain* U_λ *converges to the cardioid* $4C_0$ *in the sense of Carathéodory as* $\lambda \to 0+$.

Since U_λ is a simply connected bounded domain (indeed, a Jordan domain by Minsky's theorem), there exists a conformal homeomorphism g_λ of the unit disk $\mathbb{D}$ onto U_λ such that $g_\lambda(0) = 0$ and $g'_\lambda(0) > 0$. Recall that $g_0(t) = 2(1 - (t-1)^2)$ is a conformal map of $\mathbb{D}$ onto the cardioid $4C_0$. Then, the last theorem can be restated in the following way.

Theorem 4. *The conformal map* g_λ *converges to the map* g_0 *locally uniformly on* $\mathbb{D}$.

We illustrate the shape of the boundary of Teich(Γ_λ) for several values of λ (see Figures 3 through 6, and compare with Figure 7). We remark that the boundary curve is drawn very roughly near the "main" cusp when λ is small. This is simply because of our algorithm of computation.

Figure 3. Bers embedding for the punctured square torus ($\lambda = 1$).

Figure 4. Bers embedding for small λ.

Figure 5. Bers embedding for very small λ.

7. Proof of Theorem 4

Let $\rho_{\mathbb{C}\setminus\Omega_\lambda}(\zeta)|\mathrm{d}\zeta|$ and $\rho_{T_\lambda}([\zeta])|\mathrm{d}\zeta|$ be the hyperbolic metrics on $\mathbb{C}\setminus\Omega_\lambda$ and T_λ, respectively. The density functions (the hyperbolic densities) $\rho_{\mathbb{C}\setminus\Omega_\lambda}(\zeta)|\mathrm{d}\zeta|$ and $\rho_{T_\lambda}([\zeta])$ are characterized by the relations $\rho_{T_\lambda}([\zeta]) = \rho_{\mathbb{C}\setminus\Omega_\lambda}(\zeta)$ and $\rho_{\mathbb{C}\setminus\Omega_\lambda}(p_\lambda(z))|p'_\lambda(z)| = \rho_{\mathbb{H}}(z) = 1/(2\,\mathrm{Im}\,z)$. The following assertion can be found (in a more general form) in [26].

Lemma 5. *The hyperbolic density $\rho_{T_\lambda}([\zeta])$ of T_λ takes its minimum at the point $[a_\lambda] = [(\lambda + \mathrm{i})/2]$.*

Figure 6. Bers embedding for very very small λ.

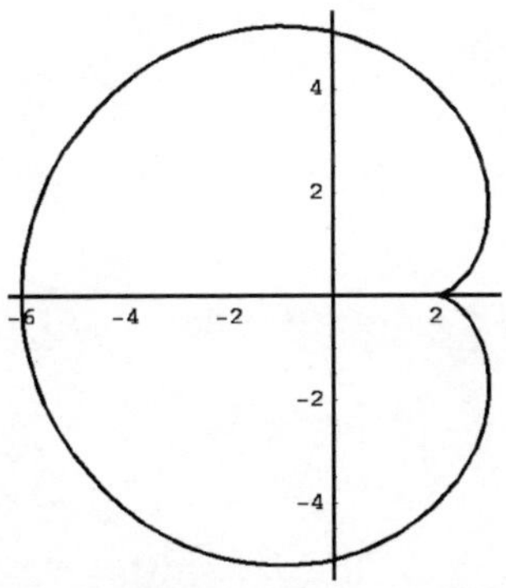

Figure 7. The cardioid $4C_0 = \{2(1-\alpha^2) : |\alpha - 1| < 1\}$.

Corollary 6. *The supremum of* $(2\,\mathrm{Im}\, z)^2|\varphi_\lambda(z)|$ *over* $z \in \mathbb{H}$ *is attained at* $z = \mathrm{i}$.

Proof. Letting $\zeta = p_\lambda(z)$, we have the relation

$$(2\,\mathrm{Im}\, z)^2|\varphi_\lambda(z)| = \rho_{\mathbb{C}\setminus\Omega_\lambda}(\zeta)^{-2},$$

which is known to take its maximum at $\zeta = a_\lambda = p_\lambda(\mathrm{i})$ by the last lemma. □

By the normalization of φ_λ, we have thus the relation $4|\varphi_\lambda(\mathrm{i})| = \|\varphi_\lambda\|_{\mathbb{H}} = 1$. On the other hand, $\tilde{\varphi}_\lambda(\mathrm{i}) = p'_\lambda(\mathrm{i})^2 < 0$. Therefore, $\varphi_\lambda(\mathrm{i}) = \tilde{\varphi}_\lambda(\mathrm{i})/\|\tilde{\varphi}_\lambda\|_{\mathbb{H}} < 0$ and thus

$$\varphi_\lambda(\mathrm{i}) = -\frac{1}{4}. \tag{7.1}$$

We also note the following fact, which will be used later.

Lemma 7. *The hyperbolic density* $\rho_{\mathbb{C}\setminus\Omega_\lambda}(\xi + \mathrm{i}/2)$ *converges to* $\pi/2$ *uniformly in* $\xi \in \mathbb{R}$ *as* $\lambda \to 0+$.

Proof. Observe that $\mathbb{C}\setminus\Omega_\lambda$ converges to the parallel strip $S = \{\zeta : 0 < \mathrm{Im}\,\zeta < 1\}$ with respect to the point $\mathrm{i}/2$ in the sense of Carathéodory when $\lambda \to 0+$. Thus, the hyperbolic density $\rho_{\mathbb{C}\setminus\Omega_\lambda}$ converges to ρ_S locally uniformly on S (cf. [8]). Since $\rho_{\mathbb{C}\setminus\Omega_\lambda}(\xi + \mathrm{i}/2)$ is a periodic function of ξ with period λ and since $\rho_S(\xi + \mathrm{i}/2) = \pi/2$, the assertion follows. □

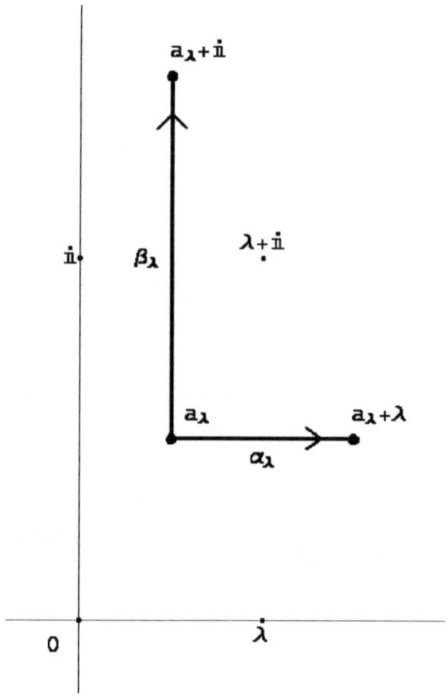

Figure 8. Geodesics on T_λ.

We now define the curves

$$\alpha_\lambda(s) = [a_\lambda + \lambda s] \; (0 \le s \le 1)$$

$$\beta_\lambda(s) = [a_\lambda + \mathrm{i}s] \; (0 \le s \le 1)$$

on T_λ, where we recall that $a_\lambda = (\lambda + \mathrm{i})/2$. Then, by the obvious symmetry of T_λ, these are simple hyperbolic geodesics of T_λ (see Figure 8).

We take the lifts $\hat{\alpha}_\lambda$ and $\hat{\beta}_\lambda$ of them starting at i via the covering projection $q_\lambda : \mathbb{H} \to T_\lambda$. We denote by A_λ and B_λ the elements of $\Gamma_\lambda \subset \mathrm{PSL}(2, \mathbb{R})$ corresponding to $\hat{\alpha}_\lambda$ and $\hat{\beta}_\lambda$, respectively. In other words, $\hat{\alpha}_\lambda(1) = A_\lambda(\hat{\alpha}_\lambda(0)) = A_\lambda(\mathrm{i})$ and $\hat{\beta}_\lambda(1) = B_\lambda(\hat{\beta}_\lambda(0)) = B_\lambda(\mathrm{i})$. See Figure 9.

Let σ and τ be the hyperbolic lengths of curves α_λ and β_λ, respectively. Then A_λ and B_λ can be represented by

$$A_\lambda = \pm \begin{pmatrix} e^\sigma & 0 \\ 0 & e^{-\sigma} \end{pmatrix} \quad \text{and} \quad B_\lambda = \pm \begin{pmatrix} \cosh\tau & \sinh\tau \\ \sinh\tau & \cosh\tau \end{pmatrix}.$$

Since the commutator $[A_\lambda, B_\lambda]$ must be parabolic, the following relation is required:

$$\sinh\sigma \sinh\tau = 1.$$

We remark that this sort of relation was already found by Hayman [7, §7].

We note the following fact.

Lemma 8. *The hyperbolic length σ of α_λ tends to* 0 *as $\lambda \to 0+$.*

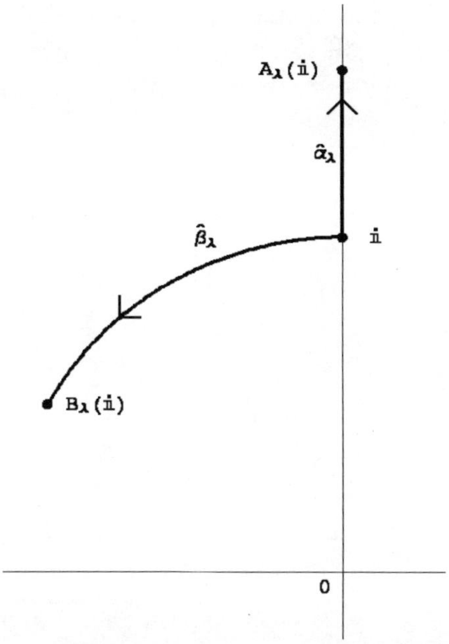

Figure 9. Lifts of curves on $\mathbb{H}$.

Proof. By definition, σ can be computed by

$$\sigma = \int_0^\lambda \rho_{T_\lambda}([\xi + \mathrm{i}/2])d\xi = \int_0^\lambda \rho_{\mathbb{C}\setminus\Omega_\lambda}(\xi + \mathrm{i}/2)d\xi.$$

By Lemma 7, we see that $\sigma \sim \pi\lambda/2$ as $\lambda \to 0+$. In particular, the assertion follows. □

A crucial result is the following variant of a proposition due to T. Nakanishi [19, Proposition 3.1].

Lemma 9 (Nakanishi's lemma). *Let G_n be a sequence of Fuchsian groups acting on $\mathbb{H}$ each of which contains a hyperbolic element of the form $z \mapsto M_n z$ such that $M_n \to 1$ as $n \to \infty$. Further let φ_n be an element of $B_2(\mathbb{H}, G_n)$ such that φ_n converges locally uniformly to a holomorphic function φ_∞ on $\mathbb{H}$. Then $\varphi_\infty(z) = c/z^2$ for some constant c.*

Proof. Write $\varphi_n(z) = P_n(z)/z^2$ and $\varphi_\infty(z) = P_\infty(z)/z^2$. Then $P_n \to P_\infty$ locally uniformly. Since $\varphi_n(M_n z)M_n^2 = \varphi_n(z)$, we have $P_n(M_n^k z) = P_n(z)$ for any $k \in \mathbb{Z}$. Letting $n \to \infty$, we can see that P_∞ is constant along the positive imaginary axis, which implies that P_∞ is constant. □

Lemma 10. *The quadratic differential $\varphi_\lambda(z)$ converges to $\varphi_0(z) = 1/(4z^2)$ locally uniformly on $\mathbb{H}$ as $\lambda \to 0+$.*

Proof. Since $\|\varphi_\lambda\|_{\mathbb{H}} = 1$, the family $\{\varphi_\lambda\}$ is locally bounded on $\mathbb{H}$ and thus normal. Let λ_n be any sequence of positive numbers tending to 0 such that φ_{λ_n} converges locally uniformly on $\mathbb{H}$. Then Lemmas 8 and 9 imply that the limit function has the form c/z^2. On the other hand, by (7.1), the constant c must be $1/4$, which is independent of the sequence λ_n. Thus a standard argument gives us the local uniform convergence of φ_λ to $1/(4z^2)$. □

Let

$$\mu_\lambda = \frac{|\varphi_\lambda|}{\varphi_\lambda}.$$

Then $t\mu_\lambda \in \mathrm{Belt}(\mathbb{H}, \Gamma_\lambda)$ for every $t \in \mathbb{D}$. Since $\Phi(-t\mu_\lambda) \in B_2(\mathbb{H}^*, \Gamma_\lambda)$, we can write $\Phi(-t\mu_\lambda) = c\varphi_\lambda^*$ for some $c \in \mathbb{C}$ for each $t \in \mathbb{D}$. We define the function $h_\lambda : \mathbb{D} \to \mathbb{C}$ by the relation

$$\Phi(-t\mu_\lambda) = h_\lambda(t)\varphi_\lambda^*.$$

We now show the following.

Lemma 11. *The function h_λ is a conformal homeomorshism of $\mathbb{D}$ onto U_λ satisfying $h_\lambda(0) = 0$, $\overline{h_\lambda(t)} = h_\lambda(\bar{t})$, and, in particular, $h'_\lambda(0) \in \mathbb{R}\backslash\{0\}$.*

Proof. Since the map $t \mapsto \Phi_{\Gamma_\lambda}(-t\mu_\lambda)$ is of the form of a Teichmüller map, h_λ is known to be a proper holomorphic injection of $\mathbb{D}$ into U_λ (see, for instance, [18]). Since $\dim \mathrm{Teich}(\Gamma_\lambda) = 1$, this map is indeed a conformal homeomorphism.

By the symmetry of Ω_λ, we have the relation $\overline{p_\lambda(-\bar{z}) - a_\lambda} + a_\lambda = p_\lambda(z)$. Therefore, $\varphi_\lambda = (p'_\lambda)^2$ satisfies $\overline{\varphi_\lambda(-\bar{z})} = \varphi_\lambda(z)$. Hence, $\overline{\mu_\lambda(-\bar{z})} = \mu_\lambda(z)$. This implies $\overline{\Phi(\bar{t}\mu_\lambda)(-\bar{z})} = \Phi(t\mu_\lambda)(z)$, which is equivalent to $\overline{h_\lambda(\bar{t})\varphi_\lambda^*(-\bar{z})} = h_\lambda(t)\varphi_\lambda^*(z)$. Since $\overline{\varphi_\lambda^*(-\bar{z})} = \varphi_\lambda^*(z)$, we have the relation $\overline{h_\lambda(t)} = h_\lambda(\bar{t})$. □

Remark. The quasiconformal deformation of T_λ corresponding to $\Phi(-t\mu_\lambda)$ can be constructed explicitly. Indeed, define a holomorphic motion H of $\mathbb{C}$ over $(\mathbb{D}, 0)$ by

$$H(t,\zeta) = H_t(\zeta) = \frac{\zeta - t\bar{\zeta}}{1+t}, \quad t \in \mathbb{D},\ \zeta \in \mathbb{C}.$$

Note that $H_t(\Omega_\lambda) = \Omega_{\frac{1-t}{1+t}\lambda}$. Thus, H_t induces a quasiconformal map $T_\lambda \to T_{\frac{1-t}{1+t}\lambda}$. Since the Beltrami coefficient of H_t is $-t$, its lift to $\mathbb{H}$ via p_λ has the Beltrami coefficient $-t\overline{p'_\lambda}/p'_\lambda = -t|\varphi_\lambda|/\varphi_\lambda = -t\mu_\lambda$.

By Lemma 10, we see that the Teichmüller differential $\mu_\lambda = |\varphi_\lambda|/\varphi_\lambda$ coverges pointwise to $\mu_0 = |\varphi_0|/\varphi_0 = z/\bar{z}$. We now use the following basic property of quasiconformal maps. It can be read from a paper of Ahlfors and Bers [1], but we give a proof for convenience.

Proposition 12. *Let μ_n be a sequence of measurable functions on $\mathbb{C}$ such that $\|\mu_n\|_\infty \leq 1$ and $\mu_n \to \mu$ a.e. Then the normalized $t\mu_n$-quasiconformal map $f^{t\mu_n}(z)$ converges to $f^{t\mu}(z)$ locally uniformly in $(t, z) \in \mathbb{D} \times \mathbb{C}$.*

Proof. We set $f_n(t,z) = f^{t\mu_n}(z)$ and $f_\infty(t,z) = f^{t\mu}(z)$. Suppose that real numbers $\varepsilon > 0, 0 < k < 1, T > 0$ are given. By Theorem 8 in [1], the function $f_n(t,z)$ is uniformly continuous in the parameter t for $n = 1, 2, \ldots, \infty$ and $|z| \leq T$. Hence, there exists a $\delta > 0$ such that $|f_n(z,t) - f_n(z,t')| < \varepsilon$ whenever $n = 1, 2, \dots, \infty$, $|z| \le T$ and $|t - t'| < \delta$. We choose finitely many points $t_1, \dots, t_m$ in the disk $|t| \le k$ so that $\min\{|t - t_j|, j = 1, \dots, m\} < \delta$ for every t with $|t| \le k$. Theorem 9 in [1] implies that $f_n(t_j, z) \to f_\infty(t_j, z)$ as $n \to \infty$ uniformly on $|z| \le T$ for each j. Thus, there exists a number N such that $|f_n(t_j, z) - f_\infty(t_j, z)| < \varepsilon$ holds whenever $j = 1, \dots, m$, $|z| \le T$ and $n \ge N$.

For each t with $|t| \le k$, z with $|z| \le T$ and $n \ge N$, choosing j so that $|t - t_j| < \delta$, we have

$$\begin{aligned}
&|f_n(t,z) - f_\infty(t,z)| \\
&\quad \le |f_n(t,z) - f_n(t_j,z)| + |f_n(t_j,z) - f_\infty(t_j,z)| + |f_\infty(t_j,z) - f_\infty(t,z)| \\
&\quad \le 3\varepsilon.
\end{aligned}$$

Thus the required assertion has been proved. □

By the above proposition, $f^{-t\mu_\lambda}(z)$ converges to $f^{-t\mu_0}(z)$ locally uniformly in $(t,z) \in \mathbb{D} \times \mathbb{C}$ as $\lambda \to 0+$. Since $f^{-t\mu_\lambda}$ is analytic in $\mathbb{H}^*$, the Weierstrass double series theorem implies that

$$\Phi(-t\mu_\lambda)(z) = S_{f^{-t\mu_\lambda}}(z) \to S_{f^{-t\mu_0}}(z) = \Phi(-t\mu_0)(z)$$

locally uniformly in $(t,z) \in \mathbb{D} \times \mathbb{H}^*$. By definition of h_λ and (5.1), the above convergence is equivalent to the locally uniform convergence

$$h_\lambda(t)\varphi_\lambda^*(z) \to g_0(t)\varphi_0^*(z).$$

Since $\varphi_\lambda^* \to \varphi_0^*$ by Lemma 10, we now see that $h_\lambda \to g_0$ locally uniformly on $\mathbb{D}$. In particular, $h_\lambda'(0) \to g_0'(0) = 4$. Since $h_\lambda'(0)$ is continuous in λ and, by Lemma 11, assumes non-zero real numbers, we also see that $h_\lambda'(0) > 0$ for all $\lambda > 0$. By the uniqueness of the Riemann mapping function and by Lemma 11, we obtain $g_\lambda = h_\lambda$. Thus Theorem 4 has been proved.

Acknowledgements

The author would like to thank the organizers of the International Workshop on Teichmüller Theory and Moduli Problems held at the Harish-Chandra Research Institute, Allahabad, India, for giving him a chance to consider the present topic seriously. Without their kind invitation to the workshop, this work would not have been carried out due to the author's busyness and laziness.

Finally, he thanks the referee for careful reading and for constructive suggestions.

References

[1] L. V. Ahlfors and L. Bers, Riemann's mapping theorem for variable metrics, *Ann. of Math. (2)*, **72** (1960), 385–404.

[2] K. Astala and G. J. Martin, Holomorphic motions, Papers on analysis, Rep. Univ. Jyväskylä Dep. Math. Stat., 83, Univ. Jyväskylä (2001) pp. 27–40.

[3] L. Bers and H. L. Royden, Holomorphic families of injections, *Acta Math.*, **157** (1986) 259–286.

[4] Lei Tan (Editor), The Mandelbrot Set, Theme and Variations, *London Math. Soc. Lecture Note Series*, vol. 274, Cambridge Univ. Press (2000).

[5] F. P. Gardiner and N. Lakic, Quasiconformal Teichmüller Theory, *Amer. Math. Soc.* (2000).

[6] D. Goodman, Spirals in the boundary of slices of quasi-Fuchsian space, *Conform. Geom. Dyn.*, **10** (2006) 136–158.

[7] W. K. Hayman, Some inequalities in the theory of functions, *Proc. Cambridge Philos. Soc.*, **44** (1948) 159–178.

[8] D. A. Hejhal, Universal covering maps for variable regions, *Math. Z.*, **137** (1974) 7–20.

[9] E. Hille, Remarks on a paper by Zeev Nehari, *Bull. Amer. Math. Soc.*, **55** (1949) 552–553.

[10] J. Hubbard, Teichmüller Theory and Applications to Geometry, Topology, and Dynamics, Vol. I: Teichmüller Theory, Matrix Editions (2006).

[11] Y. Imayoshi and M. Taniguchi, An introduction to Teichmüller spaces, Springer-Tokyo (1992).

[12] C. I. Kalme, Remarks on a paper by Lipman Bers, *Ann. of Math. (2)*, **91** (1970) 601–606.

[13] Y. Komori and T. Sugawa, Bers embedding of the Teichmüller space of a once-punctured torus, *Conform. Geom. Dyn.* **8** (2004) 115–142.

[14] Y. Komori, T. Sugawa, M. Wada and Y. Yamashita, Drawing Bers embeddings of the Teichmüller space of once-punctured tori, *Exper. Math.*, **15** (2006) 51–60.

[15] R. Mañé, P. Sad, and D. Sullivan, On dynamics of rational maps, *Ann. Sci. École Norm. Sup.*, **16** (1983) 193–217.

[16] Y. Minsky, The classification of punctured-torus groups, *Ann. of Math. (2)*, **149** (1999) 559–626.

[17] H. Miyachi, Cusps in complex boundaries of one-dimensional Teichmüller spaces, *Conform. Geom. Dyn.*, **7** (2003) 103–151.

[18] S. Nag, The Complex Analytic Theory of Teichmüller Spaces, Wiley, New York (1988).

[19] T. Nakanishi, A theorem on the outradii of Teichmüller spaces, *J. Math. Soc. Japan*, **40** (1988) 1–8.

[20] ——, The inner radii of finite-dimensional Teichmüller spaces, *Tôhoku Math. J. (2)*, **41** (1989) 679–688.

[21] T. Nakanishi and J. A. Velling, A sufficient condition for Teichmüller spaces to have smallest possible inner radii, *Ann. Acad. Sci. Fenn. A. I. Math.*, **18** (1993) 13–21.

[22] T. Nakanishi and H. Yamamoto, On the outradius of the Teichmüller space, *Comment. Math. Helvetici*, **64** (1989) 288–299.

[23] W. C. Royster, On the univalence of a certain integral, *Michigan Math. J.*, **12** (1965) 385–387.

[24] H. Sekigawa, The outradius of the Teichmüller space, *Tôhoku Math. J. (2)*, **30** (1978) 607–612.

[25] Z. Slodkowski, Holomorphic motions and polynomial hulls, *Proc. Amer. Math. Soc.*, **111** (1991) 347–355.

[26] T. Sugawa, Estimates of hyperbolic metric with applications to Teichmüller spaces, *Kyungpook Math. J.*, **42** (2002) 51–60.

[27] ——, The limiting shape of one-dimensional Teichmüller spaces, to appear in Proc. Amer. Math. Soc.

Teichmüller Theory and Moduli Problem No. 10, 2010, pp. 597–618.

Stratifying the Space of Moduli

Anthony Weaver

Department of Mathematics and Computer Science, Bronx Community College,
The City University of New York, 2155 University Avenue, Bronx, N.Y., 10453, U.S.A.
e-mail: anthony.weaver@bcc.cuny.edu; anthonyweaver@mac.com

Abstract. In the space of moduli of surfaces of genus g, the locus of surfaces with automorphisms is a complicated set which can be stratified according to topological conjugacy of the full automorphism groups of the surfaces. The strata are non-disjoint, and the intersections can contain surfaces admitting isomorphic subgroups of automorphisms which are topologically but not conformally equivalent. The paper is expository, with some explicit examples at the end.

1. Introduction

The space of moduli of compact Riemann surfaces of a given genus g has a large singular set corresponding to surfaces with automorphisms. A precise geometric understanding of the singular set requires a good deal of background material. The purpose of this expository paper is to present this material in a coherent and reasonably self-contained way. There are several possible approaches. Ours is essentially algebraic, involving Fuchsian groups whenever possible. A second purpose of the paper is to show how the distinction between topological and conformal conjugacy of group actions on surfaces is reflected in the branch locus in Teichmüller space. The main background references for the paper are [3,27,28]. Some general references for later sections are [6,31,16].

The contents of the paper are as follows. In Section 2. we define Fuchsian groups and their proper discontinuous action on the upper half plane. In Section 3. we define the Teichmüller space and modular group associated to a given Fuchsian group. In Section 4. we investigate the interaction between a quasiconformal map of the upper half plane and a given Fuchsian group, and prove, in Section 5., that every isomorphism between Fuchsian groups can be realized as conjugation by a quasiconformal map of the upper half plane. We sketch a proof of the existence and uniqueness of Teichmüller maps realizing such isomorphisms. This leads to the definition, in Section 6., of the Teichmüller metric on the Teichmüller space of a Fuchsian group. In Section 7. we show that the fixed point set of a finite subgroup of the modular group is itself a Teichmüller space, embedded in the ambient Teichmüller space on which the modular group acts. We state a version of the Nielsen-realization theorem, that every finite subgroup of the modular group has a non-empty fixed point set. In Section 8., we define the relative modular groups, which are set-wise stabilizers of fixed point sets of finite subgroups of the modular group, and the quotient groups of the relative modular groups which act effectively. Next we define the relative Riemann spaces parametrizing surfaces admitting a given symmetry group. In Section 9. we distinguish topological from conformal conjugacy of group actions on surfaces. This provides a precise way of describing the branch locus in Teichmüller space in terms of embedded Teichmüller spaces of smaller dimension. The branch locus corresponds to the singular set in the space of moduli. The stratification of this space is described in Section 10.. The intersections of the strata echo the corresponding relationships among the embedded spaces in Teichmüller space. We finish with some explicit examples in Section 11..

2. Uniformization and Fuchsian groups

According to the *uniformization theorem* (Klein, Poincaré, Koebe), a simply connected Riemann surface, up to conformal equivalence, is one of the following:

1. the complex plane;
2. the Riemann sphere;
3. the upper half plane $U = \{z \in \mathbb{C} | \mathrm{Im}(z) > 0\}$.

Each of these has a canonical complete metric of constant curvature. On U the metric is Poincarè's hyperbolic metric $|dz|/\mathrm{Im}(z)$, of constant curvature -1. A consequence of the uniformization theorem is that every Riemann surface can be represented as a quotient $\tilde{X}/\Gamma$, where $\tilde{X}$ is one of the three simply connected surfaces, and Γ is a discrete group of orientation-preserving isometries, acting *discontinuously* on $\tilde{X}$. This means that every point $x \in \tilde{X}$ is contained in an open set which does not meet any of its γ-translates, $\gamma \in \Gamma$, unless γ is the identity.

A slight weakening of the notion of discontinuity permits a covering theory of orbifolds (surfaces with cone points), analogous to the uniformization theorem for surfaces.

Definition 1. *A group of G of homeomorphisms of a topological space S acts* properly discontinuously *if*

1. *every $s \in S$ is contained in an open set $V \subset S$ such that, for $g \in G$, $gV \cap V \neq \emptyset$ implies $gs = s$;*
2. *the stabilizer of a point is finite.*

Definition 2. *A group G acts* effectively *on a set S if the (normal) subgroup $H = \{g \in G | gs = s$ for all $s \in S\}$ is trivial.*

If Γ acts effectively and properly discontinuously on $\tilde{X}$, the points in $\tilde{X}$ with nontrivial stabilizers form a discrete set $D \subset \tilde{X}$. The quotient map $\tilde{X} \to \tilde{X}/\Gamma$, restricted to $\tilde{X} - D$, is a covering which transfers the metric and conformal structures from $\tilde{X} - D$ to the quotient $(\tilde{X} - D)/\Gamma$, a surface punctured at a discrete set of points. The filled-in punctures become the cone points, and the metric and conformal structures extend in a canonical way to the resulting closed orbifold.

When $\tilde{X} = U$, Γ is a discrete subgroup of the real *Möbius group*

$$\mathcal{L} = \left\{ z \mapsto \frac{az+b}{cz+d}, \quad a, b, c, d \in \mathbb{R}, \quad ad - bc = 1 \right\} \simeq PSL(2, \mathbb{R}),$$

called a *Fuchsian group* (see [21], Chapter 5 or [22]). The boundary $\partial U = \mathbb{R} \cup \{\infty\}$ contains the *limit set* of accumulation points of Γ-orbits of points $z \in U$. A Fuchsian group is *elementary* if the limit set is finite. Here we consider only non-elementary Fuchsian groups *of the first kind*, for which the limit set is all of ∂U.

An element of $\mathcal{L}$ is *elliptic* if it has a single fixed point in U, *parabolic* if it has a single fixed point in ∂U, and *hyperbolic* if it has two fixed points in ∂U. Parabolic and hyperbolic elements have infinite order. An elliptic element can have infinite order, but, if it is contained in a Fuchsian group, by discreteness, its order must be finite. A hyperbolic element $t \in \mathcal{L}$ stabilizes a geodesic *axis* $l_t \subset U$. For $z \in l_t$ the hyperbolic distance between z and tz is an invariant of t called the *translation length*. The limits as $n \to \pm\infty$ of $t^n z$ are the endpoints of l_t on ∂U. They are called the *attracting* (+) and *repelling* (−) fixed points of t.

Lemma 1. *Non-identity elements in $\mathcal{L}$ commute if and only if they have the same fixed point set.*

Proof. [21], Theorem 5.2.4. One need only examine the centralizers of representatives of conjugacy classes. Parabolic elements are conjugate to $z \mapsto z \pm 1$, and the centralizer of these elements is $\{z \mapsto z + k | k \in \mathbb{R}\}$. All such elements fix $\{\infty\}$. Hyperbolic elements are conjugate to $z \mapsto \lambda z$, $\lambda > 0$, $\lambda \neq 1$, with centralizer $\{z \mapsto \mu z | \mu > 0\}$; the fixed point sets are $\{0, \infty\}$. After a standard conformal transformation from U to the interior of the unit disk, all elliptic elements are conjugate to rotations $w \mapsto e^{i\theta} w$, $\theta \in \mathbb{R}$, which are centralized by $\{w \mapsto e^{i\phi} w | 0 \leq \phi < 2\pi\}$. All such elements fix the origin. □

Corollary 2. *The centralizer in $\mathcal{L}$, and, in particular, the center, of a non-elementary Fuchsian group is trivial.*

Proof. The existence of a non-trivial element of $\mathcal{L}$ which commutes with every element of the group would imply that every element has the same (finite) fixed point set, making the group elementary. □

A finitely generated Fuchsian group has a fundamental polygon in $U \cup \partial U$ with finitely many sides [18]. If none of the sides is contained in ∂U, the polygon has finite hyperbolic area. If none of the vertices is contained in ∂U, Γ contains no parabolic elements, and the quotient U/Γ is compact. A Fuchsian group Γ of this type is called *co-compact*, and it has the following canonical presentation:

$$\Gamma = \left\langle a_1, b_1, \ldots, a_g, b_g, e_1, \ldots, e_n | e_1^{\nu_1} = e_2^{\nu_2} = \cdots = e_n^{\nu_n} = 1, \right.$$

$$\left. \prod_{i=1}^{g} [a_i, b_i] \prod_{j=1}^{n} e_j = 1 \right\rangle, \quad (1)$$

where the generators $\{e_i\}$ are elliptic and the other generators are hyperbolic. The $(n+1)$-tuple

$$(g; \nu_1, \nu_2, \ldots, \nu_n) \quad (2)$$

is called the *signature* of Γ, and it determines Γ uniquely if we assume, as we shall, that $\nu_1 \le \nu_2 \le \cdots \le \nu_n$. The ν_i are called the *periods* of Γ, and they form a complete set of maximal orders of elliptic elements in Γ, one for each conjugacy class. A signature of the form (2) is said to have *type* (g, n).

The presentation (1) defines a Fuchsian group if and only if the number

$$\chi(\Gamma) = \chi(g; \nu_1, \nu_2, \ldots, \nu_n) = 2\pi \left\{ 2g - 2 + \sum_{j=1}^{n} \left(1 - \frac{1}{\nu_j}\right) \right\}, \tag{3}$$

is positive. $\chi(\Gamma)$ is the hyperbolic area of a fundamental polygon for Γ acting on U. If $\chi(\Gamma) > 0$, then $\chi(\Gamma) \ge \pi/21$, with equality if and only if Γ is the group with signature (0; 2, 3, 7).

Lemma 3 (The Riemann-Hurwitz relation). *If $\Gamma \le \Gamma^*$ are co-compact Fuchsian groups, the ratio $\chi(\Gamma)/\chi(\Gamma^*)$ is finite and equal to the index $[\Gamma^* : \Gamma]$.*

Proof. See, e.g., [18], or [11], Theorem 1.2.7. □

For co-compact Γ, the compact quotient U/Γ is an *orbifold* of type (g, n), that is, a compact Riemann surface of genus g with n distinguished points, over which the quotient map $U \to U/\Gamma$, elsewhere a covering map, is ramified. All orbifolds with a finite number of distinguished points, except those of type $(g, n) = (0, 0)$, $(0, 1)$, $(0, 2)$, and $(1, 0)$, arise as quotients of co-compact Fuchsian groups acting on U. If Γ has presentation (1), the order of the isotropy subgroup of a point lying over the ith distinguished point is ν_i.

The *normalizer* of a subgroup G in a group H is the subgroup $N_H(G) = \{h \in H | hGh^{-1} = G\}$.

Lemma 4. *The normalizer $N_{\mathcal{L}}(\Gamma)$ of a co-compact Fuchsian group Γ is a co-compact Fuchsian group, and the index $[N_{\mathcal{L}}(\Gamma) : \Gamma]$ is finite.*

Proof. If $N_{\mathcal{L}}(\Gamma)$ is not discrete it contains an infinite sequence of distinct elements $\{\delta_i\}$ tending to the identity. For all $\gamma \in \Gamma$, there exists a positive integer m such that $\delta_i \gamma \delta_i^{-1} = \gamma$, $i \ge m$, otherwise $\{\delta_i \gamma \delta_i^{-1}\}$ would be an infinite sequence of distinct elements of Γ tending to γ, contradicting the discreteness of Γ. Thus, by Lemma 1, δ_i, $i \ge m$, has the same fixed point set as γ. Since Γ is nonabelian there exists $\gamma' \in \Gamma$ which does not commute with γ. Repeating the previous argument, there exists m' such that δ_i, $i \ge m'$, has the same fixed point set as γ'. But then γ, γ' have the same fixed point set and hence commute. This contradiction proves that $N_{\mathcal{L}}(\Gamma)$ is discrete. By the Riemann-Hurwitz relation, $\pi/21 \le \chi(N_{\mathcal{L}}(\Gamma)) \le \chi(\Gamma) < \infty$ whence the index $[N_{\mathcal{L}}(\Gamma) : \Gamma] < \infty$. Since a fundamental domain for $N_{\mathcal{L}}(\Gamma)$ is contained in a fundamental domain for Γ, which has no sides or vertices in ∂U, $N_{\mathcal{L}}(\Gamma)$ is co-compact. □

A co-compact Fuchsian group without elliptic elements (i.e., torsion-free) is called a *surface group*. If the signature is of type $(g, 0)$, the group is isomorphic to the fundamental group of a compact surface of genus g. We shall use Λ (Λ_g) to denote a surface group (of genus g). By the uniformization theorem, every compact surface of genus g is conformally equivalent to a quotient surface U/Λ', where Λ' is the image of an injective homomorphism $r : \Lambda_g \to \mathcal{L}$.

Theorem 5. *Surface groups Λ and Λ' of genus $g > 1$ are conjugate in $\mathcal{L}$ if and only if the surfaces U/Λ and U/Λ' are conformally equivalent.*

Proof. If $t^{-1}\Lambda t = \Lambda'$ for some $t \in \mathcal{L}$, then t induces the conformal map $\Lambda z \mapsto \Lambda' t^{-1} z$ between U/Λ and U/Λ', where Λz denotes the Λ-orbit of $z \in U$. Conversely, a conformal map $c : U/\Lambda \to U/\Lambda'$ lifts to a conformal map $\tilde{c}$ of the universal covering space U, i.e., an element $\tilde{c} \in \mathcal{L}$. Since $\tilde{c}(\Lambda z) = \Lambda'(\tilde{c}z)$, it follows that $\tilde{c}\Lambda_g\tilde{c}^{-1} = \Lambda'_g$. □

A subgroup G of a group H is called *characteristic* if all automorphisms of H preserve G. In particular, G is normal in H.

Theorem 6 (Bundgaard, Nielsen, Fox). *A co-compact Fuchsian group contains a surface group as a characteristic subgroup of finite index.*

Proof. [8,12]. The proof is based on two lemmas: (i) there exists a normal subgroup of finite index which contains no non-trivial power of any given elliptic element; (ii) the intersection of a finite number of subgroups of finite index is a subgroup of finite index. Since there are only finitely many conjugacy classes of elliptic elements, the theorem follows. □

3. Teichmüller spaces of Fuchsian groups

Henceforth Γ denotes a co-compact Fuchsian group with presentation (1).

Let $R(\Gamma)$ be the set of all injective homomorphisms $r : \Gamma \to \mathcal{L}$ such that the image $r(\Gamma)$ is Fuchsian. $R(\Gamma)$ is topologized as a subspace of the product of $2g + n$ copies of $\mathcal{L}$, by assigning to $r \in R(\Gamma)$ the point

$$(r(a_1), r(b_1), \ldots, r(a_g), r(b_g), r(e_1), \ldots r(e_n)) \in \mathcal{L}^{2g+n}.$$

The identity $\mathrm{id}_\Gamma : \Gamma \hookrightarrow \mathcal{L}$ may be taken as a base point in $R(\Gamma)$. $\mathrm{Aut}(\mathcal{L})$ acts on $R(\Gamma)$ by post-composition; since $\mathrm{Aut}(\mathcal{L}) \simeq \mathcal{L} \simeq \mathrm{Inn}(\mathcal{L})$, the action is by conjugation. $r_1, r_2 \in R(\Gamma)$ are called *equivalent* if their images are conjugate within $\mathcal{L}$. Equivalence classes $[r]$ are the points in $T(\Gamma)$, the *Teichmüller space* of Γ, which takes the quotient topology from $R(\Gamma)$.

Theorem 7 (Greenberg [17]). *Let Γ and Γ^* be co-compact Fuchsian groups. An injective homomorphism $i : \Gamma \to \Gamma^*$ induces a homeomorphism $\bar{i} : T(\Gamma^*) \to T(\Gamma)$ onto a closed subspace, defined by*

$$\bar{i} : [r] \mapsto [r \circ i]. \tag{4}$$

Proof. We sketch the proof given in [27], Theorem 7.11. Let $\delta_1, \delta_2, \ldots, \delta_h$ be a canonical finite set of generators of Γ, as in (1). Let $\gamma_1, \gamma_2, \ldots, \gamma_k \in \Gamma^*$ be a complete set of coset representatives of $i(\Gamma)$ in Γ^*. $k < \infty$ since, by the Riemann-Hurwitz relation, the inclusion $i : \Gamma \to \Gamma^*$ has finite index. It can be arranged that all the γ_i are hyperbolic elements ([27], Lemma 4.8). Γ^* is generated by the set

$$\{i(\delta_1), \ldots, i(\delta_h), \gamma_1, \ldots, \gamma_k\},$$

and $r \in R(\Gamma^*)$ corresponds to the point

$$(r \circ i(\delta_1), \ldots, r \circ i(\delta_h), r(\gamma_1), \ldots, r(\gamma_k)) \in \mathcal{L}^{h+k}.$$

For each $j = 1, \ldots, k$, there exists a smallest $n_j < \infty$ such that $\gamma_j^{n_j} \in i(\Gamma)$. Let $\gamma_j^{n_j} = i(e_j), e_j \in \Gamma$. Then $r(\gamma_j)$ is the unique n_jth root of $r \circ i(e_j)$. Call this element $l_j \in \mathcal{L}$. Uniqueness of l_j follows from uniqueness of roots in infinite cyclic groups: $r \circ i(e_j)$ belongs to the infinite cyclic group generated by the hyperbolic element $r(\gamma_j)$. Then $r \circ i \in R(\Gamma)$ corresponds to the point

$$(r \circ i(\delta_1), \ldots, r \circ i(\delta_h), l_1, \ldots, l_k) \in \mathcal{L}^{h+k},$$

and r is uniquely and continuously recoverable from $r \circ i$. Hence i induces an embedding of $R(\Gamma^*)$ into $R(\Gamma)$, whose image is a closed subset. As r varies in its Teichmüller class, so does $r \circ i$ (the images are both conjugated by an element of $\mathcal{L}$). Thus the embedding descends to an embedding of the quotient Teichmüller spaces, with respect to the quotient topology. □

Let $\mathrm{Aut}^+(\Gamma)$ be the group of automorphisms of Γ which are both *type-* and *orientation-preserving*. Type-preserving automorphisms carry elliptic elements to elliptic elements, parabolic elements to parabolic elements, etc. (For co-compact Fuchsian groups, which contain only finite-order elliptic elements and infinite-order hyperbolic elements, all automorphisms are type-preserving.) Orientation-preserving automorphisms are induced by automorphisms of the free group on the generators of (1) which carry the final relator in (1) to a conjugate of itself but not of its inverse. Inner automorphisms are type- and orientation-preserving. Let $\mathrm{Inn}(\Gamma) \subseteq \mathrm{Aut}^+(\Gamma)$ be the normal subgroup of inner automorphisms. $\alpha \in \mathrm{Aut}^+(\Gamma)$ induces a homeomorphism of $T(\Gamma)$ defined by

$$[r] \mapsto [r \circ \alpha] \tag{5}$$

If $\alpha \in \mathrm{Inn}(\Gamma)$, there exists $\delta \in \Gamma$ such that $\alpha(\gamma) = \delta^{-1}\gamma\delta$. Then $r(\alpha(\Gamma)) = c^{-1}r(\Gamma)c$, where $c = r(\delta) \in \mathcal{L}$, so that $[r \circ \alpha] = [r]$. Thus the *effective* action is by the *Teichmüller modular group* of Γ

$$\mathrm{Mod}(\Gamma) = \{[\alpha]\} \simeq \frac{\mathrm{Aut}^+(\Gamma)}{\mathrm{Inn}(\Gamma)}.$$

$[\alpha]$ denotes the class in $\mathrm{Aut}^+(\Gamma)/\mathrm{Inn}(\Gamma)$ induced by α.

Theorem 8. *The action $Mod(\Gamma) \times T(\Gamma) \to T(\Gamma)$ defined by*

$$([\alpha], [r]) \mapsto [r \circ \alpha]$$

is properly discontinuous.

Proof. This was proved by Kravetz [24] in the case where Γ is surface group. We follow the proof in [28], Theorem 7, which assumes Kravetz′ result.

By Theorem 6, Γ contains a surface group Λ as a characteristic subgroup of finite index. The inclusion $i : \Lambda \hookrightarrow \Gamma$ induces the embedding $\bar{i} : T(\Gamma) \hookrightarrow T(\Lambda)$ defined by (4). Let $[r] \in T(\Gamma)$, and suppose $[r \circ i] \in V \subset T(\Lambda)$, where V is an open set satisfying property 1 of Definition 1 for the action of $\mathrm{Mod}(\Lambda)$ on $T(\Lambda)$ (Kravetz′ result is assumed here). Let $W = V \cap \bar{i}(T(\Gamma)) = \bar{i}^{-1}(V)$. W is an open set in $T(\Gamma)$ which is non-empty, since it contains $[r]$. Since Λ is characteristic in Γ, for every $\beta \in \mathrm{Aut}^+(\Gamma)$ there exists a unique $\alpha \in \mathrm{Aut}^+(\Lambda)$ such that $\beta \circ i = i \circ \alpha$. If $\beta \in \mathrm{Aut}^+(\Gamma)$ is such that $[\beta](W) \cap W \neq \emptyset$, then $[\alpha](V) \cap V \neq \emptyset$ and by property 1 for $\mathrm{Mod}(\Lambda)$ on $T(\Lambda)$, $[\alpha]$ fixes $[r \circ i]$, that is, $[r \circ i \circ \alpha] = [r \circ i]$. Equivalently, $[r \circ \beta \circ i] = [r \circ i]$ which implies $[r \circ \beta] = [r]$. Thus $W \subset T(\Gamma)$ is an open set containing $[r]$ which satisfies property 1 for $\mathrm{Mod}(\Gamma)$ acting on $T(\Gamma)$.

Property 2 of Definition 1 for $\mathrm{Mod}(\Gamma)$ acting on $T(\Gamma)$ follows from the next theorem and Lemma 4. □

Theorem 9. *The stabilizer of a point* $[r] \in T(\Gamma)$ *is isomorphic to the (finite) subgroup* $N_{\mathcal{L}}(r(\Gamma))/r(\Gamma)$ *of* $Mod(\Gamma)$.

Proof. If $[\alpha] \in \mathrm{Mod}(\Gamma)$ fixes $[r]$, then $[r \circ \alpha] = [r]$ and there exists $t \in \mathcal{L}$ such that, for all $\gamma \in \Gamma$, $r \circ \alpha(\gamma) = tr(\gamma)t^{-1}$. It follows that $t \in N_{\mathcal{L}}(r(\Gamma))$. If $t \in r(\Gamma)$, $\alpha \in \mathrm{Inn}(\Gamma)$ and hence $[\alpha]$ is the identity in $\mathrm{Mod}(\Gamma)$. Thus the stabilizer of $[r]$ is isomorphic to a subgroup of $N_{\mathcal{L}}(r(\Gamma))/r(\Gamma)$. On the other hand, if $t \in N_{\mathcal{L}}(r(\Gamma))$, the map $\beta_t : r(\gamma) \mapsto tr(\gamma)t^{-1}$ is a type- and orientation-preserving automorphism of $r(\Gamma)$, whence $\alpha_t = r^{-1} \circ \beta_t \circ r$ is a type- and orientation-preserving automorphism of Γ. α_t is inner if and only if $t \in r(\Gamma)$. This establishes the isomorphism. □

4. Quasiconformal maps; compatibility

A homeomorphism $w : U \to U$ is called *quasiconformal*, or more precisely, *k-quasiconformal*, if there exists $k < 1$ such that $|w_{\bar{z}}/w_z| \leq k$ almost everywhere. Partial derivatives w_z and $w_{\bar{z}}$ exist in the sense of distribution theory ([3], §2), and obey the usual chain rules

$$(u \circ v)_z = (u_v \circ v)v_z + (u_{\bar{v}} \circ v)(\bar{v})_z; \tag{6}$$

$$(u \circ v)_{\bar{z}} = (u_v \circ v)v_{\bar{z}} + (u_{\bar{v}} \circ v)(\bar{v})_{\bar{z}}. \tag{7}$$

The expression $w_{\bar{z}}/w_z$ is called a *Beltrami differential.* A complex measurable function $\mu(z)$ defined on U determines a *Beltrami equation*

$$w_{\bar{z}} = \mu(z)w_z, \tag{8}$$

in which $\mu(z)$ is the *Beltrami coefficient.* If $|\mu(z)| < 1$, the Jacobian $(1 - \mu\bar{\mu})|w_z|^2$ of w is positive, and hence w is orientation-preserving.

Lemma 10.

(i) *If* $\mu(z)$ *is a Beltrami coefficient with* $|\mu(z)| \leq k < 1$ *almost everywhere, and* w *is a solution of* (8), *then* w *is* k*-quasiconformal; conversely, every* k*-quasiconformal self-homeomorphism of* U *is the solution of* (8) *for some Beltrami coefficient* $\mu(z)$, $|\mu(z)| \leq k < 1$ *almost everywhere.*

(ii) *If* w *is a solution of* (8), *and* $c \in \mathcal{L}$, *then* $w_1 = c \circ w$ *is also a solution; conversely, if* w, w_1 *are both solutions of* (8) *then there exists* $c \in \mathcal{L}$ *such that* $w_1 = c \circ w$.

Proof. [3], §4 E, §4 B. □

Lemma 11. *A quasiconformal self-homeomorphism of* U *has a unique extension to a self- homeomorphism of* $\bar{U} = U \cup \partial U$.

Proof. [3], §3 E. □

There is a definition of quasiconformality due to Ahlfors and Pfluger [1] which makes no reference to differentiability (distributional or otherwise): a conformal image of $\{(x, y) \in \mathbb{R}^2 | 0 \leq x \leq m, 0 \leq y \leq 1\}$ is called a *topological rectangle* R with *modulus* $m = m(R)$. By the Riemann mapping theorem, a continuous image of a topological rectangle is a topological rectangle (with a possibly different modulus). A homeomorphism $w : U \to U$ is k-quasiconformal if, for every topological rectangle R,

$$\frac{m(w(R))}{m(R)} \leq \frac{(1+k)}{(1-k)}. \tag{9}$$

Lemma 12. *The composition of a* k_1*-quasiconformal map with a* k_2*-quasiconformal map is* k*-quasiconformal, where*

$$\frac{1+k}{1-k} = \frac{1+k_1}{1-k_1} \cdot \frac{1+k_2}{1-k_2}.$$

Proof. An immediate consequence of (9). □

For quasiconformal w, let $k(w) < 1$ be the smallest k such that w is k-quasiconformal. The *maximal dilatation* of w is

$$K(w) = \frac{1+k(w)}{1-k(w)}.$$

We have the following consequences of the preceding definitions and lemmas. Let w_1, w_2, $w : U \to U$ be quasiconformal maps, and let $c : U \to U$ be conformal. Then

$$K(c) = 1; \tag{10}$$

$$K(w_1 \circ w_2) \leq K(w_1)K(w_2); \tag{11}$$

$$K(w) = K(w^{-1}); \tag{12}$$

$$K(w \circ c) = K(c \circ w) = K(w). \tag{13}$$

Let Γ be a Fuchsian group. A Beltrami coefficient $\mu(z)$ is Γ*-compatible* if it transforms according to the rule

$$\mu(\gamma(z))) = \mu(z)\gamma'(z)/\overline{\gamma'(z)} \quad \text{for all } \gamma \in \Gamma. \tag{14}$$

A holomorphic function $\phi(z)$ defined on U is called a Γ-2-form (automorphic form of weight 2 for Γ) if it transforms according to the rule

$$\phi(\gamma(z))) = \phi(z)/\gamma'(z)^2 \quad \text{for all } \gamma \in \Gamma. \tag{15}$$

A Γ-2-form induces an integrable *quadratic differential* on the quotient surface U/Γ, holomorphic except possibly for simple poles or removable singularities at the orbifold points. All holomorphic quadratic differentials on U/Γ arise in this way.

Lemma 13.

(i) *If ϕ is a Γ-2-form, then*

$$\mu(z) = k\overline{\phi(z)}/|\phi(z)|, \quad 0 \le k < 1, \tag{16}$$

is a Γ-compatible Beltrami coefficient. Partially conversely,

(ii) *If $\phi(z)$ is a holomorphic function such that* (16) *is a Γ-compatible Beltrami coefficient, then, up to multiplication by a real character $\chi : \Gamma \to (\mathbb{R}^+, \cdot)$, ϕ is a Γ-2-form.*

Proof. [27], Lemma 5.18. (i) is a consequence of the chain rules. (ii) A *real character* is a homomorphism of a group into the multiplicative group of positive reals. If $\phi(z)$ has modulus $r(z) > 0$ and argument $\theta(z) \in \mathbb{R}$, then $\mu(z) = k \exp(-i\theta(z))$, and, since μ is Γ-compatible, for all $\gamma \in \Gamma$,

$$-\theta(\gamma z) - \arg \gamma'(z) = -\theta(z) + \arg \gamma'(z).$$

This implies that the number $\chi(\gamma) = \phi(\gamma(z))\gamma'(z)^2/\phi(z)$ is real $(\arg(\chi(\gamma) = 0)$. In addition $\chi(\gamma) = r(\gamma(z))|\gamma'(z)|^2/r(z) > 0$. The homomorphism property $\chi(\gamma_1\gamma_2) = \chi(\gamma_1)\chi(\gamma_2)$ is easily verified. □

Theorem 14. *Let Γ_1 be a subgroup of finite index in Γ. Let ϕ be a Γ_1-2-form. Then* (16) *is Γ-compatible if and only if ϕ is a Γ-2-form.*

Proof. [27], Theorem 5.20. By Lemma 13, (16) is Γ_1-compatible; if it is also Γ-compatible, then there is a real character $\chi : \Gamma \to (\mathbb{R}^+, \cdot)$ such that

$$\phi(\gamma(z)) = \chi(\gamma) \cdot \phi(z)/\gamma'(z)^2 \quad \text{for all } \gamma \in \Gamma.$$

For some finite t, $\gamma^t \in \Gamma_1$ since the index $[\Gamma : \Gamma_1]$ is finite. It follows that $\chi(\gamma^t) = (\chi(\gamma))^t = 1$. This implies $\chi(\gamma) = 1$, since $(\mathbb{R}^+, \cdot)$ has no non-trivial element of finite order. Hence ϕ is a Γ-2-form. Conversely, if ϕ is a Γ-2-form, then (16) is Γ-compatible by Lemma 13. □

5. Geometrical realizations

$r \in R(\Gamma)$ has a *geometrical realization* if there is an orientation-preserving homeomorphism $w : U \to U$ such that, for all $\gamma \in \Gamma$, $r(\gamma) = w\gamma w^{-1}$. It is a remarkable fact, not at all obvious, that *every* $r \in R(\Gamma)$ has a quasiconformal geometrical realization.

Let $\Omega(r, \Gamma)$, $r \in R(\Gamma)$, denote the set of quasiconformal geometrical realizations of r.

Lemma 15. *If* $w \in \Omega(r, \Gamma)$*, then the Beltrami differential* $w_{\bar{z}}/w_z$ *is* Γ*-compatible. Conversely, if* $\mu(z)$ *is a* Γ*-compatible Beltrami coefficient, and* w *is a solution of* (8)*, there exists* $r \in R(\Gamma)$ *such that* $w \in \Omega(r, \Gamma)$.

Proof. [27], Theorem 5.14. Let $w \in \Omega(r, \Gamma)$. Then $w\Gamma w^{-1} = r(\Gamma)$ and for $\gamma \in \Gamma$, $z \in U$, $w(\gamma z) = r(\gamma)w(z)$. A computation using the chain rules (6), and the fact that γ and $r(\gamma)$ are conformal, shows that $w_{\bar{z}}/w_z$ satisfies (14). Conversely, if $\mu(z)$ is a Γ-compatible Beltrami differential, and w is a solution of (8), then $w \circ \gamma$ is also a solution and hence by statement (ii) of Lemma 10, there exists $c \in \mathcal{L}$ such that $w \circ \gamma = c \circ w$, or equivalently, $w\gamma w^{-1} = c \in \mathcal{L}$. Thus one can define $r : \Gamma \to \mathcal{L}$ by $r(\gamma) = w\gamma w^{-1}$ for all $\gamma \in \Gamma$. Clearly, $w \in \Omega(r, \Gamma)$. □

The next theorem is a consequence of a well-known result of Nielsen [29]: every automorphism of the fundamental group of compact surface ($= \Lambda_g$) is induced by a self-homemorphism of the surface. The homeomorphism can be taken to be piecewise linear [37]; as such it lifts to a quasiconformal homeomorphism of U. One can therefore restate Nielsen's result as follows: every automorphism $\alpha \in \mathrm{Aut}^+(\Lambda_g)$ has a quasiconformal geometrical realization.

Let $\Lambda = \Lambda_g$.

Theorem 16. $\Omega(r, \Lambda)$ *is non-empty for all* $r \in R(\Lambda)$.

Proof. Let $r \in R(\Lambda)$. There exists a piecewise linear homeomorphism $h : U/\Lambda \to U/r(\Lambda)$ which lifts to a quasiconformal homeomorphism $\tilde{h} : U \to U$ satisfying $h\{\Lambda z\} = r(\Lambda)\tilde{h}(z)$, $z \in U$. $\tilde{h}$ induces the isomorphism $r_0 : \Lambda \to r(\Lambda)$ defined by $\lambda \mapsto \tilde{h}\lambda\tilde{h}^{-1}$, $\lambda \in \Lambda$. $\alpha = r^{-1} \circ r_0 \in \mathrm{Aut}^+(\Lambda)$, so, by Nielsen's result, there is a quasiconformal map w realizing α. Hence $r = r_0 \circ \alpha^{-1}$ is geometrically realized by the quasiconformal map $\tilde{h} \circ w^{-1}$. □

Definition 3. $w : U \to U$ *is a* Teichmüller mapping *for* Γ *if it is either conformal or quasiconformal with Beltrami differential*

$$\frac{w_{\bar{z}}}{w_z} = k\frac{\overline{\phi(z)}}{|\phi(z)|}, \qquad 0 < k < 1,$$

where ϕ *is a* Γ*-2-form.*

Theorem 17 (Uniqueness of Teichmüller mappings). *Let* $w_0 \in \Omega(r, \Lambda)$*,* $r \in R(\Lambda)$*, be a Teichmüller mapping for* Λ*. If* $w \in \Omega(r, \Lambda)$*,* $w \neq w_0$*, then* $K(w_0) < K(w)$.

Proof. [3], §11, 12. Teichmüller's original proof is in [33]. □

Let the surface group $\Lambda = \Lambda_g$ be generated by hyperbolic elements

$$\{a_1, b_1, \ldots, a_g, b_g\}, \tag{17}$$

subject to the relation $a_1b_1a_1^{-1}b_1^{-1}\ldots a_gb_ga_g^{-1}b_g^{-1} = 1$. $r \in R(\Lambda)$ is (a_1, b_1)-*normalized* if the repelling and attracting fixed points of $r(a_1)$ are 0, ∞, respectively, and the repelling fixed point of $r(b_1)$ is 1. Let $Q(a_1, b_1, \Lambda) \subset R(\Lambda)$ be the subspace of (a_1, b_1)-normalized representations.

Let $\Omega(\Lambda)$ be the subspace of quasiconformal maps $w : \bar{U} \to \bar{U}$, topologized by uniform convergence, which realize some $r \in R(\Lambda)$ geometrically (i.e., such that $w\Lambda w^{-1}$ is Fuchsian) and which, in addition, fix $0, 1, \infty \in \partial U$. Let $B(\Lambda)$ be the space of Λ-compatible Beltrami coefficients $\mu(z)$ with $0 \leq |\mu(z)| < 1$, topologized by pointwise convergence at almost every $z \in U$.

Theorem 18. *$T(\Lambda)$ is homeomorphic to the open unit ball in $\mathbb{R}^{6g-6}$.*

Proof. [3], §14; [27], §6. We sketch the proof, omitting the continuity arguments. The starting point is the existence of a one-to-one correspondence between the space of Λ-2-forms and the vector space V of quadratic differentials on the quotient surface U/Λ, which has real dimension $6g - 6$, by the Riemann-Roch theorem ([11], §III.4). An inner product $(,)$ is defined on V by choosing a basis and taking it to be an orthonormal set. The continuous one-to-one map

$$\theta : \phi \mapsto \mu_\phi = (\phi, \phi)\frac{\bar{\phi}}{|\phi|}, \quad \phi \in V \tag{18}$$

has image $\theta(V) \subseteq B(\Lambda)$ by Lemma 13 and the fact that $(\phi, \phi) < 1$. By Lemmas 10 and 11, there is a (ϕ, ϕ)-quasiconformal solution to the Beltrami equation $w_{\bar{z}} = \mu_\phi(z) w_z$ whose continuous extension to ∂U fixes $0, 1, \infty$. Let $w_{[\phi]}$ be the solution; it is a Teichmüller mapping, hence, unique. The map

$$\sigma : \mu_\phi \mapsto w_{[\phi]} \tag{19}$$

from $\theta(V)$ to $\Omega(\Lambda)$ is therefore one-to-one. By the converse of Lemma 15, there exists $r_\phi \in R(\Lambda)$ defined by $r_\phi(\gamma) = w_{[\phi]}^{-1} \gamma\, w_{[\phi]}$, for all $\gamma \in \Lambda_g$. $w_{[\phi]}^{-1} 0 w_{[\phi]} = 0$ and $w_{[\phi]}^{-1} \infty w_{[\phi]} = \infty$; similarly, $w_{[\phi]}^{-1} 1 w_{[\phi]} = 1$. These are, respectively, the attracting and repelling fixed points of $r(a_1)$, and the repelling fixed point of $r(b_1)$. Thus $r_\phi \in Q(a_1, b_1, \Lambda)$. r_ϕ is unique since, by Lemma 10 (ii), another such solution would differ by a real Möbius transformation fixing $0, 1, \infty$, which must be the identity ([21], Corollary 2.5.3). A continuity argument ([27], 6.17) shows that the one-to-one map

$$\tau : w_{[\phi]} \mapsto r_\phi \tag{20}$$

is also onto $Q(a_1, b_1, \Lambda)$. This proves that $Q(a_1, b_1, \Lambda)$ is homeomorphic to the open unit ball in $\mathbb{R}^{6g-6}$. Finally, for every $r \in R(\Lambda)$ there exists a unique $c \in \mathcal{L}$ such that $c^{-1} r(\Lambda) c \in Q(a_1, b_1, \Lambda)$. Thus for every Teichmüller class $[r] \in T(\Gamma)$ there is a unique $r_0 \in Q(a_1, b_1, \Lambda)$ such that $[r] = [r_0]$. It follows that $T(\Lambda)$ is homeomorphic to $Q(a_1, b_1, \Lambda)$. □

Corollary 19. *$\Omega(r, \Lambda)$ contains a (unique) Teichmüller mapping.*

Proof. If $r \in Q(a_1, b_1, \Lambda)$, this follows immediately from the proof of Theorem 18. For the general $r \in R(\Lambda)$, there exists $c \in \mathcal{L}$ which maps the attracting and repelling fixed points of a_1, and the repelling fixed point of b, onto $0, \infty, 1$, respectively. Then the map $r' : \Lambda \to \mathcal{L}$ defined by $\lambda \mapsto c \circ r(\lambda) \circ c^{-1}$ belongs to $Q(a_1, b_1, \Lambda)$ and hence there is a unique Teichmüller mapping $t \in \Omega(r', \Lambda)$. It is not difficult to verify that $c^{-1} \circ t$ is a Teichmüller mapping in $\Omega(r, \Lambda)$, hence, the unique one. □

We now pass to the general co-compact Fuchsian group Γ.

Theorem 20. *$\Omega(r, \Gamma)$ is non-empty for all $r \in R(\Gamma)$.*

Proof. Let $r \in R(\Gamma)$. By Theorem 6, Γ contains a surface group Λ as a normal subgroup of finite index. By Theorem 16, the restriction $r|_\Lambda$ is realized by a unique Teichmuller mapping $t : U \to U$. Choose an arbitrary element $\gamma \in \Gamma$ and use it to define a map

$$t' = r(\gamma)t\gamma^{-1}, \quad t' : U \to U. \tag{21}$$

By (13), t' has the same maximal dilatation as t. For $\lambda \in \Lambda$,

$$\begin{aligned} t'\lambda t'^{-1} &= r(\gamma)t(\gamma^{-1}\lambda\gamma)t^{-1}r(\gamma^{-1}) \\ &= r(\gamma)r(\gamma^{-1}\lambda\gamma)r(\gamma^{-1}) \\ &= r(\lambda). \end{aligned}$$

Thus t' realizes $r|_\Lambda$ geometrically. Since it is also Teichmüller map for Λ, by uniqueness, $t' = t$. Then (21) shows that $t\gamma t^{-1} = r(\gamma)$. Since $\gamma \in \Gamma$ was arbitrary, $t \in \Omega(r, \Gamma)$. □

Theorem 21. *If Γ_1 is a subgroup of finite index in Γ, and t is a Teichmüller mapping for Γ_1, then t is also a Teichmüller mapping for Γ.*

Proof. Using Theorem 6 it is easily shown that there is a surface group Λ which is normal and of finite index in both Γ_1 and Γ. Let $r \in R(\Gamma)$, and let t be the unique Teichmüller mapping in $\Omega(r|_\Lambda, \Lambda)$. The proof of Theorem 20 shows that $t \in \Omega(r|_{\Gamma_1}, \Gamma_1) \cap \Omega(r, \Gamma)$. Thus t is both Γ_1- and Γ-compatible. By Theorem 14, t is a Teichmüller mapping for both Γ_1 and Γ. □

Corollary 22. *$\Omega(r, \Gamma)$ contains a (unique) Teichmüller mapping.*

Let Γ have signature (2).

Theorem 23. *$T(\Gamma)$ is homeomorphic to the open unit ball in $\mathbb{R}^{6g-6+2n}$.*

Proof. [27], Theorem 7.13. Let Λ be a surface group contained with finite index in Γ and generated by hyperbolic elements $a_1, b_1, \ldots a_h, b_h$. Define $Q(a_1, b_1, \Gamma)$ as the set of (a_1, b_1)-normalized representations of Γ. If $i : \Lambda \to \Gamma$ is the inclusion homomorphism, then $r \mapsto r \circ i$, $r \in Q(a_1, b_1, \Gamma)$, embeds $Q(a_1, b_1, \Gamma)$ as a closed subspace of $Q(a_1, b_1, \Lambda)$, by Theorem 7. Let $\psi = \tau\sigma\theta$ be the composition of the homeomorphisms (18), (19), (20). The image of the restriction of ψ^{-1} to $Q(a_1, b_1, \Gamma)$ is the open unit ball in a linear subspace, namely, the subspace of Λ-2-forms which are also Γ-2-forms. This subspace is linearly equivalent to the vector space of quadratic differentials on U/Γ having possible simple poles or removable singularities at the n orbifold points; by the Riemann-Roch theorem, the real dimension of this space is $6g - 6 + 2n$. □

6. The Teichmüller metric

If $r_1, r_2 \in R(\Gamma)$, then $r_2 \circ r_1^{-1} \in R(r_1(\Gamma))$. Let $t(r_2 \circ r_1^{-1})$ be the unique Teichmüller mapping in $\Omega(r_2 \circ r_1^{-1}, \Gamma)$.

Lemma 24. *The function*

$$d(r_1, r_2) = \log |K(t(r_2 \circ r_1^{-1}))| \tag{22}$$

defines a pseudo-metric on $R(\Gamma)$ which induces a true metric on $T(\Gamma)$.

Proof. $d(r_1, r_2) \geq 0$ since $K(t(r_2 \circ r_1^{-1})) \geq 1$. $d(r_1, r_2) = d(r_2, r_1)$ by (13). The triangle inequality follows from (12). $d(r_1, r_2) = 0$ if and only if $t(r_2 \circ r_1^{-1}) = c \in \mathcal{L}$, or, equivalently, for all $\gamma \in \Gamma$, $r_1(\gamma) = c^{-1} r_2(\gamma) c$. Thus $r_1(\Gamma)$ and $r_2(\Gamma)$ are conjugate in $\mathcal{L}$ and $[r_1] = [r_2]$ as points in $T(\Gamma)$. □

The metric on $T(\Gamma)$ is called the *Teichmüller metric*. It is not difficult to show that this metric induces the same topology on $T(\Gamma)$ as the quotient topology inherited from $R(\Gamma)$ ([27], Theorem 8.6).

Theorem 25. *With respect to the Teichmüller metrics, the embedding $\bar{i} : T(\Gamma^*) \to T(\Gamma)$ induced by the injective homomorphism $i : \Gamma \to \Gamma^*$ (cf. Theorem 7) is an isometry.*

Proof. Recall that the embedding is defined by $\bar{i} : [r] \mapsto [r \circ i]$. If d, d^* denote the Teichmüller pseudo-metrics on $R(\Gamma)$ and $R(\Gamma^*)$, respectively, and $r_1, r_2 \in R(\Gamma^*)$, then

$$d(r_1 \circ i, r_2 \circ i) = \log |K(t(r_2 \circ i \circ i^{-1} \circ r_1^{-1}))| = d^*(r_1, r_2).$$

It follows that the Teichmüller distance between $[r_1]$ and $[r_2]$ is the same as the Teichmüller distance between $[r_1 \circ i]$ and $[r_2 \circ i]$. □

Corollary 26. *$Mod(\Gamma)$ acts as a group of isometries of $T(\Gamma)$.*

Proof. This is the special case of Theorem 25 in which $\Gamma^* = \Gamma$, $d^* = d$ and $i = \alpha \in \mathrm{Aut}^+(\Gamma)$. □

There is a complex structure on $T(\Gamma)$, compatible with the Teichmüller metric, making $T(\Gamma)$ biholomorphically equivalent to a bounded domain in $\mathbb{C}^{3g-3+n}$ [2]. We shall not need this structure, so we omit the details.

7. Fixed point sets

Let H be a subgroup of $\mathrm{Mod}(\Gamma)$ and let $F(H) \subset T(\Gamma)$ be the set of points fixed by H under the action

$$\mathrm{Mod}(\Gamma) \times T(\Gamma) \to T(\Gamma). \tag{23}$$

Theorem 27 (Nielsen-Kerckhoff). *$F(H)$ is non-empty if and only if H is a finite group.*

Proof. H is necessarily finite by Theorem 9 and Lemma 4. That every finite subgroup of $\mathrm{Mod}(\Gamma)$ has a non-empty fixed point set was a long-standing conjecture, known as the *Nielsen realization problem*. It was first proved for restricted classes of groups such as cyclic and solvable, and finally in full generality by S. Kerckhoff [23] using one-parameter families of deformations of hyperbolic structures known as *earthquakes*. □

Lemma 28. *A finite subgroup $H < Mod(\Gamma)$ exists if and only if there is a short exact sequence*

$$1 \to \Gamma \overset{i}{\to} \Gamma^* \overset{\rho}{\to} H \to 1, \tag{24}$$

where $\Gamma^* \leq N_{\mathcal{L}}(i(\Gamma))$.

Proof. Let the sequence (24) be given. The natural homomorphism

$$j : N_{\mathcal{L}}(i(\Gamma)) \to \mathrm{Aut}^+(i(\Gamma)) \simeq \mathrm{Aut}^+(\Gamma) \tag{25}$$

is injective because $i(\Gamma)$ has trivial centralizer in $\mathcal{L}$; the restriction $j|_{i(\Gamma)}$ is an isomorphism between $i(\Gamma)$ and $\mathrm{Inn}(i(\Gamma))$, because $i(\Gamma)$ has trivial center (Corollary 2). Hence j induces an injective homomorphism

$$j_* : \frac{N_{\mathcal{L}}(i(\Gamma))}{i(\Gamma)} \to \mathrm{Mod}(\Gamma). \tag{26}$$

Of course, the restriction of j_* to $\Gamma^*/i(\Gamma)$ is also injective, and the image in $\mathrm{Mod}(\Gamma)$ is isomorphic to H.

Conversely, suppose H is a finite subgroup of $\mathrm{Mod}(\Gamma)$. Let $[\alpha_1], [\alpha_2], \ldots, [\alpha_k]$ be the elements of H. Assume $[\alpha_1] = 1$. By the Nielsen-Kerckhoff theorem there exists $[r] \in T(\Gamma)$ which is fixed by all of the $[\alpha_i]$, i.e., $[r \circ \alpha_i] = [r]$, for $1 \leq i \leq k$. Equivalently, there exist $c_i \in \mathcal{L}$, $1 \leq i \leq k$, such that, for all $\gamma \in \Gamma$, $r(\alpha_i(\gamma)) = c_i^{-1} r(\gamma) c_i$. Each c_i is unique because the centralizer of $r(\Gamma)$ is trivial. In particular, $c_1 = 1$. Let $\Gamma^* \leq N_{\mathcal{L}}(r(\Gamma))$ be the Fuchsian group generated by the set $r(\Gamma) \cup \{c_1, \ldots, c_k\}$. Every element of Γ^* can be written uniquely in the form $c_j r(\gamma)$, $\gamma \in \Gamma$. The group operation is $c_j r(\gamma_1) \cdot c_l r(\gamma_2) = c_j c_l r(\gamma_1)^{c_l} r(\gamma_2)$, where $r(\gamma_1)^{c_l} = c_l^{-1} r(\gamma) c_l$. The map $\rho : \Gamma^* \to H$ defined by by $c_i\{r(\gamma)\} \mapsto [\alpha_i]$, $i = 1, \ldots, k$, for all $\gamma \in \Gamma$, is an epimorphism. Thus there is a short exact sequence of the form (24), except that we have $r(\Gamma)$ in place of Γ. There is no loss of generality, since we may assume that $r = \mathrm{id}_\Gamma$. □

The next theorem characterizes the fixed point set of $H < \mathrm{Mod}(\Gamma)$ as the embedded image of Teichmüller space.

Theorem 29. *If $H < Mod(\Gamma)$ belongs to the short exact sequence* (24)*, then*

$$F(H) = \bar{i}(T(\Gamma^*)), \tag{27}$$

where $\bar{i} : T(\Gamma^*) \to T(\Gamma)$ is the embedding induced by $i : \Gamma \to \Gamma^*$.

Proof. [27], Theorem 9.11. Let $[s] \in \bar{i}(T(\Gamma^*)) \subset T(\Gamma)$. Then there exists $[\sigma] \in T(\Gamma^*)$ such that $s = \sigma \circ i$. For every $[\alpha] \in H$, there is a unique $c \in \Gamma^* \leq N_{\mathcal{L}}(\Gamma) < \mathcal{L}$ such that, for all $\gamma \in \Gamma$, $s(\alpha(\gamma)) = c^{-1} s(\gamma) c$. Thus $[s \circ \alpha] = [s]$. It follows that $s \in F(H)$. Conversely, suppose $[s] \in F(H) \subset T(\Gamma)$. We may assume $s = \mathrm{id}_\Gamma$. Let $\sigma = \mathrm{id}_{\Gamma^*}$. Clearly, $s = \sigma \circ i$ and hence, $[s] \in \bar{i}(T(\Gamma^*))$. □

Remark 1. The embedding $\bar{i} : T(\Gamma^*) \hookrightarrow T(\Gamma)$ can be a surjection even if $i(\Gamma) \neq \Gamma^*$. For this to occur, the dimensions of $T(\Gamma)$ and $T(\Gamma^*)$ (computed using Theorem 23) must be equal. This implies that $1 < \Gamma^*/i(\Gamma) \simeq H < \mathrm{Mod}(\Gamma)$ fixes *every point* in $T(\Gamma)$, so that the action of $\mathrm{Mod}(\Gamma)$ on $T(\Gamma)$ is not effective. The complete list of subgroup pairs $\Gamma < \Gamma^*$ for which the Teichmüller dimensions are equal, together with the corresponding signatures, and the indices of the inclusions, is given in [32] (see also [17]).

Let $\mathrm{Stab}_{\mathrm{Mod}(\Gamma)}(F(H))$ denote the set-wise stabilizer of $F(H)$ in $\mathrm{Mod}(\Gamma)$.

Theorem 30. $Stab_{Mod(\Gamma)}(F(H)) = N_{Mod(\Gamma)}(H)$

Proof. [28], §4. If $[\alpha]$ belongs to the normalizer $N_{\mathrm{Mod}(\Gamma)}(H)$ of H, then $[\alpha]F(H) = F(H)$, by a simple calculation. Thus

$$N_{\mathrm{Mod}(\Gamma)}(H) \subseteq \mathrm{Stab}_{\mathrm{Mod}(\Gamma)}(F(H)).$$

Conversely, let $\alpha \in \mathrm{Aut}^+(\Gamma)$, and suppose $[\alpha] \in \mathrm{Stab}_{\mathrm{Mod}(\Gamma)}(F(H))$. We may assume H belongs to the short exact sequence (24). Since $F(H) = \bar{i}(T(\Gamma^*))$, for every $[r] \in T(\Gamma^*)$, there exists $r' \in T(\Gamma^*)$ such that

$$[r \circ i \circ \alpha] = [r' \circ i].$$

This implies the existence of $\beta \in \mathrm{Aut}^+(\Gamma^*)$ such that

$$i \circ \alpha = \beta \circ i.$$

There is a unique element $c \in N_{\mathcal{L}}(\Gamma^*)$ such that

$$\beta(\delta) = c\delta c^{-1},$$

for all $\delta \in \Gamma^*$. In particular, $\alpha(i(\gamma)) = \beta(i(\gamma)) = ci(\gamma)c^{-1}$. It follows that $c \in N_{\mathcal{L}}(i(\Gamma))$ and $j_*(c) = [\alpha] \in \mathrm{Mod}(\Gamma)$, where j_* is the injective homomorphism (26). Then

$$H = j_*(\Gamma^*/i(\Gamma)) = j_*(c\Gamma^* c^{-1}/i(\Gamma)) = [\alpha]H[\alpha]^{-1},$$

which shows that $[\alpha] \in N_{\mathrm{Mod}(\Gamma)}(H)$. □

8. Relative modular groups

The action of $N_{\mathrm{Mod}(\Gamma)}(H)$ on $F(H)$ is properly discontinuous and isometric (being a restriction of (23)). Since $H < N_{\mathrm{Mod}(\Gamma)}(H)$ acts trivially on $F(H)$ by definition, the effective action is by $N_{\mathrm{Mod}(\Gamma)}(H)/H$, or, possibly, a quotient of this group.

Definition 4. $N_{Mod(\Gamma)}(H)/H$ *is called the* relative modular group of Γ^* with respect to $i(\Gamma)$.

There are other ways to define the relative modular group (see, e.g., [14,10]). One useful alternative definition is as a subgroup of $\mathrm{Mod}(\Gamma^*)$.

Lemma 31. *The subgroup*

$$Mod(\Gamma^*, i(\Gamma)) = \{[\beta] \in Mod(\Gamma^*) | \beta(i(\Gamma)) = i(\Gamma)\} \leq Mod(\Gamma^*)$$

is isomorphic to the relative modular group $N_{Mod(\Gamma)}(H)/H$.

Proof. [28], Theorem 10. Let $\mathrm{Aut}^+(\Gamma^*, i(\Gamma)) = \{\beta \in \mathrm{Aut}^+(\Gamma^*) | \beta(i(\Gamma)) = i(\Gamma)\} \leq \mathrm{Aut}^+(\Gamma^*)$. Note that $\mathrm{Inn}(\Gamma^*, i(\Gamma)) = \mathrm{Inn}(i(\Gamma))$ since $i(\Gamma)$ is normal in Γ^*. The restriction homomorphism $f : \mathrm{Aut}^+(\Gamma^*, i(\Gamma)) \to \mathrm{Aut}^+(i(\Gamma))$ defined by $f : \beta \mapsto \beta|_{i(\Gamma)}$ is injective: if $\beta'|_{i(\Gamma)} = \beta|_{i(\Gamma)}$, then, for all $\delta \in \Gamma^*$, $\beta'(\delta)^{-1}\beta(\delta)$ is in the (trivial) centralizer of $\beta(i(\Gamma))$ and therefore $\beta = \beta'$. Let $[\alpha] = [\beta|_{i(\Gamma)}]$. If $[r] \in T(\Gamma^*)$, $[r \circ i] \in \bar{i}(T(\Gamma^*))$ and

$$[\alpha][r \circ i] = [r \circ i \circ \alpha] = [r \circ \beta \circ i] = \bar{i}([r \circ \beta]) \in \bar{i}(T(\Gamma^*)).$$

Thus $[\alpha]$ stabilizes $\bar{i}(T(\Gamma^*))$, and hence, by Theorem 30, $[\alpha] \in N_{\mathrm{Mod}(\Gamma)}(H)$. $\beta \in \mathrm{Inn}(\Gamma^*, i(\Gamma))$ if and only if $[\alpha] \in H$. Thus f induces an injective homomorphism $\tilde{f} : \mathrm{Mod}(\Gamma^*, i(\Gamma)) \to N_{\mathrm{Mod}(\Gamma)}(H)/H$ defined by $[\beta] \to [\alpha]$. On the other hand, if $[\alpha] \in N_{\mathrm{Mod}(\Gamma)}(H)/H$, then $[\alpha]$ stabilizes $F(H)$ and by the proof of Theorem 30 there exists $\beta \in \mathrm{Aut}^+(\Gamma^*)$ such that $\beta|_{i(\Gamma)} = \alpha$. Thus $\beta \in \mathrm{Aut}^+(\Gamma^*, i(\Gamma))$, $\tilde{f}([\beta]) = [\alpha]$, and $\tilde{f}$ is surjective. □

If $i(\Gamma)$ is characteristic in Γ^*, $\mathrm{Mod}(\Gamma^*, i(\Gamma)) = \mathrm{Mod}(\Gamma^*)$; otherwise, the index $[\mathrm{Mod}(\Gamma^*) : \mathrm{Mod}(\Gamma^*, i(\Gamma))]$ is equal to the number of distinct images $[\beta](i(\Gamma))$, $[\beta] \in M(\Gamma^*)$. This number is finite since each $[\beta](i(\Gamma))$ is the kernel of a homomorphism from a finitely generated group (G^*) onto a finite group (H), of which there are just finitely many. (See [14].)

Even the action of the relative modular group

$$\mathrm{Mod}(\Gamma^*, i(\Gamma)) \times \bar{i}(T(\Gamma^*)) \to \bar{i}(T(\Gamma^*)), \tag{28}$$

is not always effective. Let H_1/H be the largest subgroup of $N_{\mathrm{Mod}(\Gamma)}(H)/H$ which fixes every point in $F(H)$. Then the effective action is by the group

$$\frac{N_{\mathrm{Mod}(\Gamma)}(H)/H}{H_1/H} \simeq N_{\mathrm{Mod}(\Gamma)}(H)/H_1. \tag{29}$$

9. Topological versus conformal conjugacy

The orbit space

$$\mathcal{R}(\Gamma^*, i(\Gamma)) = \bar{i}(T(\Gamma^*))/\mathrm{Mod}(\Gamma^*, i(\Gamma))$$

of the action (28) is called the *relative Riemann space* of $(\Gamma^*, i(\Gamma))$ [10], [14]. If $i(\Gamma) = \Gamma^*$ it is called simply the *Riemann space of* Γ. The Riemann space of Λ_g (a surface group) is the *space of moduli* of compact Riemann surfaces of genus g, denoted $\mathcal{R}_g$. Of course, one obtains the same relative Riemann space if one replaces $\mathrm{Mod}(\Gamma^*, i(\Gamma))$ with the *effective* relative modular group (29).

Let H_0 be a finite group. A *surface with H_0-symmetry*, or, briefly, a *surface-symmetry pair*, is a pair (X, H) in which X is a compact Riemann surface and $H \simeq H_0$ is a finite group of automorphisms of X. It is convenient to define two equivalence relations on the set of surfaces with H_0 symmetry.

Definition 5. *Surfaces with H_0 symmetry (X_1, H_1) and (X_2, H_2) are* conformally equivalent *(resp.* topologically equivalent*) if there exists a conformal map (resp. orientation-preserving homeomorphism) $t : X_1 \to X_2$ such that $H_1 = tH_2t^{-1}$. The H_i-actions are called* conformally *(resp.* topologically*) conjugate.*

Theorem 32. *If H belongs to the short exact sequence* (24)*, there is a bijection between $\mathcal{R}(\Gamma^*, i(\Gamma))$ and the set of conformal equivalence classes of surfaces with H-symmetry.*

Proof. Let $[r_1], [r_2] \in \bar{i}(T(\Gamma^*) \subset T(\Gamma)$. Then there exist $[r_j'] \in T(\Gamma^*)$, $j = 1, 2$ such that $[r_j'] = [r_j \circ i]$. Put $X_j = U/r_j(\Gamma)$, and

$$H_j = r_j'(\Gamma^*)/r_j(\Gamma) \simeq H, \quad j = 1, 2. \tag{30}$$

The H_j action on X_j is defined by

$$r_j'(\gamma)r_j(\Gamma) : r_j(\Gamma)z \mapsto r_j'(\gamma)r_j(\Gamma)z, \quad \gamma \in \Gamma^*, z \in U. \tag{31}$$

If $[r_1], [r_2]$ are in the same $\mathrm{Mod}(\Gamma^*, i(\Gamma))$- orbit, there exists $\beta \in \mathrm{Aut}^+(\Gamma^*, i(\Gamma))$ such that $[r_1' \circ \beta] = [r_2']$. Moreover, β restricts to $\alpha \in \mathrm{Aut}^+(\Gamma)$ such that $[r_1 \circ \alpha] = [r_2]$. Equivalently, there exists $c \in \mathcal{L}$ such that

$$r_1'(\beta(\gamma)) = cr_2'(\gamma)c^{-1}, \quad \gamma \in \Gamma^*. \tag{32}$$

In particular, $r_1(\Gamma) = cr_2(\Gamma)c^{-1}$. The map

$$r_1(\Gamma)z \mapsto c^{-1}r_1(\Gamma)z, \quad z \in U,$$

induces a conformal map $\tilde{c} : X_1 \to X_2$, since by (32), $c^{-1}r_1(\Gamma)z = r_2(\Gamma)c^{-1}z$.

For $\gamma \in \Gamma^*$, let $h_i(\gamma) : X_i \to X_i$ denote the map (31). To show that $\tilde{c}$ commutes with the H_j-actions, we show that the diagram

$$\begin{array}{ccc} X_1 & \overset{h_1(\gamma)}{\longrightarrow} & X_1 \\ \tilde{c}\downarrow & & \tilde{c}\downarrow \\ X_2 & \overset{h_2(\gamma)}{\longrightarrow} & X_2 \end{array} \tag{33}$$

commutes. For $z \in U$, let $p_z = r_1(\Gamma)z \in X_1$. On the one hand,

$$\tilde{c} \circ h_1(\gamma)(p_z) = r_2(\Gamma)c^{-1}r_1'(\gamma)z, \tag{34}$$

and on the other,

$$h_2(\gamma) \circ \tilde{c}(p_z) = r_2(\Gamma)r_2'(\gamma)c^{-1}z; \tag{35}$$

the diagram commutes if and only if the right-hand sides of (34) and (35) represent the same point in X_2. Using (32), the right -hand side of (34) is equal to

$$r_2(\Gamma)r_2'(\beta^{-1}(\gamma))c^{-1}z.$$

Since $r_2'(\beta^{-1}(\gamma))c^{-1}$ and $r_2'(\gamma)c^{-1}$ represent the same coset of $r_2(\Gamma)$, it follows that the right-hand sides of (34) and (35) both represent the the same point in X_2, namely, $r_2(\Gamma)c^{-1}z$.

We have shown that there is a one-to-one map from the relative Riemann space $\mathcal{R}(\Gamma^*, \Gamma)$ to the set of conformal equivalence classes of surfaces with H symmetry. The proof that this map is surjective is straightforward and we omit it. □

Suppose $i_j : \Gamma \to \Gamma^*$, $j = 1, 2$, are two different injections of Γ into Γ^* such that the image $i_j(\Gamma)$ is normal in Γ^*. Let $[r_1] \in \bar{i}_1(T(\Gamma^*)$ and $[r_2] \in \bar{i}_2(T(\Gamma^*)$. There exist $[r'_j] \in T(\Gamma^*)$, $j = 1, 2$ such that $[r_j] = [r'_j \circ i_j]$, $j = 1, 2$. Put $X_j = U/r_j(\Gamma)$ and define H_j, and H_j-actions on X_j, exactly as at (30), (31), respectively. The isomorphism

$$r'_2 \circ r_1'^{-1} : r'_1(\Gamma^*) \to r'_2(\Gamma^*)$$

has a quasiconformal geometrical realization $t : U \to U$ by Theorem 20. We may take t to be the unique Teichmüller map in $\Omega(r'_2 \circ r_1'^{-1}, \Gamma_1)$. t induces an orientation-preserving homeomorphism $\tilde{t} : X_1 \to X_2$ which (in place of $\tilde{c}$) satisfies the diagram (33). Thus the surface-symmetry pairs (X_1, H_1) and (X_2, H_2) are topologically conjugate.

One way to obtain distinct injections $i_j : \Gamma \to \Gamma^*$, $j = 1, 2$ is to precompose i_1 with $\beta \in \mathrm{Aut}^+(\Gamma)$, obtaining $i_2 = i_1 \circ \beta$. Then $[r_1] = [r_2 \circ \beta]$, i.e., $[r_1], [r_2]$ are in the same Mod(Γ)- orbit and the surfaces X_1 and X_2 are conformally equivalent. The surface-symmetry pairs (X_j, H_j), $j = 1, 2$, however, need not be, since β need not extend to an automorphism of G^*.

Under the action of Mod(Γ), the images of an embedded Teichmüller space $\bar{i}(T(\Gamma^*)) = F(H) \subset T(\Gamma)$ can intersect without coinciding (provided the embedded space has positive Teichmüller dimension). In this case, we shall argue, the intersection is composed of embedded Teichmüller spaces of strictly smaller dimension.

Let $[\beta] \in \mathrm{Mod}(\Gamma)$, $[r_1], [r_2] \in F(H) \cap [\beta]F(H)$, $[r_1] \neq [r_2]$. Suppose $[r_2] = [\beta][r_1]$, so that $[r_1]$ and $[r_2]$ project to the same point in $X \in \mathcal{R}(\Gamma)$. If $[r_1]$ and $[r_2]$ are not in the same Mod($\Gamma^*, i(\Gamma)$)-orbit, i.e., if they determine distinct points in the relative Riemann space $\mathcal{R}(\Gamma^*, i(\Gamma))$, the surface-symmetry pairs (X, H_1) and (X, H_2) are not conformally equivalent. Thus $X \in \mathcal{R}(\Gamma)$ is a surface admitting two H-actions which are topologically but not conformally conjugate. Equivalently, the full automorphism group $G = \mathrm{Aut}(X)$ contains two *non-conjugate* copies of H_1. The inclusion $i : \Gamma \to \Gamma^*$ extends to

$$\Gamma \xrightarrow{i} \Gamma^* \xrightarrow{j} \Gamma_1, \tag{36}$$

where j is a strict normal inclusion, $j \circ i : \Gamma \to \Gamma_1$ is also a normal inclusion, and $G \simeq \Gamma_1/\Gamma$. Then

$$F(G) = \bar{i} \circ \bar{j}(T(\Gamma_1)) \subseteq F(H) \cap [\beta]F(H).$$

If the dimension of $F(G)$ is positive, and there are Mod(Γ)-images of $F(G)$ which intersect without coinciding, the procedure just described can be repeated beginning with $F(G)$ instead of $F(H)$. After finitely many repetitions the procedure must terminate: we reach a finite group $G' > G$ such that all Mod(Γ)-images of $F(G')$ are either disjoint or identical. Then topological conjugacy implies conformal conjugacy and there is a bijection between the relative Riemann space and the image of $F(G')$ in $\mathcal{R}(\Gamma)$. This is always the case if the Teichmüller dimension of $F(G')$ is 0.

For further discussion of these matters from an algebraic geometric view point, see [16].

10. Stratification

Let $\mathrm{Mod}_g = \mathrm{Mod}(\Lambda_g)$. Let (H) denote the conjugacy class of a finite subgroup $H < \mathrm{Mod}_g$. There is a bijection between the set

$$\{(H)|H < \mathrm{Mod}_g, H \text{ non-trivial, finite}\}$$

and the set of topological equivalence classes of surface-symmetry pairs (X, H), where $X \in \mathcal{R}_g$ is a surface of genus g (see, e.g. [31]).

For $X \in \mathcal{R}_g$, let $\Sigma(X)$ denote the conjugacy class $(\mathrm{Aut}(X))$ in Mod_g. We call $\Sigma(X)$ the *symmetry type* of X. The (H)-*equisymmetric stratum* of $\mathcal{R}_g$ is the set

$$\mathcal{R}_g^{(H)} = \{X \in \mathcal{R}_g | \Sigma(X) \geq (H)\} \tag{37}$$

(the terminology is due to S. A. Broughton [6]). The *stratification* of $\mathcal{R}_g$ is the union of the equisymmetry strata over all conjugacy classes of non-trivial finite subgroups of Mod_g. The stratification is the image of the *branch locus* in $T(\Lambda_g) = \mathcal{T}_g$ under the action of Mod_g and is thus also the locus of surfaces with automorphisms. Each stratum is the image of an embedded Teichmüller space in $\mathcal{T}_g$. One can obtain the stratification by considering all normal inclusions of the form $i : \Lambda_g \to \Gamma^*$ where Λ_g is fixed and Γ^* varies over all co-compact Fuchsian groups containing Λ_g as a normal subgroup of finite index, and i varies over all normal inclusions $i : \Lambda_g \to \Gamma^*$. From this it is clear that the number of strata is finite: there are finitely many Γ^* such that the ratio $\chi(\Lambda_g)/\chi(\Gamma^*)$ is an integer, as required by the Riemann-Hurwtiz relation (Lemma 3), and each such Γ^*, being finitely generated, admits finitely many homomorphisms with kernel isomorphic to Λ_g.

It can happen that $\mathcal{R}_g^{(G)} = \mathcal{R}_g^{(H)}$. This occurs when $G \supset H$ and the dimensions of the covering embedded Teichmüller spaces are equal (cf. Remark 1 following Theorem 29). In this case, H is not the full automorphism group of any surface in $\mathcal{R}_g^{(H)}$. Naturally, it is preferable to associate the largest possible group with a given stratum.

It is possible to make the strata disjoint, and avoid the sort of duplication just described (allowing some empty strata), by defining

$$\tilde{\mathcal{R}}_g^{(H)} = \mathcal{R}_g^{(H)} - \bigcup_{G \supset H} \mathcal{R}_g^{(G)}.$$

On the other hand, for some purposes, it is precisely the intersection of different strata which are of most interest: the intersection and nesting relationships between the strata echo those of the covering branch locus.

The strata (37) are known to be irreducible complex algebraic varieties ([4], §7).

11. Examples

1. The equisymmetric stratification of $\mathcal{R}_2$ is simple enough to describe in words: $\mathcal{R}_2$ itself has complex dimension 3. It contains a 2-dimensional stratum $\mathcal{R}_2^{(D_2)}$, where D_2 is the Klein 4-group. This stratum contains two 1-dimensional substrata associated with the dihedral group D_4 of order 8 and the product $C_2 \times S_3$ (cyclic by symmetric). These two strata, in turn, intersect in two distinct 0-dimensional strata, one corresponding to the group $GL_2(F_3)$ of order 48 and the other to a certain group of order 24. Finally, there is an isolated 0-dimensional stratum, not contained in $\mathcal{R}_2^{(D_2)}$, associated with the cyclic group C_{10}. (See [6,26], and also, [25].)
2. For all $g > 2$, $\mathcal{R}_g$ itself is properly regarded as the stratum associated with the trivial group, but this is not the case in genus 2. In fact $\mathcal{R}_2 = \mathcal{R}_2^{(C_2)}$. The explanation is that all surfaces of genus 2 admit the hyperelliptic involution. Equivalently: there is a normal inclusion $i : \Lambda_2 \to \Gamma^*$, where Γ^* is the group with signature $(0; 2, 2, 2, 2, 2, 2)$. It is easily verified that the dimension of the Teichmüller space of Γ^* is equal to the dimension of $\mathcal{T}_2$. Thus M_2 acts ineffectively on $\mathcal{T}_2$ because $\Gamma^*/\Lambda_2 \simeq C_2 < \mathrm{Mod}_2$ fixes every point.
3. The equisymmetric stratification of $\mathcal{R}_3$ can be constructed from the information in [6]; see the author's Ph.D. dissertation [34] for a (complicated) picture.
4. Let $\Gamma^{(n)}$ denote the Fuchsian group with signature $(n; 2, \ldots, 2)$, where the number of 2's is $2g + 2 - 4n$. For $g \geq 2n - 1$ and $g > 1$, there is a short exact sequence

$$1 \to \Lambda_g \xrightarrow{i} \Gamma^{(n)} \to C_2 \to 1. \tag{38}$$

Let $\Gamma^{(n)}/\Lambda_g = \langle J_n | J_n^2 = 1\rangle \simeq C_2$. J_n is known as an n-hyperelliptic involution since it acts on the surface U/Λ_g with quotient a surface of genus n. (J_0 is the classical hyperelliptic involution.) A short argument shows that all surface-symmetry pairs $(U/\Lambda_g, \langle J_n\rangle)$ are topologically conjugate. By a theorem of Farkas and Kra [11], §V.1.9, if $g > 4n + 1$, J_n is unique (hence central) in $\mathrm{Aut}(U/\Lambda_g)$. Restated in terms of Fuchsian groups, this theorem says that whenever there is a sequence of injective homomorphisms

$$\Lambda_g \xrightarrow{i} \Gamma^{(n)} \xrightarrow{j} N_{\mathcal{L}}(\Lambda_g), \tag{39}$$

the image $j(\Gamma^{(n)})$ is normal. (The image $i(\Lambda_g)$ is normal in $\Gamma^{(n)}$ because the index is 2.) Hence any automorphism group Γ^*/Λ_g of U/Λ_g which contains $\Gamma^{(n)}/\Lambda_g = \langle J_n\rangle$, contains it as a normal subgroup, so J_n is a central element. As a consequence, all pairs $(U/\Lambda_g, \langle J_n\rangle)$ are conformally conjugate.

In the intermediate range $(g - 1)/4 \leq n < (g + 1)/2$, the theorem of Farkas and Kra does not hold and there may be $\langle J_n\rangle$ actions that are topologically but not conformally conjugate. We show this is the case when $g = 3$ and $n = 1$. In genus 3 the Klein 4-group $D_2 = \langle J_0, J_1\rangle$ acts with signature $(0; 2, 2, 2, 2, 2, 2)$. The hyperelliptic involution J_0 is unique, so the other two nontrivial elements in D_2, namely J_1 and J_0J_1, acting with signature $(1; 2, 2, 2, 2)$, are a pair of non-conjugate 1-hyperelliptic involutions.

5. (4., continued.) The theorem of Farkas and Kra has another consequence: if $g > 4n+1$, and Λ_g is contained in $\Gamma^{(n)}$, then $N_{\mathcal{L}}(\Lambda_g) \leq N_{\mathcal{L}}(\Gamma^{(n)})$. A finite subgroup $H < \mathrm{Mod}_g$ may be represented as Γ^*/Λ_g, $\Gamma^* \leq N_{\mathcal{L}}(\Lambda_g)$. By assumption, G^*/Λ_g contains $\Gamma^{(n)}/\Lambda_g = \langle J_n \rangle$ as a normal subgroup. It follows that every finite subgroup $H < \mathrm{Mod}_g$ stabilizes the fixed point set $F(\langle J_n \rangle) \subset \mathcal{T}_g$. In other words, the embedded Teichmüller space

$$\bar{i}(T(\Gamma^{(n)}) \subset \mathcal{T}_g,$$

where i is the injective homomorphism in (38), is invariant under the action of Mod_g. The n-hyperelliptic stratum $\mathcal{R}_g^{(\langle J_n \rangle)}$ is therefore disjoint from all other strata except those contained in it. This is an advantage: to determine the substrata of $\mathcal{R}_g^{(\langle J_n \rangle)}$ one need only consider the (finite) class of finite groups which are central extensions of C_2 by groups which act in genus n. See the author's paper [35] for the case $n = 0$.

References

[1] L. V. Ahlfors, On quasiconformal mappings, *Journal d'Analyse Mathematique*, **4** (1954) 1–58.

[2] L. V. Ahlfors, The complex analytic structure of the space of closed Riemann surfaces. In: Analytic Functions. Princeton University Press, Princeton, NJ (1960).

[3] L. Bers, Quasiconformal Mappings and Teichmüller's Theorem. In: Analytic Functions. Princeton University Press, Princeton, NJ (1960).

[4] L. Bers, Uniformization, moduli, and Kleinian groups, *Bull. London Math. Soc.*, **4** (1972) 257–300.

[5] L. Bers, Finite dimensional Teichmüller spaces and generalizations, *Bull. Amer. Math. Soc.*, **5(2)** (1981) 131–172.

[6] S. A. Broughton, The equisymmetric stratification of the moduli space and the Krull dimension of mapping class groups, *Topology and Appl.*, **37** (1990) 101–113.

[7] S. A. Broughton, Classifying group actions on surfaces of low genus, *J. Pure Appl. Algebra*, **69** (1990) 233–270.

[8] S. Bundgaard and J. Nielsen, On normal subgroups with finite index in F-groups, *Mat. Tidsskr.*, **B** (1951) 56–58.

[9] M. Cornalba, On the locus of curves with automorphisms, *Ann. Mat. Pura Appl.*, **149** (1987) 135–151.

[10] C. J. Earle, On the moduli of closed Riemann surfaces with symmetries. In: Advances in the theory of Riemann surfaces, *Annals of Math Studies*, 66. Princeton University Press, Princeton, NJ (1971).

[11] H. M. Farkas and I. Kra, Riemann surfaces. Springer-Verlag, New York (1991).

[12] R. H. Fox, On Fenchel's conjecture about F-groups, *Mat. Tidsskr.*, **B** (1951) 63–65.

[13] F. Gardiner, Teichmüller theory and quadratic differentials. John Wiley and Sons, New York (1987).

[14] J. Gilman, On the moduli of compact Riemann surfaces with a finite number of punctures. In: Discontinuous groups and Riemann surfaces. L. Greenberg, ed., *Annals of Math Studies*, **79**. Princeton University Press, Princeton, NJ (1974).

[15] G. González-Díez and W. J. Harvey, Moduli of Riemann surfaces with symmetry. In: Discrete Groups and Geometry. W. J. Harvey, ed., Cambridge University Press (1992).

[16] G González-Díez and R. A. Hidalgo, Conformal versus topological conjugacy of automorphisms of compact Riemann surfaces, *Bull. London Math. Soc.*, **29** (1997) 280–284.

[17] L. Greenberg, Maximal Fuchsian groups, *Bull. Amer. Math. Soc.*, **69** (1963) 569–573.

[18] L. Greenberg, Finiteness Theorems for Fuchsian and Kleinian groups. In: W. J. Harvey, ed., Discrete Groups and Automorphic Functions. Academic Press, New York (1977).

[19] W. J. Harvey, On branch loci in Teichmüller space, *Trans. Amer. Math. Soc.*, **153** (1971) 387–399.

[20] W. J. Harvey, Spaces of discrete groups. In: W. J. Harvey, ed., Discrete Groups and Automorphic Functions, Academic Press, New York (1977).

[21] G. A. Jones and D. Singerman, Complex Functions, Cambridge University Press, Cambridge (1987).

[22] S. Katok, Fuchsian groups. University of Chicago Press, Chicago (1992).

[23] S. Kerckhoff, The Nielsen realization problem, *Annals of Math.*, **117** (1983) 235–265.

[24] S. Kravetz, On the geometry of Teichmüller spaces and the structure of their modular groups, *Ann. Acad. Sci. Fenn. Ser. A I*, **278** (1959) 1–35.

[25] Kulkarni and S. Ravi, Isolated points in the branch locus of the moduli space of compact Riemann surfaces, *Ann. Acad. Sci. Fenn. Ser. A.I. Math*, **16** (1991) 71–81.

[26] T. Kuusalo, and M. Nätäänen, Geometric uniformization in genus 2, *Ann. Acad. Sci. Fenn. Ser. A.I. Math.*, **16(1)** (1991) 71–81.

[27] A. M. Macbeath and D. Singerman, Spaces of subgroups and Teichmüller space, *Proc. London Math. Soc.*, **3(31)** (1975) 211–256.

[28] C. Maclachlan and W. J. Harvey, On mapping-class groups and Teichmüller spaces, *Proc. London Math. Soc.*, **3(30)** (1975) 496–512.

[29] J. Nielsen, Untersuchungen zur Topologie der geschlossenen zwiseitigen Flachen, *Acta Math.*, **50** (1927) 189–358.

[30] H. E. Rauch, A transcendental view of the space of algebraic Riemann surfaces, *Bull. Amer. Math. Soc.*, **71** (1965) 1–39.

[31] J. F. X. Ries, Subvarieties of moduli space determined by finite groups acting on surfaces, *Trans. Amer. Math. Soc.*, **335(1)** (1993) 385–406.

[32] D. Singerman, Finitely maximal Fuchsian groups, *J. London Math. Soc. (2)*, **6** (1972) 29–38.

[33] O. Teichmüller, Extremale quasikonforme Abbildungen und quadratische differentiale, *Preuss. Akad.*, **22** (1940).

[34] A. Weaver, Moduli spaces of compact Riemann surfaces admitting automorphisms, Ph. D. Thesis, The Graduate School and University Center, City University of New York (1997).

[35] A. Weaver, Hyperelliptic surfaces and their moduli, *Geometriae Dedicata*, **103** (2004) 69–87.

[36] H. Zieschang, Über Automorphismen ebener diskontinuierlicher Gruppen, *Math. Ann.*, **166** (1966) 148–167.

[37] H. Zieschang, E. Vogt and H. D. Coldewey, Surfaces and planar discontinuous groups, Lecture Notes in Mathematics, 835 Springer-Verlag, Berlin (1980).

Made in the USA
Monee, IL
08 November 2026

68055342R00345